Student's Suite CD-ROM...
All the tools you need for your course!

When you open the Student's Suite CD-ROM you'll find:

- Lab manuals for MINITAB®, Microsoft® Excel, TI-83 Graphing Calculator, JMP® INTRO, and SPSS®

- Datasets formatted for MINITAB, Excel, SPSS, SAS®, JMP®, TI-83, and ASCII

- The applets used in the Activities found in the text, as well as links to exercises and other learning resources on the book's exclusive Companion Web Site. The text, the CD-ROM, and the Web site work in tandem to give students a visual, interactive way to learn with technology.

- InfoTrac® College Edition and Internet exercises to give students online, primary-source research assignments pertaining to specific topics in introductory statistics

- Links to Michael Larsen's guide to online statistical research, *Internet Companion for Statistics* . . . with engaging online examples and numerous helpful exercises related directly to specific topics for the course—including numerical, short-answer, and expository problems corresponding to the cited Web sites.

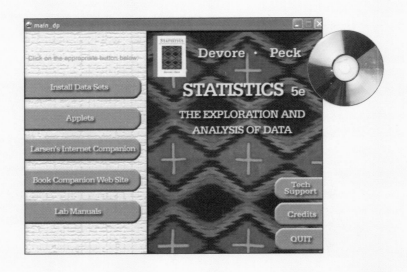

Statistics

The Exploration
and Analysis of Data

Fifth Edition

Jay Devore
California Polytechnic State University
San Luis Obispo

Roxy Peck
California Polytechnic State University
San Luis Obispo

THOMSON
™
BROOKS/COLE

Australia ▪ Canada ▪ Mexico ▪ Singapore
Spain ▪ United Kingdom ▪ United States

Statistics Editor: *Carolyn Crockett*
Assistant Editor: *Ann Day*
Editorial Assistant: *Rhonda Letts*
Technology Project Manager: *Burke Taft*
Executive Marketing Manager: *Joseph Rogove*
Marketing Assistant: *Jessica Perry*
Advertising Project Manager: *Nathaniel Michelson-Bergson*
Project Manager, Editorial Production: *Lisa Weber*
Print Buyer: *Kristine Waller*
Permissions Editor: *Sommy Ko*

Production Service: *G & S Typesetters, Inc.*
Text Designer: *Andrew Ogus ■ Book Design*
Copy Editor: *Mimi Braverman*
Illustrator: *G & S Typesetters, Inc. and Lori Heckelman*
Cover Designer: *Bill Stanton*
Cover Image: *Planet Art*
Cover Printer: *QuebecorWorld/Versailles*
Compositor: *G & S Typesetters, Inc.*
Printer: *QuebecorWorld/Versailles*

For more information about our products, contact us at:
Thomson Learning Academic Resource Center
1-800-423-0563

For permission to use material from this text, contact us by:
Phone: 1-800-730-2214   **Fax:** 1-800-730-2215
Web: http://www.thomsonrights.com

Library of Congress Control Number: 2003114200

Student Edition: ISBN 0-534-46723-7
Instructor's Edition: ISBN 0-534-49540-0

Brooks/Cole — Thomson Learning
10 Davis Drive
Belmont, CA 94002
USA

Asia
Thomson Learning
5 Shenton Way #01-01
UIC Building
Singapore 068808

Australia/New Zealand
Thomson Learning
102 Dodds Street
Southbank, Victoria 3006
Australia

Canada
Nelson
1120 Birchmount Road
Toronto, Ontario M1K 5G4
Canada

Europe/Middle East/Africa
Thomson Learning
High Holborn House
50/51 Bedford Row
London WC1R 4LR
United Kingdom

Latin America
Thomson Learning
Seneca, 53
Colonia Polanco
11560 Mexico D.F.
Mexico

Spain/Portugal
Paraninfo
Calle/Magallanes, 25
28015 Madrid, Spain

■ To our families, colleagues, and students,
for what we have learned from them.

■ About the Authors

 JAY DEVORE earned his undergraduate degree in Engineering Science from the University of California at Berkeley, spent a year at the University of Sheffield in England, and finished his Ph.D. in statistics at Stanford University. He previously taught at the University of Florida and at Oberlin College and has had visiting appointments at Stanford, Harvard, the University of Washington, and New York University. Jay is currently Professor and Chair of the Statistics Department at California Polytechnic State University, San Luis Obispo. The Statistics Department at Cal Poly has an international reputation for activities in statistics education. In addition to this book, Jay has written several widely used engineering statistics texts and is currently working on a book in applied mathematical statistics. He is the recipient of a distinguished teaching award from Cal Poly and is a Fellow of the American Statistical Association. In his spare time, he enjoys reading, cooking and eating good food, tennis, and travel to faraway places. He is especially proud of his wife, Carol, a retired elementary school teacher, his daughter Allison, who works for the Center for Women and Excellence in Boston, and his daughter Teri, who is finishing a graduate program in education at NYU.

 ROXY PECK is Associate Dean of the College of Science and Mathematics and Professor of Statistics at California Polytechnic State University, San Luis Obispo. Roxy has been on the faculty at Cal Poly since 1979, serving for six years as Chair of the Statistics Department before becoming Associate Dean. She received an M.S. in Mathematics and a Ph.D. in Applied Statistics from the University of California, Riverside. Roxy is nationally known in the area of statistics education, and in 2003 she received the American Statistical Association's Founder's Award, recognizing her contributions to K–12 and undergraduate statistics education. She is a Fellow of the American Statistical Association and an elected member of the International Statistics Institute. Roxy has recently completed five years as the Chief Reader for the Advanced Placement Statistics Exam and currently chairs the American Statistical Association's Joint Committee with the National Council of Teachers of Mathematics on Curriculum in Statistics and Probability for Grades K–12. In addition to her texts in introductory statistics, Roxy is also co-editor of *Statistical Case Studies: A Collaboration Between Academe and Industry* and a member of the editorial board for *Statistics: A Guide to the Unknown*, 4th edition. Outside the classroom and the office, Roxy likes to travel and spends her spare time reading mystery novels. She also collects Navajo rugs and heads to New Mexico whenever she can find the time.

· Brief Contents

Contents

3 ▪ Graphical Methods for Describing Data   73

4 ▪ Numerical Methods for Describing Data   139

5 ▪ Summarizing Bivariate Data   185

Preface

Statistics: The Exploration and Analysis of Data, 5th edition, is intended for use as a textbook for introductory statistics courses at two- and four-year colleges and universities. We believe that the following special features of our book distinguish it from other texts.

Features

A Traditional Structure with a Modern Flavor

The topics included in almost all introductory texts are here also. However, we have interwoven new strands that reflect current and important developments in statistical analysis. These include coverage of sampling and experimental design, the role of graphical displays as an important component of data analysis, transformations, residual analysis, normal probability plots, and simulation. The organization gives instructors considerable flexibility in deciding which of these topics to include in a course.

A Focus on Data Analysis

Students are introduced early to the idea that data analysis is a process that begins with careful planning, followed by data collection, data description using graphical and numerical summaries, data analysis, and, finally, interpretation of results. Section 2.1 describes this process in detail, and the ordering of topics in the first 10 chapters of the book is chosen to mirror this process: data collection, then data description, then statistical inference.

The Use of Real Data and the Importance of Context

Many students are skeptical of the relevance and importance of statistics. Contrived problem situations and artificial data often reinforce this skepticism. A strategy that we have used successfully to motivate students is to present examples and exercises that involve real data extracted from journal articles, newspapers, and other published sources. Most examples and exercises in the book are of this nature; they cover a wide range of disciplines and subject areas, including but not limited to health and fitness, consumer research, psychology and aging, environmental research, law and criminal justice, and entertainment.

Statistics is not about numbers; it is about data — numbers in context. It is the context that makes a problem meaningful and something worth considering. Examples and exercises with overly simple settings do not allow students to practice interpreting results in authentic situations, nor do they give students the experience necessary to be able to use statistical methods in real settings. We believe that the exercises and examples are a particular strength of this text, and we invite you to compare the examples and exercises with those in other introductory statistics texts.

■ Mathematical Level and Notational Simplicity

A good background in second-year algebra constitutes sufficient mathematical preparation for reading and understanding the material presented in this book. We want students to focus on concepts without having to grapple unnecessarily with the manipulation of formulas and symbols.

Table 5.6 ■ Power Transformation Ladder		
Power	**Transformed Value**	**Name**
3	(Original value)^3	Cube
2	(Original value)^2	Square
1	(Original value)	No transformation
$\frac{1}{2}$	$\sqrt{\text{Original value}}$	Square root
$\frac{1}{3}$	$\sqrt[3]{\text{Original value}}$	Cube root
0	Log(Original value)	Logarithm
−1	$\frac{1}{\text{Original value}}$	Reciprocal

Figure 5.31 is designed to suggest where on the ladder we should go to find an appropriate transformation. The four curved segments, labeled 1, 2, 3, and 4, represent shapes of curved scatterplots that are commonly encountered. Suppose that a scatterplot looks like the curve labeled 1. Then, to straighten the plot, we should use a power of x that is up the ladder from the no-transformation row (x^2 or x^3) and/or a power on y that is also up the ladder from the power 1. Thus, we might be led to squaring each x value, cubing each y, and plotting the transformed pairs. If the cur-

Figure 5.31 Scatterplot shapes and where to go on the transformation ladder to straighten the plot.

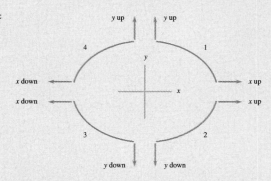

- **One-Sample t Test for a Population Mean**

Null hypothesis: $H_0: \mu = $ hypothesized value.

Test statistic: $t = \dfrac{\bar{x} - \text{hypothesized value}}{\dfrac{s}{\sqrt{n}}}$

Alternative Hypothesis:	**P-Value:**
$H_a: \mu > $ hypothesized value	Area to right of calculated t under t curve with df $= n - 1$
$H_a: \mu < $ hypothesized value	Area to the left of calculated t under t curve with df $= n - 1$
$H_a: \mu \neq $ hypothesized value	(1) 2(area to right of t) if t is positive, or (2) 2(area to left of t) if t is negative

Assumptions: 1. $\bar{x}$ and s are the sample mean and sample standard deviation, respectively, from a *random sample*.
2. The *sample size is large* (generally $n \geq 30$) or *the population distribution is at least approximately normal*.

To achieve this, we have sometimes used words and phrases in addition to and in place of symbols. For those who are apprehensive about their mathematical skills, we trust that the verbal descriptions not only are faithful to the statistical concepts but also are stepping stones to the more precise mathematical descriptions.

The Use of Technology

The computer has brought incredible statistical power to the desktop of every investigator. The wide availability of statistical computer packages (such as MINITAB, S-Plus, JMP, and SPSS) and the graphical capabilities of the modern microcomputer have transformed both the teaching and learning of statistics. To highlight the role of the computer in contemporary statistics, we have included sample outputs throughout the book.

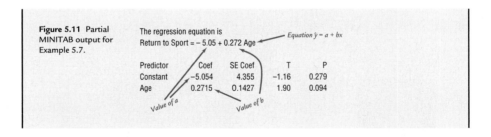

Figure 5.11 Partial MINITAB output for Example 5.7.

Most statistical computer packages and graphing calculators can calculate and report P-values for a variety of hypothesis-testing situations, including the large-sample test for a proportion. MINITAB was used to carry out the test of Example 10.10, and the resulting computer output follows (MINITAB uses p instead of π to denote the population proportion):

Test and Confidence Interval for One Proportion

Test of p = 0.5 vs p < 0.5

Sample	X	N	Sample p	95.0 % CI	Z-Value	P-Value
1	220	500	0.440000	(0.396491, 0.483509)	−2.68	0.004

From the MINITAB output, $z = -2.68$, and the associated P-value is .004. The small difference in the P-value is the result of rounding.

In addition, numerous exercises contain data that can easily be analyzed by computer, although our exposition firmly avoids a presupposition that students have access to a particular statistical package.

▪ Topic Coverage

Our book can be used in courses as short as one quarter or as long as one year. Particularly in shorter courses, an instructor will need to be selective in deciding which topics to include and which to set aside. The book divides naturally into four major sections: collecting data and descriptive methods (Chapters 1–5), probability material (Chapters 6–8), the basic one- and two-sample inferential techniques (Chapters 9–12), and more advanced inferential methods (Chapters 13–15). We include an early chapter (Chapter 5) on descriptive methods for bivariate numerical data. This early exposure raises questions and issues that should stimulate student interest in the subject; it is also advantageous for those teaching courses in which time constraints preclude covering advanced inferential material. However, this chapter can easily be postponed until the basics of inference have been covered and then combined with Chapter 13 for a unified treatment of regression and correlation.

With the possible exception of Chapter 5, Chapters 1–10 should be covered in order. We anticipate that most instructors will then continue with two-sample inference (Chapter 11) and methods for categorical data analysis (Chapter 12), although regression could be covered before either of these. Analysis of variance (Chapter 15) could be included before or after the regression material of Chapters 13 and 14.

▪ A Note on Probability

This book takes a brief and informal approach to probability, focusing on those concepts needed to understand the inferential methods covered in the later chapters. For those who prefer a more traditional approach to probability, the book *Introduction to Statistics and Data Analysis*, 2nd edition, by Roxy Peck, Chris Olsen, and Jay Devore (also published by Duxbury), may be a more appropriate choice. Except for the more formal treatment of probability and the inclusion of the optional Graphing Calculator Explorations, it parallels the material in this text.

▪ New in This Edition

There are a number of changes in the 5th edition, including the following:

- Hands-on activities have been added at the end of each chapter. These activities can be used as a chapter capstone or can be integrated into the text material of the chapter.

- **Activity 8.1:** Do Students Who Take the SATs Multiple Times Have an Advantage in College Admissions?

Technology activity: Requires use of a computer or a graphing calculator.

Background: The *Chronicle of Higher Education* (January 29, 2003) summarized an article that appeared on the *American Prospect* web site titled "College Try: Why Universities Should Stop Encouraging Applicants to Take the SAT's Over and Over Again." This paper argued that current college admission policies that permit applicants to take the SAT exam multiple times and then use the highest score for admission consideration favor students from families with higher incomes (who can afford to take the exam many times). The author proposed two alternatives that he believes would be fairer than using the highest score: (1) Use the average of all test scores, or (2) use only the most recent score.

In this activity, you will investigate the differences between the three possibilities by looking at the sampling distributions of three statistics for a test taker who takes the exam twice and for a test taker who takes the exam five times.

- The final section of most chapters is now titled "Communicating and Interpreting the Results of Statistical Analyses."

For effective communication with graphical displays, some things to remember are:

- Be sure to select a display that is appropriate for the given type of data.
- Be sure to include scales and labels on the axes of graphical displays.
- In comparative plots, be sure to include labels or a legend so that it is clear which parts of the display correspond to which samples or groups in the data set.
- Although it is sometimes a good idea to have axes that don't cross at $(0, 0)$ in a scatterplot, the vertical axis in a bar chart or a histogram should always start at 0 (see the cautions and limitations later in this section for more about this).
- Keep your graphs simple. A simple graphical display is much more effective than one that has a lot of extra "junk." Most people will not spend a great deal of time studying a graphical display, so its message should be clear and straightforward.
- Keep your graphical displays honest. People tend to look quickly at graphical displays, and so it is important that a graph's first impression is an accurate and honest portrayal of the data distribution. In addition to the graphical display itself, data analysis reports usually include a brief discussion of the features of the data distribution based on the graphical display.
- For categorical data, this discussion might be a few sentences on the relative proportion for each category, possibly pointing out categories that were either common or rare compared to other categories.
- For numerical data sets, the discussion of the graphical display usually summarizes the information that the display provides on three characteristics of the data distribution: center or location, spread, and shape.
- For bivariate numerical data, the discussion of the scatterplot would typically focus on the nature of the relationship between the two variables used to construct the plot.
- For data collected over time, any trends or patterns in the time-series plot would be described.

In addition to considering how to interpret statistical summaries found in journals and other published sources, these final sections have been expanded to include advice on how to best communicate results. Also new is a subsection titled "A Word to the Wise: Cautions and Limitations," which reminds readers of things that must be considered to ensure that statistical methods are used in reasonable and appropriate ways.

■ A Word to the Wise: Cautions and Limitations

There are several things you should watch for when conducting a hypothesis test or when evaluating a written summary of such a test.

1. The result of a hypothesis test can never show strong support for the null hypothesis. Make sure that you don't confuse "There is no reason to believe the null hypothesis is not true" with the statement "There is convincing evidence that the null hypothesis is true." These are very different statements!

2. If you have complete information for the population, don't carry out a hypothesis test! It should be obvious that no test is needed to answer questions about a population if you have complete information and don't need to generalize from a sample, but people sometimes forget this fact. For example, in an article on growth in the number of prisoners by state, the *San Luis Obispo Tribune* (August 13, 2001) reported "California's numbers showed a statistically insignificant change, with 66 fewer prisoners at the end of 2000." The use of the term "statistically insignificant" implies some sort of statistical inference, which is not appropriate when a complete accounting of the entire prison population is known. Perhaps the author confused statistical and practical significance. Which brings us to . . .

3. Don't confuse statistical significance with practical significance. When statistical significance has been declared, be sure to step back and evaluate the result in light of its practical importance. For example, we may be convinced that the proportion who respond favorably to a proposed medical treatment is greater than .4, the known proportion who respond favorably for the currently recommended treatments. But if our estimate of this proportion for the proposed treatment is .405, is this of any practical interest? It might be if the proposed treatment is less costly or has fewer side effects, but in other cases it may not be of any real interest. Results must always be interpreted in context.

- Coverage of experimental design and survey sampling has been expanded, and two new sections have been added to Chapter 2. Section 2.5 ("More on Experimental Design") discusses the use of a control group, the role of a placebo treatment, single-blind and double-blind experiments, and the use of volunteers as subjects in an experiment. Section 2.6 ("More on Observational Studies: Designing Surveys") addresses the challenges of planning a survey. In addition, Section 2.3 ("Statistical Studies: Observation and Experimentation") has been expanded to include a more complete discussion of the types of conclusions that can reasonably be drawn from statistical studies.

- Although the order of topics generally mirrors the data collection process with methods of data collection covered first, two graphical displays (dotplots and bar charts) have been moved to Chapter 1 so that these simple graphical analysis tools can be used in the conceptual development of experimental design and so that students have some tools for summarizing the data they collect through sampling and experimentation in the exercises, examples, and activities of Chapter 2.

- The coverage of scatterplots has been moved to Chapter 3, "Graphical Methods for Describing Data." This provides the necessary background for normal probability plots, even for those who choose to delay Chapter 5 ("Summarizing Bivariate Data") until after the chapters on basic inferential methods.

- Many new examples and exercises that use data from current journals and newspapers have been added. In addition, more of the exercises specifically ask students to write (e.g., explaining reasoning, interpreting results, and commenting on important features of an analysis).

■ Ancillaries

■ For Students

A Student Suite CD provided with the book includes the following items: applets; InfoTrac© College Edition and Internet Exercises; links to both the Book Companion Web Site and the Internet Companion for Statistics by Michael Larsen; data sets formatted for MINITAB, Microsoft Excel, SPSS, SAS, JMP, and ASCII; and technology manuals designed for use with MINITAB, Excel, Graphing Calculator, JMP Intro, and SPSS. Also available are a student solutions manual and a separate activities workbook.

■ For Instructors

The following supplements are available to qualified adopters. Please contact your local Thomson ■ Duxbury sales representative for details.

The Instructor's Suite CD includes everything featured on the Student Suite CD plus the Instructor's Solutions Manual for all exercises found in the text, Test Bank in Microsoft Word® format, and PowerPoint Presentations. Also available are BCA Testing and BCA Homework.

BCA Testing gives instructors the power to transform the learning and teaching experiences. BCA is fully integrated testing and course management software accessible to instructors and students anytime, anywhere. BCA uses correct statistical notation and is delivered in a browser-based format without the need for any proprietary software or plug-ins. Results flow automatically to a grade book for tracking so that instructors will be better able to assess student understanding of the material before class or before an actual test.

BCA Homework with DuxStat facilitates classroom management, finally allowing instructors to test the way they teach. DuxStat assesses students through homework, on quizzes, or on exams, in the process of doing real data analysis on the Web. Student responses are automatically graded and entered into the BCA Gradebook, making it easy for the instructor to assign and collect homework over the Web.

▪ Acknowledgments

We would like to express our thanks and gratitude to all who helped to make this book possible:

- Carolyn Crockett, our editor at Duxbury, for her good humor, patience, and, most of all, her support over the years.
- Ann Day and Rhonda Letts at Duxbury, for the development of all the ancillary materials details and for keeping us on track.
- Lisa Weber, our project manager at Brooks/Cole, and Gretchen Otto at G & S Typesetters, for artfully managing the myriad details associated with the production process.
- Mimi Braverman, for her careful copyediting.
- Mary Mortlock, for her diligence and care in producing the student and instructor solutions manuals for this book.
- Chris Olsen, Josh Tabor, and Peter Flannagan-Hyde, for producing a first-rate test bank to accompany the book.
- Beth Chance and Francisco Garcia, for producing the applet used in the confidence interval activities.
- Gary McClelland, for producing the applets from *Seeing Statistics* used in the regression activities.
- Lee Creighton, Roger Davis, Hasan Hamdan, Chris Olsen, and Charles Seiter, for producing the various technology manuals that can be found on the CD that accompanies this text.
- Gigi Williams, for checking the accuracy of the manuscript.
- David Crystal, for producing the PowerPoint presentation that can be found on the Instructor's Suite that accompanies this text.

And, as always, we thank our families, friends, and colleagues for their continued support.

Jay Devore
Roxy Peck

1 ▪ The Role of Statistics

Statistical methods for summary and analysis of data provide investigators with powerful tools for making sense out of data. Statistical techniques are being used with increasing frequency in business, medicine, agriculture, social sciences, natural sciences, and applied sciences, such as engineering. The pervasiveness of statistical analyses in such diverse fields has led to increased recognition that statistical literacy — a familiarity with the goals and methods of statistics — should be a basic component of a well-rounded educational program. In this chapter, we consider the nature and role of variability in statistical settings, introduce some basic terminology, and look at some simple graphical displays for summarizing data.

▪ 1.1 Three Reasons to Study Statistics

Because of the widespread use of statistical methods to organize, summarize, and draw conclusions from data, a familiarity with statistical techniques and statistical literacy in general is vital in today's society. It's a good idea for everyone to have a basic understanding of statistics, and many college majors require at least one course in statistics. There are three important reasons for this: (1) to be informed, (2) to understand issues and be able to make decisions, and (3) to be able to evaluate decisions that affect your life. Let's explore each reason in detail.

▪ The First Reason: Being Informed

In today's society, we are bombarded with numerical information in news, in advertisements, and even in conversation. How do we decide whether claims based on numerical information are reasonable? Consider the examples that appeared in one week's news (week of October 23, 2002).

- ▪ An analysis of data from a University of Utah study led to the conclusion that drivers engaged in cell phone conversations missed twice as many simulated traffic signals as drivers who were not talking and that cell phone users took longer to react to the signals they did detect. The researchers also found that these "driving deficits" were the same whether drivers used hands-free devices or held the phone to the ear. (*Los Angeles Times*, October 23, 2002)

- The Food and Drug Administration warned consumers about serious risks associated with the use of decorative contact lenses. These novelty lenses are imprinted with designs ranging from holiday decorations to sports logos. By analyzing data on the occurrence of corneal abrasions and other eye injuries, the FDA determined that use of these contact lenses, which generally have not been properly prescribed and fitted, can jeopardize vision. (*USA Today*, October 23, 2002)

- Results from a long-term study of cancer rates in people vaccinated for polio were reported. Between 1955 and 1963, about one-third of those vaccinated for polio received a vaccine that was contaminated with a virus called SV40. Recent experiments have shown that SV40 causes cancer and that SV40 has been detected in some types of rare cancer tumors. Researchers inferred that there has been no increase in cancer among people who received the contaminated vaccine, but they called for more research because the tumors associated with the SV40 virus are so rare that more information is needed to reach a definitive conclusion. (*USA Today*, October 23, 2002)

- Data on alcohol, tobacco, and marijuana use by junior and senior high school students in San Luis Obispo, California, were summarized. Graphs were used to show how the proportions of students who reported using each of the three substances increase from 7th to 11th grade and to confirm that these proportions did not change much between 1999 and 2001. (*San Luis Obispo Tribune*, October 25, 2002)

- In a summary of the "Human Footprint" report, the Wildlife Conservation Society and Columbia University described an analysis of the Earth's land surface and concluded that 83% of the land surface is used by humans for housing, farming, mining, or fishing. Among the few remaining wild areas are the northern forests of Alaska, Canada, and Russia, the high plateaus of Tibet, and parts of the Amazon River basin. (*USA Today*, October 23, 2002)

- A study that links hospital patient care and subsequent well-being to nurse staffing level was summarized in the *Journal of the American Medical Association*. Based on a survey of more than 10,000 nurses at 168 Pennsylvania hospitals and an analysis of the medical records of 232,342 patients who underwent routine surgery at these hospitals, the investigators concluded that surgery patients at hospitals with a severe shortage of nurses had a 31% greater risk of dying while in the hospital. (*USA Today* and *Los Angeles Times*, October 23, 2002)

- Water quality was compared for beaches in three Southern California counties. Based on data from water specimens taken at the beaches, the beaches were graded from A+ to F according to the risk of getting sick from swimming. It was reported that 70% of the 106 beaches in Orange County, 72% of the 82 beaches in Los Angeles County, and 90% of the 50 beaches in Ventura County received A grades. (*Los Angeles Times*, October 25, 2002)

- It was reported that, contrary to popular belief, both men and women tend to get more jealous over sexual rather than emotional infidelity. Previous studies had found that men tended to say it would be more upsetting to find out that their partners had been unfaithful, whereas women were more likely to say it would be more upsetting if their partner formed a strong emotional bond with someone else. However, in a new study involving 111 subjects, both men and women rated sexual infidelity as most upsetting, suggesting a change in attitudes over time. (*San Luis Obispo Tribune*, October 29, 2002)

- ▪ A study concluded that giving aspirin to heart patients soon after bypass surgery dramatically lowers the risk of death and complications. This conclusion was based on an experiment in which about 60% of 5065 patients who underwent bypass surgery received aspirin within 48 hours of surgery. The researchers found that those receiving aspirin were less likely to die in the hospital and less likely to suffer a heart attack, stroke, or kidney failure while in the hospital. (*San Luis Obispo Tribune*, October 24, 2002)

To be an informed consumer of such reports, you must be able to (1) extract information from charts and graphs, (2) follow numerical arguments, and (3) know the basics of how data should be gathered, summarized, and analyzed to draw statistical conclusions.

▪ The Second Reason: Understanding and Making Decisions

No matter what profession you choose, you will almost certainly need to understand statistical information and base decisions on it. Here are some examples:

- ▪ Almost all industries, as well as government and nonprofit organizations, use market research tools, such as consumer surveys, that are designed to provide information about who uses their products or services.

- ▪ Modern science and its applied fields, from astrophysics to zoology, rely on statistical methods for analyzing data and deciding whether various conjectures are supported by observed data. This is also true for the social sciences, such as economics and psychology. Even the liberal arts fields, such as literature and history, are beginning to use statistics as a research tool.

- ▪ In law or government, you may be called on to understand and debate statistical techniques used in another field. Class-action lawsuits can depend on a statistical analysis of whether one kind of injury or illness is more common in a particular group than in the general population. Proof of guilt in a criminal case may rest on statistical interpretation of the likelihood that DNA samples match.

Throughout your professional life, you will have to make informed decisions. To make these decisions, you must be able to do the following:

1. Decide whether existing information is adequate or whether additional information is required.
2. If necessary, collect more information in a reasonable and thoughtful way.
3. Summarize the available data in a useful and informative manner.
4. Analyze the available data.
5. Draw conclusions, make decisions, and assess the risk of an incorrect decision.

Cognitive psychologists have shown that people informally use these steps to make everyday decisions. Should you go out for a sport that involves the risk of injury? Will your college club do better by trying to raise funds with a benefit concert or with a direct appeal for donations? If you choose a particular major, what are your chances of finding a job when you graduate? How should you select a graduate program based on guidebook ratings that include information on percentage of applicants accepted, time to obtain a degree, and so on? The study of statistics formalizes the process of making decisions based on data and provides the tools for accomplishing the steps listed.

▪ The Third Reason: Evaluating Decisions That Affect Your Life

It is likely that you will need to make decisions based on data. Other people also use statistical methods to make decisions that affect you as an individual. An understanding of statistical techniques will allow you to question and evaluate decisions that affect your well-being. Some examples are:

- Insurance companies use statistical techniques to set auto insurance rates, although some states restrict the use of these techniques. Data suggest that young drivers have more accidents than older ones. Should laws or regulations limit how much more young drivers pay for insurance? What about the common practice of charging higher rates for people who live in urban areas?

- University financial aid offices survey students on the cost of going to school and collect data on family income, savings, and expenses. The resulting data are used to set criteria for deciding who receives financial aid. Are the estimates they use accurate?

- Medical researchers use statistical methods to make recommendations regarding the choice between surgical and nonsurgical treatment of such diseases as coronary heart disease and cancer. How do they weigh the risks and benefits to reach such a recommendation?

- Many companies now require drug screening as a condition of employment, but with these screening tests there is a risk of a false-positive reading (incorrectly indicating drug use) or a false-negative reading (failure to detect drug use). What are the consequences of a false result? Given the consequences, is the risk of a false result acceptable?

An understanding of elementary statistical methods can help you to evaluate whether important decisions such as the ones just mentioned are being made in a reasonable way.

We encounter data and conclusions based on data every day. **Statistics** is the scientific discipline that provides methods to help us make sense of data. Some people are suspicious of conclusions based on statistical analyses. Extreme skeptics, usually speaking out of ignorance, characterize the discipline as a subcategory of lying — something used for deception rather than for positive ends. However, we believe that statistical methods, used intelligently, offer a set of powerful tools for gaining insight into the world around us. We hope that this textbook will help you to understand the logic behind statistical reasoning, prepare you to apply statistical methods appropriately, and enable you to recognize when others are not doing so.

▪ 1.2 The Nature and Role of Variability

Statistics is the science of collecting, analyzing, and drawing conclusions from data. If we lived in a world where all measurements were identical for every individual, all three of these tasks would be simple. Imagine a population consisting of all students at a particular university. Suppose that *every* student took the same number of units, spent exactly the same amount of money on textbooks this semester, and favored increasing student fees to support expanding library services. For this population, there is *no* variability in the values of number of units, amount spent on books, or student opinion on the fee increase. A researcher studying a sample from this population to draw conclusions about these three variables would have a particularly easy task. It would not matter how many students the researcher included in the sample or how the sampled students were selected. In fact, the researcher could collect information on number of units, amount spent on books, and opinion on the fee increase by just stopping the next student who happened to walk by the library. Because there is no variability in the population, this one individual would provide complete and accurate information about the population, and the researcher could draw conclusions based on the sample with no risk of error.

The situation just described is obviously unrealistic. Populations with no variability are exceedingly rare, and they are of little statistical interest because they present no challenge! In fact, variability is almost universal. It is variability that makes life (and the life of a statistician, in particular) interesting. We need to understand variability to be able to collect, analyze, and draw conclusions from data in a sensible way. One of the primary uses of descriptive statistical methods is to increase our understanding of the nature of variability in a population.

Examples 1.1 and 1.2 illustrate how an understanding of variability is necessary to draw conclusions based on data.

▪ Example 1.1   If the Shoe Fits

The graphs in Figure 1.1 are examples of a type of graph called a histogram. (We will see how to construct histograms in Chapter 3.) Figure 1.1(a) shows the distribution of the heights of female basketball players who played at a particular university between 1990 and 1998. The height of each bar in the graph indicates how many players' heights were in the corresponding interval. For example, 40 basketball players had heights between 72 in. and 74 in., whereas only 2 players had heights between 66 in. and 68 in. Figure 1.1(b) shows the distribution of heights for members of the women's gymnastics team over the same period. Both histograms are based on the heights of 100 women.

The first histogram shows that the heights of female basketball players varied, with most heights falling between 68 in. and 76 in. In the second histogram we see that the heights of female gymnasts also varied, with most heights in the range of 60 in. to 72 in. It is also clear that there is more variation in the heights of the gymnasts than in the heights of the basketball players, because the gymnast histogram spreads out more about its center than does the basketball histogram.

Now suppose that a tall woman (5 ft 11 in.) tells you she is looking for her sister who is practicing with her team at the gym. Would you direct her to where the basketball team is practicing or to where the gymnastics team is practicing? What

Figure 1.1 Histograms of heights (in inches) of female athletes: (a) basketball players; (b) gymnasts.

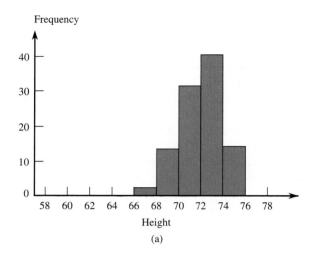

(a)

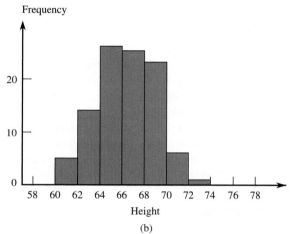

(b)

reasoning would you use to decide? What if you found a pair of size 6 tennis shoes left in the locker room? Would you first try to return them by checking with members of the basketball team or the gymnastics team?

You probably answered that you would send the woman looking for her sister to the basketball practice and that you would try to return the shoes to a gymnastics team member. To reach these conclusions, you informally used statistical reasoning that combined your own knowledge of the relationship between heights of siblings and between shoe size and height with the information about the distributions of heights presented in Figure 1.1. You might have reasoned that heights of siblings tend to be similar and that a height as great as 5 ft 11 in., although not impossible, would be unusual for a gymnast. On the other hand, a height as tall as 5 ft 11 in. would be a common occurrence for a basketball player. Similarly, you might have reasoned that tall people tend to have bigger feet and that short people tend to have smaller feet. The shoes found were a small size, so it is more likely that they belong to a gymnast than to a basketball player, because small heights and small feet are usual for gymnasts and unusual for basketball players.

■ Example 1.2   Monitoring Water Quality

As part of its regular water quality monitoring efforts, an environmental control board selects five water specimens from a particular well each day. The concentration of contaminants in parts per million (ppm) is measured for each of the five specimens, and then the average of the five measurements is calculated. The histogram in Figure 1.2 summarizes the average contamination values for 200 days.

Figure 1.2
Contaminant concentration (in parts per million) in well water.

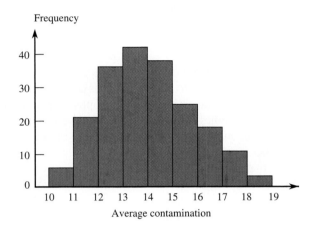

Now suppose that a chemical spill has occurred at a manufacturing plant about 1 mile from the well. It is not known whether a spill of this nature would contaminate groundwater in the area of the spill and, if so, whether a spill this distance from the well would affect the quality of well water.

One month after the spill, five water specimens are collected from the well, and the average contamination is 16 ppm. Considering the variation before the spill, would you take this as convincing evidence that the well water was affected by the spill? What if the calculated average was 18 ppm? 22 ppm? How is your reasoning related to the graph in Figure 1.2?

Before the spill, the average contaminant concentration varied from day to day. An average of 16 ppm would not have been an unusual value, and so seeing an average of 16 ppm after the spill isn't necessarily an indication that contamination has increased. On the other hand, an average as large as 18 ppm is less common, and an average as large as 22 ppm is not at all typical of the prespill values. In this case, we would probably conclude that the well contamination level has increased.

In these two examples, reaching a conclusion required an understanding of variability. Understanding variability allows us to distinguish between usual and unusual values. The ability to recognize unusual values in the presence of variability is the essence of most statistical procedures and is also what enables us to quantify the chance of being incorrect when a conclusion is based on sample data. These concepts will be developed further in subsequent chapters.

▪ 1.3 Statistics and Data Analysis

Statistical methods, used appropriately, allow us to draw reliable conclusions based on data. Data and conclusions based on data appear regularly in a variety of settings: newspapers, advertisements, magazines, and professional journals. In business, industry, and government, informed decisions are often data driven.

Statistics is the science of collecting, analyzing, and drawing conclusions from data.

Once data have been collected or once an appropriate data source has been identified, the next step in the data analysis process usually involves organizing and summarizing the information. Tables, graphs, and numerical summaries allow increased understanding and provide an effective way to present data. Methods for organizing and summarizing data make up the branch of statistics called **descriptive statistics**.

After the data have been summarized, we often wish to draw conclusions or make decisions based on the data. This usually involves generalizing from a small group of individuals or objects that we have studied to a much larger group.

For example, the admissions director at a large university might be interested in learning why some applicants who were accepted for the fall 2004 term failed to enroll at the university. The population of interest to the director consists of all accepted applicants who did not enroll in the fall 2004 term. Because this population is large and because it may be difficult to contact all the individuals, the director might be able to collect data from only 300 selected students. These 300 students constitute a sample.

▪ **Definition**

The entire collection of individuals or objects about which information is desired is called the **population** of interest. A **sample** is a subset of the population, selected for study in some prescribed manner.

The second major branch of statistics, **inferential statistics**, involves generalizing from a sample to the population from which it was selected. When we generalize in this way, we run the risk of an incorrect conclusion, because a conclusion about the population is based on incomplete information. An important aspect in the development of inferential techniques involves quantifying the chance of an incorrect conclusion.

Considering some examples will help you develop a preliminary appreciation for the scope and power of statistical methods. In Examples 1.3–1.5, we describe three problems that can be investigated using techniques presented in this textbook.

▪ Example 1.3   Student Opinion Survey

A university has recently implemented a new registration system. Students interact online with the computer to select classes for the term. To assess student opinion regarding the effectiveness of the system, a university research team wants to conduct a survey. Each student in a sample of 400 will be asked a variety of questions (such as the number of units received and the number of times the system was accessed before the registration process was completed). The survey will yield a rather large and unwieldy data set. To make sense out of the raw data and to describe student responses, the researchers must summarize the data. This would also make the results more accessible to others. Descriptive techniques can be used to accomplish this task. In addition, inferential methods can be employed to draw various conclusions about the experiences of *all* students who used the registration system.

▪ Example 1.4   Depression and Cholesterol Levels

A study linking depression to low cholesterol levels was described in an Associated Press article (*San Luis Obispo Telegram Tribune*, June 23, 1995). Researchers at a hospital in Italy compared the average cholesterol level for a sample of 331 patients who had been admitted to the hospital after a suicide attempt and who had been diagnosed with clinical depression to the average cholesterol level of 331 patients admitted to the hospital for other reasons. Statistical techniques were used to analyze the data and to show that the average cholesterol level was lower for the depressed group. The article correctly noted that because of the way in which the data were collected, it was not possible to determine from the statistical analysis alone whether a causal relationship between cholesterol level and psychological state exists — that is, whether low cholesterol levels affect psychological state or vice versa.

▪ Example 1.5   Predicting the Spread of a Forest Fire

A final example comes from the discipline of forestry. When a fire occurs in a forested area, decisions must be made about the best way to combat the fire. One possibility is to try to contain the fire by building a fire line. If building a fire line requires 4 hr, deciding where the line should be built involves making a prediction of how far the fire will spread during this period. Many factors must be taken into account, including wind speed, temperature, humidity, and time elapsed since the last rainfall. Statistical techniques make it possible to develop a model for the prediction of fire spread, using information available from past fires.

▪ Exercises 1.1–1.7

1.1 Give a brief definition of the terms *descriptive statistics* and *inferential statistics*.

1.2 Give a brief definition of the terms *population* and *sample*.

1.3 The student senate at a university with 15,000 students is interested in the proportion of students who favor a change in the grading system to allow for plus and minus grades (e.g., B+, B, B−, rather than just B). Two hundred students are interviewed to determine their attitude toward this proposed change. What is the population of interest? What group of students constitutes the sample in this problem?

1.4 The supervisors of a rural county are interested in the proportion of property owners who support the construction of a sewer system. Because it is too costly to contact all 7000 property owners, a survey of 500 owners (selected at random) is undertaken. Describe the population and sample for this problem.

1.5 Representatives of the insurance industry wished to investigate the monetary loss resulting from earthquake damage to single-family dwellings in Northridge, California, in January 1994. From the set of all single-family homes in Northridge, 100 homes were selected for inspection. Describe the population and sample for this problem.

1.6 A consumer group conducts crash tests of new model cars. To determine the severity of damage to 2003 Mazda 626s resulting from a 10-mph crash into a concrete wall, the research group tests six cars of this type and assesses the amount of damage. Describe the population and sample for this problem.

1.7 A building contractor has a chance to buy an odd lot of 5000 used bricks at an auction. She is interested in determining the proportion of bricks in the lot that are cracked and therefore unusable for her current project, but she does not have enough time to inspect all 5000 bricks. Instead, she checks 100 bricks to determine whether each is cracked. Describe the population and sample for this problem.

▪ 1.4 Types of Data and Some Simple Graphical Displays

Every discipline has its own particular way of using common words, and statistics is no exception. You will recognize some of the terminology from previous math and science courses, but much of the language of statistics will be new to you.

▪ Describing Data

The individuals or objects in any particular population typically possess many characteristics that might be studied. Consider a group of students currently enrolled in a statistics course. One characteristic of the students in the population is the brand of calculator owned (Casio, Hewlett-Packard, Sharp, Texas Instruments, and so on). Another characteristic is the number of textbooks purchased, and yet another is the distance from the university to each student's permanent residence. A **variable** is any characteristic whose value may change from one individual or object to another. For example, *calculator brand* is a variable, and so are *number of textbooks purchased* and *distance to the university*. **Data** result from making observations either on a single variable or simultaneously on two or more variables.

A **univariate data set** consists of observations on a single variable made on individuals in a sample or population. There are two types of univariate data sets: categorical and numerical. In the previous example, *calculator brand* is a categorical variable, because each student's response to the query, "What brand of calculator do you own?" is a category. The collection of responses from all these students forms a **categorical data set**. The other two attributes, *number of textbooks purchased* and *distance to the university*, are both numerical in nature. Determining the value of such a numerical variable (by counting or measuring) for each student results in a **numerical data set**.

■ **Definition**

A data set consisting of observations on a single attribute is a **univariate data set**. A univariate data set is **categorical** (or **qualitative**) if the individual observations are categorical responses. A univariate data set is **numerical** (or **quantitative**) if each observation is a number.

■ **Example 1.6**   Airline Safety Violations

The Federal Aviation Administration (FAA) monitors airlines and can take administrative actions for safety violations. Information about the fines assessed by the FAA appeared in the article "Just How Safe Is That Jet?" (*USA Today*, March 13, 2000). Violations that could lead to a fine were categorized as Security (S), Maintenance (M), Flight Operations (F), Hazardous Materials (H), or Other (O). Data for the variable *type of violation* for 20 administrative actions are given in the following list (these data are a subset of the data described in the article, but they are consistent with summary values given in the paper; for a description of the full data set, see Exercise 1.16):

S	S	M	H	M	O	S	M	S	S
F	S	O	M	S	M	S	M	S	M

Because *type of violation* is a categorical (nonnumerical) response, this is a categorical data set.

In Example 1.6, the data set consisted of observations on a single variable (*type of violation*), so this is univariate data. In some studies, attention focuses simultaneously on two different attributes. For example, both height (in inches) and weight (in pounds) might be recorded for each individual in a group. The resulting data set consists of pairs of numbers, such as (68, 146). This is called a **bivariate data set**. **Multivariate data** result from obtaining a category or value for each of two or more attributes (so bivariate data are a special case of multivariate data). For example, multivariate data would result from determining height, weight, pulse rate, and systolic blood pressure for each individual in a group. Example 1.7 illustrates a bivariate data set.

■ **Example 1.7**   Revisiting Airline Safety Violations

The same article referenced in Example 1.6 ("Just How Safe Is That Jet?" *USA Today*, March 13, 2000) gave data on both the number of violations and the average fine per violation for the period 1985–1998 for 10 major airlines. The resulting data are given in the following table:

Airline	Number of Violations	Average Fine per Violation ($)
Alaska	258	5038.760
America West	257	3112.840
American	1745	2693.410
Continental	973	5755.396
Delta	1280	3828.125
Northwest	1097	2643.573
Southwest	535	3925.234
TWA	642	2803.738
United	1110	2612.613
US Airways	891	3479.237

Each of the variables considered here is numerical (rather than categorical) in nature. This is an example of a bivariate numerical data set.

▪ Two Types of Numerical Data

With numerical data, it is useful to make a further distinction between *discrete* and *continuous* numerical data. Visualize a number line (Figure 1.3) for locating values of the numerical variable being studied. Every possible number (2, 3.125, 8.12976, etc.) corresponds to exactly one point on the number line. Now suppose that the variable of interest is the number of cylinders of an automobile engine. The possible values of 4, 6, and 8 are identified in Figure 1.4(a) by the dots at the points marked 4, 6, and 8. These possible values are isolated from one another on the line; around any possible value, we can place an interval that is small enough that no other possible value is included in the interval. On the other hand, the line segment in Figure 1.4(b) identifies a plausible set of possible values for the time it takes a car to travel one-quarter mile. Here the possible values make up an entire interval on the number line, and no possible value is isolated from the other possible values.

Figure 1.3 A number line.

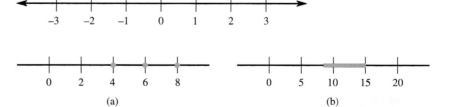

Figure 1.4 Possible values of a variable: (a) number of cylinders; (b) quarter-mile time.

▪ Definition

Numerical data are **discrete** if the possible values are isolated points on the number line. Numerical data are **continuous** if the set of possible values forms an entire interval on the number line.

Discrete data usually arise when each observation is determined by counting (e.g., the number of classes for which a student is registered or the number of petals on a certain type of flower).

■ Example 1.8   Calls to a Drug Abuse Hotline

The number of telephone calls per day to a drug abuse hotline is recorded for 12 days. The resulting data set is

$$3 \quad 0 \quad 4 \quad 3 \quad 1 \quad 0 \quad 6 \quad 2 \quad 0 \quad 0 \quad 1 \quad 2$$

Possible values for the variable *number of calls* are 0, 1, 2, 3, . . . ; these are isolated points on the number line, so we have a sample consisting of discrete numerical data.

The observations on the variable *number of violations* in Example 1.7 are also an example of discrete numerical data. However, in the same example, the variable *average fine per violation* could be 3000, 3000.1, 3000.125, 3000.12476, or any other value in an entire interval, so the observations on this variable provide an example of continuous data. Other examples of continuous data result from determining task completion times, body temperatures, and package weights.

In general, data are continuous when observations involve making measurements, as opposed to counting. In practice, measuring instruments do not have infinite accuracy, so possible measured values, strictly speaking, do not form a continuum on the number line. However, any number in the continuum *could* be a value of the variable. The distinction between discrete and continuous data will be important in our discussion of probability models.

■ Frequency Distributions and Bar Charts for Categorical Data

An appropriate graphical or tabular display of data can be an effective way to summarize and communicate information. When the data set is categorical, a common way to present the data is in the form of a table, called a *frequency distribution*.

A **frequency distribution for categorical data** is a table that displays the possible categories along with the associated frequencies or relative frequencies.

The **frequency** for a particular category is the number of times the category appears in the data set.

The **relative frequency** for a particular category is the fraction or proportion of the time that the category appears in the data set. It is calculated as

$$\text{relative frequency} = \frac{\text{frequency}}{\text{number of observations in the data set}}$$

When the table includes relative frequencies, it is sometimes referred to as a **relative frequency distribution**.

■ **Example 1.9**   Preferred Leisure Activities

Many public health efforts are directed toward increasing levels of physical activity. The article "Physical Activity in Urban White, African American, and Mexican American Women" (*Medicine and Science in Sports and Exercise* [1997]: 1608–1614) reported on physical activity patterns in urban women. The accompanying data set gives the preferred leisure-time physical activity for each of 30 Mexican American women. The following coding is used: W = walking, T = weight training, C = cycling, G = gardening, A = aerobics.

W	T	A	W	G	T	W	W	C	W
T	W	A	T	T	W	G	W	W	C
A	W	A	W	W	W	T	W	W	T

The corresponding frequency distribution is given in Table 1.1.

Table 1.1 ■ Frequency Distribution for Preferred Activity

Category	Frequency	Relative Frequency
Walking	15	.500 ← 15/30
Weight training	7	.233 ← 7/30
Cycling	2	.067
Gardening	2	.067
Aerobics	4	.133
	30	1.000

Total number of observations

Should total 1, but in some cases may be slightly off due to rounding

From the frequency distribution, we can see that 15 women indicated a preference for walking and that this was by far the most popular response. Equivalently, using the relative frequencies, we can say that .5 (half or 50%) of the women preferred walking.

A frequency distribution gives a tabular display of a data set. It is also common to display categorical data graphically. A bar chart is one of the most widely used types of graphical displays for categorical data.

■ Bar Charts

A **bar chart** is a graph of the frequency distribution of categorical data. Each category in the frequency distribution is represented by a bar or rectangle, and the picture is constructed in such a way that the *area* of each bar is proportional to the corresponding frequency or relative frequency.

▪ **Example 1.10   Revisiting Preferred Leisure Activity**

Example 1.9 gave data on preferred leisure-time physical activity for a sample of 30 Mexican American women. Figure 1.5 shows the bar chart corresponding to the frequency distribution constructed for these data (Table 1.1). Note that the bar chart in Figure 1.5 orders the categories alphabetically, whereas the frequency distribution listed the categories in a different order. When constructing a bar graph, the order in which the categories are listed generally does not matter. The three most common choices are (1) alphabetical ordering, (2) ordering by frequency, with the most commonly occurring category first and so on, and (3) ordering to match the order in an accompanying frequency distribution.

Figure 1.5 Bar chart of preferred leisure activity.

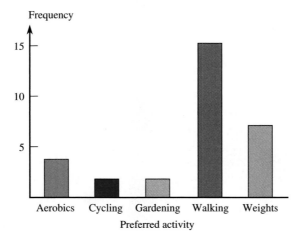

The bar chart provides a visual representation of the information in the frequency distribution. From the bar chart, it is easy to see that walking occurred most often in the data set, followed by weight training. The bar for walking is about twice as tall (and therefore has twice the area) as the bar for weight training, because approximately twice as many women preferred walking to weight training.

■ **Example 1.11**   Why Students Drop Out

The article "So Close, Yet So Far: Predictors of Attrition in College Seniors" (*Journal of College Student Development* [1998]: 343–348) examined the reasons that college seniors leave their college programs before graduating. Forty-two college seniors at a large public university who dropped out before graduation were interviewed and asked the main reason for discontinuing enrollment at the university. Data consistent with that given in the article are summarized in the following frequency distribution:

Reason for Leaving the University	Frequency
Academic problems	7
Poor advising or teaching	3
Needed a break	2
Economic reasons	11
Family responsibilities	4
To attend another school	9
Personal problems	3
Other	3

The corresponding bar chart is shown in Figure 1.6. From the bar chart, it is easy to see that more students reported leaving the university for economic reasons or to attend another school than for academic reasons.

Figure 1.6 Bar chart for the data of Example 1.11, reason for leaving university.

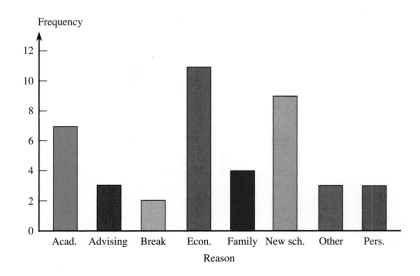

▪ Dotplots for Numerical Data

A dotplot is a simple way to display numerical data when the data set is reasonably small. Each observation is represented by a dot above the location corresponding to its value on a horizontal measurement scale. When a value occurs more than once, there is a dot for each occurrence and these dots are stacked vertically.

▪ Dotplots

When to Use
Small numerical data sets.

How to Construct

1. Draw a horizontal line and mark it with an appropriate measurement scale.

2. Locate each value in the data set along the measurement scale, and represent it by a dot. If there are two or more observations with the same value, stack the dots vertically.

What to Look For
Dotplots convey information about a representative or typical value in the data set, the extent to which the data values spread out, the nature of the distribution of values along the number line, and the presence of unusual values in the data set.

▪ Example 1.12   Graduation Rates for NCAA Division I Schools in California and Texas

The Chronicle of Higher Education (Almanac Issue, August 31, 2001) reported graduation rates for NCAA Division I schools. The rates reported are the percentages of full-time freshmen in fall 1993 who had earned a bachelor's degree by August 1999. Data from the two largest states (California, with 20 Division I schools; and Texas, with 19 Division I schools) are given in the following list:

California:	64	41	44	31	37	73	72	68	35	37
	81	90	82	74	79	67	66	66	70	63
Texas:	67	21	32	88	35	71	39	35	71	63
	12	46	35	39	28	65	25	24	22	

MINITAB, a computer software package for statistical analysis, was used to construct a dotplot of the 39 graduation rates. This dotplot is given in Figure 1.7. From the dotplot, we can see that graduation rates varied a great deal from school to school and that the graduation rates seem to form two distinguishable groups of about the same size — one group with higher graduation rates and one with lower graduation rates.

Figure 1.7 Dotplot of graduation rates.

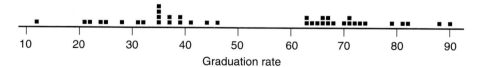

Graduation rate

Figure 1.8 Dotplot of graduation rates for California and Texas.

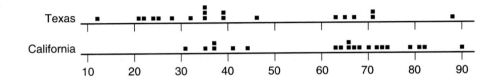

Figure 1.8 shows separate dotplots for the California and Texas schools. The dotplots are drawn using the same scale to facilitate comparisons. From the two plots, we can see that, although both states have a high group and a low group, there are only six schools in the low group for California and only six schools in the high group for Texas.

▪ Exercises 1.8–1.19

1.8 Classify each of the following attributes as either categorical or numerical. For those that are numerical, determine whether they are discrete or continuous.
a. Number of students in a class of 35 who turn in a term paper before the due date
b. Gender of the next baby born at a particular hospital
c. Amount of fluid (in ounces) dispensed by a machine used to fill bottles with soda pop
d. Thickness of the gelatin coating of a vitamin E capsule
e. Birth order classification (only child, firstborn, middle child, lastborn) of a math major

1.9 Classify each of the following attributes as either categorical or numerical. For those that are numerical, determine whether they are discrete or continuous.
a. Brand of computer purchased by a customer
b. State of birth for someone born in the United States
c. Price of a textbook
d. Concentration of a contaminant (micrograms per cubic centimeter) in a water sample
e. Zip code (Think carefully about this one.)
f. Actual weight of coffee in a 1-lb can

1.10 For the following numerical attributes, state whether each is discrete or continuous.
a. The number of insufficient-funds checks received by a grocery store during a given month
b. The amount by which a 1-lb package of ground beef decreases in weight (because of moisture loss) before purchase

c. The number of New York Yankees during a given year who will not play for the Yankees the next year
d. The number of students in a class of 35 who have purchased a used copy of the textbook
e. The length of a 1-year-old rattlesnake
f. The altitude of a location in California selected randomly by throwing a dart at a map of the state
g. The distance from the left edge at which a 12-in. plastic ruler snaps when bent sufficiently to break
h. The price per gallon paid by the next customer to buy gas at a particular station

1.11 For each of the following situations, give a set of possible data values that might arise from making the observations described.
a. The manufacturer for each of the next 10 automobiles to pass through a given intersection is noted.
b. The grade point average for each of the 15 seniors in a statistics class is determined.
c. The number of gas pumps in use at each of 20 gas stations at a particular time is determined.
d. The actual net weight of each of 12 bags of fertilizer having a labeled weight of 50 lb is determined.
e. Fifteen different radio stations are monitored during a 1-hr period, and the amount of time devoted to commercials is determined for each.

1.12 *Spider-Man* and *Star Wars: Episode II* were the top moneymakers among the summer 2002 movie releases. Box office totals for the top summer films in 2002 are given in the following table (*USA Today,* September 3, 2002):

Film	Box Office (millions of dollars)
Spider-Man	403.7
Star Wars: Episode II	300.1
Austin Powers in Goldmember	203.5
Signs	195.1
Men in Black II	189.7
Scooby-Doo	151.6
Lilo & Stitch	141.8
Minority Report	130.1
Mr. Deeds	124.0
XXX	123.9
The Sum of All Fears	118.5
The Bourne Identity	118.1
Road to Perdition	99.1
My Big Fat Greek Wedding	82.3
Spirit: Stallion of the Cimarron	73.2
Spy Kids 2: Island of Lost Dreams	69.1
Divine Secrets of the Ya-Ya Sisterhood	68.8
Insomnia	67.0
Stuart Little 2	61.9
Unfaithful	52.8

Use a dotplot to display these data. Write a few sentences commenting on the notable features of the dotplot.

1.13 Water quality ratings of 36 Southern California beaches were given in the article "How Safe Is the Surf?" (*Los Angeles Times*, October 26, 2002). The ratings, which ranged from A+ to F and which reflect the risk of getting sick from swimming at a particular beach, are given in the following list:

A+	A+	A+	F	A+	A+
A	B	A	C	C	A−
A	F	B	A+	C	A
D	F	A+	D	A+	A
D	A	A	D	A+	A+
A	A	C	F−	B	A+

a. Summarize the given ratings by constructing a relative frequency distribution and a bar chart. Comment on the interesting features of your bar chart.
b. Would it be appropriate to construct a dotplot for these data? Why or why not?

1.14 Many adolescent boys aspire to be professional athletes. The paper "Why Adolescent Boys Dream of Becoming Professional Athletes" (*Psychological Reports* [1999]: 1075–1085) examined some of the reasons. Each boy in a sample of teenage boys was asked the following question: "Previous studies have shown that more teenage boys say that they are considering becoming professional athletes than any other occupation. In your opinion, why do these boys want to become professional athletes?" The resulting data are shown in the following table:

Response	Frequency
Fame and celebrity	94
Money	56
Attract women	29
Like sports	27
Easy life	24
Don't need an education	19
Other	19

Construct a bar chart to display these data.

1.15 The article "Knee Injuries in Women Collegiate Rugby Players" (*American Journal of Sports Medicine* [1997]: 360–362) reported the following data on type of injury sustained by 13 female rugby players (MCL, ACL, and the meniscus are different sites in the knee, and the patella is the kneecap). These data are a subset of the data given in the article:

meniscus tear	patella dislocation	MCL tear
meniscus tear	ACL tear	meniscus tear
meniscus tear	MCL tear	meniscus tear
ACL tear	patella dislocation	MCL tear
MCL tear		

Summarize these data in a frequency distribution, and then construct a bar chart for this data set. Write a sentence or two describing the relative occurrence of the different types of injury.

1.16 The article "Just How Safe Is That Jet?" (*USA Today*, March 13, 2000) gave the following relative frequency distribution that summarized data on the type of violation for fines imposed on airlines by the Federal Aviation Administration:

Type of Violation	Relative Frequency
Security	.43
Maintenance	.39
Flight operations	.06
Hazardous materials	.03
Other	.09

Use this information to construct a bar chart for type of violation, and then write a sentence or two commenting on the relative occurrence of the various types of violation.

1.17 The article "Americans Drowsy on the Job and The Road" (Associated Press, March 28, 2001) summarized data from the 2001 Sleep in America poll. Each individual in a sample of 1004 adults was asked questions about his or her sleep habits. The article states that "40 percent of those surveyed say they get sleepy on the job and their work suffers at least a few days each month, while 22 percent said the problems occur a few days each week. And 7 percent say sleepiness on the job is a daily occurrence." Assuming that everyone else reported that sleepiness on the job was not a problem, summarize the given information by constructing a relative frequency bar chart.

1.18 "Ozzie and Harriet Don't Live Here Anymore" (*San Luis Obispo Tribune*, February 26, 2002) is the title of an article that looked at the changing makeup of America's suburbs. The article states that nonfamily households (e.g., homes headed by a single professional or an elderly widow) now outnumber married couples with children in suburbs of the nation's largest metropolitan areas. The article goes on to state:

> In the nation's 102 largest metropolitan areas, "nonfamilies" comprised 29 percent of households in 2000, up from 27 percent in 1990. While

the number of married-with-children homes grew too, the share did not keep pace. It declined from 28 percent to 27 percent. Married couples without children at home live in another 29 percent of suburban households. The remaining 15 percent are single-parent homes.

Use the given information on type of household in 2000 to construct a frequency distribution and a bar chart. (Be careful to extract the 2000 percentages from the given information).

1.19 Each year *U.S. News and World Report* publishes a ranking of U.S. business schools. The following data give the acceptance rates (percentage of applicants admitted) for the best 25 programs in the most recent survey:

16.3	12.0	25.1	20.3	31.9	20.7	30.1	19.5	36.2
46.9	25.8	36.7	33.8	24.2	21.5	35.1	37.6	23.9
17.0	38.4	31.2	43.8	28.9	31.4	48.9		

Construct a dotplot, and comment on the interesting features of the plot.

■ Activity 1.1: Head Sizes: Understanding Variability

Materials needed: Each team will need a measuring tape.

For this activity, you will work in teams of 6 to 10 people.

1. Designate a team leader for your team by choosing the person on your team who celebrated his or her last birthday most recently.

2. The team leader should measure and record the head size (measured as the circumference at the widest part of the forehead) of each of the other members of his or her team.

3. Record the head sizes for the individuals on your team as measured by the team leader.

4. Next, each individual on the team should measure the head size of the team leader. Do not share your measurement with the other team members until all team members have measured the team leader's head size.

5. After all team members have measured the team leader's head, record the different team leader head size measurements obtained by the individuals on your team.

6. Using the data from Step 3, construct a dotplot of the team leader's measurements of team head sizes. Then, using the same scale, construct a separate

dotplot of the different measurements of the team leader's head size (from Step 5).

Now use the available information to answer the following questions:

7. Do you think the team leader's head size changed in between measurements? If not, explain why the measurements of the team leader's head size are not all the same.

8. Which data set was more variable — head size measurements of the different individuals on your team or the different measurements of the team leader's head size? Explain the basis for your choice.

9. Consider the following scheme (you don't actually have to carry this out): Suppose that a group of 10 people measured head sizes by first assigning each person in the group a number between 1 and 10. Then person 1 measured person 2's head size, person 2 measured person 3's head size, and so on, with person 10 finally measuring person 1's head size. Do you think that the resulting head size measurements would be more variable, less variable, or show about the same amount of variability as a set of 10 measurements resulting from a single individual measuring the head size of all 10 people in the group? Explain.

▪ Activity 1.2: Estimating Sizes

1. Construct an activity sheet that consists of a table that has 6 columns and 10 rows. Label the columns of the table with the following six headings: (1) Shape, (2) Estimated Size, (3) Actual Size, (4) Difference (Est. − Actual), (5) Absolute Difference, and (6) Squared Difference. Enter the numbers from 1 to 10 in the "Shape" column.

2. Next you will be visually estimating the sizes of the shapes in Figure 1.9. Size will be described as the number of squares of this size

that would fit in the shape. For example, the shape

would be size 3, as illustrated by

You should now quickly *visually* estimate the sizes of the shapes in Figure 1.9. *Do not* draw on the figure — these are to be quick visual estimates. Record your estimates in the "Estimated Size" column of the activity sheet.

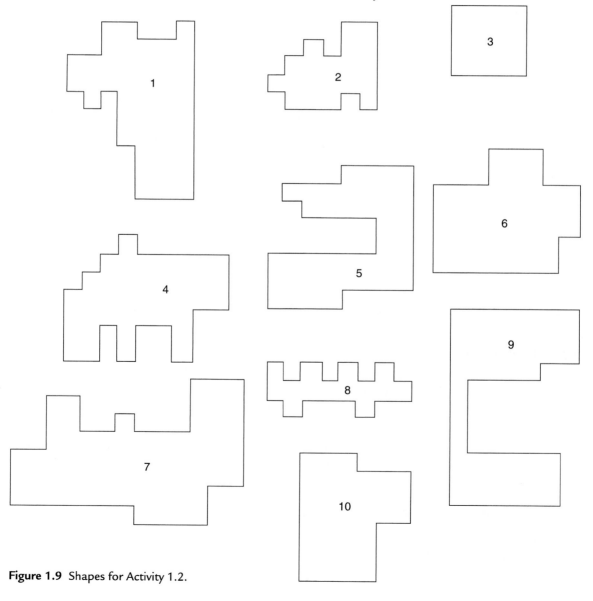

Figure 1.9 Shapes for Activity 1.2.

3. Your instructor will provide the actual sizes for the 10 shapes, which should be entered into the "Actual Size" column of the activity sheet. Now complete the "Difference" column by subtracting the actual value from your estimate for each of the 10 shapes.

4. What would cause a difference to be negative? positive?

5. Would the sum of the differences tell you if the estimates and actual values were in close agreement? Does a sum of 0 for the differences indicate that all the estimates were equal to the actual value? Explain.

6. Compare your estimates with those of another person in the class by comparing the sum of the absolute values of the differences between estimates and corresponding actual values. Who was better at estimating shape sizes? How can you tell?

7. Use the last column of the activity sheet to record the squared differences (e.g., if the difference for shape 1 was -3, the squared difference would be $(-3)^2 = 9$). Explain why the sum of the squared differences can also be used to assess how accurate your shape estimates were.

8. For this step, work with three or four other students from your class. For each of the 10 shapes, form a new size estimate by computing the average of the size estimates for that shape made by the individuals in your group. Is this new set of estimates more accurate than your own individual estimates were? How can you tell?

9. Does your answer from Step 8 surprise you? Explain why or why not.

■ Summary of Key Concepts and Formulas

Term or Formula	Comment
Descriptive statistics	Numerical, graphical, and tabular methods for organizing and summarizing data.
Population	The entire collection of individuals or measurements about which information is desired.
Sample	A part of the population selected for study.
Categorical data	Individual observations are categorical responses (nonnumerical).
Numerical data	Individual observations are numerical (quantitative) in nature.
Discrete numerical data	Possible values are isolated points along the number line.
Continuous numerical data	Possible values form an entire interval along the number line.
Bivariate and multivariate data	Each observation consists of two (bivariate) or more (multivariate) responses or values.
Frequency distribution for categorical data	A table that displays frequencies, and sometimes relative frequencies, for each of the possible values of a categorical variable.
Bar chart	A graph of a frequency distribution for a categorical data set. Each category is represented by a bar, and the area of the bar is proportional to the corresponding frequency or relative frequency.
Dotplot	A picture of numerical data in which each observation is represented by a dot on or above a horizontal measurement scale.

■ Supplementary Exercises 1.20–1.24

1.20 The paper "Profile of Sport/Leisure Injuries Treated at Emergency Rooms of Urban Hospitals" (*Canadian Journal of Sports Science* [1991]: 99–102) classified noncontact sports injuries by sport, resulting in the following table:

Sport	Number of Sport Injuries
Touch football	38
Soccer	24
Basketball	19
Baseball/softball	11
Jogging/running	11
Bicycling	11
Volleyball	17
Other	47

Calculate relative frequencies and draw the corresponding bar chart.

1.21 Nonresponse is a common problem facing researchers who rely on mail questionnaires. In the paper "Reasons for Nonresponse on the Physicians' Practice Survey" (*1980 Proceedings of the Section on Social Statistics* [Alexandria, VA: American Statistical Association, 1980]: 202), 811 doctors who did not respond to the AMA Survey of Physicians were contacted about the reason for their nonparticipation. The results are summarized in the accompanying relative frequency distribution. Draw the corresponding bar chart.

Reason	Relative Frequency
1. No time to participate	.264
2. Not interested	.300
3. Don't like surveys in general	.145
4. Don't like this particular survey	.025
5. Hostility toward the government	.054
6. Desire to protect privacy	.056
7. Other reason for refusal	.053
8. No reason given	.103

1.22 The paper "Fire Doors: A Potential Weak Link in the Protection Chain" (*Fire Technology* [1992]: 177–179) reported on a survey of companies that use fire doors to subdivide major plant facilities in case of uncontrolled fires. Four types of doors were in use: rolling steel (R), single metal-clad sliding (M), swinging with door closer (C), and single swinging (S). Suppose that the data were as shown in the following list (percentages agree with those given in the paper):

R	M	M	S	R	R	C	R	M	R
M	M	M	C	S	R	R	C	C	M
R	R	C	S	R	C	C	M	S	M
M	M	R	S	R	R	C	C	S	R
M	C	R	M	C	M	R	C	M	S

Determine the frequencies and the relative frequencies for the four categories, and display them in a relative frequency table. Construct a bar graph for this data set, and write a sentence or two commenting on the interesting features of this display.

1.23 The article "Can We Really Walk Straight?" (*American Journal of Physical Anthropology* [1992]: 19–27) reported on an experiment in which each of 20 healthy men was asked to walk as straight as possible to a target 60 m away at normal speed. Consider the following observations on cadence (number of strides per second):

0.95	0.85	0.92	1.95	0.93	1.86	1.00
0.92	0.85	0.81	0.78	0.93	0.93	1.05
0.93	1.06	1.06	0.96	0.81	0.96	

Construct a dotplot for the cadence data. Do any data values stand out as being unusual?

1.24 A sample of 50 individuals who recently joined a certain travel club yielded the responses on occupation shown in the following table (C = clerical, M = manager/executive, P = professional, R = retired, S = sales, T = skilled tradesperson, O = other):

P	R	R	M	S	R	P	R	C	P
M	R	T	O	R	S	R	S	P	M
R	P	C	R	R	S	T	P	P	C
M	S	R	R	P	R	S	R	R	P
M	P	R	R	S	C	P	R	S	M

Summarize these data using a graphical display.

■ References

Moore, David. *Statistics: Concepts and Controversies*, 4th ed. New York: W. H. Freeman, 1997. (A nice, informal survey of statistical concepts and reasoning.)

Tanur, Judith, ed. *Statistics: A Guide to the Unknown*. Belmont, CA: Duxbury Press, 1989. (Short, nontechnical articles by a number of well-known statisticians and users of statistics on the application of statistics in various disciplines and subject areas.)

Utts, Jessica. *Seeing Through Statistics*. Belmont, CA: Duxbury Press, 1999. (A nice introduction to the fundamental ideas of statistical reasoning.)

2 · The Data Analysis Process and Collecting Data Sensibly

The data analysis process can be viewed as a sequence of steps that lead from planning to data collection to informed conclusions based on the resulting data. It should come as no surprise that the correctness of such conclusions depends on the quality of the information on which they are based. The data collection step is critical to obtaining reliable information; both the type of analysis that is appropriate and the conclusions that can be drawn depend on how the data are collected. In this chapter, we first describe the data analysis process in some detail and then focus on two widely used methods of data collection: sampling and experimentation.

▪ 2.1 The Data Analysis Process

Statistics involves the collection and analysis of data. Both tasks are critical. Raw data without analysis are of little value, and even a sophisticated analysis cannot extract meaningful information from data that were not collected in a sensible way.

▪ Planning and Conducting a Study

Scientific studies are undertaken to answer questions about our world. Is a new flu vaccine effective in preventing illness? Is the use of bicycle helmets on the rise? Are injuries that result from bicycle accidents less severe for riders who wear helmets than for those who do not? How many credit cards do college students have? Do engineering students or psychology students pay more for textbooks? Data collection and analysis allow researchers to answer such questions.

The data analysis process can be organized into the following six steps:

1. **Understanding the nature of the problem.** Effective data analysis requires an understanding of the research problem. We must know the goal of the research and what questions we hope to answer. It is important to have a clear direction before gathering data to avoid being unable to answer the questions of interest using the data collected.

2. **Deciding what to measure and how to measure it.** The next step in the process is deciding what information is needed to answer the questions of interest. In some cases, the choice is obvious (e.g., in a study of the relationship between the weight of a Division I football player and position played, you would need to collect data on player weight and position), but in other cases the choice of information is not as straightforward (e.g., in a study of the relationship between preferred learning style and intelligence, how would you define learning style and measure it and what measure of intelligence would you use?). It is important to carefully define the variables to be studied and to develop appropriate methods for determining their values.

3. **Data collection.** The data collection step is crucial. The researcher must first decide whether an existing data source is adequate or whether new data must be collected. Even if a decision is made to use existing data, it is important to understand how the data were collected and for what purpose, so that any resulting limitations are also fully understood and judged to be acceptable. If new data are to be collected, a careful plan must be developed, because the type of analysis that is appropriate and the subsequent conclusions that can be drawn depend on how the data are collected.

4. **Data summarization and preliminary analysis.** After the data are collected, the next step usually involves a preliminary analysis that includes summarizing the data graphically and numerically. This initial analysis provides insight into important characteristics of the data and can provide guidance in selecting appropriate methods for further analysis.

5. **Formal data analysis.** The data analysis step requires the researcher to select and apply the appropriate statistical methods. Much of this textbook is devoted to methods that can be used to carry out this step.

6. **Interpretation of results.** Several questions should be addressed in this final step — for example, What conclusions can be drawn from the analysis? How do the results of the analysis inform us about the stated research problem or question? and How can our results guide future research? The interpretation step often leads to the formulation of new research questions, which, in turn, leads back to the first step. In this way, good data analysis is often an iterative process.

Example 2.1 illustrates the steps in the data analysis process.

▪ Example 2.1   A Proposed New Treatment for Alzheimer's Disease

The article "Brain Shunt Tested to Treat Alzheimer's" (*San Francisco Chronicle*, October 23, 2002) summarizes the findings of a study that appeared in the journal *Neurology*. Doctors at Stanford Medical Center were interested in determining whether a new surgical approach to treating Alzheimer's disease results in improved memory functioning. The surgical procedure involves implanting a thin tube, called a shunt, that is designed to drain toxins from the fluid-filled space that cushions the brain. Eleven patients had shunts implanted and were followed for a year, receiving quarterly tests of memory function. Another group of Alzheimer's patients was used as a comparison group. The patients in the comparison group received the standard care for Alzheimer's disease. After analyzing the data from this study, the investigators concluded that the "results suggested the treated patients essentially held their own in the cognitive tests while the patients in the control group steadily declined. However, the study was too small to produce conclusive statistical evidence." Based on these results, a much larger 18-month study is being planned. That study will include 256 patients at 25 medical centers around the country; the results are expected in 2005.

This study illustrates the nature of the data analysis process. A clearly defined research question and an appropriate choice for how to measure the variable of interest (the cognitive tests used to measure memory function) preceded the data collection. Assuming that a reasonable method was used to collect the data (we will see how this can be evaluated in Sections 2.4 and 2.5) and that appropriate methods of analysis were employed, the investigators reached the conclusion that the surgical procedure showed promise. However, they recognized the limitations of the study, especially those resulting from the small number of patients in the group that received surgical treatment, which in turn led to the design of a larger, more sophisticated study. As is often the case, the data analysis cycle led to further research, and the process began anew.

▪ Evaluating a Research Study

The six data analysis steps can also be used as a guide for evaluating published research studies. The following questions should be addressed as part of a study evaluation:

- What were the researchers trying to learn? What questions motivated their research?
- Was relevant information collected? Were the right things measured?
- Were the data collected in a sensible way?
- Were the data summarized in an appropriate way?
- Was an appropriate method of analysis used, given the type of data and how the data were collected?
- Are the conclusions drawn by the researchers supported by the data analysis?

Example 2.2 illustrates how these questions can guide an evaluation of a research study.

▪ Example 2.2   Spray Away the Flu

The newspaper article "Spray Away Flu" (*Omaha World-Herald*, June 8, 1998) reported on a study of the effectiveness of a new flu vaccine that is administered by nasal spray rather than by injection. The article states that the "researchers gave the spray to 1070 healthy children, 15 months to 6 years old, before the flu season two winters ago. One percent developed confirmed influenza, compared with 18 percent of the 532 children who received a placebo. And only one vaccinated child developed an ear infection after coming down with influenza. . . . Typically 30 percent to 40 percent of children with influenza later develop an ear infection." The researchers concluded that the nasal flu vaccine was effective in reducing the incidence of flu and also in reducing the number of children with flu who subsequently develop ear infections.

In this study, the researchers were trying to find out whether the nasal flu vaccine was effective in reducing the number of flu cases and in reducing the number of ear infections in children who did get the flu. The researchers recorded whether a child received the nasal vaccine or a placebo. (A **placebo** is a treatment that has the appearance of the treatment of interest but contains no active ingredients). Whether or not the child developed the flu and a subsequent ear infection was also recorded. These are appropriate measurements to make in order to answer the research question of interest. We typically cannot tell much about the data collection process from a newspaper article. As we will see in Section 2.4, to fully evaluate this study, we would also want to know how the participating children were selected, how it was determined that a particular child received the vaccine or the placebo, and how the subsequent diagnoses of flu and ear infection were made.

We will also have to delay discussion of the data analysis and the appropriateness of the conclusions because we do not yet have the necessary tools to evaluate these aspects of the study.

Other interesting examples of statistical studies can be found in Judith Tanur's *Statistics: A Guide to the Unknown* and in Roger Hock's *Forty Studies That Changed Psychology: Exploration into the History of Psychological Research* (the complete references for these two books can be found at the end of this chapter).

▪ 2.2 Sampling

There are many reasons for selecting a sample rather than obtaining information from an entire population (a **census**). The most common reason is limited resources; restrictions on available time or money usually prohibit observation of an entire population. Sometimes the process of measuring the characteristics of interest is destructive, as with measuring the breaking strength of soda bottles, the lifetime of flashlight batteries, or the sugar content of oranges, and it would be foolish to study the entire population.

Many studies are conducted in order to generalize from a sample to the corresponding population. As a result, it is important that the sample be representative of the population. To be reasonably sure of this, the researcher must carefully consider the way in which the sample is selected. It is sometimes tempting to take the easy way out and gather data in a haphazard way; but if a sample is chosen on the basis of convenience alone, it becomes impossible to interpret the resulting data with confidence. For example, it might be easy to use the students in your statistics class as a sample of students at your university. However, not all majors include a statistics course in their curriculum, and most students take statistics in their sophomore or junior year. The difficulty is that it is not clear whether or how these factors (and others that we might not be aware of) affect inferences based on information from such a sample. *There is no way to tell just by looking at a sample whether it is representative of the population from which it was drawn. Our only assurance comes from the method used to select the sample.*

▪ Bias in Sampling

Bias in sampling results in the tendency for a sample to differ from the corresponding population in some systematic way. Bias can result from the way in which the sample is selected or from the way in which information is obtained once the sample has been chosen. The most common types of bias encountered in sampling situations are selection bias, measurement or response bias, and nonresponse bias.

Selection bias is introduced when the way the sample is selected systematically excludes some part of the population of interest. This problem is sometimes also called undercoverage. For example, a researcher may wish to generalize from the results of a study to the population consisting of all residents of a particular city, but the method of selecting individuals may exclude the homeless or those without telephones. If those who are excluded from the sampling process differ in some systematic way from those who are included, the sample is guaranteed to be unrepresentative of the population. If this difference between the included and the excluded occurs on a variable that is important to the study, conclusions based on the sample data may not be valid. Selection bias also occurs if only volunteers or self-selected individuals are used in a study, because self-selected individuals (e.g., those who choose to participate in a call-in telephone poll) may well differ from those who choose not to participate.

Measurement or **response bias** occurs when the method of observation tends to produce values that systematically differ from the true value in some way. This might happen if an improperly calibrated scale is used to weigh items or if questions on a survey are worded in a way that tends to influence the response. For example, a May 1994 Gallup survey sponsored by the American Paper Institute (*Wall Street Journal*, May 17, 1994) included the following question: "It is estimated that disposable diapers account for less than 2 percent of the trash in today's landfills. In contrast, beverage containers, third-class mail and yard waste are estimated to account for about 21 percent of trash in landfills. Given this, in your opinion, would it be fair to tax or ban disposable diapers?" It is likely that the wording of this question prompted people to respond in a particular way.

Other things that might contribute to response bias are the appearance or behavior of the person asking the question, the group or organization conducting the study, and the tendency for people to not be completely honest when asked about illegal behavior or unpopular beliefs.

Although the terms *measurement bias* and *response bias* are often used interchangeably, the term *measurement bias* is usually used to describe systematic deviation from the true value as a result of a faulty measurement instrument (as with the improperly calibrated scale).

Nonresponse bias occurs when responses are not actually obtained from all individuals selected for inclusion in the sample. As with selection bias, nonresponse bias can distort results if those who respond differ in important ways from those who do not respond. Although some level of nonresponse is unavoidable in most surveys, the biasing effect on the resulting sample is lowest when the response rate is high. To minimize nonresponse bias, it is critical that a serious effort be made to follow up with individuals who do not respond to an initial request for information.

The nonresponse rate for surveys or opinion polls varies dramatically, depending on how the data are collected. Surveys are commonly conducted by mail, by phone, or by personal interview. Mail surveys are inexpensive but often have high nonresponse rates. Telephone surveys can also be inexpensive and can be implemented quickly, but they work well only for short surveys and they can also have high nonresponse rates. Personal interviews are generally expensive but tend to have better response rates. Some of the many challenges of conducting surveys are discussed in Section 2.6.

▪ **Types of Bias**

Selection Bias
Tendency for samples to differ from the corresponding population as a result of systematic exclusion of some part of the population.

Measurement or Response Bias
Tendency for samples to differ from the corresponding population because the method of observation tends to produce values that differ from the true value.

Nonresponse Bias
Tendency for samples to differ from the corresponding population because data are not obtained from all individuals selected for inclusion in the sample.

It is important to note that bias is introduced by the way in which a sample is selected or by the way in which the data are collected from the sample. Increasing the size of the sample, although possibly desirable for other reasons, does nothing to reduce bias. A good discussion of the types of bias appears in the sampling book by Lohr listed in the references at the end of this chapter.

▪ Random Sampling

Most of the inferential methods introduced in this text are based on the idea of random selection. The most straightforward of these methods is called simple random sampling. A **simple random sample** is a sample chosen using a method that ensures that each different possible sample of the desired size has an equal chance of being the one chosen. For example, suppose that we want a simple random sample of 10 employees chosen from all those who work at a large design firm. For the sample to be a simple random sample, each of the many different subsets of 10 employees must be equally likely to be the one selected. A sample taken from only full-time employees would not be a simple random sample of *all* employees, because someone who works part-time is still considered an employee but has no chance of being selected. Although a simple random sample may, by chance, include only full-time employees, it must be selected in such a way that each possible sample, and therefore *every* employee, has the same chance of inclusion in the sample. It is the selection process, not the final sample, that determines whether the sample is a simple random sample.

The letter n is used to denote sample size; it is the number of individuals or objects in the sample. For the design firm scenario of the previous paragraph, $n = 10$.

▪ **Definition**

A **simple random sample of size n** is a sample that is selected from a population in a way that ensures that every different possible sample of the desired size has the same chance of being selected.

The definition of a simple random sample implies that every individual member of the population has an equal chance of being selected. *However, the fact that every individual has an equal chance of selection, by itself, is not enough to guarantee that the sample is a simple random sample.* For example, suppose that a class is made up of 100 students, 60 of whom are female. A researcher decides to select 6 of the female students by writing all 60 names on slips of paper, mixing the slips, and then picking 6. She then selects four male students from the class using a similar procedure. Even though every student in the class has an equal chance of being included in the sample (6 of 60 females are selected and 4 of 40 males are chosen), the resulting sample is *not* a simple random sample because not all different possible samples of 10 students from the class have the same chance of selection. Many possible samples of ten students — for example, a sample of seven females and three males or a sample of all females — have no chance of being selected. The sample selection method described here is not necessarily a bad choice (in fact, it is an example of a stratified sample, to be discussed in more detail shortly), but it is not a simple random sample; this must be considered when a method is chosen for analyzing data from such a sample.

■ **Selecting a Simple Random Sample**   A number of different methods can be used to select a simple random sample. One way is to put the name or number of each member of the population on a different slip of paper; the slips are otherwise identical. The process of thoroughly mixing the slips and then selecting n slips one by one yields a random sample of size n. This method is easy to understand, but it has obvious drawbacks. The mixing must be adequate, and producing the necessary slips of paper can be extremely tedious, even for populations of moderate size.

A commonly used method for selecting a random sample is to first create a list, called a **sampling frame**, of the objects or individuals in the population. Each item on the list can then be identified by a number, and a table of random digits or a random number generator can be used to select the sample. A random number generator is a procedure that produces a sequence of numbers that satisfies all reasonable properties associated with the notion of randomness. Most statistics software packages include a random number generator, as do many calculators. A small table of random digits can be found in Appendix Table 1.

When selecting a random sample, researchers can choose to do the sampling with or without replacement. **Sampling with replacement** means that after each successive item is selected for the sample, the item is "replaced" back into the population and may therefore be selected again at a later stage. Thus, sampling with replacement allows for the possibility of having the same item or individual appear more than once in the sample. In practice, sampling with replacement is rarely used. Instead, the more common method is to not allow the same item to be included in the sample more than once. After being included in the sample, an individual or object would not be considered for further selection. Sampling in this manner is called **sampling without replacement**.

Sampling Without Replacement
Once an individual from the population is selected for inclusion in the sample, it may not be selected again in the sampling process. A sample selected without replacement includes n distinct individuals from the population.

Sampling with Replacement
After an individual from the population is selected for inclusion in the sample and the corresponding data are recorded, the individual is placed back in the population and can be selected again in the sampling process. A sample selected with replacement might include any particular individual from the population more than once.

Although these two forms of sampling are different, when the sample size *n* is small relative to the population size, as is often the case, there is little practical difference between them. In practice, the two can be viewed as equivalent if the sample size is at most 5% of the population size.

▪ Example 2.3   Selecting a Random Sample of Glass Soda Bottles

Breaking strength is an important characteristic of glass soda bottles. If the strength is too low, a bottle may burst — not a desirable outcome. Suppose that we want to measure the breaking strength of each bottle in a random sample of size *n* = 3 selected from four crates containing a total of 100 bottles (the population). Each crate contains five rows of five bottles each. We can identify each bottle with a number from 1 to 100 by numbering across the rows, starting with the top row of crate 1, as pictured:

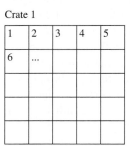
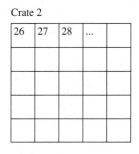
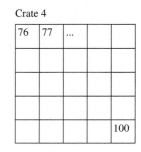

Using a random number generator from a calculator or statistical software package, we could generate three random numbers between 1 and 100 to determine which bottles would be included in our sample. This might result in bottles 15 (row 3 column 5 of crate 1), 89 (row 3 column 4 of crate 4), and 60 (row 2 column 5 of crate 3) being selected. (Alternatively, we could write the numbers from 1 to 100 on slips of paper, place them in a container, mix them well, and then select three.)

The goal of random sampling is to produce a sample that is likely to be representative of the population. Although random sampling does not *guarantee* that the sample will be representative, probability-based methods can be used to assess the risk of an unrepresentative sample. It is the ability to quantify this risk that allows us to generalize with confidence from a random sample to the corresponding population.

▪ A Note Concerning Sample Size

It is a common misconception that if the size of a sample is relatively small compared to the population size, the sample can't possibly accurately reflect the population. Critics of polls often make statements such as, "There are 14.6 million registered voters in California. How can a sample of 1000 registered voters possibly reflect public opinion when only about 1 in every 14,000 people are included in the sample?" These critics do not understand the power of random selection!

Consider a population consisting of 5000 applicants to a state university, and suppose that we are interested in math SAT scores for this population. A dotplot of the values in this population is shown in Figure 2.1(a). Figure 2.1(b) shows dot-plots of the math SAT scores for individuals in four different random samples from the population, ranging in sample size from $n = 50$ to $n = 1000$. Notice that the samples tend to reflect the distribution of scores in the population. If we were interested in using the sample to estimate the population average or to say something about the variability in SAT scores, even the smallest of the samples ($n = 50$) pictured would provide reliable information. Although it is possible to obtain a simple random sample that does not do a reasonable job of representing the population, this is likely only when the sample size is very small, and unless the population itself is small, this risk does not depend on what fraction of the population is sampled. The random selection process inherent in producing a simple random sample allows us to be confident that the sample adequately reflects the population, even when the sample consists of only a small fraction of the population.

Figure 2.1 (a) Dotplot of math SAT scores for the entire population. (b) Dotplots of math SAT scores for random samples of sizes 50, 100, 250, 500, and 1000.

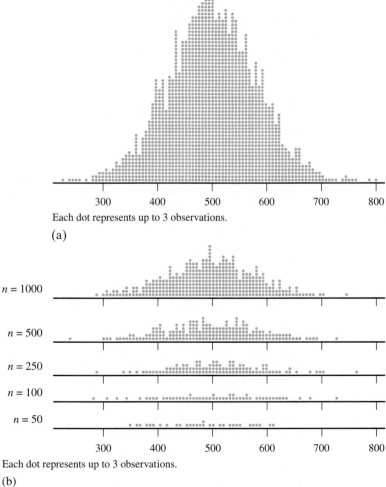

Other Sampling Methods

Simple random sampling provides researchers with a sampling method that is objective and free of selection bias. In some settings, however, alternative sampling methods may be less costly, easier to implement, or more accurate.

▪ **Stratified Random Sampling** When the entire population can be divided into a set of non-overlapping subgroups, a method known as **stratified sampling** often proves easier to implement and more cost-effective than random sampling. In stratified random sampling, separate simple random samples are independently selected from each subgroup. For example, to estimate the average cost of malpractice insurance, a researcher might find it convenient to view the population of all doctors practicing in a particular metropolitan area as being made up of four subpopulations: (1) surgeons, (2) internists and family practitioners, (3) obstetricians, and (4) a group that includes all other areas of specialization. Rather than taking a simple random sample from the population of all doctors, the researcher could take four separate random samples — one from the group of surgeons, another from the internists and family practitioners, and so on. These four samples would provide information about the four subgroups as well as information about the overall population of doctors.

When the population is divided in this way, the subgroups are called **strata**. Stratified sampling entails selecting a separate simple random sample from each stratum. Stratified sampling can be used instead of random sampling if it is important to obtain information about characteristics of the individual strata as well as of the entire population, although a stratified sample is not required to do this — subgroup estimates can also be obtained by using an appropriate subset of data from a simple random sample.

The real advantage of stratified sampling is that it often allows us to make more accurate inferences about a population than does simple random sampling. In general, it is much easier to estimate characteristics of a homogeneous group than of a heterogeneous group. For example, even with a small sample, it is possible to obtain an accurate estimate of the average grade point average (GPA) of students graduating with high honors from a university. These students are a homogeneous group with respect to GPA. The individual GPAs of group members are all quite similar, and even a sample of three or four individuals from this subpopulation should be representative. On the other hand, estimating the average GPA of *all* seniors at the university, a much more diverse group of GPAs, is a more difficult task.

If a varied population can be divided into strata, with each stratum being much more homogeneous than the population with respect to the characteristic of interest, then stratified sampling tends to produce more accurate estimates of population characteristics than simple random sampling does. This is because when the strata are relatively homogeneous, a small sample from each stratum can provide reasonably accurate information about strata characteristics. The strata information can then be used to obtain better information about the population as a whole than might have been possible using a simple random sample of the same total size.

▪ **Cluster Sampling** Sometimes it is easier to select groups of individuals from a population than it is to select individuals themselves. For example, suppose that a large urban high school has 600 senior students, all of whom are enrolled in a first period homeroom. There are 24 senior homerooms, each with approximately 25 students. If school administrators wanted to select a sample of roughly 75 seniors to participate in an evaluation of the college and career placement advising available to students, they might find it much easier to select three of the senior homerooms at random and then include all the students in the selected homerooms in the sample. In this way, an evaluation survey could be administered to all students in the selected homerooms at the same time — certainly easier logistically than randomly selecting students and then administering the survey to the 75 individual seniors.

Cluster sampling involves dividing the population of interest into nonoverlapping subgroups, called **clusters**. Clusters are then selected at random, and all individuals in the selected clusters are included in the sample.

Because whole clusters are selected, the ideal situation occurs when each cluster mirrors the characteristics of the population. When this is the case, a small number of clusters results in a sample that is representative of the population. If it is not reasonable to think that the variability present in the population is reflected in each cluster, as is often the case when the cluster sizes are small, then it becomes important to ensure that a large number of clusters is included in the sample.

Be careful not to confuse clustering and stratification. Even though both of these sampling strategies involve dividing the population into subgroups, both the way in which the subgroups are sampled and the optimal strategy for creating the subgroups are different. In stratified sampling, we sample from every stratum, whereas in cluster sampling, only selected whole clusters are included in the sample. Because of this difference, to increase the chance of obtaining a sample that is representative of the population, we want to create homogeneous (similar) groups for strata and heterogeneous (reflecting the variability in the population) groups for clusters.

■ **Systematic Sampling** **Systematic sampling** is a procedure that can be used when it is possible to view the population of interest as consisting of a list or some other sequential arrangement. A value k is specified (e.g., $k = 50$ or $k = 200$). Then one of the first k individuals is selected at random, after which every kth individual in the sequence is included in the sample. A sample selected in this way is called a **1 in k systematic sample**.

For example, a sample of faculty members at a university might be selected from the faculty phone directory. One of the first 20 faculty members listed could be selected at random, and then every 20th faculty member after that on the list would also be included in the sample. This would result in a 1 in 20 systematic sample.

The value of k for a 1 in k systematic sample is generally chosen to achieve a desired sample size. For example, in the faculty directory scenario just described, if there were 900 faculty members at the university, the 1 in 20 systematic sample described would result in a sample size of 45. If a sample size of 100 was desired, a 1 in 9 systematic sample could be used.

As long as there are no repeating patterns in the population list, systematic sampling works reasonably well. The potential danger is that if there are such patterns, systematic sampling can result in an unrepresentative sample.

■ **Convenience Sampling: Don't Go There!** It is often tempting to resort to "convenience" sampling — that is, using an easily available or convenient group to form a sample. This is a recipe for disaster! Results from such samples are rarely informative, and it is a mistake to try to generalize from a convenience sample to any larger population.

One common form of convenience sampling is sometimes called voluntary response sampling. Such samples rely entirely on individuals who volunteer to be a part of the sample, often by responding to an advertisement, calling a publicized telephone number to register an opinion, or logging on to an Internet site to complete a survey. It is extremely unlikely that individuals participating in such voluntary response surveys are representative of any larger population of interest.

■ Exercises 2.1–2.26

2.1 The psychology department at a university graduated 140 students in 2003. As part of a curriculum review, the department would like to select a simple random sample of twenty 2003 graduates to obtain information on how graduates perceived the value of the curriculum. Describe two different methods that might be used to select the sample.

2.2 During the previous calendar year, a county's small claims court processed 870 cases. A legal researcher would like to select a simple random sample of 50 cases to obtain information regarding the average award in such cases. Describe how a simple random sample of size $n = 50$ might be selected from the case files.

2.3 A petition with 500 signatures is submitted to a university's student council. The council president would like to determine the proportion of those who signed the petition who are actually registered students at the university. There is not enough time to check all 500 names with the registrar, so the council president decides to select a simple random sample of 30 signatures. Describe how this might be done.

2.4 The financial aid officers of a university wish to estimate the average amount of money that students spend on textbooks each term. They are considering taking a stratified sample. For each of the following proposed stratification schemes, discuss whether you think it would be worthwhile to stratify the university students in this manner.
a. Strata corresponding to class standing (freshman, sophomore, junior, senior, graduate student)
b. Strata corresponding to field of study, using the following categories: engineering, architecture, business, other
c. Strata corresponding to the first letter of the last name: A–E, F–K, etc.

2.5 Citrus trees are usually grown in orderly arrangements of rows to facilitate automated farming and harvesting practices. Suppose that a group of 1000 trees is laid out in 40 rows of 25 trees each. To determine the sugar content of fruit from a sample of 30 trees, researcher A suggests randomly selecting five rows and then randomly selecting six trees from each sampled row. Researcher B suggests numbering each tree on a map of the trees from 1 to 1000 and using random numbers to select 30 of the trees. Is there a reason for preferring one of these sample selection methods to the other? Explain.

2.6 For each of the situations described in what follows, state whether the sampling procedure is simple

random sampling, stratified random sampling, cluster sampling, systematic sampling, or convenience sampling.
a. All freshmen at a university are enrolled in 1 of 30 sections of a freshman seminar course. To select a sample of freshmen at this university, a researcher selects 4 sections of the freshman seminar course at random from the 30 sections and all students in the 4 selected sections are included in the sample.
b. To obtain a sample of students, faculty, and staff at a university, a researcher randomly selects 50 faculty members from a list of faculty, 100 students from a list of students, and 30 staff members from a list of staff.
c. A university researcher obtains a sample of students at his university by using the 85 students enrolled in his Psychology 101 class.
d. To obtain a sample of the seniors at a particular high school, a researcher writes the name of each senior on a slip of paper, places the slips in a box and mixes them, and then selects 10 slips. The students whose names are on the selected slips of paper are included in the sample.
e. To obtain a sample of those attending a basketball game, a researcher selects the 24th person through the door. Then, every 50th person after that is also included in the sample.

2.7 A small private college has 4500 students enrolled. Assume that the university can provide a list of the students with the students numbered from 1 to 4500. Describe the procedure you will use to select a simple random sample of 20 students, and then identify (by number) which students from the list are included in your sample.

2.8 Of the 6500 students enrolled at a community college, 3000 are part-time and the other 3500 are enrolled full time. Assume that the college can provide you with a list of students that is sorted so that all full-time students are listed first, followed by the part-time students.
a. Select a stratified random sample that uses full-time and part-time students as the two strata and that includes 10 students from each stratum. Describe the procedure you used to select the sample, and identify the students included in your sample by placement on the sorted list.
b. Does every student at this community college have the same chance of being selected for inclusion in the sample? Explain.

2.9 Give a brief explanation of why it is advisable to avoid the use of convenience samples.

2.10 Sometimes samples are composed entirely of volunteer responders. Give a brief description of the dangers of using voluntary response samples.

2.11 For no apparent reason, the authors of this textbook would like to know something about the number of words on individual pages of this book. A sample of pages from the book is to be obtained, and the number of words on each selected page will be determined. For the purposes of this exercise, equations are not counted as words and a number is counted as a word only if it is spelled out — that is, *ten* is counted as a word, but *10* is not.

a. Describe a sampling procedure that would result in a simple random sample of pages from this book.
b. Describe a sampling procedure that would result in a stratified random sample. Explain why you chose the specific strata used in your sampling plan.
c. Describe a sampling procedure that would result in a systematic sample.
d. Describe a sampling procedure that would result in a cluster sample.
e. Using the process you gave in Part (a), select a simple random sample of at least 20 pages, and record the number of words on each of the selected pages. Construct a dotplot of the resulting sample values, and write a sentence or two commenting on what it reveals about the number of words on a page.
f. Using the process you gave in Part (b), select a stratified random sample that includes a total of at least 20 selected pages, and record the number of words on each of the selected pages. Construct a dotplot of the resulting sample values, and write a sentence or two commenting on what it reveals about the number of words on a page.

2.12 In 2000, the chairman of a California ballot initiative campaign to add "none of the above" to the list of ballot options in all candidate races was quite critical of a Field poll that showed his measure trailing by 10 percentage points. The poll was based on a random sample of 1000 registered voters in California. He is quoted by the Associated Press (January 30, 2000) as saying, "Field's sample in that poll equates to one out of 17,505 voters," and he added that this was so dishonest that Field should get out of the polling business! If you worked on the Field poll, how would you respond to this criticism?

2.13 A pollster for the Public Policy Institute of California explains how the Institute selects a sample of California adults ("It's About Quality, Not Quantity," *San Luis Obispo Tribune*, January 21, 2000):

That is done by using computer-generated random residential telephone numbers with all California prefixes, and when there are no answers,

calling back repeatedly to the original numbers selected to avoid a bias against hard-to-reach people. Once a call is completed, a second random selection is made by asking for the adult in the household who had the most recent birthday. It is as important to randomize who you speak to in the household as it is to randomize the household you select. If you didn't, you'd primarily get women and older people.

Comment on this approach to selecting a sample. How does the sampling procedure attempt to minimize certain types of bias? Are there sources of bias that may still be a concern?

2.14 "More than half of California's doctors say they are so frustrated with managed care they will quit, retire early, or leave the state within three years." This statement is the lead sentence in an article titled "Doctors Feeling Pessimistic, Study Finds" (*San Luis Obispo Tribune*, July 15, 2001). This conclusion was based on a mail survey conducted by the California Medical Association. Surveys were mailed to 19,000 California doctors, and 2000 completed surveys were returned. Describe any concerns you have regarding the conclusion drawn in the first sentence of the article.

2.15 The article "I'd Like to Buy a Vowel, Drivers Say" (*USA Today*, August 7, 2001) speculates that young people prefer automobile names that consist of just numbers and/or letters that do not form a word (such as Hyundai's XG300, Mazda's 626, and BMW's 325i). The article goes on to state that Hyundai had planned to identify the car now marketed as the XG300 with the name Concerto, until they determined that consumers hated it and that they thought XG300 sounded more "technical" and deserving of a higher price. Do the students at your school feel the same way? Describe how you would go about selecting a sample to answer this question.

2.16 Based on a survey of 4113 U.S. adults, researchers at Stanford University concluded that Internet use leads to increased social isolation. The survey was conducted by an Internet-based polling company that selected its samples from a pool of 35,000 potential respondents, all of whom had been given free Internet access and WebTV hardware in exchange for agreeing to regularly participate in surveys conducted by the polling company. Two criticisms of this study were expressed in an article that appeared in the *San Luis Obispo Tribune* (February 28, 2000). The first criticism was that increased social isolation was measured by asking respondents if they were talking less to family and friends on the phone. The second criticism was that the sample was selected only from a group that was induced to participate by the offer of free Internet service, yet the results were generalized to all U.S. adults. Each of these criticisms is describing a type of bias. For each one, indicate what type of bias is being described and why it might make you question the conclusion drawn by the researchers.

2.17 "Workers Grow More Dissatisfied" is the headline for an article that appeared in the *San Luis Obispo Tribune* (August 22, 2002). The article states that "a survey of 5000 people found that while most Americans continue to find their jobs interesting, and are even satisfied with their commutes, a bare majority like their jobs." This statement was based on the fact that only 51 percent of those responding to the survey indicated that they were satisfied with their jobs. The survey was conducted by mail. Describe any potential sources of bias that you think might limit the researcher's ability to draw conclusions about working Americans based on the data collected in this survey.

2.18 *USA Today* (October 4, 2002) gave the accompanying summary data for private golf clubs in nine geographic regions of the United States in an article titled "If Women Want to Start Their Own Golf Clubs, More Power to Them." The value reported for each region is the median annual dues for the private clubs in the region that responded to the survey. The median is a middle value in the sense that half the clubs responding had dues that were lower and half had dues that were higher than the reported median. A total of 2900 private golf clubs were surveyed, with 11% responding:

Region	1	2	3	4	5
Median Dues	5200	1860	3680	2580	2950

Region	6	7	8	9
Median Dues	2680	2400	9800	2040

The article states that "private clubs in the U.S. are just that: Most decline to reveal their membership criteria or fees. But a general sense of how much their estimated 6 million members pay can be gleaned from this survey of private clubs." Do you agree that it is reasonable to use the reported values to get a sense of what the annual dues are in the nine geographic regions? Explain why or why not.

2.19 Studies linking abnormal genes to an increased risk of breast cancer are now being reevaluated. The article "Gene's Role in Cancer May Be Overstated" (*San Luis Obispo Tribune*, August 21, 2002) states that "early studies that evaluated breast cancer risk among gene mutation carriers selected women in families where sisters, mothers, and grandmothers all had breast cancer. This created a statistical bias that

skewed risk estimates for women in the general population." Is the bias described here selection bias, measurement bias, or nonresponse bias? Explain.

2.20 If liquor stores accept credit cards, does alcohol consumption change? The article "Changes in Alcohol Consumption Patterns Following the Introduction of Credit Cards in Ontario Liquor Stores" (*Journal of Studies on Alcohol* [1999]: 378–382) attempted to answer this question with data based on a telephone survey of adults. The sample was obtained by generating random phone numbers.
a. What are some potential drawbacks of using random phone numbers for sampling? Do any of these drawbacks seem critical for answering this question about alcohol and credit cards?
b. What time of day would be best for such calls?

2.21 The article "Drinking-Driving and Riding with Drunk Drivers Among Young Adults: An Analysis of Reciprocal Effects" (*Journal of Studies on Alcohol* [1999]: 615–621) concludes that although 16–24-year-olds who drive drunk are more likely to ride with drunk drivers, those who ride with drunk drivers are not more likely to drive drunk themselves. These results are based on a sample of 993 young adults from the state of New York, excluding New York City residents. The article explains the exclusion of New York City by saying that the proportion of young adults is higher but that the driving rate of young adults is far lower.
a. Describe the population of interest for this study.
b. Explain how this sample would (or would not) be representative of New York City young adults.
c. Explain how this sample would (or would not) be representative of adults from the entire United States.

2.22 The study described in Exercise 2.21 actually used a stratified sample with strata determined by age. Why do you think the stratification was desirable?

2.23 According to the article "Effect of Preparation Methods on Total Fat Content, Moisture Content, and Sensory Characteristics of Breaded Chicken Nuggets and Beef Steak Fingers" (*Family and Consumer Sciences Research Journal* [1999]: 18–27), sensory tests were conducted using 40 college student volunteers at Texas Women's University. Give three reasons, apart from the relatively small sample size, why this sample may not be ideal as the basis for generalizing to the population of all college students.

2.24 According to an article on the CNN.com web site (dated September 17, 1998) titled "Majority of U.S. Teens Are Not Sexually Active, Study Shows," 52% of surveyed teenagers had never had sexual intercourse. A very large random sample of 16,262 high school students was the source of this information. If the population of interest consists of all teenagers in the United States, are there individuals in the population who had no chance of being included in the sample? What is the name for this type of bias?

2.25 A new and somewhat controversial polling procedure that replaces the phone with the Internet is being used to conduct public opinion polls. The article "Pollsters Fear Internet Numbers Offer Warped View" (*Knight Ridder Newspapers*, October 18, 1999) states that "many pollsters say Internet users as a whole are still too upper-income, highly educated, white, and urban to produce results that accurately reflect all Americans. What's more, while 94% of

U.S. households had a telephone in 1998, only 26% had Internet access, according to a U.S. Department of Commerce study." What type of bias is being described in this statement? Do you think that this bias is a serious problem? Explain.

2.26 The article "Study Provides New Data on the Extent of Gambling by College Athletes" (*Chronicle of Higher Education*, January 22, 1999) reported that "72 percent of college football and basketball players had bet money at least once since entering college." This conclusion was based on a study in which "copies of the survey were mailed to 3000 athletes at 182 Division I institutions, 25 percent of whom responded." What types of bias might have influenced the results of this study? Explain.

▪ 2.3 Statistical Studies: Observation and Experimentation

During the week of August 10, 1998, articles with the following headlines appeared in *USA Today*:

"Prayer Can Lower Blood Pressure" (August 11, 1998)

"Wired Homes Watch 15% Less Television" (August 13, 1998)

In each of these articles, a conclusion is drawn from data. In the study on prayer and blood pressure, 2391 people, 65 years or older, were followed for 6 years. The article states that people who attended a religious service once a week and prayed or studied the Bible at least once a day were less likely to have high blood pressure. The researcher then concluded that "attending religious services lowers blood pressure."

The second article reported on a study of 5000 families, some of whom had online Internet services. The researcher concluded that TV use was lower in homes that were wired for the Internet, compared to nonwired homes. The researcher who conducted the study cautioned that the study does not answer the question of whether higher family income could be another factor in lower TV usage.

Are these conclusions reasonable? As we will see, the answer in each case depends in a critical way on how the data were collected.

▪ Observation and Experimentation

Data collection is an important step in the data analysis process. When we set out to collect information, it is important to keep in mind the questions we hope to answer on the basis of the resulting data. Sometimes we are interested in answering questions about characteristics of an existing population or in comparing two or more well-defined populations. To accomplish this, a sample is selected from each population under consideration, and the sample information is used to gain insight into characteristics of the population(s).

For example, an ecologist might be interested in estimating the average shell thickness of bald eagle eggs. A social scientist studying a rural community may want to determine whether gender and attitude toward abortion are related. A city council member in a college town may want to ascertain whether student residents of the city differ from nonstudent residents with respect to their support for various community projects. These are examples of studies that are *observational* in nature. We want to observe characteristics of members of an existing population or of several populations, and then use the resulting information to draw conclusions. In an **observational study**, it is important to obtain a sample that is representative of the corresponding population. To be reasonably certain of this, the researcher must carefully consider the way in which the sample is selected.

Sometimes the questions we are trying to answer deal with the effect of certain explanatory variables on some response and cannot be answered using data from an observational study. Such questions are often of the form, What happens when . . . ? or, What is the effect of . . . ? For example, an educator may wonder what would happen to test scores if the required lab time for a chemistry course were increased from 3 hr to 6 hr per week. To answer questions such as this one, the researcher conducts an experiment to collect relevant data. The value of some response variable (test score in the chemistry example) is recorded under different experimental conditions (3-hr lab and 6-hr lab in the chemistry example). In an experiment, the researcher manipulates one or more variables, called **factors**, to create the experimental conditions.

> A study is an **observational study** if the investigator observes characteristics of a subset of the members of one or more existing populations. The goal of an observational study is usually to draw conclusions about the corresponding population or about differences between two or more populations.
>
> A study is an **experiment** if the investigator observes how a response variable behaves when the researcher manipulates one or more factors. The usual goal of an experiment is to determine the effect of the manipulated factors on the response variable. In a well-designed experiment, the researcher determines the composition of the groups that will be exposed to different experimental conditions by random assignment to groups.

The type of conclusion that can be drawn from a research study depends on the study design. Both observational studies and experiments can be used to compare groups, but in an experiment the researcher controls who is in which group, whereas this is not the case in an observational study. This seemingly small difference is critical when it comes to the interpretation of results from the study.

A well-designed experiment can result in data that provide evidence for a cause-and-effect relationship. This is an important difference between an observational study and an experiment. In an observational study, it is impossible to draw cause-and-effect conclusions because we cannot rule out the possibility that the observed effect is due to some variable other than the factor being studied. Such variables are called **confounding variables**.

> A **confounding variable** is one that is related to both group membership and the response variable of interest in a research study.

Both of the studies reported in the *USA Today* articles described at the beginning of this section were observational studies. In the TV viewing example, the researcher merely observed the TV viewing behavior of individuals in two groups (wired and nonwired) but did not control which individuals were in which group. A possible confounding variable here is family income. Although it may be reasonable to conclude that the wired and nonwired groups differ with respect to TV viewing, it is not reasonable to conclude that the lower TV use in wired homes is caused by the fact that they have Internet service. In the prayer example, two groups were compared (those who attend a religious service at least once a week and those who do not), but the researcher did not manipulate this factor to observe the effect on blood pressure by assigning people at random to service-attending or non-service-attending groups. As a result, cause-and-effect conclusions such as "prayer can lower blood pressure" are not reasonable based on the observed data. It is possible that some other variable, such as lifestyle, is related to both religious service attendance and blood pressure. In this case, lifestyle is an example of a potential confounding variable.

▪ Drawing Conclusions from Statistical Studies

In this section, two different types of conclusions have been described. One type involves generalizing from what we have seen in a sample to some larger population, and the other involves reaching a cause-and-effect conclusion about the effect of an explanatory variable on a response. When is it reasonable to draw such conclusions? The answer depends on the way that the data were collected. Table 2.1 summarizes the types of conclusions that can be made with different study designs.

Table 2.1 ▪ Drawing Conclusions from Statistical Studies

Study Description	Reasonable to Generalize Conclusions about Group Characteristics to the Population?	Reasonable to Draw Cause-and-Effect Conclusion?
Observational study with sample selected at random from population of interest	Yes	No
Observational study based on convenience or voluntary response sample (poorly designed sampling plan)	No	No
Experiment with groups formed by random assignment of individuals or objects to experimental conditions		
Individuals or objects used in study are volunteers or not randomly selected from some population of interest	No	Yes
Individuals or objects used in study are randomly selected from some population of interest	Yes	Yes
Experiment with groups not formed by random assignment to experimental conditions (poorly designed experiment)	No	No

As you can see from Table 2.1, it is important to think carefully about the objectives of a statistical study before planning how the data will be collected. Neither type of conclusion (generalizing to a population or cause-and-effect) can be made from an observational study with a poorly designed sampling plan or from an experiment that does not create experimental groups by random assignment. Both observational studies and experiments must be carefully designed if the resulting data are to be useful. We have already considered the common sampling procedures used in observational studies. In Sections 2.4 and 2.5, we consider experimentation and explore what constitutes good practice in the design of simple experiments.

▪ Exercises 2.27–2.37

2.27 The Associated Press (December 10, 1999) reported on an investigation that concluded that women who suffer severe morning sickness early in pregnancy are more likely to have a girl. This conclusion was reached by researchers in Sweden based on a "scientific study." How do you think the researchers might have collected data that would have enabled them to reach such a conclusion? Do you think that the scientific study referred to in the article was an experiment or an observational study? Explain.

2.28 An article titled "Guard Your Kids Against Allergies: Get Them a Pet" (*San Luis Obispo Tribune*, August 28, 2002) described a study that led researchers to conclude that "babies raised with two or more animals were about half as likely to have allergies by the time they turned six."
a. Do you think this study was an observational study or an experiment? Explain.
b. Describe a potential confounding variable that illustrates why it is unreasonable to conclude that being raised with two or more animals is the cause of the observed lower allergy rate.

2.29 Researchers at the Hospital for Sick Children in Toronto compared babies born to mothers with diabetes to babies born to mothers without diabetes ("Conditioning and Hyperanalgesia in Newborns Exposed to Repeated Heel Lances," *Journal of the American Medical Association* [2002]: 857–861). Babies born to mothers with diabetes have their heels pricked numerous times during the first 36 hours of life in order to obtain blood samples to monitor blood sugar level. The researchers noted that the babies born to diabetic mothers were more likely to grimace or cry when having blood drawn than the babies born to mothers without diabetes. This led the researchers to conclude that babies who experience pain early in life become highly sensitive to pain. Comment on the appropriateness of this conclusion.

2.30 Based on a survey conducted on the Diet Smart.com web site, investigators concluded that women who regularly watched *Oprah* were only one-seventh as likely to crave fattening foods as those who watched other daytime talk shows (*San Luis Obispo Tribune*, October 14, 2000).
a. Is it reasonable to conclude that watching *Oprah* causes a decrease in cravings for fattening foods? Explain.
b. Is it reasonable to generalize the results of this survey to all women in the United States? To all women who watch daytime talk shows? Explain why or why not.

2.31 "Crime Finds the Never Married" is the conclusion drawn in an article from *USA Today* (June 29, 2001). This conclusion is based on data from the Justice Department's National Crime Victimization Survey, which estimated the number of violent crimes per 1000 people, 12 years of age or older, to be 51 for the never married, 42 for the divorced or separated, 13 for married individuals, and 8 for the widowed. Does being single cause an increased risk of violent crime? Describe a potential confounding variable that illustrates why it is unreasonable to conclude that a change in marital status causes a change in crime risk.

2.32 The paper "Prospective Randomized Trial of Low Saturated Fat, Low Cholesterol Diet During the First Three Years of Life" (*Circulation* [1996]: 1386–1393) describes an experiment to examine the effect of modifying fat intake in early childhood on blood cholesterol level. The article states that in this experiment "1062 infants were randomized to either the intervention or control group at 7 months of age. The families of the 540 intervention group children were counseled to reduce the child's intake of saturated fat and cholesterol but to ensure adequate energy intake. The control children consumed an unrestricted diet."

a. The researchers concluded that the blood cholesterol level was lower for children in the intervention group. Is it reasonable to conclude that the parental counseling and subsequent reduction in dietary fat and cholesterol are the cause of the reduction in blood cholesterol level? Explain why or why not.

b. Do you think it is reasonable to generalize the results of this experiment to all children? Explain.

2.33 *USA Today* (January 29, 2003) summarized data from a survey of Americans with household incomes of $75,000 or more. It was reported that 57% of affluent Americans would rather have more time than more money.

a. What condition on how the data were collected would make the generalization from the sample to the population of affluent Americans reasonable?

b. Would it be reasonable to generalize from the sample to say that 57% of all Americans would rather have more time than more money? Explain.

2.34 "Attending Church Found Factor in Longer Life" is the title of an article that appeared in *USA Today* (August 9, 1999). Based on a "nationally representative survey of 3617 Americans," the article concludes that "attending services extends the life span about as much as moderate exercise or not smoking." Comment on the validity of this conclusion.

2.35 An article that appeared in the *San Luis Obispo Tribune* (November 11, 1999) was titled "Study Points Out Dangerous Side to SUV Popularity: Half of All 1996 Ejection Deaths Occur in SUVs." This article states that sports utility vehicles (SUVs) have a much higher rate of passengers being thrown from a window during an accident than do automobiles. The article also states that more than half of all deaths caused by ejection involved SUVs — the basis for the conclusion that SUVs are more dangerous than cars. Later in the article, there is a comment that about 98% of those injured or killed in ejection accidents were not wearing seat belts. Comment on the conclusion that SUVs are more dangerous than cars.

2.36 Does living in the South cause high blood pressure? Data from a group of 6278 whites and blacks questioned in the Third National Health and Nutritional Examination Survey between 1988 and 1994 (see CNN.com web site article of January 6, 2000, titled "High Blood Pressure Greater Risk in U.S. South, Study Says") indicates that a greater percentage of Southerners have high blood pressure than do people in any other region of the United States. This difference in rate of high blood pressure was found in every ethnic group, gender, and age category studied. List at least two possible reasons we cannot conclude that living in the South causes high blood pressure.

2.37 Does eating broccoli reduce the risk of prostate cancer? According to an observational study from the Fred Hutchinson Cancer Research Center (see CNN.com web site article titled "Broccoli, Not Pizza Sauce, Cuts Cancer Risk, Study Finds," January 5, 2000), men who ate more cruciferous vegetables (broccoli, cauliflower, brussels sprouts, and cabbage) had a lower risk of prostate cancer. This study made separate comparisons for men who ate different levels of vegetables. According to one of the investigators, "at any given level of total vegetable consumption, as the percent of cruciferous vegetables increased, the prostate cancer risk decreased." Based on this study, is it reasonable to conclude that eating cruciferous vegetables causes a reduction in prostate cancer risk? Explain.

■ 2.4 Simple Comparative Experiments

Sometimes the questions we are trying to answer deal with the effect of certain explanatory variables on some response. Such questions are often of the form, What happens when . . . ? or, What is the effect of . . . ? For example, an educator may wonder what happens to test scores if an activity-based method of instruction is used rather than a traditional lecture method. An industrial engineer may be considering two different workstation designs and might want to know whether the choice of design affects work performance. A medical researcher may want to determine how a proposed treatment for a disease compares to a standard treatment. A nutritionist working for a commercial bakery may want to determine the effect of baking time and baking temperature on the nutritional content of bread.

To address these types of questions, the researcher conducts an experiment to collect the relevant information. The value of some response variable (test score, assembly time, bread density, etc.) is determined under different experimental conditions. Experiments must be carefully planned to obtain information that will give unambiguous answers to questions of interest.

▪ **Definition**

An **experiment** is a planned intervention undertaken to observe the effects of one or more explanatory variables, often called **factors**, on a response variable. The fundamental purpose of the intervention is to increase understanding of the nature of the relationships between the explanatory and response variables. Any particular combination of values for the explanatory variables is called an **experimental condition** or **treatment**.

The **design** of an experiment is the overall plan for conducting an experiment. A good design minimizes ambiguity in the interpretation of the results.

Suppose that we are interested in determining how student performance on a first-semester calculus exam is affected by room temperature. There are four sections of calculus being offered in the fall semester. We might design an experiment in this way: Set the room temperature (in degrees Fahrenheit) to 65° in two of the rooms and to 75° in the other two rooms on test day, and then compare the exam scores for the 65° group and the 75° group. Suppose that the average exam score for the students in the 65° group was noticeably higher than the average for the 75° group. Could we conclude that the increased temperature resulted in a lower average score? The answer is no, because many other factors might be related to exam score. Were the sections at different times of the day? Did they have the same instructor? Different textbooks? Was one section required to do more homework than the other sections? Did the sections differ with respect to the abilities of the students? Any of these other factors could provide a plausible explanation (having nothing to do with room temperature) for why the average test score was different for the two groups. It is not possible to separate the effect of temperature from the effects of these other factors. As a consequence, simply setting the room temperatures as described makes for a poorly designed experiment. A well-designed experiment requires more than just manipulating the explanatory variables; the design must also eliminate rival explanations or else the experimental results will not be conclusive.

The goal is to design an experiment that will allow us to determine the effects of the relevant factors on the chosen response variable. To do this, we must take into consideration other extraneous factors that, although not of interest in the current study, might also affect the response variable.

▪ **Definition**

An **extraneous factor** is one that is not of interest in the current study but is thought to affect the response variable.

A researcher can **directly control** some extraneous factors. In the calculus test example, the textbook used is an extraneous factor because part of the differences in test results might be attributed to this factor. We might choose to control this factor directly, by requiring that all sections participating in the study use the same textbook. If this were the case, any observed differences between temperature groups could not be explained by the use of different textbooks, because all sections would have used the same book. The extraneous factor *time of day* might also be directly controlled in this way by having all sections meet at the same time.

The effects of some extraneous factors can be filtered out by a process known as **blocking**. Extraneous factors that are addressed through blocking are called *blocking factors*. Blocking creates groups (called blocks) that are similar with respect to blocking factors; then all treatments are tried in each block. In our example, we might use *instructor* as a blocking factor. If two instructors are each teaching two sections of calculus, we would make sure that for each instructor, one section was part of the 65° group and the other section was part of the 75° group. With this design, if we see a difference in exam scores for the two temperature groups, the factor *instructor* can be ruled out as a possible explanation, because both instructors' students were present in each temperature group. If one instructor taught both 65° sections and the other taught both 75° sections, we would be unable to distinguish the effect of temperature from the effect of the instructor. These two factors (temperature and instructor) are said to be **confounded**.

Two factors are **confounded** if their effects on the response variable cannot be distinguished from one another.

We can directly control some extraneous factors by holding them constant, and we can use blocking to create groups that are similar to essentially filter out the effect of others. What about factors, such as student ability in our calculus test example, that cannot be controlled by the experimenter and that would be difficult to use as blocking factors? These extraneous factors are handled by the use of random assignment to experimental groups — a process called **randomization**. Randomization ensures that our experiment does not favor one experimental condition over any other and attempts to create "equivalent" experimental groups (groups that are as much alike as possible). For example, if the students requesting calculus could be assigned to one of the four available sections using some sort of random mechanism, we would expect the resulting groups to be similar with respect to student ability as well as with respect to other extraneous factors that we are not directly controlling or using as a basis for blocking.

To get a sense of how random assignment tends to create similar groups, suppose that 50 first-year college students are available to participate as subjects in an experiment to investigate whether completing an online review of course material before an exam improves exam performance. The 50 subjects vary quite a bit with respect to achievement, which is reflected in their math and verbal SAT scores, as shown in Figure 2.2.

If these 50 students are to be assigned to the two experimental groups (one that will complete the online review and one that will not), we want to make sure that

Figure 2.2 Dotplots of math and verbal SAT scores for 50 first-year students.

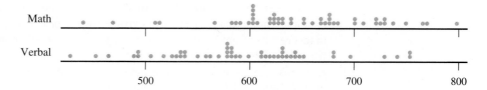

the assignment of students to groups does not favor one group over the other by tending to assign the higher achieving students to one group and the lower achieving students to the other.

Creating groups of students with similar achievement levels in a way that considers both verbal and math SAT scores simultaneously would be difficult, so we rely on random assignment. Figure 2.3(a) shows the verbal SAT scores of the students assigned to each of the two experimental groups (one shown in orange and one shown in blue) for each of three different random assignments of students to groups. Figure 2.3(b) shows the math SAT scores for the two experimental groups for each of the same three random assignments. Notice that each of the three random assignments produced groups that are similar with respect to *both* verbal and math SAT scores. So, if any of these three assignments were used and the two groups differed on exam performance, we could rule out differences in math or verbal SAT scores as possible competing explanations for the difference.

Not only will random assignment tend to create groups that are similar with respect to verbal and math SAT scores, but it will also tend to even out the groups with respect to other extraneous variables. As long as the number of subjects is not too small, we can rely on the random assignment to produce comparable experimental groups. This is the reason that randomization is a part of all well-designed experiments.

Figure 2.3 Dotplots for three different random assignments to two groups, one shown in orange and one shown in blue: (a) verbal SAT score and (b) math SAT score.

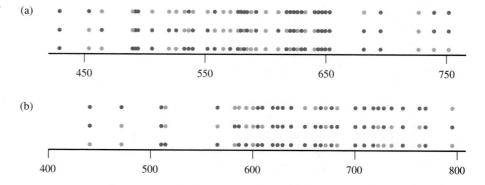

Not all experiments require the use of human subjects to evaluate different experimental conditions. For example, a researcher might be interested in comparing three different gasoline additives with respect to automobile performance, as measured by gasoline mileage. The experiment might involve using a single car with an empty tank. One gallon of gas with one of the additives will be put in the tank, and the car will be driven along a standard route at a constant speed until it runs out of gas. The total distance traveled on the gallon of gas could then be recorded. This could be repeated a number of times — ten, for example — with each additive.

The experiment just described can be viewed as consisting of a sequence of trials. Because a number of extraneous factors might have an effect on gas mileage (such as variations in environmental conditions — e.g., wind speed or humidity — and small variations in the condition of the car), it would not be a good idea to use additive 1 for the first 10 trials, additive 2 for the next 10 trials, and so on. An approach that would not unintentionally favor any one of the additives would be to randomly assign additive 1 to 10 of the 30 planned trials, and then randomly assign additive 2 to 10 of the remaining 20 trials. The resulting plan for carrying out the experiment might look as follows:

Trial	1	2	3	4	5	6	7	. . .	30
Additive	2	2	3	3	2	1	2	. . .	1

When an experiment can be viewed as a sequence of trials, randomization involves the random assignment of treatments to trials. *Just remember that random assignment — either of subjects to treatments or of treatments to trials — is a critical component of a good experiment.*

Randomization can be effective in evening out the effects of extraneous variables only if the number of subjects or observations in each treatment or experimental condition is large enough for each experimental group to reliably reflect variability in the population. For example, if there were only eight students requesting calculus, it is unlikely that we would get equivalent groups for comparison, even with random assignment to the sections. **Replication** is the design strategy of making multiple observations for each experimental condition. Together, replication and randomization allow the researcher to be reasonably confident of comparable experimental groups.

▪ Key Concepts in Experimental Design

Randomization
Random assignment (of subjects to treatments or of treatments to trials) to ensure that the experiment does not systematically favor one experimental condition over another.

Blocking
Using extraneous factors to create groups (blocks) that are similar. All experimental conditions are then tried in each block.

Direct Control
Holding extraneous factors constant so that their effects are not confounded with those of the experimental conditions.

Replication
Ensuring that there is an adequate number of observations for each experimental condition.

These four concepts — randomization, blocking, control, and replication — are the fundamental principles of experimental design. When you collect data through experimentation, you should give careful thought to the role that each of these principles plays in your design.

It is an unremarkable event when we read in our newspapers or see on the news or perhaps read in one of our textbooks that a scientist performed an experiment. The subject matter may be a new therapy for arthritis, a promising drug for asthma control, or a substance that holds promise for longer lasting tires for automobiles, but the stories are similar. A scientist, doctor, or engineer has performed an experiment, and some new or unexpected result has brought new knowledge or refined our understanding of some scientific puzzle or, in some cases, caused rethinking about a heretofore "settled" issue.

To illustrate the design of a simple experiment, consider the dilemma of Anna, a waitress in a local restaurant. Anna is saving money for college, and she depends on tips from her customers in the restaurant. She would like to increase the amount of her tips, and her strategy is simple: In addition to prompt and courteous service, she will project a positive image of herself to her customers by writing "Thank you" on the back of the check before she gives it to the patron. She will write "Thank you" on some of the checks, and on others she will write nothing. She plans to calculate the percentage of the tip as her measure of success (for instance, a 15% tip is common). She will compare the average percentage of the tip calculated from checks with and without the handwritten "Thank you."

Writing "Thank you" consumes a bit more time in her already hectic serving activities, and so she has a stake in the outcome of her experiment. If the extra time does not produce higher tips, she may try a different strategy; in any case it is of value to her to have at least a tentative answer to the question, Will writing "Thank you" produce the desired outcome of higher tips? Anna is untrained in the art of planning experiments, but already she has taken some commonsense steps in the right direction to answer her question.

1. *Identification of a specific problem.* Anna has defined a manageable problem, and collecting the appropriate data is feasible. It should be easy to gather data as a normal part of her work.

2. *Definition of the variables of interest.* Anna wonders whether writing "Thank you" on the customers' bills will have an effect on the amount of her tip. In the language of experimentation, we would refer to the writing of "Thank you" and the not writing of "Thank you" as **treatments** (the two experimental conditions to be compared in the experiment). The two treatments together are the possible values of the **explanatory variable**. The tipping percentage is the **response variable**. The idea behind this terminology is that the outcome of the experiment (the tipping percentage) is a *response* to the treatments *writing "Thank you"* or *not writing "Thank you."* Anna's experiment may be thought of as an attempt to explain the variability in the response variable in terms of its presumed cause, the variability in the explanatory variable. That is, as she manipulates the explanatory variable, she expects the response by her customers to vary in a manner consistent with her theory.

3. *Formulation of the measurement.* The explanatory variable is categorical (writing "Thank you" or not writing "Thank you"), and it can be easily recorded. The response variable, *tipping percentage*, is easily calculated from the tip and the original bill: % tip = (tip/bill total) × 100.

4. *Analysis of the resulting data.* After collecting the data, statistical methods can be used to decide whether the tip sizes tend to be higher for the "Thank you" treatment.

Anna has a good start, but now she must consider the four fundamental design principles.

Replication. Anna cannot run a successful experiment by gathering tipping information on only one person for each treatment. There is no reason to believe that any single tipping incident is representative of what would happen in other incidents (because customers vary in their tipping practices), and therefore it would be impossible to evaluate the two treatments with only two subjects. To interpret the effects of a particular treatment, she must **replicate** each treatment in the experiment.

Control and Randomization. There are a number of extraneous variables in this example — variables that might have an effect on the size of tip. Some restaurant patrons will be seated near the window with a nice view; some will have to wait for a table, whereas others may be seated immediately; and some may be on a fixed income and cannot afford a large tip. Some of these variables can be directly controlled. For example, Anna may choose to use only window tables in her experiment, thus eliminating table location as a potential confounding variable. Other variables, such as length of wait and customer income, cannot be easily controlled. As a result, it is important that Anna use randomization to decide which of the window tables will be in the "Thank you" group and which will be in the no "Thank you" group. She might do this by flipping a coin as she prepares the check for each window table. If the coin lands with the head side up, she could write "Thank you" on the bill, omitting the "Thank you" when a tail is observed.

Blocking. Suppose that Anna works on both Thursdays and Fridays. Because day of the week might affect tipping behavior, Anna should block on day of the week and make sure that observations for both treatments are made on each of the two days.

■ Evaluating an Experimental Design

The key concepts of experimental design provide a framework for evaluating an experimental design, as illustrated in Examples 2.4 and 2.5.

■ Example 2.4   Subliminal Messages

The article "The Most Powerful Manipulative Messages Are Hiding in Plain Sight" (*Chronicle of Higher Education*, January 29, 1999) reported the results of an experiment on priming — the effect of subliminal messages on how we behave. In the experiment, subjects completed a language test in which they were asked to construct a sentence using each word in a list of words. One group of subjects received a list of words related to politeness, and a second group was given a list of words related to rudeness. Subjects were told to complete the language test and then come into the hall and find the researcher so that he could explain the next part of the test. When each subject came into the hall, he or she found the researcher engaged in conversation. The researcher wanted to see whether the subject would interrupt the conversation. The researcher found that 63% of those primed with words related to rudeness interrupted the conversation, whereas only 17% of those primed with words related to politeness interrupted.

If we assume that the researcher randomly assigned the subjects to the two groups, then this study is an experiment that compares two treatments (primed with words related to rudeness and primed with words related to politeness). The response variable, *politeness*, has the values *interrupted conversation* and *did not interrupt conversation*. The experiment uses replication (many subjects in each treatment group) and randomization to control for extraneous variables that might affect the response.

Many experiments compare a group that receives a particular treatment to a **control group** that receives no treatment.

▪ Example 2.5   Health Education and Cholesterol Level

Researchers at the University of North Carolina studied 422 third- and fourth-grade children to determine whether a program of health education and exercise was effective in reducing cholesterol level (*San Luis Obispo Telegram-Tribune*, August 4, 1998). Children with high cholesterol levels were divided into three groups. One group attended classes in healthy nutrition twice a week and exercised three times a week. A second group received individualized instruction from a nurse and exercised three times a week. The third group, a control group, received no specialized instruction and participated in the school's regular physical education classes. After 2 months, cholesterol levels were measured. The researchers concluded that the classroom-based program (Group 1) was most effective in lowering cholesterol. If the researchers had randomly assigned the 422 children to the three experimental conditions (treatments), the study would have been an experiment that used both replication and randomization.

When planning an experiment or evaluating a design developed by someone else, be sure to keep the following basic principles in mind:

1. *Replication.* Design strategy of making multiple observations for each experimental treatment.
2. *Direct control.* Holding the values of extraneous variables constant so that their effect is not potentially confounded with factors in the experiment.
3. *Blocking.* Using extraneous variables to create groups (blocks) that are similar, and then making sure that all treatments are tried in each block.
4. *Randomization. A must!* The strategy for dealing with all extraneous variables not taken into account through direct control or blocking. We count on randomization to create "equivalent" experimental groups.

Before proceeding with an experiment, you should be able to give a satisfactory answer to each of the following 10 questions.

1. What is the research question that data from the experiment will be used to answer?
2. What is the response variable?
3. How will the values of the response variable be determined?
4. What are the factors (explanatory variables) for the experiment?
5. For each factor, how many different values are there, and what are these values?
6. What are the treatments for the experiment?
7. What extraneous variables might influence the response?

8. How does the design incorporate random assignment of subjects to treatments (or treatments to subjects) or random assignment of treatments to trials?

9. For each extraneous variable listed in Question 7, how does the design protect against its potential influence on the response through blocking, direct control, or randomization?

10. Will you be able to answer the research question using the data collected in this experiment?

■ Exercises 2.38–2.48

2.38 The head of the quality control department at a printing company is interested in factors that affect the strength of the binding on books that the company produces. In particular, she would like to carry out an experiment to determine which of three different glues results in the greatest binding strength. Although they are not of interest in the current investigation, other factors thought to affect binding strength are the number of pages in the book and whether the book is being bound as a paperback or a hardback.
a. What is the response variable in this experiment?
b. What factor will determine the experimental conditions?
c. What two extraneous factors are mentioned in the problem description? Can you think of other extraneous factors that might be considered?

2.39 Based on observing more than 400 drivers in the Atlanta area, two investigators at Georgia State University concluded that people exiting parking spaces did so more slowly when a driver in another car was waiting for the space than when no one was waiting ("Territorial Defense in Parking Lots: Retaliation Against Waiting Drivers," *Journal of Applied Social Psychology* [1997]: 821–834). Describe how you might design an experiment to collect information that would allow you to determine whether this phenomenon is true for your city. What is the response variable for your study? What extraneous factors might have an effect on the response variable, and how does your design control for them?

2.40 In 1993, researchers proclaimed that listening to Mozart could make you smarter. Dubbed the Mozart effect, this conclusion was based on a study that showed college students temporarily gained up to 9 IQ points after listening to a Mozart piano sonata. This research has since been criticized by a number of researchers who have been unable to confirm the result in other similar studies. Suppose that you wanted to see whether there is a Mozart effect for students at your school.

a. Describe how you might design an experiment for this purpose.
b. Does your experimental design include direct control of any extraneous variables? Explain.
c. Does your experimental design use blocking? Explain why you did or did not include blocking in your design.
d. What role does randomization play in your design?

2.41 An article from the Associated Press (May 14, 2002) led with the headline "Academic Success Lowers Pregnancy Risk." The article described an evaluation of a program that involved about 350 students at 18 Seattle schools in high crime areas. Some students took part in a program beginning in elementary school in which teachers showed children how to control their impulses, recognize the feelings of others, and get what they want without aggressive behavior. Others did not participate in the program. The study concluded that the program was effective because by the time young women in the program reached age 21, the pregnancy rate among them was 38%, compared to 56% for the women in the experiment who did not take part in the program. Explain why this conclusion is valid only if the women in the experiment were randomly assigned to one of the two experimental groups.

2.42 Do ethnic group and gender influence the type of care that a heart patient receives? The following passage is from the article "Heart Care Reflects Race and Sex, Not Symptoms" (*USA Today*, February 25, 1999, reprinted with permission):

> Previous research suggested blacks and women were less likely than whites and men to get cardiac catheterization or coronary bypass surgery for chest pain or a heart attack. Scientists blamed differences in illness severity, insurance coverage, patient preference, and health care access. The researchers eliminated those differences by videotaping actors — two black men, two black women, two white men, and two white women — describing chest pain from identical scripts. They wore

identical gowns, used identical gestures, and were taped from the same position. Researchers asked 720 primary care doctors at meetings of the American College of Physicians or the American Academy of Family Physicians to watch a tape and recommend care. The doctors thought the study focused on clinical decision-making.

Evaluate this experimental design. Do you think this is a good design or a poor design, and why? If you were designing such a study, what, if anything, would you propose to do differently?

2.43 Four new word-processing software programs are to be compared by measuring the speed with which various standard tasks can be completed. Before conducting the tests, researchers note that the level of a person's computer experience is likely to have a large influence on the test results. Discuss how you would design an experiment that fairly compares the word-processing programs while simultaneously accounting for possible differences in users' computer proficiency.

2.44 A study in Florida is examining whether health literacy classes and using simple medical instructions that include pictures and avoid big words and technical terms can keep Medicaid patients healthier (*San Luis Obispo Tribune*, October 16, 2002). Twenty-seven community health centers are participating in the study. For 2 years, half of the centers will administer standard care. The other centers will have patients attend classes and will provide special health materials that are easy to understand. Explain why it is important for the researchers to assign the 27 centers to the 2 groups (standard care and classes with simple health literature) at random.

2.45 Is status related to a student's understanding of science? The article "From Here to Equity: The Influence of Status on Student Access to and Understanding of Science" (*Culture and Comparative Studies* [1999]: 577–602) described a study on the effect of group discussions on learning biology concepts. An analysis of the relationship between status and "rate of talk" (the number of on-task speech acts per minute) during group work included gender as a blocking variable. Do you think that gender is a useful blocking variable? Explain.

2.46 In an experiment to determine whether suggesting a number of items to purchase (e.g., "limit 12 per person") increases the number of items purchased ("An Anchoring and Adjustment Model of Purchase Quantity Decision," *Journal of Marketing Research* [1999]: 71–81), data were collected at three separate stores. At one store the display of on-sale soup had a sign that said "No limit per person"; at

another store the sign read "Limit 4 per person," and at the third store the sign read "Limit 12 per person." On the second and third days of the experiment the signs were switched so that each store had each treatment for one day only.

a. Why did the researchers rotate the three treatments across the three different stores for the three days?

b. The results of this study showed that there were no day-to-day or store-to-store differences in the quantity of soup purchased. Also, in the stores with the 12-item limit, the average number of cans purchased per person was twice that of the no-limit or 4-item limit (7 instead of 3.5). This difference was large enough for the researchers to conclude that the results were not due to chance alone. If a grocery store manager asked you how she should market soup, what would you say?

c. If a grocer asked you how she should market lettuce, would you say (based on the results of this study) that a limit of 12 items per person would increase sales over a no-limit" provision? Explain.

2.47 Does alcohol consumption cause increased cravings for cigarettes? Research at Purdue University suggests this is so (see CNN.com web site article "Researchers Find Link Between Cigarette Cravings and Alcohol," dated June 13, 1997). In an experiment, 60 heavy smokers and moderate drinkers were divided into two groups. One group drank vodka tonics and the other group drank virgin tonics (tonic water alone), but all subjects were told they were drinking vodka tonics. The researchers then measured the level of nicotine cravings (by monitoring heart rate, skin conductance, etc.). Those who had consumed the vodka tonics had 35% more cravings than those who did not. Assuming that the assignment of subjects to the treatment (vodka) and control groups was made at random, do you think there are any confounding factors that would make conclusions based on this experiment questionable?

2.48 The level of lead in children's blood should be kept low, and one way to reduce a child's exposure to lead is to control dust in the home. Researchers at the University of Rochester studied the effects of families' dust control on children with low to mildly elevated blood lead levels ("A Randomized Trial of the Effect of Dust Control on Children's Blood Lead Levels," *Pediatrics* [1996]: 35–40). They designed an experiment in which they randomly assigned families whose children had elevated blood lead levels to two groups: intervention and control. The intervention group was given cleaning supplies and a demonstration of proper cleaning techniques. The control group was given a pamphlet on how to avoid lead exposure.

The researchers measured lead levels in the children's blood and in the household dust at the beginning of the study and also 7 months later.

a. Identify the population in this experiment.

b. Discuss the role and importance of randomization in this study.

c. Discuss the purpose of a control group.

d. Identify the treatments and response variable(s).

e. Discuss possible extraneous factors and how they could be controlled.

f. Discuss whether and how blocking should be used in this experiment.

■ 2.5 More on Experimental Design

The previous section covered basic principles for designing simple comparative experiments — control, blocking, randomization, and replication. The goal of an experimental design is to provide a method of data collection that (1) minimizes extraneous sources of variability in the response so that any differences in response for various experimental conditions can be more easily assessed and (2) creates experimental groups that are similar with respect to extraneous variables that cannot be controlled either directly or through blocking.

In this section, we look at some additional considerations that you may need to think about when planning an experiment.

■ Use of a Control Group

If the purpose of an experiment is to determine whether some treatment has an effect, it is important to include an experimental group that does not receive the treatment. Such a group is called a **control group**. The use of a control group allows the experimenter to assess how the response variable behaves when the treatment is not used. This provides a baseline against which the treatment groups can be compared to determine whether the treatment had an effect.

■ Example 2.6   Comparing Gasoline Additives

Suppose that an experimenter wants to know whether a gasoline additive increases fuel efficiency (miles per gallon). Such an experiment might use a single car (to eliminate car-to-car variability) and a sequence of trials in which 1 gal of gas is put in an empty tank, the car is driven around a racetrack at a constant speed, and the distance traveled on the gallon of gas is recorded.

To determine whether the additive increases gas mileage, it would be necessary to include a control group of trials where distance traveled was measured when gasoline without the additive was used. The experiment might consist of 24 trials. These trials would then be assigned *at random* to one of the two experimental conditions (additive or no additive).

Even though this experiment consists of a sequence of trials all with the same car, random assignment of trials to experimental conditions is still important because there will always be uncontrolled variability. For example, temperature or other environmental conditions might change over the sequence of trials, the physical condition of the car might change slightly from one trial to another, and so on. Random assignment of experimental conditions to trials will tend to even out the effects of these uncontrollable factors.

Although we usually think of a control group as one that receives no treatment, in experiments designed to compare a new treatment to an existing standard treatment, the term *control group* is sometimes used to describe the group that receives the current standard treatment.

Not all experiments require the use of a control group. For example, many experiments are designed to compare two or more conditions — an experiment to compare density for three different formulations of bar soap or an experiment to determine how oven temperature affects the cooking time of a particular type of cake. However, sometimes you will see a control group included even when the ultimate goal is to compare two or more different treatments. An experiment with two treatments and no control group might allow us to determine whether or not there is a difference between the two treatments and even to assess the magnitude of the difference if one exists, but it would not allow us to assess the individual effect of either treatment. For example, without a control group, we might be able to say that there is no difference in the increase in mileage for two different gasoline additives, but we wouldn't be able to tell if this was because both additives increased gas mileage by a similar amount or because neither additive had any effect on gas mileage.

▪ Use of a Placebo

In experiments that use human subjects, use of a control group may not be enough to determine whether a treatment really does have an effect. People sometimes respond merely to the power of suggestion! For example, consider a study designed to determine whether a particular herbal supplement is effective in promoting weight loss. Suppose that the study is designed with an experimental group that takes the herbal supplement and a control group that takes nothing. It is possible that those who take the herbal supplement and believe that they are taking something that will help them to lose weight may be more motivated and may unconsciously change their eating behavior or activity level, resulting in weight loss.

Although there is debate about the degree to which people respond, many studies have shown that people sometimes respond to treatments with no active ingredients, such as sugar pills or solutions that are nothing more than colored water, and that they often report that such "treatments" relieve pain or reduce symptoms such as nausea or dizziness. So, if an experiment is to enable researchers to determine whether a treatment really has an effect, comparing a treatment group to a control group may not be enough. To address the problem, many experiments use what is called a placebo.

A **placebo** is something that is identical (in appearance, taste, feel, etc.) to the treatment received by the treatment group, except that it contains no active ingredients.

For example, in the herbal supplement example, rather than using a control group that received *no* treatment, the researchers might want to include a placebo group. Individuals in the placebo group would take a pill that looked just like the herbal supplement but did not contain the herb or any other active ingredient. As long as the subjects did not know whether they were taking the herb or the placebo, the placebo group would provide a better basis for comparison and would allow the researchers to determine whether the herbal supplement had any real effect over and above the "placebo effect."

■ Single-Blind and Double-Blind Experiments

Because people often have their own personal beliefs about the effectiveness of various treatments, it is desirable to conduct experiments in such a way that subjects do not know what treatment they are receiving. For example, in an experiment comparing four different doses of a medication for relief of headache pain, someone who knows that he is receiving the medication at its highest dose may be subconsciously influenced to report a greater degree of headache pain reduction. By ensuring that subjects are not aware of which treatment they receive, we can prevent personal perception from influencing the response.

An experiment in which subjects do not know what treatment they have received is described as **single-blind**. Of course, not all experiments can be made single-blind. For example, in an experiment to compare the effect of two different types of exercise on blood pressure, it is not possible for participants to be unaware of whether they are in the swimming group or the jogging group! However, when it is possible, "blinding" the subjects in an experiment is generally a good strategy.

In some experiments, someone other than the subject is responsible for measuring the response. To ensure that the person measuring the response does not let personal beliefs influence the way in which the response is recorded, the researchers should make sure that the measurer does not know which treatment was given to any particular individual. For example, in a medical experiment to determine whether a new vaccine reduces the risk of getting the flu, doctors must decide whether or not a particular individual who is not feeling well actually has the flu or some other unrelated illness. If the doctor who had to make this assessment knew that a participant with flulike symptoms had been vaccinated with the new flu vaccine, she might be less likely to determine that the participant had the flu and more likely to interpret the symptoms as being the result of some other illness.

There are two ways in which blinding might occur in an experiment. One involves blinding the participants, and the other involves blinding the individuals who measure the response. If participants do not know which treatment was received and those measuring the response do not know which treatment was given to which participant, the experiment is described as **double-blind**. If only one of the two types of blinding is present, the experiment is single-blind.

A **double-blind** experiment is one in which neither the subjects nor the individuals who measure the response know which treatment was received.

A **single-blind** experiment is one in which the subjects do not know which treatment was received but the individuals measuring the response do know which treatment was received, or one in which the subjects do know which treatment was received but the individuals measuring the response do not know which treatment was received.

▪ Experimental Units and Replication

An **experimental unit** is the smallest unit to which a treatment is applied. In the language of experimental design, treatments are assigned at random to experimental units, and replication means that each treatment is applied to more than one experimental unit.

Replication is necessary for randomization to be an effective way to create similar experimental groups and to get a sense of the variability in the values of the response for individuals that receive the same treatment. As we will see in later chapters (Chapters 9–15), this information is important because it enables us to use statistical methods to decide whether differences in the responses in different treatment groups can be attributed to the treatment received or if they can be explained by chance variation (the natural variability seen in the responses to a single treatment).

Be careful when designing an experiment to ensure that there is replication. For example, suppose that children in two third-grade classes are available to participate in an experiment to compare two different methods for teaching arithmetic. It might at first seem reasonable to select one class at random to use one method and then assign the other method to the remaining class. But what are the experimental units here? Treatments are randomly assigned to classes, and so classes are the experimental units. Because there are only two classes, with one assigned to each treatment, this is an experiment with no replication, even though there are many children in each class. We would *not* be able to determine whether there was a difference between the two methods based on data from this experiment, because we would have only one observation per treatment.

One last note on replication: Don't confuse replication in an experimental design with replicating an experiment. When investigators talk about replicating an experiment, they mean conducting a new experiment using the same experimental design as a previous experiment. Replicating an experiment is a way of confirming conclusions based on a previous experiment, but it does not eliminate the need for replication in each of the individual experiments themselves.

▪ Using Volunteers as Subjects in an Experiment

Although the use of volunteers in a study that involves collecting data through sampling is never a good idea, it is a common practice to use volunteers as subjects in an experiment. Even though the use of volunteers limits the researcher's ability to generalize to a larger population, random assignment of the volunteers to treatments should result in comparable groups, and so treatment effects can still be assessed.

▪ Exercises 2.49–2.58

2.49 Explain why some studies include both a control group and a placebo treatment. What additional comparisons are possible if both a control group and a placebo group are included?

2.50 Give an example of an experiment for each of the following:
a. Single-blind experiment with the subjects blinded

b. Single-blind experiment with the individuals measuring the response blinded
c. Double-blind experiment
d. An experiment that is not possible to blind

2.51 Explain why blinding is a reasonable strategy in many experiments.

2.52 A novel alternative medical treatment for heart attacks seeds the damaged heart muscle with cells from the patient's thigh muscle ("Doctors Mend Damaged Hearts with Cells from Muscles," *San Luis Obispo Tribune*, November 18, 2002). Doctor Dib from the Arizona Heart Institute evaluated the approach on 16 patients with severe heart failure. The article states that "ordinarily, the heart pushes out more than half its blood with each beat. Dib's patients had such severe heart failure that their hearts pumped just 23 percent. After bypass surgery and cell injections, this improved to 36 percent, although it was impossible to say how much, if any, of the new strength resulted from the extra cells."

a. Explain why it is not reasonable to generalize to the population of all heart attack victims based on the data from these 16 patients.

b. Explain why it is not possible to say whether any of the observed improvement was due to the cell injections, based on the results of this study.

c. Describe a design for an experiment that would allow researchers to determine whether bypass surgery plus cell injections was more effective than bypass surgery alone.

2.53 An article in the *San Luis Obispo Tribune* (September 7, 1999) described an experiment designed to investigate the effect of creatine supplements on the development of muscle fibers. The article states that the researchers "looked at 19 men, all about 25 years of age and similar in weight, lean body mass, and capacity to lift weights. Ten were given creatine — 25 grams a day for the first week, followed by 5 grams a day for the rest of the study. The rest were given a fake preparation. No one was told what he was getting. All the men worked out under the guidance of the same trainer. The response variable measured was gain in fat-free mass (in percent)."

a. What extraneous variables are identified in the given statement, and what strategy did the researchers use to deal with them?

b. Do you think it was important that the men participating in the experiment were not told whether they were receiving creatine or the placebo? Explain.

c. In a double-blind experiment, not only do the subjects not know which treatment they are receiving but also the person who measures the response variable does not know which treatment the subjects are receiving. This particular experiment was not conducted in a double-blind manner. Do you think it would have been a good idea to make this a double-blind experiment? Explain.

2.54 An experiment to evaluate whether vitamins can help prevent recurrence of blocked arteries in patients who have had surgery to clear blocked arteries was described in the article "Vitamins Found to Help Prevent Blocked Arteries" (Associated Press, September 1, 2002). The study involved 205 patients who were given either a treatment consisting of a combination of folic acid, vitamin B12, and vitamin B6 or a placebo for 6 months.

a. Explain why a placebo group was used in this experiment.

b. Explain why it would be important for the researchers to have assigned the 205 subjects to the two groups (vitamin and placebo) at random.

c. Do you think it is appropriate to generalize the results of this experiment to the population of all patients who have undergone surgery to clear blocked arteries? Explain.

2.55 Pismo Beach, California, has an annual clam festival that includes a clam chowder contest. Judges rate clam chowders from local restaurants, and the judging is done in such a way that the judges are not aware of which chowder is from which restaurant. One year, much to the dismay of the seafood restaurants on the waterfront, Denny's chowder was declared the winner! (When asked what the ingredients were, the cook at Denny's said he wasn't sure — he just had to add the right amount of nondairy creamer to the soup stock that he got from Denny's distribution center!)

a. Do you think that Denny's chowder would have won the contest if the judging had not been "blind"? Explain.

b. Although this was not an experiment, your answer to Part (a) helps to explain why those measuring the response in an experiment are often blinded. Using your answer in Part (a), explain why experiments are often blinded in this way.

2.56 The *San Luis Obispo Tribune* (May 7, 2002) reported that "a new analysis has found that in the majority of trials conducted by drug companies in recent decades, sugar pills have done as well as — or better than — antidepressants." What effect is being described here? What does this imply about the design of experiments with a goal of evaluating the effectiveness of a new medication?

2.57 Researchers at the University of Pennsylvania suggest that a nasal spray derived from pheromones (chemicals emitted by animals when they are trying to attract a mate) may be beneficial in relieving symptoms of premenstrual syndrome (PMS). Early tests indicated that the spray, called PH80, eases irritability and also reduces some physical symptoms (*Los Angeles Times*, January 17, 2003).

a. Describe how you might design an experiment using 100 female volunteers who suffer from PMS to determine whether the nasal spray reduces PMS symptoms.

b. Does your design from Part (a) include a placebo treatment? Why or why not?

c. Does your design from Part (a) involve blinding? Is it single-blind or double-blind? Explain.

2.58 The article "A Debate in the Dentist's Chair" (*San Luis Obispo Tribune*, January 28, 2000) described an ongoing debate over whether newer resin fillings are a better alternative to the more traditional silver amalgam fillings. Because amalgam fillings contain mercury, there is concern that they could be mildly toxic and prove to be a health risk to those with some types of immune and kidney disorders. One experiment described in the article used sheep as subjects and reported that sheep treated with amalgam fillings had impaired kidney function.

a. In the experiment, a control group of sheep that received no fillings was used but there was no placebo group. Explain why it is not necessary to have a placebo group in this experiment.

b. The experiment compared only an amalgam filling treatment group to a control group. What would be the benefit of also including a resin filling treatment group in the experiment?

c. Why do you think the experimenters used sheep rather than human subjects?

■ 2.6 More on Observational Studies: Designing Surveys (Optional)

Designing an observational study to compare two populations on the basis of some easily measured characteristic is relatively straightforward, with attention focusing on choosing a reasonable method of sample selection. However, many observational studies attempt to measure personal opinion or attitudes using responses to a survey. In such studies, both the sampling method and the design of the survey itself are critical to obtaining reliable information.

Surveys have been around for a long time and are likely to remain a fixture in our everyday lives for the foreseeable future. They are routinely used in business and marketing, in the formulation of public policy, and even in connection with legal issues. Market research surveys lead to efficient production of desired products; the Labor Department publishes monthly unemployment statistics based on survey data; and surveys are accepted (and argued over!) in trademark infringement suits in our courts. The Gallup, Roper, and Field organizations, among others, specialize in constructing and carrying out surveys.

Survey questions may be designed for different purposes. Questions such as

Do you feel the present tax rates for married couples are too high?

To what extent do you favor implementation of the 2003 Agricultural Trade Act?

attempt to measure attitudes, opinions, preferences, beliefs, and/or values. Other survey questions are designed to elicit facts. For example, the Department of Commerce's Current Population Survey (CPS) asks factual questions such as

Have you worked at a job or business at any time during the past 12 months?

When was the month and year that you last worked?

As of the end of last week, how long had you been looking for work?

Sociologists, psychologists, criminologists (in fact, all sorts of -ologists!) gather information by asking questions of individuals and evaluating their responses. It would seem at first glance that a survey must be a simple method for acquiring information. The investigators are interested in some attitude, belief, or fact. They formulate relevant questions and administer the survey to a random sample of individuals from the population of interest. They record the responses and analyze the data. However, it turns out that designing and administering a survey is not at all easy. Great care must be taken to ensure that good information can be extracted from a survey.

▪ Survey Overview

A **survey** is a voluntary encounter between strangers in which an interviewer seeks information from a respondent by engaging in a special type of social conversation. This conversation might take place in person, over the telephone, or even in the form of a written questionnaire, and it is quite different from usual conversations. Both the interviewer and the respondent have certain roles and responsibilities. The interviewer gets to decide what is relevant to the conversation and may ask questions — possibly personal or even embarrassing questions. These questions should be both understandable and relevant to the interviewer's purpose. The respondent, in turn, may refuse to participate in the conversation and may refuse to answer any particular question. But having agreed to participate in the survey, the respondent is responsible for answering the questions truthfully.

Both parties in a survey must give some thought to the survey process. As it happens, the interviewer has an easier role in the survey conversation. The questions are already written, and from the interviewer's point of view the survey process unfolds as a sequence of scripted questions. The respondent, on the other hand, has a more difficult role. Let's consider the situation of the respondent.

▪ The Respondent's Tasks

Our understanding of the survey process has been improved in the past two decades by contributions from the field of psychology, but there is still much uncertainty about how people respond to survey questions. Survey researchers and psychologists generally agree that the respondent is confronted with a sequence of tasks when asked a question. These tasks are comprehension of the question, retrieval of information from memory, and the reporting of the response.

▪ **Task 1: Comprehension** The first task of the respondent is to understand the survey directions and each question as it is asked. Comprehension is the single most important task facing the respondent, and fortunately it is the characteristic of a survey question that is most easily controlled by the question writer. Comprehensible questions are characterized by (1) a vocabulary appropriate to the population of interest, (2) simple sentence structure, and (3) little or no ambiguity.

Vocabulary is often a problem. The investigator is usually knowledgeable about the topic of interest, and it may be difficult to remember that others do not have that special knowledge. In addition, researchers tend to be well educated and may have a more extensive vocabulary than the respondents. As a rule, it is best to use the simplest possible word that can be used without sacrificing clear meaning. A dictionary and a thesaurus are invaluable in the search for simplicity. For example:

Huh?	Oh, you mean . . .
Emulate	Imitate
Eccentric	Strange
Imbibe	Drink
Prior	Earlier

Simple sentence structure also makes it easier for the respondent to understand the question. A famous example of difficult syntax occurred in 1993 when the Roper organization created a survey related to the Holocaust, the Nazi extermination of Jews during World War II. One question in this survey was

"Does it seem possible or does it seem impossible to you that the Nazi extermination of the Jews never happened?"

The question has a complicated structure and a double negative — "impossible . . . never happened" — that could lead respondents to give an answer opposite to what they actually believed. The question was rewritten and given a year later in an otherwise unchanged survey:

"Does it seem possible to you that the Nazi extermination of the Jews never happened, or do you feel certain that it happened?"

This question wording is much clearer, and in fact the respondents' answers were quite different, as shown in the following table (the "unsure" and "no opinion" proportions have been omitted):

Original Roper Poll		Revised Roper Poll	
Impossible	65%	Certain it happened	91%
Possible	12%	Possible it never happened	1%

Keeping vocabulary and sentence structure simple is relatively easy compared to filtering out ambiguity in questions. In part, this is because precise and unambiguous language may be difficult to comprehend. Even the most innocent and seemingly clear questions can have a number of possible interpretations. For example, suppose that you are asked, "When did you move to Cedar Rapids?" This would seem to be an unambiguous question, but some possible answers might be (1) "In 1971," (2) "When I was 23," and (3) "In the summer." The respondent must decide which of these three answers, if any, is the appropriate response. It may be possible to lessen the ambiguity with more precise questions:

1. In what year did you move to Cedar Rapids?
2. How old were you when you moved to Cedar Rapids?
3. In what season of the year did you move to Cedar Rapids?

One way to find out whether or not a question is ambiguous is to field-test the question and to ask the respondents if they were unsure how to answer a question. The following table illustrates some of the ambiguities identified by field-testing questions:

Question	Ambiguity
Do you think children suffer any ill effects from watching programs with violence in them?	The word *children* was interpreted to mean everything from babies to teenagers to young adults in their early twenties.
What is the number of servings of eggs you eat in a typical day?	It was unclear to the respondents what a "serving" of eggs was, as well as what the term *typical day* meant.

Question	Ambiguity
What is the average number of days each week you consume butter?	Respondents were unclear whether margarine should count as butter.

Ambiguity can arise from the placement of questions as well as from their phrasing. Here is an example of ambiguity uncovered when the order of two questions differed in two versions of a survey on happiness. The questions were

1. Taken altogether, how would you say things are these days: Would you say that you are very happy, pretty happy, or not too happy?

2. Taking things altogether, how would you describe your marriage: Would you say that your marriage is very happy, pretty happy, or not too happy?

The proportions of responses to the general happiness question differed for the different question orders, as follows:

Response to General Happiness Question

	General Asked First	General Asked Second
Very happy	52.4%	38.1%
Pretty happy	44.2	52.8
Not too happy	3.4	9.1

If the goal in this survey was to estimate the proportion of the population that is generally happy, these numbers are quite troubling — they cannot both be right! What seems to have happened is that Question 1 was interpreted differently depending on whether it was asked first or second. When the general happiness question was asked after the marital happiness question, the respondents apparently interpreted it to be asking about their happiness in all aspects of their lives *except* their marriage. This was a reasonable interpretation, given that they had just been asked about their marital happiness, but it is a different interpretation than when the general happiness question was asked first. The troubling lesson here is that even carefully worded questions can have different interpretations in the context of the rest of the survey.

■ **Task 2: Retrieval from Memory** Once a question is understood, the respondent must retrieve relevant information from memory to answer the question. This is not always an easy task, and it is not a problem limited to questions of fact.

Psychologists do not agree completely on the nature of human memory, but most seem to believe that memory consists of stored representations of events in the lives of individuals. Some memories are particularly salient, such as wedding events and memories of where one was at the time of tragedies (e.g., a presidential assassination, a space shuttle explosion). Other events, the more typical daily memories, seem to be stored generically. For example, it is unlikely that you remember every trip to the grocery store; instead, you have a general idea of a typical trip stored in your memory. Thus, unless a survey question is about a particularly salient event, respondents reconstruct the memory by piecing together memories of typical events that are suggested by the question. For example, consider this seemingly elementary "factual" question:

How many times in the past 5 years did you visit your dentist's office?

a. 0 times

b. 1–5 times

c. 6–10 times

d. 11–15 times

e. more than 15 times

It is unlikely that many people will remember with clarity every single visit to the dentist in the most recent 5 years, especially since the advent of Novocain! But generally, people will respond to such a question with answers consistent with the memories and facts they are able to reconstruct given the time they have to respond to the question. An individual may, for example, have a sense that he usually makes about two trips a year to the dentist's office. There is no option presented as, "I think usually about two trips a year," so the respondent may extrapolate the typical year and get 10 times in 5 years. Then there may be three particularly memorable visits, say, for a root canal in the middle of winter. Thus, the best recollection is now 13, and the respondent will choose Answer (d), 11–15 times. Perhaps not exactly correct, but the best that can be reported under the circumstances.

What are the implications of this relatively fuzzy memory for those who would construct surveys about facts? First, the investigator should understand that most factual answers are going to be approximations of the truth. Second, events closer to the time of a survey are easier to recall. A question about visits to the dentist in the past year can be answered with more accuracy than a question about visits in the past 5 years. Third, memories of events are cued by the questions that are asked in a survey. The more carefully events of interest can be described in a survey question, the better the chance that the question will cue the right memories. In particular, emotional, important, and distinctive events are more easily recalled. Memories of the timing of events seem to be stored not as clock or calendar events but as events near landmark events, such as the beginning of a term or near the time that a new house was built.

Oddly, attitude and opinion questions can also be affected in significant ways by the workings of human memory. In general, the mechanism for these effects is thought to be the respondent's memory of recently asked questions. If a question is related in some way to a recently asked question, the previous question may provide fresh cues to the respondent's memory. One recent study contained a survey question asking respondents their opinion about how much they followed politics. When that question was preceded by a factual question asking whether they knew the name of the congressional representative from their district, the percentage who reported they follow politics "now and then" or "hardly ever" jumped from 21% to 39%! Respondents apparently concluded that, because they didn't know the answer to the previous knowledge question, they must not follow politics as much as they would have thought otherwise.

The manipulation of memory can sometimes affect the opinions of respondents, especially if the opinions are not strongly held. For example, suppose that a survey asks for an opinion about the degree to which the respondent believes drilling for oil should be permitted in national parks. If the question is preceded by questions about the high price of gasoline, a different response might result than if the question is preceded by questions about the environment and ecology.

■ **Task 3: Reporting the Response** The third task of the respondent in a survey is to actually formulate and report a response. It is this task that can be affected by the social aspects of the survey conversation. In general, if a respondent agrees to take a survey, he or she will be motivated to answer truthfully. Therefore, if the questions aren't too difficult (taxing the respondent's knowledge or memory) and if there aren't too many questions (taxing the respondent's patience and stamina), the answers to questions will be as accurate as possible. However, it is also true that the respondents will wish to present themselves in a favorable light. This desire leads to what is known as a social desirability bias. Sometimes this bias is a response to the particular wording in a question. In 1941, the following questions were analyzed in two different forms of a survey (emphasis added):

1. Do you think the United States should *forbid* public speeches against democracy?
2. Do you think the United States should *allow* public speeches against democracy?

It would seem logical that these questions are opposites and that the proportion who would not allow public speeches against democracy should be equal to the proportion who would forbid public speeches against democracy. But only 45% of those respondents offering an opinion on Question 1 thought the United States should "forbid," whereas 75% of the respondents offering an opinion on Question 2 thought the United States should "not allow" public speeches against democracy. Could it be that respondents thought that *forbid* had a nasty ring to it, something they would want to keep at arm's length?

Some survey questions may be sensitive or threatening, such as questions about sex, drugs, or potentially illegal behavior. In this situation a respondent not only will want to present a positive image but also will certainly think twice about admitting illegal behavior! In such cases the respondent may shade the actual truth or may even lie about particular activities and behaviors. In addition, the tendency toward positive presentation is not limited to obviously sensitive questions. For example, consider the question about general happiness previously described. Several investigators have reported higher happiness scores in face-to-face interviews than in responses to a mailed questionnaire. Presumably, a happy face presents a more positive image of the respondent to the interviewer. On the other hand, if the interviewer was a clearly unhappy person, a respondent might shade answers to the less happy side of the scale, perhaps thinking that it is inappropriate to report happiness in such a situation.

The role of the interviewer can also influence responses. Who admits to their dentist that they aren't flossing? Or suppose that English teachers are conducting a survey about the reading habits of their students. Might students suddenly develop an apparent interest in reading or report that they read for pleasure more than is the truth?

It is clear that constructing surveys and writing survey questions can be a daunting task. Keep in mind the following three things:

1. Questions should be understandable by the individuals in the population being surveyed. Vocabulary should be at an appropriate level, and sentence structure should be simple.

2. Questions should, as much as possible, recognize that human memory is fickle. Questions that are specific will aid the respondent by providing better memory cues. The limitations of memory should be kept in mind when interpreting the respondent's answers.

3. As much as possible, questions should not create opportunities for the respondent to feel threatened or embarrassed. In such cases respondents may introduce a social desirability bias, the degree of which is unknown to the interviewer. This can compromise conclusions drawn from the survey data.

Constructing good surveys is a difficult task, and we have given only a brief introduction to this topic. For a more comprehensive treatment, we recommend the book by Sudman and Bradburn listed in the references at the end of this chapter.

▪ Exercises 2.59–2.64

2.59 A tropical forest survey conducted by the group Conservation International included statements such as the following in the four pages of material that accompanied the survey:

> "A massive change is burning its way through the earth's environment."

> "The band of tropical forests that encircle the earth is being cut and burned to the ground at an alarming rate."

> "Never in history has mankind inflicted such sweeping changes on our planet as the clearing of rain forest taking place right now!"

The Tropical Forest Survey that followed included the questions given in Parts (a)–(d). For each of these questions, identify a word or phrase that might affect the response and possibly bias the results of any analysis of the questions.

a. "Did you know that the world's tropical forests are being destroyed at the rate of 80 acres per minute?"

b. "Considering what you know about vanishing tropical forests, how would you rate the problem?"

c. "Do you think we have an obligation to prevent the manmade extinction of animal and plant species?"

d. "Based on what you know now, do you think there is a link between the destruction of tropical forests and changes in the earth's atmosphere?"

2.60 Educators are increasingly concerned about the sleep habits of their students. Fast-paced lifestyles, where students balance the requirements of school, after-school activities, and jobs, are thought by some to lead to reduced sleep. One such study was described in the paper "Differences in Reported Sleep Need Among Adolescents" (*Journal of Adolescent Health* [1998]: 259–263). Suppose that you are assigned the task of designing a survey that will provide answers to the accompanying questions. Write a set of survey questions that might be used. In some cases, you may need to write more than one question to adequately address a particular issue. For example, responses might be different for weekends and school nights. You may also have to define some terms to make the questions comprehensible to the target audience, which is adolescents.

Topics to be addressed:

How much sleep do the respondents get?

Is this enough sleep?

Does sleepiness interfere with schoolwork?

If they could change the starting and ending times of the school day, what would they suggest? (Sorry, they cannot reduce the total time spent in school during the day!)

2.61 Asthma is a chronic lung condition characterized by difficulty in breathing. People with asthma have extrasensitive airways, which react by narrowing or obstructing when they become irritated, resulting in wheezing, coughing, and other symptoms. Some studies have suggested that asthma may be related to childhood exposure to some animals, especially dogs and cats, during the first year of life ("Exposure to Dogs and Cats in the First Year of Life and Risk of Allergic Sensitization at 6 to 7 Years of Age," *Journal of the American Medical Association* [2002]: 963–972). Some environmental factors that trigger an asthmatic response are (1) cold air, (2) dust, (3) strong fumes, and (4) inhaled irritants.

a. Write a set of questions that could be used in a

survey to be given to parents of young children suffering from asthma. The survey should include questions about the presence of pets in the first year of the child's life as well as questions about the presence of pets today. Also, the survey should include questions that address the four mentioned household environmental factors.

b. It is generally thought that low-income persons, who tend to be less well educated, have homes in environments where the four environmental factors are present. Mindful of the importance of comprehension, can you improve the questions in Part (a) by making your vocabulary simpler or by changing the wording of the questions?

c. One problem with the pet-related questions is the reliance on memory. That is, parents may not actually remember when they got their pets. How might you check the parents' memories about these pets?

2.62 In national surveys, parents consistently point to school safety as an important concern, and principals and teachers are constantly working to provide a safe environment for their charges. One source of violence in junior high schools is fighting ("Self-Reported Characterization of Seventh-Grade Student Fights," *Journal of Adolescent Health* [1998]: 103–109). It is not altogether clear what causes fights, and the characteristics of fights are unknown. To construct a knowledge base about student fights, a school administrator wants to give two surveys to students after fights are broken up. One of the surveys is to be given to the participants, and the other is to be given to students who witnessed the fight. You will write questions for these two surveys. Each question should include a set of possible responses from which the individual responding to the question can select. The type of information desired includes (1) the cause of the fight, (2) whether or not the fight was a continuation of a previous fight, (3) whether drugs or alcohol was a factor, (4) whether or not the fight was gang related, and (5) the role of bystanders.

a. Write a set of questions that could be used in the two surveys. For each question, indicate whether it would be used on both surveys or just on one of the two.

b. How might the tendency toward positive self-presentation affect the responses of the fighter to the survey questions you wrote for Part (a)?

c. How might the tendency toward positive self-presentation affect the responses of a bystander to the survey questions you wrote for Part (a)?

2.63 The consumption of soft drinks is a health issue of some concern to pediatricians ("Teenaged Girls, Carbonated Beverage Consumption, and Bone Fractures," *Archives of Pediatric and Adolescent Medicine* [2000]: 610–613). Doctors are concerned about young women drinking large amounts of soda and about their decreased consumption of milk in recent years. It is possible that early adolescence is a critical period in bone formation and that this changing consumption could lead to more bone fractures and enhanced risk of osteoporosis in later life. It is also possible that more physically active females also have increased risk of bone fractures. For each of the following Parts (a)–(d), construct two questions that might be included in a survey of teenage girls. Each question should include possible responses from which the respondent can select. (Note: The questions as written are vague. Your task is to clarify the questions for use in a survey, not just to change the syntax!)

a. How much "cola" beverage does the respondent consume?

b. How much milk (and milk products) is consumed by the respondent?

c. How physically active is the respondent?

d. What is the respondent's history of bone fractures?

2.64 A recent survey attempted to address psychosocial factors thought to be of importance in preventive health care for adolescents ("The Adolescent Health Review: A Brief Multidimensional Screening Instrument," *Journal of Adolescent Health* [2001]: 131–139). Some of the factors thought to be risk areas indicate a need for preventive health care. For each risk area in the following list, construct a question that would be comprehensible to students in grades 9–12 and that would provide information about the risk factor. Make your questions multiple-choice, and provide possible responses.

a. Lack of exercise

b. Poor nutrition

c. Emotional distress

d. Sexual activity

e. Cigarette smoking

f. Alcohol use

■ 2.7 Communicating and Interpreting the Results of Statistical Analyses

Statistical studies are conducted to allow investigators to answer questions about characteristics of some population of interest or about the effect of some treatment. Such questions are answered on the basis of data, and how the data are obtained determines the quality of information available and the type of conclusions that can be drawn. As a consequence, when describing a study you have conducted (or when evaluating a published study), you must consider how the data were collected.

The description of the data collection process should make it clear whether the study is an observational study or involves an experiment. For observational studies, some of the issues that should be addressed are:

1. What is the population of interest? What is the sampled population? Are these two populations the same? If the sampled population is only a subset of the population of interest, **undercoverage** limits our ability to generalize to the population of interest. For example, if the population of interest is all students at a particular university, but the sample is selected from only those students who choose to list their phone number in the campus directory, undercoverage may be a problem. We would need to think carefully about whether it is reasonable to consider the sample as representative of the population of all students at the university. For some purposes, this might be reasonable, but in other cases it may not be. **Overcoverage** results when the sampled population is actually larger than the population of interest. This would be the case if we were interested in the population of all high schools that offer Advanced Placement (AP) Statistics but sampled from a list of all schools that offered an AP class in any subject. Both undercoverage and overcoverage can be problematic.

2. How were the individuals or objects in the sample actually selected? A description of the sampling method facilitates making judgments about whether the sample can reasonably be viewed as representative of the population of interest.

3. What are potential sources of bias, and is it likely that any of these will have a substantial effect on the observed results? When describing an observational study, you should acknowledge that you are aware of potential sources of bias and explain any steps that were taken to minimize their effect. For example, in a mail survey, nonresponse can be a problem, but the sampling plan may seek to minimize its effect by offering incentives for participation and by following up one or more times with those who do not respond to the first request.

   A common misperception is that increasing the sample size is a way to reduce bias in observational studies, but this is not the case. For example, if measurement bias is present, as in the case of a scale that is not correctly calibrated and tends to weigh too high, taking 1000 measurements rather than 100 measurements cannot correct for the fact that the measured weights will be too large. Similarly, a larger sample size cannot compensate for response bias introduced by a poorly worded question.

For experiments, some of the issues that should be addressed are:

1. What is the role of randomization? All good experiments use random assignment as a means of coping with the effects of potentially confounding vari-

ables that cannot easily be directly controlled. When describing an experimental design, you should be clear about how random assignment (subjects to treatments, treatments to subjects, or treatments to trials) was incorporated into the design.

2. Were any extraneous variables directly controlled by holding them at fixed values throughout the experiment? If so, which ones and at which values?

3. Was blocking used? If so, how were the blocks created? If an experiment uses blocking to create groups of homogeneous experimental units, you should describe the criteria used to create the blocks and their rationale. For example, you might say something like "Subjects were divided into two blocks — those who exercise regularly and those who do not exercise regularly — because it was believed that exercise status might affect the responses to the diets."

Because each treatment appears at least once in each block, the block size must be at least as large as the number of treatments. Ideally, the block sizes should be equal to the number of treatments, because this presumably would allow the experimenter to create small groups of extremely homogeneous experimental units. For example, in an experiment to compare two methods for teaching calculus to first-year college students, we may want to block on previous mathematics knowledge by using math SAT scores. If 100 students are available as subjects for this experiment, rather than creating two large groups (above-average math SAT score and below-average math SAT score), we might want to create 50 blocks of two students each, the first consisting of the two students with the highest math SAT scores, the second containing the two students with the next highest scores, and so on. We would then select one student in each block at random and assign that student to teaching method 1. The other student in the block would be assigned to teaching method 2.

▪ A Word to the Wise: Cautions and Limitations

It is a cardinal sin to begin collecting data before thinking carefully about research objectives and developing a research plan. Failure to do so may result in data that do not enable the researcher to answer key questions of interest or to generalize conclusions based on the data to the desired populations of interest.

Clearly defining the objectives at the outset enables the investigator to determine whether an experiment or an observational study is the best way to proceed. Watch out for the following:

1. Drawing a cause-and-effect conclusion from an observational study. Don't do this, and don't believe it when others do it!

2. Generalizing results of an experiment that uses volunteers as subjects to a larger population without a convincing argument that the group of volunteers can reasonably be considered a representative sample from the population.

3. Generalizing conclusions based on data from a sample to some population of interest. This is sometimes a sensible thing to do, but on other occasions it is not reasonable. Generalizing from a sample to a population is justified only when there is reason to believe that the sample is likely to be representative of the population. This would be the case if the sample was a random sample from the population and there were no major potential sources of bias. If the sample was not selected at random or if potential sources of bias were

present, these issues would have to be addressed before a judgment could be made regarding the appropriateness of generalizing the study results.

For example, the Associated Press (January 25, 2003) reported on the high cost of housing in California. The median home price was given for each of the 10 counties in California with the highest home prices. Although these 10 counties are a sample of the counties in California, they were not randomly selected and (because they are the 10 counties with the highest home prices) it would not be reasonable to generalize to all California counties based on data from this sample.

4. Generalizing conclusions based on an observational study that used voluntary response or convenience sampling to a larger population. This is almost never reasonable.

▪ Activity 2.1: Designing a Sampling Plan

Background: In this activity, you will work with a partner to develop a sampling plan.

Suppose that you would like to select a sample of 50 students at your school to learn something about how many hours per week, on average, students at your school spend engaged in a particular activity (such as studying, surfing the Internet, or watching TV).

1. Discuss with your partner whether you think it would be easy or difficult to obtain a simple random sample of students at your school and to obtain the desired information from all the students selected for the sample. Summarize your discussion by writing a few sentences explaining why you think it would be easy or difficult.

2. With your partner, decide how you might go about selecting a sample of 50 students from your

school that, although it truly may not be a simple random sample, could be reasonably considered representative of the population of interest. Write a brief description of your sampling plan, and be sure to point out the aspects of your plan that you think make it reasonable to argue that it will be representative.

3. Explain your plan to another pair of students. Ask them to critique your plan, pointing out any potential flaws they see in it. Write a brief summary of the comments you received. Now reverse roles, and provide a critique of the plan devised by the other pair.

4. Based on the feedback you received in Step 3, would you modify your original sampling plan? If not, explain why this is not necessary. If so, describe how the plan would be modified.

▪ Activity 2.2: An Experiment to Test for the Stroop Effect

Background: In 1935, John Stroop published the results of his research into how people respond when presented with conflicting signals. Stroop noted that most people are able to read words quickly and that they cannot easily ignore them and focus on other attributes of a printed word, such as text color.

For example, consider the following list of words:

green   blue   red   blue   yellow   red

It is easy to quickly read this list of words. It is also easy to read the words even if the words are printed in color, and even if the text color is different from the color of the word. For example, people can read the words in the list

green   blue   red   blue   yellow   red

as quickly as they can read the list that isn't printed in color.

However, Stroop found that if people are asked to name the text colors of the words in the list (red, yellow, blue, green, red, green), it takes them longer. Psychologists believe that this is because the reader has to inhibit a natural response (reading the word) and produce a different response (naming the color of the text).

If Stroop is correct, people should be able to name colors more quickly if they do not have to inhibit the word response, as would be the case if they were shown the following:

1. Design an experiment to compare times to identify colors when they appear as text to times to identify colors when there is no need to inhibit a word response. Be sure to indicate how randomization is incorporated into your design. What is your response variable? How will you measure it? How many subjects will you use in your experiment, and how will they be chosen?

2. When you are satisfied with your experimental design, carry out the experiment. You will need to construct your list of colored words and a corresponding list of colored bars to use in the experiment. You will also need to think about how you will implement your randomization scheme.

3. Summarize the resulting data in a brief report that explains whether your findings are consistent with the Stroop effect.

■ Summary of Key Concepts and Formulas

Term or Formula	Comment
Observational study	A study that observes characteristics of an existing population.
Simple random sample of size n	A sample selected in a way that gives every different sample of size n an equal chance of being selected.
Stratified sampling	Dividing a population into subgroups (strata) and then taking a separate random sample from each stratum.
Cluster sampling	Dividing a population into subgroups (clusters) and forming a sample by randomly selecting clusters and including all individuals or objects in the selected clusters in the sample.
1 in k systematic sampling	A sample selected from an ordered arrangement of a population by choosing a starting point at random from the first k individuals on the list and then selecting every kth individual thereafter.
Confounding variable	A variable that is related both to group membership and to the response variable.
Measurement or response bias	The tendency for a sample to differ from the population because the method of observation tends to produce values that differ from the true value.
Selection bias	The tendency for a sample to differ from the population because of systematic exclusion of some part of the population.
Nonresponse bias	The tendency for a sample to differ from the population because measurements are not obtained from all individuals selected for inclusion in the sample.
Experiment	A procedure for investigating the effect of *experimental conditions* (which are manipulated by the experimenter) on a *response variable*.
Treatments	The experimental conditions imposed by the experimenter.
Extraneous factor	A variable that is not of interest in the current study but is thought to affect the response variable.
Direct control	Holding extraneous factors constant so that their effects are not confounded with those of the experimental conditions.
Blocking	Using extraneous factors to create experimental groups that are similar with respect to those factors, thereby filtering out their effect.
Randomization	Random assignment of experimental units to treatments or of treatments to trials.

Term or Formula	Comment
Replication	Ensuring that there is an adequate number of observations on each experimental treatment.
Placebo treatment	A treatment that resembles the other treatments in an experiment in all apparent ways but that has no active ingredients.
Control group	A group that receives no treatment or one that receives a placebo treatment.
Single-blind experiment	An experiment in which the subjects do not know which treatment they received but the individuals measuring the response do know which treatment was received, or an experiment in which the subjects do know which treatment they received but the individuals measuring the response do not know which treatment was received.
Double-blind experiment	An experiment in which neither the subjects nor the individuals who measure the response know which treatment was received.

■ Supplementary Exercises 2.65–2.72

2.65 The article "Heavy Drinking and Problems among Wine Drinkers" (*Journal of Studies on Alcohol* [1999]: 467–471) attempts to answer the question of whether wine drinkers tend to drink less excessively than those who drink beer and spirits. A sample of Canadians, stratified by province of residence and other socioeconomic factors, was selected.
a. Why might stratification by province be a good thing?
b. List two socioeconomic factors that would be appropriate to use for stratification. Explain how each factor would relate to the consumption of alcohol in general and of wine in particular.

2.66 According to Social Security Administration records, the most popular name for girls born in the United States in 1982 is Jennifer, with 30.6 per 1000 girls being given that name. By 1998, Jennifer had fallen in popularity to 20th place, with 8569 girls of the 1,579,399 girls born nationwide receiving the name. The 1982 results are based on a "one-percent sample of Social Security Number card applications" (required for every baby born in the United States within a year of birth). The 1998 results are based on a 100% sample of Social Security records.
a. Calculate the rate per 1000 of Jennifers in 1998.
b. What is another name for a 100% sample, such as the one of 1998 birth names?
c. The 1% sample from 1982 is "based on the last four digits of the Social Security Number, which are random and not correlated with a particular geographical region." Would you say that this sample is a simple random sample of names of girls? Explain.

2.67 The general manager of a luxury hotel decides to check the quality of housekeeping by inspecting

15 rooms himself. Discuss how the manager should determine which rooms to check. Consider that the hotel offers economy and business-class rooms as well as suites. Would a simple random sample or a stratified sample be more appropriate? Also discuss possible extraneous factors.

2.68 The size of a tip that a restaurant server receives is determined, in part, by the behavior of the server. Researchers at the University of Houston decided to test the hypothesis that servers who squat to the level of their customers would receive a larger tip ("Effect of Server Posture on Restaurant Tipping," *Journal of Applied Social Psychology* [1993]: 678–685). In the experiment, the waiter would flip a coin to determine whether he would stand or squat next to the table. The waiter would record the amount of the bill and of the tip and whether he stood or squatted.
a. Describe the treatments and the response variable.
b. Discuss possible extraneous factors and how they could be controlled.
c. Discuss whether blocking would be necessary.
d. Identify possible confounding variables.
e. Discuss the role of randomization in this experiment.

2.69 You have been asked to determine on what types of grasslands two species of birds, northern harriers and short-eared owls, build nests. The types of grasslands to be used include undisturbed native grasses, managed native grasses, undisturbed nonnative grasses, and managed nonnative grasses. You are allowed a plot of land 500 m square to study. Explain how you would determine where to plant the four types of grasses. What role would randomization play in this determination? Identify any confounding variables. Would this study be considered an observational study or an experiment? (Based on the article "Response of Northern Harriers and Short-Eared Owls to Grassland Management in Illinois," *Journal of Wildlife Management* [1999]: 517–523.)

2.70 Clay tiles are fired in a kiln. Unfortunately, some of the tiles crack during the firing process. A manufacturer of clay roofing tiles would like to investigate the effect of clay type on the proportion of cracked tiles. Two different types of clay are to be considered. One hundred tiles can be placed in the kiln at any one time. Firing temperature varies slightly at different locations in the kiln, and firing temperature may also affect cracking. Discuss the design of an experiment to collect information that could be used to decide between the two clay types. How does your proposed design deal with the extraneous factor *temperature*?

2.71 A mortgage lender routinely places advertisements in a local newspaper. The advertisements are of three different types: one focusing on low interest rates, one featuring low fees for first-time buyers, and one appealing to people who may want to refinance their homes. The lender would like to determine which advertisement format is most successful in attracting customers to call for more information. Describe an experiment that would provide the information needed to make this determination. Be sure to consider extraneous factors, such as the day of the week that the advertisement appears in the paper, the section of the paper in which the advertisement appears, or daily fluctuations in the interest rate. What role does randomization play in your design?

2.72 The headline "Men Cause More Car Accidents" appeared in the Australian newspaper *The Mercury* (September 12, 1995). The article stated that men were twice as likely as women to be involved in a car accident. This conclusion was based on a sample of 4000 insurance claims. Do you think the headline is justified based on this study? Explain.

■ References

Cobb, George. *Introduction to the Design and Analysis of Experiments*. Springer-Verlag, 1997. (An interesting and thorough introduction to the design of experiments.)

Freedman, David, Robert Pisani, and Roger Purves. *Statistics*, 3d ed. New York: W. W. Norton, 1997. (The first two chapters contain some interesting examples of both well-designed and poorly designed experimental studies.)

Hock, Roger R. *Forty Studies That Changed Psychology: Exploration into the History of Psychological Research*. New York: Prentice-Hall, 1995.

Lohr, Sharon. *Sampling Design and Analysis*. Belmont, CA: Duxbury Press, 1996. (A nice discussion of sampling and sources of bias at an accessible level.)

Moore, David. *Statistics: Concepts and Controversies*, 5th ed. New York: W. H. Freeman, 2001. (Contains an excellent chapter on the advantages and pitfalls of experimentation and another chapter in a similar vein on sample surveys and polls.)

Scheaffer, Richard L., William Mendenhall, and Lyman Ott. *Elementary Survey Sampling*, 5th ed. Belmont, CA: Duxbury Press, 1996. (An accessible yet thorough treatment of the subject.)

Sudman, Seymour, and Norman Bradburn. *Asking Questions: A Practical Guide to Questionnaire Design*. San Francisco: Jossey-Bass, 1982. (A good discussion of the art of questionnaire design.)

Tanur, Judith, ed. *Statistics: A Guide to the Unknown*. Belmont, CA: Duxbury Press, 1989. (Short, nontechnical articles by a number of well-known statisticians and users of statistics on the application of statistics in various disciplines and subject areas.)

3 · Graphical Methods for Describing Data

Most college students (and their parents) are concerned about the cost of a college education. *The Chronicle of Higher Education* (August 31, 2001) reported the average tuition and fees for 4-year public institutions in each of the 50 U.S. states for the 1999–2000 academic year. The accompanying values are given in alphabetical order by state:

2833	2855	2252	2785	2559	2775	4435	4642	2244	2524
2965	2458	4038	3646	2998	2439	2723	2430	4122	4552
4105	4538	3800	2872	3701	3011	2930	2034	6083	5255
2340	3983	2054	2990	4495	2183	3582	5610	4318	3638
3210	2698	2644	2147	6913	3733	3357	2549	3313	2416

A number of interesting questions could be posed about these data. What is a typical or representative value of average tuition and fees for the 50 states? Are observations concentrated near the typical value, or does the value of average tuition and fees differ quite a bit from state to state? Do states with low tuition and fees or states with high tuition and fees predominate, or are there roughly equal numbers of the two types? Are there any states whose average tuition and fees are somehow unusual compared to the rest of the data? What proportion or fraction of the states have average tuition and fees exceeding $3000? exceeding $5000?

Questions such as these are most easily answered if the data can be organized in a sensible manner. In this chapter, we introduce some techniques for organizing and describing data using tables and graphs.

▪ 3.1 Displaying Categorical Data: Comparative Bar Charts and Pie Charts

▪ Comparative Bar Charts

In Chapter 1 we saw that categorical data could be summarized in a frequency distribution and displayed graphically using a bar chart. Bar charts can also be used to give a visual comparison of two or more groups. This is accomplished by constructing two or more bar charts that use the same set of horizontal and vertical axes, as illustrated in Example 3.1.

■ **Example 3.1**   Perceived Risk of Smoking

The article "Most Smokers Wish They Could Quit" (*Gallup Poll Analyses*, November 21, 2002) noted that smokers and nonsmokers perceive the risks of smoking differently. The accompanying relative frequency table summarizes responses regarding the perceived harm of smoking for each of three groups: a sample of 241 smokers, a sample of 261 former smokers, and a sample of 502 nonsmokers.

Perceived Risk of Smoking	Relative Frequency		
	Smokers	Former Smokers	Nonsmokers
Very harmful	.60	.78	.86
Somewhat harmful	.30	.16	.10
Not too harmful	.07	.04	.03
Not harmful at all	.03	.02	.01

In constructing a comparative bar graph, we should use relative frequency to construct the scale on the vertical axis so that we can make meaningful comparisons when the sample sizes are not the same. The comparative bar chart for these data is shown in Figure 3.1. It is easy to see the differences among the three groups with respect to the perceived risk of smoking. The proportion believing that smoking is very harmful is noticeably smaller for smokers than for either former smokers or nonsmokers, and the proportion for former smokers is smaller than the proportion for nonsmokers.

Figure 3.1 Comparative bar chart of perceived harm of smoking.

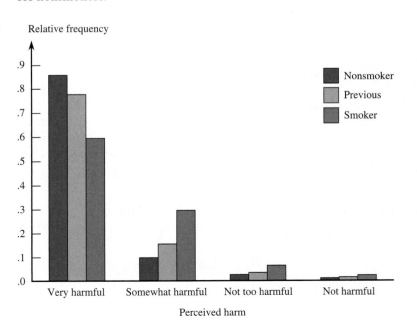

■ Example 3.2   What Could Be More Important Than Being Popular?

To see why it is important to use relative frequencies rather than frequencies to compare groups of different sizes, let's look at data from a survey of 227 boys and 251 girls in grades 4 through 6. Each student was asked what he or she thought was most important: getting good grades, being popular, or being good at sports. The resulting data, from the paper "The Role of Sport as a Social Determinant for Children" (*Research Quarterly for Exercise and Sport* [1992]: 418–424), are summarized in the accompanying table:

Most Important	Boys Frequency	Boys Relative Frequency	Girls Frequency	Girls Relative Frequency
Grades	117	.515	130	.518
Popularity	50	.220	91	.363
Sports	60	.264	30	.120
Total	**227**	**.999**	**251**	**1.001**

(Note: The relative frequencies for each group should sum to 1; here the small discrepancy is due to rounding in the relative frequencies.)

Figure 3.2(a) shows a correctly constructed comparative bar chart based on relative frequencies. Figure 3.2(b) shows an *incorrect* comparative bar chart constructed using frequencies rather than relative frequencies to determine the height of each bar. Notice that the graph of Figure 3.2(b) is misleading in that it does not accurately convey the differences between the two groups.

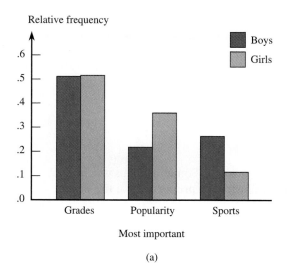

(a)

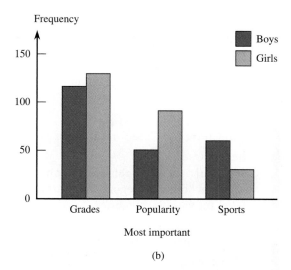

(b)

Figure 3.2 Correctly and incorrectly constructed comparative bar charts: (a) correctly constructed comparative bar chart (based on relative frequencies); (b) incorrectly constructed comparative bar chart (based on frequencies).

As you can see from Example 3.2, comparative bar charts should be constructed using relative frequencies. It is reasonable to use frequencies only if the number of observations is the same for *all* the groups being compared.

▪ Pie Charts

A categorical data set can also be summarized using a pie chart. In a pie chart, a circle is used to represent the whole data set, with "slices" of the pie representing the possible categories. The size of the slice for a particular category is proportional to the corresponding frequency or relative frequency.

▪ Example 3.3   College Choice

The Chronicle of Higher Education (August 31, 2001) published data collected in a survey of a large number of students who were college freshmen in the fall of 2001. One question asked whether the student was attending his or her first, second, or third choice of university. Fourth or higher choices were combined in a category called "other." The resulting data are summarized in the pie chart of Figure 3.3.

Figure 3.3 Pie chart of data on college choice.

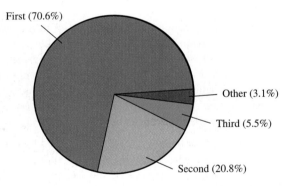

Pie charts are most effective for summarizing data sets when there are not too many different categories.

▪ Pie Chart for Categorical Data

When to Use

Categorical data with a relatively small number of possible categories. Sometimes this is achieved by using an "other" category. Pie charts are most useful for illustrating proportions of the whole data set for various categories.

How to Construct

1. Draw a circle to represent the entire data set.

2. For each category, calculate the "slice" size:

slice size = 360(category relative frequency)

(because there are 360 degrees in a circle).

3. Draw a slice of appropriate size for each category. This can be tricky, so most pie charts are generated using a graphing calculator or a statistical software package.

What to Look For

Categories that form large and small proportions of the data set.

■ **Example 3.4**   Birds That Fish

Night herons and cattle egrets are species of birds that feed on aquatic prey in shallow water. These birds wade through shallow water, stalking submerged prey and then striking rapidly and downward through the water in an attempt to catch the prey. The article "Cattle Egrets Are Less Able to Cope with Light Refraction Than Are Other Herons" (*Animal Behaviour* [1999]: 687–694) gave data on outcome when 240 cattle egrets attempted to capture submerged prey. The data are summarized in the following frequency distribution:

Outcome	Frequency	Relative Frequency
Prey caught on first attempt	103	.43
Prey caught on second attempt	41	.17
Prey caught on third attempt	2	.01
Prey not caught	94	.39

To draw a pie chart by hand, we must first compute the slice size for each category. This is done as follows:

Category	Slice Size
First attempt	(.43)(360) = 154.8°
Second attempt	(.17)(360) = 61.2°
Third attempt	(.01)(360) = 3.6°
Not caught	(.39)(360) = 140.4°

We would then draw a circle and use a protractor to mark off a slice corresponding to about 155°, as illustrated here:

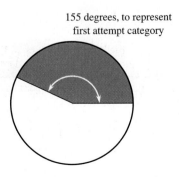

155 degrees, to represent first attempt category

Continuing to add slices in this way leads to a completed pie chart.

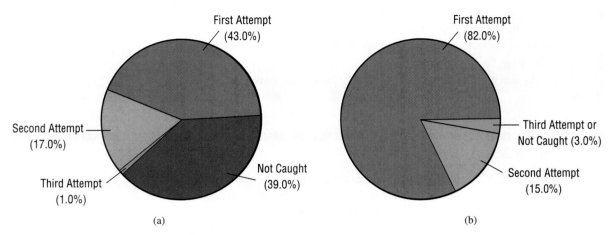

Figure 3.4 Pie charts for Example 3.4: (a) cattle egret data;
(b) night heron data.

It is much easier to use a statistical software package to create pie charts than to construct them by hand. A pie chart for the cattle egret data was created with MINITAB and is shown in Figure 3.4(a). Figure 3.4(b) shows a pie chart constructed using similar data on outcome for 180 night herons. Although some differences between night herons and cattle egrets can be seen by comparing the pie charts in Figures 3.4(a) and 3.4(b), it is difficult to actually compare category proportions using pie charts. A comparative bar chart (Figure 3.5) makes this type of comparison easier.

Figure 3.5 Comparative bar chart for the egret and heron data.

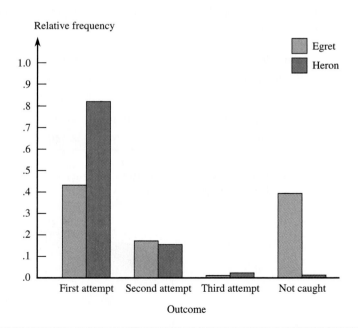

▪ A Different Type of "Pie" Chart: Segmented Bar Charts

A pie chart can be difficult to construct by hand, and the circular shape sometimes makes it difficult to compare areas for different categories, particularly when the relative frequencies for categories are similar. The **segmented bar chart** (also sometimes called a stacked bar chart) avoids these difficulties by using a rectangular bar rather than a circle to represent the entire data set. The bar is then divided into segments, with different segments representing different categories. As with pie charts, the size of the segment for a particular category is proportional to the relative frequency for that category. Example 3.5 illustrates the construction of a segmented bar graph.

▪ Example 3.5   Where Are the Men?

The paper "Community Colleges Start to Ask, Where Are the Men?" (*Chronicle of Higher Education*, June 28, 2002) gave data on gender for community college students. It was reported that 42% of students enrolled at community colleges nationwide were male and 58% were female. To construct a segmented bar graph for these data, first draw a bar of any fixed width and length. Then divide the bar into two segments, one for males and one for females. The length of the segment for males is (.42)(length of the bar) because the relative frequency for this category is .42. The segmented bar can be displayed either vertically or horizontally, as shown in Figure 3.6.

Figure 3.6 Segmented bar graphs for the gender data of Example 3.5.

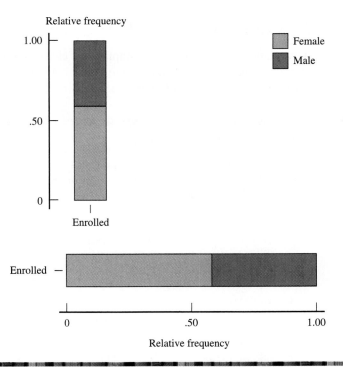

Segmented bar graphs can also be used to compare different populations on the basis of a categorical variable, as illustrated in Example 3.6.

▪ Example 3.6   Sex on TV

The segmented bar graph shown in Figure 3.7 appeared in the article "Study: More TV Shows Depicting Sexuality" (Associated Press, February 7, 2001). Based on this graph, it is easy to see how the proportion of television shows with sexual content differs for the different types of television shows studied.

Figure 3.7 Segmented bar graph for the data of Example 3.6.

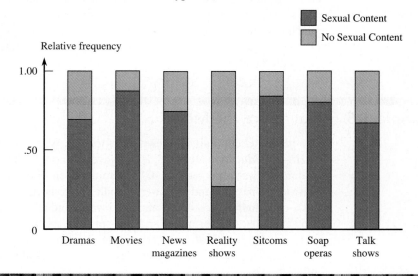

▪ Other Uses of Bar Charts and Pie Charts

As we have seen in previous examples, bar charts and pie charts can be used to summarize categorical data sets. However, they are occasionally used for other purposes, as illustrated in Example 3.7.

▪ Example 3.7   Wine Grape Production

The 1998 Grape Crush Report for San Luis Obispo and Santa Barbara counties in California gave the following information on grape production for each of seven different types of grapes used to make wine (*San Luis Obispo Tribune*, February 12, 1999):

Type of Grape	Tons Produced
Cabernet sauvignon	21,656
Chardonnay	39,582
Chenin blanc	1,601
Merlot	11,210
Pinot noir	2,856
Sauvignon blanc	5,868
Zinfandel	7,330
Total	**90,103**

Although this table is not a frequency distribution for a categorical data set, it is common to represent information of this type graphically using either a pie chart or a bar chart. A pie chart is shown in Figure 3.8(a). The pie represents the total grape production, and the slices show the proportion of the total production for each of the seven types of grapes. Figure 3.8(b) shows a bar chart representation of the grape production data.

Figure 3.8 Grape production data: (a) pie chart; (b) bar chart.

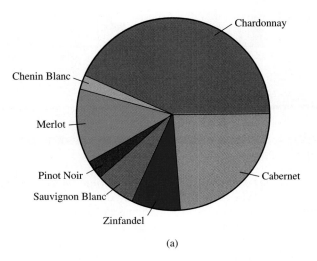

(a)

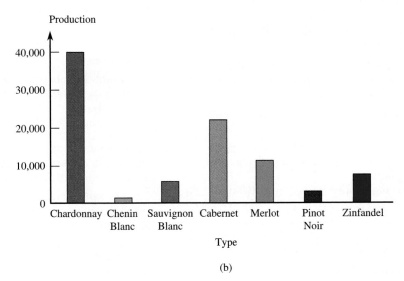

(b)

The article also gave grape production information for 1997 and commented on the effect of harsh weather on the counties' grape crop. Figure 3.9 shows a comparative bar chart of grape production in 1997 and 1998. The comparative bar chart clearly shows decreased production in 1998 for all but two of the seven types of grape.

Figure 3.9 Bar chart comparing 1997 and 1998 grape production.

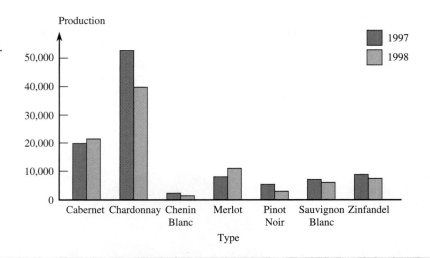

▪ Exercises 3.1–3.16

3.1 A *USA Today* poll of 1003 adults (August 12, 2002) asked people about their view of the current state of the U.S. economy. Based on his or her response, each person in the sample was classified into one of five categories:

1. Unflustered: upbeat about the economy, no significant financial worries

2. Comfortable: few financial worries, confident stock market will rebound

3. Anxious optimists: worried about maintaining standard of living but optimistic about the future

4. Strugglers: unaffected by stock market decline because they have few assets, high anxiety about retirement

5. Stressed and stretched: believe economy is in recession, worried about standard of living, view stocks as a gamble

The following relative frequency distribution shows the survey results:

Economic View	Relative Frequency
Unflustered	.31
Comfortable	.23
Anxious optimists	.10
Strugglers	.16
Stressed and stretched	.20

Construct a pie chart to show the distribution of economic view for this sample.

3.2 The web site PollingReport.com gave data from a CBS news poll conducted in December 1999. In the survey described, people were asked the follow-ing question: "All things considered, in our society today, do you think there are more advantages in being a man, more advantages in being a woman, or are there no more advantages in being one than the other?" Responses for men and for women are summarized in the following table:

Response	Relative Frequency Women	Men
Advantage in being a man	.57	.41
Advantage in being a woman	.06	.14
No advantage	.33	.40
Don't know	.04	.05

Construct a comparative bar chart for the response, and write a few sentences describing the differences in the response distribution for men and women.

3.3 The same poll described in Exercise 3.2 also asked the question, "What about salaries? These days, if a man and a woman are doing the same work, do you think the man generally earns more, the woman generally earns more, or that both earn the same amount?" The resulting data are shown in the following table:

Response	Relative Frequency Women	Men
Man earns more	.70	.59
Woman earns more	.01	.01
Both earn the same	.25	.34
Don't know	.04	.06

a. Construct a comparative bar chart that allows the responses for women and men to be compared.
b. Construct two pie charts, one summarizing the responses for women and one summarizing the response for men.
c. Is it easier to compare the responses of women and men by looking at the comparative bar chart or the two pie charts? Explain.
d. Write a brief paragraph describing the difference between women and men with respect to the way they answered this question.

3.4 The article "Online Shopping Grows, But Department, Discount Stores Still Most Popular with Holiday Shoppers" (*Gallup Poll Analyses*, December 2, 2002) included the data in the accompanying table. The data are based on two telephone surveys, one conducted in November 2000 and one in November 2002. Respondents were asked if they were very likely, somewhat likely, not too likely, or not at all likely to do Christmas shopping online.

Response	Relative Frequency	
	2000	2002
Very likely	.09	.15
Somewhat likely	.12	.14
Not too likely	.14	.12
Not at all likely	.65	.59

a. Construct a comparative bar chart. What aspects of the chart justify the claim in the title of the article that "online shopping grows"?
b. Respondents were also asked about the likelihood of doing Christmas shopping from mail order catalogues. Use the data in the accompanying table to construct a comparative bar chart. Are the changes in responses from 2000 to 2002 similar to what was observed for online shopping? Explain.

Response	Relative Frequency	
	2000	2002
Very likely	.12	.13
Somewhat likely	.22	.17
Not too likely	.20	.16
Not at all likely	.46	.54

3.5 The article "The Healthy Kids Survey: A Look at the Findings" (*San Luis Obispo Tribune*, October 25, 2002) gave the accompanying information for a sample of fifth graders in San Luis Obispo County. Responses are to the question "After school, are you home alone without adult supervision?"

Response	Percentage
Never	8
Some of the time	15
Most of the time	16
All of the time	61

a. Summarize these data using a pie chart.
b. Construct a segmented bar chart for these data.
c. Which graphing method — the pie chart or the segmented bar chart — do you think does a better job of conveying information about response? Explain.

3.6 The article "So Close, Yet So Far: Predictors of Attrition in College Seniors" (*Journal of College Student Development* [1998]: 343–348) examined the reasons that college seniors leave their college programs before graduating. Forty-two college seniors at a large public university who dropped out before graduation were interviewed and asked the main reason for discontinuing enrollment at the university. Data consistent with that given in the article are summarized in the following frequency distribution:

Reason for Leaving the University	Frequency
Academic problems	7
Poor advising or teaching	3
Needed a break	2
Economic reasons	11
Family responsibilities	4
To attend another school	9
Personal problems	3
Other	3

a. Would a bar chart or a pie chart be a better choice for summarizing this data set? Explain.
b. Based on your answer to Part (a), construct an appropriate graphical summary of the given data.
c. Write a few sentences describing the interesting features of your graphical display.

3.7 "If you were taking a new job and had your choice of a boss, would you prefer to work for a man or a woman?" That was the question posed to individuals in a sample of 576 employed adults (*Gallup at a Glance*, October 16, 2002). Responses are summarized in the following table:

Response	Frequency
Prefer to work for a man	190
Prefer to work for a woman	92
No difference	282
No opinion	12

a. Construct a pie chart to summarize this data set, and write a sentence or two summarizing how people responded to this question.

b. Summarize the given data using a segmented bar chart.

3.8 The paper "Exploring the Factors that Influence Men and Women to Form Medical Career Aspirations" (*Journal of College Student Development* [1998]: 417–421) gave estimates of the percentage of college freshmen nationwide who hoped to have a medical career. These percentages were based on an annual survey of "a nationally representative sample of all entering college students":

| | Percentage with Medical Career Aspirations | | | |
	1975	1980	1985	1990
Men	3.9	4.4	4.1	3.9
Women	2.5	2.9	3.4	3.7

Construct a comparative bar graph that shows the proportion with medical career aspirations for men and women over time. (Hint: Mark the years on the horizontal axis, and then construct a bar for each gender for each year.) Comment on any interesting features of the graph.

3.9 The paper referenced in Exercise 3.8 also gave information on parent's occupation for those who aspire to a medical career. The data are given in the following table:

Parent Occupation	Male	Female
Only mother is a physician	596	712
Only father is a physician	504	350
Both parents are physicians	734	824
Neither parent is a physician	166	114

Construct two pie charts, one for men and one for women, to display this information, and comment on the similarities and differences between the two.

3.10 The article "Physicians Accessing the Internet: The PAI Project" (*Journal of the American Medical Association* [1999]: 633–634) gave the following information on Internet usage by doctors. The data are from a survey of 324 randomly selected doctors who were asked to indicate the category that best described how often they used the Internet.

Internet Usage Pattern	Frequency
Never	31
Rarely (about 3 times per year)	15
Occasionally (about once a month)	52
Often (about once a week)	109
Daily	117

Construct a pie chart for these data, and write a brief summary that describes Internet usage for physicians who participated in this survey.

3.11 The paper "Sexual Content of Top-Grossing Motion Pictures" (*Journal of Health Education* [1998]: 354–357) gave the accompanying information on the ratings of the 10 movies that made the most money in the years 1987 and 1992:

| | Rating | | |
	G	PG/PG-13	R
1987	0	3	7
1992	1	5	4

a. Construct a pie chart to show the distribution of ratings for 1987.

b. Construct a pie chart to show the distribution of ratings for 1992, and comment on the differences between this chart and the one based on the 1987 data.

3.12 In a discussion of roadside hazards, the web site highwaysafety.com included a pie chart like the one shown:

Pie chart for Exercise 3.12

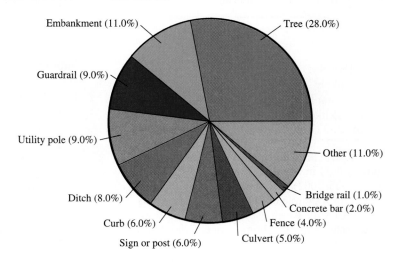

a. Do you think this is an effective use of a pie chart? Why or why not?

b. Construct a bar chart to show the distribution of deaths by object struck. Is this display more effective than the pie chart in summarizing this data set? Explain.

3.13 The percentage of U.S. gross domestic product (GDP) spent on health care over the period 1960–1995 was given in the paper "Building the Next Generation of Healthy People" (*Public Health Reports* [1999]: 213–216). Construct a bar chart for the data given in the accompanying table, and comment on the interesting features of this graph. Be sure to comment on the trend over time.

Year	Percentage of GDP Spent on Health Care
1960	15.1
1965	15.7
1970	17.1
1975	18.0
1980	18.9
1985	10.3
1990	12.2
1995	13.7

3.14 Bizrate.com reported the accompanying data on Internet sales (*San Luis Obispo Tribune*, April 26, 2002):

Year	Internet Sales (billions of dollars)
2000	28.9
2001	35.9
2002	51.5

Construct a graphical display that illustrates the trend over time. (Hint: Would a pie chart or a bar chart be the most appropriate display?)

3.15 The source referenced in Exercise 3.14 also gave data on sales (in billions of dollars) for 2002 in major retailing categories, as shown:

Category	Sales
Travel	7.0
Computer hardware	2.4
Office supplies	1.7
Apparel and accessories	1.3
Consumer electronics	0.744
Event tickets	0.581
Books	0.557
Home and garden	0.458
Health and beauty	0.308
Sports and fitness	0.258
Computer software	0.236

Construct a graphical display to summarize the data in this table. Did you choose a bar chart or a pie chart for your display? Explain your choice.

3.16 The article "Death in Roadwork Zones at Record High" (*San Luis Obispo Tribune*, July 25, 2001) included a bar chart similar to this one:

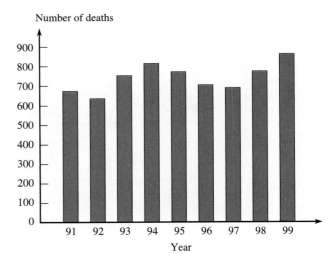

a. Comment on the trend over time in the number of people killed in highway work zones.

b. Would a pie chart have also been an effective way to summarize these data? Explain why or why not.

▪ 3.2 Displaying Numerical Data: Stem-and-Leaf Displays

A stem-and-leaf display is an effective and compact way to summarize univariate numerical data. Each number in the data set is broken into two pieces, one called the stem and the other called the leaf. The **stem** is the first part of the number and consists of the beginning digit(s). The **leaf** is the last part of the number and consists of the final digit(s). For example, the number 213 might be split into a stem of 2 and a leaf of 13 or a stem of 21 and a leaf of 3. The resulting stems and leaves are then used to construct the display.

▪ Example 3.8   Binge Drinking

The use of alcohol by college students is of great concern, not only to those in the academic community but also, because of potential health and safety consequences, to society at large. The article "Health and Behavioral Consequences of Binge Drinking in College" (*Journal of the American Medical Association* [1994]: 1672–1677) reported on a comprehensive study of heavy drinking on campuses across the country. A binge episode was defined as five or more drinks in a row for males and four or more drinks in a row for females. Figure 3.10 shows a stem-and-leaf display of 140 values of the percentage of undergraduate students who are binge drinkers at various colleges. (These values were not given in the cited article, but our display agrees with a picture of the data that did appear.)

Figure 3.10 Stem-and-leaf display for percentage of binge drinkers at each of 140 colleges.

0	4
1	1345678889
2	1223456666777889999
3	011223334455566667777788889999
4	11122222334444556666667788888999
5	001112222334556666677777888899
6	01111244455666778

Stem: Tens digit
Leaf: Ones digit

The numbers in the vertical column on the left of the display are the **stems**. Each number to the right of the vertical line is a **leaf** corresponding to one of the observations in the data set. The legend

Stem: Tens digit

Leaf: Ones digit

tells us that the observation that had a stem of 2 and a leaf of 1 corresponds to a college where 21% (as opposed to 2.1% or 0.21%) of the students were binge drinkers.

The display in Figure 3.10 suggests that a typical or representative value is in the stem 4 row, perhaps someplace in the low 40% range. The observations are not highly concentrated about this typical value, as would be the case if all values were between 20% and 49%. The display rises to a single peak as we move downward and then declines, and there are no gaps in the display. The shape of the display is not perfectly symmetric but rather appears to stretch out a bit more in the direction of low stems than in the direction of high stems. The most surprising feature of these data is that at most colleges in the sample, at least one-quarter of the students are binge drinkers.

The leaves on each line of the display in Figure 3.10 have been arranged in order from smallest to largest. Most statistical software packages order the leaves this way, but it is not necessary to do so to get an informative display that still shows many of the important characteristics of the data set, such as shape and spread.

Stem-and-leaf displays can be useful in getting a sense of a typical value for the data set, as well as how spread out the values in the data set are. It is also easy to spot data values that are unusually far from the rest of the values in the data set. Such values are called **outliers**. The stem-and-leaf display of the data on binge drinking (Figure 3.10) does not show any outliers.

■ **Definition**

An **outlier** is an unusually small or large data value. In Chapter 4 we give a precise rule for deciding when an observation is an outlier.

■ **Stem-and-Leaf Displays**

When to Use
Numerical data sets with a small to moderate number of observations (does not work well with very large data sets).

How to Construct
1. Select one or more leading digits for the stem values. The trailing digits (or sometimes just the first one of the trailing digits) become the leaves.
2. List possible stem values in a vertical column.
3. Record the leaf for every observation beside the corresponding stem value.
4. Indicate the units for stems and leaves someplace in the display.

What to Look For
The display conveys information about a representative or typical value in the data set, the extent of spread about such a value, the presence of any gaps in the data, the extent of symmetry in the distribution of values, the number and location of peaks, and the presence of any outliers.

■ **Example 3.9   Tuition at Public Universities**

The introduction to this chapter gave data on average tuition and fees at public institutions in the year 2000 for the 50 U.S. states. The observations ranged from a low value of 2054 to a high value of 6913. The data are reproduced here:

2833	2855	2252	2785	2559	2775	4435	4642	2244	2524
2965	2458	4038	3646	2998	2439	2723	2430	4122	4552
4105	4538	3800	2872	3701	3011	2930	2034	6083	5255
2340	3983	2054	2990	4495	2183	3582	5610	4318	3638
3210	2698	2644	2147	6913	3733	3357	2549	3313	2416

A natural choice for the stem is the leading (thousands) digit. This would result in a display with 5 stems (2, 3, 4, 5, 6). Using the first two digits of a number as the stem would result in 41 stems (20, 21, . . . , 60). A stem-and-leaf display with 41 stems would not be an effective summary of the data. In general, stem-and-leaf displays that use between 5 and 20 stems tend to work well.

If we choose the thousands digit as the stem, the remaining three digits (the hundreds, tens, and ones) would form the leaf. For example, for the first few values in the first column of data, we would have

$$2833 \rightarrow \text{stem} = 2, \text{leaf} = 833$$
$$2965 \rightarrow \text{stem} = 2, \text{leaf} = 965$$
$$4105 \rightarrow \text{stem} = 4, \text{leaf} = 105$$

The leaves have been entered in the display of Figure 3.11 in the order they are encountered in the data set. Commas are used to separate the leaves only when each leaf has two or more digits. Figure 3.11 shows that most states had average tuition and fees in the $2000 range and that the typical average tuition and fees is around $3000. A few states have average tuition and fees at public four-year institutions that are quite a bit higher than most other states (the four states with the highest values were New Jersey, Pennsylvania, New Hampshire, and Vermont).

```
2 | 833,855,252,785,559,775,244,524,965,458,998,439,723,430,872,930,034,340,054,990,183,698,644,147,549,416
3 | 646,800,701,011,983,582,638,210,733,357,313
4 | 435,642,038,122,552,105,538,495,318
5 | 255,610                                                      Stem:  Thousands
6 | 083,913                                                      Leaf:  Ones
```

Figure 3.11 Stem-and-leaf display of average tuition and fees.

An alternative display (Figure 3.12) results from dropping all but the first digit of the leaf. This is what most statistical computer packages do when generating a display; little information about a typical value, spread, or shape is lost in this truncation and the display is simpler and more compact.

Figure 3.12 Stem-and-leaf display of the average tuition and fees data using truncated leaves.

```
2 | 8 8 2 7 5 7 2 5 9 4 9 4 7 4 8 9 0 3 0 9 1 6 6 1 5 4
3 | 6 8 7 0 9 5 6 2 7 3 3
4 | 4 6 0 1 5 1 5 4 3
5 | 2 6                                  Stem:  Thousands
6 | 0 9                                  Leaf:  Hundreds
```

▪ Repeated Stems to Stretch a Display

Sometimes a natural choice of stems gives a display in which too many observations are concentrated on just a few stems. A more informative picture can be obtained by dividing the leaves at any given stem into two groups: those that begin with 0, 1, 2, 3, or 4 (the "low" leaves) and those that begin with 5, 6, 7, 8, or 9 (the "high" leaves). Then each stem is listed twice when constructing the display, once for the low leaves and once again for the high leaves. It is also possible to repeat a stem more than twice. For example, each stem might be repeated five times, once for each of the leaf groupings {0, 1}, {2, 3}, {4, 5}, {6, 7}, and {8, 9}.

■ Example 3.10   Protein Intake of Athletes

The accompanying data on daily protein intake (in grams of protein per kilogram of body weight) for 20 competitive athletes was obtained from a plot in the article "A Comparison of Plasma Glutamine Concentration in Athletes from Different Sports" (*Medicine and Science in Sports and Exercise* [1998]: 1693–1697):

1.4	2.2	2.7	1.5	2.3	1.7	2.3	1.5	1.8	2.8
1.8	1.9	2.0	2.3	1.5	1.9	1.7	1.8	1.6	3.0

Because each value in the data set has only two digits, we must use the first digit for the stem and the last digit for the leaf. The corresponding stem-and-leaf display is shown in Figure 3.13; the display has only three stems (1, 2, and 3), and all but one of the leaves are located at the 1 and 2 stems. A more informative display using repeated stems is shown in Figure 3.14.

Figure 3.13 Stem-and-leaf display for the protein intake data.

```
1 | 457588959786
2 | 2733803        Stem:  Ones
3 | 0              Leaf:  Tenths
```

Figure 3.14 Stem-and-leaf display for protein intake data using repeated stems.

```
1L | 4
1H | 57588959786
2L | 23303
2H | 78              Stem:  Ones
3L | 0               Leaf:  Tenths
```

■ Comparative Stem-and-Leaf Displays

Frequently, an analyst wishes to see whether two groups of data differ in some fundamental way. A comparative stem-and-leaf display, in which the leaves for one group extend to the right of the stem values and the leaves for the second group extend to the left, can provide preliminary visual impressions and insights.

■ Example 3.11   Tobacco Use in G-Rated Movies

The article "Tobacco and Alcohol Use in G-Rated Children's Animated Films" (*Journal of the American Medical Association* [1999]: 1131–1136) reported exposure to tobacco and alcohol use in all G-rated animated films released between 1937 and 1997 by five major film studios. The researchers found that tobacco use was shown in 56% of the reviewed films. Data on the total tobacco exposure time (in seconds) for films with tobacco use produced by Walt Disney, Inc., were as follows:

223	176	548	37	158	51	299	37	11	165
74	9	2	6	23	206	9			

Data for 11 G-rated animated films showing tobacco use that were produced by MGM/United Artists, Warner Brothers, Universal, and Twentieth Century Fox were also given. The tobacco exposure times (in seconds) for these films was as follows:

205   162   6   1   117   5   91   155   24   55   17

To construct a stem-and-leaf display, think of each observation as a three-digit number. For example, 2 would be 002 and 37 would be 037. We then use the first digit of each number as the stem and the remaining two digits as the leaf. For simplicity, let's truncate the leaves to one digit:

$$223 \rightarrow \text{stem} = 2, \text{leaf} = 2 \text{ (truncated from 23)}$$

$$37 \rightarrow 037 \rightarrow \text{stem} = 0, \text{leaf} = 3 \text{ (truncated from 37)}$$

$$9 \rightarrow 009 \rightarrow \text{stem} = 0, \text{leaf} = 0 \text{ (truncated from 09)}$$

The resulting comparative stem-and-leaf display, using repeated stems, is shown in Figure 3.15.

Figure 3.15
Comparative stem-and-leaf display for tobacco exposure times.

Other Studios		Disney
12000	0L	33100020
59	0H	57
1	1L	
56	1H	756
All Dogs Go to Heaven 0	2L	20 *Pinocchio, James and the Giant Peach*
	2H	9 *101 Dalmatians*
	3L	
	3H	
	4L	
	4H	
	5L	4 *The Three Caballeros*

Stem: Hundreds
Leaf: Tens

One first impression is that there are many films where the tobacco exposure time is small, both for Disney films and for those made by the other studios. Disney has more films with longer exposure to tobacco, and there is an obvious outlier in the Disney data (the 548-sec exposure in the 1945 film *The Three Caballeros*).

▪ Exercises 3.17–3.24

3.17 The stem-and-leaf display at the top of page 97 shows observations on average shower flow rate (in liters per minute) for a sample of 129 houses in Perth, Australia ("An Application of Bayes Methodology to the Analysis of Diary Records from a Water Use Study," *Journal of the American Statistical Association* [1987]: 705–711).
a. What is the smallest flow rate in the sample?
b. If one additional house yielded a flow rate of 8.9, where would this observation be placed on the display?
c. What is a typical, or representative, flow rate?
d. Does the display appear to be highly concentrated, or quite spread out?

e. Does the distribution of values in the display appear to be reasonably symmetric? If not, how would you describe the departure from symmetry?
f. Does the data set appear to contain any outliers (observations far removed from the bulk of the data)?

3.18 The Connecticut Agricultural Experiment Station conducted a study of the calorie content of different types of beer. The calorie contents (calories per 100 ml) for 26 brands of light beer are (from the web site brewery.org):

29 28 33 31 30 33 30 28 27 41 39 31 29
23 32 31 32 19 40 22 34 31 42 35 29 43

```
 2 | 23
 3 | 2344567789
 4 | 01356889
 5 | 00001114455666789
 6 | 000012222334445666778999
 7 | 00012233455555668
 8 | 02233448
 9 | 012233335666788
10 | 2344455688
11 | 2335999
12 | 37
13 | 8              Stem: Ones
14 | 36             Leaf: Tenths
15 | 0035
16 |
17 |
18 | 9
```

Stem-and-leaf display for Exercise 3.17

Construct a stem-and-leaf display using stems 1, 2, 3, and 4. Write a sentence or two describing the calorie content of light beers.

3.19 The stem-and-leaf display of Exercise 3.18 uses only four stems. Construct a stem-and-leaf display for these data using repeated stems 2L, 2H, . . . , 4L. For example, the first observation, 29, would have a stem of 2 and a leaf of 9. It would be entered into the display for the stem 2H, because it is a "high" 2 — that is, it has a leaf that is on the high end (5, 6, 7, 8, 9).

3.20 The accompanying observations are lengths (in yards) for a sample of golf courses recently listed by *Golf Magazine* as being among the most challenging in the United States. Construct a stem-and-leaf display, and explain why your choice of stems seems preferable to any of the other possible choices. The lengths are

6526	6770	6936	6770	6583	6464	7005	6927
6790	7209	7040	6850	6700	6614	7022	6506
6527	6470	6900	6605	6873	6798	6745	7280
7131	6435	6694	6433	6870	7169	7011	7168
6713	7051	6904	7105	7165	7050	7113	6890

3.21 Many states face a shortage of fully credentialed teachers. The percentages of teachers who are fully credentialed for each county in California were published in the *San Luis Obispo Tribune* (July 29, 2001) and are given in the following table:

County	Percentage Credentialed	County	Percentage Credentialed
Alameda	85.1	Butte	98.2
Alpine	100.0	Calaveras	97.3
Amador	97.3	Colusa	92.8

County	Percentage Credentialed	County	Percentage Credentialed
Contra Costa	87.7	Sacramento	95.3
Del Norte	98.8	San Benito	83.5
El Dorado	96.7	San	
Fresno	91.2	Bernardino	83.1
Glenn	95.0	San Diego	96.6
Humbolt	98.6	San Francisco	94.4
Imperial	79.8	San Joaquin	86.3
Inyo	94.4	San Luis	
Kern	85.1	Obispo	98.1
Kings	85.0	San Mateo	88.8
Lake	95.0	Santa Barbara	95.5
Lassen	89.6	Santa Clara	84.6
Los Angeles	74.7	Santa Cruz	89.6
Madera	91.0	Shasta	97.5
Marin	96.8	Sierra	88.5
Mariposa	95.5	Siskiyou	97.7
Mendicino	97.2	Solano	87.3
Merced	87.8	Sonoma	96.5
Modoc	94.6	Stanislaus	94.4
Mono	95.9	Sutter	89.3
Monterey	84.6	Tehama	97.3
Napa	90.8	Trinity	97.5
Nevada	93.9	Tulare	87.6
Orange	91.3	Tuolomne	98.5
Placer	97.8	Ventura	92.1
Plumas	95.0	Yolo	94.6
Riverside	84.5	Yuba	91.6

a. Construct a stem-and-leaf display for this data set using stems 7, 8, 9, and 10. Truncate the leaves to a single digit. Comment on the interesting features of the display.

b. Construct a stem-and-leaf display using repeated stems. Are there characteristics of the data set that are easier to see in the plot with repeated stems, or is the general shape of the two displays similar?

3.22 An article on peanut butter in *Consumer Reports* (September 1990) reported the following scores (quality ratings on a scale of 0 to 100) for various brands:

Creamy: 56 44 62 36 39 53 50 65 45 40
        56 68 41 30 40 50 56 30 22

Crunchy: 62 53 75 42 47 40 34 62 52 50
         34 42 36 75 80 47 56 62

Construct a comparative stem-and-leaf display, and discuss similarities and differences for the two types.

3.23 The article "A Nation Ablaze with Change" (*USA Today*, July 3, 2001) gave the accompanying data on percentage increase in population between 1990 and 2000 for the 50 U.S. states. Also provided in the table is a column that indicates for each state whether the state is in the eastern or western part of the United States (the states are listed in order of population size):

State	Percentage Change	East/West
California	13.8	W
Texas	22.8	W
New York	5.5	E
Florida	23.5	E
Illinois	8.6	E
Pennsylvania	3.4	E
Ohio	4.7	E
Michigan	6.9	E
New Jersey	8.9	E
Georgia	26.4	E
North Carolina	21.4	E
Virginia	14.4	E
Massachusetts	5.5	E
Indiana	9.7	E
Washington	21.1	W
Tennessee	16.7	E
Missouri	9.3	E
Wisconsin	9.6	E
Maryland	10.8	E
Arizona	40.0	W
Minnesota	12.4	E
Louisiana	5.9	E
Alabama	10.1	E
Colorado	30.6	W
Kentucky	9.7	E
South Carolina	15.1	E
Oklahoma	9.7	W
Oregon	20.4	W
Connecticut	3.6	E
Iowa	5.4	E
Mississippi	10.5	E
Kansas	8.5	W
Arkansas	13.7	E
Utah	29.6	W
Nevada	66.3	W
New Mexico	20.1	W
West Virginia	0.8	E
Nebraska	8.4	W
Idaho	28.5	W
Maine	3.9	E
New Hampshire	11.4	E
Hawaii	9.3	W
Rhode Island	4.5	E
Montana	12.9	W
Delaware	17.6	E
South Dakota	8.5	W
North Dakota	0.5	W
Alaska	14.0	W
Vermont	8.2	E
Wyoming	8.9	W

a. Construct a stem-and-leaf display for percentage growth for the data set consisting of all 50 states. Hints: Regard the observations as having two digits to the left of the decimal place. That is, think of an observation such as 8.5 as 08.5. It will also be easier to truncate leaves to a single digit; for example, a leaf of 8.5 could be truncated to 8 for purposes of constructing the display.

b. Comment on any interesting features of the data set. Do any of the observations appear to be outliers?

c. Now construct a comparative stem-and-leaf display for the eastern and western states. Write a few sentences comparing the percentage growth distributions for eastern and western states.

3.24 High school dropout rates (percentages) for the period 1997–1999 for the 50 states were given in *The Chronicle of Higher Education* (August 31, 2001) and are shown in the following table:

State	Rate	State	Rate
Alabama	10	Montana	8
Alaska	7	Nebraska	8
Arizona	17	Nevada	17
Arkansas	12	New Hampshire	7
California	9	New Jersey	6
Colorado	13	New Mexico	13
Connecticut	9	New York	9
Delaware	11	North Carolina	11
Florida	12	North Dakota	5
Georgia	13	Ohio	8
Hawaii	5	Oklahoma	9
Idaho	10	Oregon	13
Illinois	9	Pennsylvania	7
Indiana	6	Rhode Island	11
Iowa	7	South Carolina	9
Kansas	7	South Dakota	8
Kentucky	11	Tennessee	12
Louisiana	11	Texas	12
Maine	7	Utah	9
Maryland	7	Vermont	6
Massachusetts	6	Virginia	8
Michigan	9	Washington	8
Minnesota	6	West Virginia	8
Mississippi	10	Wisconsin	5
Missouri	9	Wyoming	9

Note that dropout rates range from a low of 5% to a high of 17%. In constructing a stem-and-leaf display for these data, if we regard each dropout rate as a two-digit number and use the first digit for the stem, then there are only two possible stems, 0 and 1. One solution is to use repeated stems. Consider a scheme that divides the leaf range into five parts: 0 and 1, 2 and 3, 4 and 5, 6 and 7, and 8 and 9. Then, for example, stem 1 could be repeated as

1.   with leaves 0 and 1
1t   with leaves 2 and 3
1f   with leaves 4 and 5
1s   with leaves 6 and 7

If there had been any dropout rates as large as 18 or 19 in the data set, we would also need to include a stem 1* to accommodate the leaves of 8 and 9.

Construct a stem-and-leaf display for this data set that uses stems 0f, 0s, 0*, 1., 1t, 1f, and 1s. Comment on the important features of the display.

▪ 3.3 Displaying Numerical Data: Frequency Distributions and Histograms

A stem-and-leaf display is not always an effective summary technique; it is unwieldy when the data set contains a great many observations. Frequency distributions and histograms are displays that are useful for summarizing even a large data set in a compact fashion.

▪ Frequency Distributions and Histograms for Discrete Numerical Data

Discrete numerical data almost always result from counting. In such cases, each observation is a whole number. As in the case of categorical data, a frequency distribution for discrete numerical data lists each possible value (either individually or grouped into intervals), the associated frequency, and sometimes the corresponding relative frequency. Recall that relative frequency is calculated by dividing the frequency by the total number of observations in the data set.

▪ Example 3.12   Promiscuous Raccoons!

The authors of the article "Behavioral Aspects of the Raccoon Mating System: Determinants of Consortship Success" (*Animal Behaviour* [1999]: 593–601) monitored raccoons in southern Texas during three mating seasons in an effort to describe mating behavior. Twenty-nine female raccoons were observed, and the number of male partners during the time the female was accepting partners (generally 1 to 4 days each year) was recorded for each female. The resulting data were as follows:

```
1   3   2   1   1   4   2   4   1   1   1   3   1   1   1
1   2   2   1   1   4   1   1   2   1   1   1   1   3
```

The corresponding frequency distribution is given in Table 3.1. From the frequency distribution, we can see that 18 of the female raccoons had a single partner. The corresponding relative frequency, .621, tells us that the proportion of female raccoons in the sample with a single partner was .621, or, equivalently, 62.1% of the females had a single partner. Adding the relative frequencies for the values of 1 and 2 gives

$$.621 + .172 = .793$$

indicating that 79.3% of the raccoons had 2 or fewer partners.

Table 3.1   ■   Frequency Distribution for Number of Partners

Number of Partners	Frequency	Relative Frequency	
1	18	.621	
2	5	.172	
3	3	.103	
4	3	.103	*Differs from 1 due to rounding*
	29	.999	

A histogram for discrete numerical data is a graph of the frequency distribution, and it is similar to the bar chart for categorical data. Each frequency or relative frequency is represented by a rectangle centered over the corresponding value (or range of values) and the area of the rectangle is proportional to the corresponding frequency or relative frequency.

■ **Histogram for Discrete Numerical Data**

When to Use

Discrete numerical data. Works well even for large data sets.

How to Construct

1. Draw a horizontal scale, and mark the possible values of the variable.

2. Draw a vertical scale, and mark it with either frequencies or relative frequencies.

3. Above each possible value, draw a rectangle centered at that value (so that the rectangle for 1 is centered at 1, the rectangle for 5 is centered at 5, etc.). The height of each rectangle is determined by the corresponding frequency or relative frequency. Often possible values are consecutive whole numbers, in which case the base width for each rectangle is 1.

What to Look For

Central or typical value, extent of spread or variation, general shape, location and number of peaks, and presence of gaps and outliers.

■ **Example 3.13   Revisiting Promiscuous Raccoons**

The raccoon data of Example 3.12 were summarized in a frequency distribution. The corresponding histogram is shown in Figure 3.16. Note that each rectangle in the histogram is centered over the corresponding value. When relative frequency instead of frequency is used for the vertical scale, the scale on the vertical axis is different but all essential characteristics of the graph (shape, location, spread) are unchanged.

Figure 3.16 Histogram and relative frequency histogram of raccoon data.

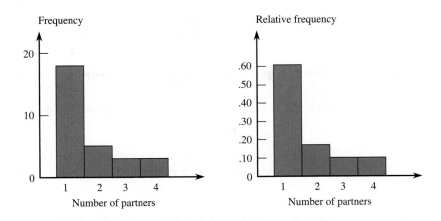

Sometimes a discrete numerical data set contains a large number of possible values and perhaps also has a few large or small values that are far away from most of the data. In this case, rather than forming a frequency distribution with a very long list of possible values, it is common to group the observed values into intervals or ranges. This is illustrated in Example 3.14.

▪ Example 3.14   Alcohol Use by College Students

Each student in a sample of 176 students at a large public university was asked about alcohol use, and the resulting data appeared in the article "Alcohol, To- bacco, and Marijuana Use: Relationships to Undergraduate Students' Creative Achievement" (*Journal of College Student Development* [1998]: 472–479). The fre- quency distribution in Table 3.2 summarizes the data on number of drinks con- sumed per week. The authors of this article chose to group the observed values (rather than list them all: 0, 1, 2, . . .), and they also created an open-ended group of "16 or more." This results in a much more compact table that still communicates one of the important features of the data: the large numbers of individuals at both the low and the high ends. Also note that we could not draw a histogram based on this frequency distribution because of the open-ended group ("16 or more"). Fur- thermore, because the number of possible values differs from group to group (2 in the "0 to 1" group, 4 in the "2 to 5" group, and 6 in the "10 to 15" group), even with- out the open-ended group, drawing a correct histogram is a bit more complicated.

Table 3.2 ▪ Frequency Distribution for Number of Drinks per Week

Drinks per Week	Frequency
0 to 1	52
2 to 5	38
6 to 9	17
10 to 15	35
16 or more	34

You cannot just use frequency or relative frequency for the vertical scale. One method of drawing a histogram that would work, as long as there are no open-ended groups, is described later in this section.

▪ Frequency Distributions and Histograms for Continuous Numerical Data

The difficulty in constructing tabular or graphical displays with continuous data, such as observations on reaction time (in seconds) or fuel efficiency (in miles per gallon), is that there are no natural categories. The way out of this dilemma is to define our own categories. For fuel efficiency data, suppose that we mark some intervals on a horizontal miles-per-gallon measurement axis, as pictured in Figure 3.17. Each data value should fall in exactly one of these intervals. If the smallest observation was 25.3 and the largest was 29.8, we might use intervals of width 0.5, with the first interval starting at 25.0 and the last interval ending at 30.0. The resulting intervals are called **class intervals**, or just **classes**. The class intervals play the same role that the categories or individual values played in frequency distributions for categorical or discrete numerical data.

Figure 3.17 Suitable class intervals for miles-per-gallon data.

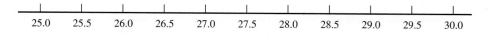

There is one further difficulty. Where should we place an observation such as 27.0, which falls on a boundary between classes? Our convention is to define intervals so that such an observation is placed in the upper rather than the lower class interval. Thus, in our frequency distribution, a typical class will be 26.5 to <27.0, where the symbol < is a substitute for the phrase *less than*. This class will contain all observations that are greater than or equal to 26.5 and less than 27.0. The observation 27.0 would then fall in the class 27.0 to <27.5.

▪ Example 3.15   Enrollments at Public Universities

States differ widely with respect to the percentage of college students who are enrolled in public institutions. The U.S. Department of Education provided the accompanying data on this percentage for the 50 U.S. states for fall 1999. Observations for a few of the states have been identified by name.

Percentage of College Students Enrolled in Public Institutions

95 (Alaska)	81	85	80	72	73	74	79
95 (Nevada)	84	89	63	91	86	89	92
87	90	83	84	89	96 (Wyoming)	87	85
76	84	75	81	73	82	81	77	75
70	55 (New York)	56 (Pennsylvania)		87	88	82
81	84	76	80	56 (Vermont)	55 (New Hampshire)
43 (Massachusetts)	52	62	80	82

The smallest observation is 43 (Massachusetts) and the largest is 96 (Wyoming). It is reasonable to start the first class interval at 40 and let each interval have a width of 10. This gives class intervals of 40 to <50, 50 to <60, 60 to <70, 70 to <80, 80 to <90, and 90 to <100.

Table 3.3 displays the resulting frequency distribution, along with the relative frequencies.

Table 3.3 ▪ Frequency Distribution for Percentage of College Students Enrolled in Public Institutions

Class Interval	Frequency	Relative Frequency
40 to <50	1	.02
50 to <60	5	.10
60 to <70	2	.04
70 to <80	11	.22
80 to <90	25	.50
90 to <100	6	.12
	50	1.00

Various relative frequencies can be combined to yield other interesting information. For example,

$$\begin{pmatrix} \text{proportion of states} \\ \text{with percentage in public} \\ \text{institutions less than 60} \end{pmatrix} = \begin{pmatrix} \text{proportion in} \\ \text{40 to} <50 \text{ class} \end{pmatrix} + \begin{pmatrix} \text{proportion in} \\ \text{50 to} <60 \text{ class} \end{pmatrix}$$

$$= .02 + .10 = .12 \,(12\%)$$

and

$$\begin{pmatrix} \text{proportion of states} \\ \text{with percentage in public} \\ \text{institutions between 60 and 90} \end{pmatrix} = \begin{pmatrix} \text{proportion in} \\ \text{60 to} <70 \text{ class} \end{pmatrix} + \begin{pmatrix} \text{proportion in} \\ \text{70 to} <80 \text{ class} \end{pmatrix}$$

$$+ \begin{pmatrix} \text{proportion in} \\ \text{80 to} <90 \text{ class} \end{pmatrix}$$

$$= .04 + .22 + .50 = .76 \,(76\%)$$

There are no set rules for selecting either the number of class intervals or the length of the intervals. Using a few relatively wide intervals will bunch the data, whereas using a great many relatively narrow intervals may spread the data over too many intervals, so that no interval contains more than a few observations. Neither type of distribution will give an informative picture of how values are distributed over the range of measurement, and interesting features of the data set may be missed. In general, with a small amount of data, relatively few intervals, perhaps between 5 and 10, should be used, whereas with a large amount of data, a distribution based on 15 to 20 (or even more) intervals is often recommended. The quantity

$$\sqrt{\text{number of observations}}$$

is often used as an estimate of an appropriate number of intervals: 5 intervals for 25 observations, 10 intervals when the number of observations is 100, and so on.

Two people making reasonable and similar choices for the number of intervals, their width, and the starting point of the first interval will usually obtain similar summaries of the data.

▪ Cumulative Relative Frequencies and Cumulative Relative Frequency Plots

Rather than wanting to know what proportion of the data fall in a particular class, we often wish to determine the proportion falling below a specified value. This is easily done when the value is a class boundary. Consider the following classes and relative frequencies:

Class	0 to <25	25 to <50	50 to <75	75 to <100	100 to <125	...
Rel. freq.	.05	.10	.18	.25	.20	...

Then

proportion of observations less than 75 = proportion in one of the first three classes
$$= .05 + .10 + .18$$
$$= .33$$

Similarly,

proportion of observations less than 100 $= .05 + .10 + .18 + .25$
$$= .33 + .25$$
$$= .58$$

Each such sum of relative frequencies is called a **cumulative relative frequency**. Notice that the cumulative relative frequency .58 is the sum of the previous cumulative relative frequency .33 and the "current" relative frequency .25. These calculations can also be done for discrete data (e.g., the proportion of observations that are at most 5, at most 6, etc.). The use of cumulative relative frequencies is illustrated in Example 3.16.

▪ Example 3.16   Strength of Aircraft Welds

The strength of welds used in aircraft construction has been of great concern to aeronautical engineers in recent years. Table 3.4 gives a frequency distribution for shear strengths (force in pounds required to break the weld) of ultrasonic spot welds ("Comparison of Properties of Joints Prepared by Ultrasonic Welding and Other Means," *Journal of Aircraft* [1983]: 552–556).

The proportion of welds with strength values less than 5400 is .85 (i.e., 85% of the observations are below 5400). What about the proportion of observations less than 4700? Because 4700 is not a class boundary, we must make an educated guess. The value 4700 is halfway between the boundaries of the 4600–4800 class; let's estimate that half of the relative frequency of .14 for this class, or .07, belongs in the 4600–4700 range. Thus,

estimate of proportion less than 4700 = .01 + .02 + .09 + .07 = .19

Table 3.4 ▪ Frequency Distribution with
Cumulative Relative Frequencies

Class Interval	Frequency	Relative Frequency	Cumulative Relative Frequency
4000 to <4200	1	.01	.01
4200 to <4400	2	.02	.03
4400 to <4600	9	.09	.12
4600 to <4800	14	.14	.26
4800 to <5000	17	.17	.43
5000 to <5200	22	.22	.65
5200 to <5400	20	.20	.85
5400 to <5600	7	.07	.92
5600 to <5800	7	.07	.99
5800 to <6000	1	.01	1.00
	100	1.00	

This proportion could also have been computed using the cumulative relative frequencies as

estimate of proportion less than $4700 = .12 + .07 = .19$

Similarly, because 5250 is one-fourth of the way from 5200 to 5400,

estimate of proportion less than $5250 = .65 + .25(.20) = .70$

A **cumulative relative frequency plot** is just a graph of the cumulative relative frequencies against the upper endpoint of the corresponding interval. The pairs

(upper endpoint of interval, cumulative relative frequency)

are plotted as points on a rectangular coordinate system, and successive points in the plot are connected by a line segment. For the weld strength data of Example 3.16, the plotted points would be

(4200, .01)   (4400, .03)   (4600, .12)   (4800, .26)   (5000, .43)
(5200, .65)   (5400, .85)   (5600, .92)   (5800, .99)   (6000, 1.00)

One additional point, the pair (lower endpoint of first interval, 0), is also included in the plot (for the weld strength data, this would be the point (4000, 0)), and then points are connected by line segments. Figure 3.18 shows the cumulative relative frequency plot for the weld strength data. The cumulative relative frequency plot can be used to obtain approximate answers to questions such as, What proportion of the observations is smaller than a particular value? and, What value separates the smallest p percent from the larger values?

For example, to determine the approximate proportion of the weld strength values that are smaller than 5700, we would follow a vertical line up from 5700 on

Figure 3.18 Cumulative relative frequency plot for the weld strength data of Example 3.16.

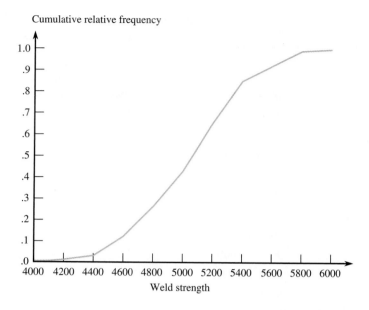

the x axis and then read across to obtain the corresponding cumulative relative frequency, as illustrated in Figure 3.19(a). Approximately .94, or 94%, of the weld strengths are smaller than 5700. Similarly, to find the weld strength value that separates the smallest 30% of the weld strengths from the larger values, start at .30 on the cumulative relative frequency axis and move across and then down to find the corresponding weld strength value, as shown in Figure 3.19(b). Approximately 30% of the weld strengths are smaller than 4850.

▪ Histograms for Continuous Numerical Data

In Example 3.15, the class intervals in the frequency distribution were all of equal width. When this is the case, it is easy to construct a histogram using the information in a frequency distribution.

▪ Histogram for Continuous Numerical Data When the Class Interval Widths Are Equal

When to Use

Continuous numerical data. Works well, even for large data sets.

How to Construct

1. Mark the boundaries of the class intervals on a horizontal axis.

2. Use either frequency or relative frequency on the vertical axis.

3. Draw a rectangle for each class directly above the corresponding interval (so that the edges are at the class boundaries). The height of each rectangle is the frequency or relative frequency of the corresponding class interval.

What to Look For

Central or typical value, extent of spread or variation, general shape, location and number of peaks, and presence of gaps and outliers.

Figure 3.19
Approximating weld strength using the cumulative relative frequency plot.
(a) Determining the approximate proportion of weld strength values less than 5700;
(b) finding the weld strength value that separates the smallest 30% of weld strengths from the larger values.

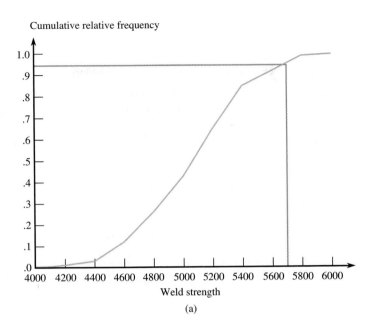

Cumulative relative frequency

Weld strength

(a)

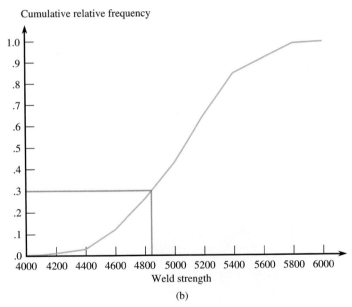

Cumulative relative frequency

Weld strength

(b)

- ■ **Example 3.17   Mercury Contamination**

Mercury contamination is a serious environmental concern. Mercury levels are particularly high in certain types of fish. Citizens of the Republic of Seychelles, a group of islands in the Indian Ocean, are among those who consume the most fish in the world. The article "Mercury Content of Commercially Important Fish of the Seychelles, and Hair Mercury Levels of a Selected Part of the Population" (*Environmental Research* [1983]: 305–312) reported the following observations on mercury content (in parts per million) in the hair of 40 fishermen:

13.26	32.43	18.10	58.23	64.00	68.20	35.35	33.92	23.94	18.28
22.05	39.14	31.43	18.51	21.03	5.50	6.96	5.19	28.66	26.29
13.89	25.87	9.84	26.88	16.81	37.65	19.63	21.82	31.58	30.13
42.42	16.51	21.16	32.97	9.84	10.64	29.56	40.69	12.86	13.80

A reasonable choice for class intervals is to start the first interval at 0 and set the interval width as 10. The resulting frequency distribution is displayed in Table 3.5, and the corresponding histogram appears in Figure 3.20. If it were not for the slight dip in the interval 50 to <60, the histogram would have a single peak; this dip might well disappear with a larger sample size. The upper or right end of the histogram is much more stretched out than the lower or left end. Typical mercury content is somewhere between 20 and 30, but the data exhibit a substantial amount of variability about the center.

Table 3.5 ▪ Frequency Distribution for Hair Mercury Content (ppm) of Seychelles Fishermen

Class Interval	Frequency	Relative Frequency
0 to <10	5	.125
10 to <20	11	.275
20 to <30	10	.250
30 to <40	9	.225
40 to <50	2	.050
50 to <60	1	.025
60 to <70	2	.050
	40	1.000

Figure 3.20 Histogram for hair mercury content of Seychelles fishermen.

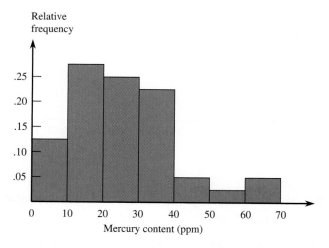

Figure 3.21 shows a histogram generated by the MINITAB statistical package. Notice that the classes are different from the ones we chose and that the centers of the intervals rather than the class boundaries are marked. There is now a gap toward the upper end of the data; this discrepancy in the two graphical displays may well be attributable to the small sample size. With larger samples, the choice of intervals has less impact on the appearance of the histogram.

Figure 3.21 Histogram for hair mercury content of Seychelles fishermen created using MINITAB.

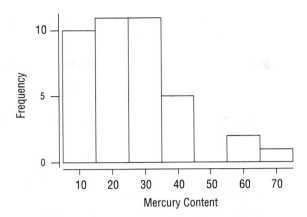

▪ **Class Intervals of Unequal Widths** Figure 3.22 shows a data set in which a great many observations are concentrated at the center of the set, with only a few outlying, or stray, values both below and above the main body of data. If a frequency distribution is based on short intervals of equal width, a great many intervals will be required to capture all observations, and many of them will contain no observations (0 frequency). On the other hand, only a few wide intervals will capture all values, but then most of the observations will be grouped into a few intervals. Neither choice yields an informative description of the distribution. In such a situation, it is best to use a combination of relatively wide class intervals where there are few data points and relatively shorter intervals where there are many data points.

Figure 3.22 Three choices of class intervals for a data set with outliers: (a) many short intervals of equal width; (b) a few wide intervals of equal width; (c) intervals of unequal width.

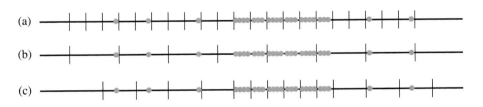

▪ **Constructing a Histogram for Continuous Data When Class Widths Are Unequal**

When class widths are unequal, frequencies or relative frequencies should *not* be used on the vertical axis of a histogram. Instead, the height of each rectangle is determined by the **density**. The density is given by

$$\text{density} = \text{rectangle height} = \frac{\text{relative frequency of class}}{\text{class width}}$$

The vertical axis is called the **density scale**; it should be marked so that each rectangle can be drawn to the calculated height.

The use of the density scale to construct the histogram ensures that the area of each rectangle in the histogram will be proportional to the corresponding relative frequency. The formula for density can also be used when class widths are identical; the denominator is then the same for each density calculation. The resulting histogram will look exactly like the one based on relative frequencies, except for the vertical scaling. When the intervals are of equal width, the extra arithmetic required to obtain the densities is unnecessary.

▪ **Example 3.18**   Misreporting Grade Point Averages

When people are asked for the values of characteristics such as age or weight, they sometimes shade the truth in their responses. The article "Self-Reports of Academic Performance" (*Social Methods and Research* [November 1981]: 165–185) focused on such characteristics as SAT scores and grade point average (GPA). For each student in a sample, the difference in GPA (reported − actual) was determined. Positive differences resulted from individuals reporting GPAs larger than the correct values. Most differences were close to 0, but there were some rather gross errors. Because of this, a frequency distribution based on unequal class widths gives an informative yet concise summary. Table 3.6 displays such a distribution based on classes with boundaries at −2.0, −0.4, −0.2, −0.1, 0, 0.1, 0.2, 0.4, and 2.0.

Table 3.6 ▪ **Frequency Distribution for Errors in Reported GPA**

Class Interval	Relative Frequency	Width	Density
−2.0 to <−0.4	.023	1.6	0.014
−0.4 to <−0.2	.055	0.2	0.275
−0.2 to <−0.1	.097	0.1	0.970
−0.1 to <0	.210	0.1	2.100
0 to <0.1	.189	0.1	1.890
0.1 to <0.2	.139	0.1	1.390
0.2 to <0.4	.116	0.2	0.580
0.4 to <2.0	.171	1.6	0.107

Figure 3.23 displays two histograms based on this frequency distribution. The histogram in Figure 3.23(a) is correctly drawn, with height equal to relative frequency divided by interval width. The histogram in Figure 3.23(b) has height equal to relative frequency and is therefore not correct. In particular, this second histogram considerably exaggerates the incidence of grossly overreported and underreported values — the areas of the two most extreme rectangles are much too large. The eye is naturally drawn to large areas, so it is important that the areas correctly represent the relative frequencies.

The formula for rectangle height given previously implies that

relative frequency = (rectangle height)(class width) = area of rectangle

That is, a histogram can always be drawn so that the area of each rectangle is the relative frequency of the corresponding class (thus the total area of all rectangles is 1).

Figure 3.23 Histograms for errors in reporting GPA: (a) a correct picture (height = density); (b) an incorrect picture (height = relative frequency).

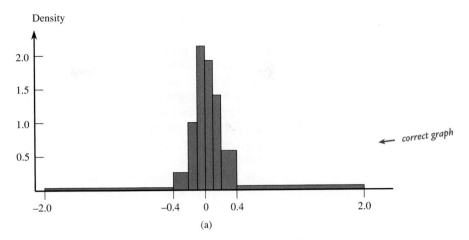

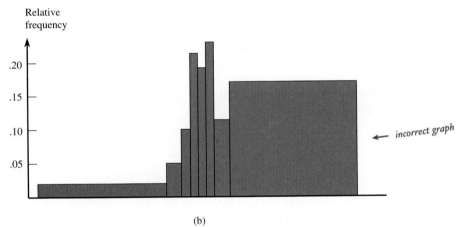

■ Histogram Shapes

The general shape of a histogram is an important characteristic. In describing various shapes, it is convenient to approximate the histogram itself with a smooth curve (called a *smoothed histogram*). This is illustrated in Figure 3.24.

Figure 3.24 Approximating a histogram with a smooth curve.

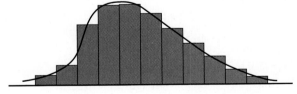

One characterization of general shape relates to the number of peaks, or **modes**. A histogram is said to be **unimodal** if it has a single peak, **bimodal** if it has two peaks, and **multimodal** if it has more than two peaks. These shapes are illustrated in Figure 3.25.

Figure 3.25 Smoothed histograms with various numbers of modes: (a) unimodal; (b) bimodal; (c) multimodal.

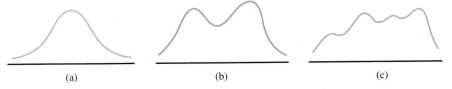

Bimodality can occur when the data set consists of observations on two quite different kinds of individuals or objects. For example, consider a large data set consisting of driving times for automobiles traveling between San Luis Obispo, California, and Monterey, California. this histogram would show two peaks, one for those cars that took the inland route (roughly 2.5 hr) and another for those cars traveling up the coast highway (3.5–4 hr). However, bimodality does not automatically follow in such situations. Bimodality will occur in the histogram of the combined groups only if the centers of the two separate histograms are far apart relative to the variability in the two data sets. Thus, a large data set consisting of heights of college students would probably not produce a bimodal histogram because the typical height for males (about 69 in.) and the typical height for females (about 66 in.) are not very far apart. Many histograms encountered in practice are unimodal, and multimodality is rather rare.

Unimodal histograms come in a variety of shapes. A unimodal histogram is **symmetric** if there is a vertical line of symmetry such that the part of the histogram to the left of the line is a mirror image of the part to the right. (Bimodal and multimodal histograms can also be symmetric in this way.) Several different symmetric smoothed histograms are shown in Figure 3.26.

Figure 3.26 Several symmetric unimodal smoothed histograms.

Proceeding to the right from the peak of a unimodal histogram, we move into what is called the **upper tail** of the histogram. Going in the opposite direction moves us into the **lower tail**. A unimodal histogram that is not symmetric is said to be **skewed**. If the upper tail of the histogram stretches out much farther than the lower tail, then the distribution of values is **positively skewed**. If, on the other hand, the lower tail is much longer than the upper tail, the histogram is **negatively skewed**. These two types of skewness are illustrated in Figure 3.27. Positive skewness is much more frequently encountered than is negative skewness. An example of positive skewness occurs in the distribution of single-family home prices in Los Angeles County; most homes are moderately priced (at least for California), whereas the relatively few homes in Beverly Hills and Malibu have much higher price tags.

Figure 3.27 Two examples of skewed smoothed histograms: (a) positive skew; (b) negative skew.

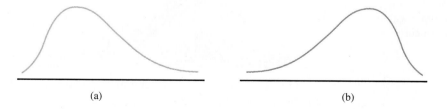

(a)   (b)

One rather specific shape, a **normal curve**, arises more frequently than any other in statistical applications. Many histograms can be well approximated by a normal curve (e.g., characteristics such as blood pressure, brain weight, adult male height, adult female height, and IQ score). Here we briefly mention several of the most important qualitative properties of such a curve, postponing a more detailed discussion until Chapter 7. A normal curve is not only symmetric but also bell

Figure 3.28 Three examples of bell-shaped histograms: (a) normal; (b) heavy-tailed; (c) light-tailed.

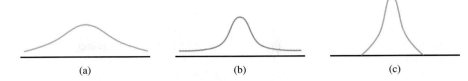

(a)   (b)   (c)

shaped; it looks like the curve in Figure 3.28(a). However, not all bell-shaped curves are normal. From the top of the bell, the height of the curve decreases at a well-defined rate when moving into either tail. (This rate of decrease is specified by a certain mathematical function.)

A curve with tails that do not decline as rapidly as the tails of a normal curve is said to specify a **heavy-tailed** distribution (compared to the normal curve). Similarly, a curve with tails that decrease more rapidly than the normal tails is called **light-tailed**. Figures 3.28(b) and 3.28(c) illustrate these possibilities. The reason that we are concerned about the tails in a distribution is that many inferential procedures that work well when the population distribution is approximately normal (i.e., they result in accurate conclusions) do poorly when the population distribution is heavy tailed.

■ Do Sample Histograms Resemble the Population Histogram?

Sample data are usually collected to make inferences about a population. The resulting conclusions may be in error if the sample is somehow unrepresentative of the population. So how similar might a histogram of sample data be to the histogram of all population values? Will the two histograms be centered at roughly the same place and spread out to about the same extent? Will they have the same number of peaks, and will these occur at approximately the same places?

A related issue concerns the extent to which histograms based on different samples from the same population resemble one another. If two different sample histograms can be expected to differ from one another in obvious ways, then at least one of them might differ substantially from the population histogram. If the sample differs substantially from the population, conclusions about the population based on the sample are likely to be incorrect. **Sampling variability** — the extent to which samples differ from one another and from the population — is a central idea in statistics. Example 3.19 illustrates such variability in histogram shapes.

■ Example 3.19   What You Should Know About Bus Drivers . . .

A sample of 708 bus drivers employed by public corporations was selected, and the number of traffic accidents in which each bus driver was involved during a 4-year period was determined ("Application of Discrete Distribution Theory to the Study of Noncommunicable Events in Medical Epidemiology," in *Random Counts in Biomedical and Social Sciences*, G. P. Patil, ed. [University Park, PA: Pennsylvania State University Press, 1970]). A listing of the 708 sample observations might look like this:

3 0 6 0 0 2 1 4 1 . . . 6 0 2

The frequency distribution (Table 3.7) shows that 117 of the 708 drivers had no accidents, a relative frequency of 117/708 = .165 (or 16.5%). Similarly, the proportion

Figure 3.29
Comparison of population and sample histograms for number of accidents.

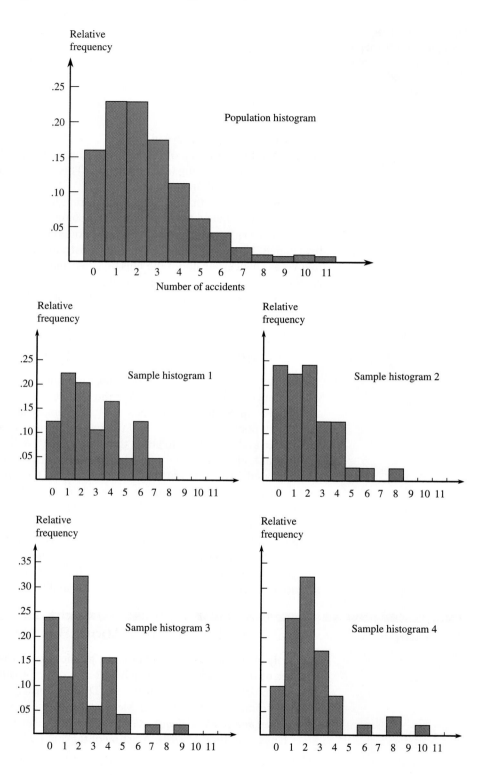

Table 3.7 ▪ Frequency Distribution for Number of Accidents by Bus Drivers

Number of Accidents	Frequency	Relative Frequency
0	117	.165
1	157	.222
2	158	.223
3	115	.162
4	78	.110
5	44	.062
6	21	.030
7	7	.010
8	6	.008
9	1	.001
10	3	.004
11	1	.001
	708	.998

of sampled drivers who had 1 accident is .222 (or 22.2%). The largest sample observation was 11.

Although the 708 observations actually constituted a sample from the population of all bus drivers, we will regard the 708 observations as constituting the entire population. The first histogram in Figure 3.29, then, represents the population histogram. The other four histograms in Figure 3.29 are based on four different samples of 50 observations each from this population. The five histograms certainly resemble one another in a general way, but some dissimilarities are also obvious. The population histogram rises to a peak and then declines smoothly, whereas the sample histograms tend to have more peaks, valleys, and gaps. Although the population data set contained an observation of 11, none of the four samples did. In fact, in the first two samples, the largest observations were 7 and 8, respectively. In Chapters 8–15 we will see how sampling variability can be described and incorporated into conclusions based on inferential methods.

▪ Exercises 3.25–3.40

3.25 People suffering from Alzheimer's disease often have difficulty performing basic activities of daily living (ADLs). In one study ("Functional Status and Clinical Findings in Patients with Alzheimer's Disease," *Journal of Gerontology* [1992]: 177–182), investigators focused on six such activities: dressing, bathing, transferring, toileting, walking, and eating. Here are data on the number of ADL impairments for each of 240 patients:

Number of impairments	0	1	2	3	4	5	6
Frequency	100	43	36	17	24	9	11

a. Determine the relative frequencies that correspond to the given frequencies.
b. What proportion of these patients had at most two impairments?

c. Use the result of Part (b) to determine what proportion of patients had more than two impairments.
d. What proportion of the patients had at least four impairments?
e. Do you notice anything especially interesting about the frequencies and relative frequencies? Note: The investigators proposed statistical models from which the number of impairments could be predicted from various patient characteristics and symptoms.

3.26 The trace element zinc is an important dietary constituent, partly because it aids in the maintenance of the immune system. The accompanying data on zinc intake (in milligrams) for a sample of 40 patients with rheumatoid arthritis were read from a graph in the article "Plasma Zinc and Copper Concentrations in Rheumatoid Arthritis: Influence of Dietary Factors and Disease Activity" (*American Journal of Clinical Nutrition* [1991]: 1082–1086):

```
8.0  12.9  13.0   8.9  10.1  17.3  11.1  10.9
6.2   8.1   8.8  10.4  15.7  13.6  19.3   9.9
8.5  11.1  10.7   8.8  10.7   6.8   7.4   4.8
11.8  13.0   9.5   8.1   6.9  11.5  11.2  13.6
4.9  18.8  15.7  10.8  10.7  11.5  16.1   9.9
```

a. Use class intervals of 3 to <6, 6 to <9, 9 to <12, 12 to <15, 15 to <18, and 18 to <21 to construct a table that includes the class intervals, frequencies, and relative frequencies.
b. Use the table constructed in Part (a) to determine the proportion of individuals whose intake is less than 12 and the proportion whose intake is between 6 and 15.
c. Construct a histogram for this data set, and comment on the key features of the histogram.

3.27 *USA Today* (July 2, 2001) gave the following information regarding cell phone use for men and women:

Average Number of Minutes Used per Month	Relative Frequency	
	Men	Women
0 to <200	.56	.61
200 to <400	.18	.18
400 to <600	.10	.13
600 to <800	.16	.08

a. Construct a relative frequency histogram for average number of minutes used per month for men. How would you describe the shape of this histogram?
b. Construct a relative frequency histogram for average number of minutes used per month for women. Is the distribution for average number of minutes used per month similar for men and women? Explain.

c. What proportion of men average less than 400 minutes per month?
d. Estimate the proportion of men that average less than 500 minutes per month.
e. Estimate the proportion of women that average 450 minutes or more per month.

3.28 The article "Associations Between Violent and Nonviolent Criminality" (*Multivariate Behavioral Research* [1981]: 237–242) reported the number of previous convictions for 283 adult males arrested for felony offenses. The following frequency distribution is a summary of the data given in the paper:

Number of Previous Convictions	Frequency
0	0
1	16
2	27
3	37
4	46
5	36
6	40
7	31
8	27
9	13
10	8
11	2

Draw the histogram corresponding to this frequency distribution, and comment on its shape.

3.29 U.S. census data for San Luis Obispo County, California, were used to construct the following frequency distribution for commute time (in minutes) of working adults (the given frequencies were read from a graph that appeared in the *San Luis Obispo Tribune* [September 1, 2002] and so are only approximate):

Commute Time	Frequency
0 to <5	5,200
5 to <10	18,200
10 to <15	19,600
15 to <20	15,400
20 to <25	13,800
25 to <30	5,700
30 to <35	10,200
35 to <40	2,000
40 to <45	2,000
45 to <60	4,000
60 to <90	2,100
90 to <120	2,200

a. Notice that not all intervals in the frequency distribution are equal in width. Why do you think that unequal width intervals were used?

b. Construct a table that adds a relative frequency and a density column to the given frequency distribution.

c. Use the densities computed in Part (b) to construct a histogram for this data set. (Note: The newspaper displayed an incorrectly drawn histogram based on frequencies rather than densities!) Write a few sentences commenting on the important features of the histogram.

d. Compute the cumulative relative frequencies, and construct a cumulative relative frequency plot.

e. Use the cumulative relative frequency plot constructed in Part (d) to answer the following questions.

i. Approximately what proportion of commute times were less than 50 min?

ii. Approximately what proportion of commute times were greater than 22 min?

iii. What is the approximate commute time value that separates the shortest 50% of commute times from the longest 50%?

3.30 The article "Determination of Most Representative Subdivision" (*Journal of Energy Engineering* [1993]: 43–55) gave data on various characteristics of subdivisions that could be used in deciding whether to provide electrical power using overhead lines or underground lines. Data on the variable x = total length of streets within a subdivision are as follows:

1280	5320	4390	2100	1240	3060	4770	1050
360	3330	3380	340	1000	960	1320	530
3350	540	3870	1250	2400	960	1120	2120
450	2250	2320	2400	3150	5700	5220	500
1850	2460	5850	2700	2730	1670	100	5770
3150	1890	510	240	396	1419	2109	

a. Construct a stem-and-leaf display for these data using the thousands digit as the stem. Comment on the various features of the display.

b. Construct a histogram using class boundaries of 0 to <1000, 1000 to <2000, and so on. How would you describe the shape of the histogram?

c. What proportion of subdivisions has total length less than 2000? between 2000 and 4000?

3.31 Student loans can add up, especially for those attending professional schools to study in such areas as medicine, law, or dentistry. Researchers at the University of Washington studied medical students and gave the following information on the educational debt of medical students on completion of their residencies (*Annals of Internal Medicine* [March 2002]: 384–398):

Educational Debt (dollars)	Relative Frequency
0 to <5000	.427
5000 to <20,000	.046
20,000 to <50,000	.109
50,000 to <100,000	.232
100,000 or more	.186

a. What are two reasons that it would be inappropriate to construct a histogram using relative frequencies to determine the height of the bars in the histogram?

b. Suppose that no student had an educational debt of $150,000 or more upon completion of his or her residency, so that the last class in the relative frequency distribution would be 100,000 to <150,000. Summarize this distribution graphically by constructing a histogram of the educational debt data. (Don't forget to use the density scale for the heights of the bars in the histogram, because the interval widths aren't all the same.)

c. Based on the histogram of Part (b), write a few sentences describing the educational debt of medical students completing their residencies.

3.32 The behavior of children watching television has been a much-studied phenomenon. The same attention has been given to children engaged in playing with toys. The paper "A Temporal Analysis of Free Toy Play and Distractibility in Young Children" (*Journal of Experimental Child Psychology* [1991]: 41–69) reported the following data on play-episode lengths (in seconds) for a particular 5-year-old boy:

Class	Frequency
0 to <5	54
5 to <10	44
10 to <15	28
15 to <20	21
20 to <40	31
40 to <60	15
60 to <90	16
90 to <120	5
120 to <180	8

a. Display this information in a histogram.

b. What proportion of episodes lasted at least 20 sec?

c. Roughly what proportion of episodes lasted between 40 and 75 sec?

3.33 An exam is given to students in an introductory statistics course. What is likely to be true of the shape of the histogram of scores if

a. the exam is quite easy?

b. the exam is quite difficult?
c. half the students in the class have had calculus, the other half have had no prior college math courses, and the exam emphasizes mathematical manipulation?
Explain your reasoning in each case.

3.34 The results of the 1990 census included a state-by-state listing of population density. The following table gives the number of people per square mile for each of the 50 states:

State	Number of People per Square Mile
Alabama	79.6
Alaska	1.0
Arizona	32.3
Arkansas	45.1
California	190.8
Colorado	31.8
Connecticut	678.4
Delaware	340.8
Florida	239.6
Georgia	111.9
Hawaii	172.5
Idaho	12.2
Illinois	205.6
Indiana	154.6
Iowa	49.7
Kansas	30.3
Kentucky	92.8
Louisiana	96.9
Maine	39.8
Maryland	489.2
Massachusetts	767.6
Michigan	163.6
Minnesota	55.0
Mississippi	54.9
Missouri	74.3
Montana	5.5
Nebraska	20.5
Nevada	10.9
New Hampshire	123.7
New Jersey	1042.0
New Mexico	12.5
New York	381.0
North Carolina	136.1
North Dakota	9.3
Ohio	264.9
Oklahoma	45.8
Oregon	29.6
Pennsylvania	265.1
Rhode Island	960.3
South Carolina	115.8

State	Number of People per Square Mile
South Dakota	9.2
Tennessee	118.3
Texas	64.9
Utah	21.0
Vermont	60.8
Virginia	156.3
Washington	73.1
West Virginia	74.5
Wisconsin	90.1
Wyoming	4.7

a. Construct a relative frequency distribution for state population density.
b. In your relative frequency distribution, did you use class intervals of equal widths? Why or why not?
c. Use the relative frequency distribution to give an approximate value for the proportion of states that have a population density of more than 100 people per square mile. Is this value close to the actual value?

3.35 The paper "Lessons from Pacemaker Implantations" (*Journal of the American Medical Association* [1965]: 231–232) gave the results of a study that followed 89 heart patients who had received electronic pacemakers. The time (in months) to the first electrical malfunction of the pacemaker was recorded:

```
24 20 16 32 14 22  2 12 24  6 10 20
 8 16 12 24 14 20 18 14 16 18 20 22
24 26 28 18 14 10 12 24  6 12 18 16
34 18 20 22 24 26 18  2 18 12 12  8
24 10 14 16 22 24 22 20 24 28 20 22
26 20  6 14 16 18 24 18 16  6 16 10
14 18 24 22 28 24 30 34 26 24 22 28
30 22 24 22 32
```

a. Summarize these data in the form of a frequency distribution, using class intervals of 0 to <6, 6 to <12, and so on.
b. Compute the relative frequencies and cumulative relative frequencies for each class interval of the frequency distribution of Part (a).
c. Show how the relative frequency for the class interval 12 to <18 could be obtained from the cumulative relative frequencies.
d. Use the cumulative relative frequencies to give approximate answers to the following:
i. What proportion of those who participated in the study had pacemakers that did not malfunction within the first year?
ii. If the pacemaker must be replaced as soon as the first electrical malfunction occurs, approximately

what proportion required replacement between 1 and 2 years after implantation?

e. Construct a cumulative relative frequency plot, and use it to answer the following questions.

i. What is the approximate time at which about 50% of the pacemakers had failed?

ii. What is the approximate time at which only about 10% of the pacemakers initially implanted were still functioning?

3.36 The clearness index was determined for the skies over Baghdad for each of the 365 days during a particular year ("Contribution to the Study of the Solar Radiation Climate of the Baghdad Environment," *Solar Energy* [1990]: 7–12). The accompanying table summarizes the resulting data:

Clearness Index	Number of Days (frequency)
0.15 to <0.25	8
0.25 to <0.35	14
0.35 to <0.45	28
0.45 to <0.50	24
0.50 to <0.55	39
0.55 to <0.60	51
0.60 to <0.65	106
0.65 to <0.70	84
0.70 to <0.75	11

a. Determine the relative frequencies and draw the corresponding histogram. (Be careful here — the intervals do not all have the same width.)

b. Cloudy days are those with a clearness index smaller than 0.35. What proportion of the days were cloudy?

c. Clear days are those for which the index is at least 0.65. What proportion of the days were clear?

3.37 How does the speed of a runner vary over the course of a marathon (a distance of 42.195 km)?

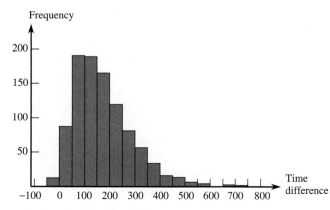

Figure for Exercise 3.37

Consider determining both the time to run the first 5 km and the time to run between the 35 km and 40 km points, and then subtracting the 5-km time from the 35–40-km time. A positive value of this difference corresponds to a runner slowing down toward the end of the race. The accompanying histogram is based on times of runners who participated in several different Japanese marathons ("Factors Affecting Runners' Marathon Performance," *Chance* [Fall, 1993]: 24–30). What are some interesting features of this histogram? What is a typical difference value? Roughly what proportion of the runners ran the late distance more quickly than the early distance?

3.38 Disparities among welfare payments by different states have been the source of much political controversy. The accompanying table reports average payment per person (in dollars) in the Aid to Families with Dependent Children Program for the 1990 fiscal year. Construct a relative frequency distribution for these data using equal interval widths. Draw the histogram corresponding to your frequency distribution.

State	Average Welfare Payment ($)
Alaska	244.90
California	218.31
Arizona	93.57
Montana	114.95
Texas	56.79
Nebraska	115.15
Minnesota	171.75
Arkansas	65.96
Alabama	39.62
Illinois	112.28
Indiana	92.43
New Hampshire	164.20
Rhode Island	179.37
New Jersey	121.99
Delaware	113.66
North Carolina	91.95
Florida	95.43
Washington	160.41
Idaho	97.93
Utah	118.36
Colorado	111.20
Oklahoma	96.98
South Dakota	95.52
Iowa	129.58
Louisiana	55.81
Tennessee	65.93

(continued)

(*continued*)

State	Average Welfare Payment ($)
Wisconsin	155.04
Ohio	115.26
Vermont	183.36
Connecticut	205.86
Pennsylvania	127.70
Maryland	132.86
South Carolina	71.91
Hawaii	187.71
Oregon	135.99
Nevada	100.25
Wyoming	113.84
New Mexico	81.87
Kansas	113.88
North Dakota	130.49
Missouri	91.93
Mississippi	40.22
Kentucky	85.21
Michigan	154.75
Maine	150.12
Massachusetts	200.99
New York	193.48
West Virginia	82.94
Virginia	97.98
Georgia	91.31

3.39 Construct a histogram corresponding to each of the five frequency distributions, I–V, given in the following table, and state whether each histogram is symmetric, bimodal, positively skewed, or negatively skewed:

Class Interval	Frequency				
	I	II	III	IV	V
0 to <10	5	40	30	15	6
10 to <20	10	25	10	25	5
20 to <30	20	10	8	8	6
30 to <40	30	8	7	7	9
40 to <50	20	7	7	20	9
50 to <60	10	5	8	25	23
60 to <70	5	5	30	10	42

3.40 Using the five class intervals 100 to <120, 120 to <140, . . . , 180 to <200, devise a frequency distribution based on 70 observations whose histogram could be described as follows:

a. symmetric
b. bimodal
c. positively skewed
d. negatively skewed

▪ 3.4 Displaying Bivariate Numerical Data

A bivariate data set consists of measurements or observations on two variables, x and y. For example, x might be distance from a highway and y the lead content of soil at that distance. When both x and y are numerical variables, each observation consists of a pair of numbers, such as (14, 5.2) or (27.63, 18.9). The first number in a pair is the value of x, and the second number is the value of y.

An unorganized list of such pairs yields little information about the distribution of either the x values or the y values separately and even less information about how the two variables are related to one another. We saw how pictures could be used to summarize univariate data. Graphical displays can help us with bivariate data also. The most important graph based on bivariate numerical data is a **scatterplot**.

In a scatterplot, each observation (pair of numbers) is represented by a point on a rectangular coordinate system, as shown in Figure 3.30(a). The horizontal axis is identified with values of x and is scaled so that any x value can be easily located. Similarly, the vertical or y axis is marked for easy location of y values. The point corresponding to any particular (x, y) pair is placed where a vertical line from the value on the x axis meets a horizontal line from the value on the y axis. Figure 3.30(b) shows the point representing the observation (4.5, 15); it is above 4.5 on the horizontal axis and to the right of 15 on the vertical axis.

Figure 3.30
Constructing a scatter-
plot: (a) rectangular
coordinate system;
(b) point corresponding
to (4.5, 15).

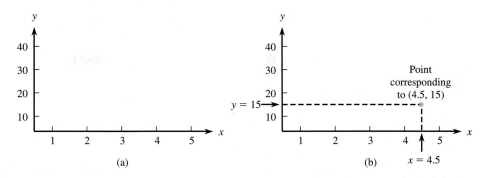

▪ **Example 3.20   You'll Pay for That Beer in More Than One Way**

The Connecticut Agricultural Experiment Station and the Excise Tax Division of
the Connecticut Department of Revenue Service conducted a study of the alcohol
and calorie content of popular beers. (Data from this study can be found on the
web site brewery.org.) Data on x = alcohol content (percent) and y = calorie content (calories per 100 ml) for 39 nonalcoholic and light beers is shown in the accompanying table:

Observation	Alcohol content, x	Calorie content, y	Observation	Alcohol content, x	Calorie content, y
1	0.10	15	21	3.39	28
2	0.44	13	22	3.55	27
3	0.48	14	23	4.53	41
4	0.50	25	24	4.52	39
5	0.10	14	25	4.61	31
6	0.20	15	26	4.40	29
7	0.10	14	27	2.41	23
8	0.10	13	28	4.81	32
9	0.50	21	29	4.70	31
10	0.05	15	30	3.82	32
11	0.10	16	31	2.50	19
12	0.73	17	32	4.49	40
13	0.10	30	33	4.32	27
14	3.74	29	34	4.07	34
15	3.96	28	35	4.28	31
16	4.40	33	36	4.91	42
17	4.12	31	37	4.45	35
18	3.56	30	38	3.56	29
19	3.88	33	39	5.44	43
20	4.36	30			

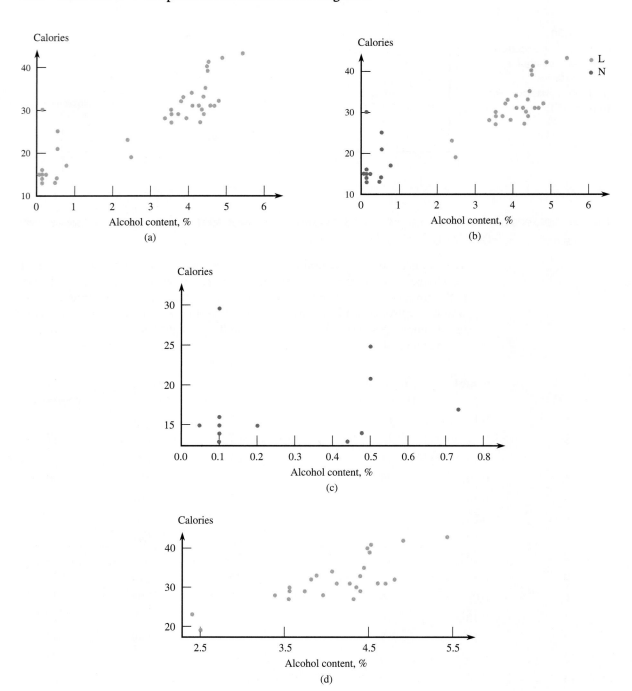

Figure 3.31 Scatterplots for the data of Example 3.20:
(a) scatterplot of data; (b) scatterplot of data with nonalcoholic
and light beers distinguished by color; (c) scatterplot for
nonalcoholic beers; (d) scatterplot for light beers.

Figure 3.31(a) gives a scatterplot of the data. Examination of the data and of the plot reveals the following:

1. Several observations have identical x values yet different y values (e.g., $x_1 = x_5 = 0.10$, but $y_1 = 15$ and $y_5 = 14$). Thus, the value of y is *not* determined *solely* by the value of x but by various other factors as well. That is, y is not a function of x.

2. There is a tendency for y to increase as x increases. That is, larger values of calorie content tend to be associated with larger values of alcohol content (a positive relationship).

3. There are two noticeable clusters of points in the scatterplot: one in the lower left-hand corner of the plot and one in the upper right-hand corner of the plot. There are also two points that stand out in the middle part of the plot.

Recall that the data set included x and y values for both nonalcoholic and light beers. The first 13 observations in the data set are for beers that are classified as nonalcoholic (note that the alcohol content of "nonalcoholic" beers is not 0!). Figure 3.31(b) shows a scatterplot of the (alcohol content, calorie content) pairs with observations for nonalcoholic beers shown in blue and observations for light beers shown in orange. We can see that the clusters of points in the scatterplot correspond to the two types of beer and that the two points that stand out in the middle of the plot are observations for light beers that have a lower alcohol and calorie content than is typical for most light beers.

Figures 3.31(c) and 3.31(d) show separate scatterplots for the nonalcoholic beers and the light beers, respectively. Note that there does not appear to be a strong relationship between $x =$ alcohol content and $y =$ calorie content for nonalcoholic beers. However, for light beers, there does appear to be a positive relationship between alcohol content and calorie content.

The horizontal and vertical axes in the scatterplot of Figure 3.31 do not intersect at the point $(0, 0)$. In many data sets, the values of x or of y or of both variables differ considerably from 0 relative to the ranges of the values in the data set. For example, a study of how air conditioner efficiency is related to maximum daily outdoor temperature might involve observations at temperatures of $80°, 82°, \ldots,$ $98°, 100°$. In such cases, the plot will be more informative if the axes intersect at some point other than $(0, 0)$ and are marked accordingly. This is illustrated in Example 3.21.

▪ Example 3.21   Vermont Sugarbushes

The growth and decline of forests is a matter of great public and scientific interest. The article "Relationships Among Crown Condition, Growth, and Stand Nutrition in Seven Northern Vermont Sugarbushes" (*Canadian Journal of Forest Research* [1995]: 386–397) included a scatterplot of $y =$ mean crown dieback (%), which is one indicator of growth retardation, and $x =$ soil pH (higher pH corresponds to less acidic soil). The following observations were read from the scatterplot:

x	y	x	y
3.3	7.3	3.9	6.6
3.4	10.8	4.0	10.0
3.4	13.1	4.1	9.2
3.5	10.4	4.2	12.4
3.6	5.8	4.3	2.3
3.6	9.3	4.4	4.3
3.7	12.4	4.5	3.0
3.7	14.9	5.0	1.6
3.8	11.2	5.1	1.0
3.8	8.0		

Figure 3.32 shows two scatterplots of these data, produced by the statistical computer package MINITAB. In Figure 3.32(a), we let MINITAB select the scale for both axes. We obtained Figure 3.32(b) by specifying minimum and maximum values for x and y such that the axes would intersect at (roughly) the point $(0, 0)$. The second plot does not make effective use of space. It is more crowded than the first plot, and such crowding can make it more difficult to ascertain the general nature of any relationship. For example, it can be more difficult to spot curvature in a crowded plot.

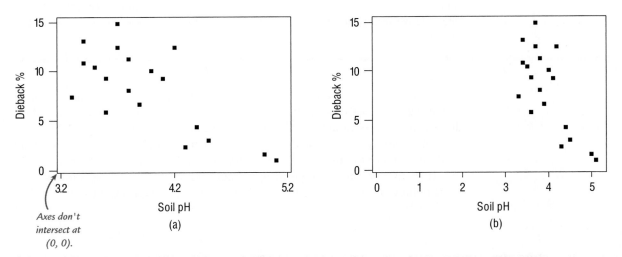

Figure 3.32 MINITAB scatterplots of data in Example 3.21: (a) scale for both axes selected by MINITAB; (b) minimum and maximum values for x and y specified such that the axes would intersect at (roughly) the point $(0, 0)$.

Large values of crown dieback tend to be associated with low soil pH, indicating a negative, or inverse, relationship. Furthermore, the two variables appear to be approximately linearly related, although the points would spread out quite a bit about any straight line drawn through the plot.

■ Example 3.22   Is Time Between Flowering and Harvesting Related to Yield of Paddy?

The focus of many agricultural experiments is to study how the yield of a crop varies with the time at which it is harvested. The accompanying data appeared in the article "Determination of Biological Maturity and Effect of Harvesting and Drying Conditions on Milling Quality of Paddy" (*Journal of Agricultural Engineering Research* [1975]: 353–361). The variables are x = time between flowering and harvesting (days) and y = yield of paddy, a type of grain farmed in India (in kilograms per hectare):

x	y	x	y	x	y	x	y
16	2508	24	3057	32	3823	40	3517
18	2518	26	3190	34	3646	42	3214
20	3304	28	3500	36	3708	44	3103
22	3423	30	3883	38	3333	46	2776

Figure 3.33 is a MINITAB scatterplot. Notice that the axes do not intersect at $(0, 0)$. This is because 0 lies well outside both the range of x values (16 to 46) and the range of y values (roughly 2500 to 3900). Had the scaling been chosen so that the axes did intersect at $(0, 0)$, all points in the plot would have been squeezed into the upper right-hand corner, obscuring the general pattern somewhat.

Figure 3.33 A scatterplot of the yield data from Example 3.22.

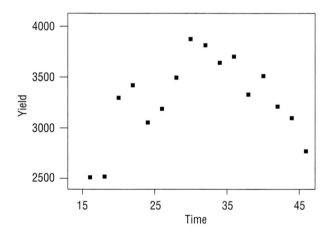

Although the pattern of the points in the plot is not linear, the scatterplot shows a fairly strong relationship between x and y: As x increases, y initially tends to increase, but after a certain point, y tends to decrease as x continues to increase. The shape of the plot is much like that of a parabola.

In Chapter 5, we will develop methods for summarizing bivariate data when the scatterplot reveals a linear pattern (as in Examples 3.20 and 3.21). A curved pattern, such as the one in Example 3.22, is a bit more complicated to analyze, and methods for summarizing such nonlinear relationships will be discussed in Section 5.4.

■ Time-Series Plots

Data sets often consist of measurements collected over time at regular intervals so that we can learn about change over time. For example, stock prices, sales figures, and other socio-economic indicators might be recorded on a weekly or monthly basis. A **time-series plot** (sometimes also called a time plot) is a simple graph of data collected over time that can be invaluable in identifying trends or patterns that might be of interest. A time-series plot can be constructed by thinking of the data set as a bivariate data set, where y is the variable observed and x is the time at which the observation was made. These (x, y) pairs are plotted as in a scatterplot. Consecutive observations are then connected by a line segment; this aids in spotting trends over time.

■ Example 3.23   Life Expectancy over Time

The article "Americans Living Longer Than Ever" (*San Luis Obispo Tribune*, September 13, 2002) included a time-series plot that showed how life expectancy at birth for people living in the United States has changed over time. The plot was based on the following data from "The Vital Statistics Report," published by the Center for Disease Control:

Year	Life Expectancy at Birth (years)
1940	62.9
1950	68.2
1960	69.7
1970	70.8
1980	73.7
1990	75.4
2000	76.9

A time-series plot of these data is shown in Figure 3.34. From this plot, the upward trend is clear, providing justification for the article headline.

Figure 3.34 Time-series plot for the life expectancy at birth data of Example 3.23.

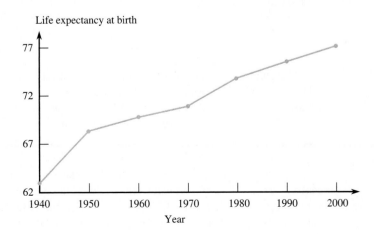

▪ **Example 3.24**   How Much Does It Cost to Get from Here to There?

The accompanying data on household expenditures on transportation for the United Kingdom appeared in "Transport Statistics for Great Britain: 2002 Edition" (in *Family Spending: A Report on the Family Expenditure Survey* [The Stationary Office, 2002]). Expenditures (in pounds per week) included costs of purchasing and maintaining any vehicles owned by members of the household and any costs associated with public transportation and leisure travel.

Year	Average Transportation Expenditure	Year	Average Transportation Expenditure
1990	247.20	1996	309.10
1991	259.00	1997	328.80
1992	271.80	1998	352.20
1993	276.70	1999	359.40
1994	283.60	2000	385.70
1995	289.90		

The time-series plot given in Figure 3.35 clearly shows the increase in transportation expenditures over time.

Figure 3.35 Time-series plot for the average transportation expenditure data of Example 3.24.

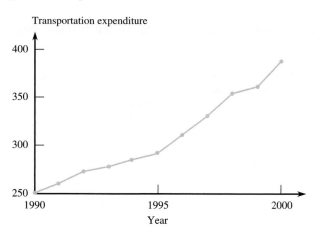

The same report also provided the accompanying data on the percentage of all household expenditures that were associated with transportation:

Year	Percentage of Household Expenditures for Transportation	Year	Percentage of Household Expenditures for Transportation
1990	16.2	1996	15.7
1991	15.3	1997	16.7
1992	15.8	1998	17.0
1993	15.6	1999	17.2
1994	15.1	2000	16.7
1995	14.9		

By comparing the time-series plot of the values for percentage of total expenditures on transportation in Figure 3.36 with the time-series plot of the transportation expenditure data of Figure 3.35, we can easily see that, although actual expenditures have been increasing, the percentage of the total household expenditures that was for transportation has remained relatively stable.

Figure 3.36 Time-series plot of the percentage of total expenditures on transportation.

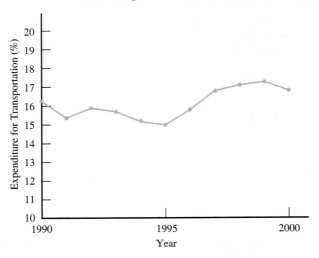

■ Exercises 3.41–3.53

3.41 Does the size of a transplanted organ matter? A study that attempted to answer this question ("Minimum Graft Size for Successful Living Donor Liver Transplantation," *Transplantation* [1999]: 1112–1116) presented a scatterplot much like the following ("graft weight ratio" is the weight of the transplanted liver relative to the ideal size liver for the recipient):

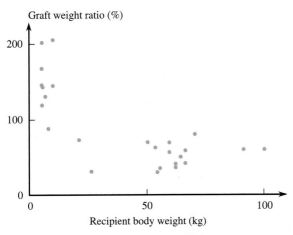

a. Discuss interesting features of this scatterplot.
b. Why do you think the overall relationship is negative?

3.42 Data on x = poverty rate (%) and y = high school dropout rate (%) for the 50 U.S. states and the District of Columbia were used to construct the following scatterplot (*Chronicle of Higher Education*, August 31, 2001):

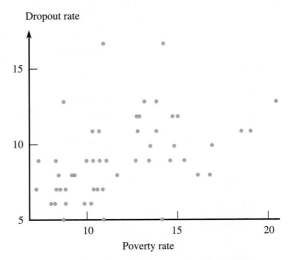

Write a few sentences commenting on this scatterplot. Would you describe the relationship between poverty rate and dropout rate as positive (y tends to increase as x increases), negative (y tends to decrease as x increases), or do you think that there is no discernible relationship between x and y?

3.43 The article "Fines and Administrative Actions Against U.S. Airlines" (*USA Today*, March 13, 2000) included the accompanying data on x = number of violations and y = total fines (millions of dollars) for the period 1994–1998 for 10 major airlines:

Violations	430	679	1038	470	528
Total Fines	1.0	2.1	2.4	1.6	1.5
Violations	516	343	302	119	129
Total Fines	1.7	1.2	0.9	0.7	0.5

a. Construct a scatterplot for these data.
b. Write a few sentences commenting on the nature of the relationship between number of violations and total fines.

3.44 A number of studies concerning muscular efficiency in humans have used competitive cyclists as subjects. It had previously been observed that variation in efficiency could not be explained by differences in cycling technique. The paper "Cycling Efficiency Is Related to the Percentage of Type I Muscle Fibers" (*Medicine and Science in Sports and Exercise* [1992]: 782–788) reported the accompanying data on x = percentage of type I (slow twitch) muscle fibers and y = cycling gross efficiency (the ratio of work accomplished per minute to caloric expenditure):

Observation	x	y
1	32	18.3
2	37	18.9
3	38	19.0
4	40	20.9
5	45	21.4
6	50	20.5
7	50	20.1
8	54	20.1
9	54	20.8
10	55	20.5
11	61	19.9
12	63	20.5
13	63	20.6
14	64	22.1
15	64	21.9
16	70	21.2
17	70	20.5
18	75	22.7
19	76	22.8

a. Construct a scatterplot of the data in which the axes intersect at the point (0, 16). How would you describe the pattern in the plot?
b. Does it appear that the value of cycling gross efficiency is determined solely by the percentage of type I muscle fibers? Explain.

3.45 The paper "Objective Measurement of the Stretchability of Mozzarella Cheese" (*Journal of Texture Studies* [1992]: 185–194) reported on an experiment to investigate how the behavior of mozzarella cheese varied with temperature. Consider the data on x = temperature and y = elongation (%) at failure of the cheese. (Note: The researchers were Italian and used *real* mozzarella, not the poor cousin widely available in the United States.)

x	59	63	68	72	74	78	83
y	118	182	247	208	197	135	132

a. Construct a scatterplot in which the axes intersect at the point (0, 0). (Mark 0, 20, 40, 60, 80, and 100 on the horizontal scale and 0, 50, 100, 150, 200, and 250 on the vertical scale.)
b. Construct a scatterplot in which the axes intersect at (55, 100) (as was done in the paper). Does this plot seem preferable to the one in Part (a)? Explain.
c. What do the plots of Parts (a) and (b) suggest about the nature of the relationship between x and y?

3.46 The article "Exhaust Emissions from Four-Stroke Lawn Mower Engines" (*Journal of the Air and Water Management Association* [1997]: 945–952) reported data from a study in which both a baseline gasoline and a reformulated gasoline were used. Consider the following observations on age (in years) and NO_x emissions (in grams per kilowatt-hour):

Engine	Age	Baseline	Reformulated
1	0	1.72	1.88
2	0	4.38	5.93
3	2	4.06	5.54
4	11	1.26	2.67
5	7	5.31	6.53
6	16	0.57	0.74
7	9	3.37	4.94
8	0	3.44	4.89
9	12	0.74	0.69
10	4	1.24	1.42

a. Construct a scatterplot of emissions versus age for the baseline gasoline.
b. Construct a scatterplot of emissions versus age for the reformulated gasoline.
c. Do age and emissions appear to be related? If so, in what way?
d. Comment on the similarities and differences in the two scatterplots of Parts (a) and (b).

3.47 A study to assess the capability of subsurface-flow wetland systems to remove biochemical oxygen demand (BOD) and various other chemical constituents resulted in the accompanying data on x = BOD mass (in kilograms per hectare per day) and y = BOD removal (in kilograms per hectare per day)

("Subsurface Flow Wetlands: A Performance Evaluation," *Water and Environmental Research* [1995]: 244–247). Construct a scatterplot of the data, and comment on any interesting features.

x	y	x	y
3	4	30	26
8	7	35	21
10	8	37	9
11	8	38	31
13	10	44	30
16	11	103	75
27	16	142	90

3.48 One factor in the development of tennis elbow, a malady that strikes fear into the hearts of all serious players of that sport (including one of the authors), is the impact-induced vibration of the racket-and-arm system at ball contact. It is well known that the likelihood of getting tennis elbow depends on various properties of the racket used. Consider the accompanying scatterplot of x = racket resonance frequency (in hertz) and y = sum of peak-to-peak accelerations (a characteristic of arm vibration, in meters per second per second) for $n = 23$ different rackets ("Transfer of Tennis Racket Vibrations into the Human Forearm," *Medicine and Science in Sports and Exercise* [1992]: 1134–1140). Discuss interesting features of the data and of the scatterplot.

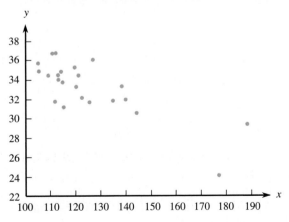

3.49 Stress can affect the physiology and behavior of animals, just as it can with humans (as many of us know all too well). The accompanying data on x = plasma cortisol concentration (in milligrams of cortisol per milliliter of plasma) and y = oxygen consumption rate (in milligrams per kilogram per hour) for juvenile steelhead after three 2-min disturbances were read from a graph in the paper "Metabolic Cost of Acute Physical Stress in Juvenile Steelhead" (*Transactions of the American Fisheries Society*

[1987]: 257–263). The paper also included data for unstressed fish.

x	25	36	48	59	62	72	80	100	100	137
y	155	184	180	220	280	163	230	222	241	350

a. Is the value of y determined solely by the value of x? Explain your reasoning.
b. Construct a scatterplot of the data.
c. Does an increase in plasma cortisol concentration appear to be accompanied by a change in oxygen consumption rate? Comment.

3.50 An article that appeared in *USA Today* (December 20, 1999) gave the following data on the 25 largest counties in the United States:

County	Population	Number of Death Row Inmates
Los Angeles, CA	8,974,105	156
Cook, IL	5,109,887	81
Harris, TX	2,958,978	140
San Diego, CA	2,576,872	27
Orange, CA	2,503,996	41
Maricopa, AZ	2,328,217	53
Kings, NY	2,283,622	0
Dade, FL	1,979,145	50
Queens, NY	1,958,299	0
Dallas, TX	1,917,501	37
King, WA	1,553,110	2
Philadelphia, PA	1,541,039	125
Santa Clara, CA	1,535,916	27
New York, NY	1,501,014	1
San Bernardino, CA	1,497,816	28
Cuyahoga, OH	1,410,021	37
Suffolk, NY	1,337,486	0
Broward, FL	1,337,001	26
Alameda, CA	1,323,155	33
Allegheny, PA	1,321,002	8
Nassau, NY	1,297,132	0
Riverside, CA	1,275,338	36
Bexar, TX	1,245,209	29
Tarrant, TX	1,227,122	25
Bronx, NY	1,198,220	0

Construct a scatterplot of these data, and comment on any unusual features. How would you describe the relationship between county population and number of death row inmates? How are the unusual features of the plot incorporated into your description of the relationship?

3.51 Example 3.23 gave data on life expectancy at birth over the time period 1940–2000. The Centers for Disease Control and Prevention also gave life expectancy at birth data separately for males and females, as shown:

Year	Life Expectancy at Birth, Males	Life Expectancy at Birth, Females
1940	60.8	65.2
1950	65.6	71.1
1960	66.6	73.1
1970	67.1	74.7
1980	70.0	77.4
1990	71.8	78.8
2000	74.1	79.6

a. Construct a time-series plot that shows the life expectancy at birth for males.
b. Add a new line to your plot from Part (a) that shows life expectancy at birth for females. Be sure to label the two lines in the plot so that it is clear which is for males and which is for females.
c. Write a few sentences describing how life expectancy at birth has changed over time for males and for females. Has the difference in life expectancy at birth for males and females changed over time? Explain.

3.52 The National Telecommunications and Information Administration published a report titled "Falling Through the Net: Toward Digital Inclusion" (U.S. Department of Commerce, October 2000) that included the following information on access to computers in the home:

Year	Percentage of Households with a Computer
1985	8.2
1990	15.0
1994	22.8
1995	24.1
1998	36.6
1999	42.1
2000	51.0

a. Construct a time-series plot for these data. Be careful — the observations are not equally spaced in time, so be sure to place them appropriately on the time (x) axis. The points in the plot should not be equally spaced along the x axis.
b. Comment on any trend over time.

3.53 The accompanying time series plot of movie box office totals (in millions of dollars) over 18 weeks of summer for both 2001 and 2002 appeared in *USA Today* (September 3, 2003):

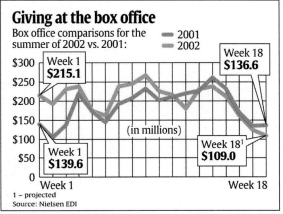

Patterns that tend to repeat on a regular basis over time are called seasonal patterns. Describe any seasonal patterns that you see in the summer box office data. Hint: Look for patterns that seem to be consistent from year to year.

▪ 3.5 Communicating and Interpreting the Results of Statistical Analyses

A graphical display, when used appropriately, can be a powerful tool for organizing and summarizing data. By sacrificing some of the detail of a complete listing of a data set, important features of the data distribution are more easily seen and more easily communicated to others.

▪ Communicating the Results of Statistical Analyses

When reporting the results of a data analysis, a good place to start is with a graphical display of the data. A well-constructed graphical display is often the best way to highlight the essential characteristics of the data distribution, such as shape and spread for numerical data sets or the nature of the relationship between the two variables in a bivariate numerical data set.

For effective communication with graphical displays, some things to remember are:

- Be sure to select a display that is appropriate for the given type of data.
- Be sure to include scales and labels on the axes of graphical displays.
- In comparative plots, be sure to include labels or a legend so that it is clear which parts of the display correspond to which samples or groups in the data set.
- Although it is sometimes a good idea to have axes that don't cross at (0, 0) in a scatterplot, the vertical axis in a bar chart or a histogram should always start at 0 (see the cautions and limitations later in this section for more about this).
- Keep your graphs simple. A simple graphical display is much more effective than one that has a lot of extra "junk." Most people will not spend a great deal of time studying a graphical display, so its message should be clear and straightforward.
- Keep your graphical displays honest. People tend to look quickly at graphical displays, and so it is important that a graph's first impression is an accurate and honest portrayal of the data distribution. In addition to the graphical display itself, data analysis reports usually include a brief discussion of the features of the data distribution based on the graphical display.
- For categorical data, this discussion might be a few sentences on the relative proportion for each category, possibly pointing out categories that were either common or rare compared to other categories.
- For numerical data sets, the discussion of the graphical display usually summarizes the information that the display provides on three characteristics of the data distribution: center or location, spread, and shape.
- For bivariate numerical data, the discussion of the scatterplot would typically focus on the nature of the relationship between the two variables used to construct the plot.
- For data collected over time, any trends or patterns in the time-series plot would be described.

▪ Interpreting the Results of Statistical Analyses

The use of graphical data displays is quite common in newspapers, magazines, and journals, so it is important to be able to extract information from such displays. For example, data on test scores for a standardized math test given to eighth-graders in 37 states, 2 territories (Guam and the Virgin Islands), and the District of Columbia were presented in the August 13, 1992, edition of *USA Today*. Figure 3.37 gives both a stem-and-leaf display and a histogram summarizing these data. Careful examination of these displays reveals the following:

1. Most of the participating states had average eighth-grade math scores between 240 and 280. We would describe the shape of this display as negatively skewed, because of the longer tail on the low end of the distribution.

2. Three of the average scores differed substantially from the others. These turn out to be 218 (Virgin Islands), 229 (District of Columbia), and 230(Guam). These three scores could be described as outliers. It is interesting to note that the three unusual values are from the areas that are not states.

3. There do not appear to be any outliers on the high side.

4. A "typical" average math score for the 37 states would be somewhere around 260.

5. There is quite a bit of variability in average score from state to state.

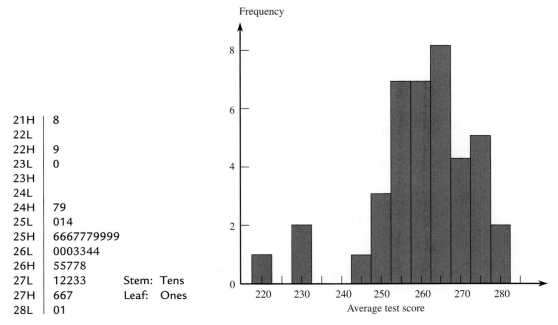

21H	8	
22L		
22H	9	
23L	0	
23H		
24L		
24H	79	
25L	014	
25H	6667779999	
26L	0003344	
26H	55778	
27L	12233	Stem: Tens
27H	667	Leaf: Ones
28L	01	

Figure 3.37 Stem-and-leaf display and histogram for math test scores.

How would the displays have been different if the two territories and the District of Columbia had not participated in the testing? The resulting histogram is shown in Figure 3.38. Note that the display is now more symmetric, with no noticeable outliers. The display still reveals quite a bit of state-to-state variability in average score, and 260 still looks reasonable as a "typical" average score.

Now suppose that the two highest values among the 37 states (Montana and North Dakota) had been even higher. The stem-and-leaf display might then look like the one given in Figure 3.39. In this stem-and-leaf display, two values stand out from the main part of the display. This would catch our attention and might cause us to look carefully at these two states to determine what factors may be related to high math scores. Researchers at Educational Testing Service found math scores to be associated with such factors as amount of time spent reading and amount of time spent watching TV.

Figure 3.38 Histogram for the modified math score data.

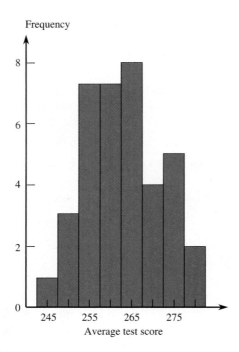

Figure 3.39 Stem-and-leaf display for modified math score data.

```
24H │ 79
25L │ 014
25H │ 6667779999
26L │ 0003344
26H │ 55778
27L │ 12233        Stem:  Tens
27H │ 667          Leaf:  Ones
28L │
28H │
29L │
29H │ 68
```

▪ What to Look for in Published Data

Here are some questions you might ask yourself when attempting to extract information from a graphical data display:

- ▪ Is the chosen display appropriate for the type of data collected?
- ▪ For graphical displays of univariate numerical data, how would you describe the shape of the distribution, and what does this say about the variable being summarized?
- ▪ Are there any outliers (noticeably unusual values) in the data set? Is there any plausible explanation for why these values differ from the rest of the data? (The presence of outliers often leads to further avenues of investigation.)
- ▪ Where do most of the data values fall? What is a typical value for the data set? What does this say about the variable being summarized?
- ▪ Is there much variability in the data values? What does this say about the variable being summarized?

Of course, you should always think carefully about how the data were collected. If the data were not gathered in a reasonable manner (based on sound sampling methods or experimental design principles), you should be cautious in formulating any conclusions based on the data.

Consider the histogram in Figure 3.40, which is based on data published by the National Center for Health Statistics. The data set summarized by this histogram consisted of 1990 infant mortality rates (deaths per 1000 live births) for the 50 states in the United States. A histogram is an appropriate way of summarizing these data (although with only 50 observations, a stem-and-leaf display would also have been reasonable). The histogram itself is slightly positively skewed, with most mortality rates between 7.5 and 12. There is quite a bit of variability in infant mortality rate from state to state — perhaps more than we might have expected. This variability might be explained by differences in economic conditions or in access to health care. We may want to look further into these issues. Although there are no obvious outliers, the upper tail is a little longer than the lower tail. The three largest values in the data set are 12.1 (Alabama), 12.3 (Georgia), and 12.8 (South Carolina) — all Southern states. Again, this may suggest some interesting questions that deserve further investigation. A typical infant mortality rate would be about 9.5 deaths per 1000 live births. This represents an improvement, because researchers at the National Center for Health Statistics stated that the overall rate for 1988 was 10 deaths per 1000 live births. However, they also point out that the United States still ranked 22d out of 24 industrialized nations surveyed, with only New Zealand and Israel having higher infant mortality rates.

Figure 3.40 Histogram of infant mortality rates.

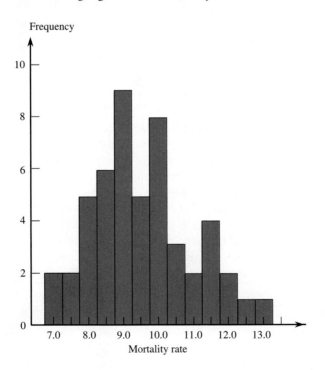

▪ A Word to the Wise: Cautions and Limitations

When constructing and interpreting graphical displays, you need to keep in mind these things:

1. Areas should be proportional to frequency, relative frequency, or the magnitude of the number being represented. The eye is naturally drawn to large ar-

eas in graphical displays, and it is natural for the observer to make informal comparisons based on area. Correctly constructed graphical displays, such as pie charts, bar charts, and histograms, are designed so that the areas of the pie slices or the bars are proportional to frequency or relative frequency. Sometimes in a effort to make graphical displays more interesting, designers lose sight of this important principle, and the resulting graphs are misleading. For example, consider the following graph (*USA Today*, October 3, 2002):

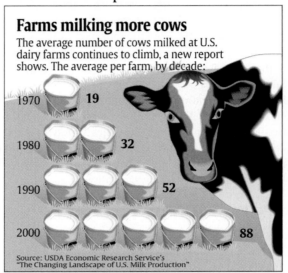

USA TODAY Snapshots®

Farms milking more cows

The average number of cows milked at U.S. dairy farms continues to climb, a new report shows. The average per farm, by decade:

1970   19

1980   32

1990   52

2000   88

Source: USDA Economic Research Service's "The Changing Landscape of U.S. Milk Production"

By Marcy E. Mullins, USA TODAY

In trying to make the graph more visually interesting by replacing the bars of a bar chart with milk buckets, areas are distorted. For example, the two buckets for 1980 represent 32 cows, whereas the one bucket for 1970 represents 19 cows. This is misleading because 32 is not twice as big as 19. Other areas are distorted as well.

Another common distortion occurs when a third dimension is added to bar charts or pie charts. For example, the pie chart for the college choice data of Example 3.3 (Figure 3.3) is reproduced here along with a three-dimensional version created by Excel. The three-dimensional version distorts the areas and, as a consequence, is much more difficult to interpret correctly.

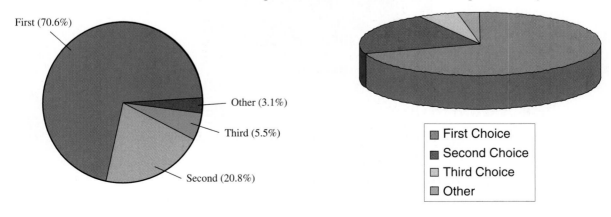

First (70.6%)

Other (3.1%)

Third (5.5%)

Second (20.8%)

First Choice
Second Choice
Third Choice
Other

2. Be cautious of graphs with broken axes. Although it is common to see scatter-plots with broken axes, be extremely cautious of time-series plots, bar charts, or histograms with broken axes. The use of broken axes in a scatterplot does not distort information about the nature of the relationship in the bivariate data set used to construct the display. On the other hand, in time-series plots, broken axes can sometimes exaggerate the magnitude of change over time. Although it is not always inadvisable to break the vertical axis in a time-series plot, it is something you should watch for, and if you see a time-series plot with a broken axis, as in the accompanying time-series plot of mortgage rates (*USA Today*, October 2, 2002), you should pay particular attention to the scale on the vertical axis and take extra care in interpreting the graph:

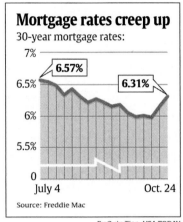

By Quin Tian, USA TODAY

In bar charts and histograms, the vertical axis (which represents frequency, relative frequency, or density) should *never* be broken. If the vertical axis is broken in this type of graph, the resulting display will violate the "propor-tional area" principle and the display will be misleading. For example, the ac-companying bar chart is similar to one appearing in an advertisement for a software product designed to help teachers raise student test scores. By start-ing the vertical axis at 50, the gain for students using the software is exagger-ated. Areas of the bars are not proportional to the magnitude of the numbers represented — the area for the rectangle representing 68 is more than three times the area of the rectangle representing 55!

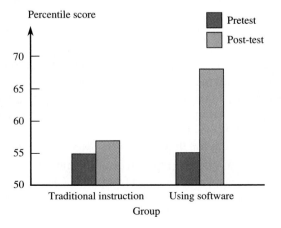

3. Watch out for unequal time spacing in time-series plots. If observations over time are not made at regular time intervals, special care must be taken in constructing the time-series plot. Consider the accompanying time-series plot, which is similar to one appearing in the *San Luis Obispo Tribune* (September 22, 2002) in an article on online banking:

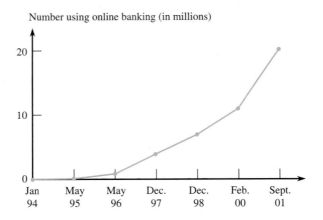

Notice that the intervals between observations are irregular, yet the points in the plot are equally spaced along the time axis. This makes it difficult to make a coherent assessment of the rate of change over time. This could have been remedied by spacing the observations differently along the time axis, as shown in the following plot:

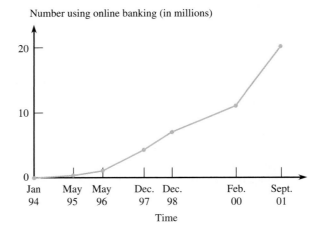

4. Be careful how you interpret patterns in scatterplots. A strong pattern in a scatterplot means that the two variables tend to vary together in a predictable way, but it does not mean that there is a cause-and-effect relationship between the two variables. We will consider this point further in Chapter 5, but in the meantime, when describing patterns in scatterplots, be careful not to use wording that implies that changes in one variable *cause* changes in the other.

5. Make sure that a graphical display creates the right first impression. For example, consider the graph from *USA Today* (June 25, 2001) at the top of page 139. Although this graph does not violate the proportional area principle, the way the "bar" for the "none" category is displayed makes this graph

USA TODAY Snapshots®

Most don't sticker their car

Number of car bumper stickers:

82%

8%

4%

1% 1% 1% 1%

None ———————→ 1   2   3   4   5-7 8-10

NO STKER

Note: More than 10 stickers are less than 1%.
Equals less than 100% due to rounding.

Source: CarMax.com   By Cindy Hall and Bob Laird, USA TODAY

Copyright June 25, 2001, USA TODAY. Reprinted with permission.

difficult to read, and a quick glance at this graph would leave the reader with an incorrect impression.

■ Activity 3.1: Locating States

Background: A newspaper article bemoaning the state of students' knowledge of geography claimed that more students could identify the island where the 2002 season of the TV show *Survivor* was filmed than could locate Vermont on a map of the United States. In this activity, you will collect data that will allow you to estimate the proportion of students that can correctly locate the states of Vermont and Nebraska.

1. Working as a class, decide how you will select a sample that you think will be representative of the students from your school.

2. Use the sampling method from Step 1 to obtain the subjects for this study. Subjects should be shown the accompanying map of the United States and asked to point out the state of Vermont. After the subject has given his or her answer, ask the subject to point

out the state of Nebraska. For each subject, record whether or not Vermont was correctly identified and whether or not Nebraska was correctly identified.

3. When the data collection process is complete, summarize the resulting data in a table like the one shown here:

Response	Frequency
Correctly identified both states	
Correctly identified Vermont but not Nebraska	
Correctly identified Nebraska but not Vermont	
Did not correctly identify either state	

4. Construct a pie chart that summarizes the data in the table from Step 3.

5. What proportion of sampled students were able to correctly identify Vermont on the map?

6. What proportion of sampled students were able to correctly identify Nebraska on the map?

7. Construct a comparative bar chart that shows the proportion correct and the proportion incorrect for each of the two states considered.

8. Which state, Vermont or Nebraska, is closer to the state in which your school is located? Based on the pie chart, do you think that the students at your school were better able to identify the state that was closer than the one that was farther away? Justify your answer.

9. Write a paragraph commenting on the level of knowledge of U.S. geography demonstrated by the students participating in this study.

10. Would you be comfortable generalizing your conclusions in Step 8 to the population of students at your school? Explain why or why not.

▪ Activity 3.2: Bean Counters!

Materials needed: A large bowl of dried beans (or marbles, plastic beads, or any other small, fairly regular objects) and a coin.

In this activity, you will investigate whether people can hold more in the right hand or in the left hand.

1. Flip a coin to determine which hand you will measure first. If the coin lands heads side up, start with the right hand. If the coin lands tails side up, start with the left hand. With the designated hand, reach into the bowl and grab as many beans as possible. Raise the hand over the bowl and count to 4. If no beans drop during the count to 4, drop the beans onto a piece of paper and record the number of beans grabbed. If any beans drop during the count, restart the count. That is, you must hold the beans for a count of 4 without any beans falling before you can determine the number grabbed. Repeat the process with the other hand, and then record the following information: (1) right-hand number, (2) left-hand number, and (3) dominant hand (left or right, depending on whether you are left- or right-handed).

2. Create a class data set by recording the values of the three variables listed in Step 1 for each student in your class.

3. Using the class data set, construct a comparative stem-and-leaf display with the right-hand counts displayed on the right and the left-hand counts displayed on the left of the stem-and-leaf display. Comment on the interesting features of the display and include a comparison of the right-hand count and left-hand count distributions.

4. Now construct a comparative stem-and-leaf display that allows you to compare dominant-hand count to nondominant-hand count. Does the display support the theory that dominant-hand count tends to be higher than nondominant-hand count?

5. For each observation in the data set, compute the difference

dominant-hand count − nondominant-hand count

Construct a stem-and-leaf display of the differences. Comment on the interesting features of this display.

6. Explain why looking at the distribution of the differences (Step 5) provides more information than the comparative stem-and-leaf display (Step 4). What information is lost in the comparative display that is retained in the display of the differences?

▪ Summary of Key Concepts and Formulas

Term or Formula	Comment
Frequency distribution	A table that displays frequencies and sometimes relative and cumulative relative frequencies for categories (categorical data), possible values (discrete numerical data), or class intervals (continuous data).

Term or Formula	Comment
Comparative bar chart	Two or more bar charts that use the same set of horizontal and vertical axes.
Pie chart	A graph of a frequency distribution for a categorical data set. Each category is represented by a slice of the pie, and the area of the slice is proportional to the corresponding frequency or relative frequency.
Segmented bar chart	A graph of a frequency distribution for a categorical data set. Each category is represented by a segment of the bar, and the size of the segment is proportional to the corresponding frequency or relative frequency.
Stem-and-leaf display	A method of organizing numerical data in which the stem values (leading digit(s) of the observations) are listed in a column and the leaf for each observation (trailing digit(s)) is listed beside the corresponding stem. Sometimes stems are repeated to stretch the display.
Histogram	A picture of the information in a frequency distribution for a numerical data set. A rectangle is drawn above each possible value (discrete data) or class interval. The rectangle's area is proportional to the corresponding frequency or relative frequency.
Histogram shapes	A (smoothed) histogram may be unimodal (single peak), bimodal (two peaks), or multimodal (more than two peaks). A unimodal histogram may be symmetric, positively skewed (a long right or upper tail), or negatively skewed (a long left or lower tail). A frequently occurring shape is the normal curve.
Cumulative relative frequency plot	A graph of a cumulative relative frequency distribution.
Scatterplot	A picture of bivariate numerical data in which each observation (x, y) is represented as a point with respect to a horizontal x axis and a vertical y axis.
Time-series plot	A picture of numerical data collected over time.

Supplementary Exercises 3.54–3.66

3.54 Each observation in the following data set is the number of housing units (homes or condominiums) sold during November 1992 in a region corresponding to a particular Orange County, California, ZIP code:

```
25  18  16   6  26  11  29   7   5  15  12  37
35  11  16  35  20  27  17  30  10  16  28  13
26  11  12   8   9  29   0  20  30  12  45  26
21  30  18  31   0  46  47  14  13  29  11  18
10  27   5  18  67  21  35  48  42  70  43   0
30  17  35  40  61  18  17  17
```

Construct a stem-and-leaf display, and comment on any interesting features.

3.55 Each murder committed in Utah during the period 1978–1990 was categorized by day of the week, resulting in the following frequencies: Sunday, 109; Monday, 73; Tuesday, 97; Wednesday, 95; Thursday, 83; Friday, 107; Saturday, 100.

a. Construct the corresponding frequency distribution.

b. What proportion of these murders were committed on a weekend day — that is, Friday, Saturday, or Sunday?

c. Do these data suggest that a murder is more likely to be committed on some days than on other days? Explain your reasoning.

3.56 An article in the *San Luis Obispo Tribune* (November 20, 2002) stated that 39% of those with critical housing needs (those who pay more than half their income for housing) lived in urban areas, 42% lived in suburban areas, and the rest lived in rural areas. Construct a pie chart that shows the distribution of type of residential area (urban, suburban, or rural) for those with critical housing needs.

3.57 Living-donor kidney transplants are becoming more common. Often a living donor has chosen to donate a kidney to a relative with kidney disease.

The following data appeared in a *USA Today* article on organ transplants ("Kindness Motivates Newest Kidney Donors," June 19, 2002):

| | Number of Kidney Transplants | |
Year	Living-Donor to Relative	Living-Donor to Unrelated Person
1994	2390	202
1995	2906	400
1996	2916	526
1997	3144	607
1998	3324	814
1999	3359	930
2000	3679	1325
2001	3879	1399

a. Construct a time-series plot for the number of living-donor kidney transplants where the donor is a relative of the recipient. Describe the trend in this plot.

b. Use the data from 1994 and 2001 to construct a comparative bar chart for the type of donation (relative or unrelated). Write a few sentences commenting on your display.

3.58 How unusual is a game with no hits in major-league baseball, and how frequently does a game have more than 10, 15, or even 20 hits? The following table is a frequency distribution for the number of hits per game for all 9-inning games that were played between 1989 and 1993:

Hits/Game	Frequency	Hits/Game	Frequency
0	20	14	569
1	72	15	393
2	209	16	253
3	527	17	171
4	1048	18	97
5	1457	19	53
6	1988	20	31
7	2256	21	19
8	2403	22	13
9	2256	23	5
10	1967	24	1
11	1509	25	0
12	1230	26	1
13	834	27	1

a. Calculate the relative frequencies, and then construct a relative frequency histogram.

b. Describe the key features of the histogram.

c. What proportion of games had at most 2 hits?

d. What proportion of games had between 5 and 10 hits?

e. What proportion of games had more than 15 hits?

3.59 Many nutritional experts have expressed concern about the high levels of sodium in prepared foods. The following data on sodium content (in milligrams) per frozen meal appeared in the article "Comparison of 'Light' Frozen Meals" (*Boston Globe*, April 24, 1991):

720	530	800	690	880	1050	340	810	760
300	400	680	780	390	950	520	500	630
480	940	450	990	910	420	850	390	600

Two histograms for these data are shown:

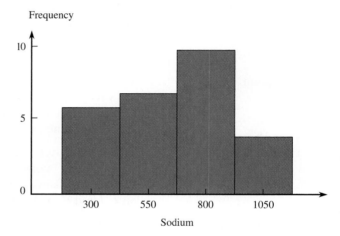

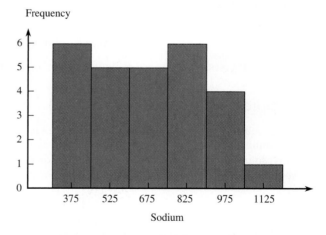

a. Do the two histograms give different impressions about the distribution of values?

b. Use each histogram to determine approximately the proportion of observations that are less than 800, and compare to the actual proportion.

3.60 The paper "Paraquat and Marijuana Risk Assessment" (*American Journal of Public Health* [1983]: 784–788) reported the results of a 1978 telephone survey on marijuana usage. The following frequency distribution gives the amount of marijuana (in grams) smoked per week for those respondents who indicated that they did use the drug:

Amount Smoked	Frequency
0 to <3	94
3 to <11	269
11 to <18	70
18 to <25	48
25 to <32	31
32 to <39	10
39 to <46	5
46 to <53	0
53 to <60	1
60 to <67	0
67 to <74	1

a. Display the information given in the frequency distribution in the form of a histogram.

b. What proportion of respondents smoked 25 g or more per week?

c. Use the histogram to estimate the proportion of respondents who smoked more than 15 g per week.

d. Calculate the cumulative relative frequencies, and use them to determine the proportion of respondents who smoked at least 25 g per week.

3.61 Data on engine emissions (hydrocarbon, HC; and carbon monoxide, CO) for 46 vehicles are given in the following table ("Determining Statistical Characteristics of Vehicle Emissions Audit Procedure," *Technometrics* [November 1980]: 483–493):

Vehicle	HC	CO	Vehicle	HC	CO
1	0.50	5.01	24	1.10	22.92
2	0.65	14.67	25	0.65	11.20
3	0.46	8.60	26	0.43	3.81
4	0.41	4.42	27	0.48	3.45
5	0.41	4.95	28	0.41	1.85
6	0.39	7.24	29	0.51	4.10
7	0.44	7.51	30	0.41	2.26
8	0.55	12.30	31	0.47	4.74
9	0.72	14.59	32	0.52	4.29
10	0.64	7.98	33	0.56	5.36
11	0.83	11.53	34	0.70	14.83
12	0.38	4.10	35	0.51	5.69
13	0.38	5.21	36	0.52	6.35
14	0.50	12.10	37	0.57	6.02
15	0.60	9.62	38	0.51	5.79
16	0.73	14.97	39	0.36	2.03
17	0.83	15.13	40	0.49	4.62
18	0.57	5.04	41	0.52	6.78
19	0.34	3.95	42	0.61	8.43
20	0.41	3.38	43	0.58	6.02
21	0.37	4.12	44	0.46	3.99
22	1.02	23.53	45	0.47	5.22
23	0.87	19.00	46	0.55	7.47

a. Construct a frequency distribution and histogram for the hydrocarbon emissions data. Do any of the observations appear to be outliers?

b. Construct a frequency distribution and histogram for the carbon monoxide data.

c. Are the hydrocarbon and carbon monoxide histograms symmetric or skewed? If they are both skewed, is the direction of the skew the same?

3.62 The following table gives data on the cost per ounce (in cents) for 31 shampoos intended for "normal" hair and 29 shampoos intended for "fine" hair (*Consumer Reports*, 1992):

Normal	79	63	19	9	37	49	20	16	55	69
	23	14	9	87	44	13	16	23	20	64
	28	18	32	81	85	47	50	8	13	21
	9									
Fine	69	9	23	22	8	12	32	12	18	74
	19	63	49	37	55	85	44	87	17	11
	23	50	65	51	35	14	20	28	8	

Use methods developed in this chapter to describe and compare the two cost distributions.

3.63 The paper "The Acid Rain Controversy: The Limits of Confidence" (*American Statistician* [1983]: 385–394) gave the following data on average SO_2 (sulfur dioxide) emission rates from utility and industrial boilers (in pounds per million Btu) for 47 states (data from Idaho, Alaska, and Hawaii were not given):

2.3 2.7 1.5 1.5 0.3 0.6 4.2 1.3 1.2 0.4 0.5
2.2 4.5 3.8 1.2 0.2 1.0 0.7 0.2 1.4 0.7 3.6
1.0 0.7 1.7 0.5 0.1 0.6 2.5 2.7 1.5 1.4 2.9
1.0 3.4 2.1 0.9 1.9 1.0 1.7 1.8 0.6 1.7 2.9
1.8 1.4 3.7

a. Summarize this data set by constructing a relative frequency distribution.

b. Draw the histogram corresponding to the frequency distribution in Part (a). Would you describe the histogram as symmetric or skewed?

c. Use the relative frequency distribution of Part (a) to compute the cumulative relative frequencies.

d. Use the cumulative relative frequencies to give the approximate proportion of states with SO_2 emission rates that

i. were below 1.0 lb/million Btu

ii. were between 1.0 and 2.0 lb/million Btu

iii. were at least 2.0 lb/million Btu

3.64 Americium 241 (^{241}Am) is a radioactive material used in the manufacture of smoke detectors. The article "Retention and Dosimetry of Injected ^{241}Am in Beagles" (*Radiation Research* [1984]: 564–575) described a study in which 55 beagles were injected with a dose of ^{241}Am (proportional to each animal's weight). Skeletal retention of ^{241}Am (in microcuries per kilogram) was recorded for each beagle, resulting in the following data:

0.196	0.451	0.498	0.411	0.324	0.190	0.489
0.300	0.346	0.448	0.188	0.399	0.305	0.304
0.287	0.243	0.334	0.299	0.292	0.419	0.236
0.315	0.447	0.585	0.291	0.186	0.393	0.419
0.335	0.332	0.292	0.375	0.349	0.324	0.301
0.333	0.408	0.399	0.303	0.318	0.468	0.441
0.306	0.367	0.345	0.428	0.345	0.412	0.337
0.353	0.357	0.320	0.354	0.361	0.329	

a. Construct a frequency distribution for these data, and draw the corresponding histogram.

b. Write a short description of the important features of the shape of the histogram.

3.65 In the paper "Reproduction in Laboratory Colonies of Bank Vole" (*Oikos* [1983]: 184), the authors presented the results of a study on litter size. (A vole is a small rodent with a stout body, blunt nose, and short ears.) As each new litter was born, the number of babies was recorded, and the following results were obtained:

Size of Litter

3	6	5	6	5	7	5	7	6	6	6	4	6
5	6	4	3	5	6	4	5	9	6	5	6	1

9	7	8	3	7	4	5	5	6	7	3	6	6
9	4	5	7	5	6	8	6	4	7	5	7	4
5	8	6	7	2	7	7	3	3	5	4	6	4
6	3	7	8	5	7	7	7	7	9	8	7	6
7	6	4	7	10	5	2	3	6	6	4	7	6
7	5	5	5	7	5	8	8	4	9	7	5	4
6	5	8	4	5	6	6	3	6	8	6	8	6
5	8	6	11	4	7	6	8	9	7	3	8	3
4	6	4	5	7	5	6	5	7	6	9	3	5
9	7	5	6	7	5	8	6	8	8	6	5	7
4	8	7	7	7	5	3	8	6				

a. Construct a relative frequency distribution for these data.

b. What proportion of the litters had more than 6 babies? between 3 and 8 babies (inclusive)?

c. Is it easier to answer questions like those posed in Part (b) using the relative frequency distribution or the raw data given in the table? Explain.

3.66 Compute the cumulative relative frequencies for the data of Exercise 3.65, and use them to answer the following questions:

a. What proportion of observations are at most 8? at least 8?

b. What proportion of litters contain between 5 and 10 (inclusive) offspring?

▪ References

Chambers, John, William Cleveland, Beat Kleiner, and Paul Tukey. *Graphical Methods for Data Analysis*. Belmont, CA: Wadsworth, 1983. (This is an excellent survey of methods, illustrated with numerous interesting examples.)

Cleveland, William. *The Elements of Graphing Data*, 2d ed. Summit, NJ: Hobart Press, 1994. (An informal and informative introduction to various aspects of graphical analysis.)

Freedman, David, Robert Pisani, and Roger Purves. *Statistics*, 3d ed. New York: W. W. Norton, 1997. (An excellent, informal introduction to concepts, with some insightful cautionary examples concerning misuses of statistical methods.)

Moore, David. *Statistics: Concepts and Controversies,* 5th ed. New York: W. H. Freeman, 2001. (A nonmathematical yet highly entertaining introduction to our discipline. Two thumbs up!)

4 · Numerical Methods for Describing Data

Exposure to various chemical substances can pose substantial health risks. Evidence suggests that a high indoor radon concentration might be linked to the development of childhood cancers. The article "Indoor Radon and Childhood Cancer" (*The Lancet* [1991]: 1537–1538) presented the accompanying data on radon concentration in two different samples of houses. The first sample consisted of houses in which a child diagnosed with cancer had been residing. Houses in the second sample had no recorded cases of childhood cancer.

Cancer

10	21	5	23	15	11	9	13	27	13	39	22
7	20	45	12	15	3	8	11	18	16	23	16
34	10	15	11	18	210	22	11	16	17	33	10
9	57	16	21	18	38						

No Cancer

9	38	11	12	29	5	7	6	8	29	24	12
17	11	3	9	33	17	55	11	29	13	24	7
11	21	6	39	29	7	8	55	9	21	9	3
85	11	14									

What is a representative or typical radon concentration value for each sample, and how do these values compare for the two samples? Do the two samples show similar amounts of variability, or are the observations in one sample substantially more spread out than those in the other sample? As we have seen, a stem-and-leaf display, a frequency distribution, or a histogram gives general impressions about where each data set is centered and how much it spreads out about its center. In this chapter we show how to calculate numerical summary measures that describe more precisely both the center and the extent of spread. In Section 4.1 we introduce the mean and the median, the two most popular measures of the center of a distribution. The variance and the standard deviation are presented in Section 4.2

as measures of variability. In later sections we discuss several techniques for using such summary measures to describe other features of the data.

▪ 4.1 Describing the Center of a Data Set

An informative way to describe the location of a numerical data set is to report a value that is representative of the observations in the set. Such a number describes roughly where the data set is located or "centered" along the number line, and it is called a measure of center. The two most popular measures of center are the *mean* and the *median*.

▪ The Mean

The **mean** of a set of numerical observations is just the familiar arithmetic average: the sum of the observations divided by the number of observations. Values in such a data set are observations on a numerical variable. The variable might be number of traffic accidents, number of pages in a book, reaction time, yield from a chemical reaction, and so on. When you work with sample data, it is helpful to have concise notation for the variable, sample size, and individual observations. Let

x = the variable for which we have sample data

n = the number of observations in the sample (the sample size)

x_1 = the first observation in the sample

x_2 = the second observation in the sample

$\vdots$

x_n = the nth (last) observation in the sample

For example, we might have a sample consisting of $n = 4$ observations on $x =$ battery lifetime (in hours):

$$x_1 = 5.9, \qquad x_2 = 7.3, \qquad x_3 = 6.6, \qquad x_4 = 5.7$$

Notice that the value of the subscript on x has no relationship to the magnitude of the observation. In this example, x_1 is just the first observation in the data set and not necessarily the smallest observation, and x_n is the last observation but not necessarily the largest.

The sum of $x_1, x_2, \ldots, x_n$ can be denoted by $x_1 + x_2 + \cdots + x_n$, but this is cumbersome. The Greek letter Σ is traditionally used in mathematics to denote summation. In particular, Σx denotes the sum of all the x values in the data set under consideration.*

▪ Definition

The **sample mean** of a numerical sample $x_1, x_2, \ldots, x_n$, denoted by $\bar{x}$, is

$$\bar{x} = \frac{\text{sum of all observations in the sample}}{\text{number of observations in the sample}} = \frac{x_1 + x_2 + \cdots + x_n}{n} = \frac{\Sigma x}{n}$$

---

*It is also common to see Σx written as Σx_i or even $\sum_{i=1}^{n} x_i$, but for simplicity we will usually omit the summation indexes.

▪ **Example 4.1**   Range of Motion After Knee Surgery

Traumatic knee dislocation often requires surgery to repair ruptured ligaments. One measure of recovery is range of motion (measured as the angle formed when, starting with the leg straight, the knee is bent as far as possible). The article "Reconstruction of the Anterior and Posterior Cruciate Ligaments After Knee Dislocation" (*American Journal of Sports Medicine* [1999]: 189–197) reported the following postsurgical range of motion for a sample of 13 patients:

Range of Motion (degrees)

$x_1 = 154$ $x_2 = 142$ $x_3 = 137$ $x_4 = 133$ $x_5 = 122$ $x_6 = 126$ $x_7 = 135$
$x_8 = 135$ $x_9 = 108$ $x_{10} = 120$ $x_{11} = 127$ $x_{12} = 134$ $x_{13} = 122$

The sum of these sample values is $154 + 142 + 137 + \cdots + 122 = 1695$, and the sample mean range of motion is

$$\bar{x} = \frac{\Sigma x}{n} = \frac{1695}{13} = 130.38$$

We would report 130.38° as a representative value of range of motion for this sample (even though there is no patient in the sample that actually had a range of motion of 130.38°).

The data values in Example 4.1 were all integers, yet the mean was given as 130.38. It is common to use more digits of decimal accuracy for the mean. This allows the value of the mean to fall between possible observable values (e.g., the average number of children per family could be 1.8, whereas no single family will have 1.8 children). It could also be argued that the mean carries with it more precision than does any single observation.

The sample mean $\bar{x}$ is computed from sample observations, so it is a characteristic of the particular sample in hand. It is customary to use Roman letters to denote sample characteristics, as we have done with $\bar{x}$. Characteristics of the population are usually denoted by Greek letters. One of the most important such characteristics is the population mean.

▪ **Definition**

The **population mean**, denoted by μ, is the average of all x values in the entire population.

For example, the true average fuel efficiency for all 600,000 cars of a certain type under specified conditions might be $\mu = 27.5$ mpg. A sample of $n = 5$ cars might yield efficiencies of 27.3, 26.2, 28.4, 27.9, and 26.5, from which we obtain $\bar{x} = 27.26$ for this particular sample (somewhat smaller than μ). However, a second sample might give $\bar{x} = 28.52$, a third $\bar{x} = 26.85$, and so on. The value of $\bar{x}$ varies from sample to sample, whereas there is just one value for μ. We illustrate in Example 4.2 how the value of $\bar{x}$ from a particular sample can differ from the value of μ and how the value of $\bar{x}$ differs from sample to sample.

▪ **Example 4.2**   County Population Sizes

The 50 states plus the District of Columbia contain 3137 counties. Let x denote the number of residents of a county. Then there are 3137 values of the variable x in the population. The sum of these 3137 values is 248,709,873 (1990 census), so the population average value of x is

$$\mu = \frac{248,709,873}{3137} = 79,282.7 \text{ residents per county}$$

We used the *World Almanac and Book of Facts* to select three different samples at random from this population of counties, with each sample consisting of five counties. The results appear in Table 4.1, along with the sample mean for each sample. Not only are the three $\bar{x}$ values different from one another — because they are based on three different samples and the value of $\bar{x}$ depends on the x values in the sample — but also none of the three values comes close to the value of the population mean, μ. If we did not know the value of μ but had only Sample 1 available, we might use $\bar{x}$ as an *estimate* of μ, but our estimate would be far off the mark.

Table 4.1 ▪ **Three Samples from the Population of All U.S. Counties (x = number of residents)**

Sample 1		Sample 2		Sample 3	
County	x Value	County	x Value	County	x Value
Fayette, TX	20,095	Stoddard, MO	28,895	Chattahoochee, GA	16,934
Monroe, IN	108,978	Johnston, OK	10,032	Petroleum, MT	519
Greene, NC	15,384	Sumter, AL	16,174	Armstrong, PA	73,478
Shoshone, ID	13,931	Milwaukee, WI	959,275	Smith, MI	14,798
Jasper, IN	24,960	Albany, WY	30,797	Benton, MO	13,859
	$\sum x = 183,348$		$\sum x = 1,045,173$		$\sum x = 119,588$
	$\bar{x} = 36,669.6$		$\bar{x} = 209,034.6$		$\bar{x} = 23,917.6$

Alternatively, we could combine the three samples into a single sample with $n = 15$ observations:

$$x_1 = 20,095, \ldots, \quad x_5 = 24,960, \ldots, \quad x_{15} = 13,859$$

$$\sum x = 1,348,109$$

$$\bar{x} = \frac{1,348,109}{15} = 89,873.9$$

This value is closer to the value of μ but is still somewhat unsatisfactory as an estimate. The problem here is that the population of x values exhibits a lot of variability (the largest value is $x = 8,863,164$ for Los Angeles County, California, and the smallest value is $x = 107$ for Loving County, Texas, which evidently few people love). Therefore, it is difficult for a sample of 15 observations, let alone just 5, to be reasonably representative of the population. But don't lose hope! Once we learn how to assess variability, we will also see how to take it into account in making accurate inferences.

One potential drawback to the mean as a measure of center for a data set is that its value can be greatly affected by the presence of even a single *outlier* (an unusually large or small observation) in the data set.

▪ **Example 4.3**   Number of Visits to a Class Web Site

Forty students were enrolled in a section of STAT 130, a general education course in statistical reasoning, during the fall quarter of 2002 at California Polytechnic State University, San Luis Obispo. The instructor made course materials, grades, and lecture notes available to students on a class web site, and course management software kept track of how often each student accessed any of the web pages on the class site. One month after the course began, the instructor requested a report that indicated how many times each student had accessed a web page on the class site. The 40 observations were:

20	37	4	20	0	84	14	36	5	331	19	0
0	22	3	13	14	36	4	0	18	8	0	26
4	0	5	23	19	7	12	8	13	16	21	7
13	12	8	42								

The sample mean for this data set is $\bar{x} = 23.10$. Figure 4.1 is a MINITAB dotplot of the data. Many would argue that 23.10 is not a very representative value for this sample, because 23.10 is larger than most of the observations in the data set — only 7 of 40 observations, or 17.5%, are larger than 23.10. The two outlying values of 84 and 331 (no, that wasn't a typo!) have a substantial impact on the value of $\bar{x}$.

Figure 4.1 A MINITAB dotplot of the data in Example 4.3.

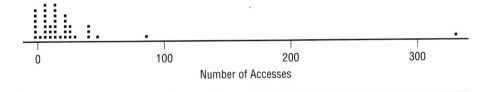

Number of Accesses

We now turn our attention to a measure of center that is not so sensitive to outliers — the median.

▪ **The Median**

The median strip of a highway divides the highway in half, and the median of a numerical data set does the same thing for a data set. Once the data values have been listed in order from smallest to largest, the **median** is the middle value in the list, and it divides the list into two equal parts. Depending on whether the sample size *n* is even or odd, the process of determining the median is slightly different. When *n* is an odd number (say, 5), the sample median is the single middle value. But when *n* is even (say, 6), there are two middle values in the ordered list, and we average these two middle values to obtain the sample median.

> ▪ **Definition**
>
> The **sample median** is obtained by first ordering the n observations from smallest to largest (with any repeated values included, so that every sample observation appears in the ordered list). Then
>
> $$\text{sample median} = \begin{cases} \text{the single middle value if } n \text{ is odd} \\ \text{the average of the middle two values if } n \text{ is even} \end{cases}$$

▪ Example 4.4   Web Site Data Revisited

The sample size for the web site access data of Example 4.3 was $n = 40$, an even number. The median is the average of the 20th and 21st values (the middle two) in the ordered list of the data. Arranging the data in order from smallest to largest produces the following ordered list:

0	0	0	0	0	0	3	4	4	4	5	5
7	7	8	8	8	12	12	13	13	13	14	14
16	18	19	19	20	20	21	22	23	26	36	36
37	42	84	331								

The median can now be determined:

$$\text{median} = \frac{13 + 13}{2} = 13$$

Looking at the dotplot (Figure 4.1), we see that this value appears to be somewhat more typical of the data than $\bar{x} = 23.10$ is.

The **population median** plays the same role for the population as the sample median plays for the sample. It is the middle value in the ordered list consisting of all population observations. We previously noted that the mean — population or sample — is sensitive to even a single value that lies far above or below the rest of the data. The value of the mean is pulled out toward such an outlying value or values. The median, on the other hand, is quite *in*sensitive to outliers. For example, the largest sample observation (331) in Example 4.3 can be increased by an arbitrarily large amount without changing the value of the median. Similarly, an increase in the second or third largest observations does not affect the median, nor would a decrease in several of the smallest observations.

This stability of the median is what sometimes justifies its use as a measure of center in some situations. For example, the July 5, 1999, issue of the *Los Angeles Times* reported that the average earnings for stockbrokers in 1997 was $158,201, whereas the median earnings was only $119,010. Income distributions are commonly summarized by reporting the median rather than the mean, because otherwise a few very high salaries could result in a mean that is not representative of a typical salary.

▪ Comparing the Mean and the Median

Figure 4.2 presents several smoothed histograms that might represent either a distribution of sample values or a population distribution. Pictorially, the median is the value on the measurement axis that separates the histogram into two parts, with .5 (50%) of the area under each part of the curve. The mean is a bit harder to visualize. If the histogram were balanced on a triangle (a fulcrum), it would tilt unless the triangle was positioned exactly at the mean. The mean is the balance point for the distribution.

Figure 4.2 The mean and the median.

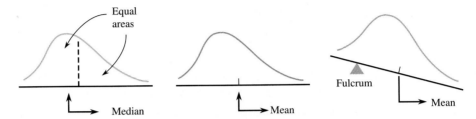

When the histogram is symmetric, the point of symmetry is both the dividing point for equal areas and the balance point, and the mean and the median are equal. However, when the histogram is unimodal (single-peaked) with a longer upper tail, the relatively few outlying values in the upper tail pull the mean up, so it generally lies above the median. For example, an unusually high exam score raises the mean but does not affect the median. Similarly, when a unimodal histogram is negatively skewed, the mean is generally smaller than the median (see Figure 4.3).

Figure 4.3 Relationship between the mean and the median.

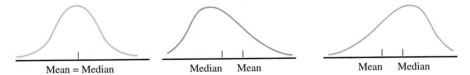

▪ Categorical Data

The natural numerical summary quantities for a categorical data set are the relative frequencies for the various categories. Each relative frequency is the proportion (fraction) of responses that are in the corresponding category. Often there are only two possible responses (a *dichotomy*)—for example, male or female, does or does not have a driver's license, did or did not vote in the last election. It is convenient in such situations to label one of the two possible responses S (for success) and the other F (for failure). As long as further analysis is consistent with the labeling, it is immaterial which category is assigned the S label. When the data set is a sample, the fraction of S's in the sample is called the *sample proportion of successes*.

▪ **Definition**

The **sample proportion of successes**, denoted p, is

$$p = \text{sample proportion of successes} = \frac{\text{number of S's in the sample}}{n}$$

where S is the label used for the response designated as success.

▪ **Example 4.5**   Tampering with Automobile Antipollution Equipment

The use of antipollution equipment on automobiles has substantially improved air quality in certain areas. Unfortunately, many car owners have tampered with smog control devices to improve performance. Suppose that a sample of $n = 15$ cars is selected and that each car is classified as S or F, according to whether or not tampering has taken place. The resulting data are:

S   F   S   S   S   F   F   S   S   F   S   S   S   F   F

This sample contains nine S's, so

$$p = \frac{9}{15} = .60$$

That is, 60% of the sample responses are S's. In 60% of the cars sampled, there has been tampering with the air pollution control devices.

The Greek letter π is used to denote the **population proportion of S's.*** We will see later how the value of p from a particular sample can be used to make inferences about π.

▪ **Trimmed Means**

The extreme sensitivity of the mean to even a single outlier and the extreme insensitivity of the median to a substantial proportion of outliers can make both of them suspect as a measure of center. Statisticians have proposed a *trimmed mean* as a compromise between these two extremes.

> ▪ **Definition**
>
> A **trimmed mean** is computed by first ordering the data values from smallest to largest, then deleting a selected number of values from each end of the ordered list, and finally averaging the remaining values.
>
> The **trimming percentage** is the percentage of values deleted from each end of the ordered list.

▪ **Example 4.6**   Alcohol Use in G-Rated Films

The film industry has been criticized for its depiction of alcohol and tobacco use in children's movies. The article "Tobacco and Alcohol Use in G-Rated Animated Films" (*Journal of the American Medical Association* [1999]: 1131–1136) gave the

---

*Note that we are using the symbol π to represent a population proportion and *not* the mathematical constant $\pi = 3.14 \ldots$. Some statistics books use the symbol p for the population proportion and $\hat{p}$ for the sample proportion.

following data on the number of seconds showing alcohol use in each of 30 animated films released between 1980 and 1997:

Alcohol Exposure (seconds)

34	414	0	0	76	123	3	0	7	0	46	38
13	73	0	72	0	5	0	0	0	0	74	0
28	0	0	0	0	39						

A MINITAB dotplot of these data is shown in Figure 4.4. Because the dotplot shows that the data distribution is not symmetric and that there is an outlier, a trimmed mean is a reasonable choice for describing where the data are centered.

Figure 4.4 A MINITAB dotplot of the data in Example 4.6.

Alcohol Exposure

Because 10% of 30 is 3, a 10% trimmed mean results from deleting the 3 largest and the 3 smallest data values and then averaging the remaining 24 data values. The ordered data values are:

| 0 | 0 | 0 | 0 | 0 | 0 | 0 | 0 | 0 | 0 | 0 | 0 | 0 | 0 | 0 |
| 3 | 5 | 7 | 13 | 28 | 34 | 38 | 39 | 46 | 72 | 73 | 74 | 76 | 123 | 414 |

Deleting three of the 0's (the smallest 10%) and 76, 123, and 414 (the largest 10%) and averaging the remaining values gives

$$10\% \text{ trimmed mean} = \frac{0 + 0 + \cdots + 74}{24} = \frac{432}{24} = 18.00$$

The mean, $\bar{x} = 34.83$, is much larger than the trimmed mean because of the unusually large value (414) in the data set. The 10% trimmed mean, which is 18.00 for this data set, falls between the mean and the median.

Sometimes the number of observations to be deleted from each end of the data set is specified. Then the corresponding trimming percentage is calculated as

$$\text{trimming percentage} = \left(\frac{\text{number deleted from each end}}{n}\right)(100)$$

In other cases, the trimming percentage is specified and then used to determine how many observations to delete from each end, with

number deleted from each end = (trimming percentage)(n)

If the number of observations to be deleted from each end resulting from this calculation is not an integer, it can be rounded to the nearest integer (which changes the trimming percentage a bit).

A trimmed mean with a small to moderate trimming percentage — between 5% and 25% — is less affected by outliers than the mean, but it is not as insensitive as the median. Trimmed means are therefore being used more and more often.

▪ Exercises 4.1–4.15

4.1 The Highway Loss Data Institute publishes data on repair costs resulting from a 5-mph crash test of a car moving forward into a flat barrier. The following table gives data for 10 midsize luxury cars tested in October 2002:

Model	Repair Cost
Audi A6	0
BMW 328i	0
Cadillac Catera	900
Jaguar X	1254
Lexus ES300	234
Lexus IS300	979
Mercedes C320	707
Saab 9-5	670
Volvo S60	769
Volvo S80	4194*

*Included cost to replace airbags, which deployed.

Compute the values of the mean and the median for this data set. Why are these values so different here? Which of the mean and median do you think is more representative of the data set, and why?

4.2 The Highway Loss Data Institute referenced in Exercise 4.1 also gave data for the same 10 midsize luxury models for a 5-mph crash test that involved backing up into a flat barrier. The following data are listed for the cars in the same order as in Exercise 4.1:

0 1039 711 739 557 418 1191 1042 322 390

a. Would the difference between the mean and the median be larger or smaller for this data set than it was for the data of Exercise 4.1? Explain why.
b. Compute the mean and the median for this data set. Write a sentence or two interpreting these two values.
c. Suppose that the first observation had been 500 rather than 0. How would the values of the mean and the median change?
d. Calculate a trimmed mean by eliminating the smallest and the largest observations. What is the corresponding trimming percentage?

4.3 The October 1, 1994, issue of the *San Luis Obispo Telegram-Tribune* reported the following monthly salaries for supervisors from six different counties: $5354 (Kern), $5166 (Monterey), $4443 (Santa Cruz), $4129 (Santa Barbara), $2500 (Placer), and $2220 (Merced). San Luis Obispo County supervisors are supposed to be paid the average of the two counties among these six in the middle of the salary range. Which measure of center determines this sal-

ary, and what is its value? Why is the other measure of center featured in this section not as favorable to these supervisors (although it might appeal to taxpayers)?

4.4 According to an article about buying Viagra over the Internet ("Direct Sale of Sildenafil (Viagra) to Consumers over the Internet," *New England Journal of Medicine* [1999]: 1389–1392), of 77 web sites offering to sell Viagra, 55 were in the United States and 11 were in the United Kingdom. Furthermore, 42 of the 77 provided information about contra-indications (reasons a patient should not use Viagra, e.g., use of nitrates).
a. What proportion of the sampled web sites were in the United Kingdom?
b. What proportion of the sampled web sites were outside the United States?
c. What proportion of the sampled web sites did not warn consumers about possible contra-indications?

4.5 A sample of 26 offshore oil workers took part in a simulated escape exercise, resulting in the accompanying data on time (in seconds) to complete the escape ("Oxygen Consumption and Ventilation During Escape from an Offshore Platform," *Ergonomics* [1997]: 281–292):

389 356 359 363 375 424 325 394 402
373 373 370 364 366 364 325 339 393
392 369 374 359 356 403 334 397

a. Construct a stem-and-leaf display of the data. How does it suggest that the sample mean and median will compare?
b. Calculate the values of the sample mean and median.
c. By how much could the largest time be increased without affecting the value of the sample median? By how much could this value be decreased without affecting the sample median?

4.6 Did Civil War veterans have different levels of wealth after the war was over than those who did not fight? The article "The Civil War Generation: Military Service and Mobility in Dubuque, Iowa, 1860–1870" (*Journal of Social History* [1999]: 791–820) discusses this issue. One bit of information presented in this article is the average property value (in dollars) of each of 123 nonsoldier sons of business-class parents and 14 soldier sons of these same parents in Dubuque, Iowa. The average property value of the 123 nonsoldier sons was $14,905.69, whereas the average property value of the 14 soldier sons was $17,678.57.

a. What is the mean property value of all 137 sons?

b. If one of the soldier sons had no property at all, what is the mean property value of the remaining 13 soldier sons?

4.7 Calculate both the 10% trimmed mean and the 20% trimmed mean for the web site access data of Example 4.3. Compare these values to $\bar{x}$ and the sample median.

4.8 The *San Luis Obispo Telegram-Tribune* (November 29, 1995) reported the values of the mean and median salary for major league baseball players for 1995. The values reported were $1,110,766 and $275,000.

a. Which of the two given values do you think is the mean and which is the median? Explain your reasoning.

b. The reported mean was computed using the salaries of all major league players in 1995. For the 1995 salaries, is the reported mean the population mean μ or the sample mean $\bar{x}$? Explain.

4.9 The cost of housing is rising in most parts of the country, and newspapers love to report on changes in typical home prices. Because some homes have selling prices that are much higher than most, the median price is usually used to describe a "typical" price for a given location. The three accompanying quotes are all from the *San Luis Obispo Tribune*, but each gives a different interpretation of the median price of a home in San Luis Obispo County. Comment on each of these statements. (Look carefully. At least one of the statements is incorrect.)

a. "So we have gone from 23 percent to 27 percent of county residents who can afford the median priced home at $278,380 in SLO County. That means that half of the homes in this county cost less than $278,380 and half cost more." (October 11, 2001)

b. "The county's median price rose to $285,170 in the fourth quarter, a 9.6 percent increase from the same period a year ago, the report said. (The median represents the midpoint of a range.)" (February 13, 2002)

c. "'Your median is going to creep up above $300,000 if there is nothing available below $300,000,' Walker said." (February 26, 2002)

4.10 Consider the following statement: More than 65% of the residents of Los Angeles earn less than the average wage for that city. Could this statement be correct? If so, how? If not, why not?

4.11 A sample consisting of four pieces of luggage was selected from among those checked at an airline counter, yielding the following data on x = weight (in pounds):

$$x_1 = 33.5, \quad x_2 = 27.3, \quad x_3 = 36.7, \quad x_4 = 30.5$$

Suppose that one more piece is selected; denote its weight by x_5. Find a value of x_5 such that $\bar{x}$ = sample median.

4.12 Suppose that 10 patients with meningitis received treatment with large doses of penicillin. Three days later, temperatures were recorded, and the treatment was considered successful if there had been a reduction in a patient's temperature. Denoting success by S and failure by F, the 10 observations are

S   S   F   S   S   S   F   F   S   S

a. What is the value of the sample proportion of successes?

b. Replace each S with a 1 and each F with a 0. Then calculate $\bar{x}$ for this numerically coded sample. How does $\bar{x}$ compare to p?

c. Suppose that it is decided to include 15 more patients in the study. How many of these would have to be S's to give $p = .80$ for the entire sample of 25 patients?

4.13 An experiment to study the lifetime (in hours) for a certain type of component involved putting 10 components into operation and observing them for 100 hr. Eight of the components failed during that period, and those lifetimes were recorded. Denote the lifetimes of the two components still functioning after 100 hr by 100+. The resulting sample observations were

48   79   100+   35   92   86   57   100+   17   29

Which of the measures of center discussed in this section can be calculated, and what are the values of those measures? (Note: The data from this experiment are said to be "censored on the right.")

4.14 An instructor has graded 19 exam papers submitted by students in a certain class of 20 students, and the average so far is 70. (The maximum possible score is 100.) How high would the score on the last paper have to be to raise the class average by 1 point? By 2 points?

4.15 "The Motley Fool," an investment column that appears in newspapers, made the following statement about mutual funds (*San Luis Obispo Tribune*, October 16, 1999):

> Funds with a great three or five or even 10-year average are likely to have that great average because of one amazing year. After all, a five-year average is just an average of five numbers. If one of them is unusually high, the average will be high. Let's consider an example. If in five consecutive years, a fund earns 8 percent, 11 percent, 4 percent, 12 percent, and 33 percent, its average annual return will be about 13 percent. That

might look respectable, but note that in reality it exceeded 13 percent in only one of five years. That 33 percent return (an "outlier" in statistical terms) has skewed the average.

Calculate the median return for the five years mentioned in the quote. Do you think that the median would be a better indication of a typical return in this case? Explain.

▪ 4.2 Describing Variability in a Data Set

Reporting a measure of center gives only partial information about a data set. It is also important to describe the spread of values about the center. The three different samples displayed in Figure 4.5 all have mean = median = 45. There is much variability in the first sample compared to the third sample. The second sample shows less variability than the first and more variability than the third; most of the variability in the second sample is due to the two extreme values being so far from the center.

Figure 4.5 Three samples with the same center and different amounts of variability.

Sample

1.   20, 40, 50, 30, 60, 70

2.   47, 43, 44, 46, 20, 70

3.   44, 43, 40, 50, 47, 46

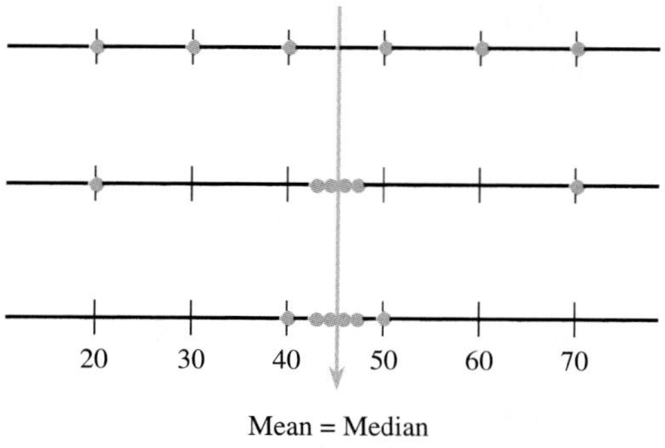

Mean = Median

The simplest numerical measure of variability is the **range**, which is defined as the difference between the largest and the smallest values. In general, more variability will be reflected in a larger range. However, variability depends on more than just the distance between the two most extreme values. Variability is a characteristic of the entire data set, and each observation contributes to variability. The first two samples plotted in Figure 4.5 both have a range of 50, but there is substantially less spread in the second sample.

▪ Deviations from the Mean

The most common measures of variability describe the extent to which the sample observations deviate from the sample mean $\bar{x}$. Subtracting $\bar{x}$ from each observation gives a set of deviations from the mean.

▪ Definition

The n **deviations from the sample mean** are the differences

$$(x_1 - \bar{x}), (x_2 - \bar{x}), \ldots, (x_n - \bar{x})$$

A particular deviation is positive if the x value exceeds $\bar{x}$ and negative if the x value is less than $\bar{x}$.

▪ Example 4.7   Acrylamide Levels in French Fries

Research by the Food and Drug Administration (FDA) shows that acrylamide (a possible cancer-causing substance) forms in high-carbohydrate foods cooked at high temperatures and that acrylamide level can vary widely even within the same brand of food (Associated Press, December 6, 2002). FDA scientists analyzed McDonald's french fries purchased at seven different locations and found the following acrylamide levels:

$$497 \quad 193 \quad 328 \quad 155 \quad 326 \quad 245 \quad 270$$

For this data set, $\sum x = 2014$ and $\bar{x} = 287.714$. Table 4.2 displays the data along with the corresponding deviations, formed by subtracting $\bar{x} = 287.714$ from each observation. Three of the deviations are positive because only three of the observations are larger than $\bar{x}$. The negative deviations correspond to observations that are smaller than $\bar{x}$. Note that some of the deviations are quite large in magnitude (209.286, −132.714), indicating observations that are far from the sample mean.

Table 4.2 ▪ Deviations from the Mean for the Acrylamide Data in Example 4.7

Observation (x)	Deviation $(x - \bar{x})$
497	209.286
193	−94.714
328	40.286
155	−132.714
326	38.286
245	−42.714
270	−17.714

In general, the greater the amount of variability in the sample, the larger the magnitudes (ignoring the signs) of the deviations. Thus, the magnitudes of the deviations for the first sample in Figure 4.5 substantially exceed those for the third sample, implying more variability in Sample 1.

We now consider how to combine the deviations into a single numerical measure. A first thought might be to calculate the average deviation, by adding the deviations together (this sum can be denoted compactly by $\sum(x - \bar{x})$) and then dividing by n. This does not work, though, because negative and positive deviations counteract one another in the summation.

Except for the effects of rounding in computing the deviations, it is always true that

$$\sum(x - \bar{x}) = 0$$

Because this sum is 0, the average deviation is always 0 and so cannot be used as a measure of variability.

As a result of rounding, the value of the sum of the seven deviations in Example 4.7 is $\sum(x - \bar{x}) = 0.002$. If we used even more decimal accuracy in computing $\bar{x}$, the sum would be even closer to 0.

▪ The Variance and Standard Deviation

The customary way to prevent negative and positive deviations from counteracting one another is to square them before combining. Then deviations with opposite signs but with the same magnitude, such as +20 and −20, make identical contributions to variability. The squared deviations are $(x_1 - \bar{x})^2, (x_2 - \bar{x})^2, \ldots, (x_n - \bar{x})^2$, and their sum is

$$(x_1 - \bar{x})^2 + (x_2 - \bar{x})^2 + \cdots + (x_n - \bar{x})^2 = \sum(x - \bar{x})^2$$

Common notation for $\sum(x - \bar{x})^2$ is S_{xx}. Dividing this sum by the sample size n gives the average squared deviation. Although this seems to be a reasonable measure of variability, we use a divisor slightly smaller than n. (The reason for this will be explained later in this section and in Chapter 9.)

▪ **Definition**

The **sample variance**, denoted by s^2, is the sum of squared deviations from the mean divided by $(n - 1)$. That is,

$$s^2 = \frac{\sum(x - \bar{x})^2}{n - 1} = \frac{S_{xx}}{n - 1}$$

The **sample standard deviation** is the positive square root of the sample variance and is denoted by s.

A large amount of variability in the sample is indicated by a relatively large value of s^2 or s, whereas a value of s^2 or s close to 0 indicates a small amount of variability. For most statistical purposes, s is the desired quantity, but s^2 must be computed first. Notice that whatever unit is used for x (such as pounds or seconds), the squared deviations and therefore s^2 are in squared units. Taking the square root gives a measure expressed in the same units as x. Thus, for a sample of heights, the standard deviation might be $s = 3.2$ in., and for a sample of textbook prices, it might be $s = \$12.43$.

▪ Example 4.8   French Fries Revisited

Let's continue using the acrylamide data and the computed deviations from the mean given in Example 4.7 to obtain the sample variance and standard deviation. Table 4.3 shows the observations, deviations from the mean, and squared deviations. Combining the squared deviations to compute the values of s^2 and s gives

$$\sum(x - \bar{x})^2 = S_{xx} = 75{,}611.429$$

and

$$s^2 = \frac{\Sigma(x - \bar{x})^2}{n - 1} = \frac{75{,}611.430}{7 - 1} = \frac{75{,}611.430}{6} = 12{,}601.905$$

$$s = \sqrt{12{,}601.905} = 112.258$$

Table 4.3 ■ Deviations and Squared Deviations for the Acrylamide Data

Observation (x)	Deviation $(x - \bar{x})$	Squared Deviation $(x - \bar{x})^2$
497	209.286	43,800.630
193	−94.714	8970.742
328	40.286	1622.962
155	−132.714	17,613.006
326	38.286	1465.818
245	−42.714	1824.486
270	−17.714	313.786
		75,611.430

The computation of s^2 can be a bit tedious, especially if the sample size is large. Fortunately, many calculators and computer software packages compute the variance and standard deviation upon request. One commonly used statistical computer package is MINITAB. The output resulting from using the MINITAB Describe command with the acrylamide data follows. MINITAB gives a variety of numerical descriptive measures, including the mean, the median, and the standard deviation.

Descriptive Statistics: acrylamide

Variable	N	Mean	Median	TrMean	StDev	SE Mean
acrylamide	7	287.7	270.0	287.7	112.3	42.4

Variable	Minimum	Maximum	Q1	Q3
acrylamide	155.0	497.0	193.0	328.0

■ Interpretation and Properties

The standard deviation can be informally interpreted as the size of a "typical" or "representative" deviation from the mean. Thus, in Example 4.8, a typical deviation from $\bar{x}$ is about 112.3; some observations are closer to $\bar{x}$ than 112.3 and others are farther away. We computed $s = 112.3$ in Example 4.8 without saying whether this value indicated a large or a small amount of variability. At this point, it is better to use s for comparative purposes than for an absolute assessment of variability. If we obtained a sample of acrylamide level values for a second set of McDonald's french fries and computed $s = 104.6$, then we would conclude that our original sample has more variability than our second sample. A particular value of s can be judged large or small only in comparison to something else.

There are measures of variability for the entire population that are analogous to s^2 and s for a sample. These measures are called the **population variance** and the **population standard deviation** and are denoted by σ^2 and σ, respectively. (We again use a lowercase Greek letter for a population characteristic.) The population standard deviation σ is expressed in the same unit of measurement as are the values in the population. As with s, the value of σ can be used for comparative purposes.

In many statistical procedures, we would like to use the value of σ, but unfortunately it is not usually available. Therefore, in its place we must use a value computed from the sample that we hope is close to σ (i.e., a good *estimate* of σ). We use the divisor $(n - 1)$ in s^2 rather than n because, on average, the resulting value tends to be a bit closer to σ^2. We will say more about this in Chapter 9.

An alternative rationale for using $(n - 1)$ is based on the property $\sum (x - \bar{x}) = 0$. Suppose that $n = 5$ and that four of the deviations are

$$x_1 - \bar{x} = -4, \qquad x_2 - \bar{x} = 6, \qquad x_3 - \bar{x} = 1, \qquad x_5 - \bar{x} = -8$$

Then, because the sum of these four deviations is -5, the remaining deviation must be $x_4 - \bar{x} = 5$ (so that the sum of all five is 0). Although there are five deviations, only four of them contain independent information about variability. More generally, once any $(n - 1)$ of the deviations are available, the value of the remaining deviation is determined. The n deviations actually contain only $(n - 1)$ independent pieces of information about variability. Statisticians express this by saying that s^2 and s are based on $(n - 1)$ *degrees of freedom* (df). Many inferential procedures encountered in later chapters are based on some appropriate number of degrees of freedom.

▪ The Interquartile Range

As with $\bar{x}$, the value of s can be greatly affected by the presence of even a single unusually small or large observation. The *interquartile range* is a measure of variability that is resistant to the effects of outliers. It is based on quantities called *quartiles*. The *lower quartile* separates the bottom 25% of the data set from the upper 75%, and the *upper quartile* separates the top 25% from the bottom 75%. The *middle quartile* is the median, and it separates the bottom 50% from the top 50%. Figure 4.6 illustrates the locations of these quartiles for a smoothed histogram.

Figure 4.6 The quartiles for a smoothed histogram.

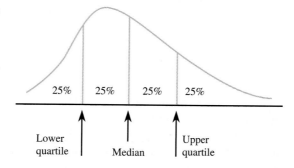

The quartiles for sample data are obtained by dividing the n ordered observations into a lower half and an upper half; if n is odd, the median is excluded from both halves. The two extreme quartiles are then the medians of the two halves.

▪ **Definition***

lower quartile = median of the lower half of the sample

upper quartile = median of the upper half of the sample

(If n is odd, the median of the entire sample is excluded from both halves.)

The **interquartile range (iqr)**, a resistant measure of variability, is given by

iqr = upper quartile − lower quartile

▪ Example 4.9   Hospital Cost-to-Charge Ratios

The Oregon Department of Health Services publishes cost-to-charge ratios for hospitals in Oregon on its web site. The cost-to-charge ratio is computed as the ratio of the actual cost of care to what the hospital actually bills for care, and the ratio is usually expressed as a percentage. A cost-to-charge ratio of 60% means that the actual cost is 60% of what was billed. The ratios for 31 hospitals in Oregon for inpatient services in 2002 were:

68	76	60	88	69	80	75	67	71	100	63	62
71	74	64	48	100	72	65	50	72	100	63	45
54	60	75	57	74	84	83					

Figure 4.7 gives a stem-and-leaf display of the data.

Figure 4.7 MINITAB stem-and-leaf display of hospital cost-to-charge ratio data in Example 4.9.

Stem-and-Leaf Display: Cost-to-Charge
Stem-and-leaf of Cost-to-Charge N = 31
Leaf Unit = 1.0
```
 4 58
 5 04
 5 7
 6 002334
 6 5789
 7 112244
 7 556
 8 034
 8 8
 9
 9
10 000
```

The sample size $n = 31$ is an odd number, so the median is excluded from both halves of the sample when computing the quartiles.

---

*There are several other sensible ways to define quartiles. Some calculators and software packages use an alternative definition.

Ordered Data

Lower half	45	48	50	54	57	60	60	**62**	63	63	64	65	67	68	69
Median								**71**
Upper half	71	72	72	74	74	75	75	**76**	80	83	84	88	100	100	100

Each half of the sample contains 15 observations. The lower quartile is just the median of the lower half of the sample (62 for this data set), and the upper quartile is the median of the upper half (76 for this data set). This gives

lower quartile = 62

upper quartile = 76

iqr = 76 − 62 = 14

The sample mean and standard deviation for this data set are 70.65 and 14.11, respectively. If we were to change the two smallest values from 45 and 48 to 25 and 28 (so that they still remain the two smallest values), the median and interquartile range would not be affected, whereas the mean and the standard deviation would change to 69.35 and 16.99, respectively.

The resistant nature of the interquartile range follows from the fact that up to 25% of the smallest sample observations and up to 25% of the largest sample observations can be made more extreme without affecting the value of the interquartile range.

The **population interquartile range** is the difference between the upper and lower population quartiles. If a histogram of the data set under consideration (whether a population or a sample) can be reasonably well approximated by a normal curve, then the relationship between the standard deviation (sd) and the interquartile range is roughly sd = iqr/1.35. A value of the standard deviation much larger than iqr/1.35 suggests a histogram with heavier (or longer) tails than a normal curve. For the hospital cost-to-charge data of Example 4.9, we had $s = 14.11$, whereas iqr/1.35 = 14/1.35 = 10.37. This suggests that the distribution of sample values is indeed heavy-tailed compared to a normal curve, as can be seen in the stem-and-leaf display.

▪ A Note Concerning Computation (Optional)

Hand calculation of s^2 using the defining formula can be a bit tedious. An alternative expression for the numerator of s^2 simplifies the arithmetic by eliminating the need to calculate the deviations. In this alternative, it is necessary to distinguish between two quantities that are similar in appearance but different in value:

1. Sum the x values, then square the sum to obtain $(\sum x)^2$

2. Square each x value, then sum the squares to obtain $\sum x^2$

If, for example, $x_1 = 3$ and $x_2 = 5$, then

$$\left(\sum x\right)^2 = (x_1 + x_2)^2 = (3 + 5)^2 = 8^2 = 64$$

$$\sum x^2 = x_1^2 + x_2^2 = 3^2 + 5^2 = 9 + 25 = 34$$

So the order in which the two operations, squaring and summation, are carried out makes a difference!

A computational formula for the sum of squared deviations is

$$S_{xx} = \sum (x - \bar{x})^2 = \sum x^2 - \frac{(\sum x)^2}{n}$$

Thus, a **computational formula for the sample variance** is

$$s^2 = \frac{S_{xx}}{n-1} = \frac{\sum x^2 - \dfrac{(\sum x)^2}{n}}{n-1}$$

According to this formula, after squaring each x value and adding these to obtain $\sum x^2$, the single quantity $\dfrac{(\sum x)^2}{n}$ is subtracted from the result. Instead of the n subtractions required to obtain the deviations, just one subtraction now suffices.

The computed value of s^2, whether obtained from the defining formula or the computational formula, can sometimes be greatly affected by rounding in $\bar{x}$ or $\dfrac{(\sum x)^2}{n}$. Protection against adverse rounding effects can almost always be achieved by using four or five digits of decimal accuracy beyond the decimal accuracy of the data values themselves.

We now illustrate the use of the computational formula with the acrylamide data of Example 4.8. For efficient computation, it is convenient to place the x values in a single column and the x^2 values in a second column. Adding the numbers in these two columns then gives $\sum x$ and $(\sum x)^2$, respectively. For example:

x	x^2
497	247,009
193	37,249
328	107,584
155	24,025
326	106,276
245	60,025
270	72,900
2014	655,068

$$\frac{(\sum x)^2}{n} = \frac{(2014)^2}{7} = \frac{4,056,196}{7} = 579,456.571$$

Then

$$s^2 = \frac{\sum x^2 - \dfrac{(\sum x)^2}{n}}{n-1} = \frac{655,068 - 579,456.571}{7-1} = \frac{75,611.429}{6} = 12,601.905$$

$$s = \sqrt{12,601.905} = 112.258$$

Note that the values of s^2 and s obtained using the computational formulas agree with those found previously in Example 4.8.

■ Exercises 4.16–4.29

4.16 A sample of $n = 5$ college students yielded the following observations on number of traffic citations for a moving violation during the previous year:

$$x_1 = 1 \quad x_2 = 0 \quad x_3 = 0 \quad x_4 = 3 \quad x_5 = 2$$

Calculate s^2 and s.

4.17 Give two sets of five numbers that have the same mean but different standard deviations, and give two sets of five numbers that have the same standard deviation but different means.

4.18 Cost-to-charge ratios (see Example 4.9 for a definition of this ratio) were reported for the 10 hospitals in California with the lowest ratios (*San Luis Obispo Tribune*, December 15, 2002). These ratios represent the 10 hospitals with the highest markup, because for these hospitals, the actual cost was only a small percentage of the amount billed. The 10 cost-to-charge values (percentages) were

    8.81 10.26 10.20 12.66 12.86 12.96 13.04
    13.14 14.70 14.84

a. Compute the variance and standard deviation for this data set.
b. If cost-to-charge data were available for all hospitals in California, would the standard deviation of this data set be larger or smaller than the standard deviation computed in Part (a) for the 10 hospitals with the lowest cost-to-charge values? Explain.
c. Explain why it would not be reasonable to use the data from the sample of 10 hospitals in Part (a) to draw conclusions about the population of all hospitals in California.

4.19 In 1997 a woman sued a computer keyboard manufacturer, charging that her repetitive stress injuries were caused by the keyboard (*Genessey v. Digital Equipment Corporation*). The jury awarded about $3.5 million for pain and suffering, but the court then set aside that award as being unreasonable compensation. In making this determination, the court identified a "normative" group of 27 similar cases and specified a reasonable award as one within 2 standard deviations of the mean of the awards in the 27 cases. The 27 award amounts were (in thousands of dollars)

    37   60   75   115   135   140   149   150
    238   290   340   410   600   750   750   750
    1050 1100 1139 1150 1200 1200 1250 1576
    1700 1825 2000

What is the maximum possible amount that could be awarded under the "2-standard deviations rule"?

4.20 The article "Genetic Factors May Be Derailing Healthful Diets" (*San Luis Obispo Tribune*, September 3, 2001) included the following statement:

> Krauss' report centered on the so-called bad cholesterol, known as LDL, that most health authorities agree is associated with coronary artery disease. When people eat diets that consist of only 10 percent fat, on average, their LDL levels drop, but the average hides a lot of individual variation, Krauss said. When looking closer at what happens to individuals, researchers find that some people on low fat diets have a real and significant drop in LDL. But others experience a change in their LDLs in such a way that their exposure to coronary artery disease probably rises, Krauss said.

Is the researcher describing a variable, LDL, that has only a little variability or a great deal of variability? What part of the statement leads you to this conclusion?

4.21 The Highway Loss Data Institute reported the following repair costs resulting from crash tests conducted in October 2002. The given data are for a 5-mph crash into a flat surface. Data are given for both a sample of 10 moderately priced midsize cars and a sample of 14 inexpensive midsize cars.

Moderately Priced	296	0	1085	148	1065
Midsize Cars	0	0	341	184	370
Inexpensive	513	719	364	295	305
Midsize Cars	335	353	156	209	288
	0	0	397	243	

a. Compute the standard deviation and the interquartile range for the repair cost of the moderately priced midsize cars.
b. Compute the standard deviation and the interquartile range for the repair cost of the inexpensive midsize cars.
c. Is there more variability in the repair cost for the moderately priced cars or for the inexpensive midsize cars? Justify your choice.
d. Compute the mean repair cost for each of the two types of cars.
e. Write a few sentences comparing repair cost for moderately priced and inexpensive midsize cars. Be sure to include information about both center and variability.

4.22 The paper "Total Diet Study Statistics on Element Results" (Food and Drug Administration, April 25, 2000) gave information on sodium content for various types of foods. Twenty-six tomato catsups were analyzed. Data consistent with summary quantities given in the paper were the following:

Sodium content (mg/kg)

12,148	10,426	10,912	9116	13,226	11,663
11,781	10,680	8457	10,788	12,605	10,591
11,040	10,815	12,962	11,644	10,047	10,478
10,108	12,353	11,778	11,092	11,673	8758
11,145	11,495				

Compute the values of the quartiles and the interquartile range.

4.23 The paper referenced in Exercise 4.22 also gave summary quantities for sodium content (in milligrams per kilogram) of chocolate pudding made from instant mix:

3099 3112 2401 2824 2682 2510 2297 3959 3068 3700

a. Use the given data to compute the mean, the standard deviation, and the interquartile range for sodium content of these chocolate puddings.
b. Based on the interquartile range, is there more or less variability in sodium content for the chocolate pudding data than for the tomato catsup data of Exercise 4.22?

4.24 Example 4.9 gave 2002 cost-to-charge ratios for inpatient services at 31 hospitals in Oregon. The same data source also provided the cost-to-charge ratios for outpatient services at these same 31 hospitals:

75 54 50 75 58 56 56 62 66 69 100 57
45 51 53 45 84 45 48 39 51 51 65 100
63 96 54 53 67 52 71

a. Use the given data to compute the quartiles and the interquartile range for the outpatient cost-to-charge ratio.
b. How does the interquartile range compare to the interquartile range for the inpatient cost-to-charge ratio from Example 4.9? Is there more variability in the inpatient or the outpatient ratio?

4.25 The paper "The Pedaling Technique of Elite Endurance Cyclists" (*International Journal of Sport Biomechanics* [1991]: 29–53) reported the following data on single-leg power at a high workload:

244 191 160 187 180 176 174 205 211 183 211 180 194 200

Calculate and interpret the sample variance and standard deviation.

4.26 Although bats are not known for their eyesight, they are able to locate prey (mainly insects) by emitting high-pitched sounds and listening for echoes. A paper appearing in *Animal Behaviour* ("The Echolocation of Flying Insects by Bats" [1960]: 141–154) gave the following distances (in centimeters) at which a bat first detected a nearby insect:

62 23 27 56 52 34 42 40 68 45 83

a. Compute the sample mean distance at which the bat first detects an insect.
b. Compute the sample variance and standard deviation for this data set. Interpret these values.

4.27 For the data in Exercise 4.26, add -10 to each sample observation. (This is the same as subtracting 10.) For the new set of values, compute the mean and the deviations from the mean. How do these deviations compare to the deviations from the mean for the original sample? How does s^2 for the new values compare to s^2 for the old values? In general, what effect does adding the same number to each observation have on s^2 and s? Explain.

4.28 For the data of Exercise 4.26, multiply each data value by 10. How does s for the new values compare to s for the original values? More generally, what happens to s if each observation is multiplied by the same positive constant c?

4.29 The standard deviation alone does not measure relative variation. For example, a standard deviation of $1 would be considered large if it is describing the variability from store to store in the price of an ice cube tray. On the other hand, a standard deviation of $1 would be considered small if it is describing store-to-store variability in the price of a particular brand of freezer. A quantity designed to give a relative measure of variability is the *coefficient of variation*. Denoted by CV, the coefficient of variation expresses the standard deviation as a percentage of the mean. It is defined by the formula $CV = 100\left(\dfrac{s}{\bar{x}}\right)$. Consider two samples. Sample 1 gives the actual weight (in ounces) of the contents of cans of pet food labeled as having a net weight of 8 oz. Sample 2 gives the actual weight (in pounds) of the contents of bags of dry pet food labeled as having a net weight of 50 lb. The weights for the two samples are:

Sample 1	8.3	7.1	7.6	8.1	7.6
	8.3	8.2	7.7	7.7	7.5
Sample 2	52.3	50.6	52.1	48.4	48.8
	47.0	50.4	50.3	48.7	48.2

a. For each of the given samples, calculate the mean and the standard deviation.
b. Compute the coefficient of variation for each sample. Do the results surprise you? Why or why not?

▪ 4.3 Summarizing a Data Set: Boxplots

In Sections 4.1 and 4.2, we looked at ways of describing the center and variability of a data set using numerical measures. It would be nice to have a method of summarizing data that gives more detail than just a measure of center and spread and yet less detail than a stem-and-leaf display or histogram. A *boxplot* is one such technique. It is compact, yet it provides information about the center, spread, and symmetry or skewness of the data. A boxplot is based on the median and interquartile range rather than on the mean and standard deviation. We consider two types of boxplots: the skeletal boxplot and the modified boxplot.

▪ Construction of a Skeletal Boxplot

1. Draw a horizontal (or vertical) measurement scale.
2. Construct a rectangular box whose left (or lower) edge is at the lower quartile and whose right (or upper) edge is at the upper quartile (so box width = iqr).
3. Draw a vertical (or horizontal) line segment inside the box at the location of the median.
4. Extend horizontal (or vertical) line segments, called whiskers, from each end of the box to the smallest and largest observations in the data set.

▪ Example 4.10 Revisiting Hospital Cost-to-Charge Ratios

Let's reconsider the cost-to-charge data for hospitals in Oregon (Example 4.9). The ordered observations are:

Ordered Data

Lower half	45	48	50	54	57	60	60	**62**	63	63	64	65	67	68	69
Median								**71**							
Upper half	71	72	72	74	74	75	75	**76**	80	83	84	88	100	100	100

To construct a boxplot of these data, we need the following information: the smallest observation, the lower quartile, the median, the upper quartile, and the largest observation. This collection of summary measures is often referred to as the **five-number summary**. For this data set (see Example 4.9), we have

smallest observation = 45

lower quartile = median of the lower half = 62

median = 16th observation in the ordered list = 71

upper quartile = median of the upper half = 76

largest observation = 100

Figure 4.8 shows the corresponding boxplot. The median line is somewhat closer to the upper edge of the box than to the lower edge, suggesting a concentration of values in the upper part of the middle half. The upper whisker is a bit longer than the lower whisker. These conclusions are consistent with the stem-and-leaf display of Figure 4.7.

Figure 4.8 Skeletal boxplot for the cost-to-charge data of Example 4.10.

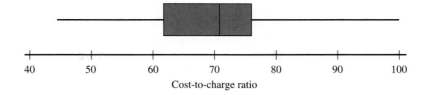

Cost-to-charge ratio

The sequence of steps used to construct a skeletal boxplot is easily modified to give information about outliers.

▪ **Definition**

An observation is an **outlier** if it is more than 1.5(iqr) away from the nearest end of the box (the nearest quartile).

An outlier is **extreme** if it is more than 3(iqr) from the nearest end of the box, and it is **mild** otherwise.

A **boxplot** represents mild outliers by shaded circles and extreme outliers by open circles.

Whiskers extend on each end to the most extreme observations that are *not* outliers.

▪ **Example 4.11   Golden Rectangles**

The accompanying data came from an anthropological study of rectangular shapes (*Lowie's Selected Papers in Anthropology*, Cora Dubios, ed. [Berkeley, CA: University of California Press, 1960], 137–142). Observations were made on the variable x = width/length for a sample of n = 20 beaded rectangles used in Shoshoni Indian leather handicrafts:

    0.553   0.570   0.576   0.601   0.606   0.606   0.609   0.611   0.615   0.628
    0.654   0.662   0.668   0.670   0.672   0.690   0.693   0.749   0.844   0.933

The quantities needed for constructing the modified boxplot follow:

    median = 0.641

    lower quartile = 0.606   upper quartile = 0.681

    iqr = 0.681 − 0.606 = 0.075

    1.5(iqr) = 0.1125   3(iqr) = 0.225

Thus,

    upper edge of box (upper quartile) + 1.5(iqr) = 0.681 + 0.1125 = 0.7935
    lower edge of box (lower quartile) − 1.5(iqr) = 0.606 − 0.1125 = 0.4935

So 0.844 and 0.933 are both outliers on the upper end (because they are larger than 0.7935), and there are no outliers on the lower end (because no observations are smaller than 0.4935). Because

    upper edge of box + 3(iqr) = 0.681 + 0.225 = 0.906

0.933 is an extreme outlier and 0.844 is only a mild outlier. The upper whisker extends to the largest observation that is not an outlier, 0.749, and the lower whisker extends to 0.553. The boxplot is presented in Figure 4.9. The median line is not at the center of the box, so there is a slight asymmetry in the middle half of the data. However, the most striking feature is the presence of the two outliers. These two x values considerably exceed the "golden ratio" of 0.618, used since antiquity as an aesthetic standard for rectangles.

Figure 4.9 Boxplot for the rectangle data in Example 4.11.

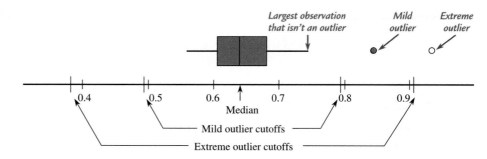

▪ Example 4.12   State Gross Products

Highway Statistics 2001 (Federal Highway Administration) gave selected descriptive measures for the 50 U.S. states and the District of Columbia. The 51 observations for gross state product (in billions of dollars) and per capita state product (in dollars) are as follows:

Gross Product

110	24	134	62	1119	142	142	34	54	419	254	40
31	426	174	85	77	107	129	32	165	239	295	161
62	163	20	52	63	41	319	48	707	236	17	341
82	105	364	30	100	21	160	646	60	16	231	193
40	158	18									

Per Capita Gross Product

25,282	39,024	28,712	24,429	34,238	35,777	43,385	45,699
103,647	28,106	33,259	33,813	28,183	35,294	29,452	29,710
29,178	27,199	29,567	25,641	32,164	38,900	30,041	34,087
22,537	29,974	22,727	31,306	36,124	34,570	39,402	27,682
38,934	31,275	26,645	30,343	34,558	31,993	30,328	30,364
26,042	28,728	29,450	32,772	28,588	27,073	34,028	33,931
22,075	30,257	37,500					

Figure 4.10(a) shows a MINITAB boxplot for gross state product. Note that the upper whisker is longer than the lower whisker and that there are three outliers on the high end (from least extreme to most extreme, they are Texas, New York, and California — not surprising, because these are the three states with the largest population sizes). Figure 4.10(b) shows a MINITAB boxplot of the per capita gross state product (state gross product divided by population size). For this boxplot, the upper and lower whiskers are about the same length and there are only two outliers on the upper end. The two outliers in this data set are Delaware and Washington, D.C., indicating that, although the three largest states have the highest gross state product, they are not the states with the highest *per capita* gross state product.

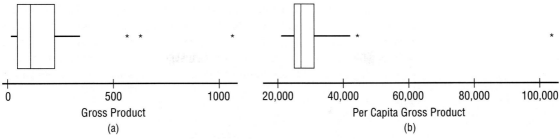

Figure 4.10 MINITAB boxplots for the data of Example 4.12: (a) gross state product; (b) per capita gross state product.

Note that MINITAB does not distinguish between mild outliers and extreme outliers in the boxplot. For the per capita gross state product data,

lower quartile = 28,183

upper quartile = 34,558

iqr = 34,558 − 28,183 = 6375

Then

1.5(iqr) = 9562.5

3(iqr) = 19,125

The observation for Delaware (45,699) would be a mild outlier, because it is more than 1.5(iqr) but less than 3(iqr) from the upper quartile. The observation for Washington, D.C. (103,647), is an extreme outlier, because it is more than 3(iqr) away from the upper quartile.

For the gross product data,

lower quartile = 41

upper quartile = 231

iqr = 231 − 41 = 190

1.5(iqr) = 285

3(iqr) = 570

so the observations for Texas (646) and New York (707) are mild outliers and the observation for California (1119) is an extreme outlier.

With two or more data sets consisting of observations on the same variable (e.g., fuel efficiencies for four types of car or weight gains for a control group and a treatment group), comparative boxplots convey initial impressions concerning similarities and differences between the data sets.

▪ Example 4.13   The Cost of Coffee

An article in the October 1994 issue of *Consumer Reports* compared various brands of brewed coffee. The study included 31 regular-roast caffeinated brands (Type 1), 9 decaffeinated brands (Type 2), and 13 dark-roast brands (Type 3). Figure 4.11 is a comparative boxplot of cost per cup for the three types. The median marker for Type 1 (above 6.0) is barely discernible at the lower end of the box, and the same is true at the higher end (above 8.0) for Type 2. The box for Type 3 is much wider than for the other two types (much larger iqr), indicating much more variability in the middle half of the data. However, there are no outliers for Type 3, whereas Type 2 has one extreme outlier and Type 1 has three extreme and four mild outliers (two of these outlier brands have the same value: 4¢ per cup).

Figure 4.11
Comparative boxplot for cost per cup of coffee data of Example 4.13.

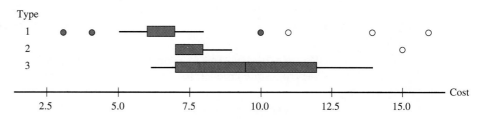

■ **Example 4.14**   Radon and Childhood Cancer

Let's revisit the radon concentration data from the chapter introduction. Recall that the first sample consisted of radon levels for houses in which a child diagnosed with cancer had been residing. Houses in the second sample had no recorded cases of childhood cancer. The data are reproduced here:

Cancer

10	21	5	23	15	11	9	13	27	13	39	22
7	20	45	12	15	3	8	11	18	16	23	16
34	10	15	11	18	210	22	11	16	17	33	10
9	57	16	21	18	38						

No Cancer

9	38	11	12	29	5	7	6	8	29	24	12
17	11	3	9	33	17	55	11	29	13	24	7
11	21	6	39	29	7	8	55	9	21	9	3
85	11	14									

A comparative boxplot for these two samples is shown in Figure 4.12. With the exception of the extreme value over 200 in the cancer sample, the difference in the radon level distributions for the two samples does not appear to be large. In fact, again with the exception of the one really large value, the radon levels for the cancer sample tend to be lower and less variable than those for the no-cancer sample.

Figure 4.12 MINITAB comparative boxplot for the radon data of Example 4.14.

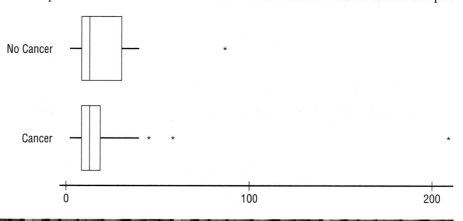

▪ Exercises 4.30–4.37

4.30 The percentage of juice lost after thawing for 19 different strawberry varieties appeared in the article "Evaluation of Strawberry Cultivars with Different Degrees of Resistance to Red Scale" (*Fruit Varieties Journal* [1991]: 12–17):

    46 51 44 50 33 46 60 41 55 46 53 53
    42 44 50 54 46 41 48

a. Are there any observations that are mild outliers? Extreme outliers?
b. Construct a boxplot, and comment on the important features of the plot.

4.31 Use the web page access data of Example 4.3 (also given here) to construct a modified boxplot. Write a few sentences relating the interesting features of the boxplot to the dotplot of the data given in Figure 4.1.

    20 37   4 20   0 84 14 36   5 331 19
     0 0 22   3 13 14 36   4   0 18 8
     0 26   4   0   5 23 19   7 12   8 13
    16 21   7 13 12   8 42

4.32 Based on a large national sample of working adults, the U.S. Census Bureau reports the following information on travel time to work for those who do not work at home:

    lower quartile = 7 min
    median = 18 min
    upper quartile = 31 min

Also given was the mean travel time, which was reported as 22.4 min.
a. Is the travel time distribution more likely to be approximately symmetric, positively skewed, or negatively skewed? Explain your reasoning based on the given summary quantities.
b. Suppose that the minimum travel time was 1 min and that the maximum travel time in the sample was 205 min. Construct a skeletal boxplot for the travel time data.
c. Were there any mild or extreme outliers in the data set? How can you tell?

4.33 The technical report "Ozone Season Emissions by State" (U.S. Environmental Protection Agency, 2002) gave the following nitrous oxide emissions (in thousands of tons) for the 48 states in the continental U.S. states:

    76 22 40   7 30   5   6 136 72 33   0
    89 136 39 92 40 13 27   1 63 33 60
    27 16 63 32 20   2 15 36 19 39 130
    40   4 85 38   7 68 151 32 34   0   6
    43 89 34   0

Use these data to construct a boxplot that shows outliers. Write a few sentences describing the important characteristics of the boxplot.

4.34 The paper "Relationship Between Blood Lead and Blood Pressure Among Whites and African Americans" (a technical report published by Tulane University School of Public Health and Tropical Medicine, 2000) gave summary quantities for blood lead level (in micrograms per deciliter) for a sample of whites and a sample of African Americans. Data consistent with the given summary quantities follow:

Whites	8.3 0.9 2.9 5.6 5.8 5.4 1.2
	1.0 1.4 2.1 1.3 5.3 8.8 6.6
	5.2 3.0 2.9 2.7 6.7 3.2
African Americans	4.8 1.4 0.9 10.8 2.4 0.4 5.0
	5.4 6.1 2.9 5.0 2.1 7.5 3.4
	13.8 1.4 3.5 3.3 14.8 3.7

a. Compute the values of the mean and the median for blood lead level for the sample of African Americans. Which of the mean or the median is larger? What characteristic of the data set explains the relative values of the mean and the median?
b. Construct a comparative boxplot for blood lead level for the two samples. Write a few sentences comparing the blood lead level distributions for the two samples.

4.35 A *Consumer Reports* article on peanut butter (September 1990) reported the following scores for various brands:

Crunchy	56 44 62 36 39 53 50 65 45
	22 40 56 68 41 30 40 50 56
	30
Creamy	62 53 75 42 47 40 34 62 52
	50 34 42 36 75 80 47 56 62

The statistical package S-Plus was used to obtain the following comparative boxplot:

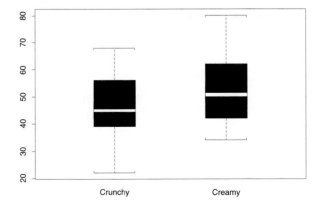

Comment on the similarities and differences in the two data sets.

4.36 The article "Compression of Single-Wall Corrugated Shipping Containers Using Fixed and Floating Test Platens" (*Journal of Testing and Evaluation* [1992]: 318–320) described an experiment in which several different types of boxes were compared with respect to compressive strength. Consider the following observations on four different types of boxes (these data are consistent with summary quantities given in the paper):

Type of Box	Compression Strength (lb)				
1	655.5	788.3	734.3	721.4	679.1	699.4
2	789.2	772.5	786.9	686.1	732.1	774.8
3	737.1	639.0	696.3	671.7	717.2	727.1
4	535.1	628.7	542.4	559.0	586.9	520.0

Construct a boxplot for each type of box. Use the same scale for all four boxplots. Discuss the similarities and differences in the boxplots.

4.37 Blood cocaine concentration (in milligrams per liter) was determined for both a sample of individuals who had died from cocaine-induced excited delirium and a sample of those who had died from a cocaine overdose without excited delirium. The accompanying data are consistent with summary values given in the paper "Fatal Excited Delirium Following Cocaine Use" (*Journal of Forensic Sciences* [1997]: 25–31):

Excited Delirium

0	0	0	0	0.1	0.1	0.1	0.1	0.2	0.2
0.3	0.3	0.3	0.4	0.5	0.7	0.7	1.0	1.5	2.7
2.8	3.5	4.0	8.9	9.2	11.7	21.0			

No Excited Delirium

0	0	0	0	0	0.1	0.1	0.1	0.1
0.2	0.2	0.2	0.3	0.3	0.3	0.4	0.5	0.5
0.6	0.8	0.9	1.0	1.2	1.4	1.5	1.7	2.0
3.2	3.5	4.1	4.3	4.8	5.0	5.6	5.9	6.0
6.4	7.9	8.3	8.7	9.1	9.6	9.9	11.0	11.5
12.2	12.7	14.0	16.6	17.8				

a. Determine the median, quartiles, and interquartile range for each of the two samples.
b. Are there any outliers in either sample? Any extreme outliers?
c. Construct a comparative boxplot, and use it as a basis for comparing and contrasting the two samples.

■ 4.4 Interpreting Center and Variability: Chebyshev's Rule, the Empirical Rule, and z Scores

The mean and standard deviation can be combined to obtain informative statements about how the values in a data set are distributed and about the relative position of a particular value in a data set. To do this, it is useful to be able to describe how far away a particular observation is from the mean in terms of the standard deviation. For example, we might say that an observation is 2 standard deviations above the mean or that an observation is 1.3 standard deviations below the mean.

■ Example 4.15 Standardized Test Scores

Consider a data set of scores on a standardized test with a mean and standard deviation of 100 and 15, respectively. We can make the following statements:

1. Because $100 - 15 = 85$, we say that a score of 85 is "1 standard deviation *below* the mean." Similarly, $100 + 15 = 115$ is "1 standard deviation *above* the mean" (see Figure 4.13).

2. Because 2 standard deviations is $2(15) = 30$, and $100 + 30 = 130$ and $100 - 30 = 70$, scores between 70 and 130 are those *within* 2 standard deviations of the mean (see Figure 4.14).

3. Because $100 + (3)(15) = 145$, scores above 145 exceed the mean by more than 3 standard deviations.

Figure 4.13 Values within 1 standard deviation of the mean (Example 4.15).

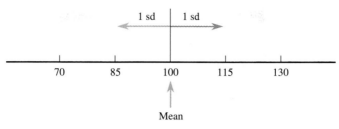

Figure 4.14 Values within 2 standard deviations of the mean (Example 4.15).

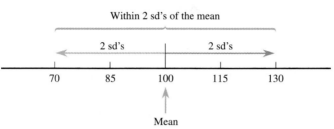

Sometimes in published articles, the mean and standard deviation are reported, but a graphical display of the data is not given. However, by using a result called Chebyshev's Rule, we can get a sense of the distribution of data values based on our knowledge of only the mean and standard deviation.

■ **Chebyshev's Rule**

Consider any number k, where $k \geq 1$. Then the percentage of observations that are within k standard deviations of the mean is at least $100\left(1 - \dfrac{1}{k^2}\right)\%$. Substituting selected values of k gives the following results:

Number of Standard Deviations, k	$1 - \dfrac{1}{k^2}$	Percentage Within k Standard Deviations of the Mean
2	$1 - \dfrac{1}{4} = .75$	at least 75%
3	$1 - \dfrac{1}{9} = .89$	at least 89%
4	$1 - \dfrac{1}{16} = .94$	at least 94%
4.472	$1 - \dfrac{1}{20} = .95$	at least 95%
5	$1 - \dfrac{1}{25} = .96$	at least 96%
10	$1 - \dfrac{1}{100} = .99$	at least 99%

■ Example 4.16 Child Care for Preschool Kids

The article "Piecing Together Child Care with Multiple Arrangements: Crazy Quilt or Preferred Pattern for Employed Parents of Preschool Children?" (*Journal of Marriage and the Family* [1994]: 669–680) examined various modes of care for preschool children. For a sample of families with one such child, it was reported that the mean and standard deviation of child care time per week were approximately 36 hr and 12 hr, respectively. Figure 4.15 displays values that are 1, 2, and 3 standard deviations from the mean.

Figure 4.15 Measurement scale for child care time (Example 4.16).

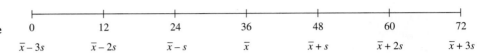

0	12	24	36	48	60	72
$\bar{x}-3s$	$\bar{x}-2s$	$\bar{x}-s$	$\bar{x}$	$\bar{x}+s$	$\bar{x}+2s$	$\bar{x}+3s$

Chebyshev's Rule allows us to assert the following:

1. At least 75% of the sample observations must be between 12 and 60 hours (within 2 standard deviations of the mean).

2. Because at least 89% of the observations must be between 0 and 72, at most 11% are outside this interval. Time cannot be negative, so we conclude that at most 11% of the observations exceed 72.

3. The values 18 and 54 are 1.5 standard deviations to either side of $\bar{x}$, so using $k = 1.5$ in Chebyshev's Rule implies that at least 55.6% of the observations must be between these two values. Thus, at most 44.4% of the observations are less than 18 — *not* at most 22.2%, because the distribution of values may not be symmetric.

Because Chebyshev's Rule is applicable to any data set (distribution), whether symmetric or skewed, we must be careful when making statements about the proportion above a particular value, below a particular value, or inside or outside an interval that is not centered at the mean. The rule must be used in a conservative fashion. There is another aspect of this conservatism. Whereas the rule states that at least 75% of the observations are within 2 standard deviations of the mean, in many data sets substantially more than 75% of the values satisfy this condition. The same sort of understatement is frequently encountered for other values of k (numbers of standard deviations).

■ Example 4.17 IQ Scores

Figure 4.16 gives a stem-and-leaf display of IQ scores of 112 children in one of the early studies that used the Stanford revision of the Binet–Simon intelligence scale (*The Intelligence of School Children*, L. M. Terman [Boston: Houghton-Mifflin, 1919]).

Summary quantities include

$$\bar{x} = 104.5 \qquad s = 16.3 \qquad 2s = 32.6 \qquad 3s = 48.9$$

Figure 4.16 Stem-and-leaf display of IQ scores used in Example 4.17.

```
 6 | 1
 7 | 25679
 8 | 0000124555668
 9 | 0000112333446666778889
10 | 00011222223333566677778899999
11 | 00001122333344444477899
12 | 01111123445669
13 | 006
14 | 26                    Stem: Tens
15 | 2                     Leaf: Ones
```

Table 4.4 shows how Chebyshev's Rule considerably understates actual percentages.

Table 4.4 ■ Summarizing the Distribution of IQ Scores

k = Number of sd's	$\bar{x} \pm ks$	Chebyshev	Actual
2	71.9 to 137.1	At least 75%	96% (108)
2.5	63.7 to 145.3	At least 84%	97% (109)
3	55.6 to 153.4	At least 89%	100% (112)

■ The Empirical Rule

The fact that statements based on Chebyshev's Rule are frequently conservative suggests that we should look for rules that are less conservative and more precise. The most useful such rule is the **Empirical Rule**, which can be applied whenever the distribution of data values can be reasonably well described by a normal curve. The word *empirical* means deriving from practical experience, and practical experience has shown that a normal curve gives a reasonable fit to many data sets.

> **■ The Empirical Rule**
>
> If the histogram of values in a data set can be reasonably well approximated by a normal curve, then
>
> Approximately 68% of the observations are within 1 standard deviation of the mean.
>
> Approximately 95% of the observations are within 2 standard deviations of the mean.
>
> Approximately 99.7% of the observations are within 3 standard deviations of the mean.

The Empirical Rule makes "approximately" instead of "at least" statements, and the percentages for $k = 1$, 2, and 3 standard deviations are much higher than those allowed by Chebyshev's Rule. Figure 4.17 illustrates the percentages given by the Empirical Rule. In contrast to Chebyshev's Rule, dividing the percentages in half is permissible, because a normal curve is symmetric.

Figure 4.17
Approximate percentages implied by the Empirical Rule.

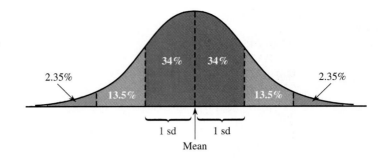

Example 4.18   Heights of Mothers and the Empirical Rule

One of the earliest articles to argue for the wide applicability of the normal distribution was "On the Laws of Inheritance in Man. I. Inheritance of Physical Characters" (*Biometrika* [1903]: 375–462). Among the data sets discussed in the article was one consisting of 1052 measurements of the heights of mothers. The mean and standard deviation were

$$\bar{x} = 62.484 \text{ in.} \quad \text{and} \quad s = 2.390 \text{ in.}$$

A normal curve did provide a good fit to the data. Table 4.5 contrasts actual percentages with those obtained from Chebyshev's Rule and the Empirical Rule. Clearly, the Empirical Rule is much more successful and informative in this case than Chebyshev's Rule.

Table 4.5 ▪ Summarizing the Distribution of Mothers' Heights

Number of sd's	Interval	Actual	Empirical Rule	Chebyshev Rule
1	60.094 to 64.874	72.1%	68%	At least 0%
2	57.704 to 67.264	96.2%	95%	At least 75%
3	55.314 to 69.654	99.2%	99.7%	At least 89%

Our detailed study of the normal distribution and areas under normal curves in Chapter 7 will enable us to make statements analogous to those of the Empirical Rule for values other than $k = 1, 2,$ or 3 standard deviations. For now, note that it is unusual to see an observation from a normally distributed population that is farther than 2 standard deviations from the mean (only 5%), and it is very surprising to see one that is more than 3 standard deviations away. If you encountered a mother whose height was 72 in., you might reasonably conclude that she was not part of the population described by the data set in Example 4.18.

▪ Measures of Relative Standing

When you obtain your score after taking an achievement test, you probably want to know how it compares to the scores of others who have taken the test. Is your score above or below the mean, and by how much? Does your score place you among the top 5% of those who took the test or only among the top 25%? Questions of this sort are answered by finding ways to measure the position of a particular value in a data set relative to all values in the set. One such measure involves calculating a *z score*.

▪ **Definition**

The ***z score*** corresponding to a particular value is

$$z\ score = \frac{value\ -\ mean}{standard\ deviation}$$

The *z* score tells us how many standard deviations the value is from the mean. It is positive or negative according to whether the value lies above or below the mean.

The process of subtracting the mean and then dividing by the standard deviation is sometimes referred to as *standardization*, and a *z* score is one example of what is called a *standardized score*.

▪ **Example 4.19   Relatively Speaking, Which Is the Better Offer?**

Suppose that two graduating seniors, one a marketing major and one an accounting major, are comparing job offers. The accounting major has an offer for $45,000 per year, and the marketing student has an offer for $43,000 per year. Summary information about the distribution of offers follows:

Accounting: mean = 46,000, standard deviation = 1500

Marketing: mean = 42,500, standard deviation = 1000

Then,

$$accounting\ z\ score = \frac{45,000\ -\ 46,000}{1500} = -0.67$$

(so $45,000 is 0.67 standard deviation below the mean), whereas

$$marketing\ z\ score = \frac{43,000\ -\ 42,500}{1000} = 0.5$$

Relative to the appropriate data sets, the marketing offer is actually more attractive than the accounting offer (although this may not offer much solace to the marketing major).

The *z* score is particularly useful when the distribution of observations is approximately normal. In this case, from the Empirical Rule, a *z* score outside the interval from -2 to $+2$ occurs in about 5% of all cases, whereas a *z* score outside the interval from -3 to $+3$ occurs only about 0.3% of the time.

A particular observation can be located even more precisely by giving the percentage of the data that fall at or below that observation. If, for example, 95% of all test scores are at or below 650, whereas only 5% are above 650, then 650 is called the *95th percentile* of the data set (or of the distribution of scores). Similarly, if 10% of all scores are at or below 400 and 90% are above 400, then the value 400 is the 10th percentile.

▪ **Definition**

For any particular number *r* between 0 and 100, the *r*th percentile is a value such that *r*% of the observations in the data set fall at or below that value.

Figure 4.18 illustrates the 90th percentile. We have already met several percentiles in disguise. The median is the 50th percentile, and the lower and upper quartiles are the 25th and 75th percentiles, respectively.

Figure 4.18 Ninetieth percentile from a smoothed histogram.

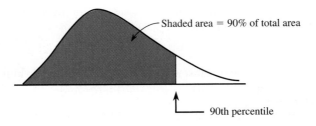

Shaded area = 90% of total area

90th percentile

▪ **Example 4.20   Variable Cost as a Percentage of Operating Revenue**

Variable cost as a percentage of operating revenue is an important indicator of a company's financial health. The article "A Credit Limit Decision Model for Inventory Floor Planning and Other Extended Trade Credit Arrangements" (*Decision Sciences* [1992]: 200–220) reported the following summary data for a sample of 350 corporations:

sample mean = 90.1%   sample median = 91.2%
5th percentile = 78.3%   95th percentile = 98.6%

Thus, half of the sampled firms had variable costs that were less than 91.2% of operating revenues because 91.2 is the 50th percentile. If the distribution of values was perfectly symmetric, the mean and the median would be identical. Further evidence against symmetry is provided by the two other given percentiles. The 5th percentile is 12.9 percentage points below the median, whereas the 95th percentile is only 7.4 percentage points above the median. The histogram given in the article had roughly the shape of the negatively skewed smoothed histogram in Figure 4.19 (the total area under the curve is 1).

Figure 4.19 The distribution of variable costs from Example 4.20.

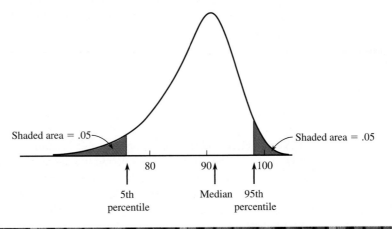

Shaded area = .05   Shaded area = .05

80   90   100

5th   Median   95th
percentile   percentile

▪ Exercises 4.38–4.52

4.38 The average playing time of compact discs in a large collection is 35 min, and the standard deviation is 5 min.
a. What value is 1 standard deviation above the mean? 1 standard deviation below the mean? What values are 2 standard deviations away from the mean?
b. Without assuming anything about the distribution of times, at least what percentage of the times are between 25 and 45 min?
c. Without assuming anything about the distribution of times, what can be said about the percentage of times that are either less than 20 min or greater than 50 min?
d. Assuming that the distribution of times is normal, approximately what percentage of times are between 25 and 45 min? less than 20 min or greater than 50 min? less than 20 min?

4.39 In a study investigating the effect of car speed on accident severity, 5000 reports of fatal automobile accidents were examined, and the vehicle speed at impact was recorded for each one. It was determined that the average speed was 42 mph and that the standard deviation was 15 mph. In addition, a histogram revealed that vehicle speed at impact could be described by a normal curve.
a. Roughly what proportion of vehicle speeds were between 27 and 57 mph?
b. Roughly what proportion of vehicle speeds exceeded 57 mph?

4.40 The U.S. Census Bureau (2000 census) reported the following relative frequency distribution for travel time to work for a large sample of adults who did not work at home:

Travel Time (minutes)	Relative Frequency
0 to <5	.04
5 to <10	.13
10 to <15	.16
15 to <20	.17
20 to <25	.14
25 to <30	.05
30 to <35	.12
35 to <40	.03
40 to <45	.03
45 to <60	.06
60 to <90	.05
90 or more	.02

a. Draw the histogram for the travel time distribution. In constructing the histogram, assume that the last interval in the relative frequency distribution (90 or more) ends at 200; so the last interval is 90 to <200. Be sure to use the density scale to determine the heights of the bars in the histogram because not all the intervals have the same width.
b. Describe the interesting features of the histogram from Part (a), including center, shape, and spread.
c. Based on the histogram from Part (a), would it be appropriate to use the Empirical Rule to make statements about the travel time distribution? Explain why or why not.
d. The approximate mean and standard deviation for the travel time distribution are 27 min and 24 min, respectively. Based on this mean and standard deviation and the fact that travel time cannot be negative, explain why the travel time distribution could not be well approximated by a normal curve.
e. Use the mean and standard deviation given in Part (d) and Chebyshev's Rule to make a statement about
i. the percentage of travel times that were between 0 and 75 min
ii. the percentage of travel times that were between 0 and 47 min
f. How well do the statements in Part (e) based on Chebyshev's Rule agree with the actual percentages for the travel time distribution? (Hint: You can estimate the actual percentages from the given relative frequency distribution.)

4.41 Mobile homes are tightly constructed for energy conservation. This can lead to a buildup of indoor pollutants. The paper "A Survey of Nitrogen Dioxide Levels Inside Mobile Homes" (*Journal of the Air Pollution Control Association* [1988]: 647–651) discussed various aspects of NO_2 concentration in these structures.
a. In one sample of mobile homes in the Los Angeles area, the mean NO_2 concentration in kitchens during the summer was 36.92 ppb, and the standard deviation was 11.34. Making no assumptions about the shape of the histogram, what can be said about the percentage of observations between 14.24 and 59.60?
b. Inside what interval is it guaranteed that at least 89% of the concentration observations will lie?
c. In a sample of non–Los Angeles mobile homes, the average kitchen NO_2 concentration during the winter was 24.76 ppb, and the standard deviation was 17.20. Do these values suggest that the histogram of

sample observations did not closely resemble a normal curve? (Hint: What is $\bar{x} - 2s$?)

4.42 The article "Taxable Wealth and Alcoholic Beverage Consumption in the United States" (*Psychological Reports* [1994]: 813–814) reported that the mean annual adult consumption of wine was 3.15 gal and that the standard deviation was 6.09 gal. Would you use the Empirical Rule to approximate the proportion of adults who consume more than 9.24 gal (i.e., the proportion of adults whose consumption value exceeds the mean by more than 1 standard deviation)? Explain your reasoning.

4.43 A student took two national aptitude tests in the course of applying for admission to colleges. The national average and standard deviation were 475 and 100, respectively, for the first test and 30 and 8, respectively, for the second test. The student scored 625 on the first test and 45 on the second test. Use z scores to determine on which exam the student performed better.

4.44 Suppose that your younger sister is applying for entrance to college and has taken the SATs. She scored at the 83d percentile on the verbal section of the test and at the 94th percentile on the math section of the test. Because you have been studying statistics, she asks you for an interpretation of these values. What would you tell her?

4.45 A sample of concrete specimens of a certain type is selected, and the compressive strength of each specimen is determined. The mean and standard deviation are calculated as $\bar{x} = 3000$ and $s = 500$, and the sample histogram is found to be well approximated by a normal curve.

a. Approximately what percentage of the sample observations are between 2500 and 3500, and what result justifies your assertion?

b. Approximately what percentage of sample observations are outside the interval from 2000 to 4000?

c. What can be said about the approximate percentage of observations between 2000 and 2500?

d. Why would you not use Chebyshev's Rule to answer the questions posed in Parts (a)–(c)?

4.46 The paper "Modeling and Measurements of Bus Service Reliability" (*Transportation Research* [1978]: 253–256) studied various aspects of bus service and presented data on travel times from several different routes. We give here a frequency distribution for bus travel times from origin to destination on one particular route in Chicago during peak morning traffic periods:

Class	Frequency	Relative Frequency
15 to <16	4	.02
16 to <17	0	.00
17 to <18	26	.13
18 to <19	99	.49
19 to <20	36	.18
20 to <21	8	.04
21 to <22	12	.06
22 to <23	0	.00
23 to <24	0	.00
24 to <25	0	.00
25 to <26	16	.08

a. Construct the corresponding histogram.
b. Compute (approximately) the following percentiles:
i. 86th   ii. 15th   iii. 90th   iv. 95th   v. 10th

4.47 An advertisement for the "30-in. Wonder" that appeared in the September 1983 issue of the journal *Packaging* claimed that the 30-in. Wonder weighs cases and bags up to 110 lb and provides accuracy to within 0.25 oz. Suppose that a 50-oz weight was repeatedly weighed on this scale and the weight readings recorded. The mean value was 49.5 oz, and the standard deviation was 0.1. What can be said about the proportion of the time that the scale actually showed a weight that was within 0.25 oz of the true value of 50 oz? (Hint: Try to make use of Chebyshev's Rule.)

4.48 Suppose that your statistics professor returned your first midterm exam with only a z score written on it. She also told you that a histogram of the scores was closely described by a normal curve. How would you interpret each of the following z scores?
a. 2.2   b. −0.4   c. −1.8   d. 1.0   e. 0

4.49 The paper "Answer Changing on Multiple-Choice Tests" (*Journal of Experimental Education* [1980]: 18–21) reported that for a group of 162 college students, the average number of responses changed from the correct answer to an incorrect answer on a test containing 80 multiple-choice items was 1.4. The corresponding standard deviation was reported to be 1.5. Based on this mean and standard deviation, what can you tell about the shape of the distribution of the variable *number of answers changed from right to wrong*? What can you say about the number of students who changed at least six answers from correct to incorrect?

4.50 The average reading speed of students completing a speed-reading course is 450 words per minute (wpm). If the standard deviation is 70 wpm, find the z score associated with each of the following reading speeds.

a. 320 wpm **b.** 475 wpm **c.** 420 wpm **d.** 610 wpm

4.51 The following data values are 1989 per capita expenditures on public libraries for each of the 50 states (New York is the highest and Arkansas is the lowest):

```
29.48  24.45  23.64  23.34  22.10  21.16  19.83
18.01  17.95  17.23  16.53  16.29  15.89  15.85
13.64  13.37  13.16  13.09  12.66  12.37  11.93
10.99  10.55  10.24  10.06   9.84   9.65   8.94
 7.70   7.56   7.46   7.04   6.58   5.98  19.81
19.25  19.18  18.62  14.74  14.53  14.46  13.83
11.85  11.71  11.53  11.34   8.72   8.22   8.13
 8.01
```

a. Summarize this data set with a frequency distribution. Construct the corresponding histogram.

b. Use the histogram in Part (a) to find approximate values of the following percentiles:
i. 50th   **ii.** 70th   **iii.** 10th   **iv.** 90th   **v.** 40th

4.52 The accompanying table gives the mean and standard deviation of reaction times (in seconds) for each of two different stimuli:

	Stimulus 1	Stimulus 2
Mean	6.0	3.6
Standard deviation	1.2	0.8

If your reaction time is 4.2 sec for the first stimulus and 1.8 sec for the second stimulus, to which stimulus are you reacting (compared to all other individuals) relatively more quickly?

▪ 4.5 Communicating and Interpreting the Results of Statistical Analyses

As was the case with the graphical displays of Chapter 3, the primary function of the descriptive tools introduced in this chapter is to help us better understand the variables under study. If we have collected data on the amount of money students spend on textbooks at a particular university, most likely we did so because we wanted to learn about the distribution of this variable (amount spent on textbooks) for the population of interest (in this case, students at the university). Numerical measures of center and spread and boxplots help to enlighten us, and they also allow us to communicate to others what we have learned from the data.

▪ Communicating the Results of Statistical Analyses

When reporting the results of a data analysis, it is common to start with descriptive information about the variables of interest. It is always a good idea to start with a graphical display of the data, and, as we saw in Chapter 3, graphical displays of numerical data are usually described in terms of center, variability, and shape. The numerical measures of this chapter can help you to be more specific in describing the center and spread of a data set.

When describing center and spread, you must first decide which measures to use. Common choices are to use either the sample mean and standard deviation or the sample median and interquartile range (and maybe even a boxplot) to describe center and spread. Because the mean and standard deviation can be sensitive to extreme values in the data set, they are best used when the distribution shape is approximately symmetric and when there are few outliers. If the data set is noticeably skewed or if there are outliers, then the observations are more spread out in one part of the distribution than in the others. In this situation, a five-number summary or a boxplot conveys more information than the mean and standard deviation do.

▪ Interpreting the Results of Statistical Analyses

It is relatively rare to find raw data in published sources. Typically, only a few numerical summary quantities are reported. We must be able to interpret these values and understand what they tell us about the underlying data set.

For example, a university recently conducted an investigation of the amount of time required to enter the information contained in an application for admission into the university computer system. One of the individuals who performs this task was asked to note starting time and completion time for 50 randomly selected application forms. The resulting entry times (in minutes) were summarized using the mean, median, and standard deviation:

$\bar{x} = 7.854$

median $= 7.423$

$s = 2.129$

What do these summary values tell us about entry times? The average time required to enter admissions data was 7.854 min, but the relatively large standard deviation suggests that many observations differ substantially from this mean. The median tells us that half of the applications required less than 7.423 min to enter. The fact that the mean exceeds the median suggests that some unusually large values in the data set affected the value of the mean. This last conjecture is confirmed by the stem-and-leaf display of the data given in Figure 4.20.

Figure 4.20 Stem-and-leaf display of data entry times.

```
 4 8
 5 0234579
 6 00001234566779
 7 223556688
 8 23334
 9 002
10 011168
11 134
12 2              Stem: Ones
13               Leaf: Tenths
14 3
```

The administrators conducting the data-entry study looked at the outlier 14.3 min and at the other relatively large values in the data set; they found that the five largest values came from applications that were entered before lunch. After talking with the individual who entered the data, the administrators speculated that morning entry times might differ from afternoon entry times because there tended to be more distractions and interruptions (phone calls, etc.) during the morning hours, when the admissions office generally was busier. When morning and afternoon entry times were separated, the following summary statistics resulted:

Morning (based on $n = 20$ applications): $\bar{x} = 9.093$,   median $= 8.7$, $s = 2.329$

Afternoon (based on $n = 30$ applications): $\bar{x} = 7.027$,   median $= 6.737$, $s = 1.529$

Clearly, the average entry time is higher for applications entered in the morning; also, the individual entry times differ more from one another in the mornings than in the afternoons (because the standard deviation for morning entry times, 2.329, is about 1.5 times as large as 1.529, the standard deviation for afternoon entry times).

▪ What to Look for in Published Data

Here are a few questions to ask yourself when you interpret numerical summary measures.

- ▪ Is the chosen summary measure appropriate for the type of data collected? In particular, watch for inappropriate use of the mean and standard deviation with categorical data that has simply been coded numerically.

- ▪ If both the mean and the median are reported, how do the two values compare? What does this suggest about the distribution of values in the data set? If only the mean or the median was used, was the appropriate measure selected?

- ▪ Is the standard deviation large or small? Is the value consistent with your expectations regarding variability? What does the value of the standard deviation tell you about the variable being summarized?

- ▪ Can anything of interest be said about the values in the data set by applying Chebyshev's Rule or the Empirical Rule?

For example, consider a study on smoking and lactation. The journal article "Smoking During Pregnancy and Lactation and Its Effects on Breast Milk Volume" (*American Journal of Clinical Nutrition* [1991]: 1011–1016) reported the following summary values for data collected on daily breast milk volume (in grams) for two groups of nursing mothers:

Group	Mean	Standard Deviation	Median
Smokers	693	110	589.5
Nonsmokers	961	120	946

Because milk volume is a numerical variable, these descriptive measures reasonably summarize the data set. For nonsmoking mothers, the mean and the median are similar, indicating that the milk volume distribution is approximately symmetric. The average milk volume for smoking mothers is 693, whereas for nonsmoking mothers the mean milk volume is 961 — quite a bit higher than for the smoking mothers. The standard deviations for the two groups are similar, suggesting that, although the average milk volume may be lower for smokers, the variability in mothers' milk volume is about the same. Because the median for the smoking group is smaller than the corresponding mean, we suspect that the distribution of milk volume for smoking mothers is positively skewed.

▪ A Word to the Wise: Cautions and Limitations

When computing or interpreting numerical descriptive measures, you need to keep in mind the following:

1. **Measures of center don't tell all.** Although measures of center, such as the mean and the median, do give us a sense of what might be considered a typical value for a variable, this is only one characteristic of a data set. Without additional information about variability and distribution shape, we don't really know much about the behavior of the variable.

2. Data distributions with different shapes can have the same mean and standard deviation. For example, consider the following two histograms:

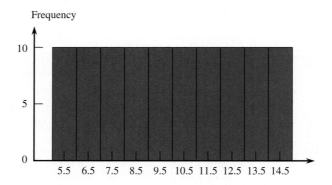

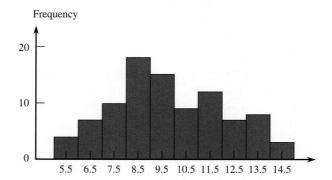

Both histograms summarize data sets that have a mean of 10 and a standard deviation of 2, yet they have different shapes.

3. Both the mean and the standard deviation are sensitive to extreme values in a data set, especially if the sample size is small. If a data distribution is skewed or if the data set has outliers, the median and the interquartile range may be a better choice for describing center and spread.

4. Measures of center and variability describe the values of the variable studied, not the frequencies in a frequency distribution or the heights of the bars in a histogram. For example, consider the following two frequency distributions and histograms:

Frequency Distribution A		Frequency Distribution B	
Value	**Frequency**	**Value**	**Frequency**
1	10	1	5
2	10	2	10
3	10	3	20
4	10	4	10
5	10	5	5

There is more variability in the data summarized by Frequency Distribution and Histogram A than in the data summarized by Frequency Distribution

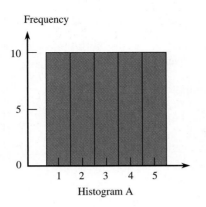

Histogram A

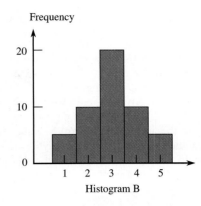

Histogram B

and Histogram B. This is because the values of the variable described by Histogram and Frequency Distribution B are more concentrated near the mean than are the values for the variable described by Histogram and Frequency Distribution A. Don't be misled by the fact that there is no variability in the frequencies in Frequency Distribution A or the heights of the bars in Histogram A.

5. Be careful with boxplots based on small sample sizes. Boxplots convey information about center, variability, and shape, but when the sample size is small, you should be hesitant to overinterpret shape information. It is really not possible to decide whether a data distribution is symmetric or skewed if only a small sample of observations from the distribution is available.

6. Not all distributions are normal (or even approximately normal). Be cautious in applying the Empirical Rule in situations where you are not convinced that the data distribution is at least approximately normal. Using the Empirical Rule in such situations can lead to incorrect statements.

7. Watch out for outliers! Unusual observations in a data set often provide important information about the variable under study, so it is important to consider outliers in addition to describing what is typical. Outliers can also be problematic — both because the values of some descriptive measures are influenced by outliers and because some of the methods for drawing conclusions from data may not be appropriate if the data set has outliers.

▪ Activity 4.1: Collecting and Summarizing Numerical Data

In this activity, you will work in groups to collect data that will provide information about how many hours per week, on average, students at your school spend engaged in a particular activity. You will use the sampling plan designed in Activity 2.1 to collect the data.

1. With your group, pick one of the following activities to be the focus of your study:
   i. Surfing the web
   ii. Studying or doing homework
   iii. Watching TV
   iv. Exercising
   v. Sleeping

or you may choose a different activity, *subject to the approval of your instructor*.

2. Use the plan developed in Activity 2.1 to collect data on the variable you have chosen for your study.

3. Summarize the resulting data using both numerical and graphical summaries. Be sure to address both center and variability.

4. Write a short article for your school paper summarizing your findings regarding student behavior. Your article should include both numerical and graphical summaries.

▪ Activity 4.2: Boxplot Shapes

In this activity, you will investigate the relationship between boxplot shapes and the corresponding five-number summary.

The accompanying figure shows four boxplots, labeled A–D. Also given are four five-number summaries, labeled I–IV. Match each five-number summary to the appropriate boxplot. Note that scales are not included on the boxplots, so you will have to think about what the five-number summary implies about characteristics of the boxplot.

Five-Number Summaries

	I	II	III	IV
Minimum	4.3	40.0	0.3	4.9
Lower quartile	5.2	46.6	0.7	21.0
Median	9.3	51.3	1.5	27.5
Upper quartile	11.9	59.2	2.4	52.4
Maximum	14.9	75.7	6.8	83.0

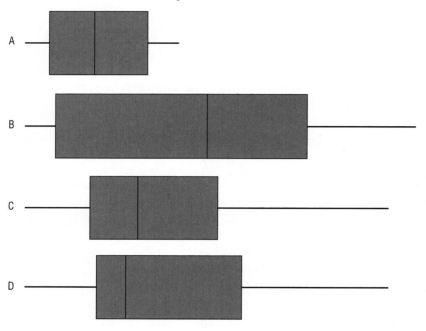

▪ Summary of Key Concepts and Formulas

Term or Formula	Comment
$x_1, x_2, \ldots, x_n$	Notation for sample data consisting of observations on a variable x, where n is the sample size.
Sample mean, $\bar{x}$	The most frequently used measure of center of a sample. It can be sensitive to the presence of even a single outlier (an unusually large or small observation).
Population mean, μ	The average x value in an entire population.
Sample median	The middle value in an ordered list of sample observations. (If n is even, the median is the average of the two middle values.) The median is insensitive to outliers.
Trimmed mean	A measure of center in which the observations are first ordered from smallest to largest, then one or more observations are deleted from each end, and finally the remaining observations are averaged. In terms of sensitivity to outliers, the trimmed mean is a compromise between the mean and the median.

Term or Formula	Comment
Deviations from the mean: $x_1 - \bar{x}, x_2 - \bar{x}, \ldots, x_n - \bar{x}$	Quantities used to assess variability in a sample. Except for rounding effects, $\sum (x - \bar{x}) = 0$.
The sample variance $$s^2 = \frac{\sum (x - \bar{x})^2}{n - 1}$$ and standard deviation $s = \sqrt{s^2}$	The most frequently used measures of variability for sample data.
$$S_{xx} = \sum x^2 - \frac{(\sum x)^2}{n}$$	The computing formula for S_{xx}, the numerator of s^2.
The population variance σ^2 and standard deviation σ	Measures of variability for an entire population.
Quartiles and the interquartile range	The lower quartile separates the smallest 25% of the data from the remaining 75%, and the upper quartile separates the largest 25% from the smallest 75%. The interquartile range (iqr), a measure of variability less sensitive to outliers than s, is the difference between the upper and the lower quartiles.
Chebyshev's Rule	For any number $k \geq 1$, *at least* $100 \left(1 - \frac{1}{k^2} \right) \%$ of the observations in *any* data set are within k standard deviations of the mean. The percentage value is typically conservative in that the actual percentages often considerably exceed the stated lower bound.
Empirical Rule	This rule gives the approximate percentage of observations within 1 standard deviation (68%), 2 standard deviations (95%), and 3 standard deviations (99.7%) of the mean when the histogram is well approximated by a normal curve.
z score	The distance between an observation and the mean expressed as a certain number of standard deviations. The z score is positive (negative) if the observation lies above (below) the mean.
rth percentile	The value such that $r\%$ of the observations in the data set fall at or below that value.
Five-number summary	A summary of a data set that includes the minimum, the lower quartile, the median, the upper quartile, and the maximum.
Boxplot	A picture that conveys information about the most important features of a data set: center, spread, extent of skewness, and presence of outliers.

■ Supplementary Exercises 4.53–4.68

4.53 Cardiac output and maximal oxygen capacity typically decrease with age in sedentary individuals, but these decreases are at least partially arrested in middle-aged individuals who engage in a substantial amount of physical exercise. To understand better the effects of exercise and aging on various circulatory functions, the article "Cardiac Output in Male Middle-Aged Runners" (*Journal of Sports Medicine*

[1982]: 17–22) presented data from a study of 21 middle-aged male runners. The following data set gives values of oxygen capacity values (in milliliters per kilogram per minute) while the participants pedaled at a specified rate on a bicycle ergometer:

12.81 14.95 15.83 15.97 17.90 18.27 18.34
19.82 19.94 20.62 20.88 20.93 20.98 20.99
21.15 22.16 22.24 23.16 23.56 35.78 36.73

a. Compute the median and the quartiles for this data set.

b. What is the value of the interquartile range? Are there outliers in this data set?

c. Draw a modified boxplot, and comment on the interesting features of the plot.

4.54 The risk of developing iron deficiency is especially high during pregnancy. Detecting such a deficiency is complicated by the fact that some methods for determining iron status can be affected by the state of pregnancy itself. Consider the following data on transferrin receptor concentration for a sample of women with laboratory evidence of overt iron-deficiency anemia ("Serum Transferrin Receptor for the Detection of Iron Deficiency in Pregnancy," *American Journal of Clinical Nutrition* [1991]: 1077–1081):

15.2 9.3 7.6 11.9 10.4 9.7 20.4 9.4 11.5
16.2 9.4 8.3

Compute the values of the sample mean and median. Why are these values different here? Which one do you regard as more representative of the sample, and why?

4.55 The paper "The Pedaling Technique of Elite Endurance Cyclists" (*International Journal of Sport Biomechanics* [1991]: 29–53) reported the following data on single-leg power at a high workload:

244 191 160 187 180 176 174 205 211
183 211 180 194 200

a. Calculate and interpret the sample mean and median.

b. Suppose that the first observation had been 204, not 244. How would the mean and median change?

c. Calculate a trimmed mean by eliminating the smallest and the largest sample observations. What is the corresponding trimming percentage?

d. Suppose that the largest observation had been 204 rather than 244. How would the trimmed mean in Part (c) change? What if the largest value had been 284?

4.56 The paper cited in Exercise 4.55 also reported values of single-leg power for a low workload. The sample mean for $n = 13$ observations was $\bar{x} = 119.8$ (actually 119.7692), and the 14th observation, some-

what of an outlier, was 159. What is the value of $\bar{x}$ for the entire sample?

4.57 The paper "Sodium–Calcium Exchange Equilibria in Soils as Affected by Calcium Carbonate and Organic Matter" (*Soil Science* [1984]: 109) gave 10 observations on soil pH. The following data resulted from analysis of 10 samples of soil from the Central Soil Salinity Research experimental farm:

8.53 8.52 8.01 7.99 7.93 7.89 7.85 7.82
7.80 7.72

a. Compute the upper quartile, lower quartile, and interquartile range.

b. Construct a skeletal boxplot, and comment on the interesting features of the plot.

4.58 The amount of aluminum contamination (in parts per million) in plastic of a certain type was determined for a sample of 26 plastic specimens, resulting in the following data ("The Log Normal Distribution for Modeling Quality Data When the Mean Is Near Zero," *Journal of Quality Technology* [1990]: 105–110):

30   30   60   63   70   79   87   90   101
102 115 118 119 119 120 125 140 145
172 182 183 191 222 244 291 511

Construct a boxplot that shows outliers, and comment on the interesting features of this plot.

4.59 Reconsider the radon concentration data given in the chapter introduction.

a. Construct a comparative stem-and-leaf display with 210 shown separately as a high value, and comment on similarities and differences.

b. $\Sigma x = 958$ for the cancer sample and $\Sigma x = 747$ for the no-cancer sample. Calculate the two sample means and medians, and comment.

c. $S_{xx} = 41,084.476$ for the cancer sample and $S_{xx} = 10,969.077$ for the no-cancer sample. What do the values of the two sample standard deviations suggest about variability in the two samples?

d. Determine the value of the interquartile range for each sample. What do these values tell you about variability in the two samples? Is the conclusion here different from the conclusion in Part (c)? Explain. (Hint: Without 210 and 85, cancer $s = 11.4$ and no-cancer $s = 13.3$.)

e. Construct a comparative boxplot.

4.60 The article "Can We Really Walk Straight?" (*American Journal of Physical Anthropology* [1992]: 19–27) reported on an experiment in which each of 20 healthy men was asked to walk as straight as possible to a target 60 m away at normal speed. Consider the following data on cadence (number of strides per second):

0.95	0.85	0.92	0.95	0.93	0.86	1.00	0.92
0.85	0.81	0.78	0.93	0.93	1.05	0.93	1.06
1.06	0.96	0.81	0.96

Use the methods developed in this chapter to summarize the data; include an interpretation or discussion wherever appropriate. (Note: The author of the paper used a rather sophisticated statistical analysis to conclude that people cannot walk in a straight line and suggested several explanations for this.)

4.61 The *San Luis Obispo Telegram-Tribune* (December 18, 1992) reported that for the 1992 season, the average major league baseball salary was just over $1 million, whereas only 267 of the 772 players made at least this much money. What does this imply about the median salary and the shape of a histogram of salaries?

4.62 The article "Comparing the Costs of Major Hotel Franchises" (*Real Estate Review* [1992]: 46–51) gave the following data on franchise cost as a percentage of total room revenue for chains of three different types:

Budget	2.7 2.8 3.8 3.8 4.0 4.1 5.5
	5.9 6.7 7.0 7.2 7.2 7.5 7.5
	7.7 7.9 7.9 8.1 8.2 8.5
Midrange	1.5 4.0 6.6 6.7 7.0 7.2 7.2
	7.4 7.8 8.0 8.1 8.3 8.6 9.0
First-class	1.8 5.8 6.0 6.6 6.6 6.6 7.1
	7.2 7.5 7.6 7.6 7.8 7.8 8.2
	9.6

Construct a boxplot for each type of hotel, and comment on interesting features, similarities, and differences.

4.63 In recent years, many teachers have been subject to increased levels of stress, contributing to disenchantment with teaching as a profession. The paper "Professional Burnout Among Public School Teachers" (*Public Personnel Management* [1988]: 167–189) looked at various aspects of this problem. Consider the following information on total psychological effects scores for a sample of 937 teachers:

Burnout Level	Range of Scores	Frequency
Low	10–21	554
Moderate	22–43	342
High	44–65	41

$\bar{x} = 19.93$, and $s = 12.89x$.

a. Draw a histogram.
b. Mark the value of $\bar{x}$ on the measurement scale, and use the histogram to locate the approximate value of the median. Why are the two measures of center not identical?

c. In what interval of values are we guaranteed to find at least 75% of the scores, irrespective of the histogram shape? From the histogram, roughly what percentage of the scores actually fall in this interval?

4.64 The amount of radiation received at a greenhouse plays an important role in determining the rate of photosynthesis. The following observations on incoming solar radiation were read from a graph in the paper "Radiation Components over Bare and Planted Soils in a Greenhouse" (*Solar Energy* [1990]: 1–6):

6.3	6.4	7.1	7.7	8.4	8.5	8.8	8.9
9.0	9.1	10.0	10.1	10.2	10.6	10.6	10.7
10.7	10.8	10.9	11.1	11.2	11.2	11.4	11.9
11.9	12.2	13.1

Use some of the methods discussed in this chapter and in Chapter 3 to describe these data.

4.65 The accompanying data on milk volume (in grams per day) was taken from the paper "Smoking During Pregnancy and Lactation and Its Effects on Breast Milk Volume" (*American Journal of Clinical Nutrition* [1991]: 1011–1016):

| Smoking mothers | 621 793 593 545 753 655 895 767 714 598 693 |
| Nonsmoking mothers | 947 945 1086 1202 973 981 930 745 903 899 961 |

Compare and contrast the two samples.

4.66 The *Los Angeles Times* (July 17, 1995) reported that in a sample of 364 lawsuits in which punitive damages were awarded, the sample median damage award was $50,000, and the sample mean was $775,000. What does this suggest about the distribution of values in the sample?

4.67 Age at diagnosis for each of 20 patients under treatment for meningitis was given in the paper "Penicillin in the Treatment of Meningitis" (*Journal of the American Medical Association* [1984]: 1870–1874). The ages (in years) were as follows:

| 18 | 18 | 25 | 19 | 23 | 20 | 69 | 18 | 21 | 18 | 20 | 18 |
| 18 | 20 | 18 | 19 | 28 | 17 | 18 | 18 |

a. Calculate the values of the sample mean and the standard deviation.
b. Calculate the 10% trimmed mean. How does the value of the trimmed mean compare to that of the sample mean? Which would you recommend as a measure of location? Explain.
c. Compute the upper quartile, the lower quartile, and the interquartile range.
d. Are there any mild or extreme outliers present in this data set?
e. Construct the boxplot for this data set.

4.68 Suppose that the distribution of scores on an exam is closely described by a normal curve with mean 100. The 16th percentile of this distribution is 80.
a. What is the 84th percentile?
b. What is the approximate value of the standard deviation of exam scores?
c. What z score is associated with an exam score of 90?
d. What percentile corresponds to an exam score of 140?
e. Do you think there were many scores below 40? Explain.

▪ References

See the references at the end of Chapter 3.

5 · Summarizing Bivariate Data

How can the cost-effectiveness of schools be assessed? People inside and outside the education community have been arguing over this issue for years, and the debate has become more heated as spending for education has come under increasing scrutiny. The article "Cost-Effectiveness in Public Education" (*Chance* [1995]: 38–41) considered data from various school districts in New Jersey. Part of these data consisted of observations on the two variables x = dollars spent per pupil and y = average SAT scores for 44 districts having only grade levels 7–12 or 9–12. Some representative observations (as read from a graph) are as follows:

x	7750	9900	10,870	12,080 . . .
y	878	893	966	950 . . .

In general, what can be said about the general nature of the relationship between expenditure per pupil and average SAT score? In particular, are higher scores associated with high expenditure levels? Can average SAT score for a district be predicted once expenditure per pupil for the district is known? How accurate are predictions of average SAT score based only on expenditure per pupil?

In this chapter, we introduce methods for describing relationships between two numerical variables and for assessing the strength of a relationship. These methods allow us to answer questions such as the ones just posed regarding the relationship between school expenditures and test scores. In Chapter 13, methods of statistical inference for drawing conclusions from data consisting of observations on two variables are developed. The techniques introduced in this chapter are also important stepping stones for analyzing data consisting of observations on three or more variables, the topic of Chapter 14.

▪ 5.1 Correlation

An investigator is often interested in how two or more attributes of individuals or objects in a population are related to one another. For example, an environmental researcher might wish to know how the lead content of soil varies with distance from a major highway. Researchers in early childhood education might investigate how vocabulary size is related to age. College admissions officers, who must try to predict whether an applicant will succeed in college, might use a model relating college grade point average to high school grades, ACT or SAT scores, and various personal and family characteristics.

Recall that a scatterplot of bivariate numerical data gives a visual impression of how strongly x values and y values are related. However, to make precise statements and draw conclusions from data, we must go beyond pictures. A **correlation coefficient** (from *co-* and *relation*) is a quantitative assessment of the strength of relationship between the x and y values in a set of (x, y) pairs. In this section, we introduce the most commonly used correlation coefficient.

Figure 5.1 displays several scatterplots that show different relationships between the x and y values. The plot in Figure 5.1(a) suggests a strong *positive relationship* between x and y; for every pair of points in the plot, the one with the larger x value also has the larger y value. That is, an increase in x is inevitably paired with an increase in y. The plot in Figure 5.1(b) shows a strong *tendency* for y to increase as x does, but there are a few exceptions. For example, the x and y values of the two points in the extreme upper right-hand corner of the plot go in opposite directions (x increases but y decreases). Nevertheless, a plot such as this would again indicate a rather strong positive relationship. Figure 5.1(c) suggests that x and y are *negatively related*—as x increases, y tends to decrease. Obviously, the negative relationship in this plot is not as strong as the positive relationship in Figure 5.1(b), although both plots show a well-defined linear pattern. The plot of Figure 5.1(d) indicates no strong relationship between x and y; there is no tendency for y either to increase or to decrease as x increases. Finally, as illustrated in Figure 5.1(e), a

Figure 5.1 Scatterplots illustrating various types of relationships: (a) positive linear relation; (b) another positive linear relation; (c) negative linear relation; (d) no relation; (e) curved relation.

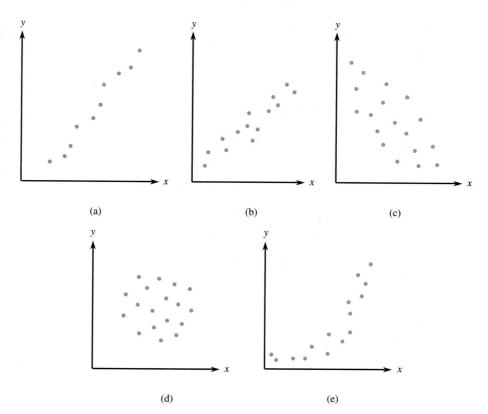

scatterplot can show evidence of a strong relationship through a pattern that is curved rather than linear.

■ Pearson's Sample Correlation Coefficient

Let $(x_1, y_1), (x_2, y_2), \ldots, (x_n, y_n)$ denote a sample of (x, y) pairs. Consider replacing each x value by the corresponding z score, z_x (by subtracting $\bar{x}$ and then dividing by s_x) and similarly replacing each y value by its z score. Note that x values that are larger than the mean x value have a positive z score and that x values that are smaller than $\bar{x}$ have negative z scores. The same is true for the y values: y values larger than $\bar{y}$ have positive z scores and y values smaller than $\bar{y}$ have negative z scores. A common measure of the strength of linear relationship, called Pearson's sample correlation coefficient, is based on $\sum z_x z_y$—that is, the sum of the product of z_x and z_y for each observation in the bivariate data set.

The scatterplot in Figure 5.2(a) indicates a substantial positive relationship. A vertical line through $\bar{x}$ and a horizontal line through $\bar{y}$ divide the plot into four

Figure 5.2 Viewing a scatterplot according to the signs of z_x and z_y: (a) a positive relation; (b) a negative relation; (c) no strong relation.

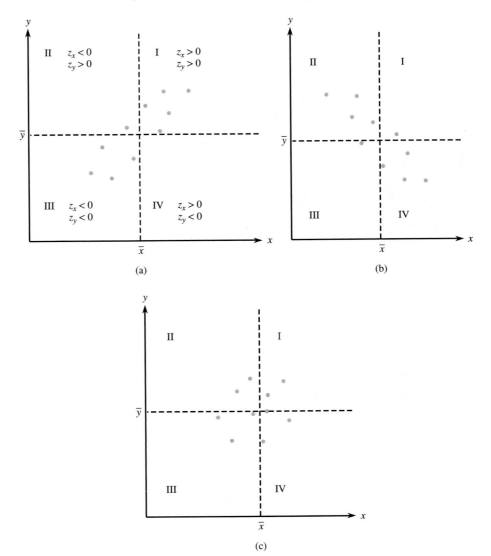

regions. In Region I, both x and y exceed their mean values, so the z score for x and the z score for y are both positive numbers. It follows that $z_x z_y$ is positive. The product of the z scores is also positive for any point in Region III, because both z scores are negative there and multiplying two negative numbers gives a positive number. In each of the other two regions, one z score is positive and the other is negative, so $z_x z_y$ is negative. But because the points generally fall in Regions I and III, the products of z scores tend to be positive. Thus, the *sum* of the products is a large positive number.

Similar reasoning for the data displayed in Figure 5.2(b), which exhibits a strong negative relationship, implies that $\sum z_x z_y$ is a large negative number. When there is no strong relationship, as in Figure 5.2(c), positive and negative products tend to counteract one another, producing a value of $\sum z_x z_y$ that is close to 0. In summary, $\sum z_x z_y$ seems to be a reasonable measure of the degree of association between x and y; it can be a large positive number, a large negative number, or a number close to 0, depending on whether there is a strong positive, a strong negative, or no strong linear relationship.

Pearson's sample correlation coefficient, denoted r, is obtained by dividing $\sum z_x z_y$ by $(n - 1)$.

■ **Definition**

Pearson's sample correlation coefficient r is given by

$$r = \frac{\sum z_x z_y}{n - 1}$$

Although there are several different correlation coefficients, Pearson's correlation coefficient is by far the most commonly used, and so the name *Pearson's* is often omitted and it is referred to as simply the correlation coefficient.

Hand calculation of the correlation coefficient is illustrated in Example 5.1; such a calculation can be quite tedious. Fortunately, all statistical software packages and most scientific calculators can compute r once the x and y values have been input.

■ **Example 5.1   Crisis Management and Family Strength**

The article "Parents' and Adolescents' Perceptions of a Strong Family" (*Psychological Reports* [1999]: 1219–1224) examined the relationship between family strength and other characteristics such as commitment, time spent together, and crisis management skills. Adolescents in what the researchers called intact families (families with both parents present in the home) rated their families with respect to the degree that the family exhibited effective crisis management and with respect to family strength. Data consistent with findings that appeared in the article are as follows:

Observation	Crisis management score	Family strength score
1	20	50
2	13	60
3	27	67
4	18	57
5	19	49
6	21	72
7	0	52
8	21	68
9	21	60
10	11	58

Figure 5.3 is a scatterplot of these data.

Figure 5.3 MINITAB scatterplot for the data of Example 5.1.

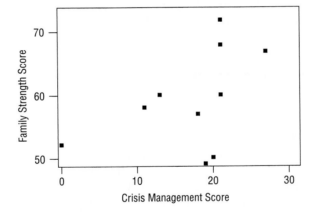

Let x denote the crisis management score and y denote the family strength score. It is easy to verify that

$$\bar{x} = 17.10 \qquad s_x = 7.48 \qquad \bar{y} = 59.30 \qquad s_y = 7.82$$

To illustrate the calculation of the correlation coefficient, we begin by computing z scores for each (x, y) pair in the data set. For example, the first observation is $(20, 50)$. The corresponding z scores are

$$z_x = \frac{20 - 17.10}{7.48} = 0.39$$

$$z_y = \frac{50 - 59.30}{7.82} = -1.19$$

The following table shows the z scores and the product $z_x z_y$ for each observation:

x	y	z_x	z_y	$z_x z_y$
20	50	0.39	−1.19	−0.46
13	60	−0.55	0.09	−0.05
27	67	1.32	0.98	1.29
18	57	0.12	−0.29	−0.03
19	49	0.25	−1.32	−0.33
21	72	0.52	1.62	0.84
0	52	−2.29	−0.93	2.13
21	68	0.52	1.11	0.58
21	60	0.52	0.09	0.05
11	58	−0.82	−0.17	0.14
				$4.16 = \Sigma z_x z_y$

Then, with $n = 10$,

$$r = \frac{\Sigma z_x z_y}{n - 1} = \frac{4.16}{9} = 0.46$$

Based on the scatterplot and the properties of the correlation coefficient presented in the discussion that follows this example, we conclude that there is only a weak positive linear relationship between crisis management score and family strength score.

▪ Properties of r

1. *The value of r does not depend on the unit of measurement for either variable.*
   For example, if x is height, the corresponding z score is the same whether height is expressed in inches, meters, or miles, and thus the value of the correlation coefficient is not affected. The correlation coefficient measures the inherent strength of the linear relationship between two numerical variables.

2. *The value of r does not depend on which of the two variables is considered x.*
   Thus, if we had let x = family strength score and y = crisis management score in Example 5.1, the same value, $r = 0.46$, would have resulted.

3. *The value of r is between −1 and +1.* A value near the upper limit, +1, indicates a substantial positive relationship, whereas an r close to the lower limit, −1, suggests a substantial negative relationship. Figure 5.4 shows a useful way to describe the strength of relationship based on r. It may seem surprising that a value of r as extreme as −.5 or .5 should be in the weak category; an explanation for this is given later in the chapter. Even a weak correlation can indicate a meaningful relationship.

Figure 5.4 Describing the strength of a linear relationship.

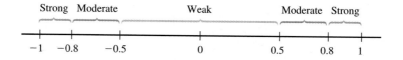

4. *The correlation coefficient r = 1 only when all the points in a scatterplot of the data lie exactly on a straight line that slopes upward. Similarly, r = −1 only when all the points lie exactly on a downward-sloping line.* Only when there is a perfect linear relationship between x and y in the sample does r take on one of its two possible extreme values.

5. *The value of r is a measure of the extent to which x and y are linearly related—* that is, the extent to which the points in the scatterplot fall close to a straight line. A value of r close to 0 does not rule out *any* strong relationship between x and y; there could still be a strong relationship but one that is not linear.

▪ **Example 5.2   How Strong Is the Relationship Between Hours Worked and GPA?**

College administrators and instructors have long puzzled over what factors affect student performance. The article "The Impact of Athletics, Part-Time Employment, and Other Activities on Academic Achievement" (*Journal of College Student Development* [1992]: 447–453) summarized the results of one large-scale study.

With x = grade point average and y = time spent working at a job (in hours per week), it was reported that $r = -.08$ (based on $n = 528$). It appears that the linear relationship between time spent working at a job and grade point average is extremely weak: There is a very slight tendency for those who work more to have lower grades. If time spent working was re-expressed in minutes per week, r would still be $-.08$.

▪ **Example 5.3   The Misery Index and Suicide**

Economists sometimes use what is called the Misery Index to reflect the gloominess of the U.S. population. The Misery Index is defined as the sum of the inflation rate and the unemployment rate. The article "The Misery Index and Suicide" (*Psychological Reports* [1999]: 1086) examined the relationship between suicide rate and both the Misery Index and a proposed revised misery index (the sum of the inflation rate and 2 times the unemployment rate). The revised index was proposed based on the belief that "unemployment inflicts twice the pain of inflation on the general public."

Using inflation, unemployment, and suicide rates for 1958–1992, the researchers reported that "Pearson correlations between the Misery Indices and suicide rate were .97 for the original Misery Index and .61 for the revised index." Based on the correlation coefficient, which measures the strength of a linear relationship, it appears that, although there is a positive relationship between suicide rate and both indexes, the relationship is much stronger for the original index than for the revised index.

▪ **Example 5.4   Is Foal Weight Related to Mare Weight?**

Foal weight at birth is an indicator of health, so it is of interest to breeders of thoroughbred horses. Is foal weight related to the weight of the mare (mother)? The accompanying data are from the article "Suckling Behaviour Does Not Measure Milk Intake in Horses" (*Animal Behaviour* [1999]: 673–678):

Observation	Mare weight (x, in kg)	Foal weight (y, in kg)	Observation	Mare weight (x, in kg)	Foal weight (y, in kg)
1	556	129	9	556	104
2	638	119	10	616	93.5
3	588	132	11	549	108.5
4	550	123.5	12	504	95
5	580	112	13	515	117.5
6	642	113.5	14	551	128
7	568	95	15	594	127.5
8	642	104			

MINITAB was used to compute the value of the correlation coefficient (Pearson's) for these data, with the following result:

Correlations (Pearson)

Correlation of Mare Weight and Foal Weight = 0.001

A correlation coefficient this close to 0 indicates no linear relationship between mare weight and foal weight. A scatterplot of the data (Figure 5.5) supports the conclusion that mare weight and foal weight are unrelated. From the correlation coefficient alone, we can conclude only that there is no *linear* relationship. We cannot rule out a more complicated curved relationship without also examining the scatterplot.

Figure 5.5 A MINITAB scatterplot of the mare and foal weight data of Example 5.4.

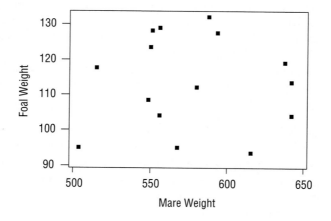

▪ Example 5.5   Paddy Data Revisited

Let's reconsider the data on x = time between planting and harvesting and y = yield for the grain data first given in Example 3.22. The *scatterplot* for these data is reproduced here as Figure 5.6, and it shows a strong curved pattern.

Figure 5.6 Scatterplot of y = yield and x = time between flowering and harvesting for the data of Example 5.5.

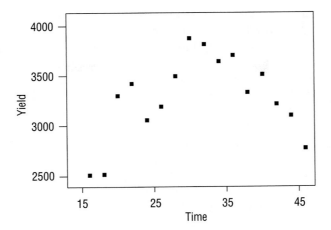

Using MINITAB to compute Pearson's correlation coefficient between time and yield results in the following:

Correlations (Pearson)
Correlation of Time and Yield = 0.274

This example shows the importance of interpreting r as a measure of the strength of a linear association. Here, r is not large, but there is a strong nonlinear relationship between time and yield. We should not conclude that there is no relationship whatsoever just because the value of r is small.

▪ The Population Correlation Coefficient

The sample correlation coefficient r measures how strongly the x and y values in a *sample* of pairs are linearly related to one another. There is an analogous measure of how strongly x and y are related in the entire population of pairs from which the sample was obtained. It is called the **population correlation coefficient** and is denoted ρ. (Notice again the use of a Greek letter for a population characteristic and a Roman letter for a sample characteristic.) We will never have to calculate ρ from an entire population of pairs, but it is important to know that ρ satisfies properties paralleling those of r:

1. ρ is a number between -1 and $+1$ that does not depend on the unit of measurement for either x or y, or on which variable is labeled x and which is labeled y.

2. $\rho = +1$ or -1 if and only if all (x, y) pairs in the population lie exactly on a straight line, so ρ measures the extent to which there is a linear relationship in the population.

In Chapter 13, we show how the sample characteristic r can be used to make an inference concerning the population characteristic ρ. In particular, r can be used to decide whether it is plausible that $\rho = 0$ (no linear relationship in the population).

▪ Correlation and Causation

A value of r close to 1 indicates that relatively large values of one variable tend to be associated with relatively large values of the other variable. This is far from saying that a large value of one variable *causes* the value of the other variable to be large. Correlation measures the extent of association, but *association does not imply causation.* It frequently happens that two variables are highly correlated not because one is causally related to the other but because they are both strongly related to a third variable. Among all elementary school children, the relationship between the number of cavities in a child's teeth and the size of his or her vocabulary is strong and positive. Yet no one advocates eating foods that result in more cavities to increase vocabulary size (or working to decrease vocabulary size to protect against cavities). Number of cavities and vocabulary size are both strongly related to age, so older children tend to have higher values of both variables than do younger ones. In the ABCNews.com series "Who's Counting?" (February 1, 2001), John Paulos reminded readers that correlation does not imply causation and gives the following example: Consumption of hot chocolate is negatively correlated with crime rate (high values of hot chocolate consumption tend to be paired with lower crime rates), but both are responses to cold weather.

Scientific experiments can frequently make a strong case for causality by carefully controlling the values of all variables that might be related to the ones under study. Then, if *y* is observed to change in a "smooth" way as the experimenter changes the value of *x*, the most plausible explanation would be a causal relationship between *x* and *y*. In the absence of such control and ability to manipulate values of one variable, we must admit the possibility that an unidentified underlying third variable is influencing both the variables under investigation. A high correlation in many uncontrolled studies carried out in different settings can marshal support for causality — as in the case of cigarette smoking and cancer — but proving causality is an elusive task.

▪ Exercises 5.1–5.16

5.1 For each of the following pairs of variables, indicate whether you would expect a positive correlation, a negative correlation, or a correlation close to 0. Explain your choice.
a. Maximum daily temperature and cooling costs
b. Interest rate and number of loan applications
c. Incomes of husbands and wives when both have full-time jobs
d. Height and IQ
e. Height and shoe size
f. Score on the math section of the SAT exam and score on the verbal section of the same test
g. Time spent on homework and time spent watching television during the same day by elementary school children
h. Amount of fertilizer used per acre and crop yield (Hint: As the amount of fertilizer is increased, yield tends to increase for a while but then tends to start decreasing.)

5.2 Is the following statement correct? Explain why or why not.

A correlation coefficient of 0 implies that no relationship exists between the two variables under study.

5.3 Draw two scatterplots, one for which $r = 1$ and a second for which $r = -1$.

5.4 The newspaper article "That's Rich: More You Drink, More You Earn" (*Calgary Herald*, April 16, 2002) described research that showed that light drinkers and people who don't drink earn about 10% less than moderate and heavy drinkers. It was reported that there was a positive correlation between alcohol consumption and income. Is it reasonable to conclude that increasing alcohol consumption will increase income? Give at least two reasons or examples to support your answer.

5.5 The paper "A Cross-National Relationship Between Sugar Consumption and Major Depression?" (Depression and Anxiety [2002]:118–120) concluded that there was a correlation between refined sugar consumption (calories per person per day) and annual rate of major depression (cases per 100 people) based on data from 6 countries. The following data were read from a graph that appeared in the paper:

Country	Sugar Consumption	Depression Rate
Korea	150	2.3
United States	300	3.0
France	350	4.4
Germany	375	5.0
Canada	390	5.2
New Zealand	480	5.7

a. Compute the correlation coefficient for this data set. Is the correlation positive or negative? Weak, moderate, or strong?

b. Based on the value of the correlation coefficient from Part (a), is it reasonable to conclude that increasing sugar consumption leads to higher rates of depression? Explain.

c. Do you have any concerns about this study that would make you hesitant to generalize these conclusions to other countries?

5.6 Cost-to-charge ratios (the percentage of the amount billed that represents the actual cost) in October 2002 for 11 Oregon hospitals of similar size were given by the Oregon Department of Health Services. Cost-to-charge ratio was reported separately for inpatient and outpatient services. The data are:

	Cost-to-Charge Ratio	
Hospital	Inpatient	Outpatient
Blue Mountain	80	62
Curry General	76	66
Good Shepherd	75	63
Grande Ronde	62	51
Harney District	100	54
Lake District	100	75
Pioneer	88	65
St. Anthony	64	56
St. Elizabeth	50	45
Tillamook	54	48
Wallowa Memorial	83	71

a. Does there appear to be a strong linear relationship between the cost-to-charge ratio for inpatient and outpatient services? Justify your answer based on the value of the correlation coefficient and examination of a scatterplot of the data.

b. Are there any unusual features of the data that are evident in the scatterplot?

c. Suppose that the observation for Harney District was removed from the data set. Would the correla-

tion coefficient for the new data set be greater than or less than the one computed in Part (a)? Explain.

5.7 The following time-series plot is based on data from the article "Bubble Talk Expands: Corporate Debt Is Latest Concern Turning Heads" (*San Luis Obispo Tribune*, September 13, 2002) and shows how household debt (in trillions of dollars) and corporate debt (in trillions of dollars) have changed over the time period from 1991 (year 1 in the graph) to 2002:

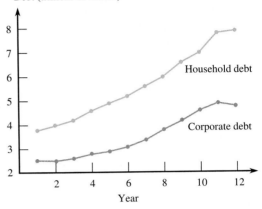

Based on the time-series plot, would the correlation coefficient between household debt and corporate debt be positive or negative? Weak or strong? What aspect of the time-series plot supports your answer?

5.8 The Federal Reserve Board (Household Debt-Service Burden, 2002) provided data on how consumers in the United States spend their money. Data on the percentage of disposable personal income required to meet consumer loan payments and mortgage payments for selected years are shown in the following table:

Consumer Debt	Household Debt
7.88	6.22
7.91	6.14
7.65	5.95
7.61	5.83
7.48	5.83
7.49	5.85
7.37	5.81
6.57	5.79
6.24	5.73
6.09	5.95
6.32	6.09
6.97	6.28
7.38	6.08
7.52	5.79
7.84	5.81

a. What is the value of the correlation coefficient for this data set?

b. The value of the correlation coefficient from Part (a) indicates a weak linear relationship between consumer debt and mortgage debt. Is it reasonable to conclude in this case that there is no strong relationship between the variables (linear or otherwise)? Use a graphical display to support your answer.

5.9 The accompanying data were read from graphs that appeared in the article "Bush Timber Proposal Runs Counter to the Record" (*San Luis Obispo Tribune*, September 22, 2002). The variables shown are the number of acres burned in forest fires in the Western United States and timber sales.

Year	Thousands of Acres Burned	Timber Sales (billions of board feet)
1945	200	2.0
1950	250	3.7
1955	260	4.4
1960	380	6.8
1965	80	9.7
1970	450	11.0
1975	180	11.0
1980	240	10.2
1985	440	10.0
1990	400	11.0
1995	180	3.8

a. Is there a correlation between timber sales and acres burned in forest fires? Compute and interpret the value of the correlation coefficient.

b. The article concludes that "heavier logging led to large forest fires." Do you think this conclusion is justified based on the given data? Explain.

5.10 Peak heart rate (beats per minute) was determined both during a shuttle run and during a 300-yard run for a sample of $n = 10$ individuals afflicted with Down syndrome ("Heart Rate Responses to Two Field Exercise Tests by Adolescents and Young Adults with Down Syndrome," *Adapted Physical Activity Quarterly* [1995]: 43–51), resulting in the following data:

Shuttle	168	168	188	172	184	176	192
300-yd	184	192	200	192	188	180	182

Shuttle	172	188	180
300-yd	188	196	196

a. A scatterplot of the data is as follows:

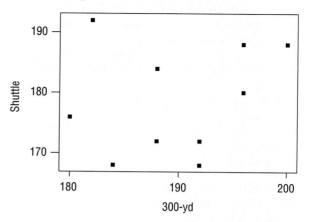

What does the scatterplot suggest about the nature of the relationship between the two variables?

b. With x = shuttle run peak rate and y = 300-yd run peak rate, calculate r. Is the value of r consistent with your answer in Part (a)?

c. With x = 300-yd peak rate and y = shuttle run peak rate, how does the value of r compare to what you calculated in Part (b)?

5.11 The detection of anemia requires an accurate method for determining hemoglobin level. The accompanying scatterplot is for observations on hemoglobin level, determined both by the standard spectrophotometric method (y) and by a new, simpler method based on a color scale (x) ("A Simple and Reliable Method for Estimating Hemoglobin," *Bulletin of the World Health Organization* [1995]: 369–373):

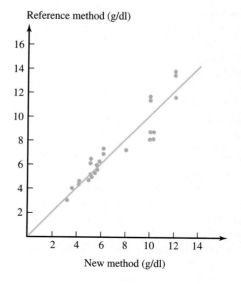

a. Does it appear that x and y are highly correlated?
b. The paper reported that $r = .9366$. How would you describe the relationship between the two variables?
c. The line pictured in the scatterplot has a slope of 1 and passes through $(0, 0)$. If x and y were always identical, all points would lie exactly on this line. The authors of the paper claimed that perfect correlation $(r = 1)$ would result in this line. Do you agree? Explain your reasoning.

5.12 Is there a correlation between test anxiety and exam score performance? Data on x = score on a measure of test anxiety and y = exam score for a sample of $n = 9$ students are the following (the data are consistent with summary quantities given in the paper "Effects of Humor on Test Anxiety and Performance," *Psychological Reports* [1999]: 1203–1212):

x	23	14	14	0	17	20	20	15	21
y	43	59	48	77	50	52	46	51	51

Higher values for x indicate higher levels of anxiety.
a. Construct a scatterplot, and comment on the features of the plot.
b. Does there appear to be a linear relationship between the two variables? Based on the scatterplot, would you characterize the relationship as positive or negative? Strong or weak?
c. Compute the value of the correlation coefficient. Is the value of r consistent with your answer to Part (b)?
d. Based on the value of the correlation, is it reasonable to conclude that test anxiety caused poor exam performance? Explain.

5.13 According to the article "First-Year Academic Success: A Prediction Combining Cognitive and Psychosocial Variables for Caucasian and African American Students" (*Journal of College Student Development* [1999]: 599–605), there is a mild correlation between high school GPA (x) and first-year college GPA (y). The data, collected from a "Southeastern public research university," can be summarized as follows:

$$n = 2600 \qquad \sum x = 9620 \qquad \sum y = 7436$$

$$\sum xy = 27{,}918 \qquad \sum x^2 = 36{,}168 \qquad \sum y^2 = 23{,}145$$

An alternative formula for computing the correlation coefficient that is based on raw data and is algebraically equivalent to the one given in the text is

$$r = \frac{\sum xy - \dfrac{(\sum x)(\sum y)}{n}}{\sqrt{\sum x^2 - \dfrac{(\sum x)^2}{n}}\sqrt{\sum y^2 - \dfrac{(\sum y)^2}{n}}}$$

Use this formula to compute the value of the correlation coefficient, and interpret this value.

5.14 An employee of an auction house has a list of 25 recently sold paintings. Eight artists were represented in these sales. The sale price of each painting appears on the list. Would the correlation coefficient be an appropriate way to summarize the relationship between artist (x) and sale price (y)? Why or why not?

5.15 Each individual in a sample was asked to indicate on a quantitative scale how willing he or she was to spend money on the environment and also how strongly he or she believed in God ("Religion and Attitudes Toward the Environment," *Journal for the Scientific Study of Religion* [1993]: 19–28). (This article was written by Andrew Greeley, who also writes entertaining mysteries with religious themes.) The resulting value of the sample correlation coefficient was $r = -.085$. Would you agree with the stated conclusion that stronger support for environmental spending is associated with a weaker degree of belief in God? Explain your reasoning.

5.16 A sample of automobiles traversing a certain stretch of highway is selected. Each one travels at roughly a constant rate of speed, although speed does vary from auto to auto. Let x = speed and y = time needed to traverse this segment of highway. Would the sample correlation coefficient be closest to .9, .3, −.3, or −.9? Explain.

■ 5.2 Linear Regression: Fitting a Line to Bivariate Data

Given two variables x and y, the general objective of *regression analysis* is to use information about x to draw some sort of conclusion concerning y. Often an investigator wants to predict the y value that would result from making a single observation at a specified x value — for example, to predict product sales y during a given period when amount spent on advertising is $x = \$10,000$. The different roles played by the two variables are reflected in standard terminology: y is called the **dependent** or **response variable**, and x is referred to as the **independent, predictor**, or **explanatory variable**.

A scatterplot of y versus x (i.e., of the (x, y) pairs in a sample) frequently exhibits a linear pattern. It is natural in such cases to summarize the relationship between the variables by finding a line that is as close as possible to the points in the plot. Before doing so, let's review some elementary facts about lines and linear relationships.

Suppose that a car dealership advertises that a particular model of car can be rented for a flat fee of \$25 plus an additional \$0.30 per mile. If this type of car is rented and driven for 100 miles, the dealer's revenue y is

$$y = 25 + (0.30)(100) = 25 + 30 = 55$$

More generally, if x denotes distance driven (in miles), then

$$y = 25 + 0.30x$$

That is, x and y are linearly related.

The general form of a linear relation between x and y is $y = a + bx$. A particular relation is specified by choosing values of a and b. Thus, one such relationship is $y = 10 + 2x$; another is $y = 100 - 5x$. If we choose some x values and compute $y = a + bx$ for each value, the points in the plot of the resulting (x, y) pairs will fall exactly on a straight line.

▪ Definition

The relationship

$$\overset{\text{Intercept}}{\underset{\text{Slope}}{y = a + bx}}$$

is the equation of a straight line. The value of b, called the **slope** of the line, is the amount by which y increases when x increases by 1 unit. The value of a, called the **intercept** (or sometimes the **y intercept or vertical intercept**) of the line, is the height of the line above the value $x = 0$.

The equation $y = 10 + 2x$ has slope $b = 2$, so each 1-unit increase in x is paired with an increase of 2 in y. When $x = 0$, $y = 10$, and the height at which the line crosses the vertical axis (where $x = 0$) is 10. This is illustrated in Figure 5.7(a). The slope of the line determined by $y = 100 - 5x$ is -5, so y increases by -5 (i.e., decreases by 5) when x increases by 1. The height of the line above $x = 0$ is $a = 100$. The resulting line is pictured in Figure 5.7(b).

It is easy to draw the line corresponding to any particular linear equation. First choose any two x values and substitute them into the equation to obtain the corresponding y values. Then plot the resulting two (x, y) pairs as two points. The desired

Figure 5.7 Graphs of two lines: (a) slope $b = 2$, intercept $a = 10$; (b) slope $b = -5$, intercept $a = 100$.

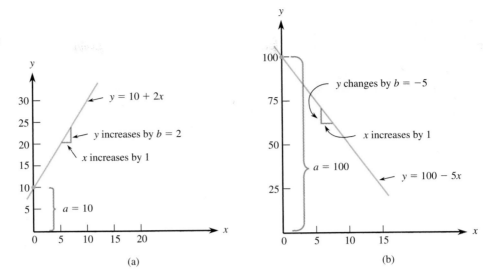

(a)   (b)

line is the one passing through these points. For the equation $y = 10 + 2x$, substituting $x = 5$ yields $y = 20$, whereas using $x = 10$ gives $y = 30$. The two points are then $(5, 20)$ and $(10, 30)$. The line in Figure 5.7(a) does indeed pass through these points.

▪ Fitting a Straight Line: The Principle of Least Squares

Figure 5.8 shows a scatterplot with two lines superimposed on the plot. The line that gives the most effective summary of the approximate linear relationship is the one that in some sense is the best-fit line, the one closest to the sample data. Line II clearly gives a better fit to the data than does Line I. In order to measure the extent to which a particular line provides a good fit, we focus on the vertical deviations from the line. For example, Line II in Figure 5.8 has equation $y = 10 + 2x$, and the third and fourth points from the left in the scatterplot are $(15, 44)$ and $(20, 45)$. For these two points, the vertical deviations from this line are

$$\text{3d deviation} = y_3 - \text{height of the line above } x_3$$
$$= 44 - [10 + 2(15)]$$
$$= 4$$

and

$$\text{4th deviation} = 45 - [10 + 2(20)] = -5$$

A positive vertical deviation results from a point that lies above the chosen line, and a negative deviation results from a point that lies below this line. A particular line is said to be a good fit to the data if the deviations from the line are small in magnitude. Line I in Figure 5.8 fits poorly, because all deviations from that line are larger in magnitude (some are much larger) than the corresponding deviations from Line II.

We now need a way to combine the n deviations into a single measure of fit. The standard approach is to square the deviations (to obtain nonnegative numbers) and sum these squared deviations.

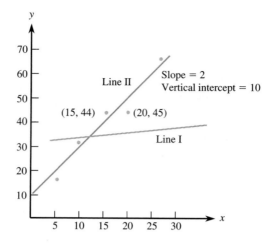

Figure 5.8 Line I gives a poor fit and Line II gives a good fit to the data.

▪ Definition

The most widely used criterion for measuring the goodness of fit of a line $y = a + bx$ to bivariate data $(x_1, y_1), \ldots, (x_n, y_n)$ is the sum of the squared deviations about the line

$$\sum [y - (a + bx)]^2 = [y_1 - (a + bx_1)]^2 + [y_2 - (a + bx_2)]^2 + \cdots + [y_n - (a + bx_n)]^2$$

The line that gives the best fit to the data is the one that minimizes this sum; it is called the **least-squares line** or the **sample regression line**.

Fortunately, the equation of the least-squares line can be obtained without having to calculate deviations from any particular line. This is because mathematical techniques can be applied to obtain relatively simple formulas for the slope and intercept of the least-squares line. The formula for b involves $\sum (x - \bar{x})(y - \bar{y})$, which is the sum of products of the x and y deviations from their respective means.

The slope of the least-squares line is

$$b = \frac{\sum (x - \bar{x})(y - \bar{y})}{\sum (x - \bar{x})^2}$$

and the y intercept is

$$a = \bar{y} - b\bar{x}$$

We write the equation of the least-squares line as

$$\hat{y} = a + bx$$

where the ^ above y indicates that $\hat{y}$ (read as y-hat) is a prediction of y resulting from the substitution of a particular x value into the equation.

Statistical software packages and many calculators can compute the slope and intercept of the least-squares line. If the slope and intercept are to be computed by hand, the following computational formula can be used to reduce the amount of time required to perform the calculations.

■ **Calculating Formula for the Slope of the Least-Squares Line**

$$b = \frac{\sum xy - \dfrac{(\sum x)(\sum y)}{n}}{\sum x^2 - \dfrac{(\sum x)^2}{n}}$$

■ **Example 5.6**   Time to Defibrillator Shock and Heart Attack Survival Rate

Studies have shown that people who suffer sudden cardiac arrest (SCA) have a better chance of survival if a defibrillator shock is administered very soon after cardiac arrest. How is survival rate related to the time between when cardiac arrest occurs and when the defibrillator shock is delivered? This question is addressed in the paper "Improving Survival from Sudden Cardiac Arrest: The Role of Home Defibrillators" (by J. K. Stross, University of Michigan, February 2002; available at www.heartstarthome.com). The accompanying data give y = survival rate (percent) and x = mean call-to-shock time (minutes) for a cardiac rehabilitation center (where cardiac arrests occurred while victims were hospitalized and so the call-to-shock time tended to be short) and for four communities of different sizes:

Mean call-to-shock time, x	2	6	7	9	12
Survival rate, y	90	45	30	5	2

A scatterplot of these data (Figure 5.9) shows that the relationship between survival rate and mean call-to-shock time for times in the range 2–12 min could reasonably be summarized by a straight line.

Figure 5.9 MINITAB scatterplot for the data of Example 5.6.

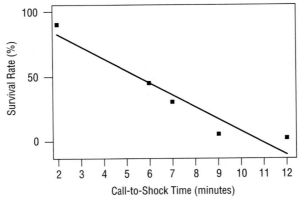

The summary quantities necessary to compute the equation of the least-squares line are

$$\sum x = 36 \qquad \sum x^2 = 314 \qquad \sum xy = 729$$
$$\sum y = 172 \qquad \sum y^2 = 11{,}054$$

From these quantities, we compute

$$\bar{x} = 7.20 \qquad \bar{y} = 34.4$$

$$b = \frac{\sum xy - \frac{(\sum x)(\sum y)}{n}}{\sum x^2 - \frac{(\sum x)^2}{n}} = \frac{729 - \frac{(36)(172)}{5}}{314 - \frac{(36)^2}{5}} = \frac{-509.40}{54.80} = -9.30$$

and

$$a = \bar{y} - b\bar{x} = 34.4 - (-9.30)(7.20) = 34.4 + 66.96 = 101.36$$

The least-squares line is then

$$\hat{y} = 101.36 - 9.30x$$

This line is also shown in the scatterplot of Figure 5.9.

If we wanted to predict SCA survival rate for a community with a mean call-to-shock time of 5 min, we could use the point on the least-squares line above $x = 5$:

$$\hat{y} = 101.36 - 9.30(5) = 101.36 - 46.50 = 54.86$$

Predicted survival rates for communities with different mean call-to-shock times could be obtained in a similar way.

However, the least-squares line should not be used to predict survival rate for communities much outside the range 2–12 min (the range of x values in the data set) because we do not know whether the linear pattern observed in the scatterplot continues outside this range. This is sometimes referred to as the **danger of extrapolation**.

In this example, we can see that using the least-squares line to predict survival rate for communities with mean call-to-shock times much above 12 min leads to nonsensical predictions. For example, if the mean call-to-shock time is 15 min, the predicted survival rate is negative:

$$\hat{y} = 101.36 - 9.30(15) = -38.14$$

Because it is impossible for survival rate to be negative, this is a clear indication that the pattern observed for x values in the 2–12 range does not continue outside this range. Nonetheless, the least-squares line can be a useful tool for making predictions for x values within this range.

Calculations involving the least-squares line can obviously be tedious, and fitting a function to multivariate data even more so. This is where the computer comes to our rescue. All the standard statistical packages can fit a straight line to bivariate data.

▪ Example 5.7 Is Age Related to Recovery Time for Injured Athletes?

How quickly can athletes return to their sport following injuries requiring surgery? The paper "Arthroscopic Distal Clavicle Resection for Isolated Atraumatic Osteolysis in Weight Lifters" (*American Journal of Sports Medicine* [1998]: 191–195) gave the following data on x = age and y = days after arthroscopic shoulder surgery before being able to return to their sport, for 10 weight lifters:

x	33	31	32	28	33	26	34	32	28	27
y	6	4	4	1	3	3	4	2	3	2

Figure 5.10 is a scatterplot of the data. The predominant pattern is linear, although the points in the plot spread out quite a bit about *any* line (even the least-squares line).

Figure 5.10 Scatterplot and least-squares line for the data of Example 5.7.

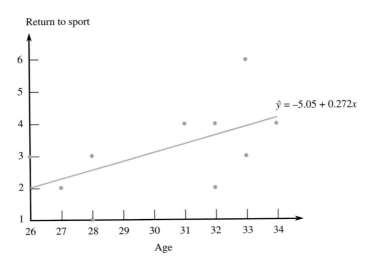

MINITAB was used to fit the least-squares line, and Figure 5.11 shows part of the resulting output. Instead of x and y, the variable labels "Return to sport" and "Age" are used. The equation at the top is that of the least-squares line. In the rectangular table just below the equation, the first row gives information about the intercept, a, and the second row gives information concerning the slope, b. In particular, the coefficient column labeled "Coef" contains the values of a and b using more significant figures than in the rounded values that appear in the equation.

Figure 5.11 Partial MINITAB output for Example 5.7.

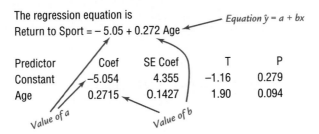

The regression equation is
Return to Sport = −5.05 + 0.272 Age   ← *Equation $\hat{y} = a + bx$*

Predictor	Coef	SE Coef	T	P
Constant	−5.054	4.355	−1.16	0.279
Age	0.2715	0.1427	1.90	0.094

Value of a   *Value of b*

The least-squares line should not be used to predict the number of days before returning to practice for weight lifters with ages such as $x = 21$ or $x = 48$. These x values are well outside the range of the data, and there is no evidence to support extrapolation of the linear relationship.

■ Regression

The least-squares line is often called the **sample regression line**. This terminology comes from the relationship between the least-squares line and Pearson's correlation coefficient. To understand this relationship, we first need alternative expressions for the slope b and the equation of the line itself. With s_x and s_y denoting the sample standard deviations of the x's and y's, respectively, a bit of algebraic manipulation gives

$$b = r\left(\frac{s_y}{s_x}\right)$$

$$\hat{y} = \bar{y} + r\left(\frac{s_y}{s_x}\right)(x - \bar{x})$$

You do not need to use these formulas in any computations, but several of their implications are important for appreciating what the least-squares line does.

1. When $x = \bar{x}$ is substituted in the equation of the line, $\hat{y} = \bar{y}$ results. That is, the least-squares line passes through the *point of averages* $(\bar{x}, \bar{y})$.

2. Suppose for the moment that $r = 1$, so that all points lie exactly on the line whose equation is

$$\hat{y} = \bar{y} + \frac{s_y}{s_x}(x - \bar{x})$$

Now substitute $x = \bar{x} + s_x$ which is 1 standard deviation above $\bar{x}$:

$$\hat{y} = \bar{y} + \frac{s_y}{s_x}(\bar{x} + s_x - \bar{x}) = \bar{y} + s_y$$

That is, with $r = 1$, when x is 1 standard deviation above its mean, we predict that the associated y value will be 1 standard deviation above its mean. Similarly, if $x = \bar{x} - 2s_x$ (2 standard deviations below its mean), then

$$\hat{y} = \bar{y} + \frac{s_y}{s_x}(\bar{x} - 2s_x - \bar{x}) = \bar{y} - 2s_y$$

which also is 2 standard deviations below the mean. If $r = -1$, then $x = \bar{x} + s_x$ results in $\hat{y} = \bar{y} - s_y$, so the predicted y is also 1 standard deviation from its mean but on the opposite side of $\bar{y}$ from where x is relative to $\bar{x}$. In general, if x and y are perfectly correlated, the predicted y value associated with a given x value will be the same number of standard deviations (of y) from its mean, $\bar{y}$, as x is from its mean, $\bar{x}$.

3. Now suppose that x and y are not perfectly correlated. For example, suppose $r = .5$, so that the least-squares line has the equation

$$\hat{y} = \bar{y} + .5\left(\frac{s_y}{s_x}\right)(x - \bar{x})$$

Then, substituting $x = \bar{x} + s_x$ gives

$$\hat{y} = \bar{y} + .5\left(\frac{s_y}{s_x}\right)(\bar{x} + s_x - \bar{x}) = \bar{y} + .5s_y$$

That is, for $r = .5$, when x lies 1 standard deviation above its mean, we predict that y will be only 0.5 standard deviation above its mean.

Similarly, we can predict y when r is negative. If $r = -.5$, then the predicted y value will be only half the number of standard deviations from $\bar{y}$ that x is from $\bar{x}$, but x and the predicted y will now be on opposite sides of their respective means.

Consider using the least-squares line to predict the value of y associated with an x value some specified number of standard deviations away from $\bar{x}$. Then the predicted y value will be only r times this number of standard deviations from $\bar{y}$. In terms of standard deviations, except when $r = 1$ or -1, the predicted y will always be closer to $\bar{y}$ than x is to $\bar{x}$.

Using the least-squares line for prediction results in a predicted y that is pulled back in, or regressed, toward the mean of y compared to where x is relative to the mean of x. This regression effect was first noticed by Sir Francis Galton (1822–1911), a famous biologist, when he was studying the relationship between the heights of fathers and their sons. He found that predicted heights of sons whose fathers were above average in height were also above average (because r is positive here) but not by as much as the father's height; he found a similar relationship for fathers whose heights were below average. This regression effect has led to the term **regression analysis** for the collection of methods involving the fitting of lines, curves, and more complicated functions to bivariate and multivariate data.

The alternative form of the regression (least-squares) line emphasizes that predicting y from knowledge of x is not the same problem as predicting x from knowledge of y. The slope of the least-squares line for predicting x is $r\left(\dfrac{s_x}{s_y}\right)$ rather than $r\left(\dfrac{s_y}{s_x}\right)$ and the intercepts of the lines are almost always different. For purposes of prediction, it makes a difference whether y is regressed on x, as we have done, or x is regressed on y. The regression line of y on x should not be used to predict x, because it is not the line that minimizes the sum of squared x deviations.

■ Exercises 5.17–5.31

5.17 The article "Air Pollution and Medical Care Use by Older Americans" (*Health Affairs* [2002]: 207–214) concluded that elderly people who live in areas with high levels of pollution are more likely to require medical treatment. The following data give a measure of pollution (in micrograms of particulate matter per cubic meter of air) and the cost of medical care per person over age 65 for six geographical regions of the United States:

Region	Pollution	Cost of Medical Care
North	30.0	915
Upper South	31.8	891
Deep South	32.1	968
West South	26.8	972
Big Sky	30.4	952
West	40.0	899

a. Construct a scatterplot of the data. Write a sentence or two describing any interesting features of the scatterplot.
b. Find the equation of the least-squares line describing the relationship between y = medical cost and x = pollution.
c. Is the slope of the least-squares line positive or negative? Is this consistent with your description of the relationship in Part (a)?
d. Do the scatterplot and the equation of the least-squares line support the researchers' conclusion that elderly people who live in more polluted areas have higher medical costs? Explain.

5.18 In the article "Reproductive Biology of the Aquatic Salamander *Amphiuma tridactylum* in Louisiana" (*Journal of Herpetology* [1999]: 100–105), 14 female salamanders were studied. Using regression, the researchers predicted y = clutch size (number of salamander eggs) from x = snout-vent length (in centimeters) as follows:

$$\hat{y} = -147 + 6.175x$$

For the salamanders in the study, the range of snout-vent lengths was approximately 30 to 70 cm. Because the scatterplot looked linear, the researchers thought a simple linear regression was a reasonable way to describe the relationship between x and y.

a. What is the value of the y intercept of the least-squares line? What is the value of the slope of the least-squares line? Give an interpretation of the slope in the context of this problem.

b. Would you be reluctant to predict the clutch size when snout-vent length is 22 cm? Explain.

5.19 Percentages of public school students in fourth grade in 1996 and in eighth grade in 2000 who were at or above the proficient level in mathematics were given in the article "Mixed Progress in Math" (*USA Today*, August 3, 2001). The following data for eight Western states were given:

State	4th grade (1996)	8th grade (2000)
Arizona	15	21
California	11	18
Hawaii	16	16
Montana	22	37
New Mexico	13	13
Oregon	21	32
Utah	23	26
Wyoming	19	25

a. Construct a scatterplot, and comment on the interesting features of the scatterplot.

b. Find the equation of the least-squares line that summarizes the relationship between x = 1996 fourth-grade math proficiency percentage and y = 2000 eighth-grade math proficiency percentage.

c. Nevada, a Western state not included in the data set, had a 1996 fourth-grade math proficiency of 14%. What would you predict for Nevada's 2000 eighth-grade math proficiency percentage? How does your prediction compare to the actual eighth-grade value of 20 for Nevada?

5.20 According to the article "First-Year Academic Success: A Prediction Combining Cognitive and Psychosocial Variables for Caucasian and African American Students" (*Journal of College Student Development* [1999]: 599–605), high school GPA (x) and first-year college GPA (y) are mildly correlated. The data, which were collected from a Southeastern public research university, can be summarized as follows:

$$n = 2600 \qquad \sum x = 9620 \qquad \sum y = 7436$$

$$\sum xy = 27,918 \qquad \sum x^2 = 36,168 \qquad \sum y^2 = 23,145$$

a. Find the equation of the least-squares regression line.

b. Interpret the value of b, the slope of the least-squares line, in the context of this problem.

c. What first-year GPA would you predict for a student with a 4.0 high school GPA?

5.21 Corrosion of the steel frame is the most important factor affecting the durability of reinforced concrete buildings. It is believed that carbonation of the concrete (caused by a chemical reaction that lowers the pH of the concrete) leads to corrosion of the steel frame and thus reduces the strength of the concrete. Representative data on x = carbonation depth (in millimeters) and y = strength of concrete (in megapascals) for a sample of core specimens taken from a particular building were read from a plot in the article "The Carbonation of Concrete Structures in the Tropical Environment of Singapore" (*Magazine of Concrete Research* [1996]: 293–300):

Depth, x	8.0	20.0	20.0	30.0	35.0
Strength, y	22.8	17.1	21.1	16.1	13.4

Depth, x	40.0	50.0	55.0	65.0
Strength, y	12.4	11.4	9.7	6.8

a. Construct a scatterplot. Does the relationship between carbonation depth and strength appear to be linear?

b. Find the equation of the least-squares line.

c. What would you predict for strength when carbonation depth is 25 mm?

d. Explain why it would not be reasonable to use the least-squares line to predict strength when carbonation depth is 100 mm.

5.22 The data given in Example 5.6 on x = call-to-shock time (in minutes) and y = survival rate (percent) were used to compute the equation of the least-squares line, which was

$$\hat{y} = 101.36 - 9.30x$$

The newspaper article "FDA OKs Use of Home Defibrillators" (*San Luis Obispo Tribune*, November 13, 2002) reported that "every minute spent waiting for paramedics to arrive with a defibrillator lowers the chance of survival by 10 percent." Is this statement consistent with the given least-squares line? Explain.

5.23 An article on the cost of housing in California that appeared in the *San Luis Obispo Tribune* (March 30, 2001) included the following statement: "In Northern California, people from the San Francisco Bay area pushed into the Central Valley, benefiting from home prices that dropped on average $4000 for every mile traveled east of the Bay area." If this statement is correct, what is the slope of the least-squares regression line, $\hat{y} = a + bx$, where y = house price (in dollars) and x = distance east of the Bay (in miles)? Explain.

5.24 The following data on sale price, size, and land-to-building ratio for 10 large industrial properties appeared in the paper "Using Multiple Regression Analysis in Real Estate Appraisal" (*Appraisal Journal* [2002]: 424–430):

Property	Sale Price (millions of dollars)	Size (thousands of sq. ft.)	Land-to-Building Ratio
1	10.6	2166	2.0
2	2.6	751	3.5
3	30.5	2422	3.6
4	1.8	224	4.7
5	20.0	3917	1.7
6	8.0	2866	2.3
7	10.0	1698	3.1
8	6.7	1046	4.8
9	5.8	1108	7.6
10	4.5	405	17.2

a. Calculate and interpret the value of the correlation coefficient between sale price and size.
b. Calculate and interpret the value of the correlation coefficient between sale price and land-to-building ratio.
c. If you wanted to predict sale price and you could use either size or land-to-building ratio as the basis for making predictions, which would you use? Explain.
d. Based on your choice in Part (c), find the equation of the least-squares regression line you would use for predicting y = sale price.

5.25 The article "Effect of Cattle Treading on Erosion from Hill Pasture: Modeling Concepts and Analysis of Rainfall Simulator Data" (*Australian Journal of Soil Research* [2002]: 963–977) summarized results of a study of the effect of rainfall runoff on land already damaged as a result of cattle grazing. The effect of rainfall runoff was measured by determining the concentration of sediment in runoff water. Representative data (read from a plot that appeared in the paper) on sediment concentration for plots with varying amounts of grazing damage, measured by the percentage of bare ground in the plot, are shown for gradually sloped plots and for steeply sloped plots:

Gradually Sloped Plots

Bare ground (%)	5	10	15	25
Concentration	50	200	250	500

Bare ground (%)	30	40		
Concentration	600	500		

Steeply Sloped Plots

Bare ground (%)	5	5	10	15
Concentration	100	250	300	600

Bare ground (%)	20	25	20	30
Concentration	500	500	900	800

Bare ground (%)	35	40	35	
Concentration	1100	1200	1000	

a. Using the data for steeply sloped plots, find the equation of the least-squares line for predicting y = runoff sediment concentration using x = percentage of bare ground.
b. What would you predict runoff sediment concentration to be for a steeply sloped plot with 18% bare ground?
c. Would you recommend using the least-squares equation from Part (a) to predict runoff sediment concentration for gradually sloped plots? If so, explain why it would be appropriate to do so. If not, provide an alternative way to make predictions of runoff sediment concentrations for gradually sloped plots.

5.26 Shells of mollusks function as part of the skeletal system and as protective armor. It has been argued that many features of these shells were the result of natural selection in the constant battle against predators. The paper "Postmortem Changes in Strength of Gastropod Shells" (*Paleobiology* [1992]: 367–377) included scatterplots of data on x = shell height (in centimeters) and y = breaking strength (in newtons). The least-squares line for a sample of $n = 38$ hermit crab shells was $\hat{y} = -275.1 + 244.9x$.
a. What are the slope and the intercept of this line?
b. When shell height increases by 1 cm, by how much does breaking strength tend to change?
c. What breaking strength would you predict when shell height is 2 cm?
d. Does this approximate linear relationship appear to hold for shell heights as small as 1 cm? Explain your reasoning.

5.27 It has been observed that Andean high-altitude natives have larger chest dimensions and lung volumes than do sea-level residents. Is this also true of lifelong Himalayan residents? The paper "Increased Vital and Total Lung Capacities in Tibetan Compared to Han Residents of Lhasa" (*American Journal of Physical Anthropology* [1991]: 341–351) reported on the results of an investigation into this question.

Included in the paper was a plot of vital capacity (y) versus chest circumference (x) for a sample of 16 Tibetan natives, from which the following data were read:

x	79.4	81.8	81.8	82.3	83.7	84.3	84.3	85.2
y	4.3	4.6	4.8	4.7	5.0	4.9	4.4	5.0
x	87.0	87.3	87.7	88.1	88.1	88.6	89.0	89.5
y	6.1	4.7	5.7	5.7	5.2	5.5	5.0	5.3

a. Construct a scatterplot. What does it suggest about the nature of the relationship between x and y?
b. The summary quantities are

$$\sum x = 1368.1 \qquad \sum y = 80.9 \qquad \sum x^2 = 117{,}123.85$$

$$\sum y^2 = 412.81 \qquad \sum xy = 6933.48$$

Verify that the equation of the least-squares line is $\hat{y} = -4.54 + 0.1123x$, and draw this line on your scatterplot.
c. On average, roughly what change in vital capacity is associated with a 1-cm increase in chest circumference? with a 10-cm increase?
d. What vital capacity would you predict for a Tibetan native whose chest circumference is 85 cm?
e. Is vital capacity completely determined by chest circumference? Explain.

5.28 Explain why it can be dangerous to use the least-squares line to obtain predictions for x values that are substantially larger or smaller than those contained in the sample.

5.29 The sales manager of a large company selected a random sample of $n = 10$ salespeople and determined for each one the values of x = years of sales experience and y = annual sales (in thousands of dollars). A scatterplot of the resulting (x, y) pairs showed a marked linear pattern.
a. Suppose that the sample correlation coefficient is $r = .75$ and that the average annual sales is $\bar{y} = 100$.

If a particular salesperson is 2 standard deviations above the mean in terms of experience, what would you predict for that person's annual sales?
b. If a particular person whose sales experience is 1.5 standard deviations below the average experience is predicted to have an annual sales value that is 1 standard deviation below the average annual sales, what is the value of r?

5.30 Explain why the slope b of the least-squares line always has the same sign (positive or negative) as does the sample correlation coefficient r.

5.31 The accompanying data resulted from an experiment in which weld diameter x and shear strength y (in pounds), were determined for five different spot welds on steel. A scatterplot shows a pronounced linear pattern. With $\sum (x - \bar{x})^2 = 1000$ and $\sum (x - \bar{x})(y - \bar{y}) = 8577$, the least-squares line is $\hat{y} = -936.22 + 8.577x$.

x	200.1	210.1	220.1	230.1	240.0
y	813.7	785.3	960.4	1118.0	1076.2

a. Because 1 lb = 0.4536 kg, strength observations can be re-expressed in kilograms through multiplication by this conversion factor: new y = 0.4536(old y). What is the equation of the least-squares line when y is expressed in kilograms?
b. More generally, suppose that each y value in a data set consisting of n (x, y) pairs is multiplied by a conversion factor c (which changes the units of measurement for y). What effect does this have on the slope b (i.e., how does the new value of b compare to the value before conversion), on the intercept a, and on the equation of the least-squares line? Verify your conjectures by using the given formulas for b and a. (Hint: Replace y with cy, and see what happens — and remember, this conversion will affect $\bar{y}$.)

▪ 5.3 Assessing the Fit of a Line

Once the least-squares regression line has been obtained, it is natural to examine how effectively the line summarizes the relationship between x and y. Important questions to consider are:

1. Is a line an appropriate way to summarize the relationship between the two variables?

2. Are there any unusual aspects of the data set that we need to consider before proceeding to use the regression line to make predictions?

3. If we decide that it is reasonable to use the regression line as a basis for pre-diction, how accurate can we expect predictions based on the regression line to be?

In this section, we look at graphical and numerical methods that will allow us to answer these questions. Most of these methods are based on the vertical deviations of the data points from the regression line. These vertical deviations are called *residuals*, and each represents the difference between an actual y value and the corresponding predicted value, $\hat{y}$, that would result from using the regression line to make a prediction.

▪ Predicted Values and Residuals

If the x value for the first observation is substituted into the equation for the least-squares line, the result is $a + bx_1$, the height of the line above x_1. The point (x_1, y_1) in the scatterplot also lies above x_1, so the difference

$$y_1 - (a + bx_1)$$

is the vertical deviation from this point to the line (see Figure 5.12). A point lying above the line gives a positive deviation, and a point lying below the line results in a negative deviation. The remaining vertical deviations come from repeating this process for $x = x_2$, then $x = x_3$, and so on.

Figure 5.12 Positive and negative deviations (residuals) from the least-squares line.

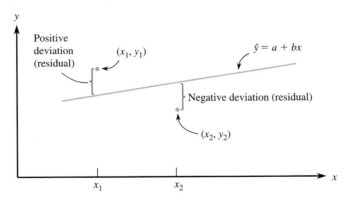

▪ Definition

The **predicted** or **fitted values** result from substituting each sample x value in turn into the equation for the least-squares line. This gives

$\hat{y}_1 = $ first predicted value $= a + bx_1$

$\hat{y}_2 = $ second predicted value $= a + bx_2$

$\vdots$

$\hat{y}_n = $ nth predicted value $= a + bx_n$

The residuals from the least-squares line are the n quantities

$$y_1 - \hat{y}_1, y_2 - \hat{y}_2, \ldots, y_n - \hat{y}_n$$

Each residual is the difference between an observed y value and the corresponding predicted y value.

■ **Example 5.8**   How Does Range of Motion After
Knee Surgery Change with Age?

One measure of the success of knee surgery is postsurgical range of motion for the
knee joint. Postsurgical range of motion was recorded for 12 patients who had surgery following a knee dislocation. The age of each patient was also recorded ("Reconstruction of the Anterior and Posterior Cruciate Ligaments After Knee Dislocation," *American Journal of Sports Medicine* [1999]: 189–194). The data are given
in Table 5.1.

Table 5.1   ■   Predicted Values and Residuals
for the Data of Example 5.8

Patient	Age (x)	Range of Motion (y)	Predicted Range of Motion ($\hat{y}$)	Residual $y - \hat{y}$
1	35	154	138.07	15.93
2	24	142	128.49	13.51
3	40	137	142.42	−5.42
4	31	133	134.58	−1.58
5	28	122	131.97	−9.97
6	25	126	129.36	−3.36
7	26	135	130.23	4.77
8	16	135	121.52	13.48
9	14	108	119.78	−11.78
10	20	120	125.00	−5.00
11	21	127	125.87	1.13
12	30	122	133.71	−11.71

MINITAB was used to fit the least-squares regression line. A partial computer
output follows:

Regression Analysis
The regression equation is
Range of motion = 108 + 0.871(Age)

Predictor	Coef	StDev	T	P
Constant	107.58	11.12	9.67	0.000
Age	0.8710	0.4146	2.10	0.062

$s = 10.42$ R-Sq = 30.6%   R-Sq(adj) = 23.7%

The resulting least-squares regression line is $\hat{y} = 107.58 + 0.871x$. A scatterplot
that also includes the regression line is shown in Figure 5.13. The residuals for this
data set are the vertical distances from the points to the regression line.

For the youngest patient (patient 9 with $x_9 = 14$ and $y_9 = 108$), the corresponding predicted value and residual are

predicted value = $\hat{y}_9 = 107.58 + 0.871(14) = 119.78$

residual = $y_9 - \hat{y}_9 = 108 - 119.78 = -11.78$

Figure 5.13 Scatter-plot for the data of Example 5.8.

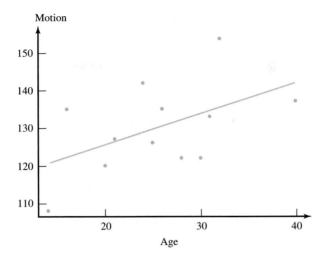

The other predicted values and residuals are computed in a similar manner and are included in Table 5.1.

Computing the predicted values and residuals by hand can be tedious, but MINITAB and other statistical software packages, as well as many graphing calculators, include them as part of the output, as shown in Figure 5.14. The predicted values and residuals can be found in the table at the bottom of the MINITAB output in the columns labeled "Fit" and "Residual," respectively.

Figure 5.14 MINITAB output for the data of Example 5.8.

The regression equation is

Range of Motion = 108 + 0.871 Age

Predictor	Coef	StDev	T	P
Constant	107.58	11.12	9.67	0.000
Age	0.8710	0.4146	2.10	0.062

s = 10.42   R-Sq = 30.6%   R-Sq(adj) = 23.7%

Analysis of Variance

Source	DF	SS	MS	F	P
Regression	1	479.2	479.2	4.41	0.062
Residual Error	10	1085.7	108.6		
Total	11	1564.9			

Obs	Age	Range of	Fit	SE Fit	Residual	St Resid
1	35.0	154.00	138.07	4.85	15.93	1.73
2	24.0	142.00	128.49	3.10	13.51	1.36
3	40.0	137.00	142.42	6.60	−5.42	−0.67
4	31.0	133.00	134.58	3.69	−1.58	−0.16
5	28.0	122.00	131.97	3.14	−9.97	−1.00
6	25.0	126.00	129.36	3.03	−3.36	−0.34
7	26.0	135.00	130.23	3.01	4.77	0.48
8	16.0	135.00	121.52	5.07	13.48	1.48
9	14.0	108.00	119.78	5.75	−11.78	−1.36
10	20.0	120.00	125.00	3.86	−5.00	−0.52
11	21.0	127.00	125.87	3.61	1.13	0.12
12	30.0	122.00	133.71	3.47	−11.71	−1.19

▪ **Plotting the Residuals** A careful look at residuals can reveal many potential problems. A *residual plot* is a good place to start when assessing the appropriateness of the regression line.

▪ **Definition**

A **residual plot** is a scatterplot of the (*x*, residual) pairs.

Isolated points or a pattern of points in the residual plot indicate potential problems. A desirable plot is one that exhibits no particular pattern, such as curvature. Curvature in the residual plot is an indication that the relationship between *x* and *y* is not linear and that a curve would be a better choice than a line for describing the relationship between *x* and *y*. This is sometimes easier to see in a residual plot than in a scatterplot of *y* versus *x*, as illustrated in Example 5.9.

▪ **Example 5.9**   **Heights and Weights of American Women**

Consider the accompanying data on x = height (in inches) and y = average weight (in pounds) for American females, age 30–39 (from *The World Almanac and Book of Facts*). The scatterplot displayed in Figure 5.15(a) appears rather straight. However, when the residuals from the least-squares line ($\hat{y} = 98.23 + 3.59x$) are plotted, substantial curvature is apparent (even though $r \approx .99$). It is not accurate to say that weight increases in direct proportion to height (linearly with height). Instead, average weight increases somewhat more rapidly for relatively large heights than it does for relatively small heights.

x	58	59	60	61	62	63	64	65	66	67	68	69	70	71	72
y	113	115	118	121	124	128	131	134	137	141	145	150	153	159	164

Figure 5.15 Plots for the data of Example 5.9: (a) scatterplot; (b) residual plot.

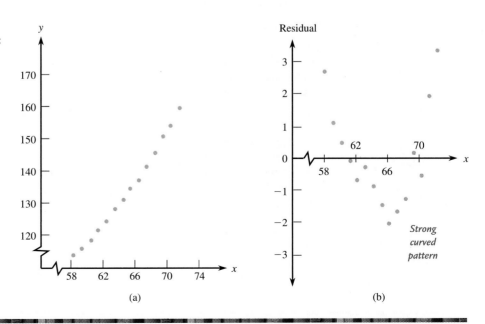

(a)

(b)

There is another common type of residual plot — one that plots the residuals versus the corresponding $\hat{y}$ values rather than versus the x values. Because $\hat{y} = a + bx$ is simply a linear function of x, the only real difference between the two types of residual plots is the scale on the horizontal axis. The pattern of points in the residual plots will be the same, and it is this pattern of points that is important, not the scale. Thus the two plots give equivalent information, as can be seen in Figure 5.16, which gives both plots for the data of Example 5.8.

Figure 5.16 (a) Plot of residuals versus x; (b) plot of residuals versus $\hat{y}$.

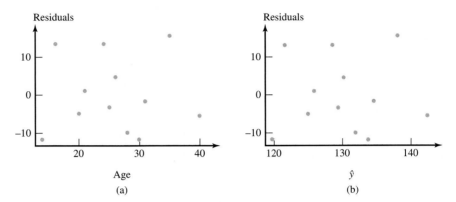

It is also important to look for unusual values in the scatterplot or in the residual plot. A point falling far above or below the horizontal line at height 0 corresponds to a large residual, which may indicate some type of unusual behavior, such as a recording error, a nonstandard experimental condition, or an atypical experimental subject. A point whose x value differs greatly from others in the data set may have exerted excessive influence in determining the fitted line. One method for assessing the impact of such an isolated point on the fit is to delete it from the data set, recompute the best-fit line, and evaluate the extent to which the equation of the line has changed.

▪ Example 5.10   Tennis Elbow

One factor in the development of tennis elbow is the impact-induced vibration of the racket and arm at ball contact. Tennis elbow is thought to be related to various properties of the tennis racket used. The following data are a subset of those analyzed in the article "Transfer of Tennis Racket Vibrations into the Human Forearm" (*Medicine and Science in Sports and Exercise* [1992]: 1134–1140). Measurements on x = racket resonance frequency (in hertz) and y = sum of peak-to-peak accelerations (a characteristic of arm vibration in meters per seconds squared) are given for $n = 14$ different rackets:

Racket	Resonance (x)	Acceleration (y)	Racket	Resonance (x)	Acceleration (y)
1	105	36.0	8	114	33.8
2	106	35.0	9	114	35.0
3	110	34.5	10	119	35.0
4	111	36.8	11	120	33.6
5	112	37.0	12	121	34.2
6	113	34.0	13	126	36.2
7	113	34.2	14	189	30.0

A scatterplot and a residual plot are shown in Figures 5.17(a) and 5.17(b), respectively. One observation in the data set is far to the right of the other points in the scatterplot. Because the least-squares line minimizes the sum of squared residuals, the least-squares line is pulled down toward this discrepant point. This single observation plays a big role in determining the slope of the least-squares line, and it is therefore called an *influential observation*. Notice that an influential observation is not necessarily the one with the largest residual, because the least-squares line actually passes near this point.

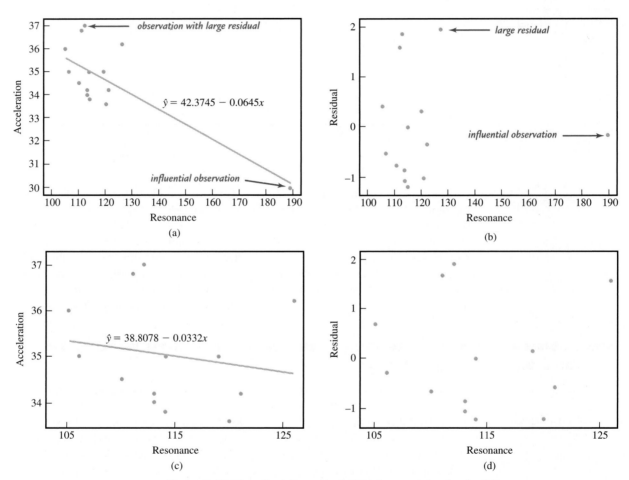

Figure 5.17 Plots from Example 5.10: (a) scatterplot for the full sample; (b) residual plot for the full sample; (c) scatterplot when the influential observation is deleted; (d) residual plot when the influential observation is deleted.

Figure 5.17(c) and 5.17(d) show what happens when the influential observation is removed from the sample. Both the slope and the intercept of the least-squares line are quite different from the slope and intercept of the line with this observation included.

Deletion of the observation corresponding to the largest residual (1.955 for observation 13) also changes the values of the slope and intercept of the least-squares line, but these changes are not profound. For example, the least-squares line for the data set consisting of the 13 points that remain when observation 13 is deleted is $\hat{y} = 42.5 - 0.0670x$. This observation does not appear to be all that influential.

Careful examination of a scatterplot and a residual plot can help us determine the appropriateness of a line for summarizing a relationship. If we decide that a line is appropriate, the next step is to think about assessing the accuracy of predictions based on the least-squares line and whether these predictions (based on the value of x) are better in general than those made without knowledge of the value of x. Two numerical measures that are helpful in this assessment are the coefficient of determination and the standard deviation about the regression line.

■ Coefficient of Determination

Suppose that we would like to predict the price of homes in a particular city. A random sample of 20 homes that are for sale is selected, and y = price and x = size (in square feet) are recorded for each house in the sample. Undoubtedly, there will be variability in house price (the houses will differ with respect to price), and it is this variability that makes accurate prediction of price a challenge. How much of the variability in house price can be explained by the fact that price is related to house size and that houses differ in size? If differences in size account for a large proportion of the variability in price, a price prediction that takes house size into account is a big improvement over prediction that is not based on size.

The **coefficient of determination** is a measure of the proportion of variability in the y variable that can be "explained" by a linear relationship between x and y.

■ **Definition**

The **coefficient of determination**, denoted by r^2, gives the proportion of variation in y that can be attributed to an approximate linear relationship between x and y; $100r^2$ is the percentage of variation in y that can be attributed to an approximate linear relationship between x and y.

Variation in y can effectively be explained by an approximate straight-line relationship when the points in the scatterplot fall close to the least-squares line — that is, when the residuals are small in magnitude. A natural measure of variation about the least-squares line is the sum of the squared residuals. (Squaring before combining prevents negative and positive residuals from counteracting one another.) A second sum of squares assesses the total amount of variation in observed y values.

■ **Definition**

The **total sum of squares**, denoted by **SSTo**, is defined as

$$\text{SSTo} = (y_1 - \bar{y})^2 + (y_2 - \bar{y})^2 + \cdots + (y_n - \bar{y})^2 = \sum (y - \bar{y})^2$$

(*continued*)

The **residual sum of squares** (sometimes referred to as the error sum of squares), denoted by **SSResid**, is defined as

$$\text{SSResid} = (y_1 - \hat{y}_1)^2 + (y_2 - \hat{y}_2)^2 + \cdots + (y_n - \hat{y}_n)^2 = \sum (y - \hat{y})^2$$

These sums of squares can be found as part of the regression output from most standard statistical packages or can be obtained using the following computational formulas:

$$\text{SSTo} = \sum y^2 - \frac{(\sum y)^2}{n}$$

$$\text{SSResid} = \sum y^2 - a \sum y - b \sum xy$$

▪ Example 5.11   Range of Motion Revisited

Figure 5.18 displays part of the MINITAB output that results from fitting the least-squares line to the data on range of motion and age from Example 5.8. From the output,

SSTo = 1564.9 and SSResid = 1085.7

Notice that SSResid is relatively large compared to SSTo.

Figure 5.18 MINITAB output for the data of Example 5.11.

Regression Analysis

The regression equation is
Range of Motion = 108 + 0.871 Age

Predictor	Coef	StDev	T	P
Constant	107.58	11.12	9.67	0.000
Age	0.8710	0.4146	2.10	0.062

S = 10.42   R-Sq = 30.6%   R-Sq(adj) = 23.7%

Analysis of Variance

Source	DF	SS	MS	F	P
Regression	1	479.2	479.2	4.41	0.062
Residual Error	10	1085.7	108.6		
Total	11	1564.9			

SSResid ↖ 1085.7   SSTo ↗ 1564.9

The residual sum of squares is the sum of squared vertical deviations from the least-squares line. As Figure 5.19 illustrates, SSTo is also a sum of squared vertical deviations from a line — the horizontal line at height $\bar{y}$. The least-squares line is, by definition, the one having the smallest sum of squared deviations. It follows that SSResid ≤ SSTo. The two sums of squares are equal only when the least-squares line *is* the horizontal line.

Figure 5.19 Interpreting sums of squares: (a) SSResid = sum of squared vertical deviations from the least-squares line; (b) SSTo = sum of squared vertical deviations from the horizontal line at height $\bar{y}$.

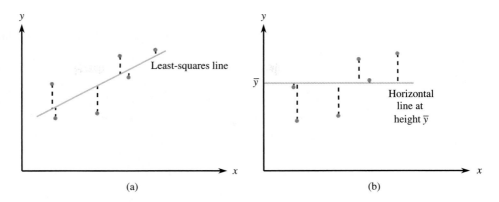

(a)   (b)

SSResid is often referred to as a measure of unexplained variation — the amount of variation in y that cannot be attributed to the linear relationship between x and y. The more the points in the scatterplot deviate from the least-squares line, the larger the value of SSResid and the greater the amount of y variation that cannot be explained by the approximate linear relationship. Similarly, SSTo is interpreted as a measure of total variation. The larger the value of SSTo, the greater the amount of variability in $y_1, y_2, \ldots, y_n$. The ratio SSResid/SSTo is the fraction or proportion of total variation that is unexplained by a straight-line relation. Subtracting this ratio from 1 gives the proportion of total variation that *is* explained:

The coefficient of determination can be computed as

$$r^2 = 1 - \frac{\text{SSResid}}{\text{SSTo}}$$

Multiplying r^2 by 100 gives the percentage of y variation attributable to the approximate linear relationship. The closer this percentage is to 100%, the more successful is the relationship in explaining variation in y.

▪ **Example 5.12** r^2 **for Age and Range of Motion Data**

For the data on range of motion and age from Example 5.11, we found SSTo = 1564.9 and SSResid = 1085.7. Thus

$$r^2 = 1 - \frac{\text{SSResid}}{\text{SSTo}} = 1 - \frac{1085.7}{1564.9} = .306$$

This means that only 30.6% of the observed variability in postsurgical range of motion can be explained by an approximate linear relationship between range of motion and age. Note that the r^2 value can be found in the MINITAB output of Figure 5.18, labeled "R-Sq."

The symbol r was used in Section 5.1 to denote Pearson's sample correlation coefficient. It is not coincidental that r^2 is used to represent the coefficient of determination. The notation suggests how these two quantities are related:

$$(\text{correlation coefficient})^2 = \text{coefficient of determination}$$

Thus, if $r = .8$ or $r = -.8$, then $r^2 = .64$, so 64% of the observed variation in the dependent variable can be explained by the linear relationship. Because the value of r does not depend on which variable is labeled x, the same is true of r^2. The coefficient of determination is one of the few quantities computed in a regression analysis whose value remains the same when the role of dependent and independent variables are interchanged. When $r = .5$, we get $r^2 = .25$, so only 25% of the observed variation is explained by a linear relation. This is why a value of r between $-.5$ and $.5$ is not considered evidence of a strong linear relationship.

▪ Standard Deviation About the Least-Squares Line

The coefficient of determination measures the extent of variation about the best-fit line *relative* to overall variation in y. A high value of r^2 does not by itself promise that the deviations from the line are small in an absolute sense. A typical observation could deviate from the line by quite a bit, yet these deviations might still be small relative to overall y variation. Recall that in Chapter 4 the sample standard deviation

$$s = \sqrt{\frac{\sum (x - \bar{x})^2}{n - 1}}$$

was used as a measure of variability in a single sample; roughly speaking, s is the typical amount by which a sample observation deviates from the mean. There is an analogous measure of variability when a least-squares line is fit.

▪ **Definition**

The **standard deviation about the least-squares line** is given by

$$s_e = \sqrt{\frac{\text{SSResid}}{n - 2}}$$

Roughly speaking, s_e is the typical amount by which an observation deviates from the least-squares line. Justification for division by $(n - 2)$ and the use of the subscript e is given in Chapter 13.

▪ Example 5.13   Commuting Distances and Commuting Times

The values of x = commuting distance and y = commuting time were determined for workers in samples from three different regions. The data are given in Table 5.2; the three scatterplots are displayed in Figure 5.20.

For Region 1, a rather small proportion of variation in y can be attributed to an approximate linear relationship, and a typical deviation from the least-squares line is roughly 4. The amount of variability about the line for Region 2 is the same as for Region 1, but the value of r^2 is much higher, because y variation is much greater overall in Region 2 than in Region 1. Region 3 yields roughly the same high value of r^2 as does Region 2, but the typical deviation from the line for Region 3 is only half that for Region 2. For a complete picture of variation, both r^2 and s_e should be computed.

Table 5.2 ■ Data for Example 5.13

Region 1		Region 2		Region 3	
x	y	x	y	x	y
15	42	5	16	5	8
16	35	10	32	10	16
17	45	15	44	15	22
18	42	20	45	20	23
19	49	25	63	25	31
20	46	50	115	50	60

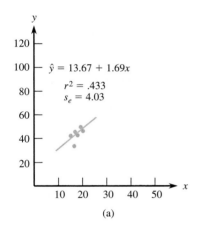

(a)

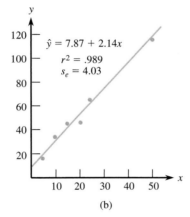

(b)

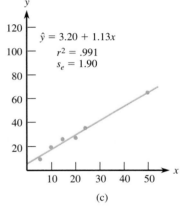

(c)

Figure 5.20 Scatterplots for the data of Example 5.13:
(a) Region 1; (b) Region 2; (c) Region 3.

■ Exercises 5.32–5.46

5.32 The following table gives the number of organ transplants performed in the United States each year from 1990 to 1999 (The Organ Procurement and Transplantation Network, 2003):

Year	Number of Transplants (in thousands)
1 (1990)	15.0
2	15.7
3	16.1
4	17.6
5	18.3
6	19.4
7	20.0
8	20.3
9	21.4
10 (1999)	21.8

a. Construct a scatterplot of these data, and then find the equation of the least-squares regression line that describes the relationship between y = number of transplants performed and x = year. Write a few sentences describing how the number of transplants performed has changed over time from 1990 to 1999.
b. Compute the 10 residuals, and construct a residual plot. Are there any features of the residual plot that indicate that the relationship between year and number of transplants performed would be better described by a curve rather than a line? Explain.

5.33 Anthropologists often study soil composition for clues to how the land was used during different periods. The following data on x = soil depth (in centimeters) and y = percentage of montmorillonite in the soil was taken from a scatterplot in the paper "Ancient Maya Drained Field Agriculture: Its

Possible Application Today in the New River Flood-plain, Belize, C.A." (*Agricultural Ecosystems and Environment* [1984]: 67–84):

x	40	50	60	70	80	90	100
y	58	34	32	30	28	27	22

a. Draw a scatterplot of y versus x.
b. The equation of the least-squares line is $\hat{y} = 64.50 - 0.45x$. Draw this line on your scatter-plot. Do there appear to be any large residuals?
c. Compute the residuals, and construct a residual plot. Are there any unusual features in the plot?

5.34 Data on pollution and cost of medical care for elderly people were given in Exercise 5.17 and are also shown here. The following data give a measure of pollution (micrograms of particulate matter per cubic meter of air) and the cost of medical care per person over age 65 for six geographic regions of the United States:

Region	Pollution	Cost of Medical Care
North	30.0	915
Upper South	31.8	891
Deep South	32.1	968
West South	26.8	972
Big Sky	30.4	952
West	40.0	899

The equation of the least-squares regression line for this data set is $\hat{y} = 1082.2 - 4.691x$, where y = medical cost and x = pollution.
a. Compute the six residuals.
b. What is the value of the correlation coefficient for this data set? Does the value of r indicate that the linear relationship between pollution and medical cost is strong, moderate, or weak? Explain.
c. Construct a residual plot. Are there any unusual features of the plot?
d. Because the observation for the West, (40.0, 899), has an x value that is far removed from the other x values in the sample, this observation might be influential in determining the values of the slope and/or intercept of the least-squares line. Is this observation influential? Justify your answer.

5.35 There have been numerous studies on the effects of radiation. The following data on the relationship between degree of exposure to ^{242}Cm alpha particles (x) and the percentage of exposed cells without aberrations (y) appeared in the paper "Chromosome Aberrations Induced in Human Lymphocytes by D-T Neutrons" (*Radiation Research* [1984]: 561–573):

x	0.106	0.193	0.511	0.527
y	98	95	87	85
x	1.08	1.62	1.73	2.36
y	75	72	64	55
x	2.72	3.12	3.88	4.18
y	44	41	37	40

Summary quantities are

$$n = 12 \qquad \sum x = 22.027 \qquad \sum y = 793$$

$$\sum x^2 = 62.600235 \qquad \sum xy = 1114.5$$

$$\sum y^2 = 57{,}939$$

a. Obtain the equation of the least-squares line.
b. Construct a residual plot, and comment on any interesting features.

5.36 Cost-to-charge ratio (the percentage of the amount billed that represents the actual cost) for in-patient and outpatient services at 11 Oregon hospitals is shown in the following table (Oregon Department of Health Services, 2002):

Hospital	Cost-to-Charge Ratio	
	Outpatient Care	Inpatient Care
1	62	80
2	66	76
3	63	75
4	51	62
5	75	100
6	65	88
7	56	64
8	45	50
9	48	54
10	71	83
11	54	100

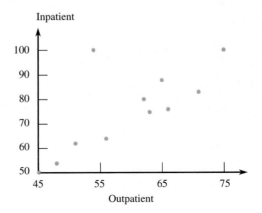

A scatterplot of the data is also shown:

The least-squares regression line with y = inpatient cost-to-charge ratio and x = outpatient cost-to-charge ratio is $\hat{y} = -1.1 + 1.29x$.

a. Is the observation for Hospital 11 an influential observation? Justify your answer.

b. Is the observation for Hospital 11 an outlier? Explain.

c. Is the observation for Hospital 5 an influential observation? Justify your answer.

d. Is the observation for Hospital 5 an outlier? Explain.

5.37 The article "Examined Life: What Stanley H. Kaplan Taught Us About the SAT" (*The New Yorker* [December 17, 2001]: 86–92) included a summary of findings regarding the use of SAT I scores, SAT II scores, and high school grade point average (GPA) to predict first-year college GPA. The article states that "among these, SAT II scores are the best predictor, explaining 16 percent of the variance in first-year college grades. GPA was second at 15.4 percent, and SAT I was last at 13.3 percent."

a. If the data from this study were used to fit a least-squares line with y = first-year college GPA and x = high school GPA, what would the value of r^2 have been?

b. The article stated that SAT II was the best predictor of first-year college grades. Do you think that predictions based on a least-squares line with y = first-year college GPA and x = SAT II score would have been very accurate? Explain why or why not.

5.38 Exercise 5.18 gave the least-squares regression line for predicting y = clutch size from x = snout-vent length ("Reproductive Biology of the Aquatic Salamander *Amphiuma tridactylum* in Louisiana," *Journal of Herpetology* [1999]: 100–105). The paper also reported $r^2 = .7664$ and SSTo = 43,951.

a. Interpret the value of r^2.

b. Find and interpret the value of s_e (the sample size was $n = 14$).

5.39 The article "Characterization of Highway Runoff in Austin, Texas, Area" (*Journal of Environmental Engineering* [1998]: 131–137) gave a scatterplot, along with the least-squares line for x = rainfall volume (in cubic meters) and y = runoff volume (in cubic meters), for a particular location. The following data were read from the plot in the paper:

x	5	12	14	17	23	30	40	47
y	4	10	13	15	15	25	27	46

x	55	67	72	81	96	112	127
y	38	46	53	70	82	99	100

a. Does a scatterplot of the data suggest a linear relationship between x and y?

b. Calculate the slope and intercept of the least-squares line.

c. Compute an estimate of the average runoff volume when rainfall volume is 80.

d. Compute the residuals, and construct a residual plot. Are there any features of the plot that indicate that a line is not an appropriate description of the relationship between x and y? Explain.

5.40 The article cited in the chapter introduction reported that for a regression of y = average SAT score on x = expenditure per pupil, based on data from $n = 44$ New Jersey school districts, $a = 766$, $b = 0.015$, $r^2 = .160$, and $s_e = 53.7$.

a. One observation in the sample was (9900, 893). What average SAT score would you predict for this district, and what is the corresponding residual?

b. Interpret the value of s_e.

c. How effectively do you think the least-squares line summarizes the relationship between x and y? Explain your reasoning.

5.41 The decline of salmon fisheries along the Columbia River in Oregon has caused great concern among commercial and recreational fishermen. The paper "Feeding of Predaceous Fishes on Out-Migrating Juvenile Salmonids in John Day Reservoir, Columbia River" (*Transactions of the American Fisheries Society* [1991]: 405–420) gave the following data on y = maximum size of salmonids consumed by a northern squawfish (the most abundant salmonid predator) and x = squawfish length, both in millimeters:

x	218	246	270	287	318	344
y	82	85	94	127	141	157

x	375	386	414	450	468
y	165	216	219	238	249

Use the accompanying output from MINITAB to answer the following questions.

The regression equation is
size = −89.1 + 0.729 length

Predictor	Coef	Stdev	t ratio	p
Constant	−89.09	16.83	−5.29	0.000
length	0.72907	0.04778	15.26	0.000

$s = 12.56$ R-sq = 96.3%   R-sq(adj) = 95.9%

Analysis of Variance

SOURCE	DF	SS	MS	F	p
Regression	1	36736	36736	232.87	0.000
Error	9	1420	158		
Total	10	38156			

a. What maximum size would you predict for a squawfish whose length is 375 mm, and what is the residual corresponding to the observation (375, 165)?

b. What proportion of observed variation in y can be attributed to the approximate linear relationship between the two variables?

5.42 The paper "Crop Improvement for Tropical and Subtropical Australia: Designing Plants for Difficult Climates" (*Field Crops Research* [1991]: 113–139) gave the following data on x = crop duration (in days) for soybeans and y = crop yield (in tons per hectare):

x	92	92	96	100	102
y	1.7	2.3	1.9	2.0	1.5
x	102	106	106	121	143
y	1.7	1.6	1.8	1.0	0.3

$$\sum x = 1060 \qquad \sum y = 15.8 \qquad \sum xy = 1601.1$$

$$\sum x^2 = 114{,}514 \qquad \sum y^2 = 27.82$$

$$a = 5.20683380 \qquad b = -0.3421541$$

a. Construct a scatterplot of the data. Do you think the least-squares line will give accurate predictions? Explain.
b. The largest x value in the sample greatly exceeds the remaining ones. Delete the corresponding observation from the sample, and recalculate the equation of the least-squares line. Does this observation greatly affect the equation of the line?
c. What effect does the deletion suggested in Part (b) have on the value of r^2? Can you explain why this is so?

5.43 A study was carried out to investigate the relationship between the hardness of molded plastic (y, in Brinell units) and the amount of time elapsed since termination of the molding process (x, in hours). Summary quantities include $n = 15$, SSResid = 1235.470, and SSTo = 25,321.368. Calculate and interpret the coefficient of determination.

5.44 There has been an ongoing debate in recent years as to whether various species of wolves are endangered and what steps should be taken to prevent their demise. Disease can certainly pose a serious threat to such wildlife populations. The paper "Effects of Canine Parvovirus (CPV) on Gray Wolves in Minnesota" (*Journal of Wildlife Management* [1995]: 565–570) summarized a regression of y = percentage of pups in a capture on x = percentage of CPV prevalence among adults and pups. The equation of the least-squares line, based on $n = 10$ observations, was $\hat{y} = 62.9476 - 0.54975x$, with $r^2 = .57$.

a. One observation was (25, 70). What is the corresponding residual?
b. What is the value of the sample correlation coefficient?
c. Suppose that SSTo = 2520.0 (this value was not given in the paper). What is the value of s_e?

5.45 Both r^2 and s_e are used to assess the fit of a line.
a. Is it possible that both r^2 and s_e could be large for a bivariate data set? Explain. (A picture might be helpful.)
b. Is it possible that a bivariate data set could yield values of r^2 and s_e that are both small? Explain. (Again, a picture might be helpful.)
c. Explain why it is desirable to have r^2 large and s_e small if the relationship between two variables x and y is to be described using a straight line.

5.46 Some straightforward but slightly tedious algebra shows that

$$\text{SSResid} = (1 - r^2) \sum (y - \bar{y})^2$$

from which it follows that

$$s_e = \sqrt{\frac{n-1}{n-2}} \sqrt{1 - r^2} s_y$$

Unless n is quite small, $\dfrac{n-1}{n-2} \approx 1$, so $s_e \approx \sqrt{1 - r^2} s_y$.

a. For what value of r is s_e as large as s_y? What is the least-squares line in this case?
b. For what values of r will s_e be much smaller than s_y?
c. A study by the Berkeley Institute of Human Development (see the book *Statistics* by Freedman et al., listed in the references for Chapter 2) reported the following summary data for a sample of $n = 66$ California boys:

> $r \approx .80$
> At age 6, average height $\approx$ 46 in., standard deviation $\approx$ 1.7 in.
> At age 18, average height $\approx$ 70 in., standard deviation $\approx$ 2.5 in.

What would s_e be for the least-squares line used to predict 18-year-old height from 6-year-old height?
d. Referring to Part (c), suppose that you wanted to predict the past value of 6-year-old height from knowledge of 18-year-old height. Find the equation for the appropriate least-squares line. What is the corresponding value of s_e?

▪ 5.4 Nonlinear Relationships and Transformations

As we have seen in previous sections, when the points in a scatterplot exhibit a linear pattern and the residual plot does not reveal any problems with the linear fit, the least-squares line is a sensible way to summarize the relationship between x and y. A linear relationship is easy to interpret, departures from the line are easily detected, and using the line to predict y from our knowledge of x is straightforward. Often, though, a scatterplot or residual plot exhibits a curved pattern, indicating a more complicated relationship between x and y. In this case, finding a curve that fits the observed data well is a more complex task. In this section, we consider two common approaches to fitting nonlinear relationships: polynomial regression and transformations.

▪ Polynomial Regression

Let's reconsider the data first introduced in Example 3.22 on $x =$ time between flowering and harvesting and $y =$ yield of paddy (a type of grain):

x	16	18	20	22	24	26	28	30
y	2508	2518	3304	3423	3057	3190	3500	3883

x	32	34	36	38	40	42	44	46
y	3823	3646	3708	3333	3517	3214	3103	2776

The scatterplot of these data is reproduced here as Figure 5.21. Because this plot shows a marked curved pattern, it is clear that no straight line can do a reasonable job of describing the relationship between x and y. However, the relationship can be described by a curve, and in this case the curved pattern in the scatterplot looks like a parabola (the graph of a quadratic function). This suggests trying to find a quadratic function of the form

$$\hat{y} = a + b_1 x + b_2 x^2$$

that would reasonably describe the relationship. That is, the values of the coefficients a, b_1, and b_2 in this equation must be selected to obtain a good fit to the data.

Figure 5.21 ▪ Scatterplot for the paddy data.

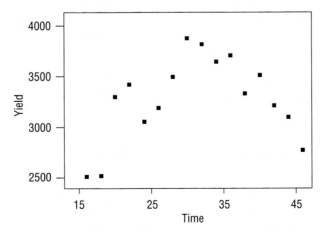

What are the best choices for the values of a, b_1, and b_2? In fitting a line to data, we used the principle of least squares to guide our choice of slope and intercept.

Least squares can be used to fit a quadratic function as well. The deviations, $y - \hat{y}$, are still represented by vertical distances in the scatterplot, but now they are vertical distances from the points to a parabolic curve (the graph of a quadratic equation) rather than to a line, as shown in Figure 5.22. We then choose values for the coefficients in the quadratic equation so that the sum of squared deviations is as small as possible.

Figure 5.22 Deviation for a quadratic function.

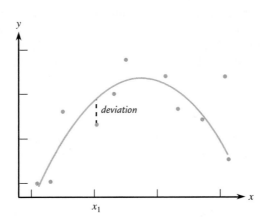

For a quadratic regression, the least-squares estimates of a, b_1, and b_2 are those values that minimize the sum of squared deviations $\sum (y - \hat{y})^2$, where $\hat{y} = a + b_1 x + b_2 x^2$.

For quadratic regression, a measure that is useful for assessing fit is

$$R^2 = 1 - \frac{\text{SSResid}}{\text{SSTo}}$$

where $\text{SSResid} = \sum (y - \hat{y})^2$. The measure R^2 is defined in a way similar to r^2 for simple linear regression and is interpreted in a similar fashion. The notation r^2 is used only with linear regression to emphasize the relationship between r^2 and the correlation coefficient, r, in the linear case.

The general expressions for computing the least-squares estimates are somewhat complicated, so we rely on a statistical software package or graphing calculator to do the computations for us.

▪ Example 5.14   Paddy Yield Data Revisited: Fitting a Quadratic Model

For the paddy yield data, the scatterplot (see Figure 5.21) showed a marked curved pattern. If the least-squares line is fit to these data, it is no surprise that the line does not do a good job of describing the relationship ($r^2 = .075$ or 7.5% and $s_e = 416.7$), and the residual plot shows a distinct curved pattern as well (Figure 5.23).

Part of the MINITAB output from fitting a quadratic function to these data is as follows:

Figure 5.23 Plots for the paddy data of Example 5.14: (a) fitted line plot; (b) residual plot.

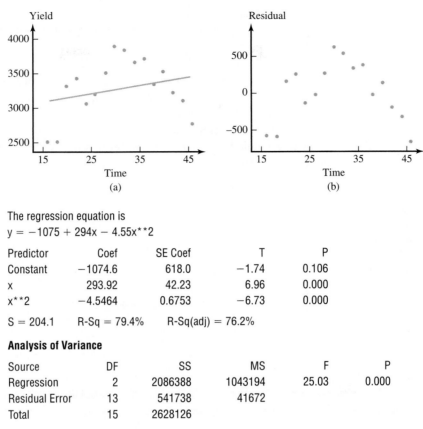

The regression equation is
$$y = -1075 + 294x - 4.55x^{**}2$$

Predictor	Coef	SE Coef	T	P
Constant	−1074.6	618.0	−1.74	0.106
x	293.92	42.23	6.96	0.000
x**2	−4.5464	0.6753	−6.73	0.000

$S = 204.1$ R-Sq = 79.4%   R-Sq(adj) = 76.2%

Analysis of Variance

Source	DF	SS	MS	F	P
Regression	2	2086388	1043194	25.03	0.000
Residual Error	13	541738	41672		
Total	15	2628126			

The least-squares coefficients are $a = -1074.63$, $b_1 = 293.924$, and $b_2 = -4.54644$, and the least-squares quadratic is

$$\hat{y} = -1074.63 + 293.924x - 4.54644x^2$$

A plot showing the curve and the corresponding residual plot for the quadratic regression are given in Figure 5.24. Notice that there is no strong pattern in the residual plot for the quadratic case as there was in the linear case. For the quadratic regression, $R^2 = .794$ (as opposed to .075 for the least-squares line), which means that 79.4% of the variability in yield can be explained by an approximate quadratic relationship between yield and time between flowering and harvesting.

Figure 5.24 Quadratic regression of Example 5.14: (a) scatterplot; (b) residual plot.

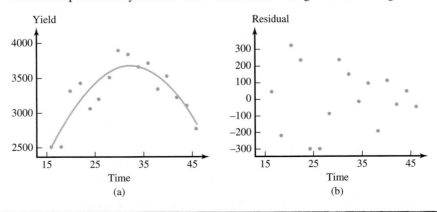

Linear and quadratic regression are special cases of polynomial regression. A polynomial regression curve is described by a function of the form

$$\hat{y} = a + b_1 x + b_2 x^2 + b_3 x^3 + \cdots + b_k x^k$$

which is called a kth-degree polynomial. The case of $k = 1$ results in linear regression ($\hat{y} = a + b_1 x$) and $k = 2$ yields a quadratic regression ($\hat{y} = a + b_1 x + b_2 x^2$). A quadratic curve has only one bend (see Figures 5.25(a) and 5.25(b)). A less frequently encountered special case is for $k = 3$, where $\hat{y} = a + b_1 x + b_2 x^2 + b_3 x^3$, which is called a cubic regression curve. Cubic curves have two bends, as shown in Figure 5.25(c).

Figure 5.25 Polynomial regression curves: (a) quadratic curve with $b_2 < 0$; (b) quadratic curve with $b_2 > 0$; (c) cubic curve.

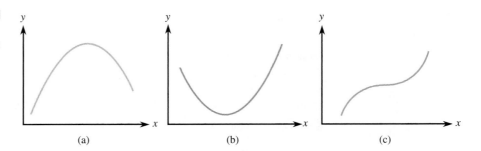

(a)   (b)   (c)

▪ Example 5.15   Nonlinear Relationship Between Cloud Cover Index and Sunshine Index

Researchers have examined a number of climatic variables in an attempt to understand the mechanisms that govern rainfall runoff. The article "The Applicability of Morton's and Penman's Evapotranspiration Estimates in Rainfall-Runoff Modeling" (*Water Resources Bulletin* [1991]: 611–620) reported on a study that examined the relationship between x = cloud cover index and y = sunshine index.

Suppose that the cloud cover index can have values between 0 and 1. Consider the accompanying data, which are consistent with some of the summary quantities in the article:

Cloud Cover Index (x)	Sunshine Index (y)
0.2	10.98
0.5	10.94
0.3	10.91
0.1	10.94
0.2	10.97
0.4	10.89
0	10.88
0.4	10.92
0.3	10.86

The authors of the article used a cubic regression to describe the relationship between cloud cover and sunshine. Using the given data, MINITAB was instructed to fit a cubic regression, resulting in the following output:

Polynomial Regression

The regression equation is

y = 10.8768 + 1.46036x − 7.25901x**2 + 9.23423x**3

S = 0.0315265   R-Sq = 62.0%   R-Sq (adj) = 39.3%

Analysis of Variance

Source	DF	SS	MS	F	P
Regression	3	0.0081193	0.0027064	2.72299	0.154
Error	5	0.0049696	0.0009939		
Total	8	0.0130889			

The least-squares cubic regression is then

$$\hat{y} = 10.8768 + 1.460636x - 7.25901x^2 + 9.23423x^3$$

To predict the sunshine index for a day when the cloud cover index is 0.45, we use the regression equation to obtain

$$\hat{y} = 10.8768 + 1.46036(0.45) - 7.25901(0.45)^2 + 9.23423(0.45)^3 = 10.91$$

The least-squares cubic regression and the corresponding residual plot are shown in Figure 5.26. The residual plot does not reveal any troublesome patterns that would suggest a choice other than the cubic regression.

Figure 5.26 Plots for the data of Example 5.15: (a) least-squares cubic regression; (b) residual plot.

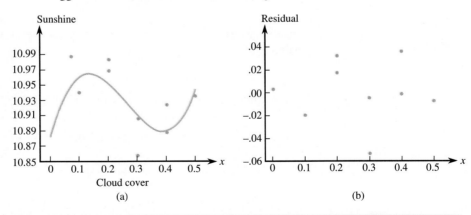

▪ Transformations

An alternative to finding a curve to fit the data is to find a way to transform the x values and/or y values so that a scatterplot of the transformed data has a linear appearance. A **transformation** (sometimes called a reexpression) involves using a simple function of a variable in place of the variable itself. For example, instead of trying to describe the relationship between x and y, it might be easier to describe the relationship between $\sqrt{x}$ and y or between x and log(y). And, if we can describe the relationship between, say, $\sqrt{x}$ and y, we will still be able to predict the value of y for a given x value. Common transformations involve taking square roots, logarithms, or reciprocals.

▪ **Example 5.16** River Water Velocity and Distance from Shore

As fans of white-water rafting know, a river flows more slowly close to its banks (because of friction between the river bank and the water). To study the nature of the relationship between water velocity and the distance from the shore, data were gathered on velocity (in centimeters per second) of a river at different distances (in meters) from the bank. Suppose that the resulting data were as follows:

Distance	0.5	1.5	2.5	3.5	4.5	5.5	6.5	7.5	8.5	9.5
Velocity	22.00	23.18	25.48	25.25	27.15	27.83	28.49	28.18	28.50	28.63

A graph of the data exhibits a curved pattern, as seen in both the scatterplot and the residual plot from a linear fit (see Figures 5.27(a) and 5.27(b)).

Figure 5.27 Plots for the data of Example 5.16: (a) scatterplot of the river data; (b) residual plot from linear fit.

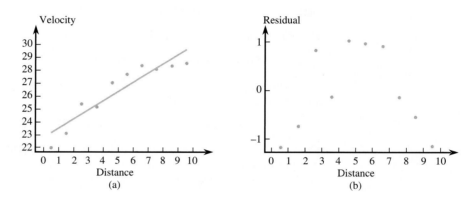

Let's try transforming the x values by replacing each x value by its square root. We define

$$x' = \sqrt{x}$$

The resulting transformed data are given in Table 5.3.

Table 5.3 ▪ **Original and Transformed Data of Example 5.16**

Original Data		Transformed Data	
x	y	x'	y
0.5	22.00	0.7071	22.00
1.5	23.18	1.2247	23.18
2.5	25.48	1.5811	25.48
3.5	25.25	1.8708	25.25
4.5	27.15	2.1213	27.15
5.5	27.83	2.3452	27.83
6.5	28.49	2.5495	28.49
7.5	28.18	2.7386	28.18
8.5	28.50	2.9155	28.50
9.5	28.63	3.0822	28.63

Figure 5.28 Plots for the transformed data of Example 5.16: (a) scatterplot of y versus x'; (b) residual plot resulting from a linear fit to the transformed data.

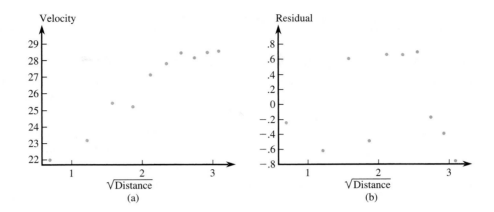

(a)   (b)

Figure 5.28(a) shows a scatterplot of y versus x' (or equivalently y versus $\sqrt{x}$). The pattern of points in this plot looks linear, and so we can fit a least-squares line using the transformed data. The MINITAB output from this regression is as follows:

Regression Analysis

The regression equation is
velocity = 20.1 + 3.01 sqrt distance

Predictor	Coef	StDev	T	P
Constant	20.1102	0.6097	32.99	0.000
sqrt dis	3.0085	0.2726	11.03	0.000

S = 0.6292   R-Sq = 93.8%   R-Sq(adj) = 93.1%

Analysis of Variance

Source	DF	SS	MS	F	P
Regression	1	48.209	48.209	121.76	0.000
Residual Error	8	3.168	0.396		
Total	9	51.376			

The residual plot in Figure 5.28(b) shows no indication of a pattern. The resulting regression equation is

$$\hat{y} = 20.1 + 3.01x'$$

or, equivalently,

$$\hat{y} = 20.1 + 3.01\sqrt{x}$$

The values of r^2 and s_e (see the MINITAB output) indicate that a line is a reasonable way to describe the relationship between y and x'. To predict velocity of the river at a distance of 9 m from shore, we first compute $x' = \sqrt{x} = \sqrt{9} = 3$ and then use the sample regression line to obtain a prediction of y:

$$\hat{y} = 20.1 + 3.01x' = 20.1 + (3.01)(3) = 29.13$$

In Example 5.16, transforming the x values using the square root function worked well. In general, how can we choose a transformation that will result in a

Table 5.4   ▪   Commonly Used Transformations

Transformation	Mathematical Description	Try This Transformation When
No transformation	$\hat{y} = a + bx$	The change in y is constant as x changes. A 1-unit increase in x is associated with, on average, an increase of b in the value of y.
Square root of x	$\hat{y} = a + b\sqrt{x}$	The change in y is not constant. A 1-unit increase in x is associated with smaller increases or decreases in y for larger x values.
Log[a] of x	$\hat{y} = a + b\log_{10}(x)$ or $\hat{y} = a + b\ln(x)$	The change in y is not constant. A 1-unit increase in x is associated with smaller increases or decreases in the value of y for larger x values.
Reciprocal of x	$\hat{y} = a + b\left(\dfrac{1}{x}\right)$	The change in y is not constant. A 1-unit increase in x is associated with smaller increases or decreases in the value of y for larger x values. In addition, y has a limiting value of a as x increases.
Log[a] of y (exponential growth or decay)	$\log(\hat{y}) = a + bx$ or $\ln(\hat{y}) = a + bx$	The change in y associated with a 1-unit change in x is proportional to x.

[a]The values of a and b in the regression equation will depend on whether $\log_{10}$ or ln is used, but the $\hat{y}$'s and r^2 values will be identical.

linear pattern? Table 5.4 gives some guidance and summarizes some of the properties of the most commonly used transformations.

■ **Example 5.17   Tortilla Chips: Relationship Between Frying Time and Moisture Content**

No tortilla chip lover likes soggy chips, so it is important to find characteristics of the production process that produce chips with an appealing texture. The following data on x = frying time (in seconds) and y = moisture content (%) appeared in the article "Thermal and Physical Properties of Tortilla Chips as a Function of Frying Time" (*Journal of Food Processing and Preservation* [1995]: 175–189):

Frying time, x	5	10	15	20	25	30	45	60
Moisture content, y	16.3	9.7	8.1	4.2	3.4	2.9	1.9	1.3

Figure 5.29(a) shows a scatterplot of the data. The pattern in this plot is typical of exponential decay, with the change in y as x increases much smaller for large x values than for small x values. You can see that a 5-sec change in frying time is associated with a much larger change in moisture content in the part of the plot where the x values are small than in the part of the plot where the x values are large. Table 5.4 suggests transforming the y values (moisture content in this example) by taking their logarithms.

Figure 5.29 Plots for the data of Example 5.17: (a) scatterplot of the tortilla chip data; (b) scatterplot of transformed data with $y' = \log(y)$; (c) scatterplot of transformed data with $y' = \ln(y)$.

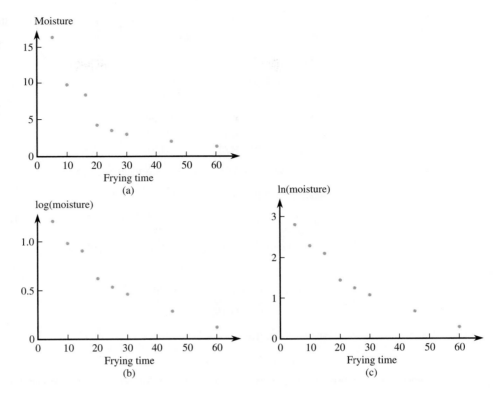

Two standard logarithmic functions are commonly used for such transformations — the common logarithm (log base 10, denoted by log or $\log_{10}$) and the natural logarithm (log base e, denoted ln). Either the common or the natural logarithm can be used; the only difference in the resulting scatterplots is the scale of the transformed y variable. This can be seen in Figures 5.29(b) and 5.29(c). These two scatterplots show the same pattern, and it looks like a line would be appropriate to describe this relationship.

Table 5.5 displays the original data along with the transformed y values using $y' = \log(y)$.

Table 5.5 ▪ Transformed Data from Example 5.17

Frying Time x	Moisture y	Log(moisture) y'
5	16.3	1.21219
10	9.7	0.98677
15	8.1	0.90849
20	4.2	0.62325
25	3.4	0.53148
30	2.9	0.46240
45	1.9	0.27875
60	1.3	0.11394

The following MINITAB output shows the result of fitting the least-squares line to the transformed data:

Regression Analysis

The regression equation is
log(moisture) = 1.14 − 0.0192 frying time

Predictor	Coef	StDev	T	P
Constant	1.14287	0.08016	14.26	0.000
frying t	−0.019170	0.002551	−7.52	0.000

S = 0.1246   R-Sq = 90.4%   R-Sq(adj) = 88.8%

Analysis of Variance

Source	DF	SS	MS	F	P
Regression	1	0.87736	0.87736	56.48	0.000
Residual Error	6	0.09320	0.01553		
Total	7	0.97057			

The resulting regression equation is

$$y' = 1.14 - 0.0192x$$

or, equivalently,

$$\log(y) = 1.14 - 0.0192x$$

▪ **Fitting a Curve Using Transformations** The objective of a regression analysis is usually to describe the approximate relationship between x and y with an equation of the form $y =$ some function of x.

If we have transformed only x, fitting a least-squares line to the transformed data results in an equation of the desired form, for example,

$$\hat{y} = 5 + 3x' = 5 + 3\sqrt{x}, \quad \text{where } x' = \sqrt{x}$$

or

$$\hat{y} = 4 + 0.2x' = 4 + 0.2\frac{1}{x}, \quad \text{where } x' = \frac{1}{x}$$

These functions specify lines when graphed using y and x', and they specify curves when graphed using y and x, as illustrated in Figure 5.30 for the square root transformation.

Figure 5.30 (a) A plot of $\hat{y} = 5 + 3x'$, where $x' = \sqrt{x}$; (b) plot of $\hat{y} = 5 + 3\sqrt{x}$.

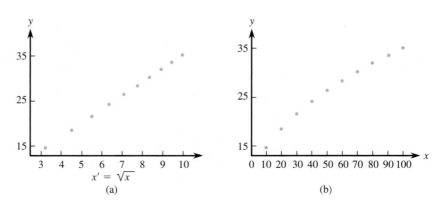

If the y values have been transformed, after obtaining the least-squares line, the transformation can be undone to yield an expression of the form y = some function of x (as opposed to y' = some function of x). For example, to reverse a logarithmic transformation $[y' = \log(y)]$, we can take the antilogarithm of each side of the equation. To reverse a square root transformation $(y' = \sqrt{y})$, we can square both sides of the equation, and to reverse a reciprocal transformation $\left(y' = \dfrac{1}{y}\right)$, we can take the reciprocal of each side of the equation. This is illustrated in Example 5.18.

▪ **Example 5.18   Revisiting the Tortilla Chip Data**

For the tortilla chip data of Example 5.17, $y' = \log(y)$, and the least-squares line relating y' and x was

$$y' = 1.14 - 0.0192x$$

or, equivalently,

$$\log(y) = 1.14 - 0.0192x$$

To reverse this transformation, we take the antilogarithm of both sides of the equation:

$$10^{\log(y)} = 10^{1.14 - 0.0192x}$$

Using properties of logarithms and exponents, that is,

$$10^{\log(y)} = y$$

and

$$10^{1.14 - 0.0192x} = (10^{1.14})(10^{-0.0192x})$$

we get

$$\hat{y} = (10^{1.14})(10^{-0.0192x}) = (13.8038)(10^{-0.0192x})$$

This equation can now be used to predict the y value (moisture content) for a given x (frying time). For example, the predicted moisture content when frying time is 35 sec is

$$\hat{y} = (13.8038)(10^{-0.0192x}) = (13.8038)(0.2128) = 2.9374$$

It should be noted that the process of transforming data, fitting a line to the transformed data, and then undoing the transformation to get an equation for a curved relationship between x and y usually results in a curve that provides a reasonable fit to the sample data, but it is not the least-squares curve for the data. For example, in Example 5.18, a transformation was used to fit the curve $\hat{y} = (13.8038)(10^{-0.0192x})$. However, there may be another equation of the form $\hat{y} = a(10^{bx})$ that has a smaller sum of squared residuals for the *original* data than the one we obtained using transformations. Finding the least-squares estimates for a and b in an equation of this form is complicated. Fortunately, the curves found using transformations usually provide reasonable predictions of y.

▪ **Power Transformations** Frequently, an appropriate transformation is suggested by the data. One type of transformation that statisticians have found useful for straightening a plot is a **power transformation**. A power (exponent) is first selected, and each original value is raised to that power to obtain the corresponding transformed value. Table 5.6 displays a "ladder" of the most frequently used power transformations. The power 1 corresponds to no transformation at all. Using the power 0 would transform every value to 1, which is certainly not informative, so statisticians use the logarithmic transformation in its place in the ladder of transformations. Other powers intermediate to or more extreme than those listed can be used, of course, but they are less frequently needed than those on the ladder. Notice that all the transformations previously presented are included in this ladder.

Table 5.6 ▪ Power Transformation Ladder

Power	Transformed Value	Name
3	$(\text{Original value})^3$	Cube
2	$(\text{Original value})^2$	Square
1	(Original value)	No transformation
$\frac{1}{2}$	$\sqrt{\text{Original value}}$	Square root
$\frac{1}{3}$	$\sqrt[3]{\text{Original value}}$	Cube root
0	$\text{Log}(\text{Original value})$	Logarithm
-1	$\dfrac{1}{\text{Original value}}$	Reciprocal

Figure 5.31 is designed to suggest where on the ladder we should go to find an appropriate transformation. The four curved segments, labeled 1, 2, 3, and 4, represent shapes of curved scatterplots that are commonly encountered. Suppose that a scatterplot looks like the curve labeled 1. Then, to straighten the plot, we should use a power of x that is up the ladder from the no-transformation row (x^2 or x^3) and/or a power on y that is also up the ladder from the power 1. Thus, we might be led to squaring each x value, cubing each y, and plotting the transformed pairs. If the cur-

Figure 5.31 Scatterplot shapes and where to go on the transformation ladder to straighten the plot.

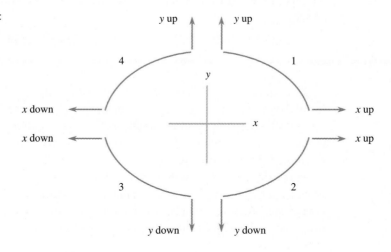

vature looks like curved segment 2, a power up the ladder from no transformation for x and/or a power down the ladder for y (e.g., $\sqrt{y}$ or $\log(y)$) should be used.

The scatterplot for the tortilla chip data (Figure 5.29(a)) has the pattern of segment 3 in Figure 5.31. This suggests going down the ladder of transformations for x and/or for y. We found that transforming the y values in this data set by taking logarithms worked well, and this is consistent with the suggestion of going down the ladder of transformations for y.

■ **Example 5.19**   A Nonlinear Model: Iron Absorption and Polyphenol Content of Foods

In many parts of the world, a typical diet consists mainly of cereals and grains, and many individuals suffer from a substantial iron deficiency. The article "The Effects of Organic Acids, Phytates, and Polyphenols on the Absorption of Iron from Vegetables" (*British Journal of Nutrition* [1983]: 331–342) reported the data in Table 5.7 on x = proportion of iron absorbed and y = polyphenol content (in milligrams per gram) when a particular food is consumed.

Table 5.7 ■ Original and Transformed Data for Example 5.19

Food	x	y	$\sqrt{x}$	$\sqrt{y}$
Wheat germ	0.007	6.4	0.083666	2.52982
Aubergine	0.007	3.0	0.083666	1.73205
Butter beans	0.012	2.9	0.109545	1.70294
Spinach	0.014	5.8	0.118322	2.40832
Brown lentils	0.024	5.0	0.154919	2.23607
Beetroot greens	0.024	4.3	0.154919	2.07364
Green lentils	0.032	3.4	0.178885	1.84391
Carrot	0.096	0.7	0.309839	0.83666
Potato	0.115	0.2	0.339116	0.44721
Beetroot	0.185	1.5	0.430116	1.22474
Pumpkin	0.206	0.1	0.453872	0.31623
Tomato	0.224	0.3	0.473286	0.54772
Broccoli	0.260	0.4	0.509902	0.63246
Cauliflower	0.263	0.7	0.512835	0.83666
Cabbage	0.320	0.1	0.565685	0.31623
Turnip	0.327	0.3	0.571839	0.54772
Sauerkraut	0.327	0.2	0.571839	0.44721

The scatterplot of the data in Figure 5.32(a) shows a clear curved pattern, which resembles the curved segment 3 in Figure 5.31. This suggests that x and/or y should be transformed by a power transformation down the ladder from 1. The authors of the article applied a square root transformation to both x and y. The resulting scatterplot in Figure 5.32(b) is reasonably straight.

Figure 5.32 Scatterplot of data of Example 5.19: (a) original data; (b) data transformed by taking square roots.

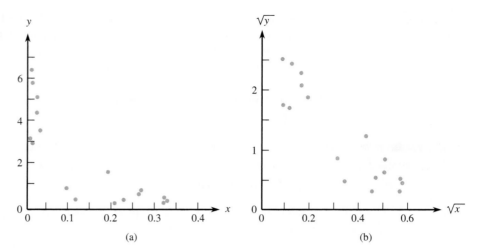

(a)   (b)

Using MINITAB to fit a least-squares line to the transformed data ($y' = \sqrt{y}$ and $x' = \sqrt{x}$) results in the following output:

Regression Analysis

The regression equation is
sqrt y = 2.45 − 3.72 sqrt x

Predictor	Coef	StDev	T	P
Constant	2.4483	0.1833	13.35	0.000
sqrt x	−3.7247	0.4836	−7.70	0.000

S = 0.3695   R-Sq = 79.8%   R-Sq(adj) = 78.5%

The corresponding least-squares line is

$$\hat{y}' = 2.45 - 3.72x'$$

This transformation can be reversed by squaring both sides to obtain an equation of the form y = some function of x:

$$\hat{y}'^2 = (2.45 - 3.72x')^2$$

Because $y'^2 = y$ and $x' = \sqrt{x}$, we get

$$\hat{y} = (2.45 - 3.72\sqrt{x})^2$$

▪ Exercises 5.47–5.54

5.47 The following data on x = frying time (in seconds) and y = moisture content (%) appeared in the paper "Thermal and Physical Properties of Tortilla Chips as a Function of Frying Time" (*Journal of Food Processing and Preservation* [1995]: 175–189):

x	5	10	15	20	25	30	45	60
y	16.3	9.7	8.1	4.2	3.4	2.9	1.9	1.3

a. Construct a scatterplot of the data. Would a straight line provide an effective summary of the relationship?
b. Here are the values of $x' = \log(x)$ and $y' = \log(y)$:

x'	0.70	1.00	1.18	1.30	1.40	1.48	1.65	1.78
y'	1.21	0.99	0.91	0.62	0.53	0.46	0.28	0.11

Construct a scatterplot of these transformed data, and comment on the pattern.

c. Based on the accompanying MINITAB output, does the least-squares line effectively summarize the relationship between y' and x'?

The regression equation is
log(moisture) = 2.01 − 1.05 log(time)

Predictor	Coef	Stdev	t-ratio	p
Constant	2.01444	0.09166	21.98	0.000
log(time)	−1.04920	0.06786	−15.46	0.000

s = 0.06293   R-sq = 97.6%   R-sq(adj) = 97.1%

Analysis of Variance

SOURCE	DF	SS	MS	F	p
Regression	1	0.94680	0.94680	239.06	0.000
Error	6	0.02376	0.00396		
Total	7	0.97057			

d. Use the MINITAB output in Part (c) to predict moisture content when frying time is 35 sec.
e. Do you think that predictions of moisture content using the model in Part (c) will be better than those using the model fit in Example 5.17, which used transformed y values but did not transform x? Explain.

5.48 The paper "Aspects of Food Finding by Wintering Bald Eagles" (*The Auk* [1983]: 477–484) examined the relationship between the time that eagles spend aerially searching for food (indicated by the percentage of eagles soaring) and relative food availability. The accompanying data were taken from a scatterplot that appeared in this paper. Let x denote salmon availability and y denote the percentage of eagles in the air.

x	0	0	0.2	0.5	0.5	1.0
y	28.2	69.0	27.0	38.5	48.4	31.1
x	1.2	1.9	2.6	3.3	4.7	6.5
y	26.9	8.2	4.6	7.4	7.0	6.8

a. Draw a scatterplot for this data set. Would you describe the plot as linear or curved?
b. One possible transformation that might lead to a straighter plot involves taking the square root of both the x and y values. Use Figure 5.31 to explain why this might be a reasonable transformation.
c. Construct a scatterplot using the variables $\sqrt{x}$ and $\sqrt{y}$. Is this scatterplot straighter than the scatterplot in Part (a)?
d. Using Table 5.6, suggest another transformation that might be used to straighten the original plot.

5.49 Data on salmon availability (x) and the percentage of eagles in the air (y) were given in the previous exercise.
a. Calculate the correlation coefficient for these data.
b. Because the scatterplot of the original data appeared curved, transforming both the x and y values by taking square roots was suggested. Calculate the correlation coefficient for the variables $\sqrt{x}$ and $\sqrt{y}$. How does this value compare with that calculated in Part (a)? Does this indicate that the transformation was successful in straightening the plot?

5.50 The article "Organ Transplant Demand Rises Five Times as Fast as Existing Supply" (*San Luis Obispo Tribune*, February 23, 2001) included a graph that showed the number of people waiting for organ transplants each year from 1990 to 1999. The following data are approximate values and were read from the graph in the article:

Year	Number Waiting for Transplant (in thousands)
1 (1990)	22
2	25
3	29
4	33
5	38
6	44
7	50
8	57
9	64
10 (1999)	72

a. Construct a scatterplot of the data with y = number waiting for transplant and x = year. Write a few sentences describing how the number of people waiting for transplants has changed over time from 1990 to 1999.
b. The scatterplot in Part (a) is shaped like segment 2 in Figure 5.31. Use this information to find a transformation of x and/or y that straightens the plot. Construct a scatterplot for your transformed variables.
c. Using the transformed variables from Part (b), fit a least-squares line and use it to predict the number waiting for an organ transplant in 2000 (Year 11).
d. The prediction made in Part (c) involves prediction for an x value that is outside the range of the x values in the sample. What assumption must you be willing to make for this to be reasonable? Do you think this assumption is reasonable in this case? Would your answer be the same if the prediction had been for the year 2010 rather than 2000? Explain.

5.51 Example 5.6 examined the relationship between survival rate of cardiac arrest patients and the time between when the cardiac arrest occurred and an initial shock from a defibrillator. Another study, described in the paper "Prediction of Defibrillation Success from a Single Defibrillation Threshold Measurement" (*Circulation* [1988]: 1144–1149), investigated the relationship between defibrillation success and the energy of the defibrillation shock (expressed as a multiple of the defibrillation threshold) and presented the following data:

Energy of Shock	Success (%)
0.5	33.3
1.0	58.3
1.5	81.8
2.0	96.7
2.5	100.0

a. Construct a scatterplot of y = success and x = energy of shock. Does the relationship between success and energy of shock appear to be linear or nonlinear?
b. Fit a least-squares line to the given data, and construct a residual plot. Does the residual plot support your conclusion in Part (a)? Explain.
c. Consider transforming the data by leaving y unchanged and using either $x' = \sqrt{x}$ or $x'' = \log(x)$. Which of these transformations would you recommend? Justify your choice by appealing to appropriate graphical displays.
d. Using the transformation you recommended in Part (c), find the equation of the least-squares line that describes the relationship between y and the transformed x.
e. What would you predict success to be when the energy of shock is 1.75 times the threshold level? When it is 0.8 times the threshold level?

5.52 Penicillin was administered orally to five horses, and the concentration of penicillin in the blood was determined after five different lengths of time (a different horse was used each time). The following data appeared in the paper "Absorption and Distribution Patterns of Oral Phenoxymethyl Penicillin in the Horse" (*Cornell Veterinarian* [1983]: 314–323):

x (time elapsed, hr)

1	2	3	6	8

y (penicillin concentration, mg/ml)

1.8	1.0	0.5	0.1	0.1

Construct scatterplots using the following variables. Which transformation, if any, would you recommend?
a. x and y
b. $\sqrt{x}$ and y
c. x and $\sqrt{y}$
d. $\sqrt{x}$ and $\sqrt{y}$
e. x and $\log(y)$ (values of $\log(y)$ are 0.26, 0, −0.30, −1, and −1)

5.53 The paper "Population Pressure and Agricultural Intensity" (*Annals of the Association of American Geographers* [1977]: 384–396) reported a positive association between population density and agricultural intensity. The following data consist of measures of population density (x) and agricultural intensity (y) for 18 different subtropical locations:

x	1.0	26.0	1.1	101.0	14.9	134.7
y	9	7	6	50	5	100
x	3.0	5.7	7.6	25.0	143.0	27.5
y	7	14	14	10	50	14
x	103.0	180.0	49.6	140.6	140.0	233.0
y	50	150	10	67	100	100

a. Construct a scatterplot of y versus x. Is the scatterplot compatible with the statement of positive association made in the paper?
b. The scatterplot in Part (a) is curved upward like segment 2 in Figure 5.31, suggesting a transformation that is up the ladder for x or down the ladder for y. Try a scatterplot that uses y and x^2. Does this transformation straighten the plot?
c. Draw a scatterplot that uses $\log(y)$ and x. The $\log(y)$ values, given in order corresponding to the y values, are 0.95, 0.85, 0.78, 1.70, 0.70, 2.00, 0.85, 1.15, 1.15, 1.00, 1.70, 1.15, 1.70, 2.18, 1.00, 1.83, 2.00, and 2.00. How does this scatterplot compare with that of Part (b)?
d. Now consider a scatterplot that uses transformations on both x and y: $\log(y)$ and x^2. Is this effective in straightening the plot? Explain.

5.54 Determining the age of an animal can sometimes be a difficult task. One method of estimating the age of harp seals is based on the width of the pulp canal in the seal's canine teeth. To investigate the relationship between age and the width of the pulp canal, researchers measured age and canal width in seals of known age. The following data on x = age (in years) and y = canal length (in millimeters) are a portion of a larger data set that appeared in the paper "Validation of Age Estimation in the Harp Seal Using Dentinal Annuli" (*Canadian Journal of Fisheries and Aquatic Science* [1983]: 1430–1441):

x	0.25	0.25	0.50	0.50	0.50	0.75	0.75	1.00
y	700	675	525	500	400	350	300	300
x	1.00	1.00	1.00	1.00	1.25	1.25	1.50	1.50
y	250	230	150	100	200	100	100	125
x	2.00	2.00	2.50	2.75	3.00	4.00	4.00	5.00
y	60	140	60	50	10	10	10	10
x	5.00	5.00	5.00	6.00	6.00
y	15	10	10	15	10

Construct a scatterplot for this data set. Would you describe the relationship between age and canal length as linear? If not, suggest a transformation that might straighten the plot.

■ 5.5 Communicating and Interpreting the Results of Statistical Analyses

Using either a least-squares line to summarize a linear relationship or a correlation coefficient to describe the strength of a linear relationship is common in investigations that focus on more than a single variable. In fact, the methods described in this chapter are among the most widely used of all statistical tools. When numerical bivariate data are analyzed in journal articles and other published sources, it is common to find a scatterplot of the data and a least-squares line or a correlation coefficient.

■ Communicating the Results of Statistical Analyses

When reporting the results of a data analysis involving bivariate numerical data, it is important to include graphical displays as well as numerical summaries. Including a scatterplot and providing a description of what the plot reveals about the form of the relationship between the two variables under study establish the context in which numerical summary measures, such as the correlation coefficient or the equation of the least-squares line, can be interpreted.

In general, the goal of an analysis of bivariate data is to give a quantitative description of the relationship, if any, between the two variables. If there is a relationship, you can describe how strong or weak the relationship is or model the relationship in a way that allows various conclusions to be drawn. If the goal of the study is to describe the strength of the relationship and the scatterplot shows a linear pattern, you can report the value of the correlation coefficient or the coefficient of determination as a measure of the strength of the linear relationship. When you interpret the value of the correlation coefficient, it is a good idea to relate the interpretation to the pattern observed in the scatterplot. This is especially important before making a statement of no relationship between two variables, because a correlation coefficient near 0 does not necessarily imply that there is no relationship of any form. Similarly, a correlation coefficient near 1 or -1, by itself, does not guarantee that the relationship is linear. A curved pattern, such as the one we saw in Figure 5.15, can produce a correlation coefficient that is near 1.

If the goal of a study is prediction, then, when you report the results of the study, you should not only give a scatterplot and the equation of the least-squares line but also address how well the linear prediction model fits the data. At a minimum, you should include both the values of s_e (the standard deviation about the regression line) and r^2 (the coefficient of determination). Including a residual plot can also provide support for the appropriateness of the linear model for describing the relationship between the two variables.

■ What to Look for in Published Data

Here are a few things to consider when you read an article that includes an analysis of bivariate data:

- What two variables are being studied? Are they both numerical? Is a distinction made between a dependent variable and an independent variable?
- Does the article include a scatterplot of the data? If so, does there appear to be a relationship between the two variables? Can the relationship be described

as linear, or is some type of nonlinear relationship a more appropriate description?

■ Does the relationship between the two variables appear to be weak or strong? Is the value of a correlation coefficient reported?

■ If the least-squares line is used to summarize the relationship between the dependent and independent variables, is any measure of goodness of fit reported, such as r^2 or s_e? How are these values interpreted, and what do they imply about the usefulness of the least-squares line?

■ If a correlation coefficient is reported, is it interpreted properly? Be cautious of interpretations that claim a causal relationship.

The article "Rubbish Regression and the Census Undercount" (*Chance* [1992]: 33) describes work done by the Garbage Project at the University of Arizona. Project researchers had analyzed different categories of garbage for a number of households. They were asked by the Census Bureau to see whether any of the garbage data variables were related to household size. They reported that "the weight data for different categories of garbage were plotted on graphs against data on household size, dwelling by dwelling, and the resulting scatterplots were analyzed to see in which categories the weight showed a steady, monotonic rise relative to household size."

The researchers determined that the strongest linear relationship appeared to be that between the amount of plastic discarded and household size. The line used to summarize this relationship was stated to be $\hat{y} = 0.2815x$, where y = household size and x = weight (in pounds) of plastic during a 5-week collection. Note that this line has an intercept of 0. Scientists at the Census Bureau believed that this relationship would extend to entire neighborhoods, and so the amount of plastic discarded by a neighborhood could be measured (rather than having to measure house to house) and then used to approximate the neighborhood size.

An example of the use of r^2 and the correlation coefficient is found in the article "Affective Variables Related to Mathematics Achievement Among High-Risk College Freshmen" (*Psychological Reports* [1991]: 399–403). The researchers wrote that "an examination of the Pearson correlations indicated that the two scales, Confidence in Learning Mathematics and Mathematics Anxiety, were highly related, with values ranging from .87 to .91 for the men, women, and for the total group. Such a high relationship indicates that the two scales are measuring essentially the same concept, so for further analysis, one of the two scales could be eliminated."

The reported correlation coefficients (from .87 to .91) indicate a strong positive linear relationship between scores on the two scales. Students who scored high on the Confidence in Learning Mathematics scale tended also to score high on the Mathematics Anxiety scale. (The researchers explain that a high score on the Mathematics Anxiety scale corresponds to low anxiety toward mathematics.) We should be cautious of the statement that the two scales are measuring essentially the same concept. Although that interpretation may make sense in this context, given what the researchers know about these two scales, it does not follow solely from the fact that the two are highly correlated. (For example, there might be a correlation between number of years of education and income, but it would be hard to argue that these two variables are measuring the same thing!)

The same article also states that Mathematics Anxiety score accounted for 27% of the variability in first-quarter math grades. The reported value of 27% is the value of r^2 converted to a percentage. Although a value of 27% ($r^2 = .27$) may not seem particularly large, it is in fact surprising that the variable *Math Anxiety score* would be able to explain more than one-fourth of the student-to-student variability in first-quarter math grades.

▪ A Word to the Wise: Cautions and Limitations

There are a number of ways to get into trouble when analyzing bivariate numerical data! Here are some of the things you need to keep in mind when conducting your own analyses or when reading reports of such analyses:

1. Correlation does not imply causation. A common media blunder is to infer a cause-and-effect relationship between two variables just because there is a strong correlation between them. Don't fall into this trap! A strong correlation implies only that the two variables tend to vary together in a predictable way, but there are many possible explanations for why this is occurring besides one variable causing changes in the other.

   For example, the article "Ban Cell Phones? You May as Well Ban Talking Instead" (*USA Today*, April 27, 2000) gave data that showed a strong negative correlation between the number of cell phone subscribers and traffic fatality rates. During the years from 1985 to 1998, the number of cell phone subscribers increased from 200,000 to 60,800,000, and the number of traffic deaths per 100 million miles traveled decreased from 2.5 to 1.6 over the same period. However, based on this correlation alone, the conclusion that cell phone use improves road safety is not reasonable!

   Similarly, the *Calgary Herald* (April 16, 2002) reported that heavy and moderate drinkers earn more than light drinkers or those who do not drink. Based on the correlation between number of drinks consumed and income, the author of the study concluded that moderate drinking "causes" higher earnings. This is obviously a misleading statement, but at least the article goes on to state that "there are many possible reasons for the correlation. It could be because better-off men simply choose to buy more alcohol. Or it might have something to do with stress: Maybe those with stressful jobs tend to earn more after controlling for age, occupation, etc. and maybe they also drink more in order to deal with the stress."

2. A correlation coefficient near 0 does not necessarily imply that there is no relationship between two variables. Before such an interpretation can be given, it is important to examine a scatterplot of the data carefully. Although it may be true that the variables are unrelated, there may in fact be a strong but non-linear relationship.

3. The least-squares line for predicting y from x is not the same line as the least-squares line for predicting x from y. The least-squares line is, by definition, the line that has the smallest possible sum of squared deviations of points from the line *in the y direction* (it minimizes $\sum(y - \hat{y})^2$). The line that minimizes the sum of squared deviations in the y direction is not generally the same as the line that minimizes the sum of the squared deviations in the x direction. So, for example, it is not appropriate to fit a line to data using $y =$ house price and $x =$ house size and then use the resulting least-squares line Price $= a + b$(Size) to predict the size of a house by substituting in a price and then solving for size. Make sure that the dependent and independent variables are clearly identified and that the appropriate line is fit.

4. Beware of extrapolation. It is dangerous to assume that a linear model fit to data is valid over a wider range of x values. Using the least-squares line to make predictions outside the range of x values in the data set often leads to poor predictions.

5. Be careful in interpreting the value of the slope and intercept in the least-squares line. In particular, in many instances interpreting the intercept as the value of y that would be predicted when $x = 0$ is equivalent to extrapolating way beyond the range of the x values in the data set, and this should be avoided unless $x = 0$ is within the range of the data.

6. Remember that the least-squares line may be the "best" line (in that it has a smaller sum of squared deviations than any other line), but that doesn't necessarily mean that the line will produce good predictions. Be cautious of predictions based on a least-squares line without any information about the adequacy of the linear model, such as s_e and r^2.

7. It is not enough to look at just r^2 or just s_e when assessing a linear model. These two measures address different aspects of the fit of the line. In general, we would like to have a small value for s_e (which indicates that deviations from the line tend to be small) and a large value for r^2 (which indicates that the linear relationship explains a large proportion of the variability in the y values). It is possible to have a small s_e combined with a small r^2 or a large r^2 combined with a large s_e. Remember to consider both values.

8. The value of the correlation coefficient as well as the values for the intercept and slope of the least-squares line can be sensitive to influential observations in the data set, particularly if the sample size is small. Because potentially influential observations are those whose x values are far away from most of the x values in the data set, it is important to look for such observations when examining the scatterplot. (Another good reason for *always* starting with a plot of the data!)

9. If the relationship between two variables is nonlinear, it is preferable to model the relationship using a nonlinear model rather than fitting a line to the data. A plot of the residuals from a linear fit is particularly useful in determining whether a nonlinear model would be a more appropriate choice.

▪ Activity 5.1: Exploring Correlation and Regression Technology Activity (Applets)

Open the applet (on the CD that accompanies this textbook) called CorrelationPoints. You should see a screen like the one shown to the right.

1. Using the mouse, you can click to add points to form a scatterplot. The value of the correlation coefficient for the data in the plot at any given time (once you have two or more points in the plot) will be displayed at the top of the screen. Try to add points to obtain a correlation coefficient that is close to −.8. Briefly describe the factors that influenced where you placed the points.

2. Reset the plot (by clicking on the Reset bar in the lower left corner of the screen) and add points trying to produce a data set with a correlation that is close

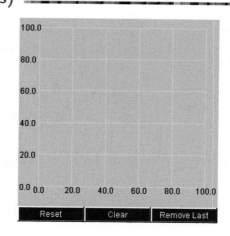

to +.4. Briefly describe the factors that influenced where you placed the points.

Now open the applet called RegDecomp. You should see a screen that looks like the one shown here:

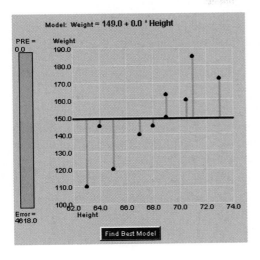

The blue points in the plot represent the data set, and the blue line represents a possible line that might be used to describe the relationship between the two variables shown. The pink lines in the graph represent the deviations of points from the line, and the pink bar on the left-hand side of the display depicts the sum of squared errors for the line shown.

3. Using your mouse, you can move the line and see how the deviations from the line and the sum of squared errors change as the line changes. Try to move the line into a position that you think is close to the least-squares regression line. When you think you are close to the line that minimizes the sum of squared errors, you can click on the bar that says "Find Best Model" to see how close you were to the actual least-squares line.

■ Activity 5.2: Age and Flexibility

Materials needed: Yardsticks.

In this activity you will investigate the relationship between age and a measure of flexibility. Flexibility will be measured by asking a person to bend at the waist as far as possible, extending his or her arms toward the floor. Using a yardstick, measure the distance from the floor to the fingertip closest to the floor.

1. Age and the measure of flexibility just described will be measured for a group of individuals. Our goal is to determine whether there is a relationship between age and this measure of flexibility. What are two reasons why it would not be a good idea to use just the students in your class as the subjects for your study?

2. Working as a class, decide on a reasonable way to collect data on the two variables of interest.

3. After your class has collected appropriate data, use it to construct a scatterplot. Comment on the interesting features of the plot. Does it look like there is a relationship between age and flexibility?

4. If there appears to be a relationship between age and flexibility, fit a model that is appropriate for describing the relationship.

5. *In the context of this activity*, write a brief description of the danger of extrapolation.

■ Summary of Key Concepts and Formulas

Term or Formula	Comment
Scatterplot	A picture of bivariate numerical data in which each observation (x, y) is represented as a point located with respect to a horizontal x axis and a vertical y axis.
Pearson's sample correlation coefficient $$r = \frac{\sum z_x z_y}{n-1}$$	A measure of the extent to which sample x and y values are linearly related; $-1 \le r \le 1$, so values close to 1 or -1 indicate a strong linear relationship.
Principle of least squares	The method used to select a line that summarizes an approximate linear relationship between x and y. The least-squares line is the line that minimizes the sum of the squared vertical deviations from the points in the scatterplot.

Term or Formula	Comment
$b = \dfrac{\sum(x - \bar{x})(y - \bar{y})}{\sum(x - \bar{x})^2} = \dfrac{\sum xy - \dfrac{(\sum x)(\sum y)}{n}}{\sum x^2 - \dfrac{(\sum x)^2}{n}}$	The slope of the least-squares line.
$a = \bar{y} - b\bar{x}$	The intercept of the least-squares line.
Predicted (fitted) values $\hat{y}_1, \hat{y}_2, \ldots, \hat{y}_n$	Obtained by substituting the x value for each observation into the least-squares line: $\hat{y}_1 = a + bx_1, \ldots,$ $\hat{y}_n = a + bx_n$
Residuals	The vertical deviations from the least-squares line. Residuals are obtained by subtracting each predicted value from the corresponding observed y value: $y_1 - \hat{y}_1, \ldots, y_n - \hat{y}_n$.
Residual plot	A scatterplot of the $(x, \text{residual})$ pairs. Isolated points or a pattern of points in a residual plot is indicative of potential problems.
Residual (error) sum of squares, $\text{SSResid} = \sum(y - \hat{y})^2$	The sum of the squared residuals is a measure of y variation that cannot be attributed to an approximate linear relationship (unexplained variation).
Total sum of squares, $\text{SSTo} = \sum(y - \bar{y})^2$	The sum of squared deviations from the sample mean is a measure of total variation in the observed y values.
Coefficient of determination, $r^2 = 1 - \dfrac{\text{SSResid}}{\text{SSTo}}$	The proportion of variation in observed y's that can be explained by an approximate linear relationship.
Standard deviation about the least-squares line, $s_e = \sqrt{\dfrac{\text{SSResid}}{n - 2}}$	The size of a "typical" deviation from the least-squares line.
Transformation	A simple function of the x and/or y variable that is then used in a regression.
Power transformation	An exponent, or power, p, is first specified, and then new (transformed) data values are calculated as transformed value = (original value)^p. A logarithmic transformation is identified with $p = 0$. When the scatterplot of original data exhibits curvature, a power transformation of x and/or y often results in a scatterplot that has a linear appearance.

▪ Supplementary Exercises 5.55–5.67

5.55 The accompanying data resulted from an experiment in which x was the amount of catalyst added to accelerate a chemical reaction and y was the resulting reaction time:

x	1	2	3	4	5
y	49	46	41	34	25

a. Calculate r. Does the value of r suggest a strong linear relationship?
b. Construct a scatterplot. From the plot, does the word *linear* really provide the most effective description of the relationship between x and y? Explain.

5.56 A number of studies have shown that lichens (certain plants composed of an alga and a fungus) are excellent bioindicators of air pollution. The article "The Epiphytic Lichen *Hypogymnia physodes* as a Bioindicator of Atmospheric Nitrogen and Sulphur Deposition in Norway" (*Environmental Monitoring and Assessment* [1993]: 27–47) gives the following data (read from a graph in the paper) on x = NO_3 wet deposition (in grams per cubic meter) and y = lichen (% dry weight):

x	0.05	0.10	0.11	0.12	0.31
y	0.48	0.55	0.48	0.50	0.58
x	0.37	0.42	0.58	0.68	0.68
y	0.52	1.02	0.86	0.86	1.00
x	0.73	0.85	0.92		
y	0.88	1.04	1.70		

a. What is the equation of the least-squares regression line?

b. Predict lichen dry weight percentage for an NO_3 deposition of 0.5 g/m³.

5.57 Athletes competing in a triathlon participated in a study described in the paper "Myoglobinemia and Endurance Exercise" (*American Journal of Sports Medicine* [1984]: 113–118). The following data on finishing time x (in hours) and myoglobin level y (in nanograms per milliliter) were read from a scatterplot in the paper:

x	4.90	4.70	5.35	5.22	5.20
y	1590	1550	1360	895	865
x	5.40	5.70	6.00	6.20	6.10
y	905	895	910	700	675
x	5.60	5.35	5.75	5.35	6.00
y	540	540	440	380	300

a. Obtain the equation of the least-squares line.

b. Interpret the value of b.

c. What happens if the line in Part (a) is used to predict the myoglobin level for a finishing time of 8 hr? Is this reasonable? Explain.

5.58 The paper "Root Dentine Transparency: Age Determination of Human Teeth Using Computerized Densitometric Analysis" (*American Journal of Physical Anthropology* [1991]: 25–30) reported on an investigation of methods for age determination based on tooth characteristics. With y = age (in years) and x = percentage of root with transparent dentine, a regression analysis for premolars using an image-analyzer method to determine x gave n = 36, SSResid = 5987.16, and SSTo = 17,409.60. Calculate and interpret the values of r^2 and s_e.

5.59 The paper cited in Exercise 5.58 gave a scatterplot in which x values were for anterior teeth. Consider the following representative subset of the data:

x	15	19	31	39	41
y	23	52	65	55	32
x	44	47	48	55	65
y	60	78	59	61	60

$$\sum x = 404 \qquad \sum x^2 = 18,448 \qquad \sum y = 545$$

$$\sum y^2 = 31,993 \qquad \sum xy = 23,198$$

$$a = 32.080888 \qquad b = 0.554929$$

a. Calculate the predicted values and residuals.

b. Use the results of Part (a) to obtain SSResid and r^2.

c. Does the least-squares line appear to give accurate predictions? Explain your reasoning.

5.60 There have been numerous studies on the effects of radiation. The following data on the relationship between degree of exposure to ^{242}Cm alpha particles (x) and the percentage of exposed cells without aberrations (y) appeared in the paper "Chromosome Aberrations Induced in Human Lymphocytes by D-T Neutrons" (*Radiation Research* [1984]: 561–573):

x	0.106	0.193	0.511	0.527
y	98	95	87	85
x	1.08	1.62	1.73	2.36
y	75	72	64	55
x	2.72	3.12	3.88	4.18
y	44	41	37	40

Summary quantities are

$$n = 12 \qquad \sum x = 22.027 \qquad \sum y = 793$$

$$\sum x^2 = 62.600235 \qquad \sum xy = 1114.5$$

$$\sum y^2 = 57,939$$

a. Obtain the equation of the least-squares line.

b. Calculate SSResid and SSTo.

c. What percentage of observed variation in y can be explained by the approximate linear relationship between the two variables?

d. Calculate and interpret the value of s_e.

e. Using just the results of Parts (a) and (c), what is the value of Pearson's sample correlation coefficient?

5.61 The article "Reduction in Soluble Protein and Chlorophyll Contents in a Few Plants as Indicators of Automobile Exhaust Pollution" (*International Journal of Environmental Studies* [1983]: 239–244) reported the following data on x = distance from a highway (in meters) and y = lead content of soil at that distance (in parts per million):

x	0.3	1	5	10	15	20
y	62.75	37.51	29.70	20.71	17.65	15.41
x	25	30	40	50	75	100
y	14.15	13.50	12.11	11.40	10.85	10.85

a. Use a statistical computer package to construct scatterplots of y versus x, y versus $\log(x)$, $\log(y)$ versus $\log(x)$, and $\dfrac{1}{y}$ versus $\dfrac{1}{x}$.

b. Which transformation considered in Part (a) does the best job of producing an approximately linear relationship? Use the selected transformation to predict lead content when distance is 25 m.

5.62 The sample correlation coefficient between annual raises and teaching evaluations for a sample of $n = 353$ college faculty was found to be $r = .11$ ("Determination of Faculty Pay: An Agency Theory Perspective," *Academy of Management Journal* [1992]: 921–955).

a. Interpret this value.

b. If a straight line were fit to the data using least squares, what proportion of variation in raises could be attributed to the approximate linear relationship between raises and evaluations?

5.63 An accurate assessment of oxygen consumption provides important information for determining energy expenditure requirements for physically demanding tasks. The paper "Oxygen Consumption During Fire Suppression: Error of Heart Rate Estimation" (*Ergonomics* [1991]: 1469–1474) reported on a study in which $x =$ oxygen consumption (in milliliters per kilogram per minute) during a treadmill test was determined for a sample of 10 firefighters. Then $y =$ oxygen consumption at a comparable heart rate was measured for each of the 10 individuals while they performed a fire-suppression simulation. This resulted in the following data and scatterplot:

Firefighter	1	2	3	4	5
x	51.3	34.1	41.1	36.3	36.5
y	49.3	29.5	30.6	28.2	28.0
Firefighter	6	7	8	9	10
x	35.4	35.4	38.6	40.6	39.5
y	26.3	33.9	29.4	23.5	31.6

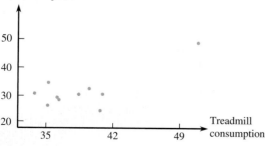

a. Does the scatterplot suggest an approximate linear relationship?

b. The investigators fit a least-squares line. The resulting MINITAB output is given in the following:

The regression equation is
fi recon = −11.4 + 1.09 treadcon

Predictor	Coef	Stdev	t-ratio	p
Constant	−11.37	12.46	−0.91	0.388
treadcon	1.0906	0.3181	3.43	0.009

$s = 4.70$ R-sq = 59.5%   R-sq(adj) = 54.4%

Predict fire-simulation consumption when treadmill consumption is 40.

c. How effectively does a straight line summarize the relationship?

d. Delete the first observation, (51.3, 49.3), and calculate the new equation of the least-squares line and the value of r^2. What do you conclude? (Hint: For the original data, $\sum x = 388.8$, $\sum y = 310.3$, $\sum x^2 = 15{,}338.54$, $\sum xy = 12{,}306.58$, and $\sum y^2 = 10{,}072.41$.)

5.64 The relationship between the depth of flooding and the amount of flood damage was examined in the paper "Significance of Location in Computing Flood Damage" (*Journal of Water Resources Planning and Management* [1985]: 65–81). The following data on $x =$ depth of flooding (feet above first-floor level) and $y =$ flood damage (as a percentage of structure value) was obtained using a sample of flood insurance claims:

x	1	2	3	4	5	6	7
y	10	14	26	28	29	41	43
x	8	9	10	11	12	13	
y	44	45	46	47	48	49	

a. Obtain the equation of the least-squares line.

b. Construct a scatterplot, and draw the least-squares line on the plot. Does it look as though a straight line provides an adequate description of the relationship between y and x? Explain.

c. Predict flood damage for a structure subjected to 6.5 ft of flooding.

d. Would you use the least-squares line to predict flood damage when depth of flooding is 18 ft? Explain.

5.65 Consider the four (x, y) pairs $(0, 0)$, $(1, 1)$, $(1, −1)$, and $(2, 0)$.

a. What is the value of the sample correlation coefficient r?

b. If a fifth observation is made at the value $x = 6$, find a value of y for which $r > .5$.

c. If a fifth observation is made at the value $x = 6$, find a value of y for which $r < −.5$.

5.66 The paper "Biomechanical Characteristics of the Final Approach Step, Hurdle, and Take-Off of Elite American Springboard Divers" (*Journal of*

Human Movement Studies [1984]: 189–212) gave the following data on y = judge's score and x = length of final step (in meters) for a sample of seven divers performing a forward pike with a single somersault:

y	7.40	9.10	7.20	7.00	7.30	7.30	7.90
x	1.17	1.17	0.93	0.89	0.68	0.74	0.95

a. Construct a scatterplot.
b. Calculate the slope and intercept of the least-squares line. Draw this line on your scatterplot.
c. Calculate and interpret the value of Pearson's sample correlation coefficient.

5.67 The following data on y = concentration of penicillin-G in pig's blood plasma (units per milliliter) and x = time (in minutes) from administration of a dose of penicillin (22 mg/kg body weight) appeared in the paper "Calculation of Dosage Regimens of Antimicrobial Drugs for Surgical Prophylaxis" (*Journal of the American Veterinary Medicine Association* [1984]: 1083–1087):

x	5	15	45	90	180
y	32.6	43.3	23.1	16.7	5.7
x	240	360	480	1440	
y	6.4	9.2	0.4	0.2	

a. Construct a scatterplot for these data.
b. Using the ladder of transformations of Section 5.4 (see Figure 5.31 and Table 5.4), suggest a transformation that might straighten the plot. Give reasons for your choice of transformation.

■ Reference

Neter, John, William Wasserman, and Michael Kutner. *Applied Linear Statistical Models*, 4th ed. New York: McGraw-Hill, 1996. (The first half of this book gives a comprehensive treatment of regression analysis without overindulging in mathematical development; a highly recommended reference.)

6 · Probability

Every day, each of us makes decisions based on uncertainty. Should you buy an extended warranty for your new DVD player? It depends on the likelihood that it will fail during the warranty period. Should you allow 45 min to get to your 8 A.M. class, or is 35 min enough? From experience, you may know that most mornings you can drive to school and park in 25 min or less. Most of the time, the walk from your parking space to class is 5 min or less. But how often will the drive to school or the walk to class take longer than you expect? How often will both take longer? When it takes longer than usual to drive to campus, is it more likely that it will also take longer to walk to class? less likely? Or are the driving and walking times unrelated? Some questions involving uncertainty are more serious: If an artificial heart has four key parts, how likely is each one to fail? How likely is it that at least one will fail? If a satellite has a backup solar power system, how likely is it that both the main and the backup components will fail?

We can answer questions such as these using the ideas and methods of probability, the systematic study of uncertainty. From its roots in the analysis of games of chance, probability has evolved into a science that enables us to make important decisions with confidence. In this chapter we introduce the basic rules of probability that are most widely used in statistics, and we explain two ways to estimate probabilities when it is difficult to derive them analytically.

▪ 6.1 Interpreting Probabilities and Basic Probability Rules

In many situations, it is uncertain what outcome will occur. For example, when a ticketed passenger shows up at the airport, she may be denied a seat on the flight as a result of overbooking by the airline. There are two possible outcomes of interest for the passenger: (1) She is able to take the flight, or (2) she is denied a seat on the plane and must take a later flight. Before she arrives at the airport, there is uncertainty about which outcome will occur. Based on past experience with this particular flight, the passenger may know that one outcome is more likely than the other. She may believe that the chance of being denied a seat is quite small; she may not be overly concerned about this possibility. Although this outcome is possible, she views it as unlikely. Assigning a probability to such an outcome is an attempt to quantify what is meant by "likely" or "unlikely." We do this by using a number between 0 and 1 to indicate the likelihood of occurrence of some outcome.

There is more than one way to interpret such numbers. One is a subjective interpretation, in which a probability is interpreted as a personal measure of the strength of belief that the outcome will occur. A probability of 1 represents a belief that the outcome will certainly occur. A probability of 0 then represents a belief that the outcome will certainly *not* occur — that is, that it is impossible. Other probabilities are intermediate between these two extremes. This interpretation is common in ordinary speech. For example, we say, "There's about a 50–50 chance," or "My chances are nil." In the airline flight example, suppose that the passenger reported her subjective assessment of the probability of being denied a seat to be .01. Because this probability is close to 0, she believes it is highly unlikely that she will be denied a seat. The subjective interpretation, however, presents some difficulties. For one thing, different people may assign different probabilities to the same outcome, because each person could have a different subjective belief.

Wherever possible, we use instead the objective *relative frequency* interpretation of probability. In this interpretation, a probability is the *long-run proportion* of the time that an outcome will occur, given many repetitions under identical circumstances. A probability of 1 corresponds to an outcome that occurs 100% of the time — that is, a certain outcome. A probability of 0 corresponds to an outcome that occurs 0% of the time — that is, an impossible outcome.

▪ Relative Frequency Interpretation of Probability

A **probability** is a number between 0 and 1 that reflects the likelihood of occurrence of some outcome.

The **probability of an outcome**, denoted by **P(outcome)**, is interpreted as the long-run proportion of the time that the outcome occurs when the experiment is performed repeatedly under identical conditions.

To illustrate the relative frequency interpretation of probability, consider the following situation. A package delivery service promises 2-day delivery between two cities in California but is often able to deliver packages in just 1 day. Suppose that the company reports that there is a 50–50 chance that a package will arrive in 1 day. Alternatively, they might report that the probability of next-day delivery is .5, implying that in the long run, 50% of all packages arrive in 1 day.

Suppose that we track the delivery of packages shipped with this company. With each new package shipped, we could compute the relative frequency of packages shipped so far that arrived in one day:

number of packages that arrived in one day
───────────────────────────────────────
total number of packages shipped

The results for the first 10 packages might be as follows:

Package number	1	2	3	4	5	6	7	8	9	10
Arrived next day	N	Y	Y	Y	N	N	Y	Y	N	N
Relative frequency delivered next day	0	.5	.667	.75	.6	.5	.571	.625	.556	.5

Figure 6.1 illustrates how the relative frequency of packages arriving in 1 day fluctuates during a sample sequence of 50 shipments. As the number of packages in the sequence increases, the relative frequency does not continue to fluctuate wildly but instead stabilizes and approaches some fixed number, called the *limiting value*. This limiting value is the true probability. Figure 6.2 illustrates this process of stabilization for a sample sequence of 1000 shipments.

Figure 6.1 The fluctuation of relative frequency.

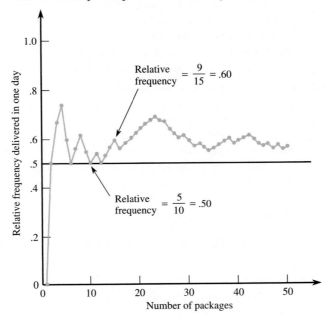

Figure 6.2 The stabilization of relative frequency.

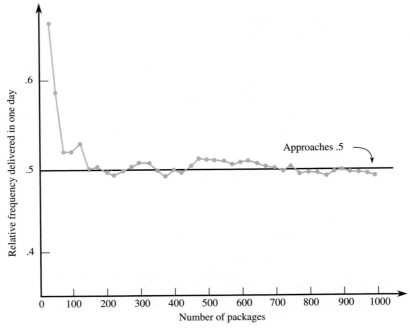

The stabilization of the relative frequency of occurrence illustrated in Figure 6.2 is not unique to this particular example. The general phenomenon of stabilization makes the relative frequency interpretation of probabilities possible.

▪ Some Basic Properties

The relative frequency interpretation of probability makes it easy to understand the following basic properties of probability.

1. *The probability of any outcome is a number between 0 and 1.* A relative frequency, which is the number of times the outcome occurs divided by the total number of repetitions, cannot be less than 0 (you cannot have a negative number of occurrences) or greater than 1 (the outcome cannot occur more often than the total number of repetitions).

2. *If outcomes cannot occur simultaneously, then the probability that any one of them will occur is the sum of the outcome probabilities.* For example, a hotel near Disney World offers three different rental options to its guests. Package A includes hotel room only; Package B includes hotel room and meals; and Package C includes hotel room, meals, and admission to Disney World. From past experience, the hotel knows that about 30% of those who stay at the hotel choose Package A and about 25% choose Package B. Interpreting these percentages, which are really long-run relative frequencies, as probabilities, we can say that

$$P(\text{next person calling to make a reservation will choose Package A}) = .30$$

and

$$P(\text{next person calling to make a reservation will choose Package B}) = .25$$

Because no one can choose more than one package for a single visit, the probability that the next person calling for a reservation will choose either Package A or Package B is

$$P(\text{Package A } or \text{ Package B}) = .30 + .25 = .55$$

3. *The probability that an outcome will not occur is equal to 1 minus the probability that the outcome will occur.* According to this property, the probability that the next person calling the hotel for a reservation will not choose either Package A or Package B is

$$P(not \text{ (Package A } or \text{ Package B)}) = 1 - P(\text{Package A } or \text{ Package B})$$
$$= 1 - .55$$
$$= .45$$

Because there are only three packages available, if the next reservation is not for Package A or B, it must be for Package C. So, the probability that the next person calling for a reservation chooses package C is

$$P(C) = P(not \text{ (Package A } or \text{ Package B)}) = .45$$

Remember that, in probability, 1 represents an outcome that is certain — that is, an outcome that will occur 100% of the time. At this hotel, it is certain that a reservation will be for one of the three packages, because these are the only options available. So

$$P(\text{Package A } or \text{ B } or \text{ C}) = .30 + .25 + .45 = 1$$

Similarly, it is certain that any event either will or will not occur. Thus,

$$P(\text{Package A } or \text{ B}) + P(not \text{ (Package A } or \text{ B)}) = .55 + .45 = 1$$

▪ Independence

Suppose that the incidence rate for a particular disease in a certain population is known to be 1 in 1000. Then the probability that a randomly selected individual from this population has the disease is .001. A diagnostic test for the disease is given to the selected individual, and the test result is positive. However, suppose that the diagnostic test is not infallible; it returns a false-positive result about 10% of the time. That is, about 90% of those without the disease get a correct (negative) response, but the other 10% of people without the disease incorrectly test positive. When we learn that the outcome *positive test result* has occurred, we obviously want to revise the probability that the selected individual has the disease upward from .001. Knowing that one outcome has occurred (the selected person has a positive test result) can change the probability of another outcome (the selected person has the disease).

As another example, consider a university's course registration system, which consists of 12 priority groups. Overall, only 10% of all students receive all requested classes, but 75% of those in the first priority group receive all requested classes. Interpreting these figures as probabilities, we say that the probability that a randomly selected student at this university received all requested classes is .10. However, if we know that the selected student is in the first priority group, we revise the probability that the selected student received all requested classes to .75. Knowing that the outcome *selected person is in the first priority group* has occurred changes the probability of the outcome *selected person received all requested classes*. These two outcomes are said to be **dependent**.

It is often the case, however, that the likelihood of one outcome is not affected by the occurrence of another outcome. For example, suppose that you purchase a computer system with a separate monitor and keyboard. Two possible outcomes of interest are

Outcome 1:   The monitor needs service while under warranty.

Outcome 2:   The keyboard needs service while under warranty.

Because the two components operate independently of one another, learning that the monitor has needed warranty service should not affect our assessment of the likelihood that the keyboard will need repair. If we know that 1% of all keyboards need repair while under warranty, we say that the probability that a keyboard needs warranty service is .01. Knowing that the monitor has needed warranty service does not affect this probability. We say that *monitor failure* (Outcome 1) and *keyboard failure* (Outcome 2) are **independent** outcomes.

> ▪ **Definition**
>
> **Independent outcomes:** Two outcomes are said to be **independent** if the chance that one outcome occurs is not affected by knowledge of whether the other outcome has occurred. If there are more than two outcomes under consideration, they are independent if knowledge that some of the outcomes have occurred does not change the probabilities that any of the other outcomes have occurred.
>
> **Dependent outcomes:** If the occurrence of one outcome changes the probability that the other outcome occurs, the outcomes are **dependent**.

In the previous examples, the outcomes *has disease* and *tests positive* are dependent; the outcomes *receives all requested classes* and *is in the first priority group* are dependent; but the outcomes *monitor needs warranty service* and *keyboard needs warranty service* are independent.

▪ Multiplication Rule for Independent Outcomes

An individual who purchases a computer system might wonder how likely it is that *both* the monitor and the keyboard will need service while under warranty. A student who must take a chemistry or a physics course to fulfill a science requirement might be concerned about the chance that both courses are full. In each of these cases, interest is centered on the probability that two different outcomes occur together. For independent events, a simple multiplication rule relates the individual outcome probabilities to the probability that the outcomes occur together.

> ▪ **Multiplication Rule for Independent Outcomes**
>
> If two outcomes are independent, the probability that *both* outcomes occur is the product of the individual outcome probabilities. Denoting the outcomes as A and B, we write
>
> $$P(A \text{ and } B) = P(A)P(B)$$
>
> More generally, if there are k independent outcomes, the probability that all the outcomes occur is simply the product of all the individual outcome probabilities.

▪ Example 6.1 Nuclear Power Plant Warning Sirens

The Diablo Canyon nuclear power plant in Avila Beach, California, has a warning system that includes a network of sirens in the vicinity of the plant. Sirens are located approximately 0.5 mi from each other. When the system is tested, individual sirens sometimes fail. The sirens operate independently of one another. That is, knowing that a particular siren has failed does not change the probability that any other siren fails.

Imagine that you live near Diablo Canyon (as do the authors of this book) and that there are two sirens that can be heard from your home. We will call these Siren 1 and Siren 2. You might be concerned about the probability that *both* Siren 1 *and* Siren 2 fail. Suppose that when the siren system is activated, about 5% of the individual sirens fail. Then

P(Siren 1 fails) = .05

P(Siren 2 fails) = .05

and the two outcomes *Siren 1 fails* and *Siren 2 fails* are independent. Using the multiplication rule for independent outcomes, we obtain

$$P(\text{Siren 1 fails } and \text{ Siren 2 fails}) = P(\text{Siren 1 fails})P(\text{Siren 2 fails})$$
$$= (.05)(.05)$$
$$= .0025$$

Even though the probability that any individual siren will fail is .05, the probability that *both* of the two sirens will fail is much smaller. This would happen in the long run only about 25 times in 10,000 times that the system is activated.

In Example 6.1, the two outcomes *Siren 1 fails* and *Siren 2 fails* are independent outcomes, so using the multiplication rule for independent outcomes is justified. It is not so easy to calculate the probability that both outcomes occur when two outcomes are dependent, and we won't go into the details here. Be careful to use this multiplication rule only when outcomes are independent!

▪ Example 6.2   DNA Testing

DNA testing has become prominent in the arsenal of criminal detection techniques. It illustrates a somewhat controversial application of the multiplication rule for independent outcomes. Let's focus on three different characteristics (such as blue eye color or blood type B), each of which might or might not occur in a given DNA specimen. Define the following three outcomes:

Outcome 1:   The first characteristic occurs in the specimen.

Outcome 2:   The second characteristic occurs in the specimen.

Outcome 3:   The third characteristic occurs in the specimen.

Suppose that the long-run frequency of occurrence is 1 in 10 people for the first characteristic, 1 in 20 for the second characteristic, and 1 in 5 for the third characteristic. *Assuming independence of the three characteristics*, we calculate the probability that all three characteristics appear in the specimen (all three outcomes occur together) to be

P(1st *and* 2d *and* 3d characteristics all occur) = (.10)(.05)(.20) = .001

That is, in the long run, only 1 in every 1000 individuals will yield DNA specimens with all three characteristics. For more than three characteristics, the probability that all are present would be quite small. So if an accused individual has all characteristics present in a DNA specimen found at the scene of a crime, it is extremely unlikely that another person would have such a genetic profile.

The assumption of independence in this situation makes computing probabilities in this manner controversial. Suppose, for example, that the three characteristics had been as follows:

Outcome 1:   The gene for blue eyes occurs in the DNA specimen.
Outcome 2:   The gene for blond hair occurs in the DNA specimen.
Outcome 3:   The gene for fair skin occurs in the DNA specimen.

Is it reasonable to believe that knowing that Outcome 2 (gene for blond hair) has occurred would not affect your assessment of the probability of Outcome 1 (gene for blue eyes)? If these outcomes are in fact dependent, we are not justified in using the multiplication rule for independent outcomes to compute the probability that all three outcomes occur together.

▪ Using Probability Rules

The probability rules introduced so far are summarized in the accompanying box.

▪ **Probability Rules**

Addition Rule for Outcomes That Cannot Occur at the Same Time:
   $P(\text{Outcome 1 } or \text{ Outcome 2}) = P(\text{Outcome 1}) + P(\text{Outcome 2})$

Complement Rule:
   $P(not \text{ outcome}) = 1 - P(\text{outcome})$

Multiplication Rule for Independent Outcomes:
   $P(\text{Outcome 1 } and \text{ Outcome 2}) = P(\text{Outcome 1})P(\text{Outcome 2})$

In the chapter introduction we posed several questions that can be addressed by using these probability rules. These questions are considered in Examples 6.3 and 6.4.

▪ Example 6.3   Satellite Backup Systems

A satellite has both a main and a backup solar power system, and these systems function independently of one another. (It is common to design such redundancy into products to increase their reliability.) Suppose that the probability of failure during the 10-yr lifetime of the satellite is .05 for the main power system and .08 for the backup system. What is the probability that both systems will fail? Because whether or not one of these systems functions properly has no effect on whether or not the other one does, it is reasonable to assume that the following two outcomes are independent:

Outcome 1:   Main system fails.

Outcome 2:   Backup system fails.

It is then appropriate to use the multiplication rule for independent outcomes to compute

P(main system fails *and* backup system fails)

   $= P$(main system fails)P(backup system fails)

   $= (.05)(.08)$

   $= .004$

The probability that the satellite has at least one workable power system is, using the complement rule,

$P(not$(main system fails *and* backup system fails))

   $= 1 - P$(main system fails *and* backup system fails)

   $= 1 - .004$

   $= .996$

■ Example 6.4   Artificial Heart Components

A certain type of artificial heart has four independent, critical components. Failure of any of these four components is a serious problem. Suppose that 5-year failure rates (expressed as the proportion that fail within 5 years) for these components are known to be

Component 1:   .01

Component 2:   .06

Component 3:   .04

Component 4:   .02

What is the probability that an artificial heart functions for 5 years? For the heart to function for 5 years, all four components must not fail during that time. Let's define the following outcomes:

Outcome 1:   Component 1 does not fail in the first 5 years.

Outcome 2:   Component 2 does not fail in the first 5 years.

Outcome 3:   Component 3 does not fail in the first 5 years.

Outcome 4:   Component 4 does not fail in the first 5 years.

These outcomes are independent (because the components function independently of one another), so

P(no components fail)

   $= P$(1 does not fail *and* 2 does not fail *and* 3 does not fail *and* 4 does not fail)

   $= P$(1 does not fail)P(2 does not fail)P(3 does not fail) P(4 does not fail)

To compute P(Component 1 does not fail), we use the complement rule:

$$P(\text{Component 1 does not fail}) = 1 - P(\text{Component 1 does fail})$$
$$= 1 - .01$$
$$= .99$$

The corresponding probabilities for the other three components are computed in a similar fashion. Then the desired probability is

$$P(\text{no components fail}) = (.99)(.94)(.96)(.98) = .8755$$

This probability, interpreted as a long-run relative frequency, tells us that in the long run, 87.55% of these artificial hearts will function for 5 years with no failures of critical components.

▪ Exercises 6.1–6.14

6.1 An article in the *New York Times* (March 2, 1994) reported that people who suffer cardiac arrest in New York City have only a 1 in 100 chance of survival. Using probability notation, an equivalent statement would be

$P(\text{survival}) = .01$ for people who suffer a cardiac arrest in New York City

(The article attributed this poor survival rate to factors common in large cities: traffic congestion and the difficulty of finding victims in large buildings. Similar studies in smaller cities showed higher survival rates.)
a. Give a relative frequency interpretation of the given probability.
b. The research that was the basis for the *New York Times* article was a study of 2329 consecutive cardiac arrests in New York City. To justify the "1 in 100 chance of survival" statement, how many of the 2329 cardiac arrest sufferers do you think survived? Explain.

6.2 Many fire stations handle emergency calls for medical assistance as well as calls requesting fire-fighting equipment. A particular station says that the probability that an incoming call is for medical assistance is .85. This can be expressed as $P(\text{call is for medical assistance}) = .85$.
a. Give a relative frequency interpretation of the given probability.
b. What is the probability that a call is not for medical assistance?
c. Assuming that successive calls are independent of one another (i.e., knowing that one call is for medical assistance doesn't influence our assessment of the probability that the next call will be for medical as-

sistance), calculate the probability that both of two successive calls will be for medical assistance.
d. Still assuming independence, calculate the probability that for two successive calls, the first is for medical assistance and the second is not for medical assistance.
e. Still assuming independence, calculate the probability that exactly one of the next two calls will be for medical assistance. (Hint: There are two different possibilities that you should consider. The one call for medical assistance might be the first call, or it might be the second call.)
f. Do you think it is reasonable to assume that the requests made in successive calls are independent? Explain.

6.3 A Gallup survey of 2002 adults found that 46% of women and 37% of men experience pain daily (*San Luis Obispo Tribune*, April 6, 2000). Suppose that this information is representative of U.S. adults. If a U.S. adult is selected at random, are the outcomes *selected adult is male* and *selected adult experiences pain daily* independent or dependent? Explain.

6.4 "N.Y. Lottery Numbers Come Up 9-1-1 on 9/11" was the headline of an article that appeared in the *San Francisco Chronicle* (September 13, 2002). More than 5600 people had selected the sequence 9-1-1 on that date, many more than is typical for that sequence. A professor at the University of Buffalo is quoted as saying, "I'm a bit surprised, but I wouldn't characterize it as bizarre. It's randomness. Every number has the same chance of coming up. People tend to read into these things. I'm sure that whatever numbers come up tonight, they will have some special meaning to someone, somewhere." The New

York state lottery uses balls numbered 0–9 circulating in 3 separate bins. To select the winning sequence, one ball is chosen at random from each bin. What is the probability that the sequence 9-1-1 would be the one selected on any particular day?

6.5 The Associated Press (*San Luis Obispo Telegram-Tribune*, August 23, 1995) reported on the results of mass screening of schoolchildren for tuberculosis (TB). It was reported that for Santa Clara County, California, the proportion of all tested kindergartners who were found to have TB was .0006. The corresponding proportion for recent immigrants (thought to be a high-risk group) was .0075. Suppose that a Santa Clara County kindergartner is to be selected at random. Are the outcomes *selected student is a recent immigrant* and *selected student has TB* independent or dependent outcomes? Justify your answer using the given information.

6.6 The article "Men, Women at Odds on Gun Control" (*Cedar Rapids Gazette*, September 8, 1999) included the following statement: "The survey found that 56 percent of American adults favored stricter gun control laws. Sixty-six percent of women favored the tougher laws, compared with 45 percent of men." These figures are based on a large telephone survey conducted by Associated Press Polls. If an adult is selected at random, are the outcomes *selected adult is female* and *selected adult favors stricter gun control* independent or dependent outcomes? Explain.

6.7 The Australian newspaper *The Mercury* (May 30, 1995) reported that, based on a survey of 600 reformed and current smokers, 11.3% of those who had attempted to quit smoking in the previous 2 years had used a nicotine aid (such as a nicotine patch). It also reported that 62% of those who quit smoking without a nicotine aid began smoking again within 2 weeks and 60% of those who used a nicotine aid began smoking again within 2 weeks. If a smoker who is trying to quit smoking is selected at random, are the outcomes *selected smoker who is trying to quit uses a nicotine aid* and *selected smoker who has attempted to quit begins smoking again within 2 weeks* independent or dependent outcomes? Justify your answer using the given information.

6.8 In a small city, approximately 15% of those eligible are called for jury duty in any one calendar year. People are selected for jury duty at random from those eligible, and the same individual cannot be called more than once in the same year. What is the probability that an eligible person in this city is selected 2 years in a row? 3 years in a row?

6.9 Jeanie is a bit forgetful, and if she doesn't make a "to do" list, the probability that she forgets something she is supposed to do is .1. Tomorrow she intends to run three errands, and she fails to write them on her list.

a. What is the probability that Jeanie forgets all three errands? What assumptions did you make to calculate this probability?

b. What is the probability that Jeanie remembers at least one of the three errands?

c. What is the probability that Jeanie remembers the first errand but not the second or third?

6.10 Approximately 30% of the calls to an airline reservation phone line result in a reservation being made.

a. Suppose that an operator handles 10 calls. What is the probability that none of the 10 calls results in a reservation?

b. What assumption did you make in order to calculate the probability in Part (a)?

c. What is the probability that at least one call results in a reservation being made?

6.11 The following case study is reported in the article "Parking Tickets and Missing Women," which appears in the book *Statistics: A Guide to the Unknown*, Judith Tanur, ed. (see Chapter 1 references). In a Swedish trial on a charge of overtime parking, a police officer testified that he had noted the position of the two air valves on the tires of a parked car: To the closest hour, one valve was at the 1 o'clock position and the other was at the 6 o'clock position. After the allowable time for parking in that zone had passed, the policeman returned, noted that the valves were in the same position, and ticketed the car. The owner of the car claimed that he had left the parking place in time and had returned later. The valves just happened by chance to be in the same positions. An "expert" witness computed the probability of this occurring as $(1/12)(1/12) = 1/144$.

a. What reasoning did the expert use to arrive at the probability of 1/144?

b. Can you spot the error in the reasoning that leads to the stated probability of 1/144? What effect does this error have on the probability of occurrence? Do you think that 1/144 is larger or smaller than the correct probability of occurrence?

6.12 Three friends (A, B, and C) will participate in a round-robin tournament in which each one plays both of the others. Suppose that $P(A$ beats $B) = .7$, $P(A$ beats $C) = .8$, and $P(B$ beats $C) = .6$ and that the outcomes of the three matches are independent of one another.

a. What is the probability that A wins both her matches and that B beats C?

b. What is the probability that A wins both her matches?

c. What is the probability that A loses both her matches?

d. What is the probability that each person wins one match? (Hint: There are two different ways for this to happen. Calculate the probability of each separately, and then add.)

6.13 Consider a system consisting of four components, as pictured in the following diagram:

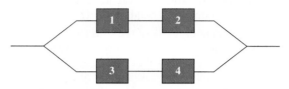

Components 1 and 2 form a series subsystem, as do Components 3 and 4. The two subsystems are connected in parallel. Suppose that $P(1 \text{ works}) = .9$, $P(2 \text{ works}) = .9$, $P(3 \text{ works}) = .9$, and $P(4 \text{ works}) = .9$ and that these four outcomes are independent (the four components work independently of one another).

a. The 1–2 subsystem works only if both components work. What is the probability of this happening?

b. What is the probability that the 1–2 subsystem doesn't work? that the 3–4 subsystem doesn't work?

c. The system won't work if the 1–2 subsystem doesn't work and if the 3–4 subsystem also doesn't work. What is the probability that the system won't work? that it will work?

d. How would the probability of the system working change if a 5–6 subsystem was added in parallel with the other two subsystems?

e. How would the probability that the system works change if there were three components in series in each of the two subsystems?

6.14 *USA Today* (March 15, 2001) introduced a measure of racial and ethnic diversity called the Diversity Index. The Diversity Index is supposed to approximate the probability that two randomly selected individuals are racially or ethnically different. The equation used to compute the Diversity Index after the 1990 census was

$$1 - [P(W)^2 + P(B)^2 + P(AI)^2 + P(API)^2] \cdot [P(H)^2 + P(not\ H)^2]$$

where W is the event that a randomly selected individual is white, B is the event that a randomly selected individual is black, AI is the event that a randomly selected individual is American Indian, API is the event that a randomly selected individual is Asian or Pacific Islander, and H is the event that a randomly selected individual is Hispanic. The explanation of this index stated that

1. $[P(W)^2 + P(B)^2 + P(AI)^2 + P(API)^2]$ is the probability that two randomly selected individuals are the same race

2. $[P(H)^2 + P(not\ H)^2]$ is the probability that two randomly selected individuals are either both Hispanic or both not Hispanic

3. The calculation of the Diversity Index treats Hispanic ethnicity as if it were independent of race

a. What additional assumption about race must be made to justify use of the addition rule in the computation of $[P(W)^2 + P(B)^2 + P(AI)^2 + P(API)^2]$ as the probability that two randomly selected individuals are of the same race?

b. Three different probability rules are used in the calculation of the Diversity Index: the Complement Rule, the Addition Rule, and the Multiplication Rule. Describe the way in which each is used.

■ 6.2 Probability as a Basis for Making Decisions

The concepts of probability play a critical role in developing statistical methods that allow us to draw conclusions based on available information, as we can see in Examples 6.5 and 6.6.

■ Example 6.5   Age and Gender of College Students

Table 6.1 shows the proportions of the student body who fall in various age–gender combinations at a university. These numbers can be interpreted as probabilities if we think of selecting a student at random from the student body. For example, the probability (long-run proportion of the time) that a male, age 21–24, will be selected is .16.

Table 6.1 ▪ Age and Gender Distribution

Gender	Age					
	Under 17	17–20	21–24	25–30	31–39	40 and Older
Male	.006	.15	.16	.08	.04	.01
Female	.004	.18	.14	.10	.04	.09

Suppose that you are told that a 43-year-old student from this university is waiting to meet you. You are asked to decide whether the student is male or female. How would you respond? What if the student had been 27 years old? What about 33 years old? Are you equally confident in all three of your choices?

A reasonable response to these questions could be based on the probability information in Table 6.1. We would decide that the 43-year-old student was female. We cannot be certain that this is correct, but we can see that a male is much less likely to be in the 40-and-older age group than a female is. We would also decide that the 27-year-old was female. However, we would be less confident in our conclusion than we were for the 43-year-old student. For the age group 31–39, males and females constitute the same proportion of the student body, so we would think it equally likely that a 33-year-old student would be male or female. We could decide in favor of male (or female), but with little confidence in our conclusion; in other words, there is a good chance of being incorrect.

▪ Example 6.6   Can You Pass by Guessing?

A professor planning to give a quiz that consists of 20 true–false questions is interested in knowing how someone who answers by guessing would do on such a test. To investigate, he asks the 500 students in his introductory psychology course to write the numbers from 1 to 20 on a piece of paper and then to arbitrarily write T or F next to each number. The students are forced to guess at the answer to each question, because they are not even told what the questions are. These answer sheets are then collected and graded using the key for the quiz. The results are summarized in Table 6.2.

Because probabilities are long-run proportions, an entry in the right-hand column of Table 6.2 can be considered an estimate of the probability of correctly guessing a specific number of responses. For example, a proportion of .122 (or 12.2%) of the 500 students got 12 of the 20 correct when guessing. We then estimate the long-run proportion of guessers who would get 12 correct to be .122; we say that the probability that a student who is guessing will get 12 correct is (approximately) .122.

Let's use the information in Table 6.2 to answer the following questions.

1. Would you be surprised if someone who is guessing on a 20-question true–false quiz got only 3 correct? The approximate probability of a guesser getting 3 correct is .002. This means that, in the long run, only about 2 in 1000 guessers would score exactly 3 correct. This would be an unlikely outcome, and we would consider its occurrence surprising.

Table 6.2 ▪ Quiz "Guessing" Distribution

Number of Correct Responses	Number of Students	Proportion of Students	Number of Correct Responses	Number of Students	Proportion of Students
0	0	.000	11	79	.158
1	0	.000	12	61	.122
2	1	.002	13	39	.078
3	1	.002	14	18	.036
4	2	.004	15	7	.014
5	8	.016	16	1	.002
6	18	.036	17	1	.002
7	37	.074	18	0	.000
8	58	.116	19	0	.000
9	81	.162	20	0	.000
10	88	.176			

2. If a score of 15 or more correct is required to receive a passing grade on the quiz, is it likely that someone who is guessing will be able to pass? The long-run proportion of guessers who would pass is the sum of the proportions for all the passing scores (15, 16, 17, 18, 19, and 20). Thus,

probability of passing ≈ .014 + .002 + .002 + .000 + .000 + .000 = .018

It would be unlikely that a student who is guessing would be able to pass.

3. The professor actually gives the quiz, and a student scores 16 correct. Do you think that the student was just guessing? Let's begin by assuming that the student was guessing and determine whether a score as high as 16 is a likely or an unlikely occurrence. Table 6.2 tells us that the approximate probability of getting a score at least as high as this student's score is

probability of scoring 16 or higher ≈ .002 + .002 + .000 + .000 + .000

= .004

That is, in the long run, only about 4 times in 1000 would a guesser score 16 or higher. This would be a rare outcome. There are two possible explanations for the observed score: (1) The student was guessing and was really lucky, or (2) the student was not just guessing. Given that the first explanation is highly unlikely, a more plausible choice is the second explanation. We would conclude that the student was not just guessing at the answers. Although we cannot be certain that we are correct in this conclusion, the evidence is compelling.

4. What score on the quiz would it take to convince you that a student was not just guessing? We would be convinced that a student was not just guessing if his or her score was high enough that it was unlikely that a guesser would have been able to do as well. Consider the following approximate probabilities (computed from the entries in Table 6.2):

Score	Approximate Probability
20	.000
19 or better	.000 + .000 = .000
18 or better	.000 + .000 + .000 = .000
17 or better	.002 + .000 + .000 + .000 = .002
16 or better	.002 + .002 + .000 + .000 + .000 = .004
15 or better	.014 + .002 + .002 + .000 + .000 + .000 = .018
14 or better	.036 + .014 + .002 + .002 + .000 + .000 + .000 = .054
13 or better	.078 + .036 + .014 + .002 + .002 + .000 + .000 + .000 = .132

We might say that a score of 14 or higher is reasonable evidence that some-one is not guessing, because the approximate probability that a guesser would score this high is only .054. Of course, if we conclude that a student is not guessing based on a quiz score of 14 or higher, there is a risk that we are incorrect (about 1 in 20 guessers would score this high by chance). About 13.2% of the time, a guesser will score 13 or more correct. This would happen by chance often enough that most people would not rule out the student being a guesser.

Examples 6.5 and 6.6 show how probability information can be used to make a decision. This is a primary goal of statistical inference. Later chapters look more formally at the problem of drawing a conclusion based on available but often incomplete information and then assessing the reliability of such a conclusion.

▪ Exercises 6.15–6.18

6.15 Is ultrasound a reliable method for determining the gender of an unborn baby? Consider the following data on 1000 births, which are consistent with summary values that appeared in the online *Journal of Statistics Education* ("New Approaches to Learning Probability in the First Statistics Course" [2001]):

	Ultrasound Predicted Female	Ultrasound Predicted Male
Actual gender is female	432	48
Actual gender is male	130	390

Do you think ultrasound is equally reliable for predicting gender for boys and for girls? Explain, using the information in the table to calculate estimates of any probabilities that are relevant to your conclusion.

6.16 Researchers at UCLA were interested in whether working mothers were more likely to suffer workplace injuries than women without children. They studied 1400 working women, and a summary of their findings was reported in the *San Luis Obispo Telegram-Tribune* (February 28, 1995). The information in the following table is consistent with summary values reported in the article:

	No Children	Children Under 6	Children, but None Under 6	Total
Injured on the Job in 1989	32	68	56	**156**
Not Injured on the Job in 1989	368	232	644	**1244**
Total	**400**	**300**	**700**	**1400**

The researchers drew the following conclusion: Women with children younger than age 6 are much more likely to be injured on the job than childless women or mothers with older children. Provide a justification for the researchers' conclusion. Use the information in the table to calculate estimates of any probabilities that are relevant to your justification.

6.17 A Gallup Poll conducted in November 2002 examined how people perceived the risks associated with smoking. The following table summarizes data on smoking status and perceived risk of smoking that is consistent with summary quantities published by Gallup:

Smoking Status	Perceived Risk			
	Very Harmful	Somewhat Harmful	Not Too Harmful	Not at All Harmful
Current Smoker	60	30	5	1
Former Smoker	78	16	3	2
Never Smoked	86	10	2	1

Assume that it is reasonable to consider these data representative of the U.S. adult population. Consider the following conclusion: Current smokers are less likely to view smoking as very harmful than either former smokers or those who have never smoked.

Provide a justification for this conclusion. Use the information in the table to calculate estimates of any probabilities that are relevant to your justification.

6.18 Students at a particular university use a telephone registration system to select their courses for the next term. There are four different priority groups, with students in Group 1 registering first, followed by those in Group 2, and so on. Suppose that the university provided the accompanying information on registration for the fall semester. The entries in the table represent the proportion of students falling into each of the 20 priority–unit combinations.

Priority Group	Number of Units Secured During First Call to Registration System				
	0–3	4–6	7–9	10–12	More than 12
1	.01	.01	.06	.10	.07
2	.02	.03	.06	.09	.05
3	.04	.06	.06	.06	.03
4	.04	.08	.07	.05	.01

a. What proportion of students at this university got 10 or more units during the first call?
b. Suppose that a student reports receiving 11 units during the first call. Is it more likely that he or she is in the first or the fourth priority group?
c. If you are in the third priority group next term, is it likely that you will get more than 9 units during the first call? Explain.

▪ 6.3 Estimating Probabilities Empirically and by Using Simulation

In the examples presented so far, reaching conclusions required knowledge of the probabilities of various outcomes. In some cases, this is reasonable, and we know the true long-run proportion of the time that each outcome will occur. In other situations, these probabilities are not known and must be determined. Sometimes probabilities can be determined analytically, by using mathematical rules and probability properties, including the basic ones introduced in this chapter.

In this section we change gears a bit and focus on an empirical approach to probability. When an analytical approach is impossible, impractical, or just beyond the limited probability tools of the introductory course, we can *estimate* probabilities empirically through observation or by using simulation.

▪ Estimating Probabilities Empirically

It is fairly common practice to use observed long-run proportions to estimate probabilities. The process used to estimate probabilities is simple:

1. Observe a large number of chance outcomes under controlled circumstances.
2. By appealing to the interpretation of probability as a long-run relative frequency, estimate the probability of an event by using the observed proportion of occurrence.

This process is illustrated in Examples 6.7 and 6.8.

▪ **Example 6.7   Fair Hiring Practices**

The biology department at a university plans to recruit a new faculty member and intends to advertise for someone with a Ph.D. in biology and at least 10 years of college-level teaching experience. A member of the department expresses the belief that requiring at least 10 years of teaching experience will exclude most potential applicants and will exclude far more female applicants than male applicants. The biology department would like to determine the probability that someone with a Ph.D. in biology who is looking for an academic position would be eliminated from consideration because of the experience requirement.

A similar university just completed a search in which there was no requirement for prior teaching experience but the information about prior teaching experience was recorded. The 410 applications yielded the following data:

	Number of Applicants		
	Less Than 10 Years of Experience	10 Years of Experience or More	Total
Male	178	112	**290**
Female	99	21	**120**
Total	**277**	**133**	**410**

Let's assume that the populations of applicants for the two positions can be regarded as the same. We will use the available information to approximate the probability that an applicant will fall into each of the four gender–experience combinations. The estimated probabilities (obtained by dividing the number of applicants for each gender–experience combination by 410) are given in Table 6.3. From Table 6.3, we calculate

estimate of P(candidate excluded) $= .4341 + .2415 = .6756$

We can also assess the impact of the experience requirement separately for male applicants and for female applicants. From the given information, we calculate that the proportion of male applicants who have less than 10 years of experience is

Table 6.3 ▪ Estimated Probabilities for Example 6.7

	Less than 10 Years of Experience	10 Years of Experience or More
Male	.4341	.2732
Female	.2415	.0512

178/290 = .6138, whereas the corresponding proportion for females is 99/120 = .8250. Therefore, approximately 61% of the male applicants would be eliminated by the experience requirement, and about 83% of the female applicants would be eliminated.

These subgroup proportions — .6138 for males and .8250 for females — are examples of *conditional probabilities*. As discussed in Section 6.1, outcomes are dependent if the occurrence of one outcome changes our assessment of the probability that the other outcome will occur. A conditional probability shows how the original probability changes in light of new information. In this example, the probability that a potential candidate has less than 10 years experience is .6756, but this probability changes to .8250 if we know that a candidate is female. These probabilities can be expressed as an unconditional probability

P(less than 10 years of experience) = .6756

and a conditional probability

read as "given"

P(less than 10 years of experience | female) = .8250

■ Example 6.8   Who Has the Upper Hand?

Men and women frequently express intimacy through the act of touching. A common instance of mutual touching is the simple act of holding hands. Some researchers have suggested that touching in general, and hand-holding in particular, not only might be an expression of intimacy but also might communicate status differences. For two people to hold hands, one must assume an "overhand" grip and one an "underhand" grip. Research in this area has shown that it is predominantly the male who assumes the overhand grip. In the view of some investigators, the overhand grip is seen to imply status or superiority. The authors of the paper "Men and Women Holding Hands: Whose Hand is Uppermost?" (*Perceptual and Motor Skills* [1999]: 537–549) investigated an alternative explanation: Perhaps the positioning of hands is a function of the heights of the individuals. Because men, on average, tend to be taller than women, maybe comfort, not status, dictates the positioning. Investigators at two separate universities observed hand-holding male–female pairs, resulting in the following data:

Number of Hand-Holding Couples

	Sex of Person with Uppermost Hand		
	Men	Women	Total
Man taller	2149	299	**2448**
Equal height	780	246	**1026**
Woman taller	241	205	**446**
Total	**3170**	**750**	3920

Assuming that these hand-holding couples are representative of hand-holding couples in general, we can use the available information to estimate various probabilities. For example, if a hand-holding couple is selected at random, then

$$\text{estimate of } P(\text{man's hand uppermost}) = \frac{3170}{3920} = 0.809$$

For a randomly selected hand-holding couple, if the man is taller, then the probability that the male has the uppermost hand is 2149/2448 = .878. On the other hand —so to speak—if the woman is taller, the probability that the female has the uppermost hand is 205/446 = .460. Notice that these last two estimates are estimates of the conditional probabilities P(male uppermost *given* male taller) and P(female uppermost *given* female taller), respectively. Also, because P(male uppermost *given* male taller) is not equal to P(male uppermost), the events *male uppermost* and *male taller* are not independent events. But, even when the female is taller, the male is still more likely to have the upper hand!

■ Estimating Probabilities by Using Simulation

Simulation provides a means of estimating probabilities when we are unable (or do not have the time or resources) to determine probabilities analytically and when it is impractical to estimate them empirically by observation. Simulation is a method that generates "observations" by performing a chance experiment that is as similar as possible in structure to the real situation of interest.

To illustrate the idea of simulation, consider the situation in which a professor wishes to estimate the probabilities of different possible scores on a 20-question true–false quiz when students are merely guessing at the answers. Observations could be collected by having 500 students actually guess at the answers to 20 questions and then scoring the resulting papers. However, obtaining the probability estimates in this way requires considerable time and effort; simulation provides an alternative approach.

Because each question on the quiz is a true–false question, a person who is guessing should be equally likely to answer correctly or incorrectly on any given question. Rather than asking a student to select true or false and then comparing the choice to the correct answer, an equivalent process would be to pick a ball at random from a box that contains half red balls and half blue balls, with a blue ball representing a correct answer. Making 20 selections from the box (replacing each ball selected before picking the next one) and then counting the number of correct choices (the number of times a blue ball is selected) is a physical substitute for an observation from a student who has guessed at the answers to 20 true–false questions. The number of blue balls in 20 selections, with replacement, from a box that contains half red and half blue balls should have the same probability as the number of correct responses to the quiz when a student is guessing.

For example, 20 selections of balls might yield the following results (R, red ball; B, blue ball):

Selection	1	2	3	4	5	6	7	8	9	10
Result	R	R	B	R	B	B	R	R	R	B

Selection	11	12	13	14	15	16	17	18	19	20
Result	R	R	B	R	R	B	B	R	R	B

These data would correspond to a quiz with eight correct responses, and they provide us with one observation for estimating the probabilities of interest. This process is then repeated a large number of times to generate additional observations. For example, we might find the following:

Repetition	Number of "Correct" Responses
1	8
2	11
3	10
4	12
⋮	⋮
1000	11

The 1000 simulated quiz scores could then be used to construct a table of estimated probabilities.

Taking this many balls out of a box and writing down the results would be cumbersome and tedious. The process can be simplified by using random digits to substitute for drawing balls from the box. For example, a single digit could be selected at random from the 10 digits 0, 1, 2, 3, 4, 5, 6, 7, 8, 9. When using random digits, each of the 10 possibilities is equally likely to occur, so we can use the even digits (including 0) to indicate a correct response and the odd digits to indicate an incorrect response. This would maintain the important property that a correct response and an incorrect response are equally likely, because correct and incorrect are each represented by 5 of the 10 digits.

To aid in carrying out such a simulation, tables of random digits (such as Appendix Table 1) or computer-generated random digits can be used. The numbers in Appendix Table 1 were generated using a computer's random number generator. You can think of the table as being produced by repeatedly drawing a chip from a box containing 10 chips numbered 0, 1, ..., 9. After each selection, the result is recorded, the chip returned to the box, and the chips mixed. Thus, any of the digits is equally likely to occur on any of the selections.

To see how a table of random numbers can be used to carry out a simulation, let's reconsider the quiz example. We use a random digit to represent the guess on a single question, with an even digit representing a correct response. A series of 20 digits represents the answers to the 20 quiz questions. We pick an arbitrary starting point in Appendix Table 1. Suppose that we start at Row 10 and take the 20 digits in a row to represent one quiz. The first five "quizzes" and the corresponding number correct are

Quiz	Random Digits	Number Correct
	* * * * * * * * * * * * *	
1	9 4 6 0 6 9 7 8 8 2 5 2 9 6 0 1 4 6 0 5	13
2	6 6 9 5 7 4 4 6 3 2 0 6 0 8 9 1 3 6 1 8	12
3	0 7 1 7 7 7 2 9 7 8 7 5 8 8 6 9 8 4 1 0	9
4	6 1 3 0 9 7 3 3 6 6 0 4 1 8 3 2 6 7 6 8	11
5	2 2 3 6 2 1 3 0 2 2 6 6 9 7 0 2 1 2 5 8	13

denotes correct response →

This process is repeated to generate a large number of observations, which are then used to construct a table of estimated probabilities.

The method for generating observations must preserve the important characteristics of the actual process being considered if simulation is to be successful. For example, it would be easy to adapt the simulation procedure for the true–false quiz to one for a multiple-choice quiz. Suppose that each of the 20 questions on the quiz has 5 possible responses, only 1 of which is correct. For any particular question, we would expect a student to be able to guess the correct answer only one-fifth of the time in the long run. To simulate this situation, we could select at random from a box that contained four red balls and only one blue ball (or, more generally, four times as many red balls as blue balls). If we are using random digits for the simulation, we could use 0 and 1 to represent a correct response and 2, 3, . . . , 9 to represent an incorrect response.

▪ **Using Simulation to Approximate a Probability**

1. Design a method that uses a random mechanism (such as a random number generator or table, the selection of a ball from a box, or the toss of a coin) to represent an observation. Be sure that the important characteristics of the actual process are preserved.

2. Generate an observation using the method from Step 1, and determine whether the outcome of interest has occurred.

3. Repeat Step 2 a large number of times.

4. Calculate the estimated probability by dividing the number of observations for which the outcome of interest occurred by the total number of observations generated.

The simulation process is illustrated in Examples 6.9–6.11.

▪ Example 6.9   Building Permits

Many California cities limit the number of building permits that are issued each year. Because of limited water resources, one such city plans to issue permits for only 10 dwelling units in the upcoming year. The city will decide who is to receive permits by holding a lottery. Suppose that you are 1 of 39 individuals who apply for permits. Thirty of these individuals are requesting permits for a single-family home, eight are requesting permits for a duplex (which counts as two dwelling units), and one person is requesting a permit for a small apartment building with eight units (which counts as eight dwelling units). Each request will be entered into the lottery. Requests will be selected at random one at a time, and if there are enough permits remaining, the request will be granted. This process will continue until all 10 permits have been issued. If your request is for a single-family home, what are your chances of receiving a permit? Let's use simulation to estimate this probability. (It is not easy to determine analytically.)

To carry out the simulation, we can view the requests as being numbered from 1 to 39 as follows:

1–30:	Requests for single-family homes
31–38:	Requests for duplexes
39:	Request for 8-unit apartment

For ease of discussion, let's assume that your request is Request 1.

One method for simulating the permit lottery consists of these three steps:

1. Choose a random number between 1 and 39 to indicate which permit request is selected first, and grant this request.

2. Select another random number between 1 and 39 to indicate which permit request is considered next. Determine the number of dwelling units for the selected request. Grant the request only if there are enough permits remaining to satisfy the request.

3. Repeat Step 2 until permits for 10 dwelling units have been granted.

We used MINITAB to generate random numbers between 1 and 39 to imitate the lottery drawing. (The random number table in Appendix Table 1 could also be used, by selecting two digits and ignoring 00 and any value over 39.) For example, the first sequence generated by MINITAB is:

Random Number	Type of Request	Total Number of Units So Far
25	Single-family home	1
07	Single-family home	2
38	Duplex	4
31	Duplex	6
26	Single-family home	7
12	Single-family home	8
33	Duplex	10

We would stop at this point, because permits for 10 units would have been issued. In this simulated lottery, Request 1 was not selected, so you would not have received a permit.

The next simulated lottery (using MINITAB to generate the selections) is as follows:

Random Number	Type of Request	Total Number of Units So Far
38	Duplex	2
16	Single-family home	3
30	Single-family home	4
39	Apartment, not granted, because there are not 8 permits remaining	4
14	Single-family home	5
26	Single-family home	6
36	Duplex	8
13	Single-family home	9
15	Single-family home	10

Again, Request 1 was not selected, so you would not have received a permit in this simulated lottery.

Now that a strategy for simulating a lottery has been devised, the tedious part of the simulation begins. We now have to simulate a large number of lottery drawings, determining for each whether Request 1 is granted. We simulated 500 such drawings and found that Request 1 was selected in 85 of the lotteries. Thus,

$$\text{estimated probability of receiving a building permit} = \frac{85}{500} = .17$$

■ **Example 6.10   One-Boy Family Planning**

Suppose that couples who wanted children were to continue having children until a boy is born. Assuming that each newborn child is equally likely to be a boy or a girl, would this behavior change the proportion of boys in the population? This question was posed in an article that appeared in *The American Statistician* (1994: 290–293), and many people answered the question incorrectly. We use simulation to estimate the long-run proportion of boys in the population if families were to continue to have children until they have a boy. This proportion is an estimate of the probability that a randomly selected child from this population is a boy. Note that every sibling group would have exactly one boy.

We use a single-digit random number to represent a child. The odd digits (1, 3, 5, 7, 9) represent a male birth, and the even digits represent a female birth. An observation is constructed by selecting a sequence of random digits. If the first random number obtained is odd (a boy), the observation is complete. If the first selected number is even (a girl), another digit is chosen. We continue in this way until an odd digit is obtained. For example, reading across Row 15 of the random number table (Appendix Table 1), we find that the first 10 digits are

0 7 1 7 4 2 0 0 0 1

Using these numbers to simulate sibling groups, we get

Sibling group 1	0 7	girl, boy
Sibling group 2	1	boy
Sibling group 3	7	boy
Sibling group 4	4 2 0 0 0 1	girl, girl, girl, girl, girl, boy

Continuing along Row 15 of the random number table, we get

Sibling group 5	3	boy
Sibling group 6	1	boy
Sibling group 7	2 0 4 7	girl, girl, girl, boy
Sibling group 8	8 4 1	girl, girl, boy

After simulating 8 sibling groups, we have 8 boys among 19 children. The proportion of boys is 8/19, which is close to .5. Continuing the simulation to obtain a large number of observations suggests that the long-run proportion of boys in the population would still be .5, which is indeed the case.

■ **Example 6.11   ESP?**

Can a close friend read your mind? Try the following chance experiment. Write the word *blue* on one piece of paper and the word *red* on another, and place the two slips of paper in a box. Select one slip of paper from the box, look at the word written on it, and then try to convey the word by sending a mental message to a friend who is seated in the same room. Ask your friend to select either red or blue, and record whether the response is correct. Repeat this 10 times and total the number of correct responses. How did your friend do? Is your friend receiving your mental messages or just guessing?

Let's investigate this issue by using simulation to get the approximate probabilities of the various possible numbers of correct responses for someone who is guessing. Someone who is guessing should have an equal chance of responding correctly (C) or incorrectly (X). We can use a random digit to represent a response, with an even digit representing a correct response and an odd digit representing an incorrect response. A sequence of 10 digits can be used to simulate one execution of the chance experiment.

For example, using the last 10 digits in Row 25 of the random number table (Appendix Table 1) gives

5	2	8	3	4	3	0	7	3	5
X	C	C	X	C	X	C	X	X	X

which is a simulated chance experiment resulting in four correct responses. We used MINITAB to generate 150 sequences of 10 random digits and obtained the following results:

Sequence Number	Digits	Number Correct
1	3996285890	5
2	1690555784	4
3	9133190550	2
⋮	⋮	⋮
149	3083994450	5
150	9202078546	7

Table 6.4 summarizes the results of our simulation. The estimated probabilities in Table 6.4 are based on the assumption that a correct and an incorrect response are equally likely (guessing). Evaluate your friend's performance in light of the information in Table 6.4. Is it likely that someone who is guessing would have been able to get as many correct as your friend did? Do you think your friend was receiving your mental messages? How are the estimated probabilities in Table 6.4 used to support your answer?

Table 6.4 ▪ Estimated Probabilities for Example 6.11

Number Correct	Number of Sequences	Estimated Probability
0	0	.0000
1	1	.0067
2	8	.0533
3	16	.1067
4	30	.2000
5	36	.2400
6	35	.2333
7	17	.1133
8	7	.0467
9	0	.0000
10	0	.0000
Total	**150**	**1.0000**

■ Exercises 6.19–6.27

6.19 The *Los Angeles Times* (June 14, 1995) reported that the U.S. Postal Service is getting speedier, with higher overnight on-time delivery rates than in the past. Postal Service standards call for overnight delivery within a zone of about 60 mi for any first-class letter deposited by the last collection time posted on a mailbox. Two-day delivery is promised within a 600-mi zone, and three-day delivery is promised for distances over 600 mi. The Price Waterhouse accounting firm conducted an independent audit by "seeding" the mail with letters and recording on-time delivery rates for these letters. Suppose that the results of the Price Waterhouse study were as follows (these numbers are fictitious, but they are compatible with summary values given in the article):

	Number of Letters Mailed	Number of Letters Arriving on Time
Los Angeles	500	425
New York	500	415
Washington, D.C.	500	405
Nationwide	6000	5220

Use the given information to estimate the following probabilities:
a. The probability of an on-time delivery in Los Angeles
b. The probability of late delivery in Washington, D.C.
c. The probability that both of two letters mailed in New York are delivered on time
d. The probability of on-time delivery nationwide

6.20 Five hundred first-year students at a state university were classified according to both high school grade point average (GPA) and whether they were on academic probation at the end of their first semester. The data are summarized in the table at the top of the next column.

	High School GPA			
Probation	2.5 to <3.0	3.0 to <3.5	3.5 and Above	Total
Yes	50	55	30	**135**
No	45	135	185	**365**
Total	**95**	**190**	**215**	**500**

a. Construct a table of the estimated probabilities for each GPA–probation combination by dividing the number of students in each of the 6 cells of the table by 500.
b. Use the table constructed in Part (a) to approximate the probability that a randomly selected first-year student at this university will be on academic probation at the end of the first semester.
c. What is the estimated probability that a randomly selected first-year student at this university had a high school GPA of 3.5 or above?
d. Are the two outcomes *selected student has a GPA of 3.5 or above* and *selected student is on academic probation at the end of the first semester* independent outcomes? How can you tell?
e. Estimate the proportion of first-year students with high school GPAs between 2.5 and 3.0 who are on academic probation at the end of the first semester.
f. Estimate the proportion of those first-year students with high school GPAs 3.5 and above who are on academic probation at the end of the first semester.

6.21 The table below describes (approximately) the distribution of students by gender and college at a midsize public university in the West. If we were to randomly select one student from this university:
a. What is the probability that the selected student is a male?
b. What is the probability that the selected student is in the College of Agriculture?
c. What is the probability that the selected student is a male in the College of Agriculture?

Table for Exercise 6.21

	College						
Gender	Education	Engineering	Liberal Arts	Science and Math	Agriculture	Business	Architecture
Male	200	3200	2500	1500	2100	1500	200
Female	300	800	1500	1500	900	1500	300

d. What is the probability that the selected student is a male who is not from the College of Agriculture?

6.22 On April 1, 2000, the Bureau of the Census in the United States attempted to count every U.S. citizen and every resident. When counting such a large number of people, however, mistakes of various sorts are likely to arise. In 2000, many people were employed by the Bureau of the Census who were, in essence, trying to verify some of the information obtained by other census workers. Suppose that the counts in the table below are obtained for four counties in one region:

a. If a census fact-checker selects one person at random from this region, what is the probability that the selected person is from Ventura County?
b. If one person is selected at random from Ventura County, what is the probability that the selected person is Hispanic?
c. If one Hispanic person is selected at random from this region, what is the probability that the selected individual is from Ventura?
d. If one person is selected at random from this region, what is the probability that the selected person is an Asian from San Luis Obispo County?
e. If one person is selected at random from this region, what is the probability that the person is either Asian or from San Luis Obispo County?
f. If one person is selected at random from this region, what is the probability that the person is Asian or from San Luis Obispo County but not both?
g. If two people are selected at random from this region, what is the probability that both are Caucasians?
h. If two people are selected at random from this region, what is the probability that neither is Caucasian?
i. If two people are selected at random from this region, what is the probability that exactly one is a Caucasian?
j. If two people are selected at random from this region, what is the probability that both are residents of the same county?
k. If two people are selected at random from this region, what is the probability that both are from different racial/ethnic groups?

6.23 A medical research team wishes to evaluate two different treatments for a disease. Subjects are selected two at a time, and then one of the pair is assigned to each of the two treatments. The treatments are applied, and each is either a success (S) or a failure (F). The researchers keep track of the total number of successes for each treatment. They plan to continue the chance experiment until the number of successes for one treatment exceeds the number of successes for the other treatment by 2. For example, they might observe the results in the table on the next page. The chance experiment would stop after the sixth pair, because Treatment 1 has two more successes than Treatment 2. The researchers would conclude that Treatment 1 is preferable to Treatment 2.

Suppose that Treatment 1 has a success rate of .7 (i.e., $P(\text{success}) = .7$ for Treatment 1) and that Treatment 2 has a success rate of .4. Use simulation to estimate the probabilities listed in Parts (a) and (b). (Hint: Use a pair of random digits to simulate one pair of subjects. Let the first digit represent Treatment 1 and use 1–7 as an indication of a success and 8, 9, and 0 to indicate a failure. Let the second digit represent Treatment 2, with 1–4 representing a success. For example, if the two digits selected to represent a pair were 8 and 3, you would record failure for Treatment 1 and success for Treatment 2. Continue to select pairs, keeping track of the cumulative number of successes for each treatment. Stop the trial as soon as the number of successes for one treatment exceeds that for the other by 2. This would complete one trial. Now repeat this whole process until you have results for at least 20 trials [more is better]. Finally, use the simulation results to estimate the desired probabilities.)

a. What is the probability that more than 5 pairs must be treated before a conclusion can be reached? (Hint: $P(\text{more than } 5) = 1 - P(5 \text{ or fewer})$.)
b. What is the probability that the researchers will incorrectly conclude that Treatment 2 is the better treatment?

Table for Exercise 6.22

| County | Race/Ethnicity | | | | |
	Caucasian	Hispanic	Black	Asian	American Indian
Monterey	163,000	139,000	24,000	39,000	4000
San Luis Obispo	180,000	37,000	7000	9000	3000
Santa Barbara	230,000	121,000	12,000	24,000	5000
Ventura	430,000	231,000	18,000	50,000	7000

Table for Exercise 6.23

Pair	Treatment 1	Treatment 2	Cumulative Number of Successes for Treatment 1	Cumulative Number of Successes for Treatment 2
1	S	F	1	0
2	S	S	2	1
3	F	F	2	1
4	S	S	3	2
5	F	F	3	2
6	S	F	4	2

6.24 Many cities regulate the number of taxi licenses, and there is a great deal of competition for both new and existing licenses. Suppose that a city has decided to sell 10 new licenses for $25,000 each. A lottery will be held to determine who gets the licenses, and no one may request more than three licenses. Twenty individuals and taxi companies have entered the lottery. Six of the 20 entries are requests for 3 licenses, 9 are requests for 2 licenses, and the rest are requests for a single license. The city will select requests at random, filling as much of the request as possible. For example, the city might fill requests for 2, 3, 1, and 3 licenses and then select a request for 3. Because there is only one license left, the last request selected would receive a license, but only one.
a. An individual who wishes to be an independent driver has put in a request for a single license. Use simulation to approximate the probability that the request will be granted. Perform *at least* 20 simulated lotteries (more is better!).
b. Do you think that this is a fair way of distributing licenses? Can you propose an alternative procedure for distribution?

6.25 Four students must work together on a group project. They decide that each will take responsibility for a particular part of the project, as follows:

Person	Maria	Alex	Juan	Jacob
Task	Survey design	Data collection	Analysis	Report writing

Because of the way the tasks have been divided, one student must finish before the next student can begin work. To ensure that the project is completed on time, a timeline is established, with a deadline for each team member. If any one of the team members is late, the timely completion of the project is jeopardized. Assume the following probabilities:
1. The probability that Maria completes her part on time is .8.

2. If Maria completes her part on time, the probability that Alex completes on time is .9, but if Maria is late, the probability that Alex completes on time is only .6.
3. If Alex completes his part on time, the probability that Juan completes on time is .8, but if Alex is late, the probability that Juan completes on time is only .5.
4. If Juan completes his part on time, the probability that Jacob completes on time is .9, but if Juan is late, the probability that Jacob completes on time is only .7.

Use simulation (with at least 20 trials) to estimate the probability that the project is completed on time. Think carefully about this one. For example, you might use a random digit to represent each part of the project (four in all). For the first digit (Maria's part), 1–8 could represent *on time* and 9 and 0 could represent *late*. Depending on what happened with Maria (late or on time), you would then look at the digit representing Alex's part. If Maria was on time, 1–9 would represent *on time* for Alex, but if Maria was late, only 1–6 would represent *on time*. The parts for Juan and Jacob could be handled similarly.

6.26 In Exercise 6.25, the probability that Maria completes her part on time was .8. Suppose that this probability is really only .6. Use simulation (with at least 20 trials) to estimate the probability that the project is completed on time.

6.27 Refer to Exercises 6.25 and 6.26. Suppose that the probabilities of timely completion are as in Exercise 6.25 for Maria, Alex, and Juan but that Jacob has a probability of completing on time of .7 if Juan is on time and .5 if Juan is late.
a. Use simulation (with at least 20 trials) to estimate the probability that the project is completed on time.
b. Compare the probability from Part (a) to the one computed in Exercise 6.26. Which decrease in the probability of on-time completion (Maria's or Jacob's) made the biggest change in the probability that the project is completed on time?

▪ Activity 6.1: Kisses

Background: The paper "What is the Probability of a Kiss? (It's Not What You Think)" (found in the online *Journal of Statistics Education* [2002]) posed the following question: What is the probability that a Hershey's Kiss will land on its base (as opposed to its side) if it is flipped onto a table? Unlike flipping a coin, there is no reason to believe that this probability is .5.

Working as a class, develop a plan for a simulation that would enable you to estimate this probability.

Once you have an acceptable plan, carry out the simulation and use it to produce an estimate of the desired probability.

Do you think that a Hershey's Kiss is equally likely to land on its base or its side? Explain.

▪ Activity 6.2: A Crisis for European Sports Fans?

Background: The *New Scientist* (January 4, 2002) reported on a controversy surrounding the new Euro coins that have been introduced as a common currency across most of Europe. Each country mints its own coins, but these coins are accepted in any of the countries that have adopted the Euro as its currency.

A group in Poland claims that the Belgium-minted Euro does not have an equal chance of landing heads or tails. This claim was based on 250 tosses of the Belgium Euro, of which 140 (56%) came up heads. Should this be cause for alarm for European sports fans, who know that "important" decisions are made by the flip of a coin?

In this activity, we will investigate whether this difference should be cause for alarm by examining whether observing 140 heads out of 250 tosses is an unusual outcome if the coin is fair.

1. For this first step, you can either (a) flip a U.S. penny 250 times, keeping a tally of the number of heads and tails observed (this won't take as long as you think) or (b) simulate 250 coins tosses by using your calculator or a statistics software package to generate random numbers (if you choose this option, give a brief description of how you carried out the simulation).

2. For your sequence of 250 tosses, calculate the proportion of heads observed.

3. Form a data set that consists of the values for proportion of heads observed in 250 tosses of a fair coin for the entire class. Summarize this data set by constructing a graphical display.

4. Working with a partner, write a paragraph explaining why European sports fans should or should not be worried by the results of the Polish experiment. Your explanation should be based on the observed proportion of heads from the Polish experiment and the graphical display constructed in Step 3.

▪ Activity 6.3: The "Hot Hand" in Basketball

Background: Consider a mediocre basketball player who has consistently made only 50% of his free throws over several seasons. If we were to examine his free throw record over the last 50 free throw attempts, is it likely that we would see a "streak" of 5 in a row where he is successful in making the free throw? In this activity, we will investigate this question. We will assume that the outcomes of successive free throw attempts are independent and that the probability that the player is successful on any particular attempt is .5.

1. Begin by simulating a sequence of 50 free throws for this player. Because this player has a probability of success = .5 for each attempt and the attempts are independent, we can model a free throw by tossing a coin. Using heads to represent a successful free throw and tails to represent a missed free throw, simulate 50 free throws by tossing a coin 50 times, recording the outcome of each toss.

2. For your sequence of 50 tosses, identify the longest streak by looking for the longest string of heads in your sequence. Determine the length of this longest streak.

3. Combine your longest streak value with those from the rest of the class, and construct a histogram or dotplot of these longest streak values.

4. Based on the graph from Step 3, does it appear likely that a player of this skill level would have a streak of 5 or more successes sometime during a sequence of 50 free throw attempts? Justify your answer based on the graph from Step 3.

5. Use the combined class data to estimate the probability that a player of this skill level has a streak of at least 5 somewhere in a sequence of 50 free throw attempts.

6. Using basic probability rules, we can calculate that the probability that a player of this skill level is successful on the *next* 5 free throw attempts is

$$P(\text{SSSSS}) = \left(\frac{1}{2}\right)\left(\frac{1}{2}\right)\left(\frac{1}{2}\right)\left(\frac{1}{2}\right)\left(\frac{1}{2}\right)$$

$$= \left(\frac{1}{2}\right)^5 = .031$$

which is relatively small. At first, this value might seem inconsistent with your answer in Step 5, but the estimated probability from Step 5 and the computed

probability of .031 are really considering different situations. Explain why it is plausible that both probabilities are correct.

7. Do you think that the assumption that the outcomes of successive free throws are independent is reasonable? Explain. (This is a hotly debated topic among both sports fans and statisticians!)

■ Summary of Key Concepts and Formulas

Term or Formula	Comment
Probability	A number between 0 and 1 that reflects the likelihood of occurrence of some outcome.
Independent outcomes	Two outcomes are independent if the chance that one outcome occurs is not affected by knowledge of whether or not the other outcome occurs.
$P(\text{not outcome}) = 1 - P(\text{outcome})$	Complement rule.
$P(\text{Outcome 1 or Outcome 2})$ $= P(\text{Outcome 1}) + P(\text{Outcome 2})$	Addition rule for outcomes that cannot occur at the same time.
$P(\text{Outcome 1 and Outcome 2})$ $= P(\text{Outcome 1})P(\text{Outcome 2})$	Multiplication rule for independent outcomes.
Simulation	A technique for estimating probabilities that generates observations by performing an experiment that is similar in structure to the real situation of interest.

■ Supplementary Exercises 6.28–6.34

6.28 Of the 10,000 students at a certain university, 7000 have Visa cards, 6000 have MasterCards, and 5000 have both. Suppose that a student is randomly selected.
a. What is the probability that the selected student has a Visa card?
b. What is the probability that the selected student has both cards?
c. Suppose that you learn that the selected individual has a Visa card (was one of the 7000 with such a card). Now what is the probability that this student has both cards?
d. Are the outcomes *has a Visa card* and *has a MasterCard* independent? Explain.
e. Answer the question posed in Part (d) if only 4200 of the students have both cards.

6.29 The article "Baseball: Pitching No-Hitters" (*Chance* [summer 1994]: 24–30) gives information on the number of hits per team per game for all nine-inning major league games played between 1989 and 1993. Each game resulted in two observations, one for each team. No distinction was made

between the home team and the visiting team. The data are summarized in the following table:

Hits per team per game	Number of observations
0	20
1	72
2	209
3	527
4	1048
5	1457
6	1988
7	2256
8	2403
9	2256
10	1967
11	1509
12	1230
13	843
14	569
15	393
>15	633

a. If one of these games is selected at random, what is the probability that the visiting team got fewer than 3 hits?

b. What is the probability that the home team got more than 13 hits?

c. Assume that the following outcomes are independent:

Outcome 1: Home team got 10 or more hits.
Outcome 2: Visiting team got 10 or more hits.

What is the probability that both teams got 10 or more hits?

d. Calculation of the probability in Part (c) required that we assume independence of Outcomes 1 and 2. If the outcomes are independent, knowing that one team had 10 or more hits would not change the probability that the other team had 10 or more hits. Do you think that the independence assumption is reasonable? Explain.

6.30 Consider the five outcomes (shown in the table below) for an experiment in which the type of ice cream purchased by the next customer at a certain store is noted.

a. What is the probability that Dreyer's ice cream is purchased?

b. What is the probability that Von's brand is not purchased?

c. What is the probability that the size purchased is larger than a pint?

6.31 A radio station that plays classical music has a "by request" program each Saturday evening. The percentages of requests for composers on a particular night are as follows:

Bach	5%
Mozart	21%
Beethoven	26%
Schubert	12%
Brahms	9%
Schumann	7%
Dvorak	2%
Tchaikovsky	14%
Mendelssohn	3%
Wagner	1%

Suppose that one of these requests is randomly selected.

a. What is the probability that the request is for one of the three B's?

b. What is the probability that the request is not for one of the two S's?

c. Neither Bach nor Wagner wrote any symphonies. What is the probability that the request is for a composer who wrote at least one symphony?

6.32 Suppose that the following information on births in the United States over a given period of time is available to you:

Type of Birth	Number of Births
Single birth	41,500,000
Twins	500,000
Triplets	5,000
Quadruplets	100

Use this information to approximate the probability that a randomly selected pregnant woman who reaches full term

a. Delivers twins

b. Delivers quadruplets

c. Gives birth to more than a single child

6.33 Two individuals, A and B, are finalists for a chess championship. They will play a sequence of games, each of which can result in a win for A, a win for B, or a draw. Suppose that the outcomes of successive games are independent, with $P(A$ wins game$) = .3$, $P(B$ wins game$) = .2$, and $P($draw$) = .5$. Each time a player wins a game, he earns one point and his opponent earns no points. The first player to win 5 points wins the championship. For the sake of simplicity, assume that the championship will end in a draw if both players obtain 5 points at the same time.

a. What is the probability that A wins the championship in just five games?

b. What is the probability that it takes just five games to obtain a champion?

c. If a draw earns a half-point for each player, describe how you would perform a simulation experiment to estimate $P(A$ wins the championship$)$.

d. If neither player earns any points from a draw, would the simulation requested in Part (c) take longer to perform? Explain your reasoning.

Table for Exercise 6.30

	Brand				
Size of container	Steve's Pint	Ben and Jerry's Pint	Dreyer's Quart	Dreyer's Half-gallon	Von's Half-gallon
Probability	.10	.15	.20	.25	.30

6.34 A single-elimination tournament with four players is to be held. A total of three games will be played. In Game 1, the players seeded (rated) first and fourth play. In Game 2, the players seeded second and third play. In Game 3, the winners of Games 1 and 2 play, with the winner of Game 3 declared the tournament winner. Suppose that the following probabilities are given:

P(Seed 1 defeats Seed 4) = .8
P(Seed 1 defeats Seed 2) = .6
P(Seed 1 defeats Seed 3) = .7
P(Seed 2 defeats Seed 3) = .6
P(Seed 2 defeats Seed 4) = .7
P(Seed 3 defeats Seed 4) = .6

a. Describe how you would use a selection of random digits to simulate Game 1 of this tournament.
b. Describe how you would use a selection of random digits to simulate Game 2 of this tournament.
c. How would you use a selection of random digits to simulate the third game in the tournament? (This will depend on the outcomes of Games 1 and 2.)
d. Simulate one complete tournament, giving an explanation for each step in the process.
e. Simulate 10 tournaments, and use the resulting information to estimate the probability that the first seed wins the tournament.
f. Ask four classmates for their simulation results. Along with your own results, this should give you information on 50 simulated tournaments. Use this information to estimate the probability that the first seed wins the tournament.
g. Why do the estimated probabilities from Parts (e) and (f) differ? Which do you think is a better estimate of the true probability? Explain.

■ Reference

Peck, Roxy, Chris Olsen, and Jay Devore. *Introduction to Statistics and Data Analysis*, 2d ed. Belmont, CA: Duxbury, 2004. (This textbook contains a more comprehensive treatment of probability than what appears here, but it is still at a modest mathematical level.)

7 · Population Distributions

This chapter is the first of two that together link the basic ideas of probability explored in Chapter 6 with the techniques of statistical inference. Chapter 6 used probability to describe the long-run frequency of occurrence of various types of outcomes. Here, we introduce probability models that can be used to describe the distribution of characteristics of individuals in a population. In Chapter 8 we will see how such models help us reach conclusions based on a sample from the population.

In this chapter we begin by distinguishing between categorical and numerical variables and between discrete and continuous numerical variables. We show how a continuous numerical variable can be described by a probability distribution curve, which can also be used to make probability statements about values of the variable. Finally, one particular probability model, the normal distribution, is presented in detail.

▪ 7.1 Describing the Distribution of Values in a Population

In Chapter 1, we described statistical inference as the branch of statistics that involves generalizing from a sample to the population from which it was selected. A population is the entire collection of individuals or objects about which information is desired. Interest usually centers on the value of one or more variables. For example, if one wants to obtain information about the performance of a telephone registration system for selecting college courses, the population would consist of all students at the university who use the phone registration system. The variable of interest (a characteristic of each individual member of the population) might be *time to complete registration*.

A variable can be either categorical or numerical, depending on the possible values of the variable that occur in the population. The variable *time to complete registration* is numerical, because it associates a number with each individual in the population. If interest instead had centered on whether a student was able to complete registration for classes with a single phone call, rather than on the time required to complete registration, the variable of interest, which we might name *first call*, would be categorical. This variable associates a categorical response (successful completion or unsuccessful) with each individual in the population. The variable *class standing* is also categorical, with the four categories freshman, sophomore, junior, and senior.

▪ **Definition**

A **variable** associates a value with each individual or object in a population. A variable can be either **categorical** or **numerical**, depending on its possible values.

The distribution of categories or values in a population provides important information about the population. For example, if we knew something about the distribution of registration completion times for the population of all students, we might be able to give students an idea of how much time to allow (e.g., most students require 8 to 13 min to complete the registration process) or determine a reasonable amount of time after which a student should be automatically disconnected from the system. The distribution of all the values of a numerical variable or all the categories of a categorical variable is called a **population distribution**.

▪ Categorical Variables

Categorical variables are often dichotomous (having only two possible categories). For example, each individual in the population of students at a state university might be classified as a resident of the state or as a nonresident, or each owner of a Mazda automobile might be classified according to whether he or she would consider purchasing a Mazda the next time around. Each of these variables (*residence status* for the college population or *future purchase* for the population of Mazda owners) is a categorical variable with two possible categories.

▪ Example 7.1   Residence Status

Consider the variable *residence status* for the population of students at a state university. This variable associates a category (resident or nonresident) with each individual in the population. The population distribution for this variable can be summarized in a bar chart, with a rectangle for each possible category. The height of each rectangle corresponds to the relative frequency (proportion) of the corresponding value in the population. Figure 7.1 shows a possible population distribution for the variable *residence status*. If an individual is randomly selected from this population, the two category relative frequencies can be interpreted as the probabilities of observing each of the two possible categories of residence status. Thus, in the long run, a resident will be selected about 73% of the time, and a nonresident will be selected about 27% of the time.

Figure 7.1 Population distribution of the variable *residence status* for Example 7.1.

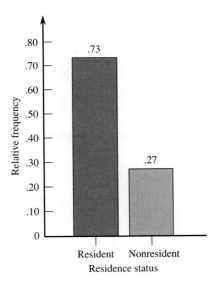

Example 7.2   Mode of Transportation

In a study of factors related to air quality, monitors were posted at every entrance to a California university campus on October 10, 2003. From 6 A.M. to 10 P.M., monitors recorded the mode of transportation for each person entering the campus. Based on the information collected, the population distribution of the variable

   x = mode of transportation

was constructed; it is shown in Figure 7.2.

Figure 7.2 Population distribution of the variable *mode of transportation* for Example 7.2.

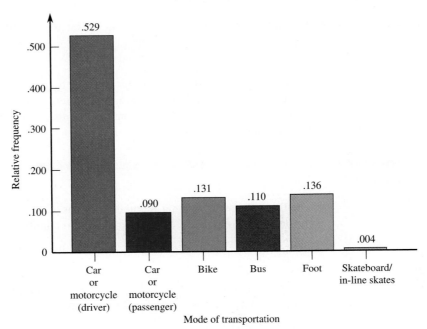

In this example, the population consists of all individuals entering campus on October 10, 2003. The variable x is categorical, with the following possible categories:

car or motorcycle (driver), car or motorcycle (passenger), bike, bus, foot, and skateboard or in-line skates. (This was in California!)

If an individual is selected at random from those who were on campus on October 10, 2003, the probability that the selected individual arrived as the driver of a car or motorcycle is .529. The probability that the selected individual arrived by car or motorcycle (either as the driver or as a passenger) is .529 + .090 = .619. The probability that the selected individual arrived on foot is .136.

It is common practice to use special notation when writing probability statements, such as those at the end of Example 7.2. The probability of an outcome is denoted by P(outcome). The outcome can be described with words or by using a variable name. Thus, continuing Example 7.2 where the outcome of interest is mode of transportation, we write

P(selected individual arrives by bus) = .110   or $P(x = \text{bus}) = .110$

P(selected individual arrives by bike) = .131   or $P(x = \text{bike}) = .131$

▪ Numerical Variables

Before considering examples of numerical variables, we must distinguish between two types of such variables.

▪ **Definition**

Numerical variables can be either discrete or continuous.

A **discrete numerical variable** is one whose possible values are isolated points along the number line.

A **continuous numerical variable** is one whose possible values form an interval along the number line.

The population distribution for a discrete numerical variable can be summarized by a relative frequency histogram, whereas a density histogram is used to summarize the distribution of a continuous numerical variable. The examples that follow demonstrate how this is done.*

▪ Example 7.3   Pet Ownership

The Department of Animal Regulation released information on pet ownership for the population of all households in a particular county. The variable considered was

x = number of licensed dogs or cats for a household

This variable associates a numerical value with each household in the population; possible x values are 0, 1, 2, 3, 4, and 5 (county regulations prohibit more than 5 dogs or cats per household). Because possible values of x are isolated points along the number line, x is a discrete numerical variable.

*One frequently encountered discrete distribution, the **binomial distribution**, is discussed in the Appendix at the end of this textbook.

Figure 7.3 Population distribution of the variable *number of licensed dogs or cats* for Example 7.3.

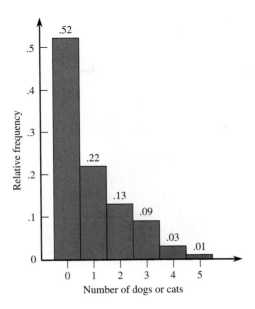

One way to summarize the population distribution for a discrete numerical variable is to use a relative frequency histogram. The population distribution for the variable x = number of licensed dogs or cats is shown in Figure 7.3.

The most common value of x in the population (from Figure 7.3) is $x = 0$, and $P(x = 0) = .52$. Only 1% of all households have five licensed dogs or cats. Therefore if a household was selected at random from this population, it would be unusual to observe five licensed dogs or cats. The probability of observing a household with three or more licensed dogs or cats is

$$P(3 \text{ or more licensed dogs or cats}) = P(x \geq 3) = .09 + .03 + .01 = .13$$

▪ Example 7.4   Birth Weights

Birth weight was recorded for all full-term babies born during 2002 in a semirural county. The variable

x = birth weight for a full-term baby

for the population of all full-term babies in this county is an example of a continuous numerical variable. One way to describe the population distribution of x values for this population is to construct a density histogram. Recall from Chapter 3 that, in a density histogram, the measurement scale (here, the range that includes all possible birth weights) is divided into class intervals. Each value in the data set (here, each birth weight) is classified into one of the intervals. For each interval, the resulting relative frequency is used to compute

$$\text{density} = \frac{\text{relative frequency}}{\text{interval width}}$$

The density histogram has a rectangle for each interval; the height of the rectangle is determined by the corresponding density. (Review Section 3.3 of Chapter 3 for a more detailed description of density histograms.)

Figure 7.4 Population distribution of birth-weight values for Example 7.4.

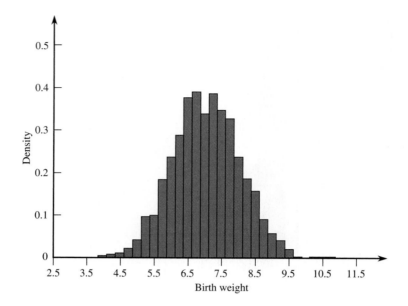

Figure 7.4 is a density histogram of the birth-weight values. This density histogram shows the distribution of birth weights for all full-term babies in this county, and so it can be viewed as the population distribution of the variable x = birth weight.

From Figure 7.4, we can see that most birth weights are between 5 and 9 lb and that it would be unusual for a full-term baby born in this county to have a birth weight over 10 lb. In the next section, we will see how this distribution can be used to calculate various probabilities regarding the birth weight of a randomly selected child.

The population distribution for a numerical variable is useful for describing the distribution of its values in the population. Such a population distribution can also be summarized by a mean and a standard deviation. The mean describes where the distribution of values is centered, and the standard deviation describes how much the distribution spreads out about this central value.

▪ **Definition**

The **mean value of a numerical variable** x, denoted by μ, describes where the population distribution of x is centered.

The **standard deviation of a numerical variable** x, denoted by σ, describes variability in the population distribution. When σ is close to 0, the values of x in the population tend to be close to the mean value (little variability). When the value of σ is large, there is more variability in the population of x values.

Figure 7.5(a) shows two discrete distributions with the same standard deviation (spread) but different means (center). One distribution has a mean of $\mu = 6$; the other has $\mu = 11$. Which is which? Figure 7.5(b) shows two continuous distributions that have the same mean but different standard deviations. Smooth curves have been used to approximate the shape of the underlying density histograms. Which distribution, (i) or (ii), has the larger standard deviation? Finally, Figure 7.5(c) shows three continuous distributions with different means and standard deviations. Which of the three distributions has the largest mean? Which has a mean of about 5? Which distribution has the smallest standard deviation? (Check your answers to the previous questions: Figure 7.5(a)(ii) has a mean of 6, and Figure 7.5(a)(i) has a mean of 11; Figure 7.5(b)(ii) has the larger standard deviation; Figure 7.5(c)(iii) has the largest mean, Figure 7.5(c)(ii) has a mean of about 5, and Figure 7.5(c)(iii) has the smallest standard deviation.)

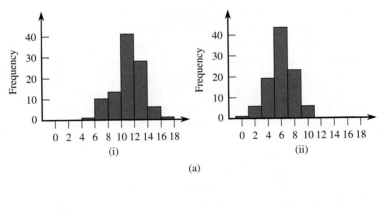

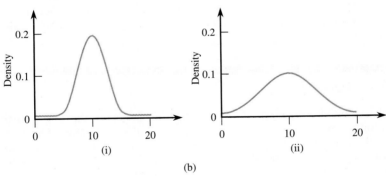

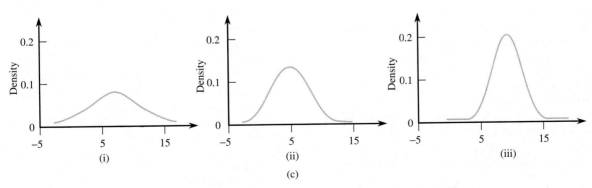

Figure 7.5 (a) Different values of μ with the same value of σ; (b) different values of σ with the same value of μ; (c) different values of μ and σ.

We will not concern ourselves with how to compute the values of μ and σ (it is handled differently for discrete and for continuous variables), but we will look at how these values are interpreted. For the population distribution of Example 7.3, the mean value of x turns out to be $\mu = .92$. This value is interpreted as the average number of licensed dogs or cats for the population of households in the county considered. Notice that μ is not a possible value of x. If another county (another population) had a mean of $\mu = .5$, we would know that the average number of licensed pets per household was smaller for the second county than for the first county, and so the distribution for the second county would be centered to the left of the first county's distribution.

For the population distribution in Example 7.4, $\mu = 7$ and $\sigma = 1$. The value of μ tells us that the average (mean) birth weight of full-term babies in the population is 7 lb. The value of the standard deviation provides information on the extent to which individual birth weights vary in the population. Because the population distribution is roughly bell shaped and symmetric, the Empirical Rule tells us that about 68% of the birth weights fall between 6 and 8 lb and that it would be extremely rare for a full-term baby to have a birth weight under 4 lb (or over 10 lb). If the birth-weight distribution for a different population (e.g., an urban county) had $\mu = 7.4$ and $\sigma = 1.3$, we would know that the average birth weight for the urban population was higher and that individual birth weights varied more than for the semirural area considered in Example 7.4 (i.e., the urban distribution is more spread out than the semirural distribution).

If the population distribution for a variable is known, it is possible to determine the values of μ and σ. Most often, however, the population distribution is not fully known, and the mean and standard deviation are estimated using sample data. In Chapter 9 we will use $\overline{x}$ and s for this purpose.

▪ Exercises 7.1–7.10

7.1 State whether each of the following numerical variables is discrete or continuous:
a. The number of defective tires on a car
b. The body temperature of a hospital patient
c. The number of pages in a book
d. The number of checkout lines operating at a large grocery store
e. The lifetime of a light bulb

7.2 Classify each of the following numerical variables as either discrete or continuous:
a. The fuel efficiency (in miles per gallon) of an automobile
b. The amount of rainfall at a particular location during the next year
c. The distance that a person throws a baseball
d. The number of questions asked during a 1-hr lecture
e. The tension (in pounds per square inch) at which a tennis racket is strung
f. The amount of water used by a household during a given month
g. The number of traffic citations issued by the highway patrol in a particular county on a given day

7.3 Consider the variable x = earthquake insurance status for the population of homeowners in an earthquake-prone California county. This variable associates a category (insured or not insured) with each individual in the population.
a. Construct a relative frequency bar chart that represents the population distribution for x for the case where 60% of the county homeowners have earthquake insurance.
b. If an individual is randomly selected from this population, what is the probability that the selected homeowner does not have earthquake insurance?

7.4 Based on past history, a fire station reports that 25% of the calls to the station are false alarms, 60% are for small fires that can be handled by station personnel without outside assistance, and 15% are for major fires that require outside help.
a. Construct a relative frequency bar chart that represents the distribution of the variable x = type of call, where *type of call* has three categories: false alarm, small fire, and major fire. What is the underlying population for this variable?

b. Based on the given information, we can write $P(x = \text{false alarm}) = .25$. Use the other two relative frequencies shown in the bar chart from Part (a) to write two other probability statements.

7.5 Suppose that fund-raisers at a university call recent graduates to request donations for campus outreach programs. They report the following information for last year's graduates:

Size of donation	$0	$10	$25	$50
Proportion of calls	.45	.30	.20	.05

Three attempts were made to contact each graduate; a donation of $0 was recorded both for those who were contacted but who declined to make a donation and for those who were not reached in three attempts. Consider the variable x = amount of donation for the population of last year's graduates of this university.
a. Construct a relative frequency histogram to represent the population distribution of this variable.
b. What is the most common value of x in this population?
c. What is $P(x \geq 25)$? **d.** What is $P(x > 0)$?

7.6 A pizza shop sells pizzas in four different sizes. The 1000 most recent orders for a single pizza gave the following proportions for the various sizes:

Size	12 in.	14 in.	16 in.	18 in.
Proportion	.20	.25	.50	.05

With x denoting the size of a pizza in a single-pizza order, the given table is an approximation to the population distribution of x.
a. Construct a relative frequency histogram to represent the approximate distribution of this variable.
b. Approximate $P(x < 16)$.
c. Approximate $P(x \leq 16)$.
d. It can be shown that the mean value of x is approximately 14.8 in. What is the approximate probability that x is within 2 in. of this mean value?

7.7 Airlines sometimes overbook flights. Suppose that for a plane with 100 seats, an airline takes 110 reservations. Define the variable x as the number of people who actually show up for a sold-out flight. From past experience, the population distribution of x is given in the following table:

x	Proportion	x	Proportion
95	.05	103	.03
96	.10	104	.02
97	.12	105	.01
98	.14	106	.005
99	.24	107	.005
100	.17	108	.005
101	.06	109	.0037
102	.04	110	.0013

a. What is the probability that the airline can accommodate everyone who shows up for the flight?
b. What is the probability that not all passengers can be accommodated?
c. If you are trying to get a seat on such a flight and you are number 1 on the standby list, what is the probability that you will be able to take the flight? What if you are number 3?

7.8 Homicide rate (homicides per 100,000 population) for each of the 50 states appeared in the *San Luis Obispo Telegram-Tribune* (February 2, 1995). A frequency distribution constructed from the 50 observations is shown in the following table:

Homicide Rate	Frequency
0 to <3	5
3 to <6	16
6 to <9	9
9 to <12	9
12 to <15	9
15 to <18	1
18 to <21	1

a. Calculate the relative frequency and density for each of the seven intervals in the frequency distribution. Use the computed densities to construct a density histogram for the variable x = homicide rate for the population consisting of the 50 states.
b. Is the population distribution symmetric or skewed?
c. Use the population distribution to determine the following probabilities.
i. $P(x \geq 15)$ **ii.** $P(x < 9)$ **iii.** $P(12 \leq x < 18)$

7.9 A company receives light bulbs from two different suppliers. Define the variables x and y by

x = lifetime of a bulb from Supplier 1
y = lifetime of a bulb from Supplier 2

Five hundred bulbs from each supplier are tested, and the lifetime of each bulb is recorded. The density histograms on the next page are constructed from these two sets of observations. Although these histograms are constructed using data from only 500 bulbs, they can be considered approximations to the corresponding population distributions.
a. Which population distribution has the larger mean?
b. Which population distribution has the larger standard deviation?
c. Assuming that the cost of the light bulbs is the same for both suppliers, which supplier would you recommend? Explain.
d. One of the two distributions pictured has a mean of approximately 1000, and the other has a mean of

Figure for Exercise 7.9

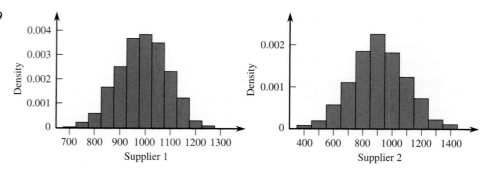

Supplier 1

Supplier 2

about 900. What is the mean of the distribution for the variable x (lifetime for a bulb made by Supplier 1)?

e. One of the two distributions pictured has a standard deviation of approximately 100, and the other has a standard deviation of about 175. What is the standard deviation of the distribution for the variable x (lifetime for a bulb made by Supplier 1)?

7.10 A certain basketball player makes 70% of his free throws. Assume that results of successive free throws are independent of one another. At the end of a particular practice, the coach tells the player to begin shooting free throws and to stop only when he has made two consecutive shots. Let x denote the number of shots until the player can stop. Describe how you would carry out a simulation experiment to approximate the distribution of x.

▪ 7.2 Population Models for Continuous Numerical Variables

In Example 7.4 of Section 7.1, we saw how a density histogram could be used to summarize a population distribution when the variable of interest is numerical and continuous. When this is done, the exact shape of the histogram and any approximate probability statements that we may make based on the population distribution depend somewhat on the number and location of intervals used in constructing the density histogram.

For example, let's look again at the distribution of birth weights for the population of all full-term babies born in 2002 in a particular county (see Example 7.4). Suppose that 2000 full-term babies were born. Then the population consists of 2000 individuals, and each individual has an associated value of the variable $x =$ birth weight. Figure 7.6(a) shows a density histogram for the population distribution of x based on 7 intervals: 3.5 to <4.5, 4.5 to <5.5, and so on.

The area of any rectangle in a density histogram such as Figure 7.6(a) can be interpreted as the probability of observing a variable value in the corresponding interval, if an individual is selected at random from the population. This follows from the fact that, for any interval,

$$\text{density} = \frac{\text{relative frequency}}{\text{interval width}}$$

By construction, the corresponding rectangle has height equal to density, so the area of the rectangle is

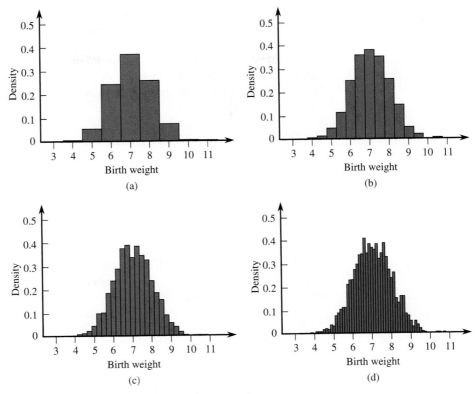

Figure 7.6 Density histograms for birth weight.

$$\text{area} = (\text{height})(\text{interval width})$$
$$= (\text{density})(\text{interval width})$$
$$= \left(\frac{\text{relative frequency}}{\text{interval width}}\right)(\text{interval width})$$
$$= \text{relative frequency}$$

Thus, the area of the rectangle above each interval is equal to the relative frequency of values that fall in the interval. Because the area of a rectangle in the density histogram specifies the proportion of the population values that fall in the corresponding interval, it can be interpreted as the long-run proportion of time that a value in the interval would occur if individual after individual were randomly selected from the population.

Thus, for the interval 4.5 to <5.5 in Figure 7.6(a), the probability that the weight of a randomly selected individual falls in the interval is

$$P(4.5 < x < 5.5) = \text{area of rectangle above the interval from 4.5 to 5.5}$$
$$= (\text{density})(1)$$
$$= (.05)(1)$$
$$= .05$$

Similarly,

$$P(7.5 < x < 8.5) = .255$$

The probability of observing a value in an interval other than those used to construct the density histogram can be approximated. For example, to approximate the probability of observing a birth weight between 7 and 8 lb, we could add half the area of the rectangle for the 6.5–7.5 interval and half the area for the 7.5–8.5 interval. This gives only an approximate probability, because the area of each rectangle in the density histogram is equal to the proportion of the population falling in the corresponding interval. So

$$P(7 < x < 8) \approx \frac{1}{2}(\text{area of rectangle for } 6.5\text{--}7.5)$$

$$+ \frac{1}{2}(\text{area of rectangle for } 7.5\text{--}8.5)$$

$$= \frac{1}{2}(.370)(1) + \frac{1}{2}(.255)(1)$$

$$= .3125$$

The approximation of probabilities can be improved by increasing the number of intervals on which the density histogram is based. As Figure 7.6(a) shows, a density histogram based on a small number of intervals can be quite jagged. Figures 7.6(b)–(d) show density histograms based on 14, 28, and 56 intervals, respectively. As the number of intervals increases, the rectangles in the density histogram become much narrower and the histogram appears smoother.

Two important ideas were presented in the foregoing discussion. First, when summarizing a population distribution with a density histogram, the area of any rectangle in the histogram can be interpreted as the probability of observing a variable value in the corresponding interval when an individual is selected at random from the population. The second important idea is that when a density histogram based on a small number of intervals is used to summarize a population distribution for a continuous numerical variable, the histogram can be quite jagged. However, when the number of intervals is increased, the resulting histograms become much smoother in appearance. (You can see this in the histograms of Figure 7.6.)

It is often useful to represent a population distribution for a continuous variable by using a simple smooth curve that approximates the actual population distribution. For example, Figure 7.7 shows a smooth curve superimposed over the density histogram of Figure 7.6(d). Such a curve is called a **continuous probability distribution**. Because the total area of the rectangles in a density histogram is equal to 1, we consider only smooth curves for which the total area under the curve is equal to 1.

Figure 7.7 A smooth curve specifies a continuous distribution for birth weight.

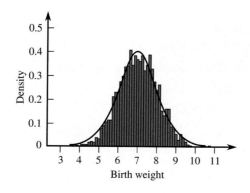

A continuous probability distribution is an abstract but simplified description of the population distribution that preserves important population characteristics (general shape, center, spread, etc.). Thus, it can serve as a model for the distribution of values in the population. Because the area under the curve approximates the areas of rectangles in the density histogram, the area under the curve and above any particular interval can be interpreted as the approximate probability of observing a value in that interval when an individual is selected at random from the population.

A **continuous probability distribution** is a smooth curve, called a **density curve**, that serves as a model for the population distribution of a continuous variable.

Properties of continuous probability distributions are:

1. The total area under the curve is equal to 1.
2. The area under the curve and above any particular interval is interpreted as the (approximate) probability of observing a value in the corresponding interval when an individual or object is selected at random from the population.

Examples 7.5–7.7 show how a continuous probability distribution can be used to make probability statements about a variable.

■ **Example 7.5**   Departure Delays

A morning commuter train never leaves before its scheduled departure time. The length of time that elapses between the scheduled departure time and the actual departure time is recorded on 200 occasions. The resulting observations are summarized in the density histogram shown in Figure 7.8(a). This histogram can serve as an approximation to the population distribution of the variable x = elapsed time (in minutes).

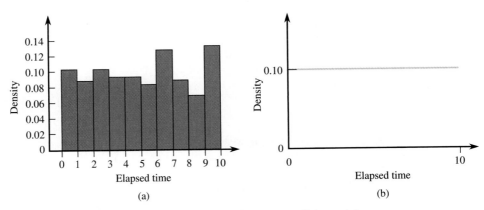

Figure 7.8 Graphs for Example 7.5: (a) histogram of elapsed time values; (b) continuous probability distribution for elapsed time.

Because the histogram in Figure 7.8(a) is fairly flat, a reasonable model (smooth curve) for the population distribution is the probability distribution "curve" shown in Figure 7.8(b). This model is sometimes referred to as a *uniform distribution*. The height of the curve (density = 0.1) is chosen so that the total area under the density curve is equal to 1.

The model can be used to approximate probabilities involving the variable x. For example, the probability that between 5 and 7 min elapse between scheduled and actual departure time is the area under the density curve and above the interval from 5 to 7, as shown in the accompanying illustration:

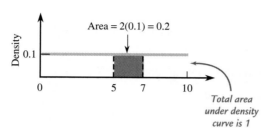

So

$$P(\text{elapsed time is between 5 and 7}) = P(5 < x < 7) = (2)(.1) = .2$$

Other probabilities are determined in a similar fashion. For example,

$P(\text{elapsed time is less than 2.5})$

$\quad = P(x < 2.5)$

$\quad =$ area under curve and above interval from 0 to 2.5

$\quad = .25$

For continuous numerical variables, probabilities are represented by an area under a probability distribution curve and above an interval. The area above an interval is not changed by including the interval endpoints, because there is no area above a single point. In Example 7.5, we found that $P(x < 2.5) = .25$. It is also true that $P(x \leq 2.5) = .25$.

For continuous numerical variables and any particular numbers a and b,

$$P(x \leq a) = P(x < a)$$
$$P(x \geq b) = P(x > b)$$
$$P(a < x < b) = P(a \leq x \leq b)$$

▪ Example 7.6   Priority Mail Package Weights

Two hundred packages shipped using the Priority Mail rate for packages under 2 lb were weighed, resulting in a sample of 200 observations of the variable

$x =$ package weight (in pounds)

from the population of all Priority Mail packages under 2 lb. A density histogram constructed from the 200 weights is shown in Figure 7.9(a). Because the histogram is based on a sample of 200 packages, it provides only an approximation to the population histogram. However, the shape of the sample density histogram does suggest that a reasonable model for the population might be the triangular distribution shown in Figure 7.9(b).

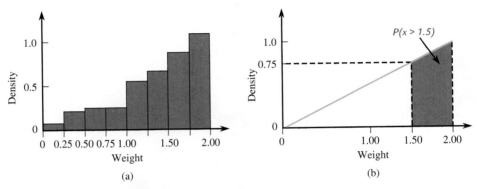

Figure 7.9 Graphs for Example 7.6: (a) histogram of package weight values; (b) continuous probability distribution for package weight.

Note that the total area under the probability distribution curve (the density curve) is equal to

$$\text{total area of triangle} = \frac{1}{2}(\text{base})(\text{height}) = \frac{1}{2}(2)(1) = 1$$

The probability model can be used to compute the proportion of packages over 1.5 lb, $P(x > 1.5)$. This corresponds to the area of the shaded trapezoid in Figure 7.9(b). In this case, it is easier to compute the area of the unshaded region (which corresponds to $P(x \leq 1.5)$), because this is just the area of a triangle:

$$P(x \leq 1.5) = \frac{1}{2}(1.5)(.75) = .5625$$

Because the total area under a probability density curve is 1,

$$P(x > 1.5) = 1 - .5625 = .4375$$

It is also the case that

$$P(x \geq 1.5) = .4375$$

and that

$$P(x = 1.5) = 0$$

The last probability is a consequence of the fact that there is 0 area under the density curve above a single x value.

▪ Example 7.7   Service Times

An airline's toll-free reservation number recorded the length of time required to provide service to each of 500 callers. This resulted in 500 observations of the continuous numerical variable

x = service time

A density histogram is shown in Figure 7.10(a).

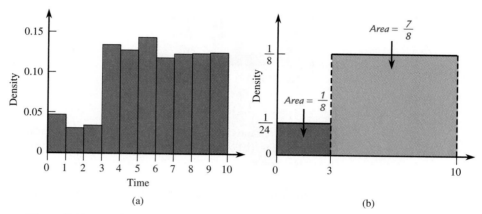

Figure 7.10 Graphs for Example 7.7: (a) histogram of service times; (b) continuous distribution of service times.

The population of interest is all callers to the reservation line. After studying the density histogram, we might think that a model for the population distribution would be flat over the interval from 0 to 3 and higher but also flat over the interval from 3 to 10. This type of model was thought to be reasonable, because service requests were usually one of two types: (1) requests to make a flight reservation and (2) requests to cancel a reservation. Canceling a reservation, which accounted for about one-eighth of the calls to the reservation line, could usually be accomplished fairly quickly, whereas making a reservation (seven-eighths of the calls) required more time.

Figure 7.10(b) shows the probability distribution curve proposed as a model for the variable x = service time. The height of the curve for each of the two segments was chosen so that the total area under the curve would be 1 and so that $P(x \le 3) = 1/8$ (these were thought to be cancellation calls) and $P(x > 3) = 7/8$.

Once the model has been developed, it can be used to compute probabilities. For example,

$P(x > 8)$ = area under curve and above interval from 8 to 10

$$= 2\left(\frac{1}{8}\right) = \frac{2}{8} = \frac{1}{4}$$

In the long run, one-fourth of all service requests will require more than 8 min.

Similarly,

$P(2 < x < 4)$ = area under curve and above interval from 2 to 4

= (area under curve and above interval from 2 to 3)
 + (area under curve and above interval from 3 to 4)

$$= 1\left(\frac{1}{24}\right) + 1\left(\frac{1}{8}\right)$$

$$= \frac{1}{24} + \frac{3}{24}$$

$$= \frac{4}{24}$$

$$= \frac{1}{6}$$

In each of the previous examples, the continuous probability distribution used as a model for the population distribution was simple enough that we were able to calculate probabilities (evaluate areas under the curve) using simple geometry. Example 7.8 shows that this is not always the case.

■ Example 7.8   Telephone Registration Times

Students at a university use a telephone registration system to register for courses. The variable

   x = length of time required for a student to register

was recorded for a large number of students using the system, and the resulting values were used to construct the density histogram of Figure 7.11. The general form of the density histogram can be described as bell shaped and symmetric, and a smooth curve has been superimposed. This smooth curve serves as a reasonable model for the population distribution represented by the density histogram. Although this is a common population model (there are many variables whose distri-

Figure 7.11 Histogram and continuous probability distribution for time to register for Example 7.8.

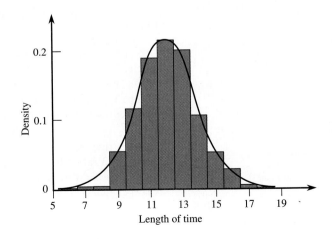

butions are described by curves of this sort), it is not obvious how we could use such a model to calculate probabilities, because at this point it is not clear how to find areas under such a curve.

The probability model of Example 7.8 is an example of a type of symmetric bell-shaped distribution known as a *normal probability distribution*. Normal distributions have many and varied applications, and they are investigated in more detail in the next section.

▪ Exercises 7.11–7.15

7.11 Consider the population of batteries made by a particular manufacturer. The following density curve represents the probability distribution for the variable x = lifetime (in hours):

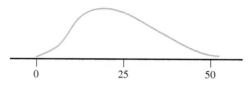

Shade the region under the curve corresponding to each of the following probabilities (draw a new curve for each part):
a. $P(10 < x < 25)$
b. $P(10 \le x \le 25)$
c. $P(x < 30)$
d. The probability that the lifetime is at least 25 hr
e. The probability that the lifetime exceeds 30 hr

7.12 A particular professor never dismisses class early. Let x denote the amount of time past the hour (in minutes) that elapses before the professor dismisses class. Suppose that the density curve shown in the following figure is an appropriate model for the probability distribution of x:

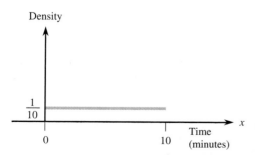

a. What is the probability that at most 5 min elapse before dismissal?

b. What is the probability that between 3 and 5 min elapse before dismissal?
c. What do you think is the value of the mean of this distribution?

7.13 Consider the population that consists of all soft contact lenses made by a particular manufacturer, and define the variable x = thickness (in millimeters). Suppose that a reasonable model for the population distribution is the one shown in the following figure:

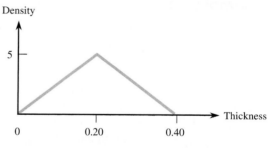

a. Verify that the total area under the density curve is equal to 1. (Hint: The area of a triangle is equal to 0.5(base)(height).)
b. What is the probability that x is less than .20? less than .10? more than .30?
c. What is the probability that x is between .10 and .20? (Hint: First find the probability that x is *not* between .10 and .20.)
d. Because the density curve is symmetric, the mean of the distribution is .20. What is the probability that thickness is within 0.05 mm of the mean thickness?

7.14 A delivery service charges a special rate for any package that weighs less than 1 lb. Let x denote the weight of a randomly selected parcel that qualifies for this special rate. The probability distribution of x is specified by the following density curve:

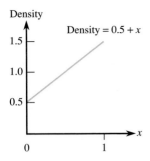

Use the fact that the area of a trapezoid = (base)(average of two side lengths) to answer each of the following questions.
a. What is the probability that a randomly selected package of this type weighs at most 0.5 lb?
b. What is the probability that a randomly selected package of this type weighs between 0.25 lb and 0.5 lb?
c. What is the probability that a randomly selected package of this type weighs at least 0.75 lb?

7.15 Let x denote the time (in seconds) necessary for an individual to react to a certain stimulus. The probability distribution of x is specified by the following density curve:

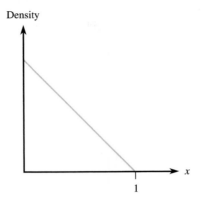

a. What is the height of the density curve above $x = 0$? (Hint: Total area $= 1$.)
b. What is the probability that reaction time exceeds 0.5 sec?
c. What is the probability that reaction time is at most 0.25 sec?

▪ 7.3 Normal Distributions

Normal distributions formalize the notion of mound-shaped histograms introduced in Chapter 4. Normal distributions are widely used for two reasons. First, they provide a reasonable approximation to the distribution of many different variables. They also play a central role in many of the inferential procedures that will be discussed in Chapters 9–11. Normal distributions are continuous probability distributions that are bell shaped and symmetric, as shown in Figure 7.12. Normal distributions are also referred to as *normal curves*.

Figure 7.12 A normal distribution.

There are many different normal distributions, and they are distinguished from one another by their mean μ and standard deviation σ. The mean μ of a normal distribution describes where the corresponding curve is centered, and the standard deviation σ describes how much the curve spreads out around that center. As with all continuous probability distributions, the total area under any normal curve is equal to 1. Three normal distributions are shown in Figure 7.13. Notice that the smaller the standard deviation, the taller and narrower the corresponding curve. Recall that areas under a continuous probability distribution curve represent probabilities; therefore, when the standard deviation is small, a larger area is concen-

Figure 7.13 Three normal distributions.

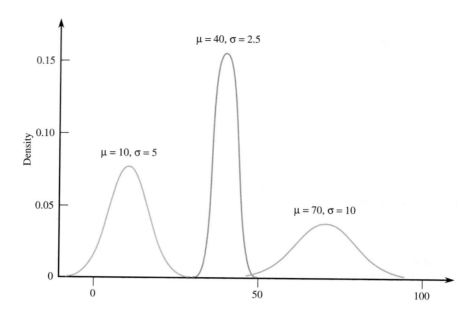

trated near the center of the curve, and the chance of observing a value near the mean is much greater (because μ is at the center).

The value of μ is the number on the measurement axis lying directly below the top of the bell. The value of σ can also be ascertained from a picture of the curve. Consider the normal curve in Figure 7.14. Starting at the top of the bell (above $\mu = 100$) and moving to the right, the curve turns downward until it is above the value 110. After that point, it continues to decrease in height but is turning upward rather than downward. Similarly, to the left of $\mu = 100$, the curve turns downward until it reaches 90 and then begins to turn upward. The curve changes from turning downward to turning upward at a distance of 10 on either side of μ, so $\sigma = 10$. In general, σ is the distance to either side of μ at which a normal curve changes from turning downward to turning upward.

Figure 7.14 μ and σ for a normal curve.

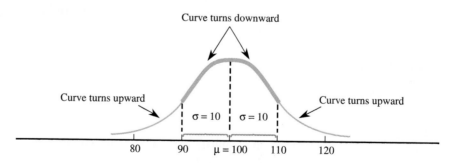

If a particular normal distribution is to be used as a population model, a mean and a standard deviation must be specified. For example, a normal distribution with mean 7 and standard deviation 1 might be used as a model for the distribution of $x =$ birth weight from Section 7.2. If this model is a reasonable description of the probability distribution, we could use areas under the normal curve with $\mu = 7$ and $\sigma = 1$ to approximate various probabilities related to birth weight. The probability that a birth weight is over 8 lb (expressed symbolically as $P(x > 8)$) corresponds to

Figure 7.15 Normal distribution for birth weight: (a) shaded area = $P(x > 8)$; (b) shaded area = $P(6.5 < x < 8)$.

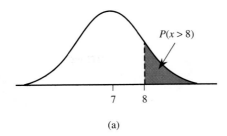

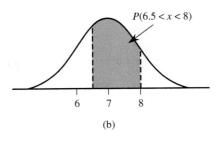

(a)

(b)

the shaded area in Figure 7.15(a). The shaded area in Figure 7.15(b) is the (approximate) probability $P(6.5 < x < 8)$ of a birth weight falling between 6.5 and 8 lb.

Unfortunately, direct computation of such probabilities (areas under a normal curve) is not simple. To overcome this difficulty, we rely on a table of areas for a reference normal distribution called the *standard normal distribution*.

> ▪ **Definition**
>
> The **standard normal distribution** is the normal distribution with $\mu = 0$ and $\sigma = 1$. The corresponding density curve is called the *standard normal curve*. It is customary to use the letter z to represent a variable whose distribution is described by the standard normal curve. The term z *curve* is often used in place of standard normal curve.

Few naturally occurring variables have distributions that are well described by the standard normal distribution, but this distribution is important because it is also used in probability calculations for other normal distributions. When we are interested in finding a probability based on some other normal curve, we first translate the problem into an "equivalent" problem that involves finding an area under the standard normal curve. A table for the standard normal distribution is then used to find the desired area. To be able to do this, we must first learn to work with the standard normal distribution.

▪ The Standard Normal Distribution

In working with normal distributions, we need two general skills:

1. We must be able to use the normal distribution to compute probabilities, which are areas under a normal curve and above given intervals.

2. We must be able to characterize extreme values in the distribution, such as the largest 5%, the smallest 1%, and the most extreme 5% (which would include the largest 2.5% and the smallest 2.5%).

Let's begin by looking at how to accomplish these tasks when the distribution of interest is the standard normal distribution.

The standard normal or z curve is shown in Figure 7.16(a). It is centered at $\mu = 0$, and the standard deviation, $\sigma = 1$, is a measure of the extent to which it spreads out about its mean (in this case, 0). Note that this picture is consistent with the Empirical Rule of Chapter 4: About 95% of the area (probability) is associated

Figure 7.16
(a) A standard
normal (z) curve;
(b) a cumulative area.

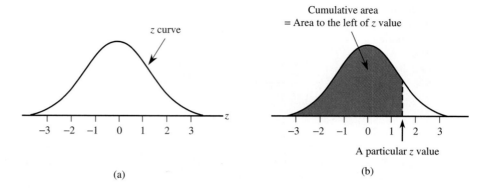

(a)

(b)

with values that are within 2 standard deviations of the mean (between −2 and 2) and almost all of the area is associated with values that are within 3 standard deviations of the mean (between −3 and 3).

Appendix Table 2 tabulates cumulative z curve areas of the sort shown in Figure 7.16(b) for many different values of z. The smallest value for which the cumulative area is given is −3.89, a value far out in the lower tail of the z curve. The next smallest value for which the area appears is −3.88, then −3.87, then −3.86, and so on in increments of 0.01, terminating with the cumulative area to the left of 3.89.

■ Using the Table of Standard Normal Curve Areas

For any number z^* between −3.89 and 3.89 and rounded to two decimal places, Appendix Table 2 gives

(area under z curve to the left of z^*) = $P(z < z^*)$ = $P(z \leq z^*)$

where the letter z is used to represent a variable whose distribution is the standard normal distribution.

To find this probability, locate the following:

1. The row labeled with the sign of z^* and the digit to either side of the decimal point (e.g., −1.7 or 0.5)

2. The column identified with the second digit to the right of the decimal point in z^* (e.g., .06 if $z^* = -1.76$)

The number at the intersection of this row and column is the desired probability, $P(z < z^*)$.

A portion of the table of standard normal curve areas appears in Figure 7.17. To find the area under the z curve to the left of 1.42, look in the row labeled 1.4 and the column labeled .02 (the highlighted row and column in Figure 7.17). From the table, the corresponding cumulative area is .9222. So

z curve area to the left of 1.42 = .9222

We can also use the table to find the area to the right of a particular value. Because the total area under the z curve is 1, it follows that

z curve area to the right of $1.42 = 1 - (z$ curve area to the left of $1.42)$

$$= 1 - .9222$$

$$= .0778$$

These probabilities can be interpreted to mean that in a long sequence of observations, roughly 92.22% of the observed z values will be smaller than 1.42 and 7.78% will be larger than 1.42.

Figure 7.17 Portion of Appendix Table 2 (standard normal curve areas).

z^*	.00	.01	.02	.03	.04	.05
0.0	.5000	.5040	.5080	.5120	.5160	.5199
0.1	.5398	.5438	.5478	.5517	.5557	.5596
0.2	.5793	.5832	.5871	.5910	.5948	.5987
0.3	.6179	.6217	.6255	.6293	.6331	.6368
0.4	.6554	.6591	.6628	.6664	.6700	.6736
0.5	.6915	.6950	.6985	.7019	.7054	.7088
0.6	.7257	.7291	.7324	.7357	.7389	.7422
0.7	.7580	.7611	.7642	.7673	.7704	.7734
0.8	.7881	.7910	.7939	.7967	.7995	.8023
0.9	.8159	.8186	.8212	.8238	.8264	.8289
1.0	.8413	.8438	.8461	.8485	.8508	.8531
1.1	.8643	.8665	.8686	.8708	.8729	.8749
1.2	.8849	.8869	.8888	.8907	.8925	.8944
1.3	.9032	.9049	.9066	.9082	.9099	.9115
1.4	.9192	.9207	.9222	.9236	.9251	.9265
1.5	.9332	.9345	.9357	.9370	.9382	.9394
1.6	.9452	.9463	.9474	.9484	.9495	.9505
1.7	.9554	.9564	.9573	.9582	.9591	.9599
1.8	.9641	.9649	.9656	.9664	.9671	.9678

$P(z < 1.42)$

▪ **Example 7.9** Finding Standard Normal Curve Areas

The probability $P(z < -1.76)$ is found at the intersection of the -1.7 row and the .06 column of the z table. The result is

$$P(z < -1.76) = .0392$$

as shown in the following figure:

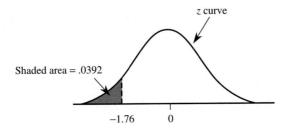

z curve

Shaded area $= .0392$

-1.76 0

In other words, in a long sequence of observations, roughly 3.9% of the observed z values will be smaller than -1.76. Similarly,

$$P(z \leq 0.58) = \text{entry in 0.5 row and .08 column of Appendix Table 2} = .7190$$

as shown in the following figure:

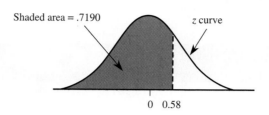

Now consider $P(z < -4.12)$. This probability does not appear in Appendix Table 2; there is no -4.1 row. However, it must be less than $P(z < -3.89)$, the smallest z value in the table, because -4.12 is farther out in the lower tail of the z curve. Because $P(z < -3.89) = .0000$ (i.e., 0 to four decimal places), it follows that

$$P(z < -4.12) \approx 0$$

Similarly,

$$P(z < 4.18) > P(z < 3.89) = 1.0000$$

from which we conclude that

$$P(z < 4.18) \approx 1$$

As illustrated in Example 7.9, we can use the cumulative areas tabulated in Appendix Table 2 to calculate other probabilities involving z. The probability that z is larger than a value c is

$$P(z > c) = \text{area under the } z \text{ curve to the right of } c = 1 - P(z \leq c)$$

In other words, the area to the right of a value (a right-tail area) is 1 minus the corresponding cumulative area. This is illustrated in Figure 7.18.

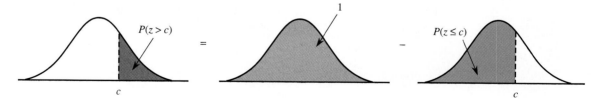

Figure 7.18 The relationship between an upper-tail area and a cumulative area.

Similarly, the probability that z falls in the interval between a lower limit a and an upper limit b is

$$P(a < z < b) = \text{area under the } z \text{ curve and above the interval from } a \text{ to } b$$
$$= P(z < b) - P(z < a)$$

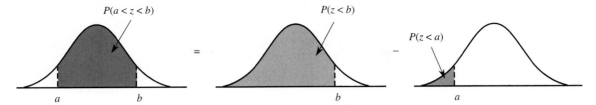

Figure 7.19 $P(a < z < b)$ as a difference of two cumulative areas.

That is, $P(a < z < b)$ is the difference between two cumulative areas, as illustrated in Figure 7.19.

▪ **Example 7.10**   More About Standard Normal Curve Areas

The probability that z is between -1.76 and 0.58 is

$$P(-1.76 < z < 0.58) = P(z < 0.58) - P(z < -1.76)$$
$$= .7190 - .0392$$
$$= .6798$$

as shown in the following figure:

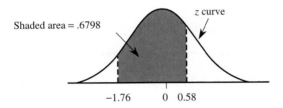

The probability that z is between -2 and $+2$ (within 2 standard deviations of its mean, because $\mu = 0$ and $\sigma = 1$) is

$$P(-2.00 < z < 2.00) = P(z < 2.00) - P(z < -2.00)$$
$$= .9772 - .0228$$
$$= .9544$$
$$\approx .95$$

as shown in the following figure:

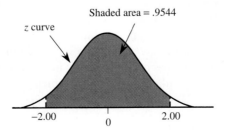

This last probability is the basis for one part of the Empirical Rule, which states that when a histogram is well approximated by a normal curve, roughly 95% of the values are within 2 standard deviations of the mean.

The probability that the value of z exceeds 1.96 is

$$P(z > 1.96) = 1 - P(z \le 1.96)$$
$$= 1 - .9750$$
$$= .0250$$

as shown in the following figure:

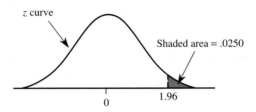

That is, 2.5% of the area under the z curve lies to the right of 1.96 in the upper tail. Similarly,

$$P(z > -1.28) = \text{area to the right of } -1.28$$
$$= 1 - P(z \le -1.28)$$
$$= 1 - .1003$$
$$= .8997$$
$$\approx .90$$

▪ **Identifying Extreme Values** Suppose that we want to describe the values included in the smallest 2% of a distribution or the values making up the most extreme 5% (which includes the largest 2.5% and the smallest 2.5%). Let's see how we can identify extreme values in the distribution by working through Examples 7.11 and 7.12.

▪ Example 7.11   Identifying Extreme Values

Suppose that we want to describe the values that make up the smallest 2% of the standard normal distribution. Symbolically, we are trying to find a value (call it z^*) such that

$$P(z < z^*) = .02$$

This is illustrated in Figure 7.20, which shows that the cumulative area for z^* is .02. Therefore we look for a cumulative area of .0200 in the body of Appendix Table 2. The closest cumulative area in the table is .0202, in the -2.0 row and .05 column; so we use $z^* = -2.05$, the best approximation from the table. Variable values less than -2.05 make up the smallest 2% of the standard normal distribution.

Now suppose that we had been interested in the largest 5% of all z values. We would then be trying to find a value of z^* for which

$$P(z > z^*) = .05$$

as illustrated in Figure 7.21.

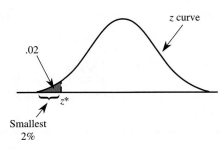

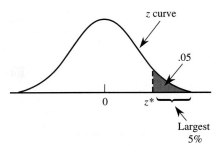

Figure 7.20 The smallest 2% of the standard normal distribution.

Figure 7.21 The largest 5% of the standard normal distribution.

Because Appendix Table 2 always works with cumulative area (area to the left), the first step is to determine

area to the left of $z^* = 1 - .05 = .95$

Looking for the cumulative area closest to .95 in Appendix Table 2, we find that .95 falls exactly halfway between .9495 (corresponding to a z value of 1.64) and .9505 (corresponding to a z value of 1.65). Because .9500 is exactly halfway between the two areas, we use a z value that is halfway between 1.64 and 1.65. (If one value had been closer to .9500 than the other, we would just use the z value corresponding to the closest area). This gives

$$z^* = \frac{1.64 + 1.65}{2} = 1.645$$

Values greater than 1.645 make up the largest 5% of the standard normal distribution. By symmetry, -1.645 separates the smallest 5% of all z values from the others.

■ **Example 7.12   More Extremes**

Sometimes we are interested in identifying the most extreme (unusually large *or* small) values in a distribution. Consider describing the values that make up the most extreme 5% of the standard normal distribution. That is, we want to separate the middle 95% from the extreme 5%. This is illustrated in Figure 7.22.

Figure 7.22 The most extreme 5% of the standard normal distribution.

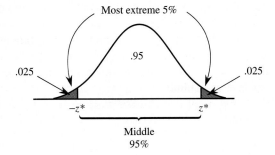

Because the standard normal distribution is symmetric, the most extreme 5% is equally divided between the high side and the low side of the distribution, resulting in an area of .025 for each of the tails of the z curve. Symmetry about 0 implies that if z^* denotes the value that separates the largest 2.5%, the value that separates the smallest 2.5% is simply $-z^*$.

To find z^*, we must first determine the cumulative area for z^*, which is

area to the left of $z^* = .95 + .025 = .975$

The cumulative area .9750 appears in the 1.9 row and .06 column of Appendix Table 2, so $z^* = 1.96$. For the standard normal distribution, 95% of the variable values fall between -1.96 and 1.96; the most extreme 5% are those values that are either greater than 1.96 or less than -1.96.

▪ Other Normal Distributions

We now show how z curve areas can be used to calculate probabilities and to describe extreme values for any normal distribution. Remember that the letter z is reserved for those variables that have a standard normal distribution; the letter x is used more generally for any variable whose distribution is described by a normal curve with mean μ and standard deviation σ.

Suppose that we want to compute $P(a < x < b)$, the probability that the variable x lies in a particular range. This probability corresponds to an area under a normal curve and above the interval from a to b, as shown in Figure 7.23(a).

Figure 7.23 Equality of nonstandard and standard normal curve areas.

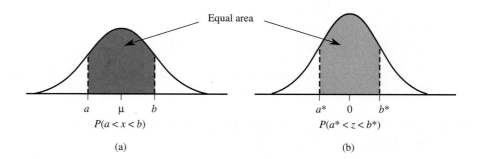

$P(a < x < b)$

(a)

$P(a^* < z < b^*)$

(b)

Our strategy for obtaining this probability is to find an "equivalent" problem involving the standard normal distribution. Finding an equivalent problem means determining an interval (a^*, b^*) that has the same probability for z (same area under the z curve) as does the interval (a, b) in our original normal distribution (see Figure 7.23). The asterisk notation is used to distinguish a and b, the values from the original normal distribution with mean μ and standard deviation σ, from a^* and b^*, the corresponding values from the z curve. To find a^* and b^*, we simply calculate z scores for the endpoints of the interval for which a probability is desired. This process is called **standardizing** the endpoints. For example, suppose that the variable x has a normal distribution with mean $\mu = 100$ and standard deviation $\sigma = 5$. To calculate

$P(98 < x < 107)$

we first translate this problem into an equivalent problem for the standard normal distribution. Recall from Chapter 4 that a z score, or standardized score, tells how many standard deviations away from the mean a value lies; the z score is calculated by first subtracting the mean and then dividing by the standard deviation. Converting the lower endpoint $a = 98$ to a z score gives

$$a^* = \frac{98 - 100}{5} = \frac{-2}{5} = -0.40$$

and converting the upper endpoint yields

$$b^* = \frac{107 - 100}{5} = \frac{7}{5} = 1.40$$

Then

$$P(98 < x < 107) = P(-0.40 < z < 1.40)$$

The probability $P(-0.40 < z < 1.40)$ can now be evaluated using Appendix Table 2.

▪ **Finding Probabilities**

To calculate probabilities for any normal distribution, standardize the relevant values and then use the table of z curve areas. More specifically, if x is a variable whose behavior is described by a normal distribution with mean μ and standard deviation σ, then

$$P(x < b) = P(z < b^*)$$
$$P(a < x) = P(a^* < z) \quad \text{(equivalently, } P(x > a) = P(z > a^*))$$
$$P(a < x < b) = P(a^* < z < b^*)$$

where z is a variable whose distribution is standard normal and

$$a^* = \frac{a - \mu}{\sigma} \qquad b^* = \frac{b - \mu}{\sigma}$$

▪ **Example 7.13   Children's Heights**

In poor countries, the growth of children can be an important indicator of general levels of nutrition and health. Data from the article "The Osteological Paradox: Problems in Inferring Prehistoric Health from Skeletal Samples" (*Current Anthropology* [1992]: 343–370) suggest that a reasonable model for the probability distribution of the continuous numerical variable x = height of a randomly selected 5-year-old child is a normal distribution with a mean of $\mu = 100$ cm and standard deviation $\sigma = 6$ cm. What proportion of the heights is between 94 and 112 cm?

To answer this question, we must find

$$P(94 < x < 112)$$

First, we translate the endpoints to an equivalent problem for the standard normal distribution:

$$a^* = \frac{a - \mu}{\sigma} = \frac{94 - 100}{6} = -1.00$$

$$b^* = \frac{b - \mu}{\sigma} = \frac{112 - 100}{6} = 2.00$$

Then

$$P(94 < x < 112) = P(-1.00 < z < 2.00)$$
$$= (z \text{ curve area to the left of } 2.00)$$
$$\quad - (z \text{ curve area to the left of } -1.00)$$
$$= .9772 - .1587$$
$$= .8185$$

The probabilities for x and z are shown in Figure 7.24. If heights were observed for many children from this population, about 82% of them would fall between 94 and 112 cm.

Figure 7.24
$P(94 < x < 112)$ and corresponding z curve area for Example 7.13.

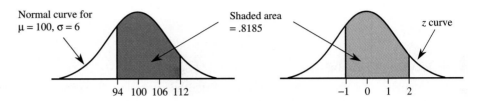

What is the probability that a randomly chosen child will be taller than 110 cm? To evaluate $P(x > 110)$, we first compute

$$a^* = \frac{a - \mu}{\sigma} = \frac{110 - 100}{6} = 1.67$$

Then (see Figure 7.25)

$$P(x > 110) = P(z > 1.67)$$
$$= z \text{ curve area to the right of } 1.67$$
$$= 1 - (z \text{ curve area to the left of } 1.67)$$
$$= 1 - .9525$$
$$= .0475$$

Figure 7.25 $P(x > 110)$ and corresponding z curve area for Example 7.13.

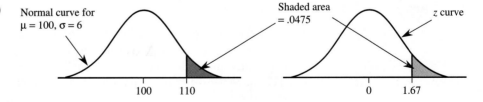

▪ Example 7.14   IQ Scores

Although there is some controversy regarding the appropriateness of IQ scores as a measure of intelligence, IQ scores are commonly used for a variety of purposes. One commonly used IQ scale has a mean of 100 and a standard deviation of 15, and scores are approximately normally distributed. (IQ score is actually a discrete variable [because it is based on the number of correct responses on a test], but its population distribution closely resembles a normal curve.) If we define

x = IQ score of a randomly selected individual

then x has approximately a normal distribution with $\mu = 100$ and $\sigma = 15$.

One way to become eligible for membership in Mensa, an organization purportedly for those of high intelligence, is to have a Stanford–Binet IQ score above 130. What proportion of the population would qualify for Mensa membership? An answer to this question requires evaluating $P(x > 130)$. This probability is shown in Figure 7.26.

Figure 7.26 Normal distribution and desired proportion for Example 7.14.

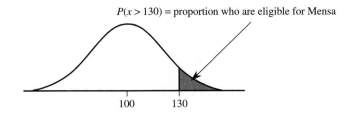

$P(x > 130) = $ proportion who are eligible for Mensa

100   130

With $a = 130$,

$$a^* = \frac{a - \mu}{\sigma} = \frac{130 - 100}{15} = 2.00$$

Therefore (see Figure 7.27)

$$P(x > 130) = P(z > 2.00)$$
$$= z \text{ curve area to the right of } 2.00$$
$$= 1 - (z \text{ curve area to the left of } 2.00)$$
$$= 1 - .9772$$
$$= .0228$$

Only 2.28% of the population would qualify for Mensa membership.

Figure 7.27 $P(x > 130)$ and corresponding z curve area for Example 7.14.

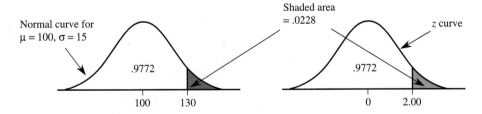

Normal curve for $\mu = 100$, $\sigma = 15$

.9772

100   130

Shaded area $= .0228$

z curve

.9772

0   2.00

Suppose that we are interested in the proportion of the population with IQ scores below 80 — that is, $P(x < 80)$. With $b = 80$,

$$b^* = \frac{b - \mu}{\sigma} = \frac{80 - 100}{15} = -1.33$$

Therefore

$$P(x < 80) = P(z < -1.33)$$
$$= z \text{ curve area to the left of } -1.33$$
$$= .0918$$

as shown in Figure 7.28. This probability (.0918) tells us that just a little over 9% of the population has an IQ score below 80.

Figure 7.28 $P(x < 80)$ and corresponding z curve area for Example 7.14.

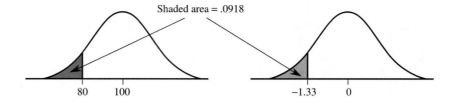

Shaded area = .0918

Now consider the proportion of the population with IQs between 75 and 125. Using $a = 75$ and $b = 125$, we obtain

$$a^* = \frac{75 - 100}{15} = -1.67$$

$$b^* = \frac{125 - 100}{15} = 1.67$$

Therefore

$$
\begin{aligned}
P(75 < x < 125) &= P(-1.67 < z < 1.67) \\
&= z \text{ curve area between } -1.67 \text{ and } 1.67 \\
&= (z \text{ curve area to the left of } 1.67) \\
&\quad - (z \text{ curve area to the left of } -1.67) \\
&= .9525 - .0475 \\
&= .9050
\end{aligned}
$$

This probability is illustrated in Figure 7.29. The calculation tells us that 90.5% of the population has an IQ score between 75 and 125. Of the 9.5% whose IQ score is not between 75 and 125, half of them (4.75%) have scores over 125, and the other half have scores below 75.

Figure 7.29 $P(75 < x < 125)$ and corresponding z curve area for Example 7.14.

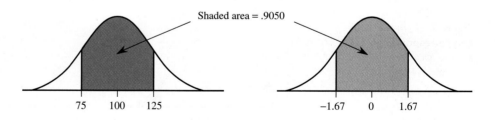

Shaded area = .9050

When we translate from a problem involving a normal distribution with mean μ and standard deviation σ to a problem involving the standard normal distribution, we convert to z scores:

$$z = \frac{x - \mu}{\sigma}$$

Because a z score can be interpreted as giving the distance of an x value from the mean in units of the standard deviation, a z score of 1.4 corresponds to an x value that is 1.4 standard deviations above the mean, and a z score of -2.1 corresponds to an x value that is 2.1 standard deviations below the mean.

Suppose that we are trying to evaluate $P(x < 60)$ for a variable whose distribution is normal with $\mu = 50$ and $\sigma = 5$. Converting the endpoint 60 to a z score gives

$$z = \frac{60 - 50}{5} = 2$$

which tells us that the value 60 is 2 standard deviations above the mean. We then have

$$P(x < 60) = P(z < 2)$$

where z is a standard normal variable. Notice that for the standard normal distribution, the value 2 is 2 standard deviations above the mean, because the mean is 0 and the standard deviation is 1. The value $z = 2$ is located the same distance (measured in standard deviations) from the mean of the standard normal distribution as the value $x = 60$ is from the mean in the normal distribution with $\mu = 50$ and $\sigma = 5$. This is why the translation using z scores results in an "equivalent" problem involving the standard normal distribution.

▪ Describing Extreme Values in a Normal Distribution

To describe the extreme values for a normal distribution with mean μ and standard deviation σ, we first solve the corresponding problem for the standard normal distribution and then translate our answer into one for the normal distribution of interest. This process is illustrated in Example 7.15.

▪ Example 7.15   Registration Times

Data on the length of time required to complete registration for classes using a telephone registration system suggests that the distribution of the variable

x = time to register

for students at a particular university can be well approximated by a normal distribution with mean $\mu = 12$ min and standard deviation $\sigma = 2$ min. (The normal distribution might not be an appropriate model for x = time to register at another university. Many factors influence the shape, center, and spread of such a distribution.) Because some students do not sign off properly, the university would like to disconnect students automatically after some amount of time has elapsed. It is decided to choose this time such that only 1% of the students are disconnected while they are still attempting to register. To determine the amount of time that should be allowed before disconnecting a student, we need to describe the largest 1% of the distribution of time to register. These are the individuals who will be mistakenly disconnected. This is illustrated in Figure 7.30(a). To determine the value of x^*, we first solve the analogous problem for the standard normal distribution, as shown in Figure 7.30(b).

By looking in Appendix Table 2 for a cumulative area of .99, we find the closest entry (.9901) in the 2.3 row and the .03 column, from which $z^* = 2.33$. For the standard normal distribution, the largest 1% of the distribution is made up of those values greater than 2.33. An equivalent statement is that the largest 1% are those with z scores greater than 2.33. This implies that in the distribution of time to register x (or any other normal distribution), the largest 1% are those values with

Figure 7.30 Capturing the largest 1% in a normal distribution for Example 7.15.

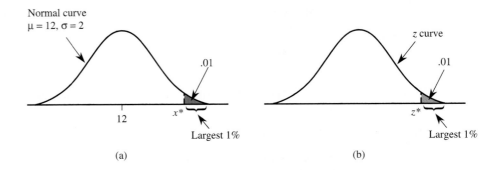

z scores greater than 2.33 or, equivalently, those x values that are more than 2.33 standard deviations above the mean. Here the standard deviation is 2, so 2.33 standard deviations is 2.33(2), and it follows that

$$x^* = 12 + 2.33(2) = 12 + 4.66 = 16.66$$

The largest 1% of the distribution for time to register is made up of values that are greater than 16.66 min. If the university system was set to disconnect students after 16.66 min, only 1% of the students registering would be disconnected before completing their registration.

A general formula for converting a z score back to an x value results from solving $z^* = \dfrac{x^* - \mu}{\sigma}$ for x^*, as shown in the accompanying box.

To convert a z score z^* back to an x value, use

$$x^* = \mu + z^*\sigma$$

▪ Example 7.16   Motor Vehicle Emissions

The Environmental Protection Agency (EPA) has developed a testing program to monitor vehicle emission levels of several pollutants. The article "Determining Statistical Characteristics of a Vehicle Emissions Audit Procedure" (*Technometrics* [1980]: 483–493) described the program, which measures pollutants from various types of vehicles, all following a fixed driving schedule. Data from the article suggest that the emissions of nitrogen oxides, which are major constituents of smog, can be plausibly modeled using a normal distribution. Let x denote the amount of this pollutant emitted by a randomly selected vehicle. The distribution of x can be described by a normal distribution with $\mu = 1.6$ and $\sigma = 0.4$.

Suppose that the EPA wants to offer some sort of incentive to get the worst polluters off the road. What emission levels constitute the worst 10% of the vehicles? The worst 10% would be the 10% with the highest emissions level, as shown in the following illustration:

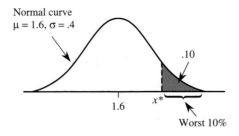

Normal curve
$\mu = 1.6$, $\sigma = .4$

.10

1.6 x^*

Worst 10%

For the standard normal distribution, the largest 10% are those with z values greater than $z^* = 1.28$ (from Appendix Table 2, based on a cumulative area of .90). Then

$$x^* = \mu + z^*\sigma$$
$$= 1.6 + 1.28(0.4)$$
$$= 1.6 + 0.512$$
$$= 2.112$$

In the population of vehicles of the type considered, about 10% would have oxide emission levels greater than 2.112.

■ Exercises 7.16–7.32

7.16 Determine the following standard normal (z) curve areas:
a. The area under the z curve to the left of 1.75
b. The area under the z curve to the left of -0.68
c. The area under the z curve to the right of 1.20
d. The area under the z curve to the right of -2.82
e. The area under the z curve between -2.22 and 0.53
f. The area under the z curve between -1 and 1
g. The area under the z curve between -4 and 4

7.17 Determine each of the areas under the standard normal (z) curve
a. To the left of -1.28 **b.** To the right of 1.28
c. Between -1 and 2   **d.** To the right of 0
e. To the right of -5 **f.** Between -1.6 and 2.5
g. To the left of 0.23

7.18 Let z denote a variable that has a standard normal distribution. Determine each of the following probabilities:
a. $P(z < 2.36)$ **b.** $P(z \leq 2.36)$
c. $P(z < -1.23)$ **d.** $P(1.14 < z < 3.35)$
e. $P(-0.77 \leq z \leq -0.55)$ **f.** $P(z > 2)$
g. $P(z \geq -3.38)$ **h.** $P(z < 4.98)$

7.19 Let z denote a variable having a normal distribution with $\mu = 0$ and $\sigma = 1$. Determine each of the following probabilities:

a. $P(z < 0.10)$ **b.** $P(z < -0.10)$
c. $P(0.40 < z < 0.85)$ **d.** $P(-0.85 < z < -0.40)$
e. $P(-0.40 < z < 0.85)$ **f.** $P(z > -1.25)$
g. $P(z < -1.50 \; or \; z > 2.50)$

7.20 Let z denote a variable that has a standard normal distribution. Determine the value z^* to satisfy the following conditions:
a. $P(z < z^*) = .025$ **b.** $P(z < z^*) = .01$
c. $P(z < z^*) = .05$ **d.** $P(z > z^*) = .02$
e. $P(z > z^*) = .01$
f. $P(z > z^* \; or \; z < -z^*) = .20$

7.21 Determine the value z^* that
a. Separates the largest 3% of all z values from the others
b. Separates the largest 1% of all z values from the others
c. Separates the smallest 4% of all z values from the others
d. Separates the smallest 10% of all z values from the others

7.22 Determine the value of z^* such that
a. z^* and $-z^*$ separate the middle 95% of all z values from the most extreme 5%
b. z^* and $-z^*$ separate the middle 90% of all z values from the most extreme 10%

c. z^* and $-z^*$ separate the middle 98% of all z values from the most extreme 2%

d. z^* and $-z^*$ separate the middle 92% of all z values from the most extreme 8%

7.23 Because $P(z < 0.44) = .67$, 67% of all z values are less than 0.44, and 0.44 is the 67th percentile of the standard normal distribution. Determine the value of each of the following percentiles for the standard normal distribution (Hint: If the cumulative area that you must look for does not appear in the z table, use the closest entry):

a. The 91st percentile (Hint: Look for area .9100.)
b. The 77th percentile
c. The 50th percentile
d. The 9th percentile
e. What is the relationship between the 70th z percentile and the 30th z percentile?

7.24 Consider the population of all 1-gal cans of dusty rose paint manufactured by a particular paint company. Suppose that a normal distribution with mean $\mu = 5$ ml and standard deviation $\sigma = 0.2$ ml is a reasonable model for the distribution of the variable $x =$ amount of red dye in the paint mixture. Use the normal distribution model to calculate the following probabilities.

a. $P(x < 5.0)$ **b.** $P(x < 5.4)$
c. $P(x \leq 5.4)$ **d.** $P(4.6 < x < 5.2)$
e. $P(x > 4.5)$ **f.** $P(x > 4.0)$

7.25 Consider babies born in the "normal" range of 37–43 weeks gestational age. Extensive data support the assumption that for such babies born in the United States, birth weight is normally distributed with mean 3432 g and standard deviation 482 g ("Are Babies Normal," *The American Statistician* [1999]: 298–302). (The investigators who wrote the article analyzed data from a particular year. For a sensible choice of class intervals, the histogram did not look at all normal, but after further investigations the researchers determined that this was due to some hospitals measuring weight in grams and others measuring to the nearest ounce and then converting to grams. A modified choice of class intervals that allowed for this measurement difference gave a histogram that was well described by a normal distribution.)

a. What is the probability that the birth weight of a randomly selected baby of this type exceeds 4000 g? is between 3000 and 4000 g?

b. What is the probability that the birth weight of a randomly selected baby of this type is either less than 2000 g or greater than 5000 g?

c. What is the probability that the birth weight of a randomly selected baby of this type exceeds 7 lb? (Hint: 1 lb = 453.59 g.)

d. How would you characterize the most extreme 0.1% of all birth weights?

e. If x is a variable with a normal distribution and a is a numerical constant $(a \neq 0)$, then $y = ax$ also has a normal distribution. Use this formula to determine the distribution of birth weight expressed in pounds (shape, mean, and standard deviation), and then recalculate the probability from Part (c). How does this compare to your previous answer?

7.26 A machine that cuts corks for wine bottles operates in such a way that the distribution of the diameter of the corks produced is well approximated by a normal distribution with mean 3 cm and standard deviation 0.1 cm. The specifications call for corks with diameters between 2.9 and 3.1 cm. A cork not meeting the specifications is considered defective. (A cork that is too small leaks and causes the wine to deteriorate; a cork that is too large doesn't fit in the bottle.) What proportion of corks produced by this machine are defective?

7.27 Refer to Exercise 7.26. Suppose that there are two machines available for cutting corks. The machine described in the preceding problem produces corks with diameters that are approximately normally distributed with mean 3 cm and standard deviation 0.1 cm. The second machine produces corks with diameters that are approximately normally distributed with mean 3.05 cm and standard deviation 0.01 cm. Which machine would you recommend? (Hint: Which machine would produce the fewest defective corks?)

7.28 A gasoline tank for a certain car is designed to hold 15 gal of gas. Suppose that the variable $x =$ actual capacity of a randomly selected tank has a distribution that is well approximated by a normal curve with mean 15.0 gal and standard deviation 0.1 gal.

a. What is the probability that a randomly selected tank will hold at most 14.8 gal?

b. What is the probability that a randomly selected tank will hold between 14.7 and 15.1 gal?

c. If two such tanks are independently selected, what is the probability that both tanks hold at most 15 gal?

7.29 The time that it takes a randomly selected job applicant to perform a certain task has a distribution that can be approximated by a normal distribution with a mean value of 120 sec and a standard deviation of 20 sec. The fastest 10% are to be given advanced training. What task times qualify individuals for such training?

7.30 A machine that produces ball bearings has initially been set so that the true average diameter of the bearings it produces is 0.500 in. A bearing is acceptable if its diameter is within 0.004 in. of this target value. Suppose, however, that the setting has changed during the course of production, so that the distribution of the diameters produced is well approximated by a normal distribution with mean 0.499 in. and standard deviation 0.002 in. What percentage of the bearings produced will not be acceptable?

7.31 Suppose that the distribution of net typing rate in words per minute (wpm) for experienced typists can be approximated by a normal curve with mean 60 wpm and standard deviation 15 wpm. The paper "Effects of Age and Skill in Typing" (*Journal of Experimental Psychology* [1984]: 345–371) described how net rate is obtained from gross rate by using a correction for errors.
a. What is the probability that a randomly selected typist's net rate is at most 60 wpm? less than 60 wpm?
b. What is the probability that a randomly selected typist's net rate is between 45 and 90 wpm?
c. Would you be surprised to find a typist in this population whose net rate exceeded 105 wpm? (Note: The largest net rate in a sample described in the paper cited is 104 wpm.)

d. Suppose that two typists are independently selected. What is the probability that both their typing rates exceed 75 wpm?
e. Suppose that special training is to be made available to the slowest 20% of the typists. What typing speeds would qualify individuals for this training?

7.32 Consider the variable x = time required for a college student to complete a standardized exam. Suppose that for the population of students at a particular university, the distribution of x is well approximated by a normal curve with mean 45 min and standard deviation 5 min.
a. If 50 min is allowed for the exam, what proportion of students at this university would be unable to finish in the allotted time?
b. How much time should be allowed for the exam if we wanted 90% of the students taking the test to be able to finish in the allotted time?
c. How much time is required for the fastest 25% of all students to complete the exam?

▪ 7.4 Checking for Normality and Normalizing Transformations

Some of the most frequently used statistical methods are valid only when a sample $x_1, x_2, \ldots, x_n$, has come from a population distribution that is at least approximately normal. One way to see whether an assumption of population normality is plausible is to construct a **normal probability plot** of the data. One version of this plot uses certain quantities called **normal scores**. The values of the normal scores depend on the sample size n. For example, the normal scores when $n = 10$ are as follows:

$$-1.539 \quad -1.001 \quad -0.656 \quad -0.376 \quad -0.123$$
$$0.123 \quad 0.376 \quad 0.656 \quad 1.001 \quad 1.539$$

To interpret these numbers, think of selecting sample after sample from a standard normal distribution, each one consisting of $n = 10$ observations. Then -1.539 is the long-run average of the smallest observation from each sample, -1.001 is the long-run average of the second smallest observation from each sample, and so on. In other words, -1.539 is the mean value of the smallest observation in a sample of size 10 from the z distribution, -1.001 is the mean value of the second smallest observation, and so on.

Extensive tabulations of normal scores for many different sample sizes are available. Alternatively, many software packages (such as MINITAB and SAS) and some graphing calculators can compute these scores on request and then construct a normal probability plot. Not all calculators and software packages use the same algorithm to compute normal scores. However, this does not change the overall character of a normal probability plot, so either the tabulated values or those given by the computer or calculator can be used.

After ordering the sample observations from smallest to largest, the smallest normal score is paired with the smallest observation, the second smallest normal score with the second smallest observation, and so on. The first number in a pair is the normal score, and the second number in the pair is the observed data value. A normal probability plot is just a scatterplot of the (normal score, observed value) pairs.

If the sample has been selected from a *standard* normal distribution, the second number in each pair should be reasonably close to the first number (ordered observation ≈ corresponding mean value). Then the n plotted points fall near a line with slope equal to 1 (a 45° line) passing through $(0, 0)$. When the sample has been obtained from *some* normal population distribution, the plotted points should be close to *some* straight line.

■ **Definition**

A **normal probability plot** is a scatterplot of the (normal score, observed value) pairs.

A substantial linear pattern in a normal probability plot suggests that population normality is plausible. On the other hand, a systematic departure from a straight-line pattern (such as curvature in the plot) casts doubt on the legitimacy of assuming a normal population distribution.

■ **Example 7.17   Window Widths**

The following 10 observations are widths of contact windows in integrated circuit chips:

| 3.21 | 2.49 | 2.94 | 4.38 | 4.02 | 3.62 | 3.30 | 2.85 | 3.34 | 3.81 |

The 10 pairs for the normal probability plot are then

$(-1.539, 2.49)$	$(0.123, 3.34)$
$(-1.001, 2.85)$	$(0.376, 3.62)$
$(-0.656, 2.94)$	$(0.656, 3.81)$
$(-0.376, 3.21)$	$(1.001, 4.02)$
$(-0.123, 3.30)$	$(1.539, 4.38)$

The normal probability plot is shown in Figure 7.31. The linearity of the plot supports the assumption that the window width distribution from which these observations were drawn is normal.

Figure 7.31 A normal probability plot for Example 7.17.

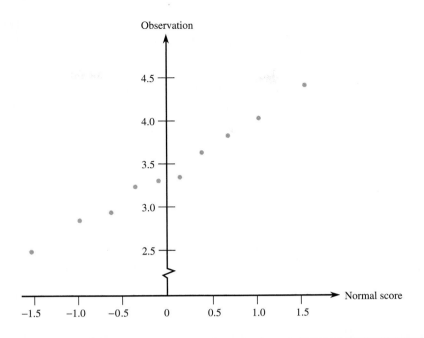

The decision as to whether or not a plot shows a substantial linear pattern is somewhat subjective. Particularly when n is small, normality should not be ruled out unless the departure from linearity is clear-cut. Figure 7.32 displays several plots that suggest a nonnormal population distribution.

Figure 7.32 Plots suggesting nonnormality: (a) indication that the population distribution is skewed; (b) indication that the population distribution has heavier tails than a normal curve; (c) presence of an outlier.

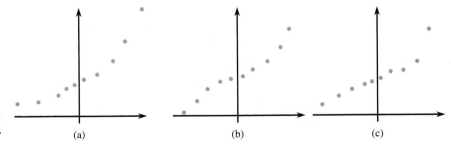

(a)   (b)   (c)

■ Using the Correlation Coefficient to Check Normality

The correlation coefficient r was introduced in Chapter 5 as a quantitative measure of the extent to which the points in a scatterplot fall close to a straight line. Let's denote the n (normal score, observed value) pairs as follows:

(smallest normal score, smallest observation)

$\vdots$

(largest normal score, largest observation)

Then the correlation coefficient can be computed as discussed in Chapter 5. The normal probability plot always slopes upward (because it is based on values ordered from smallest to largest), so r is a positive number. A value of r quite close to

1 indicates a very strong linear relationship in the normal probability plot. If r is too much smaller than 1, normality of the underlying distribution is questionable.

How far below 1 does r have to be before we begin to seriously doubt the plausibility of normality? The answer depends on the sample size n. If n is small, an r value somewhat below 1 is not surprising even when the distribution is normal, but if n is large, only an r value very close to 1 supports the assumption of normality. For selected values of n, Table 7.1 gives critical values to which r can be compared in checking for normality. If your sample size is between two tabulated values of n, use the critical value for the larger sample size. (For example, if $n = 46$, use the value of .966 for sample size 50.)

Table 7.1 ▪ Values to Which r Can Be Compared to Check for Normality

n	5	10	15	20	25	30	40	50	60	75
Critical r	.832	.880	.911	.929	.941	.949	.960	.966	.971	.976

*Source: *MINITAB User's Manual.*

If

$r <$ critical r for corresponding n

considerable doubt is cast on the assumption of population normality.

How were the critical values in Table 7.1 obtained? Consider the critical value .941 for $n = 25$. Suppose that the underlying distribution is actually normal. Consider obtaining a large number of different samples, each one consisting of 25 observations, and computing the value of r for each one. Then it can be shown that only 1% of the samples result in an r value less than the critical value .941. That is, .941 was chosen to guarantee a 1% error rate: In only 1% of all cases will we judge normality implausible when the distribution is really normal. The other critical values are also chosen to yield a 1% error rate for the corresponding sample sizes. It might have occurred to you that another type of error is possible: obtaining a large value of r and concluding that normality is a reasonable assumption when the distribution is actually nonnormal. This type of error is more difficult to control than the type mentioned previously, but the procedure we have described generally does a good job in both respects.

▪ Example 7.18   Window Widths Continued

The sample size for the contact window width data of Example 7.17 is $n = 10$. The critical r from Table 7.1 is then .880. The correlation coefficient calculated using the (normal score, observed value) pairs is $r = .995$. Because r is larger than the critical r for a sample of size 10, it is plausible that the population distribution of window widths from which this sample was drawn is approximately normal.

■ Transforming Data to Obtain a Distribution That Is Approximately Normal

Many of the most frequently used statistical methods are valid only when the sample is selected at random from a population whose distribution is at least approximately normal. When a sample histogram shows a distinctly nonnormal shape, it is common to use a transformation or reexpression of the data. By *transforming* data, we mean applying some specified mathematical function (such as the square root, logarithm, or reciprocal) to each data value to produce a set of transformed data. We can then study and summarize the distribution of these transformed values using methods that require normality. We saw in Chapter 5 that, with bivariate data, one or both of the variables can be transformed in an attempt to find two variables that are linearly related. With univariate data, a transformation is usually chosen to yield a distribution of transformed values that is more symmetric and more closely approximated by a normal curve than was the original distribution.

■ Example 7.19   Rainfall Data

Data that have been used by several investigators to introduce the concept of transformation (e.g., "Exploratory Methods for Choosing Power Transformations," *Journal of the American Statistical Association* [1982]: 103–108) consist of values of March precipitation for Minneapolis–St. Paul over a period of 30 years. These values are given in Table 7.2 along with the square root of each value. Histograms of both the original and the transformed data appear in Figure 7.33. The distribution

Table 7.2 ■ Original and Square-Root-Transformed Values of March Precipitation in Minneapolis–St. Paul over a 30-Year Period

Year	Precipitation	$\sqrt{\text{Precipitation}}$	Year	Precipitation	$\sqrt{\text{Precipitation}}$
1	0.77	0.88	16	1.62	1.27
2	1.74	1.32	17	1.31	1.14
3	0.81	0.90	18	0.32	0.57
4	1.20	1.10	19	0.59	0.77
5	1.95	1.40	20	0.81	0.90
6	1.20	1.10	21	2.81	1.68
7	0.47	0.69	22	1.87	1.37
8	1.43	1.20	23	1.18	1.09
9	3.37	1.84	24	1.35	1.16
10	2.20	1.48	25	4.75	2.18
11	3.00	1.73	26	2.48	1.57
12	3.09	1.76	27	0.96	0.98
13	1.51	1.23	28	1.89	1.37
14	2.10	1.45	29	0.90	0.95
15	0.52	0.72	30	2.05	1.43

Figure 7.33 Histograms of precipitation data for Example 7.19: (a) untransformed data; (b) square-root-transformed data.

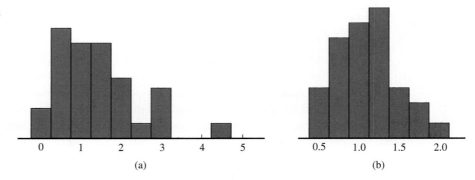

of the original data is clearly skewed, with a long upper tail. The square-root transformation has resulted in a substantially more symmetric distribution, with a typical (i.e., central) value near the 1.25 boundary between the third and fourth class intervals.

Logarithmic transformations are also common and, as with bivariate data, either the natural logarithm or the base 10 logarithm can be used. A logarithmic transformation is usually applied to data that are positively skewed (a long upper tail). This affects values in the upper tail substantially more than values in the lower tail, yielding a more symmetric — and often more normal — distribution.

▪ Example 7.20   Beryllium Exposure

Exposure to beryllium is known to produce adverse effects on lungs as well as on other tissues and organs in both laboratory animals and humans. The article "Time Lapse Cinematographic Analysis of Beryllium: Lung Fibroblast Interactions" (*Environmental Research* [1983]: 34–43) reported the results of experiments designed to study the behavior of certain individual cells that had been exposed to beryllium. An important characteristic of such an individual cell is its interdivision time (IDT). IDTs were determined for a large number of cells under both exposed (treatment) and unexposed (control) conditions. The authors of the article state, "The IDT distributions are seen to be skewed, but the natural logs do have an approximate normal distribution." The same property holds for $\log_{10}$-transformed data. We give representative IDT data in Table 7.3 and the resulting histograms in Figure 7.34, which are in agreement with the authors' statement.

The sample size for the IDT data is $n = 40$. The correlation coefficient for the (normal score, untransformed data) pairs is .950, which is less than the critical r for $n = 40$ (critical $r = .960$). The correlation coefficient using the transformed data is .998, which is much larger than the critical r, supporting the assertion that $\log_{10}(IDT)$ has approximately a normal distribution. Figure 7.35 displays MINITAB normal probability plots for the original data and for the transformed data. The plot for the transformed data is clearly more linear in appearance than the plot for the original data.

Table 7.3 ▪ Original and $\log_{10}$-Transformed IDT Values

IDT	$\log_{10}(IDT)$	IDT	$\log_{10}(IDT)$	IDT	$\log_{10}(IDT)$	IDT	$\log_{10}(IDT)$
28.1	1.45	15.5	1.19	26.2	1.42	21.1	1.32
46.0	1.66	38.4	1.58	17.4	1.24	60.1	1.78
34.8	1.54	21.4	1.33	55.6	1.75	21.4	1.33
17.9	1.25	40.9	1.61	21.0	1.32	32.0	1.51
31.9	1.50	31.2	1.49	36.3	1.56	38.8	1.59
23.7	1.37	25.8	1.41	72.8	1.86	25.5	1.41
26.6	1.42	62.3	1.79	20.7	1.32	22.3	1.35
43.5	1.64	19.5	1.29	13.7	1.14	19.1	1.28
30.6	1.49	28.9	1.46	16.8	1.23	48.9	1.69
52.1	1.72	18.6	1.27	28.0	1.45	57.3	1.76

Figure 7.34 Histograms of IDT data for Example 7.20: (a) untransformed data; (b) $\log_{10}$-transformed data.

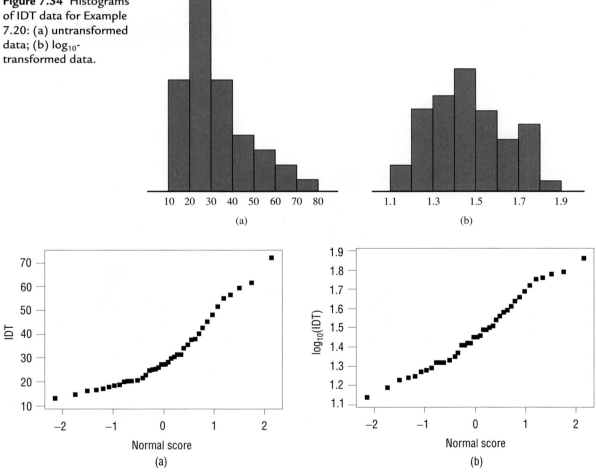

Figure 7.35 MINITAB-generated normal probability plots for Example 7.20: (a) original IDT data; (b) log-transformed IDT data.

▪ Selecting a Transformation

Occasionally, a particular transformation can be dictated by some theoretical argument, but often this is not the case and you may wish to try several different transformations to find one that is satisfactory. Figure 7.36, from the article "Distribution of Sperm Counts in Suspected Infertile Men" (*Journal of Reproduction and Fertility* [1983]: 91–96), shows what can result from such a search. Other investigators in this field had previously used all three of the transformations illustrated.

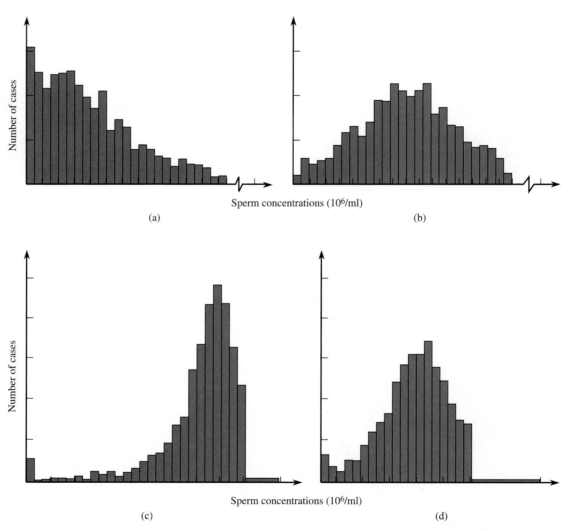

Figure 7.36 Histograms of sperm concentrations for 1711 suspected infertile men: (a) untransformed data (highly skewed); (b) log-transformed data (reasonably symmetric); (c) square-root-transformed data; (d) cube-root-transformed data.

▪ Exercises 7.33–7.44

7.33 Ten measurements of the steam rate (in pounds per hour) of a distillation tower were used to construct the following normal probability plot ("A Self-Descaling Distillation Tower," *Chemical Engineering Process* [1968]: 79–84):

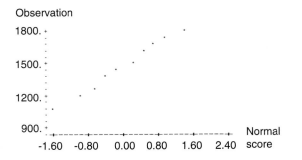

Based on the plot, do you think it is reasonable to assume that the normal distribution provides an adequate description of the steam rate distribution? Explain.

7.34 The following normal probability plot was constructed using part of the data appearing in the paper "Trace Metals in Sea Scallops" (*Environmental Concentration and Toxicology* 19: 1326–1334):

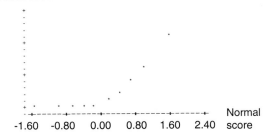

The variable under study was the amount of cadmium in North Atlantic scallops. Do the sample data suggest that the cadmium concentration distribution is not normal? Explain.

7.35 Consider the following 10 observations on the lifetime (in hours) for a certain type of component:

152.7 172.0 172.5 173.3 193.0
204.7 216.5 234.9 262.6 422.6

Construct a normal probability plot, and comment on the plausibility of a normal distribution as a model for component lifetime.

7.36 The paper "The Load-Life Relationship for M50 Bearings with Silicon Nitride Ceramic Balls"

(*Lubrication Engineering* [1984]: 153–159) reported the following data on bearing load life (in millions of revolutions); the corresponding normal scores are also given:

x	Normal Score	x	Normal Score
47.1	−1.867	240.0	0.062
68.1	−1.408	240.0	0.187
68.1	−1.131	278.0	0.315
90.8	−0.921	278.0	0.448
103.6	−0.745	289.0	0.590
106.0	−0.590	289.0	0.745
115.0	−0.448	367.0	0.921
126.0	−0.315	385.9	1.131
146.6	−0.187	392.0	1.408
229.0	−0.062	395.0	1.867

Construct a normal probability plot. Is normality plausible?

7.37 The following observations are DDT concentrations in the blood of 20 people:

24 26 30 35 35 38 39 40 40 41
42 52 56 58 61 75 79 88 102 42

Use the normal scores from Exercise 7.36 to construct a normal probability plot, and comment on the appropriateness of a normal probability model.

7.38 Consider the following sample of 25 observations on the diameter x (in centimeters) of a disk used in a certain system.

16.01 16.08 16.13 15.94 16.05 16.27 15.89
15.84 15.95 16.10 15.92 16.04 15.82 16.15
16.06 15.66 15.78 15.99 16.29 16.15 16.19
16.22 16.07 16.13 16.11

The 13 largest normal scores for a sample of size 25 are 1.965, 1.524, 1.263, 1.067, 0.905, 0.764, 0.637, 0.519, 0.409, 0.303, 0.200, 0.100, and 0. The 12 smallest scores result from placing a negative sign in front of each of the given nonzero scores. Construct a normal probability plot. Does it appear plausible that disk diameter is normally distributed? Explain.

7.39 Example 7.19 examined rainfall data for Minneapolis–St. Paul. The square-root transformation was used to obtain a distribution of values that was more symmetric than the distribution of the original data. Another transformation that has been suggested by meteorologists is the cube root: transformed value = (original value)$^{1/3}$. The original values and their cube roots (the transformed values) are given in the following table:

Original	Transformed	Original	Transformed
0.32	0.68	1.51	1.15
0.47	0.78	1.62	1.17
0.52	0.80	1.74	1.20
0.59	0.84	1.87	1.23
0.77	0.92	1.89	1.24
0.81	0.93	1.95	1.25
0.81	0.93	2.05	1.27
0.90	0.97	2.10	1.28
0.96	0.99	2.20	1.30
1.18	1.06	2.48	1.35
1.20	1.06	2.81	1.41
1.20	1.06	3.00	1.44
1.31	1.09	3.09	1.46
1.35	1.11	3.37	1.50
1.43	1.13	4.75	1.68

Construct a histogram of the transformed data. Compare your histogram to those given in Figure 7.33. Which of the cube-root and the square-root transformations appears to result in the more symmetric histogram(s)?

7.40 The following data are a sample of survival times (in days from diagnosis) for patients suffering from chronic leukemia of a certain type (*Statistical Methodology for Survival Time Studies*, Bethesda, MD: National Cancer Institute, 1986):

7	47	58	74	177	232	273	285
317	429	440	445	455	468	495	497
532	571	579	581	650	702	715	779
881	900	930	968	1077	1109	1314	1334
1367	1534	1712	1784	1877	1886	2045	2056
2260	2429	2509	2045	2056			

a. Construct a relative frequency distribution for this data set, and draw the corresponding histogram.
b. Would you describe this histogram as having a positive or a negative skew?
c. Would you recommend transforming the data? Explain.

7.41 In a study of warp breakage during the weaving of fabric (*Technometrics* [1982]: 59–65), 100 pieces of yarn were tested. The number of cycles of strain to breakage was recorded for each yarn sample. The resulting data are given in the following table:

86	146	251	653	98	249	400	292	131	176
76	264	15	364	195	262	88	264	42	321
180	198	38	20	61	121	282	180	325	250
196	90	229	166	38	337	341	40	40	135
597	246	211	180	93	571	124	279	81	186
497	182	423	185	338	290	398	71	246	185
188	568	55	244	20	284	93	396	203	829
239	236	277	143	198	264	105	203	124	137
135	169	157	224	65	315	229	55	286	350
193	175	220	149	151	353	400	61	194	188

a. Construct a frequency distribution using the class intervals 0 to <100, 100 to <200, and so on.
b. Draw the histogram corresponding to the frequency distribution in Part (a). How would you describe the shape of this histogram?
c. Find a transformation for these data that results in a more symmetric histogram than what you obtained in Part (b).

7.42 The article "The Distribution of Buying Frequency Rates" (*Journal of Marketing Research* [1980]: 210–216) reported the results of a $3\frac{1}{2}$-yr study of dentifrice purchases. The investigators conducted their research using a national sample of 2071 households and recorded the number of toothpaste purchases for each household participating in the study. The results are given in the following frequency distribution:

Number of Purchases	Number of Households (Frequency)	Number of Purchases	Number of Households (Frequency)
10 to <20	904	90 to <100	13
20 to <30	500	100 to <110	9
30 to <40	258	110 to <120	7
40 to <50	167	120 to <130	6
50 to <60	94	130 to <140	6
60 to <70	56	140 to <150	3
70 to <80	26	150 to <160	0
80 to <90	20	160 to <170	2

a. Draw a histogram for this frequency distribution. Would you describe the histogram as positively or negatively skewed?
b. Does the square-root transformation result in a histogram that is more symmetric than that of the original data? (Be careful! This one is a bit tricky because you don't have the raw data; transforming the endpoints of the class intervals results in class intervals that are not necessarily of equal widths, so the histogram of the transformed values has to be drawn with this in mind.)

7.43 Ecologists have long been interested in factors that might explain how far north or south particular animal species are found. As part of one such study, the paper "Temperature and the Northern Distributions of Wintering Birds" (*Ecology* [1991]: 2274–2285) gave the following body masses (in grams) for 50 different bird species that had previously been thought to have northern boundaries corresponding to a particular isotherm:

7.7	10.1	21.6	8.6	12.0	11.4	16.6	9.4	11.5	
9.0	8.2	20.2	48.5	21.6	26.1	6.2	19.1	21.0	
28.1	10.6	31.6	6.7	5.0	68.8	23.9	19.8	20.1	
6.0	99.6	19.8	16.5	9.0	448.0	21.3	17.4	36.9	
34.0	41.0	15.9	12.5	10.2	31.0	21.5	11.9	32.5	
9.8	93.9	10.9	19.6	14.5					

a. Construct a stem-and-leaf display in which 448.0 is shown beside the display as an outlier value, the stem of an observation is the tens digit, the leaf is the ones digit, and the tenths digit is suppressed (e.g., 21.5 has stem 2 and leaf 1). What do you perceive as the most prominent feature of the display?

b. Draw a histogram based on the class intervals 5 to <10, 10 to <15, 15 to <20, 20 to <25, 25 to <30, 30 to <40, 40 to <50, 50 to <100, and 100 to <500. Is a transformation of the data desirable? Explain.

c. Use a calculator or statistical computer package to calculate logarithms of these observations and construct a histogram. Is the logarithmic transformation satisfactory?

d. Consider transformed value $= \dfrac{1}{\sqrt{\text{original value}}}$ and construct a histogram of the transformed data. Does it appear to resemble a normal curve?

7.44 The following figure appeared in the paper "EDTA-Extractable Copper, Zinc, and Manganese in Soils of the Canterbury Plains" (*New Zealand Journal of Agricultural Research* [1984]: 207–217):

Figure for Exercise 7.44

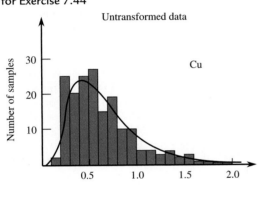

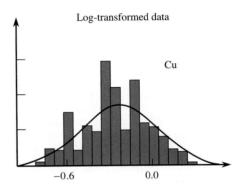

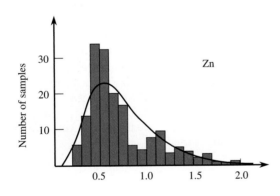

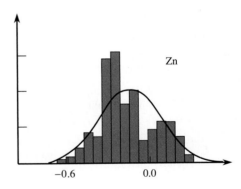

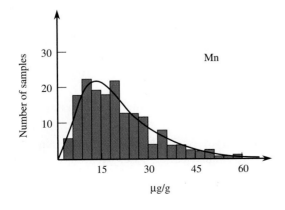

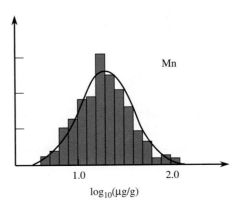

A large number of topsoil samples were analyzed for manganese (Mn), zinc (Zn), and copper (Cu), and the resulting data were summarized using histograms. The investigators transformed each data set using logarithms in an effort to obtain more symmetric distributions of values. Do you think the transformations were successful? Explain.

▪ Activity 7.1: Rotten Eggs?

Background: The Salt Lake Tribune (October 11, 2002) printed the following account of an exchange between a restaurant manager and a health inspector:

> The recipe calls for four fresh eggs for each quiche. A Salt Lake County Health Department inspector paid a visit recently and pointed out that research by the Food and Drug Administration indicates that one in four eggs carries salmonella bacterium, so restaurants should never use more than three eggs when preparing quiche. The manager on duty wondered aloud if simply throwing out three eggs from each dozen and using the remaining nine in four-egg quiches would serve the same purpose.

1. Working in a group or as a class, discuss the folly of the above statement!

2. Suppose the following argument is made for three-egg quiches rather than four-egg quiches: Let x = number of eggs that carry salmonella. Then

$$p(0) = P(x = 0) = (.75)^3 = .422$$

for three-egg quiches and

$$p(0) = P(x = 0) = (.75)^4 = .316$$

for four-egg quiches. What assumption must be made to justify these probability calculations? Do you think this is reasonable or not? Explain.

3. Suppose that a carton of one dozen eggs does happen to have exactly three eggs that carry salmonella and that the manager does as he proposes: selects three eggs at random and throws them out, then uses the remaining nine eggs in four-egg quiches. Let x = number of eggs that carry salmonella among four eggs selected at random from the remaining nine.

Working with a partner, conduct a simulation to approximate the distribution of x by carrying out the following sequence of steps:

a. Take 12 identical slips of paper and write "Good" on 9 of them and "Bad" on the remaining 3. Place the slips of paper in a paper bag or some other container.

b. Mix the slips and then select three at random and remove them from the bag.

c. Mix the remaining slips and select four "eggs" from the bag.

d. Note the number of bad eggs among the four selected. (This is an observed x value.)

e. Replace all slips, so that the bag now contains all 12 "eggs."

f. Repeat Steps (b)–(d) at least 10 times, each time recording the observed x value.

4. Combine the observations from your group with those from the other groups. Use the resulting data to approximate the distribution of x. Comment on the resulting distribution in the context of the risk of salmonella exposure if the manager's proposed procedure is used.

▪ Summary of Key Concepts and Formulas

Term or Formula	Comment
Population distribution	The distribution of all the values of a numerical variable or categories of a categorical variable.
Discrete numerical variable	A numerical variable whose possible values are isolated points along the number line.
Continuous numerical variable	A numerical variable whose possible values form an interval along the number line.
Continuous probability distribution (density curve)	A smooth curve that serves as a model for the population distribution of a continuous numerical variable. Areas under this curve are interpreted as probabilities.

Term or Formula	Comment
μ, σ	The mean and standard deviation, respectively, of a probability distribution. The mean describes the center, and the standard deviation describes the spread of the probability distribution.
Normal distribution	A continuous probability distribution that is specified by a particular bell-shaped density curve.
Standard normal distribution (z curve)	The normal distribution with $\mu = 0$ and $\sigma = 1$.
$z = \dfrac{x - \mu}{\sigma}$	The formula for a z score. When x has a normal distribution, z has a standard normal distribution.
Normal probability plot	A scatterplot of the (normal score, observed value) pairs. A linear pattern in the plot suggests that it is plausible that the data are from a normal population distribution. Curvature in the plot suggests that the population distribution is not normal.

▪ Supplementary Exercises 7.45–7.52

7.45 A machine producing vitamin E capsules operates in such a way that the distribution of x = actual amount of vitamin E in a capsule can be modeled by a normal curve with mean 5 mg and standard deviation 0.05 mg. What is the probability that a randomly selected capsule contains less than 4.9 mg of vitamin E? at least 5.2 mg?

7.46 Accurate labeling of packaged meat is difficult because of weight decrease resulting from moisture loss (defined as a percentage of the package's original net weight). Suppose that the normal distribution with mean value 4.0% and standard deviation 1.0% is a reasonable model for the variable x = moisture loss for a package of chicken breasts. (This model is suggested in the paper "Drained Weight Labeling for Meat and Poultry: An Economic Analysis of a Regulatory Proposal," *Journal of Consumer Affairs* [1980]: 307–325.)
a. What is the probability that x is between 3.0% and 5.0%?
b. What is the probability that x is at most 4.0%?
c. What is the probability that x is at least 7%?
d. Describe the largest 10% of the moisture loss distribution.
e. Describe the most extreme 10% of the moisture loss distribution. (Hint: The most extreme 10% consists of the largest 5% and the smallest 5%.)

7.47 The *Wall Street Journal* (February 15, 1972) reported that General Electric was being sued in Texas for sex discrimination because of a minimum height requirement of 5 ft 7 in. The suit claimed that this restriction eliminated more than 94% of adult females from consideration. Let x represent the height of a randomly selected adult woman. Suppose that the probability distribution of x is approximately normal with mean 66 in. and standard deviation 2 in.
a. Is the claim that 94% of all women are shorter than 5 ft 7 in. correct?
b. What proportion of adult women would be excluded from employment because of the height restriction?

7.48 Your statistics professor tells you that the distribution of scores on a midterm exam was approximately normal with a mean of 78 and a standard deviation of 7. The top 15% of all scores have been designated A's. Your score was 89. Did you receive an A? Explain.

7.49 Suppose that the distribution of pH readings for soil samples taken in a certain geographic region can be approximated by a normal distribution with mean 6.00 and standard deviation 0.10. The pH of a randomly selected soil sample from this region is to be determined.
a. What is the probability that the resulting pH is between 5.90 and 6.15?
b. What is the probability that the resulting pH exceeds 6.10?
c. What is the probability that the resulting pH is at most 5.95?
d. Describe the largest 5% of the pH distribution.

7.50 The light bulbs used to provide exterior lighting for a large office building have an average lifetime of 700 hr. If the distribution of the variable x = length of bulb life can be modeled as a normal distribution with a standard deviation of 50 hr, how often should all the bulbs be replaced so that only 20% of the bulbs will have already burned out?

7.51 Let x denote the duration of a randomly selected pregnancy (the time elapsed between conception and birth). Accepted values for the mean value and standard deviation of x are 266 days and 16 days, respectively. Suppose that a normal distribution is an appropriate model for the probability distribution of x.

a. What is the probability that the duration of pregnancy is between 250 and 300 days?

b. What is the probability that the duration of pregnancy is at most 240 days?

c. What is the probability that the duration of pregnancy is within 16 days of the mean duration?

d. A *Dear Abby* column dated January 20, 1973, contained a letter from a woman who stated that the duration of her pregnancy was exactly 310 days. (She wrote that the last visit with her husband, who was in the navy, occurred 310 days before the birth.) What is the probability that the duration of a pregnancy is at least 310 days? Does this probability make you a bit skeptical of the claim?

e. Some insurance companies will pay the medical expenses associated with childbirth only if the insurance has been in effect for more than 9 months

(275 days). This restriction is designed to ensure that the insurance company has to pay benefits only for those pregnancies for which conception occurred during coverage. Suppose that conception occurred 2 weeks after coverage began. What is the probability that the insurance company will refuse to pay benefits because of the 275-day insurance requirement?

7.52 Let x denote the systolic blood pressure of an individual selected at random from a certain population. Suppose that the normal distribution with mean $\mu = 120$ mm Hg and standard deviation $\sigma = 10$ mm Hg is a reasonable model for describing the population distribution of x. (The article "Oral Contraceptives, Pregnancy, and Blood Pressure," *Journal of the American Medical Association* [1972]: 1507–1510, reported on the results of a large study in which a sample histogram of blood pressures among women of similar ages was found to be well approximated by a normal curve.)

a. Calculate $P(110 < x < 140)$. How does this compare to $P(110 \leq x \leq 140)$, and why?

b. Calculate $P(x < 75)$.

c. Calculate $P(x < 95 \text{ or } x > 145)$, the probability that x is more than 2.5 standard deviations from its mean value.

▪ Reference

McClave, James, and Terry Sincich. *Statistics*, 8th ed. Englewood Cliffs, NJ: Prentice-Hall, 2000. (Chapters 4 and 5 present a good exposition of discrete and continuous distributions at a modest mathematical level.)

8 · Sampling Variability and Sampling Distributions

Many investigations and research projects try to draw conclusions about how the values of some variable x are distributed in a population. Often, attention is focused on a single characteristic of that distribution. Examples include the following:

1. x = fat content (in grams) of a quarter-pound hamburger, with interest centered on the mean fat content μ of all such hamburgers

2. x = fuel efficiency (in miles per gallon) for a 2003 Honda Accord, with interest focused on the variability in fuel efficiency as described by σ, the standard deviation for the fuel efficiency population distribution

3. x = time to first recurrence of skin cancer for a patient treated using a particular therapy, with attention focused on the proportion π of such individuals whose first recurrence is within 5 years of treatment

4. x = marital status (a categorical variable) of a student at a particular university, with interest centered on π, the proportion of all students at the university who are married

The inferential methods presented in Chapters 9–15 use information contained in a sample to reach conclusions about one or more characteristics of the population from which the sample was selected. For example, let μ denote the true mean fat content of quarter-pound hamburgers marketed by a national fast-food chain. To learn something about μ, we might obtain a sample of $n = 50$ hamburgers and determine the fat content for each burger in the sample. The sample data might produce a mean of $\bar{x} = 28.4$ g. How close is this sample mean to the population mean, μ? If we selected another sample of 50 quarter-pound burgers and computed the fat content, would this second $\bar{x}$ value be near 28.4, or might it be quite different?

These questions can be addressed by studying what is called the *sampling distribution* of $\bar{x}$. Just as the distribution of a numerical variable describes its long-run behavior, the sampling distribution of $\bar{x}$ provides information about the long-run behavior of $\bar{x}$ when sample after sample is selected.

Later in the text, in Chapters 9 and 10, we will use properties of the sampling distribution of $\bar{x}$ to develop inferential procedures for drawing conclusions about μ. In a similar manner, we will study the sampling distribution of a sample proportion (the fraction of individuals or objects in a sample that have some characteristic of interest). This will enable us to describe procedures for making inferences about the corresponding population proportion π.

▪ 8.1 Statistics and Sampling Variability

The usual way to obtain information regarding the value of a population characteristic is by selecting a sample from the population. For example, to gain insight about the mean value of the fat content distribution for the population of quarter-pound hamburgers from a specific fast-food chain, we might select a sample of 50 hamburgers for analysis. Each hamburger in the sample could be tested to yield a value of x = fat content. As in Chapter 7, we could construct a histogram of the 50 sample x values, and we could view this histogram as a rough approximation of the population distribution of x. In a similar way, we could view the sample mean $\bar{x}$ (the mean of the sample of n values) as an approximation of μ, the mean of the population distribution. It would be nice if the value of $\bar{x}$ were equal to the value of μ. Although it could happen that $\bar{x} = \mu$, this would be an unusual occurrence because $\bar{x}$ is only an *estimate* of μ. An important question addressed later in this chapter is, How good is this estimate?

Not only does the value of $\bar{x}$ for a particular sample from a population usually differ from μ, but also the $\bar{x}$ values from different samples typically differ from one another. (For example, 2 different samples of 50 hamburgers will usually result in different $\bar{x}$ values.) This sample-to-sample variability makes it challenging to generalize from a sample to the population from which it was selected. To meet this challenge, we must understand sample-to-sample variability.

▪ Example 8.1   Constructing a Sampling Distribution

When we sample from a population, any one of many different samples might result. Consider a small population consisting of the board of directors of a day care center (also listed are the values of x = number of children for each board member):

Board member	Jay	Carol	Allison	Teresa	Lygia	Bob	Roxy	Kyle
Number of children	2	2	0	0	2	2	0	3

The average number of children for the entire group of eight (the population) is

$$\mu = \frac{2 + 2 + 0 + 0 + 2 + 2 + 0 + 3}{8} = \frac{11}{8} = 1.38$$

There are 28 different possible outcomes when a sample of size 2 is selected (without replacement) from the board; these are shown in Table 8.1. If sample 1 (Jay, Carol) is selected, the associated sample mean is $\bar{x}_1 = \frac{2 + 2}{2} = 2$; if sample 14 (Allison, Teresa) is selected, the associated sample mean is $\bar{x}_{14} = \frac{0 + 0}{2} = 0$. Each of the different possible samples has an $\bar{x}$ value associated with it. The $\bar{x}$ values not only differ from μ (neither $\bar{x}_1$ nor $\bar{x}_{14}$ is equal to 1.38), but they can also differ from sample to sample (e.g., $\bar{x}_1 \neq \bar{x}_{14}$). Table 8.1 gives the value of $\bar{x}$ for each of the 28 different possible samples, and a dotplot of the $\bar{x}$ values in the table is shown in Figure 8.1.

Table 8.1 ■ **All Possible Samples of Size 2 and Their Respective Means for Example 8.1**

Sample	$\bar{x}$	Sample	$\bar{x}$
1 Jay, Carol	2	15 Allison, Lygia	1
2 Jay, Allison	1	16 Allison, Bob	1
3 Jay, Teresa	1	17 Allison, Roxy	0
4 Jay, Lygia	2	18 Allison, Kyle	1.5
5 Jay, Bob	2	19 Teresa, Lygia	1
6 Jay, Roxy	1	20 Teresa, Bob	1
7 Jay, Kyle	2.5	21 Teresa, Roxy	0
8 Carol, Allison	1	22 Teresa, Kyle	1.5
9 Carol, Teresa	1	23 Lygia, Bob	2
10 Carol, Lygia	2	24 Lygia, Roxy	1
11 Carol, Bob	2	25 Lygia, Kyle	2.5
12 Carol, Roxy	1	26 Bob, Roxy	1
13 Carol, Kyle	2.5	27 Bob, Kyle	2.5
14 Allison, Teresa	0	28 Roxy, Kyle	1.5

Figure 8.1 A dotplot of the $\bar{x}$ values in Example 8.1.

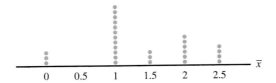

Figure 8.1 provides some insight into the behavior of the mean of a sample of size 2 from this population. We see that the value $\bar{x} = 1$ is more common than any other value. However, $\bar{x}$ values can differ greatly not only from one sample to another but also from the value of the population mean, $\mu = 1.38$. If we were to select a random sample of size $n = 2$ (so that each different sample has the same chance of being selected), we could not count on the resulting value of $\bar{x}$ being very close to the population mean $\mu = 1.38$.

To formalize these ideas, we begin by introducing some terminology.

■ **Definition**

Any quantity computed from values in a sample is called a **statistic**.

The observed value of a statistic depends on the particular sample selected from the population; typically, it varies from sample to sample. This variability is called **sampling variability**.

The statistic considered in Example 8.1 was $\bar{x}$, the sample mean. Values of statistics such as $\bar{x}$, the sample median, the sample interquartile range, and so on, are our primary sources of information regarding various population characteristics.

▪ Sampling Distributions

It is possible to view a statistic as a random variable and to use a distribution to describe its behavior. Before examining how this is done, let's review how a population distribution is used to describe the distribution of the values of some variable in a particular population. For example, suppose that we are studying the population of all mail carriers in Los Angeles County. Each carrier has an assigned route, and associated with each carrier is a value of the variable x = length of route (in miles). One carrier might have a 3.8-mi route ($x = 3.8$), whereas for another carrier we might find that $x = 2.4$. If we knew the x value for every individual in the population, we could construct a density histogram that would represent the population distribution for the variable x. The distribution might look like the histogram in Figure 8.2. The population distribution tells us about the distribution of individual x values in the population. The population distribution could be used to evaluate the probabilities, such as the probability that a randomly selected carrier has a route that is more than 4 mi, $P(x > 4)$, or a route that is between 2 and 3 mi, $P(2 < x < 3)$.

Figure 8.2 Histogram of x = length of route.

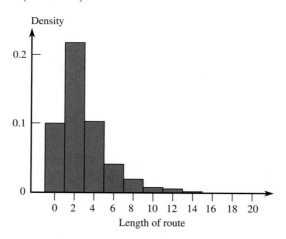

Now suppose that we are interested in μ, the mean value of x (the average route length) for this population. If the population distribution is known, the value of μ can be determined. For the distribution in Figure 8.2, $\mu \approx 3$.

What if the population distribution is not known? Then we must estimate μ using information in a sample. Suppose that we plan to take a random sample of size $n = 100$ from the population of all mail carriers in Los Angeles County. There are many, many different possible samples that might result (many more than the 28 from the day care example). We now define a hypothetical population, which consists of all the different possible samples of a given size n. That is, we consider a *population of samples*. (This is a bit abstract, but it is an important idea!) The population of samples is viewed as a population because it consists of *every* different sample; it is the complete collection of all possible samples.

Just as a variable associates a value with every individual or object in a population and can be described by its distribution, a statistic associates a value with each individual sample in the population of samples. Thus, a statistic can also be described by a distribution.

▪ Definition

The distribution of the values of a statistic is called its **sampling distribution**.

▪ Example 8.2   More on Constructing a Sampling Distribution

In Example 8.1, we considered a small population consisting of the board of directors of a day care center. The variable of interest was x = number of children, and the statistic of interest was the sample mean $\bar{x}$. Figure 8.1 showed a dotplot of the $\bar{x}$ values for the population of all 28 samples of size $n = 2$. The corresponding density histogram appears in Figure 8.3; this is the sampling distribution of the statistic $\bar{x}$ when a random sample of size 2 is selected from the original population of 8 people.

Figure 8.3 Sampling distribution of $\bar{x}$ for Example 8.2.

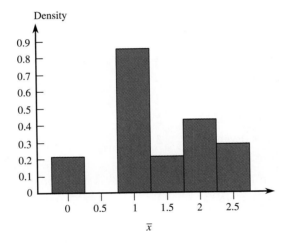

Looking at the sampling distribution of $\bar{x}$ depicted in Figure 8.3 helps us to understand what happens when a random sample of size 2 is selected from this population. For example, we can compute the probability of observing a sample mean that is within 0.5 of $\mu = 1.38$, the mean number of children in the population. This probability is

$$P(\bar{x} \text{ is within 0.5 of } \mu) = P(\bar{x} = 1 \text{ or } \bar{x} = 1.5)$$
$$= \frac{12}{28} + \frac{3}{28}$$
$$= .4286 + .1071$$
$$= .5357$$

The probabilities $P(\bar{x} = 1) = 12/28$ and $P(\bar{x} = 1.5) = 3/28$ were computed using the information in Figure 8.1. Note that they are also the areas of the corresponding rectangles in the density histogram of Figure 8.3.

The sampling distribution of $\bar{x}$ can also be displayed in tabular form in this case (because there are only a few different possible values), as shown in the following:

$\bar{x}$	0	1	1.5	2	2.5
$p(\bar{x})$	3/28	12/28	3/28	6/28	4/28

The sampling distribution of statistics other than $\bar{x}$ can be constructed in a similar manner. For example, if we had been interested in the maximum (rather than the mean) number of children for the individuals in the samples of size 2 in the day care board setting of this example, we could proceed in a similar manner. We would consider each possible sample of size 2, as in Table 8.1, but rather than computing the mean for each sample, we would determine the maximum. This would result in the sample maximums given in Table 8.2.

Table 8.2   ■   Values of the Maximum for All Possible Samples of Size 2 in Example 8.2

Sample	Maximum	Sample	Maximum
1 Jay, Carol	2	15 Allison, Lygia	2
2 Jay, Allison	2	16 Allison, Bob	2
3 Jay, Teresa	2	17 Allison, Roxy	0
4 Jay, Lygia	2	18 Allison, Kyle	3
5 Jay, Bob	2	19 Teresa, Lygia	2
6 Jay, Roxy	2	20 Teresa, Bob	2
7 Jay, Kyle	3	21 Teresa, Roxy	0
8 Carol, Allison	2	22 Teresa, Kyle	3
9 Carol, Teresa	2	23 Lygia, Bob	2
10 Carol, Lygia	2	24 Lygia, Roxy	2
11 Carol, Bob	2	25 Lygia, Kyle	3
12 Carol, Roxy	2	26 Bob, Roxy	2
13 Carol, Kyle	3	27 Bob, Kyle	3
14 Allison, Teresa	0	28 Roxy, Kyle	3

Then the sampling distribution of the sample maximum is

Max	0	2	3
$p(\text{Max})$	3/28	17/28	7/28

The sampling distribution of a statistic provides important information about the statistic's behavior. In some cases, finding the sampling distribution by considering all possible samples is not feasible, but the sampling distribution can be approximated using simulation, as illustrated in Example 8.3.

■ **Example 8.3**   Approximating a Sampling Distribution Using Simulation

Consider a small population consisting of the 20 students enrolled in an upper division class. The students are numbered from 1 to 20, and the amount of money (in dollars) each student spent on textbooks for the current semester is shown in the following table:

Student	Amount Spent on Books	Student	Amount Spent on Books
1	267	11	319
2	258	12	263
3	342	13	265
4	261	14	262
5	275	15	333
6	295	16	184
7	222	17	231
8	270	18	159
9	278	19	230
10	168	20	323

For this population,

$$\mu = \frac{(267 + 258 + \cdots + 323)}{20} = 260.25$$

Suppose that the value of the population mean is unknown to us and that we want to estimate μ by taking a random sample of five students and computing the sample mean amount spent on textbooks, $\bar{x}$. Is this a reasonable thing to do? Will the estimate that results be close to the true population mean? To answer these questions, we can perform a simple experiment that allows us to look at the behavior of the statistic $\bar{x}$ when random samples of size 5 are taken from the population. (Note that this is not a realistic scenario. If a population consisted of only 20 individuals, we would probably conduct a census rather than select a sample. However, this small population size is easier to work with as we develop the ideas of sampling distributions.)

Let's select a random sample of size 5 from this population. This can be done by writing the numbers from 1 to 20 on identical slips of paper, mixing them well, and then selecting 5 slips without replacement. The numbers on the slips selected identify which of the 20 students will be included in our sample. Alternatively, either a table of random digits or a random number generator can be used to determine which 5 students should be selected. We used MINITAB to obtain 5 numbers between 1 and 20, resulting in 17, 20, 7, 11, and 9, and the following sample of amounts spent on books:

231   323   222   319   278

For this sample,

$$\bar{x} = \frac{1373}{5} = 274.60$$

The sample mean is larger than the population mean of 260.25 by about $15. Is this difference typical, or is this sample mean unusually far away from μ? Taking additional samples could provide some insight.

Four more random samples (Samples 2–5) from this same population are shown here.

Sample 2		Sample 3		Sample 4		Sample 5	
Student	x	Student	x	Student	x	Student	x
4	261	15	333	20	323	18	159
15	333	12	263	16	184	8	270
12	263	3	342	19	230	9	278
1	267	7	222	1	267	7	222
18	159	18	159	8	270	14	262
$\bar{x}$	256.60		263.80		254.80		238.20

Because $\mu = 260.25$, we can see the following:

1. The value of $\bar{x}$ differs from one random sample to another (sampling variability).

2. Some samples produced $\bar{x}$ values larger than μ (Samples 1 and 3), whereas others produced $\bar{x}$ values smaller than μ (Samples 2, 4, and 5).

3. Samples 2, 3, and 4 produced $\bar{x}$ values that were fairly close to the population mean, but Sample 5 resulted in an $\bar{x}$ value that was $22 below the population mean.

Continuing with the experiment, we selected 45 additional random samples (each of size 5). The resulting sample means are as follows:

Sample	$\bar{x}$	Sample	$\bar{x}$	Sample	$\bar{x}$
1	274.6	16	255.0	31	253.4
2	256.6	17	307.2	32	279.6
3	263.8	18	280.0	33	252.6
4	254.8	19	277.4	34	242.2
5	238.2	20	241.2	35	262.6
6	275.6	21	216.0	36	215.4
7	279.2	22	270.0	37	266.2
8	241.6	23	301.0	38	261.4
9	255.4	24	247.0	39	275.0
10	263.8	25	273.8	40	301.4
11	239.6	26	282.8	41	237.0
12	248.2	27	220.4	42	287.4
13	330.8	28	213.6	43	249.2
14	288.8	29	287.6	44	236.8
15	252.8	30	214.8	45	264.6

Figure 8.4, a MINITAB density histogram of the 45 sample means, provides us with a lot of information about the behavior of the statistic $\bar{x}$. Most samples resulted in sample means that were reasonably near $\mu = 260.25$, falling between 235 and 295. A few samples, however, produced $\bar{x}$ values that were far from μ. If we

were to take a sample of size 5 from this population and use $\bar{x}$ as an estimate of the population mean μ, we should not expect $\bar{x}$ to be close to μ.

Figure 8.4 $\bar{x}$ values from 45 random samples for Example 8.3.

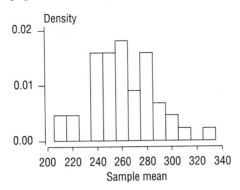

If we had continued sampling indefinitely, the resulting density histogram of $\bar{x}$ values would represent the sampling distribution of $\bar{x}$ for samples of size 5 from this population. The information provided by this sampling distribution enables us to evaluate the behavior of the statistic $\bar{x}$.

The sampling distribution of $\bar{x}$ provides important information about the behavior of the statistic $\bar{x}$ and how it relates to μ. However, you may have noticed that obtaining the sampling distribution of $\bar{x}$, even for a small population and a small sample size, was a lot of work. For realistic examples with larger population and sample sizes, the situation becomes worse because there are so many possible samples. Fortunately, as we look at more examples, patterns emerge that enable us to describe some important aspects of the sampling distribution of the statistic $\bar{x}$ without actually having to look at all possible samples. These general results are described in Section 8.2.

■ Exercises 8.1–8.13

8.1 Explain the difference between a population characteristic and a statistic.

8.2 What is the difference between $\bar{x}$ and μ? between s and σ?

8.3 For each of the following statements, identify the number that appears in boldface type as the value of either a population characteristic or a statistic:

a. A department store reports that **84**% of all customers who use the store's credit plan pay their bills on time.

b. A sample of 100 students at a large university had a mean age of **24.1** years.

c. The Department of Motor Vehicles reports that **22**% of all vehicles registered in a particular state are imports.

d. A hospital reports that based on the 10 most recent cases, the mean length of stay for surgical patients is **6.4** days.

e. A consumer group, after testing 100 batteries of a certain brand, reported an average life of **63** hr of use.

8.4 Consider the following population: $\{1, 2, 3, 4\}$. Note that the population mean is

$$\mu = \frac{1 + 2 + 3 + 4}{4} = 2.5$$

a. Suppose that a random sample of size 2 is to be selected without replacement from this population. There are 12 possible samples (provided that the order in which observations are selected is taken into account):

| 1, 2 | 1, 3 | 1, 4 | 2, 1 | 2, 3 | 2, 4 |
| 3, 1 | 3, 2 | 3, 4 | 4, 1 | 4, 2 | 4, 3 |

Compute the sample mean for each of the 12 possible samples. Use this information to construct the sampling distribution of $\bar{x}$. (Display the sampling distribution in a table.)

b. Suppose that a random sample of size 2 is to be selected, but this time sampling will be done with replacement. Using a method similar to that of Part (a), construct the sampling distribution of $\bar{x}$. (Hint: There are 16 different possible samples in this case.)
c. In what ways are the two sampling distributions of Parts (a) and (b) similar? In what ways are they different?

8.5 Simulate sampling from the population of Exercise 8.4 by using four slips of paper individually marked 1, 2, 3, and 4. Select a sample of size 2 without replacement, and compute $\bar{x}$. Repeat this process 50 times, and construct a relative frequency distribution of the 50 $\bar{x}$ values. How does the relative frequency distribution compare to the sampling distribution of $\bar{x}$ derived in Exercise 8.4, Part (a)?

8.6 Use the method of Exercise 8.4 to find the sampling distribution of the sample range when a random sample of size 2 is to be selected with replacement from the population {1, 2, 3, 4}. (Recall that sample range = largest observation − smallest observation.)

8.7 On four different occasions, you have borrowed money from a friend in the amounts of $1, $5, $10, and $20. The friend has four IOU slips with each of these amounts written on a different slip. He will randomly select two slips from among the four and ask that those IOUs be repaid immediately. Let t denote the amount that you must repay immediately.
a. List all possible outcomes and the value of t for each, and then determine the sampling distribution of t.
b. Calculate μ_t, the mean value of t. How is μ_t related to the population mean μ?

8.8 Refer to Exercise 8.7. Suppose that instead of asking that the total amount on the two selected slips be repaid immediately, the maximum m of the two selected amounts is to be repaid immediately (e.g., $m = 10$ if the $5 and $10 slips are drawn). Obtain the sampling distribution of m if the two slips are selected
a. Without replacement
b. With replacement

8.9 Consider the following population: {2, 3, 3, 4, 4}. The value of μ is 3.2, but suppose that this is not known to an investigator, who therefore wants to estimate μ from sample data. Three possible statistics for estimating μ are

  Statistic 1: the sample mean, $\bar{x}$
  Statistic 2: the sample median
  Statistic 3: the average of the largest and the
      smallest values in the sample

A random sample of size 3 will be selected without replacement. Provided that we disregard the order in which the observations are selected, there are

10 possible samples that might result (writing 3 and 3* , 4 and 4* to distinguish the two 3's and the two 4's in the population):

2, 3, 3*	2, 3, 4	2, 3, 4*	2, 3*, 4	2, 3*, 4*
2, 4, 4*	3, 3*, 4	3, 3*, 4*	3, 4, 4*	3*, 4, 4*

For each of these 10 samples, compute Statistics 1, 2, and 3. Construct the sampling distribution of each of these statistics. Which statistic would you recommend for estimating μ? Explain the reasons for your choice.

8.10 On your shelf, you have four books that you are planning to read in the near future. Two are works of fiction, containing 212 and 379 pages, respectively, and the other two are nonfiction, with 350 and 575 pages, respectively.
a. Suppose that you plan to randomly select two books from among these four to take on a one-week ski trip (in case you are injured). Let $\bar{x}$ denote the sample average number of pages for the two books selected. Obtain the sampling distribution of $\bar{x}$ for this sampling method, and determine the mean of the $\bar{x}$ distribution.
b. Suppose that you plan to randomly select one of the two fiction books and also randomly select one of the two nonfiction books. Determine the sampling distribution of $\bar{x}$, and then calculate the mean of the $\bar{x}$ distribution.

8.11 Consider a population consisting of the following five values, which represent the number of video rentals during the academic year for each of five housemates:

  8   14   16   10   11

a. Compute the mean of this population.
b. Select a random sample of size 2 by writing the numbers on slips of paper, mixing them, and then selecting two. Compute the mean of your sample.
c. Repeatedly select samples of size 2, and compute the associated $\bar{x}$ values until you have the results of 25 samples.
d. Construct a density histogram using the 25 $\bar{x}$ values. Are most of the $\bar{x}$ values near the population mean? Do the $\bar{x}$ values differ a lot from sample to sample, or do they tend to be similar?

8.12 Select 10 additional random samples of size 5 from the population of 20 students given in Example 8.3, and compute the mean amount spent on books for each of the 10 samples. Are the $\bar{x}$ values consistent with the results of the sampling experiment summarized in Figure 8.4?

8.13 Suppose that the sampling experiment described in Example 8.3 had used samples of size 10 rather than size 5. If 45 samples of size 10 were selected, $\bar{x}$ computed, and a density histogram constructed, how do you think this histogram would differ from the density histogram constructed for samples of size 5 (Figure 8.4)? In what way would it be similar?

▪ 8.2 The Sampling Distribution of a Sample Mean

When the objective of a statistical investigation is to make an inference about the population mean μ, it is natural to consider the sample mean $\bar{x}$. To understand how inferential procedures based on $\bar{x}$ work, we must first study how sampling variability causes $\bar{x}$ to vary in value from one sample to another. The behavior of $\bar{x}$ is described by its sampling distribution. The sample size n and characteristics of the population — its shape, mean value μ, and standard deviation σ — are important in determining properties of the sampling distribution of $\bar{x}$.

It is helpful first to consider the results of some sampling experiments. In Examples 8.4 and 8.5, we start with a specified x population distribution, fix a sample size n, and select 500 different random samples of this size. We then compute $\bar{x}$ for each sample and construct a sample histogram of these 500 $\bar{x}$ values. Because 500 is reasonably large (a reasonably long sequence of samples), the histogram of the $\bar{x}$ values should closely resemble the true sampling distribution of $\bar{x}$ (which would be obtained from an unending sequence of $\bar{x}$ values). We repeat the experiment for several different values of n to see how the choice of sample size affects the sampling distribution. Careful examination of these histograms will aid in understanding the general properties to be stated shortly.

▪ Example 8.4   Blood Platelet Size

The article "Changes in Platelet Size and Count in Unstable Angina Compared to Stable Angina or Non-Cardiac Chest Pain" (*European Heart Journal* [1998]: 80–84) presented data suggesting that the distribution of platelet size for patients with noncardiac chest pain is approximately normal with mean $\mu = 8.25$ and standard deviation $\sigma = 0.75$. Figure 8.5 shows the corresponding normal curve centered at 8.25, the mean value of platelet size. The value of the population standard deviation, 0.75, determines the extent to which the x distribution spreads out about its mean value.

Figure 8.5 Normal distribution of platelet size x with $\mu = 8.25$ and $\sigma = 0.75$ for Example 8.4.

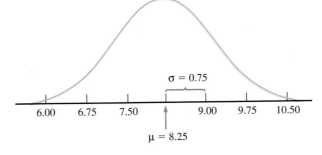

We first used MINITAB to select 500 random samples from this normal distribution, with each sample consisting of $n = 5$ observations. A histogram of the resulting 500 $\bar{x}$ values appears in Figure 8.6(a). This procedure was repeated for

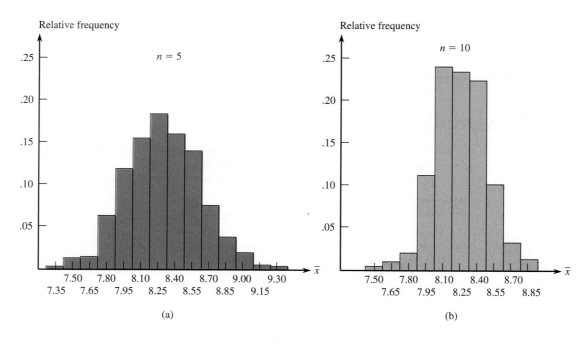

(a)

(b)

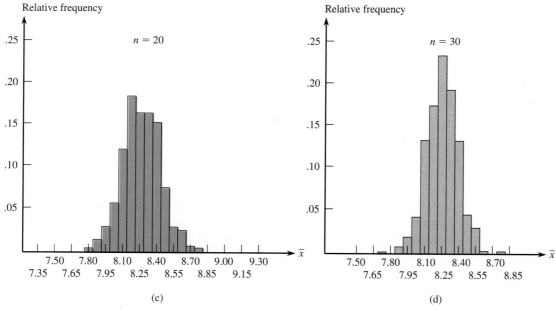

(c)

(d)

Figure 8.6 Histograms for $\bar{x}$ based on 500 samples, each
consisting of n observations, for Example 8.4: (a) $n = 5$;
(b) $n = 10$; (c) $n = 20$; (d) $n = 30$.

samples of size $n = 10$, $n = 20$, and $n = 30$. The resulting sample histograms of
$\bar{x}$ values are displayed in Figures 8.6(b)–(d).

    The first thing to notice about the histograms is their shape. To a reasonable approximation, each of the four histograms looks like a normal curve. The resemblance would be even more striking if each histogram had been based on many more than 500 $\bar{x}$ values. Second, each histogram is centered approximately at 8.25, the mean of the population being sampled. Had the histograms been based on an unending sequence of $\bar{x}$ values, their centers would have been exactly the population mean, 8.25.

The final aspect of the histograms to note is their spread relative to one another. The smaller the value of n, the greater the extent to which the sampling distribution spreads out about the population mean value. This is why the histograms for $n = 20$ and $n = 30$ are based on narrower class intervals than those for the two smaller sample sizes. For the larger sample sizes, most of the $\bar{x}$ values are quite close to 8.25. This is the effect of averaging. When n is small, a single unusual x value can result in an $\bar{x}$ value far from the center. With a larger sample size, any unusual x values, when averaged with the other sample values, still tend to yield an $\bar{x}$ value close to μ. Combining these insights yields a result that should appeal to your intuition: $\bar{x}$ *based on a large n will tend to be closer to μ than will $\bar{x}$ based on a small n.*

■ **Example 8.5** **Length of Overtime Period in Hockey**

Now consider properties of the $\bar{x}$ distribution when the population is quite skewed (and thus very unlike a normal distribution). The Winter 1995 issue of *Chance* magazine gave data on the length of the overtime period for all 251 National Hockey League play-off games between 1970 and 1993 that went into overtime. In hockey, the overtime period ends as soon as one of the teams scores a goal. Figure 8.7 displays a histogram of the data. The histogram has a long upper tail, indicating that although most overtime periods lasted less than 20 min, a few games had long overtime periods.

Figure 8.7 The population distribution for Example 8.5 ($\mu = 9.841$).

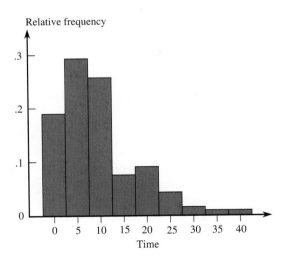

If we think of the 251 values as a population, the histogram in Figure 8.7 shows the distribution of values in that population. The skewed shape makes identification of the mean value from the picture more difficult than for a normal distribution. We found the average of these 251 values to be $\mu = 9.841$, so that is the balance point for the population histogram. The median value for the population, 8.000, is less than μ as a result of the distribution's positively skewed nature.

For each of the sample sizes $n = 5$, 10, 20, and 30, we selected 500 random samples of size n. This was done with replacement to approximate more nearly the usual situation in which the sample size n is only a small fraction of the population

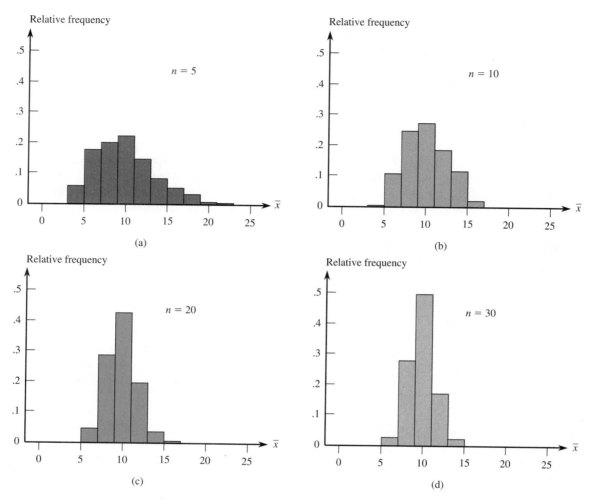

Figure 8.8 Four histograms of 500 $\bar{x}$ values for Example 8.5.

size. We then constructed a histogram of the 500 $\bar{x}$ values for each of the four sample sizes. These histograms are displayed in Figure 8.8.

As with samples from a normal population, the averages of the 500 $\bar{x}$ values for the four different sample sizes are all close to the population mean $\mu = 9.841$. If each histogram had been based on an unending sequence of sample means rather than just 500 of them, each one would have been centered at exactly 9.841. Comparison of the four $\bar{x}$ histograms in Figure 8.8 also shows that as n increases, the histogram's spread about its center decreases. This was also true of increasing sample sizes from a normal population: $\bar{x}$ is less variable for a large sample size than it is for a small sample size.

One aspect of these histograms distinguishes them from the distribution of $\bar{x}$ based on a sample from a normal population. They are skewed and differ in shape more, but they become progressively more symmetric as the sample size increases. The four histograms in Figure 8.8 were drawn using the same horizontal scale, to make it easy to see the decrease in spread as n increases. However, this makes it harder to see how the histograms' shapes change with sample size. Figure 8.9 is a histogram based on narrower class intervals for the $\bar{x}$ values from samples of size 30. This figure shows that for $n = 30$, the histogram has a shape much like a normal curve. Again this is the effect of averaging. Even when n is large, one of the few large x values in the population appears infrequently in the sample. When one does appear, its contribution to $\bar{x}$ is swamped by the contributions of more typical sample values. The normal-curve shape of the histogram for $n = 30$ is what is predicted by the Central Limit Theorem, which will be introduced shortly. According to this theorem, even if the population distribution bears no resemblance whatsoever to a normal curve, the $\bar{x}$ sampling distribution is approximately normal when the sample size n is reasonably large.

Figure 8.9 Histogram of 500 $\bar{x}$ values for Example 8.5 when $n = 30$.

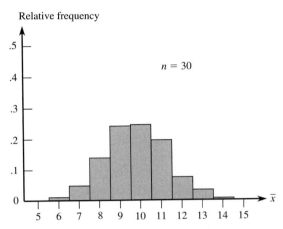

▪ General Properties of the Sampling Distribution of $\bar{x}$

Examples 8.4 and 8.5 suggest that for any n, the center of the $\bar{x}$ distribution (the mean value of $\bar{x}$) coincides with the mean of the population being sampled and that the spread of the $\bar{x}$ distribution decreases as n increases, indicating that the standard deviation of $\bar{x}$ is smaller for large n than for small n. The sample histograms also suggest that in some cases, the $\bar{x}$ distribution is approximately normal in shape. These observations are stated more formally in the following general rules.

▪ General Properties of the Sampling Distribution of $\bar{x}$

Let $\bar{x}$ denote the mean of the observations in a random sample of size n from a population having mean μ and standard deviation σ. Denote the mean value of the $\bar{x}$ distribution by $\mu_{\bar{x}}$ and the standard deviation of the $\bar{x}$ distribution by $\sigma_{\bar{x}}$. Then the following rules hold:

Rule 1: $\mu_{\bar{x}} = \mu$.

Rule 2: $\sigma_{\bar{x}} = \dfrac{\sigma}{\sqrt{n}}$. This rule is exact if the population is infinite and is approximately correct when the population is finite if no more than 10% of the population is included in the sample.

Rule 3: When the population distribution is normal, the sampling distribution of $\bar{x}$ is also normal for any sample size n.

Rule 4 (Central Limit Theorem): When n is sufficiently large, the sampling distribution of $\bar{x}$ is well approximated by a normal curve, even when the population distribution is not itself normal.

Rule 1, $\mu_{\bar{x}} = \mu$, states that the sampling distribution of $\bar{x}$ is always centered at the mean of the population sampled. Rule 2, $\sigma_{\bar{x}} = \dfrac{\sigma}{\sqrt{n}}$, not only states that the spread of the sampling distribution of $\bar{x}$ decreases as n increases but also gives a precise relationship between the standard deviation of the $\bar{x}$ distribution and the population standard deviation. When $n = 4$, for example,

$$\sigma_{\bar{x}} = \frac{\sigma}{\sqrt{n}} = \frac{\sigma}{\sqrt{4}} = \frac{\sigma}{2}$$

so the $\bar{x}$ distribution has a standard deviation only half as large as the population standard deviation. Rules 3 and 4 specify circumstances under which the $\bar{x}$ distribution is normal (when the population is normal) or approximately normal (when the sample size is large). Figure 8.10 illustrates these rules by showing several $\bar{x}$ distributions superimposed over a graph of the population distribution.

Figure 8.10 Population distribution and sampling distributions of $\bar{x}$: (a) symmetric population; (b) skewed population.

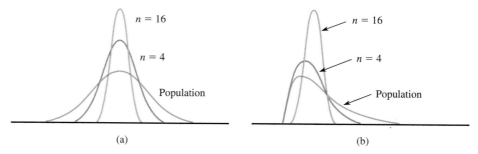

The Central Limit Theorem of Rule 4 states that when n is sufficiently large, the $\bar{x}$ distribution is approximately normal, no matter what the population distribution looks like. This result has enabled statisticians to develop procedures for making inferences about a population mean μ using a large sample, even when the shape of the population distribution is unknown.

Recall that a variable is standardized by subtracting the mean value and then dividing by its standard deviation. Using Rules 1 and 2 to standardize $\bar{x}$ gives an important consequence of the last two rules.

If n is large or if the population distribution is normal, then the standardized variable

$$z = \frac{\bar{x} - \mu_{\bar{x}}}{\sigma_{\bar{x}}} = \frac{\bar{x} - \mu}{\dfrac{\sigma}{\sqrt{n}}}$$

has (at least approximately) a standard normal (z) distribution.

Application of the Central Limit Theorem in specific situations requires a rule of thumb for deciding whether n is indeed sufficiently large. Such a rule is not as easy to come by as you might think. Look back at Figure 8.8, which shows the approximate sampling distribution of $\bar{x}$ for $n = 5, 10, 20$, and 30 when the population distribution is quite skewed. Certainly the histogram for $n = 5$ is not well described by a normal curve, and this is still true of the histogram for $n = 10$, particularly in the tails of the histogram (far away from the mean value). Among the four histograms, only the histogram for $n = 30$ has a reasonably normal shape.

On the other hand, when the population distribution is normal, the sampling distribution of $\bar{x}$ is normal for any n. If the population distribution is somewhat skewed but not to the extent of Figure 8.7, we might expect the $\bar{x}$ sampling distribution to be a bit skewed for $n = 5$ but quite well fit by a normal curve for n as small as 10 or 15. How large an n is needed for the $\bar{x}$ distribution to approximate a normal curve depends on how much the population distribution differs from a normal distribution. The closer the population distribution is to being normal, the smaller the value of n necessary for the Central Limit Theorem approximation to be accurate.

Many statisticians recommend the following conservative rule:

The Central Limit Theorem can safely be applied if n exceeds 30.

If the population distribution is believed to be reasonably close to a normal distribution, an n of 15 or 20 is often large enough for $\bar{x}$ to have approximately a normal distribution. At the other extreme, we can imagine a distribution with a much longer tail than that of Figure 8.7, in which case even $n = 40$ or 50 would not suffice for approximate normality of $\bar{x}$. In practice, however, few population distributions are likely to be this badly behaved.

■ **Example 8.6   Bluethroat Song Duration**

Male bluethroats have a complex song, which is thought to be used to attract female birds. Let x denote the duration of a randomly selected song (in seconds) for a male bluethroat. Suppose that the mean value of song duration is $\mu = 13.8$ sec and that the standard deviation of song duration is $\sigma = 11.8$ sec. (These values are suggested in the paper "Response of Male Bluethroats *Luscinia svecica* to Song Playback," *Ornis Fennica* [2000]: 43–47.) The authors of the paper noted that the song length distribution is not normal (which could also be deduced from the relative values of the mean and standard deviation and the fact that song length cannot be negative). The sampling distribution of $\bar{x}$ based on a random sample of $n = 25$ song durations then also has mean value

$$\mu_{\bar{x}} = 13.8 \text{ sec}$$

That is, the sampling distribution of $\bar{x}$ is centered at 13.8. The standard deviation of $\bar{x}$ is

$$\sigma_{\bar{x}} = \frac{\sigma}{\sqrt{n}} = \frac{11.8}{\sqrt{25}} = 2.36$$

which is only one-fifth as large as the population standard deviation σ. Because the population distribution is not normal and because the sample size is not larger than 30, we cannot assume that the sampling distribution of $\bar{x}$ is normal in shape.

■ **Example 8.7   Soda Volumes**

A soft-drink bottler claims that, on average, cans contain 12 oz of soda. Let x denote the actual volume of soda in a randomly selected can. Suppose that x is normally distributed with $\sigma = 0.16$ oz. Sixteen cans are to be selected, and the soda volume will be determined for each one; let $\bar{x}$ denote the resulting sample mean soda volume. Because the x distribution is normal, the sampling distribution of $\bar{x}$ is also normal. *If the bottler's claim is correct*, the sampling distribution of $\bar{x}$ has a mean value of $\mu_{\bar{x}} = \mu = 12$ and a standard deviation of

$$\sigma_{\bar{x}} = \frac{\sigma}{\sqrt{n}} = \frac{0.16}{\sqrt{16}} = 0.04$$

To calculate a probability involving $\bar{x}$, we simply standardize by subtracting the mean value, 12, and dividing by the standard deviation (of $\bar{x}$), which is 0.04. For example, the probability that the sample mean soda volume is between 11.96 oz and 12.08 oz is the area between 11.96 and 12.08 under the normal density curve with mean 12 and standard deviation 0.04, as shown in the following figure:

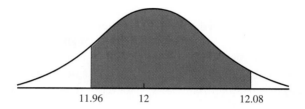

This area is calculated by first standardizing the interval limits:

Lower limit: $a^* = \dfrac{11.96 - 12}{0.04} = -1.0$

Upper limit: $b^* = \dfrac{12.08 - 12}{0.04} = 2.0$

Then

$$
\begin{aligned}
P(11.96 \leq \bar{x} \leq 12.08) &= \text{area under the } z \text{ curve between } -1.0 \text{ and } 2.0 \\
&= (\text{area to the left of } 2.0) - (\text{area to the left of } -1.0) \\
&= .9772 - .1587 \\
&= .8185
\end{aligned}
$$

The probability that the sample mean soda volume is at most 11.9 oz is

$$
\begin{aligned}
P(\bar{x} \leq 11.9) &= P\left(z \leq \frac{11.9 - 12}{0.04} = -2.5 \right) \\
&= (\text{area under the } z \text{ curve to the left of } -2.5) \\
&= .0062
\end{aligned}
$$

If the x distribution is as described and the claim is correct, a sample mean soda volume based on 16 observations will be less than 11.9 oz for less than 1% of all such samples. Thus, observation of an $\bar{x}$ value that is smaller than 11.9 oz would cast doubt on the bottler's claim that the average soda volume is 12 oz.

▪ **Example 8.8   Fat Content of Hot Dogs**

A hot dog manufacturer asserts that one of its brands of hot dogs has an average fat content of $\mu = 18$ g per hot dog. Consumers of this brand would probably not be disturbed if the mean was less than 18 but would be unhappy if it exceeded 18. Let x denote the fat content of a randomly selected hot dog, and suppose that σ, the standard deviation of the x distribution, is 1.

An independent testing organization is asked to analyze a random sample of 36 hot dogs. Let $\bar{x}$ be the average fat content for this sample. The sample size, $n = 36$, is large enough to rely on the Central Limit Theorem and to regard the $\bar{x}$ distribution as approximately normal. The standard deviation of the $\bar{x}$ distribution is

$$\sigma_{\bar{x}} = \frac{\sigma}{\sqrt{n}} = \frac{1}{\sqrt{36}} = 0.1667$$

If the manufacturer's claim is correct, we know that $\mu_{\bar{x}} = \mu = 18$ g. Suppose that the sample resulted in a mean of $\bar{x} = 18.4$ g. Does this result indicate that the manufacturer's claim is incorrect?

We can answer this question by looking at the sampling distribution of $\bar{x}$. Because of sampling variability, even if $\mu = 18$, we know that $\bar{x}$ typically deviates somewhat from this value. Is it likely that we would see a sample mean at least as large as 18.4 when the population mean is really 18? If the company's claim is correct,

$$P(\bar{x} \geq 18.4) \approx P\left(z \geq \frac{18.4 - 18}{0.1667} \right)$$

$$= P(z \geq 2.4)$$

$$= \text{area under the } z \text{ curve to the right of 2.4}$$

$$= .0082$$

The value $\bar{x} = 18.4$ exceeds 18 by enough to cast substantial doubt on the manufacturer's claim. Values of $\bar{x}$ at least as large as 18.4 will be observed only approximately 0.82% of the time when a random sample of size 36 is taken from a population with mean 18 and standard deviation 1.

▪ Other Cases

We now know a great deal about the sampling distribution of $\bar{x}$ in two cases: for a normal population distribution and for a large sample size. What happens when the population distribution is not normal and n is small? Although it is still true that $\mu_{\bar{x}} = \mu$ and $\sigma_{\bar{x}} = \frac{\sigma}{\sqrt{n}}$, unfortunately there is no general result about the shape of the distribution. When the objective is to make an inference about the center of such a population, one way to proceed is to replace the normality assumption with some other assumption about the shape of the distribution. Statisticians have proposed and studied a number of such models. Theoretical methods or simulation can be used to describe the $\bar{x}$ distribution corresponding to the assumed model. An alternative strategy is to use one of the transformations presented in Chapter 7 to create a data set that more closely resembles a sample from a normal population and then to base inferences on the transformed data. Yet another path is to use an inferential procedure based on a statistic other than $\bar{x}$.

▪ Exercises 8.14–8.26

8.14 A random sample is selected from a population with mean $\mu = 100$ and standard deviation $\sigma = 10$. Determine the mean and standard deviation of the $\bar{x}$ sampling distribution for each of the following sample sizes:

a. $n = 9$ **b.** $n = 15$ **c.** $n = 36$
d. $n = 50$ **e.** $n = 100$ **f.** $n = 400$

8.15 For which of the sample sizes given in Exercise 8.14 would it be reasonable to think that the $\bar{x}$ sampling distribution is approximately normal in shape?

8.16 Explain the difference between σ and $\sigma_{\bar{x}}$ and between μ and $\mu_{\bar{x}}$.

8.17 Suppose that a random sample of size 64 is to be selected from a population with mean 40 and standard deviation 5.
a. What are the mean and standard deviation of the $\bar{x}$ sampling distribution? Describe the shape of the $\bar{x}$ sampling distribution.
b. What is the approximate probability that $\bar{x}$ will be within 0.5 of the population mean μ?
c. What is the approximate probability that $\bar{x}$ will differ from μ by more than 0.7?

8.18 The time that a randomly selected individual waits for an elevator in an office building has a uniform distribution over the interval from 0 to 1 min. It can be shown that for this distribution $\mu = 0.5$ and $\sigma = 0.289$.
a. Let $\bar{x}$ be the sample average waiting time for a random sample of 16 individuals. What are the mean value and standard deviation of the sampling distribution of $\bar{x}$?
b. Answer Part (a) for a random sample of 50 individuals. In this case, sketch a picture of a good approximation to the actual $\bar{x}$ distribution.

8.19 Let x denote the time (in minutes) that it takes a fifth-grade student to read a certain passage. Suppose that the mean value and standard deviation of x are $\mu = 2$ min and $\sigma = 0.8$ min, respectively.
a. If $\bar{x}$ is the sample average time for a random sample of $n = 9$ students, where is the $\bar{x}$ distribution centered, and how much does it spread out about the center (as described by its standard deviation)?
b. Repeat Part (a) for a sample of size of $n = 20$ and again for a sample of size $n = 100$. How do the centers and spreads of the three $\bar{x}$ distributions compare to one another? Which sample size would be most likely to result in an $\bar{x}$ value close to μ, and why?

8.20 In the library on a university campus, there is a sign in the elevator that indicates a limit of 16 per-

sons. Furthermore, there is a weight limit of 2500 lb. Assume that the average weight of students, faculty, and staff on campus is 150 lb, that the standard deviation is 27 lb, and that the distribution of weights of individuals on campus is approximately normal. If a random sample of 16 persons from the campus is to be taken:
a. What is the expected value of the sample mean of their weights?
b. What is the standard deviation of the sampling distribution of the sample mean weight?
c. What average weights for a sample of 16 people will result in the total weight exceeding the weight limit of 2500 lb?
d. What is the chance that a random sample of 16 persons on the elevator will exceed the weight limit?

8.21 Suppose that the mean value of interpupillary distance (the distance between the pupils of the left and right eyes) for adult males is 65 mm and that the population standard deviation is 5 mm.
a. If the distribution of interpupillary distance is normal and a sample of $n = 25$ adult males is to be selected, what is the probability that the sample average distance $\bar{x}$ for these 25 will be between 64 and 67 mm? at least 68 mm?
b. Suppose that a sample of 100 adult males is to be obtained. Without assuming that interpupillary distance is normally distributed, what is the approximate probability that the sample average distance will be between 64 and 67 mm? at least 68 mm?

8.22 Suppose that a sample of size 100 is to be drawn from a population with standard deviation 10.
a. What is the probability that the sample mean will be within 2 of the value of μ?
b. For this example ($n = 100$, $\sigma = 10$), complete each of the following statements by computing the appropriate value:
i. Approximately 95% of the time, $\bar{x}$ will be within _____ of μ.
ii. Approximately 0.3% of the time, $\bar{x}$ will be farther than _____ from μ.

8.23 A manufacturing process is designed to produce bolts with a 0.5-in. diameter. Once each day, a random sample of 36 bolts is selected and the diameters recorded. If the resulting sample mean is less than 0.49 in. or greater than 0.51 in., the process is shut down for adjustment. The standard deviation for diameter is 0.02 in. What is the probability that the manufacturing line will be shut down unnecessarily? (Hint: Find the probability of observing an $\bar{x}$

in the shutdown range when the true process mean really is 0.5 in.)

8.24 College students with a checking account typically write relatively few checks in any given month, whereas full-time residents typically write many more checks during a month. Suppose that 50% of a bank's accounts are held by students and that 50% are held by full-time residents. Let x denote the number of checks written in a given month by a randomly selected bank customer.

a. Give a sketch of what the probability distribution of x might look like.

b. Suppose that the mean value of x is 22.0 and that the standard deviation is 16.5. If a random sample of $n = 100$ customers is to be selected and $\bar{x}$ denotes the sample average number of checks written during a particular month, where is the sampling distribution of $\bar{x}$ centered, and what is the standard deviation of the $\bar{x}$ distribution? Sketch a rough picture of the sampling distribution.

c. Referring to Part (b), what is the approximate probability that $\bar{x}$ is at most 20? at least 25?

8.25 An airplane with room for 100 passengers has a total baggage limit of 6000 lb. Suppose that the total weight of the baggage checked by an individual passenger is a random variable x with a mean value of 50 lb and a standard deviation of 20 lb. If 100 passengers will board a flight, what is the approximate probability that the total weight of their baggage will exceed the limit? (Hint: With $n = 100$, the total weight exceeds the limit when the average weight $\bar{x}$ exceeds 6000/100.)

8.26 The thickness (in millimeters) of the coating applied to disk drives is a characteristic that determines the usefulness of the product. When no unusual circumstances are present, the thickness (x) has a normal distribution with a mean of 3 mm and a standard deviation of 0.05 mm. Suppose that the process will be monitored by selecting a random sample of 16 drives from each shift's production and determining $\bar{x}$, the mean coating thickness for the sample.

a. Describe the sampling distribution of $\bar{x}$ (for a sample of size 16).

b. When no unusual circumstances are present, we expect $\bar{x}$ to be within $3\sigma_{\bar{x}}$ of 3 mm, the desired value. An $\bar{x}$ value farther from 3 than $3\sigma_{\bar{x}}$ is interpreted as an indication of a problem that needs attention. Compute $3 \pm 3\sigma_{\bar{x}}$. (A plot over time of $\bar{x}$ values with horizontal lines drawn at the limits $\mu \pm 3\sigma_{\bar{x}}$ is called a process control chart.)

c. Referring to Part (b), what is the probability that a sample mean will be outside $3 \pm 3\sigma_{\bar{x}}$ just by chance (i.e., when there are no unusual circumstances)?

d. Suppose that a machine used to apply the coating is out of adjustment, resulting in a mean coating thickness of 3.05 mm. What is the probability that a problem will be detected when the next sample is taken? (Hint: This will occur if $\bar{x} > 3 + 3\sigma_{\bar{x}}$ or $\bar{x} < 3 - 3\sigma_{\bar{x}}$ when $\mu = 3.05$.)

▪ 8.3 The Sampling Distribution of a Sample Proportion

The objective of many statistical investigations is to draw a conclusion about the proportion of individuals or objects in a population that possess a specified property — for example, Maytag washers that don't require service during the warranty period, Europeans who favor the deployment of a certain type of missile, or coffee drinkers who regularly drink decaffeinated coffee. Traditionally, any individual or object that possesses the property of interest is labeled a success (S), and one that does not possess the property is termed a failure (F). The letter π denotes the proportion of successes in the population. The value of π is a number between 0 and 1, and 100π is the percentage of successes in the population. If $\pi = .75$, 75% of the population members are successes, and if $\pi = .01$, the population contains only 1% successes and 99% failures.

The value of π is usually unknown to an investigator. When a random sample of size n is selected from this type of population, some of the individuals in the sample are successes, and the remaining individuals in the sample are failures. The statistic that provides a basis for making inferences about π is p, the **sample proportion of successes**:

$$p = \frac{\text{number of successes in the sample}}{n}$$

For example, if $n = 5$ and three successes result, then $p = 3/5 = .6$.

Just as making inferences about μ requires knowing something about the sampling distribution of the statistic $\bar{x}$, making inferences about π requires first learning about properties of the sampling distribution of the statistic p. For example, when $n = 5$, the six possible values of p are 0, .2 (from 1/5), .4, .6, .8, and 1. The sampling distribution of p gives the probability of each of these six possible values, the long-run proportion of the time that each value would occur if samples with $n = 5$ were selected over and over again. Similarly, when $n = 100$, the 101 possible values of p are 0, .01, .02, ... , .98, .99, 1.

As we did for the distribution of the sample mean, we will look at some simulation experiments to develop an intuitive understanding of the distribution of the sample proportion before stating general rules. In each example, 500 random samples (each of size n) are selected from a population having a specified value of π. We compute p for each sample and then construct a sample histogram of the 500 values.

■ Example 8.9   Gender of College Students

In the fall of 1999, there were 16,250 students enrolled at California Polytechnic State University, San Luis Obispo. Of these students, 6825 (42%) were female. To illustrate properties of the sampling distribution of a sample proportion, we decided to simulate sampling from Cal Poly's student population. With S denoting a female student and F a male student, the proportion of S's in the population is $\pi = .42$. A statistical software package was used to select 500 samples of size $n = 5$, then 500 samples of size $n = 10$, then 500 samples with $n = 25$, and finally 500 samples with $n = 50$. The sample histograms of the 500 values of p for the four sample sizes are displayed in Figure 8.11.

The most noticeable feature of the histogram shapes is the progression toward the shape of a normal curve as n increases. The histogram for $n = 10$ is more bell shaped than the histogram for samples of size 5, although it is still slightly skewed. The histograms for $n = 25$ and $n = 50$ look much like normal curves.

Although the skewness of the first two histograms makes it a bit difficult to locate their centers, all four histograms appear to be centered at roughly .42, the value of π for the population sampled. Had the histograms been based on an unending sequence of samples, each histogram would have been centered at exactly .42. Finally, as was the case with the sampling distribution of $\bar{x}$, the histograms spread out more for small n than for large n. The value of p based on a large sample size tends to be closer to π, the population proportion of successes, than does p from a small sample.

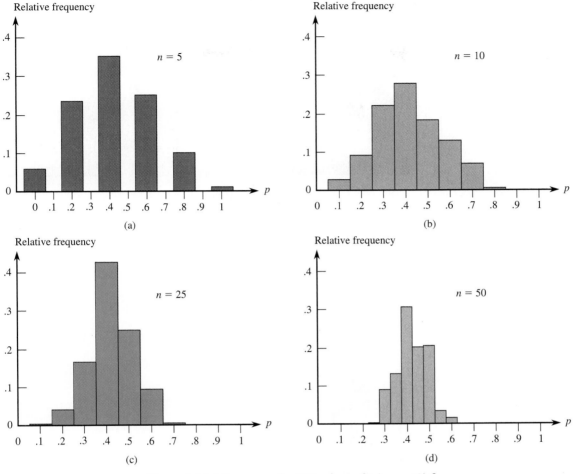

Figure 8.11 Histograms for 500 values of p ($\pi = .42$) for
Example 8.9: (a) $n = 5$; (b) $n = 10$; (c) $n = 25$; (d) $n = 50$.

▪ **Example 8.10**   Contracting Hepatitis from Blood Transfusion

The development of viral hepatitis subsequent to a blood transfusion can cause serious complications for a patient. The article "Lack of Awareness Results in Poor
Autologous Blood Transfusion" (*Health Care Management*, May 15, 2003) reported
that hepatitis occurs in 7% of patients who receive multiple blood transfusions during heart surgery. Here, we simulate sampling from the population of blood recipients, with S denoting a recipient who contracts hepatitis (not the sort of characteristic one usually thinks of as a success, but the S–F labeling is arbitrary), so
$\pi = .07$. Figure 8.12 displays histograms of 500 values of p for the four sample sizes
$n = 10, 25, 50,$ and 100.

As was the case in Example 8.9, all four histograms are centered at approximately the value of π for the population being sampled. (The average values of p
for these simulations are .0690, .0677, .0707, and .0694.) If the histograms had been
based on an unending sequence of samples, they would all have been centered at

Figure 8.12 Histograms of 500 values of p ($\pi = .07$) for Example 8.10: (a) $n = 10$; (b) $n = 25$; (c) $n = 50$; (d) $n = 100$.

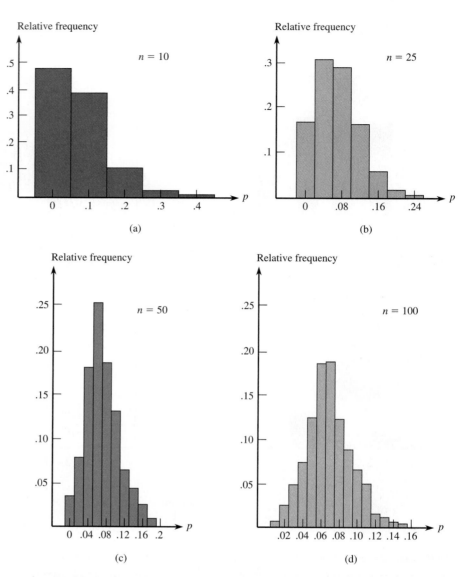

exactly $\pi = .07$. Again, the spread of a histogram based on a large n is smaller than the spread of a histogram resulting from a small sample size. The larger the value of n, the closer the sample proportion p tends to be to the value of the population proportion π.

Furthermore, there is a progression toward the shape of a normal curve as n increases. However, the progression is much slower here than in the previous example, because the value of π is so extreme. (The same thing would happen for $\pi = .93$, except that the histograms would be negatively rather than positively skewed.) The histograms for $n = 10$ and $n = 25$ exhibit substantial skew, and the skew of the histogram for $n = 50$ is still moderate (compare Figure 8.12(c) with Figure 8.11(d)). Only the histogram for $n = 100$ is reasonably well fit by a normal curve. It appears that whether a normal curve provides a good approximation to the sampling distribution of p depends on the values of both n and π. Knowing only that $n = 50$ is not enough to guarantee that the shape of the histogram is approximately normal.

▪ General Properties of the Sampling Distribution of p

Examples 8.9 and 8.10 suggest that the sampling distribution of p depends on both n, the sample size, and π, the proportion of successes in the population. Key results are stated more formally in the following general rules.

▪ General Properties of the Sampling Distribution of p

Let p be the proportion of successes in a random sample of size n from a population whose proportion of successes is π. Denote the mean value of p by μ_p and the standard deviation by σ_p. Then the following rules hold:

Rule 1: $\mu_p = \pi$.

Rule 2: $\sigma_p = \sqrt{\dfrac{\pi(1 - \pi)}{n}}$. This rule is exact if the population is infinite and is approximately correct if the population is finite and no more than 10% of the population is included in the sample.

Rule 3: When n is large and π is not too near 0 or 1, the sampling distribution of p is approximately normal.

Thus, the sampling distribution of p is always centered at the value of the population success proportion π, and the extent to which the distribution spreads out about π decreases as the sample size n increases.

Examples 8.9 and 8.10 indicate that both π and n must be considered in judging whether the sampling distribution of p is approximately normal.

The farther the value of π is from .5, the larger n must be for a normal approximation to the sampling distribution of p to be accurate. A conservative rule of thumb is that if both $n\pi \geq 10$ and $n(1 - \pi) \geq 10$, then it is safe to use a normal approximation.

A sample size of $n = 100$ is not by itself sufficient to justify the use of a normal approximation. If $\pi = .01$, the distribution of p is extremely positively skewed, so a bell-shaped curve does not give a good approximation. Similarly, if $n = 100$ and $\pi = .99$ (so that $n(1 - \pi) = 1 < 10$), the distribution of p has a substantial negative skew. The conditions $n\pi \geq 10$ and $n(1 - \pi) \geq 10$ ensure that the sampling distribution of p is not too skewed. If $\pi = .5$, the normal approximation can be used for n as small as 20, whereas for $\pi = .05$ or .95, n should be at least 200.

▪ Example 8.11   Blood Transfusions Continued

The proportion of all cardiac patients receiving blood transfusions who contract hepatitis was given as .07 in the article referenced in Example 8.10. Suppose that a new blood screening procedure is believed to reduce the incidence rate of hepatitis. Blood screened using this procedure is given to $n = 200$ blood recipients. Only

6 of the 200 patients contract hepatitis. This appears to be a favorable result, because $p = 6/200 = .03$. The question of interest to medical researchers is, Does this result indicate that the true (long-run) proportion of patients who contract hepatitis when the new screening procedure is used is less than .07, or could this result be plausibly attributed to sampling variability (i.e., to the fact that p typically deviates from its mean value, π)? If the screening procedure is ineffective,

$$\mu_p = \pi = 0.07$$

$$\sigma_p = \sqrt{\frac{\pi(1 - \pi)}{200}} = \sqrt{\frac{(.07)(.93)}{200}} = 0.018$$

Furthermore, because

$$n\pi = 200(.07) = 14 \geq 10$$

and

$$n(1 - \pi) = 200(.93) = 186 \geq 10$$

the sampling distribution of p is approximately normal. Then, if the screening procedure is ineffective,

$$P(p \leq .03) = P\left(z \leq \frac{.03 - .07}{.018}\right)$$
$$= P(z \leq -2.22)$$
$$= .0132$$

Thus, it is unlikely that a sample proportion .03 or smaller would be observed if the screening procedure really was ineffective. The new screening procedure appears to yield a smaller incidence rate for hepatitis.

▪ Exercises 8.27–8.35

8.27 A random sample is to be selected from a population that has a proportion of successes $\pi = .65$. Determine the mean and standard deviation of the sampling distribution of p for each of the following sample sizes:

a. $n = 10$ b. $n = 20$ c. $n = 30$
d. $n = 50$ e. $n = 100$ f. $n = 200$

8.28 For which of the sample sizes given in Exercise 8.27 would the sampling distribution of p be approximately normal if $\pi = .65$? if $\pi = .2$?

8.29 The article "Unmarried Couples More Likely to Be Interracial" (*San Luis Obispo Tribune*, March 13, 2002) reported that 7% of married couples in the United States are mixed racially or ethnically. Consider the population consisting of all married couples in the United States.

a. A random sample of $n = 100$ couples will be selected from this population and p, the proportion of couples that are mixed racially or ethnically, will be computed. What are the mean and standard deviation of the sampling distribution of p?
b. Is it reasonable to assume that the sampling distribution of p is approximately normal for random samples of size $n = 100$? Explain.
c. Suppose that the sample size is $n = 200$ rather than $n = 100$, as in Part (b). Does the change in sample size change the mean and standard deviation of the sampling distribution of p? If so, what are the new values for the mean and standard deviation? If not, explain why not.
d. Is it reasonable to assume that the sampling distribution of p is approximately normal for random samples of size $n = 200$? Explain.

e. When $n = 200$, what is the probability that the proportion of couples in the sample who are racially or ethnically mixed will be greater than .10?

8.30 The article referenced in Exercise 8.29 reported that for unmarried couples living together, the proportion that are racially or ethnically mixed is .15. Answer the questions posed in Parts (a)–(e) of Exercise 8.29 for the population of unmarried couples living together.

8.31 A certain chromosome defect occurs in only 1 out of 200 adult Caucasian males. A random sample of $n = 100$ adult Caucasian males is to be obtained.
a. What is the mean value of the sample proportion p, and what is the standard deviation of the sample proportion?
b. Does p have approximately a normal distribution in this case? Explain.
c. What is the smallest value of n for which the sampling distribution of p is approximately normal?

8.32 The article "Should Pregnant Women Move? Linking Risks for Birth Defects with Proximity to Toxic Waste Sites" (*Chance* [1992]: 40–45) reported that in a large study carried out in the state of New York, approximately 30% of the study subjects lived within 1 mi of a hazardous waste site. Let π denote the proportion of all New York residents who live within 1 mi of such a site, and suppose that $\pi = .3$.
a. Would p based on a random sample of only 10 residents have approximately a normal distribution? Explain why or why not.
b. What are the mean value and standard deviation of p based on a random sample of size 400?
c. When $n = 400$, what is $P(.25 \le p \le .35)$?
d. Is the probability calculated in Part (c) larger or smaller than would be the case if $n = 500$? Answer without actually calculating this probability.

8.33 The article "Thrillers" (*Newsweek*, April 22, 1985) stated, "Surveys tell us that more than half of America's college graduates are avid readers of mystery novels." Let π denote the actual proportion of college graduates who are avid readers of mystery novels. Consider p to be based on a random sample of 225 college graduates.
a. If $\pi = .5$, what are the mean value and standard deviation of p? Answer this question for $\pi = .6$. Does p have approximately a normal distribution in both cases? Explain.
b. Calculate $P(p \ge .6)$ for both $\pi = .5$ and $\pi = .6$.
c. Without doing any calculations, how do you think the probabilities in Part (b) would change if n were 400 rather than 225?

8.34 Suppose that a particular candidate for public office is in fact favored by 48% of all registered voters in the district. A polling organization will take a random sample of 500 voters and will use p, the sample proportion, to estimate π. What is the approximate probability that p will be greater than .5, causing the polling organization to incorrectly predict the result of the upcoming election?

8.35 A manufacturer of computer printers purchases plastic ink cartridges from a vendor. When a large shipment is received, a random sample of 200 cartridges is selected, and each cartridge is inspected. If the sample proportion of defective cartridges is more than .02, the entire shipment is returned to the vendor.
a. What is the approximate probability that a shipment will be returned if the true proportion of defective cartridges in the shipment is .05?
b. What is the approximate probability that a shipment will not be returned if the true proportion of defective cartridges in the shipment is .10?

▪ Activity 8.1: Do Students Who Take the SATs Multiple Times Have an Advantage in College Admissions?

Technology activity: Requires use of a computer or a graphing calculator.

Background: The *Chronicle of Higher Education* (January 29, 2003) summarized an article that appeared on the *American Prospect* web site titled "College Try: Why Universities Should Stop Encouraging Applicants to Take the SAT's Over and Over Again." This paper argued that current college admission policies that permit applicants to take the SAT exam multiple times and then use the highest score for admission consideration favor students from families with higher incomes (who can afford to take the exam many times). The author proposed two alternatives that he believes would be fairer than using the highest score: (1) Use the average of all test scores, or (2) use only the most recent score.

In this activity, you will investigate the differences between the three possibilities by looking at the sampling distributions of three statistics for a test taker who takes the exam twice and for a test taker who takes the exam five times. The three statistics are

Max = maximum score
Mean = average score
Recent = most recent score

An individual's score on the SAT exam fluctuates between test administrations. Suppose that a particular student's "true ability" is reflected by an SAT score of 1200 but, because of chance fluctuations, the test score on any particular administration of the exam can be considered a random variable that has a distribution that is approximately normal with mean 1200 and standard deviation 30. If we select a sample from this normal distribution, the resulting set of observations can be viewed as a collection of test scores that might have been obtained by this student.

Part 1: Begin by considering what happens if this student takes the exam twice. You will use simulation to generate samples of two test scores, Score1 and Score2, for this student. Then you will compute the values of Max, Mean, and Recent for each pair of scores. The resulting values of Max, Mean, and Recent will be used to construct approximations to the sampling distributions of the three statistics.

The instructions that follow assume the use of MINITAB. If you are using a different software package or a graphing calculator, your instructor will provide alternative instructions.
a. Obtain 500 sets of 2 test scores by generating observations from a normal distribution with mean 1200 and standard deviation 30.

> MINITAB:　　Calc → Random Data → Normal
> 　　　　Enter 500 in the Generate box (to get 500 sets of scores)
> 　　　　Enter C1–C2 in the Store in Columns box (to get two test scores in each set)
> 　　　　Enter 1200 in the Mean box (because we want scores from a normal distribution with mean 1200)
> 　　　　Enter 30 in the Standard Deviation box (because we want scores from a normal distribution with standard deviation 30)
> 　　　　Click on OK

b. Looking at the MINITAB worksheet, you should now see 500 rows of values in each of the first two columns. The two values in any particular row can be regarded as the test scores that might be observed when the student takes the test twice. For each pair of test scores, we now calculate the values of Max, Mean, and Recent.
i. Recent is just the last test score, so the values in C2 are the values of Recent. Name this column recent2 by typing the name into the gray box at the top of C2.

ii. Compute the maximum test score (Max) for each pair of scores, and store the values in C3, as follows:

> MINITAB:　　Calc → Row statistics
> 　　　　Click the button for maximum
> 　　　　Enter C1-C2 in the Input variables box
> 　　　　Enter C3 in the Store Result In box.
> 　　　　Click on OK

You should now see the maximum value for each pair in C3. Name this column max2.
iii. Compute the average test score (Mean) for each pair of scores, and store the values in C4, as follows:

> MINITAB:　　Calc → Row statistics
> 　　　　Click the button for mean
> 　　　　Enter C1-C2 in the Input Variables box
> 　　　　Enter C4 in the Store Result In box.
> 　　　　Click on OK

You should now see the average for each pair in C4. Name this column mean2.
c. Construct density histograms for each of the three statistics (these density histograms approximate the sampling distributions of the three statistics), as follows:

> MINITAB:　　Graph → Histogram
> 　　　　Enter max2, mean2, and recent2 into the first three rows of the Graph Variables box
> 　　　　Click on the Options button. Select Density. Click on OK. (This will produce histograms that use the density scale rather than the frequency scale.)
> 　　　　Click on the Frame drop-down menu, and select Multiple Graphs. Select Same X and Same Y. (This will cause MINITAB to use the same scales for all three histograms, so that they can be easily compared.
> 　　　　Click on OK.

Part 2: Now you will produce approximate sampling distributions for these same three statistics, but for the case of a student who takes the exam five times. Follow the same steps as in Part 1, with the following modifications:
a. Obtain 500 sets of 5 test scores, and store these values in columns C11–C15.
b. Recent will just be the values in C15; name this column recent5. Compute the Max and Mean values, and store them in columns C16 and C17. Name these columns max5 and mean5.
c. Construct density histograms for max5, mean5, and recent5.

Part 3: Now use the approximate sampling distributions constructed in Parts 1 and 2 to answer the following questions.
a. The statistic that is the average of the test scores is just a sample mean (for a sample of size 2 in Part 1

and for a sample of size 5 in Part 2). How do the sampling distributions of mean2 and mean5 compare to what is expected based on the general properties of the $\bar{x}$ distribution given in Section 8.2? Explain.

b. Based on the three distributions from Part 1, for a two-time test taker, describe the advantage of using the maximum score compared to using either the average score or the most recent score.

c. Now consider the approximate sampling distributions of the maximum score for two-time and for five-time test takers. How do these two distributions compare?

d. Does a student who takes the exam five times have a big advantage over a student of equal ability who takes the exam only twice if the maximum score is used for college admission decisions? Explain.

e. If you were writing admission procedures for a selective university, would you recommend using the maximum test score, the average test score, or the most recent test score in making admission decisions? Write a paragraph explaining your choice.

■ Summary of Key Concepts and Formulas

Term or Formula	Comment
Statistic	Any quantity whose value is computed from sample data.
Sampling distribution	The probability distribution of a statistic: The sampling distribution describes the long-run behavior of the statistic.
Sampling distribution of $\bar{x}$	The probability distribution of the sample mean $\bar{x}$ based on a random sample of size n. Properties of the $\bar{x}$ sampling distribution: $\mu_{\bar{x}} = \mu$ and $\sigma_{\bar{x}} = \dfrac{\sigma}{\sqrt{n}}$ (where μ and σ are the population mean and standard deviation, respectively). In addition, when the population distribution is normal or the sample size is large, the sampling distribution of $\bar{x}$ is (approximately) normal.
Central Limit Theorem	This important theorem states that when n is sufficiently large, the $\bar{x}$ distribution is approximately normal. The standard rule of thumb is that the theorem can safely be applied when n exceeds 30.
Sampling distribution of p	The probability distribution of the sample proportion p, based on a random sample of size n. When the sample size is sufficiently large, the sampling distribution of p is approximately normal, with $\mu_p = \pi$ and $\sigma_p = \sqrt{\dfrac{\pi(1 - \pi)}{n}}$ (where π is the true population proportion).

■ Supplementary Exercises 8.36–8.42

8.36 The nicotine content in a single cigarette of a particular brand has a distribution with mean 0.8 mg and standard deviation 0.1 mg. If 100 of these cigarettes are analyzed, what is the probability that the resulting sample mean nicotine content will be less than 0.79? less than 0.77?

8.37 Let $x_1, x_2, \ldots, x_{100}$ denote the actual net weights (in pounds) of 100 randomly selected bags of fertilizer. Suppose that the weight of a randomly selected bag has a distribution with mean 50 lb and variance 1 lb^2. Let $\bar{x}$ be the sample mean weight ($n = 100$).

a. Describe the sampling distribution of $\bar{x}$.

b. What is the probability that the sample mean is between 49.75 lb and 50.25 lb?

c. What is the probability that the sample mean is less than 50 lb?

8.38 Suppose that 20% of the subscribers of a cable television company watch the shopping channel at least once a week. The cable company is trying to decide whether to replace this channel with a new local station. A survey of 100 subscribers will be undertaken. The cable company has decided to keep the shopping channel if the sample proportion is greater than .25. What is the approximate probability that the cable company will keep the shopping channel, even though the true proportion who watch it is only .20?

8.39 Although a lecture period at a certain university lasts exactly 50 min, the actual lecture time of a statistics instructor on any particular day has a distribution with mean value 52 min and standard deviation 2 min. Suppose that times of different lectures are independent of one another. Let $\bar{x}$ represent the mean of 36 randomly selected lecture times.
a. What are the mean value and standard deviation of the sampling distribution of $\bar{x}$?
b. What is the probability that the sample mean exceeds 50 min? 55 min?

8.40 Water permeability of concrete is an important characteristic in assessing suitability for various applications. Permeability can be measured by letting water flow across the surface and determining the amount lost (in inches per hour). Suppose that the permeability index x for a randomly selected concrete specimen of a particular type is normally distributed with mean value 1000 and standard deviation 150.

a. How likely is it that a single randomly selected specimen will have a permeability index between 850 and 1300?
b. If the permeability index is to be determined for each specimen in a random sample of size 10, how likely is it that the sample average permeability index will be between 950 and 1100? between 850 and 1300?

8.41 *Newsweek* (November 23, 1992) reported that 40% of all U.S. employees participate in "self-insurance" health plans ($\pi = .40$).
a. In a random sample of 100 employees, what is the approximate probability that at least half of those in the sample participate in such a plan?
b. Suppose you were told that at least 60 of the 100 employees in a sample from your state participated in such a plan. Would you think $\pi = .40$ for your state? Explain.

8.42 The amount of money spent by a customer at a discount store has a mean of $100 and a standard deviation of $30. What is the probability that a randomly selected group of 50 shoppers will spend a total of more than $5300? (Hint: The total will be more than $5300 when the sample average exceeds what value?)

▪ References

Freedman, David, Robert Pisani, Roger Purves. *Statistics*, 3d ed. New York: W. W. Norton, 1997.

(This book gives an excellent informal discussion of sampling distributions.)

9 · Estimation Using a Single Sample

Affirmative action in university admissions is a controversial topic. Some believe that affirmative action programs are no longer needed, whereas others argue that using race and ethnicity as factors in university admissions is necessary to achieve diverse student populations. To assess public opinion on this issue, investigators conducted a survey of 1013 randomly selected U.S. adults. The results are summarized in the article "Poll Finds Sharp Split on Affirmative Action" (*San Luis Obispo Tribune*, March 8, 2003). The investigators wanted to use the survey data to estimate the true proportion of U.S. adults who believed that programs that give "advantages and preferences to Blacks, Hispanics, and other minorities in hiring, promotions, and college admissions" should be continued. The methods introduced in this chapter will be used to produce the desired estimate. Because the estimate is based only on a sample rather than on a census of all U.S. adults, it is important that this estimate be constructed in a way that also conveys information about the anticipated accuracy.

The objective of inferential statistics is to use sample data to decrease our uncertainty about the corresponding population. Often, data are collected to obtain information that allows the investigator to estimate the value of some population characteristic, such as a population mean μ or a population proportion π. One way to accomplish this uses the sample data to arrive at a single number that represents a plausible value for the characteristic of interest. Alternatively, an entire range of plausible values for the characteristic can be reported. These two estimation techniques, *point estimation* and *interval estimation*, are introduced in this chapter.

▪ 9.1 Point Estimation

The simplest approach to estimating a population characteristic involves using sample data to compute a single number that can be regarded as a plausible value of the characteristic. For example, sample data might suggest that 1000 hr is a plausible value for μ, the true mean lifetime for light bulbs of a particular brand. In a different setting, a sample survey of students at a particular university might lead to the statement that .41 is a plausible value for π, the true proportion of students who favor a fee for recreational facilities.

▪ Definition

A **point estimate** of a population characteristic is a single number that is based on sample data and that represents a plausible value of the characteristic.

In the examples just given, 1000 is a point estimate of μ and .41 is a point estimate of π. The adjective *point* reflects the fact that the estimate corresponds to a single point on the number line.

A point estimate is obtained by first selecting an appropriate statistic. The estimate is then the value of the statistic for the given sample. For example, the computed value of the sample mean $\bar{x}$ provides a point estimate of a population mean μ.

▪ Example 9.1   Support for Affirmative Action

One of the purposes of the survey on affirmative action described in the chapter introduction was to estimate the proportion of the U.S. population who believe that affirmative action programs should be continued. The article reported that 537 of the 1013 people surveyed believed that affirmative action programs should be continued. Let's use this information to estimate π, where π is the true proportion of all U.S. adults who favor continuing affirmative action programs. With success identified as a person who believes that affirmative action programs should continue, π is then just the population proportion of successes. The statistic

$$p = \frac{\text{number of successes in the sample}}{n}$$

which is the sample proportion of successes, is an obvious choice for obtaining a point estimate of π. Based on the reported information, the point estimate of π is

$$p = \frac{537}{1013} = .530$$

That is, based on this random sample, we estimate that 53% of the adults in the United States believe that affirmative action programs should be continued.

For purposes of estimating a population proportion π, there is no obvious alternative to the statistic p. In other situations, such as the one illustrated in Example 9.2, there may be several statistics that can be used to obtain an estimate.

▪ Example 9.2   Internet Use by College Students

The article "Online Extracurricular Activity" (*USA Today*, March 13, 2000) reported the results of a study of college students conducted by a polling organization called The Student Monitor. One aspect of computer use examined in this study was the number of hours per week spent on the Internet. Suppose that the following observations represent the number of Internet hours per week reported by 20 college students (these data are compatible with summary values given in the article):

| 4.00 | 5.00 | 5.00 | 5.25 | 5.50 | 6.25 | 6.25 | 6.50 | 6.50 | 7.00 |
| 7.25 | 7.75 | 8.00 | 8.00 | 8.00 | 8.25 | 8.50 | 8.50 | 9.50 | 10.50 |

A dotplot of the data is shown here:

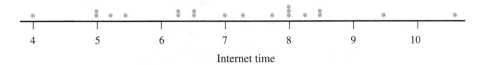

Internet time

Suppose further that a point estimate of μ, the true mean Internet time per week for college students, is desired. An obvious choice of a statistic for estimating μ is the sample mean $\bar{x}$. However, there are other possibilities. We might consider using a trimmed mean or even the sample median, because the data set exhibits some symmetry. (If the corresponding population distribution is symmetric, the population mean μ and the population median are equal).

The three statistics and the resulting estimates of μ calculated from the data are

$$\text{sample mean} = \bar{x} = \frac{\Sigma x}{n} = \frac{141.50}{20} = 7.075$$

$$\text{sample median} = \frac{7.0 + 7.25}{2} = 7.125$$

$$10\% \text{ trimmed mean} = (\text{average of middle 16 observations}) = \frac{112.5}{16} = 7.031$$

The estimates of the mean Internet time per week for college students differ somewhat from one another. The choice from among them should depend on which statistic tends, on average, to produce an estimate closest to the true value of μ. The following subsection discusses criteria for choosing among competing statistics.

■ Choosing a Statistic for Computing an Estimate

The point of Example 9.2 is that more than one statistic may be reasonable to use to obtain a point estimate of a specified population characteristic. Loosely speaking, the statistic used should be one that tends to yield an accurate estimate — that is, an estimate close to the value of the population characteristic. Information about the accuracy of estimation for a particular statistic is provided by the statistic's sampling distribution. Figure 9.1 displays the sampling distributions of three different statistics. The value of the population characteristic, which we refer to as the *true value*, is marked on the measurement axis.

The distribution in Figure 9.1(a) is that of a statistic unlikely to yield an estimate close to the true value. The distribution is centered to the right of the true value, making it very likely that an estimate (a value of the statistic for a particular sample) will be larger than the true value. If this statistic is used to compute an estimate based on a first sample, then another estimate based on a second sample,

Figure 9.1 Sampling distributions of three different statistics for estimating a population characteristic.

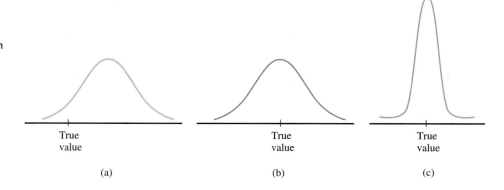

(a)   (b)   (c)

and another estimate based on a third sample, and so on, the long-run average value of these estimates will exceed the true value.

The sampling distribution of Figure 9.1(b) is centered at the true value. Thus, although one estimate may be smaller than the true value and another may be larger, when this statistic is used many times over with different samples, there will be no long-run tendency to over- or underestimate the true value. Note that even though the sampling distribution is correctly centered, it spreads out quite a bit about the true value. Because of this, some estimates resulting from the use of this statistic will be far above or far below the true value, even though there is no systematic tendency to underestimate or overestimate the true value.

In contrast, the mean value of the statistic with the distribution shown in Figure 9.1(c) is exactly the true value of the population characteristic (implying no systematic error in estimation), and the statistic's standard deviation is relatively small. Estimates based on this third statistic will almost always be quite close to the true value — certainly more often than estimates resulting from the statistic with the sampling distribution shown in Figure 9.1(b).

■ Definition

A statistic whose mean value is equal to the value of the population characteristic being estimated is said to be an **unbiased statistic**. A statistic that is not unbiased is said to be **biased**.

As an example of a statistic that is biased, consider using the sample range as an estimate of the population range. Because the range of a population is defined as the difference between the largest value in the population and the smallest value, the range for a sample tends to underestimate the population range. This is true because the largest value in a sample can be at most as large as the largest value in the population and the smallest sample value must be at least as large as the smallest value in the population. The sample range equals the population range *only* if the sample includes both the largest and the smallest values in the population; in all other instances, the sample range is smaller than the population range. Thus, $\mu_{\text{sample range}} <$ population range, implying bias.

Let $x_1, x_2, \ldots, x_n$ represent the values in a random sample. One of the general results concerning the sampling distribution of $\overline{x}$, the sample mean, is that $\mu_{\overline{x}} = \mu$. This result says that the $\overline{x}$ values from all possible random samples of size n center around μ, the population mean. For example, if $\mu = 100$, the $\bar{x}$ distribution is centered at 100, whereas if $\mu = 5200$, then the $\bar{x}$ distribution is centered at 5200. Therefore, $\bar{x}$ is an unbiased statistic for estimating μ. Similarly, because $\mu_p = \pi$, the sampling distribution of p is centered at π, and it follows that p is an unbiased statistic for estimating a population proportion.

Using an unbiased statistic that also has a small standard deviation ensures that there will be no systematic tendency to under- or overestimate the value of the population characteristic *and* that estimates will almost always be relatively close to the true value.

Given a choice between several unbiased statistics that could be used for estimating a population characteristic, the best statistic to use is the one with the smallest standard deviation.

Consider the problem of estimating a population mean μ. The obvious choice of statistic for obtaining a point estimate of μ is the sample mean $\bar{x}$, an unbiased statistic for this purpose. However, when the population distribution is symmetric, $\bar{x}$ is not the only choice. Other unbiased statistics for estimating μ in this case include the sample median and any trimmed mean (with the same number of observations trimmed from each end of the ordered sample). Which statistic should be used? The following facts may be helpful in making a choice.

1. If the population distribution is normal, then $\bar{x}$ has a smaller standard deviation than any other unbiased statistic for estimating μ. However, in this case, a trimmed mean with a small trimming percentage (such as 10%) performs almost as well as $\bar{x}$.

2. When the population distribution is symmetric with heavy tails compared to the normal curve, a trimmed mean is a better statistic than $\bar{x}$ for estimating μ.

When the population distribution is unquestionably normal, the choice is clear: Use $\bar{x}$ to estimate μ. However, with a heavy-tailed distribution, a trimmed mean gives protection against one or two outliers in the sample that might otherwise have a large effect on the value of the estimate.

Now consider estimating another population characteristic, the population variance σ^2. The sample variance

$$s^2 = \frac{\Sigma (x - \bar{x})^2}{n - 1}$$

is a good choice for obtaining a point estimate of the population variance σ^2. It can be shown that s^2 is an unbiased statistic for estimating σ^2; that is, whatever the value of σ^2, the sampling distribution of s^2 is centered at that value. It is precisely for this reason — to obtain an unbiased statistic — that the divisor $(n - 1)$ is used. An alternative statistic is the average squared deviation

$$\frac{\Sigma (x - \bar{x})^2}{n}$$

which one might think has a more natural divisor than s^2. However, the average squared deviation is biased, with its values tending to be smaller, on average, than σ^2.

▪ **Example 9.3** Airborne Times for Flight 448

The Bureau of Transportation Statistics provides data on U.S. airline flights. The airborne times (in minutes) for United Airlines flight 448 from Albuquerque to Denver on 10 randomly selected days between January 1, 2003, and March 31, 2003, are

$$57 \quad 54 \quad 55 \quad 51 \quad 56 \quad 48 \quad 52 \quad 51 \quad 59 \quad 59$$

For these data $\sum x = 542$, $\sum x^2 = 29{,}498$, $n = 10$, and

$$\sum (x - \overline{x})^2 = \sum x^2 - \frac{(\sum x)^2}{n}$$

$$= 29{,}498 - \frac{(542)^2}{10}$$

$$= 121.6$$

Let σ^2 denote the true variance in airborne time for flight 448. Using the sample variance s^2 to provide a point estimate of σ^2 yields

$$s^2 = \frac{\sum (x - \overline{x})^2}{n - 1} = \frac{121.6}{9} = 13.51$$

Using the average squared deviation (with divisor $n = 10$), we obtain the resulting point estimate:

$$\frac{\sum (x - \overline{x})^2}{n} = \frac{121.6}{10} = 12.16$$

Because s^2 is an unbiased statistic for estimating σ^2, most statisticians would recommend using the point estimate 13.51.

An obvious choice of a statistic for estimating the population standard deviation σ is the sample standard deviation s. For the data given in Example 9.3,

$$s = \sqrt{13.51} = 3.68$$

Unfortunately, the fact that s^2 is an unbiased statistic for estimating σ^2 does not imply that s is an unbiased statistic for estimating σ. The sample standard deviation tends to underestimate slightly the true value of σ. However, unbiasedness is not the only criterion by which a statistic can be judged, and there are other good reasons for using s to estimate σ. In what follows, whenever we need to estimate σ based on a single random sample, we use the statistic s to obtain a point estimate.

▪ **Exercises 9.1–9.10**

9.1 Three different statistics are being considered for estimating a population characteristic. The sampling distributions of the three statistics are shown in the following illustration:

Figure for Exercise 9.1

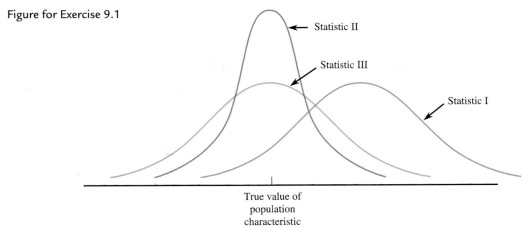

Statistic II

Statistic III

Statistic I

True value of
population
characteristic

Which statistic would you recommend? Explain your choice.

9.2 Why is an unbiased statistic generally preferred over a biased statistic for estimating a population characteristic? Does unbiasedness alone guarantee that the estimate will be close to the true value? Explain. Under what circumstances might you choose a biased statistic over an unbiased statistic if two statistics are available for estimating a population characteristic?

9.3 Radon, an odorless gas that can cause lung cancer, is released in the decay of radium, which is naturally present in soil and rocks. The Environmental Protection Agency (EPA) considers radon the second leading cause of lung cancer (behind cigarette smoking). One radon hot spot has been identified on the Spokane Indian reservation in Washington. The Associated Press (June 10, 1991) reported that of 270 randomly selected homes on the reservation, 68 showed radon readings above 4 pCi (the level that the EPA considers hazardous). Estimate π, the proportion of all homes on the reservation with a radon level above 4 pCi.

9.4 Despite protests from civil libertarians and gay rights activists, many people favor mandatory AIDS testing of certain at-risk groups, and some people even believe that all citizens should be tested. What proportion of the adults in the United States favor mandatory testing for all citizens? To assess public opinion on this issue, researchers conducted a survey of 1014 randomly selected adult U.S. citizens ("Large Majorities Continue to Back AIDS Testing," *Gallup Poll Monthly* [1991]: 25–28). The article reported that 466 of the 1014 people surveyed believed that all citizens should be tested. Use this information to estimate π, the true proportion of all U.S. adults who favor AIDS testing of all citizens.

9.5 The article "Sensory and Mechanical Assessment of the Quality of Frankfurters" (*Journal of Texture Studies* [1990]: 395–409) reported the following salt content (percentage by weight) for 10 frankfurters:

2.26 2.11 1.64 1.17 1.64 2.36 1.70 2.10 2.19 2.40

a. Use the given data to produce a point estimate of μ, the true mean salt content for frankfurters.
b. Use the given data to produce a point estimate of σ^2, the variance of salt content for frankfurters.
c. Use the given data to produce an estimate of σ, the standard deviation of salt content. Is the statistic you used to produce your estimate unbiased?

9.6 A study reported in *Newsweek* (December 23, 1991) involved a sample of 935 smokers. Each individual received a nicotine patch, which delivers nicotine to the bloodstream but at a much slower rate than cigarettes do. Dosage was decreased to 0 over a 12-week period. Suppose that 245 of the subjects were still not smoking 6 months after treatment (this figure is consistent with information given in the article). Estimate the percentage of all smokers who, when given this treatment, would refrain from smoking for at least 6 months.

9.7 The following data on gross efficiency (ratio of work accomplished per minute to calorie expenditure per minute) for trained endurance cyclists were given in the article "Cycling Efficiency Is Related to the Percentage of Type I Muscle Fibers" (*Medicine and Science in Sports and Exercise* [1992]: 782–788):

18.3 18.9 19.0 20.9 21.4 20.5 20.1 20.1
20.8 20.5 19.9 20.5 20.6 22.1 21.9 21.2
20.5 22.6 22.6

a. Assuming that the distribution of gross energy in the population of all endurance cyclists is normal, give a point estimate of μ, the population mean gross efficiency.

b. Making no assumptions about the shape of the population distribution, estimate the proportion of all such cyclists whose gross efficiency is at most 20.

9.8 A random sample of $n = 12$ four-year-old red pine trees was selected, and the diameter (in inches) of each tree's main stem was measured. The resulting observations are as follows:

> 11.3 10.7 12.4 15.2 10.1 12.1 16.2 10.5
> 11.4 11.0 10.7 12.0

a. Compute a point estimate of σ, the population standard deviation of main stem diameter. What statistic did you use to obtain your estimate?

b. Making no assumptions whatsoever about the shape of the population distribution of diameters, give a point estimate for the population median diameter (i.e., for the middle diameter value in the entire population of four-year-old red pine trees). What statistic did you use to obtain the estimate?

c. Suppose that the population distribution of diameter is symmetric but with heavier tails than the normal distribution. Give a point estimate of the population mean diameter based on a statistic that gives some protection against the presence of outliers in the sample. What statistic did you use?

d. Suppose that the diameter distribution is normal. Then the 90th percentile of the diameter distribution is $\mu + 1.28\sigma$ (so 90% of all trees have diameters less than this value). Compute a point estimate for this percentile. (Hint: First compute an estimate of μ in this case; then use it along with your estimate of σ from Part (a).)

9.9 Each person in a random sample of 20 students at a particular university was asked whether he or she is registered to vote. The responses (R = registered, N = not registered) are given here:

> R R N R N N R R R N
> R R R R R N R R R N

Use these data to estimate π, the true proportion of all students at the university who are registered to vote.

9.10 A random sample of 10 houses in a particular area, each of which is heated with natural gas, is selected, and the amount of gas (in therms) used during the month of January is determined for each house. The resulting observations are as follows:

> 103 156 118 89 125 147 122 109 138 99

a. Let μ_J denote the average gas usage during January by all houses in this area. Compute a point estimate of μ_J.

b. Suppose that 10,000 houses in this area use natural gas for heating. Let τ denote the total amount of gas used by all of these houses during January. Estimate τ using the data of Part (a). What statistic did you use in computing your estimate?

c. Use the data in Part (a) to estimate π, the proportion of all houses that used at least 100 therms.

d. Give a point estimate of the population median usage (the middle value in the population of all houses) based on the sample of Part (a). Which statistic did you use?

▪ 9.2 Large-Sample Confidence Interval for a Population Proportion

In Section 9.1 we saw how to use a statistic to produce a point estimate of a population characteristic. The value of a point estimate depends on which sample, out of all the possible samples, happens to be selected. Different samples usually yield different estimates as a result of chance differences from one sample to another. Because of sampling variability, rarely is the point estimate from a sample exactly equal to the true value of the population characteristic. We hope that the chosen statistic produces an estimate that is close, on average, to the true value. Although a point estimate may represent our best single-number guess for the value of the population characteristic, it is not the only plausible value. These considerations suggest the need to indicate in some way a range of plausible values for the population characteristic. A point estimate by itself does not provide this information.

As an alternative to a point estimate, suppose that we report not just a single credible value for the population characteristic but an entire interval of reasonable values based on the sample data. For example, we might be confident that for all calls made from AT&T pay phones, the proportion π of calls that are billed to a credit card is in the interval from .53 to .57. The narrowness of this interval implies that we have rather precise information about the value of π. If, with the same high degree of confidence, we could only state that π was between .32 and .74, it would be clear that we had relatively imprecise knowledge of the value of π.

■ **Definition**

A **confidence interval (CI)** for a population characteristic is an interval estimate of plausible values for the characteristic. It is constructed so that, with a chosen degree of confidence, the value of the characteristic is captured between the lower and upper endpoints of the interval.

Associated with each confidence interval is a *confidence level*. The confidence level provides information on how much "confidence" we can have in the *method* used to construct the interval estimate (*not* our confidence in any one particular interval). Usual choices for confidence levels are 90%, 95%, and 99%, although other levels are also possible. If we were to construct a 95% confidence interval using the technique to be described shortly, we would be using a method that is "successful" 95% of the time. That is, if this method was used to generate an interval estimate over and over again with different samples, in the long run 95% of the resulting intervals would capture the true value of the characteristic being estimated. Similarly, a 99% confidence interval is one that is constructed using a method that is, in the long run, successful in capturing the true value of the population characteristic 99% of the time.

■ **Definition**

The **confidence level** associated with a confidence interval estimate is the success rate of the *method* used to construct the interval.

Many factors influence the choice of confidence level. These factors will be discussed after we develop the method for constructing confidence intervals. We first consider a large-sample confidence interval for a population proportion π.

Often an investigator wishes to make an inference about the proportion of individuals or objects in a population that possesses a particular property of interest. For example, a university administrator might be interested in the proportion of students who prefer a new web-based computer registration system to the previous registration method. In a different setting, a quality control engineer might be concerned about the proportion of defective parts manufactured using a particular process.

Let π be the proportion of the population that possesses the property of interest. In this section, we consider the problem of estimating π using information in a random sample of size n from the population. Previously, we used the sample proportion

$$p = \frac{\text{number in the sample that possess the property of interest}}{n}$$

to calculate a point estimate of π. We can also use p to form a confidence interval for π.

Although a small-sample confidence interval for π can be obtained, our focus is on the large-sample case. The justification for the large-sample interval rests on properties of the sampling distribution of the statistic p:

1. The sampling distribution of p is centered at π; that is, $\mu_p = \pi$. Therefore, p is an unbiased statistic for estimating π.

2. The standard deviation of p is $\sigma_p = \sqrt{\dfrac{\pi(1 - \pi)}{n}}$.

3. As long as n is large (i.e., $n\pi \geq 10$ and $n(1 - \pi) \geq 10$) and as long as the sample size is less than 10% of the population size, the sampling distribution of p is well approximated by a normal curve.

The accompanying box summarizes these properties.

When n is large, the statistic p has a sampling distribution that is approximately normal with mean π and standard deviation

$$\sqrt{\frac{\pi(1 - \pi)}{n}}$$

The development of a confidence interval for π is easier to follow if we select a particular confidence level. For a confidence level of 95%, Appendix Table 2, the table of standard normal (z) curve areas, can be used to determine a value z^* such that a central area of .95 falls between $-z^*$ and z^*. In this case, the remaining area of .05 is divided equally between the two tails, as shown in Figure 9.2. The total area to the left of the desired z^* is .975 (.95 central area + .025 area below $-z^*$). By locating .9750 in the body of Appendix Table 2, we find that the corresponding z critical value is 1.96.

Figure 9.2 Capturing a central area of .95 under the z curve.

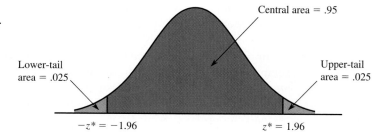

Generalizing this result to normal distributions other than the standard normal distribution tells us that for *any* normal distribution, about 95% of the values are within 1.96 standard deviations of the mean. Because (for large random samples) the sampling distribution of p is approximately normal with mean $\mu_p = \pi$ and standard deviation $\sigma_p = \sqrt{\dfrac{\pi(1 - \pi)}{n}}$, we get the following result.

> When n is large, approximately 95% of all samples of size n will result in a value of p that is within $1.96\sigma_p = 1.96\sqrt{\dfrac{\pi(1 - \pi)}{n}}$ of the true population proportion π.

If p is within $1.96\sqrt{\dfrac{\pi(1 - \pi)}{n}}$ of π, this means that the interval

$$p - 1.96\sqrt{\frac{\pi(1 - \pi)}{n}} \quad \text{to} \quad p + 1.96\sqrt{\frac{\pi(1 - \pi)}{n}}$$

captures π (and this happens for 95% of all possible samples). However, if p is farther away from π than $1.96\sqrt{\dfrac{\pi(1 - \pi)}{n}}$ (which happens for about 5% of all possible samples), the interval will not include the true value of π. This is shown in Figure 9.3.

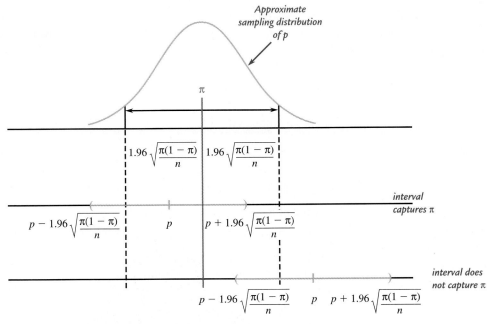

Figure 9.3 The population proportion π is captured in the interval from $p - 1.96\sqrt{\dfrac{\pi(1 - \pi)}{n}}$ to $p + 1.96\sqrt{\dfrac{\pi(1 - \pi)}{n}}$ when p is within $1.96\sqrt{\dfrac{\pi(1 - \pi)}{n}}$ of π.

Because p is within $1.96\sigma_p$ approximately 95% of the time, this implies that in repeated sampling, 95% of the time the interval

$$p - 1.96\sqrt{\frac{\pi(1-\pi)}{n}} \quad \text{to} \quad p + 1.96\sqrt{\frac{\pi(1-\pi)}{n}}$$

will contain π. Because π is unknown, $\sqrt{\frac{\pi(1-\pi)}{n}}$ must be estimated. As long as the sample size is large, the value of $\sqrt{\frac{p(1-p)}{n}}$ should be close to $\sqrt{\frac{\pi(1-\pi)}{n}}$, and this value can be used in its place.

When n is large, a 95% confidence interval for π is

$$\left(p - 1.96\sqrt{\frac{p(1-p)}{n}}, \quad p + 1.96\sqrt{\frac{p(1-p)}{n}} \right)$$

An abbreviated formula for the interval is

$$p \pm 1.96\sqrt{\frac{p(1-p)}{n}}$$

where $p + 1.96\sqrt{\frac{p(1-p)}{n}}$ gives the upper endpoint of the interval and $p - 1.96\sqrt{\frac{p(1-p)}{n}}$ gives the lower endpoint of the interval. The interval can be used as long as (1) $np \geq 10$ and $n(1-p) \geq 10$, (2) the sample size is less than 10% of the population size if sampling is without replacement, and (3) the sample can be regarded as a random sample from the population of interest.

▪ Example 9.4   Affirmative Action Continued

Let's return to the information from the survey on attitudes towards affirmative action (see the chapter introduction and Example 9.1):

Total number of people surveyed: 1013

Number who believe affirmative action programs should be continued: 537

A point estimate of π, the true proportion of U.S. adults who believe that affirmative action programs should be continued, is

$$p = \frac{537}{1013} = .530$$

Because $np = (1013)(.53) = 537 \geq 10$ and $n(1-p) = (1013)(.47) = 476 \geq 10$ and the adults in the sample were randomly selected from a large population, the large-sample interval can be used. For a 95% confidence level, a confidence interval for π is

$$p \pm 1.96\sqrt{\frac{p(1-p)}{n}} = .530 \pm 1.96\sqrt{\frac{(.530)(.470)}{1013}}$$

$$= .530 \pm (1.96)(.016)$$

$$= .530 \pm .031$$

$$= (.499, .561)$$

Based on this sample, we can be 95% confident that π, the true proportion who believe that affirmative action programs should continue, is between .499 and .561. We used a *method* to construct this estimate that in the long run will successfully capture the true value of π 95% of the time.

The 95% confidence interval for π calculated in Example 9.4 is (.499, .561). It is tempting to say that there is a "probability" of .95 that π is between .499 and .561. *Do not yield to this temptation!* The 95% refers to the percentage of *all* possible samples resulting in an interval that includes π. In other words, if we take sample after sample from the population and use each one separately to compute a 95% confidence interval, in the long run roughly 95% of these intervals will capture π. Figure 9.4 illustrates this concept for intervals generated from 100 different random samples; 93 of the intervals include π, whereas 7 do not. Any specific interval,

Figure 9.4 One hundred 95% confidence intervals for π computed from 100 different random samples (asterisks identify intervals that do not include π).

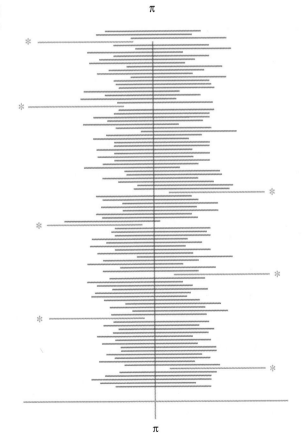

and our interval (.499, .561) in particular, either includes π or it does not (remember, the value of π is fixed but not known to us). We cannot make a chance (probability) statement concerning this particular interval. *The confidence level 95% refers to the method used to construct the interval rather than to any particular interval, such as the one we obtained.*

The formula given for a 95% confidence interval can easily be adapted for other confidence levels. The choice of a 95% confidence level led to the use of the z value 1.96 (chosen to achieve a central area of .95 under the standard normal curve) in the formula. Any other confidence level can be obtained by using an appropriate z critical value in place of 1.96. For example, suppose that we wanted to achieve a confidence level of 99%. To obtain a central area of .99, the approximate z critical value would have a cumulative area (area to the left) of .995, as illustrated in Figure 9.5. From Appendix Table 2, we find that the corresponding z critical value is 2.58. A 99% confidence interval for π is then obtained by using 2.58 in place of 1.96 in the formula for the 95% confidence interval.

Figure 9.5 Finding the z critical value for a 99% confidence level.

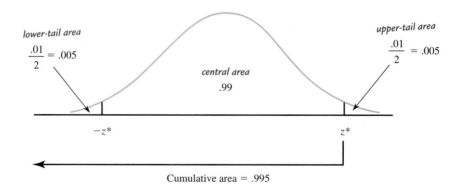

▪ Large-Sample Confidence Interval for π

The general formula for a confidence interval for a population proportion π when

1. p is the sample proportion from a **random sample**, and
2. the sample size **n is large** ($np \geq 10$ and $n(1 - p) \geq 10$), and
3. **the sample size is small relative to the population size** if the sample is selected without replacement (i.e., n is at most 10% of the population size)

is

$$p \pm (z \text{ critical value})\sqrt{\frac{p(1 - p)}{n}}$$

The desired confidence level determines which z critical value is used. The three most commonly used confidence levels, 90%, 95%, and 99%, use z critical values of 1.645, 1.96, and 2.58, respectively.

Note: This interval is not appropriate for small samples. It is possible to construct a confidence interval in the small-sample case, but this is beyond the scope of this textbook.

Why settle for 95% confidence when 99% confidence is possible? Because the higher confidence level comes with a price tag. The resulting interval is wider than the 95% interval. The width of the 95% interval is $2\left(1.96\sqrt{\dfrac{p(1-p)}{n}}\right)$, whereas the 99% interval has width $2\left(2.58\sqrt{\dfrac{p(1-p)}{n}}\right)$. The higher *reliability* of the 99% interval (where "reliability" is specified by the confidence level) entails a loss in precision (as indicated by the wider interval). In the opinion of many investigators, a 95% interval is a reasonable compromise between reliability and precision.

▪ Example 9.5   Violent Behavior in the Workplace

An Associated Press article on potential violent behavior reported the results of a survey of 750 workers who were employed full time (*San Luis Obispo Tribune*, September 7, 1999). Of those surveyed, 125 indicated that they were so angered by a coworker during the past year that he or she felt like hitting the person (but didn't). Assuming that it is reasonable to regard this sample of 750 as a random sample from the population of full-time workers, we can use this information to construct an estimate of π, the true proportion of full-time workers so angered in the last year that they wanted to hit a colleague.

For this sample

$$p = \frac{125}{750} = .167$$

Because $np = 125$ and $n(1-p) = 625$ are both greater than or equal to 10, the sample size is large enough to use the formula for a large-sample confidence interval. A 90% confidence interval for π is then

$$p \pm (z \text{ critical value})\sqrt{\frac{p(1-p)}{n}} = .167 \pm 1.645\sqrt{\frac{(.167)(.833)}{750}}$$
$$= .167 \pm (1.645)(.014)$$
$$= .167 \pm .023$$
$$= (.144, .190)$$

Based on these sample data, we can be 90% confident that the true proportion of full-time workers who have been angry enough in the last year to consider hitting a coworker is between .144 and .190. We have used a *method* to construct this interval estimate that has a 10% error rate.

The confidence level for the z confidence interval for a population proportion is only approximate. That is, when we report a 95% confidence interval for a population proportion, the 95% confidence level implies that we have used a method that produces an interval that includes the actual value of the population proportion 95% of the time in repeated sampling. In fact, because the normal distribution is only an approximation to the sampling distribution of p, the true confidence level may differ somewhat from the reported value. If the conditions (1) $np \geq 10$ and

$n(1 - p) \geq 10$ and (2) n is at most 10% of the population size if sampling without replacement are met, the normal approximation is reasonable and the actual confidence level is usually quite close to the reported level; this is why it is important to check these conditions before computing and reporting a z confidence interval for a population proportion.

What should you do if these conditions are not met? If the sample size is too small to satisfy the np and $n(1 - p)$ greater than or equal to 10 condition, an alternative procedure can be used. Consult a statistician or a more advanced textbook in this case. If the condition that the sample size is less than 10% of the population size when sampling without replacement is not satisfied, the z confidence interval tends to be conservative (i.e., it tends to be wider than is necessary to achieve the desired confidence level). In this case, a finite population correction factor can be used to obtain a more precise interval. Again, it would be wise to consult a statistician or a more advanced textbook.

▪ An Alternative to the Large-Sample z Interval

Investigators have shown that in some instances, even when the sample size conditions of the large sample z confidence interval for a population proportion are met, the actual confidence level associated with the method may be noticeably different from the reported confidence level. A modified interval that has an actual confidence level that is closer to the reported confidence level is based on a modified sample proportion, p_{mod}, the proportion of successes after adding two successes and two failures to the sample. Then p_{mod} is

$$p_{mod} = \frac{\text{number of successes} + 2}{n + 4}$$

p_{mod} is used in place of p in the usual confidence interval formula. Properties of this modified confidence interval are investigated in Activity 9.2 at the end of the chapter.

▪ General Form of a Confidence Interval

Many confidence intervals have the same general form as the large-sample z interval for π just considered. We started with a statistic p, from which a point estimate for π was obtained. The standard deviation of this statistic is $\sqrt{\frac{\pi(1 - \pi)}{n}}$. This resulted in a confidence interval of the form

$$\left(\begin{array}{c} \text{point estimate using} \\ \text{a specified statistic} \end{array} \right) \pm (\text{critical value}) \left(\begin{array}{c} \text{standard deviation} \\ \text{of the statistic} \end{array} \right)$$

Because π was unknown, we estimated the standard deviation of the statistic by $\sqrt{\frac{p(1 - p)}{n}}$, which yielded the interval

$$\left(\begin{array}{c} \text{point estimate using} \\ \text{a specified statistic} \end{array} \right) \pm (\text{critical value}) \left(\begin{array}{c} \text{estimated standard deviation} \\ \text{of the statistic} \end{array} \right)$$

For a population characteristic other than π, a statistic for estimating the characteristic is selected. Then (drawing on statistical theory) a formula for the standard deviation of the statistic is given. In practice, it is almost always necessary to estimate this standard deviation (using something analogous to $\sqrt{\dfrac{p(1-p)}{n}}$ rather than $\sqrt{\dfrac{\pi(1-\pi)}{n}}$, for example), so that the second interval is the prototype confidence interval. It is common practice to refer to both the standard deviation of a statistic and the *estimated* standard deviation of a statistic as the *standard error*. In this textbook, when we use the term *standard error*, we mean the estimated standard deviation of a statistic.

▪ **Definition**

The **standard error** of a statistic is the estimated standard deviation of the statistic.

The 95% confidence interval for π is based on the fact that, for approximately 95% of all random samples, p is within $1.96\sqrt{\dfrac{\pi(1-\pi)}{n}}$ of π. The quantity $1.96\sqrt{\dfrac{\pi(1-\pi)}{n}}$ is sometimes called the *bound on the error of estimation* associated with a 95% confidence level — we have 95% confidence that the point estimate p is no farther than this quantity from π.

▪ **Definition**

If the sampling distribution of a statistic is (at least approximately) normal, the **bound on the error of estimation B** associated with a 95% confidence interval is (1.96)(standard error of the statistic).

▪ Choosing the Sample Size

Before collecting any data, an investigator may wish to determine a sample size for which a particular value of the bound on the error is achieved. For example, with π representing the true proportion of students at a university who purchase textbooks over the Internet, the objective of an investigation may be to estimate π to within .05 with 95% confidence. The value of n necessary to achieve this is obtained by equating .05 to $1.96\sqrt{\dfrac{\pi(1-\pi)}{n}}$ and solving for n.

In general, suppose that we wish to estimate π to within an amount B (the specified bound on the error of estimation) with 95% confidence. Finding the necessary sample size requires solving the equation

$$B = 1.96\sqrt{\frac{\pi(1-\pi)}{n}}$$

Solving this equation for n results in

$$n = \pi(1 - \pi)\left(\frac{1.96}{B}\right)^2$$

Unfortunately, the use of this formula requires the value of π, which is unknown. One possible way to proceed is to carry out a preliminary study and use the resulting data to get a rough estimate of π. In other cases, prior knowledge may suggest a reasonable estimate of π. If there is no reasonable basis for estimating π and a preliminary study is not feasible, a conservative solution follows from the observation that $\pi(1 - \pi)$ is never larger than .25 (its value when $\pi = .5$). Replacing $\pi(1 - \pi)$ with .25, the maximum value, yields

$$n = .25\left(\frac{1.96}{B}\right)^2$$

Using this formula to obtain n gives us a sample size for which we can be 95% confident that p will be within B of π, no matter what the value of π.

The sample size required to estimate a population proportion π to within an amount B with 95% confidence is

$$n = \pi(1 - \pi)\left(\frac{1.96}{B}\right)^2$$

The value of π may be estimated using prior information. In the absence of any such information, using $\pi = .5$ in this formula gives a conservatively large value for the required sample size (this value of π gives a larger n than would any other value).

■ Example 9.6   Doctor-Assisted Suicide

The 1991 publication of the book *Final Exit*, which includes chapters on doctor-assisted suicide, caused a great deal of controversy in the medical community. The Society for the Right to Die and the American Medical Association quoted very different figures regarding the proportion of primary-care physicians who have participated in some form of doctor-assisted suicide for terminally ill patients (*USA Today*, July 1991). Suppose that a survey of physicians is to be designed to estimate this proportion to within .05 with 95% confidence. How many primary-care physicians should be included in a random sample?

Using a conservative value of $\pi = .5$ in the formula for required sample size gives

$$n = \pi(1 - \pi)\left(\frac{1.96}{B}\right)^2 = .25\left(\frac{1.96}{.05}\right)^2 = 384.16$$

Thus, a sample of at least 385 doctors should be used. Note that in sample size calculations, we always round up.

▪ Exercises 9.11–9.27

9.11 For each of the following choices, explain which would result in a wider large-sample confidence interval for π:
a. 90% confidence level or 95% confidence level
b. $n = 100$ or $n = 400$

9.12 The formula used to compute a large-sample confidence interval for π is

$$p \pm (z \text{ critical value})\sqrt{\frac{p(1-p)}{n}}$$

What is the appropriate z critical value for each of the following confidence levels?
a. 95%   **b.** 90%   **c.** 99%
d. 80%   **e.** 85%

9.13 The use of the interval

$$p \pm (z \text{ critical value})\sqrt{\frac{p(1-p)}{n}}$$

requires a large sample. For each of the following combinations of n and p, indicate whether the given interval would be appropriate.
a. $n = 50$ and $p = .30$ **b.** $n = 50$ and $p = .05$
c. $n = 15$ and $p = .45$ **d.** $n = 100$ and $p = .01$
e. $n = 100$ and $p = .70$ **f.** $n = 40$ and $p = .25$
g. $n = 60$ and $p = .25$ **h.** $n = 80$ and $p = .10$

9.14 Discuss how each of the following factors affects the width of the confidence interval for π:
a. The confidence level
b. The sample size
c. The value of p

9.15 The article "Doctors Cite Burnout in Mistakes" (*San Luis Obispo Tribune*, March 5, 2002) reported that many doctors who are completing their residency have financial struggles that could interfere with training. In a sample of 115 residents, 38 reported that they worked moonlighting jobs and 22 reported a credit card debt of more than $3000. Suppose that it is reasonable to consider this sample of 115 as a random sample of all medical residents in the United States.
a. Construct and interpret a 95% confidence interval for the proportion of U.S. medical residents who work moonlighting jobs.
b. Construct and interpret a 90% confidence interval for the proportion of U.S. medical residents who have a credit card debt of more than $3000.
c. Give two reasons why the confidence interval in Part (a) is wider than the confidence interval in Part (b).

9.16 Girls younger than 18 seeking services at Planned Parenthood family planning clinics in Wisconsin were surveyed to determine whether mandatory parental notification would cause them to stop using sexual health services. Of the 118 girls surveyed, 55 indicated that they would stop using all services if their parents were informed they were seeking prescription contraceptives. Construct a 95% confidence interval for the proportion who would stop using services if parents were informed. To what population would it be reasonable to generalize this confidence interval estimate?

9.17 The National Geographic Society conducted a study that included 3000 respondents, age 18 to 24, in 9 different countries (*San Luis Obispo Tribune*, November 21, 2002). The Society found that 10% of the participants could not identify their own country on a blank world map.
a. Construct a 90% confidence interval for the proportion who can identify their own country on a blank world map.
b. What assumptions are necessary for the confidence interval in Part (a) to be valid?
c. To what population would it be reasonable to generalize the confidence interval estimate from Part (a)?

9.18 "Tongue Piercing May Speed Tooth Loss, Researchers Say" is the headline of an article that appeared in the *San Luis Obispo Tribune* (June 5, 2002). The article describes a study of 52 young adults with pierced tongues. The researchers found receding gums, which can lead to tooth loss, in 18 of the participants. Construct a 95% confidence interval for the proportion of young adults with pierced tongues who have receding gums. What assumptions must be made for use of the z confidence interval to be appropriate?

9.19 *USA Today* (October 14, 2002) reported that 36% of adult drivers admit that they often or sometimes talk on a cell phone when driving. This estimate was based on data from a sample of 1004 adult drivers, and a bound on the error of estimation of 3.1% was reported. Explain how the given bound on the error can be justified.

9.20 The article "Consumers Show Increased Liking for Diesel Autos" (*USA Today*, January 29, 2003) reported that 27% of U.S. consumers would opt for a diesel car if it ran as cleanly and performed as well as a car with a gas engine. Suppose that you suspect that the proportion might be different in your area and that you want to conduct a survey to estimate this proportion for the adult residents of your city. What is the required sample size if you want to estimate this proportion to within .05 with 95% confidence? Compute the required sample size first using .27 as a preliminary estimate of π and then using the conservative value of .5. How do the two sample sizes compare? What sample size would you recommend for this study?

9.21 The Gallup Organization conducts an annual survey on crime victimizations. It was reported that 25% of all households experienced some sort of crime during the past year. This estimate was based on a sample of 1002 randomly selected adults. The report states, "One can say with 95% confidence that the margin of sampling error is ± 3 percentage points." Explain how this statement can be justified.

9.22 The article "Ban on Smoking Shown to Improve Lung Health" (*Los Angeles Times*, December 9, 1998) reported the results of a study of San Francisco bartenders. Of 39 bartenders who reported respiratory problems before a ban on smoking in bars, 21 were symptom-free 2 months after the ban began. Suppose that it is reasonable to regard this group of 39 bartenders as a random sample of bartenders with reported respiratory problems. Assuming that there are more than 400 bartenders with respiratory problems, use this information to construct a 95% confidence interval for the proportion of bartenders with respiratory problems who were symptom-free 2 months after the smoking ban.

9.23 The *Princeton Metro Times* (September 25, 1999) reported that 48% of a random sample of 369 students at the College of New Jersey indicated that they were "binge drinkers." Binge drinking was defined as consuming five to six drinks in one sitting for men and four to five drinks in one sitting for women. Construct and interpret a 90% confidence interval for π, the proportion of students at the College of New Jersey who are binge drinkers.

9.24 According to a 1998 survey of 4000 randomly selected public school teachers, many full-time public school teachers feel inadequately prepared for various classroom situations. Of those surveyed, 21% indicated that they felt unprepared to address the needs of students with disabilities and 20% stated that they felt unprepared to integrate technology into the grade or subject they taught.

a. Construct a 99% confidence interval for the true proportion of teachers who feel unprepared to integrate technology into the classroom. Is your interval relatively narrow or wide? What aspect of this problem explains the interval width?
b. Construct and interpret a 95% confidence interval for the proportion of teachers who feel unprepared to address the needs of students with disabilities.
c. Because a higher percentage of the sample reported that they felt unprepared to address the needs of students with disabilities (21%) than reported that they felt unprepared to integrate technology (20%), is it reasonable to conclude that this proportion is also higher for the entire population of teachers? Use the results of Parts (a) and (b) to justify your answer.

9.25 In the article "Fluoridation Brushed Off by Utah" (Associated Press, August 24, 1998), it was reported that a small but vocal minority in Utah has been successful in keeping fluoride out of Utah water supplies despite evidence that fluoridation reduces tooth decay and despite the fact that a clear majority of Utah residents favor fluoridation. To support this statement, the article included the result of a survey of Utah residents that found 65% to be in favor of fluoridation. Suppose that this result was based on a random sample of 150 Utah residents. Construct and interpret a 90% confidence interval for π, the true proportion of Utah residents who favor fluoridation. Is this interval consistent with the statement that fluoridation is favored by a clear majority of residents?

9.26 In a study of 1710 schoolchildren in Australia (*Herald Sun*, October 27, 1994), 1060 children indicated that they normally watch TV before school in the morning. (Interestingly, only 35% of the parents said their children watched TV before school!) Construct a 95% confidence interval for the true proportion of Australian children who say they watch TV before school. What assumption about the sample must be true for the method used to construct the interval to be valid?

9.27 A consumer group is interested in estimating the proportion of packages of ground beef sold at a particular store that have an actual fat content exceeding the fat content stated on the label. How many packages of ground beef should be tested to estimate this proportion to within .05 with 95% confidence?

■ 9.3 Confidence Interval for a Population Mean

Often the purpose of an investigation is to estimate the population mean μ. In this section, we consider how to use information from a random sample to construct a confidence interval estimate of a population mean.

We begin by considering the case in which (1) σ, **the population standard deviation, is known** (not realistic, but we will see shortly how to handle the more common situation where σ is unknown) and (2) **the sample size n is large** enough for the Central Limit Theorem to apply. In this case, the following three properties about the sampling distribution of $\bar{x}$ hold:

1. The sampling distribution of $\bar{x}$ is centered at μ, so $\bar{x}$ is an unbiased statistic for estimating μ ($\mu_{\bar{x}} = \mu$).

2. The standard deviation of $\bar{x}$ is $\sigma_{\bar{x}} = \dfrac{\sigma}{\sqrt{n}}$.

3. As long as n is large (generally $n \geq 30$), the sampling distribution of $\bar{x}$ is approximately normal, even when the population distribution itself is not normal.

The same reasoning that was used to develop the large-sample confidence interval for a population proportion π can be used to obtain a confidence interval estimate for μ.

■ **One-Sample z Confidence Interval for μ**

The general formula for a confidence interval for a population mean μ when

1. $\bar{x}$ is the sample mean from a **random sample**,

2. the **sample size n is large** (generally $n \geq 30$), and

3. σ, **the population standard deviation, is known**

is

$$\bar{x} \pm (z \text{ critical value})\left(\frac{\sigma}{\sqrt{n}}\right)$$

■ Example 9.7   Radiation Exposure

Public Citizens, a consumer group, ranks the 111 U.S. nuclear reactors according to employee exposure to radiation. Based on a sample of workers at Diablo Canyon Nuclear Power Plant, Public Citizens rated Diablo's Unit 2 reactor as the 28th worst in the country. They reported a mean annual radiation exposure of 0.481 rem for a sample of Unit 2 workers (*San Luis Obispo Telegram-Tribune*, April 11, 1990). Suppose that this mean was based on a random sample of 100 workers.

Let μ denote the true mean radiation exposure for Unit 2 workers at Diablo Canyon. Although σ, the true population standard deviation, is not usually known, suppose for illustrative purposes that $\sigma = 0.35$. Then a 95% confidence interval for μ is

$$\bar{x} \pm (z \text{ critical value})\left(\frac{\sigma}{\sqrt{n}}\right) = 0.481 \pm (1.96)\left(\frac{0.35}{\sqrt{100}}\right)$$

$$= 0.481 \pm 0.069$$

$$= (0.412, 0.550)$$

Based on this sample, *plausible* values of μ, the true mean annual radiation exposure for Diablo's Unit 2 workers, are between 0.412 and 0.550 rem. A 95% confidence level is associated with the method used to produce this interval estimate.

The confidence interval just introduced is appropriate when σ is known and n is large, and it can be used regardless of the shape of the population distribution. This is because this confidence interval is based on the Central Limit Theorem, which says that when n is sufficiently large, the sampling distribution of $\bar{x}$ is approximately normal for any population distribution. When n is small, the Central Limit Theorem cannot be used to justify the normality of the $\bar{x}$ sampling distribution, so the interval developed previously should not be used. One way to proceed in the small-sample case is to make a specific assumption about the shape of the population distribution and then to use a method of estimation that is valid only under this assumption.

The one instance where this is easy to do is when it is reasonable to believe that the population distribution is normal in shape. Recall that for a normal population distribution the sampling distribution of $\bar{x}$ is normal even for small sample sizes. So, if n is small but the population distribution is normal, the same confidence interval formula just introduced can still be used.

> If n is small (generally $n < 30$) but it is reasonable to believe that the distribution of values in the population is normal, a confidence interval for μ (when σ is known) is
>
> $$\bar{x} \pm (z \text{ critical value})\left(\frac{\sigma}{\sqrt{n}}\right)$$

There are several ways that sample data can be used to assess the plausibility of normality. The two most common ways are to look at a normal probability plot of the sample data (looking for a plot that is reasonably straight) and to construct a boxplot of the data (looking for approximate symmetry and no outliers).

▪ Confidence Interval for μ When σ Is Unknown

The confidence intervals just developed have an obvious drawback: To compute the interval endpoints, σ must be known. Unfortunately, this is rarely the case in practice. We now turn our attention to the situation when σ is unknown. The development of the confidence interval in this instance depends on the assumption that the population distribution is normal. This assumption is not critical if the sample size is large, but it is important when the sample size is small.

To understand the derivation of this confidence interval, it is instructive to begin by taking another look at the previous 95% confidence interval. We know that

$\mu_{\bar{x}} = \mu$ and $\sigma_{\bar{x}} = \dfrac{\sigma}{\sqrt{n}}$. Also, when the population distribution is normal, the $\bar{x}$ distribution is normal. These facts imply that the standardized variable

$$z = \frac{\bar{x} - \mu}{\dfrac{\sigma}{\sqrt{n}}}$$

has approximately a standard normal distribution. Because the interval from -1.96 to 1.96 captures an area of .95 under the z curve, approximately 95% of all samples result in an $\bar{x}$ value that satisfies

$$-1.96 < \frac{\bar{x} - \mu}{\dfrac{\sigma}{\sqrt{n}}} < 1.96$$

Manipulating these inequalities to isolate μ in the middle results in the equivalent inequalities:

$$\bar{x} - 1.96\left(\frac{\sigma}{\sqrt{n}}\right) < \mu < \bar{x} + 1.96\left(\frac{\sigma}{\sqrt{n}}\right)$$

The term $\bar{x} - 1.96\left(\dfrac{\sigma}{\sqrt{n}}\right)$ is the lower endpoint of the 95% large-sample confidence interval for μ, and $\bar{x} + 1.96\left(\dfrac{\sigma}{\sqrt{n}}\right)$ is the upper endpoint.

However, if σ is unknown, we must use the sample data to estimate σ. If we use the sample standard deviation as our estimate, the result is a different standardized variable denoted by t:

$$t = \frac{\bar{x} - \mu}{\dfrac{s}{\sqrt{n}}}$$

The value of s may not be all that close to σ, especially when n is small. As a consequence, the use of s in place of σ introduces extra variability, so the distribution of t is more spread out than the standard normal (z) curve. (The value of z varies from sample to sample, because different samples generally result in different $\bar{x}$ values. There is even more variability in t, because different samples may result in different values of both $\bar{x}$ and s.)

To develop an appropriate confidence interval, we must investigate the probability distribution of the standardized variable t for a sample from a normal population. This requires that we first learn about probability distributions called t *distributions*.

▪ t Distributions

Just as there are many different normal distributions, there are also many different t distributions. Although normal distributions are distinguished from one another by their mean μ and standard deviation σ, t distributions are distinguished by a positive whole number called the number of *degrees of freedom* (df). There is a t distribution with 1 df, another with 2 df, and so on.

> ▪ **Important Properties of *t* Distributions**
> 1. The *t* curve corresponding to any fixed number of degrees of freedom is bell shaped and is centered at 0 (just like the standard normal (*z*) curve).
> 2. Each *t* curve is more spread out than the *z* curve.
> 3. As the number of degrees of freedom increases, the spread of the corresponding *t* curve decreases.
> 4. As the number of degrees of freedom increases, the corresponding sequence of *t* curves approaches the *z* curve.

The properties discussed in the preceding box are illustrated in Figure 9.6, which shows two *t* curves along with the *z* curve.

Figure 9.6 Comparison of the *z* curve and *t* curves for 12 df and 4 df.

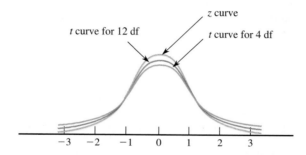

Appendix Table 3 gives selected critical values for various *t* distributions. The central areas for which values are tabulated are .80, .90, .95, .98, .99, .998, and .999. To find a particular critical value, go down the left margin of the table to the row labeled with the desired number of degrees of freedom. Then move over in that row to the column headed by the desired central area. For example, the value in the 12-df row under the column corresponding to central area .95 is 2.18, so 95% of the area under the *t* curve with 12 df lies between −2.18 and 2.18. Moving over two columns, we find the critical value for central area .99 (still with 12 df) to be 3.06 (see Figure 9.7). Moving down the .99 column to the 20-df row, we see the critical value is 2.85, so the area between −2.85 and 2.85 under the *t* curve with 20 df is .99.

Figure 9.7 *t* critical values illustrated.

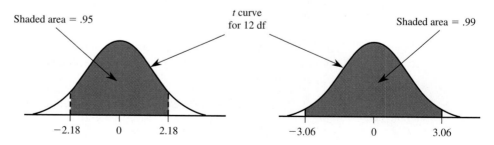

Notice that the critical values increase from left to right in each row of Appendix Table 3. This makes sense because as we move to the right, we capture larger central areas. In each column, the critical values decrease as we move downward, reflecting decreasing spread for *t* distributions with larger degrees of freedom.

The larger the number of degrees of freedom, the more closely the *t* curve resembles the *z* curve. To emphasize this, we have included the *z* critical values as the last row of the t table. Furthermore, once the number of degrees of freedom exceeds 30, the critical values change little as the number of degrees of freedom increases. For this reason, Appendix Table 3 jumps from 30 df to 40 df, then to 60 df, then to 120 df, and finally to the row of z critical values. If we need a critical value for a number of degrees of freedom between those tabulated, we just use the critical value for the closest df. For df > 120, we use the z critical values. Many graphing calculators calculate t critical values for any number of degrees of freedom; so, if you are using such a calculator, it is not necessary to approximate the t critical values as described.

■ One-Sample t Confidence Interval

The fact that the sampling distribution of $\dfrac{\bar{x} - \mu}{\sigma/\sqrt{n}}$ is approximately the z (standard normal) distribution when n is large led to the z interval when σ is known. In the same way, the following proposition provides the key to obtaining a confidence interval when the population distribution is normal and σ is unknown.

Let $x_1, x_2, \ldots, x_n$ constitute a random sample from a normal population distribution. Then the probability distribution of the standardized variable

$$t = \frac{\bar{x} - \mu}{\dfrac{s}{\sqrt{n}}}$$

is the t distribution with $(n - 1)$ df.

To see how this result leads to the desired confidence interval, consider the case $n = 25$. We use the t distribution with df = 24 $(n - 1)$. From Appendix Table 3, the interval between -2.06 and 2.06 captures a central area of $.95$ under the t curve with 24 df. Then 95% of all samples (with $n = 25$) from a normal population result in values of $\bar{x}$ and s for which

$$-2.06 < \frac{\bar{x} - \mu}{\dfrac{s}{\sqrt{n}}} < 2.06$$

Algebraically manipulating these inequalities to isolate μ yields

$$\bar{x} - 2.06\left(\frac{s}{\sqrt{25}}\right) < \mu < \bar{x} + 2.06\left(\frac{s}{\sqrt{25}}\right)$$

The 95% confidence interval for μ in this situation extends from the lower endpoint $\bar{x} - 2.06\left(\dfrac{s}{\sqrt{25}}\right)$ to the upper endpoint $\bar{x} + 2.06\left(\dfrac{s}{\sqrt{25}}\right)$. This interval can be written

$$\bar{x} \pm 2.06\left(\frac{s}{\sqrt{25}}\right)$$

The major difference between this interval and the interval when σ is known is the use of the t critical value 2.06 rather than the z critical value 1.96. The extra uncertainty that results from estimating σ causes the t interval to be wider than the z interval.

If the sample size is something other than 25 or if the desired confidence level is something other than 95%, a different t critical value (obtained from Appendix Table 3) is used in place of 2.06.

▪ **One-Sample t Confidence Interval for μ**

The general formula for a confidence interval for a population mean μ based on a sample of size n when

1. $\bar{x}$ is the sample mean from a **random sample**,
2. the **population distribution is normal**, *or* the **sample size n is large** (generally $n \geq 30$), and
3. σ, **the population standard deviation, is unknown**

is

$$\bar{x} \pm (t \text{ critical value})\left(\frac{s}{\sqrt{n}}\right)$$

where the t critical value is based on $(n-1)$ df. Appendix Table 3 gives critical values appropriate for each of the confidence levels 90%, 95%, and 99% as well as several other less frequently used confidence levels.

If n is large (generally $n \geq 30$), the normality of the population distribution is not critical. *However, this confidence interval is appropriate for small n only when the population distribution is (at least approximately) normal.* If this is not the case, as might be suggested by a normal probability plot or boxplot, another estimation method should be used.

▪ Example 9.8   Executive Salaries

"Executives Whose Wives Stay at Home Earn More" was the headline for an article that appeared in the *San Luis Obispo Telegram-Tribune* (October 14, 1994). To support this claim, the article presented data from a random sample of 231 married men who received MBA degrees in the late 1970s. For this sample, the mean salary in 1993 for married men in two-income families was $95,140, whereas the mean salary for the married men who were the sole source of family income was $124,510. The article did not give the sample standard deviations, but suppose for the purposes of this example that they are as given in the following table:

	n	Mean Salary	Standard Deviation
Two-income family	140	95,140	15,000
Sole source of income	91	124,510	18,000

If we had access to the raw data (the $140 + 91 = 231$ salary observations), we might begin by looking at boxplots. Data consistent with the given summary quantities were used to generate the boxplots of Figure 9.8. The boxplots show that salaries for the one-income group tend to be higher than those for the two-income group. There is some overlap, and the smallest salary in each group is similar. The outliers in the boxplots indicate that the income distributions may not be normal, but because the sample sizes are large, we can still use the *t* confidence interval.

Figure 9.8 Boxplots for Example 9.8.

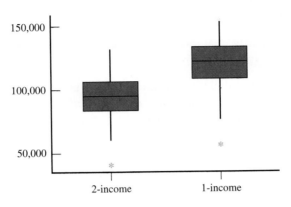

As a next step, we can use the confidence interval of this section to estimate the true mean salary of each group. Let's first focus on the sample of men from two-income families. For this group,

sample size = n = 140

sample mean salary = $\bar{x}$ = 95,140

sample standard deviation = s = 15,000

The newspaper article generalized from this sample to the population of "executives whose wives work." Is this appropriate? Remember that everyone in the sample received an MBA degree in the late 1970s. A more cautious approach would be to consider this group a sample from the population of married men who received MBAs in the late 1970s and whose wives also work. Then, with μ denoting the mean salary for this population, we can estimate μ using a 90% confidence interval.

From Appendix Table 3, we use *t* critical value = 1.645 (from the *z* critical value row because df = 139 > 120, the largest number of degrees of freedom in the table). The 90% confidence interval for μ is

$$\bar{x} \pm (t \text{ critical value})\left(\frac{s}{\sqrt{n}}\right) = 95{,}140 \pm (1.645)\left(\frac{15{,}000}{\sqrt{140}}\right)$$

$$= 95{,}140 \pm 2085.42$$

$$= (\$93{,}054.58, \$97{,}225.42)$$

Based on this sample, we are 90% confident that μ is between $93,054.58 and $97,225.42. This interval is fairly narrow, given the context of this problem; the interval spans a range of only about $4200. This indicates that our information about the value of μ is relatively precise.

A 90% confidence interval estimate for the mean of the population of married men who received MBAs in the late 1970s and who are the sole source of family income is ($121,406.03, $127,613.97). This interval is wider than the first interval for two reasons: The sample size is smaller (91 rather than 140), and the sample standard deviation is larger (18,000 rather than 15,000).

Based on these two interval estimates, it does appear that the mean salary for the two-income family group is lower than the mean for the one-income group. However, this analysis does not prove the existence of a cause-and-effect relationship between wife's work status and increased salary. The cited article falls into this trap by speculating that "the upper ranks of corporate America favor men whose wives stay at home" and that stay-at-home wives "actually further their husbands careers." Can you see the flaw in this argument?

■ **Example 9.9**   Walking a Straight Line

A study of the ability of individuals to walk in a straight line ("Can We Really Walk Straight?" *American Journal of Physical Anthropology* [1992]: 19–27) reported the following data on cadence (strides per second) for a sample of $n = 20$ randomly selected healthy men:

| 0.95 | 0.85 | 0.92 | 0.95 | 0.93 | 0.86 | 1.00 | 0.92 | 0.85 | 0.81 |
| 0.78 | 0.93 | 0.93 | 1.05 | 0.93 | 1.06 | 1.06 | 0.96 | 0.81 | 0.96 |

Figure 9.9 is a normal probability plot of these data. The plot is reasonably straight, so it seems plausible that the population distribution is approximately normal. Calculation of a confidence interval for the population mean cadence requires $\bar{x}$ and s:

$$\bar{x} = 0.926$$
$$s^2 = 0.006552$$
$$s = \sqrt{0.006552} = 0.0809$$

The t critical value for a 99% confidence interval based on 19 df is 2.86. The interval is

$$\bar{x} \pm (t \text{ critical value})\left(\frac{s}{\sqrt{n}}\right) = 0.926 \pm (2.86)\left(\frac{0.0809}{\sqrt{20}}\right)$$
$$= 0.926 \pm 0.052$$
$$= (0.874, 0.978)$$

With 99% confidence, we estimate the population mean cadence to be between 0.874 and 0.978 stride per second. Remember that the 99% confidence level implies that if the same formula is used to calculate intervals for sample after sample randomly selected from the population, in the long run 99% of these intervals will capture μ between the lower and upper confidence limits.

Figure 9.9 Normal probability plot for Example 9.9.

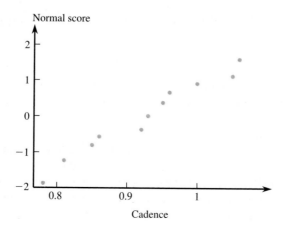

Any of the commercially available statistical computing packages can produce a one-sample t confidence interval upon request. The following is an example of output from MINITAB. The slight discrepancy between the previous interval and MINITAB's interval occurs because MINITAB uses more decimal accuracy in $\bar{x}, s,$ and t critical values.

	N	MEAN	STDEV	SE MEAN	99.0 PERCENT C.I.
CADENCE	20	0.9255	0.0809	0.0181	(0.8737, 0.9773)

■ **Example 9.10**   Housework

How much time do school-age children spend helping with housework? The article "The Three Corners of Domestic Labor: Mothers', Fathers', and Children's Weekday and Weekend Housework" (*Journal of Marriage and the Family* [1994]: 657–668) gave information on the number of minutes per weekday spent on housework. The following mean and standard deviation are for a random sample of 26 girls in two-parent families where both parents work full-time: $\bar{x} = 14.0$ and $s = 8.6$.

The authors of the article analyzed these data using methods designed for population distributions that are approximately normal. This assumption appears a bit questionable based on the reported mean and standard deviation (it is impossible to spend less than 0 min per day on housework, so the smallest possible value, 0, is only 1.63 standard deviations below the mean). However, because the authors reported that there were no outliers in the data and because n is relatively close to 30, we use the t confidence interval formula of this section to compute a 95% confidence interval.

Because $n = 26$, df $= 25$ and the appropriate t critical value is 2.06. The confidence interval is then

$$\bar{x} \pm (t \text{ critical value})\left(\frac{s}{\sqrt{n}}\right) = 14.0 \pm (2.06)\left(\frac{8.6}{\sqrt{26}}\right)$$

$$= 14.0 \pm 3.5$$

$$= (10.5, 17.5)$$

Based on the sample data, we believe that the true mean time per weekday spent on housework is between 10.5 and 17.5 min for girls in two-parent families where both parents work. We used a method that has a 5% error rate to construct this interval. We should be somewhat cautious in interpreting this confidence interval because of the concern expressed about the normality of the population distribution.

■ **Choosing the Sample Size**

When estimating μ using a large sample or a small sample from a normal population with known σ, the bound B on the error of estimation associated with a 95% confidence interval is

$$B = 1.96\left(\frac{\sigma}{\sqrt{n}}\right)$$

Before collecting any data, an investigator may wish to determine a sample size for which a particular value of the bound is achieved. For example, with μ representing the average fuel efficiency (in miles per gallon, mpg) for all cars of a certain type, the objective of an investigation may be to estimate μ to within 1 mpg with 95% confidence. The value of n necessary to achieve this is obtained by setting $B = 1$ and then solving $1 = 1.96\left(\dfrac{\sigma}{\sqrt{n}}\right)$ for n.

In general, suppose that we wish to estimate μ to within an amount B (the specified bound on the error of estimation) with 95% confidence. Finding the necessary sample size requires solving the equation $B = 1.96\left(\dfrac{\sigma}{\sqrt{n}}\right)$ for n. The result is

$$n = \left(\frac{1.96\sigma}{B}\right)^2$$

Notice that, in general, a large value of σ forces n to be large, as does a small value of B. Use of the sample-size formula requires that σ be known, but this is rarely the case in practice. One possible strategy for estimating σ is to carry out a preliminary study and use the resulting sample standard deviation (or a somewhat larger value, to be conservative) to determine n for the main part of the study. Another possibility is simply to make an educated guess about the value of σ and to use that value to calculate n. For a population distribution that is not too skewed, dividing the range (the difference between the largest and the smallest values) by 4 often gives a rough idea of the value of the standard deviation.

The sample size required to estimate a population mean μ to within an amount B with 95% confidence is

$$n = \left(\frac{1.96\sigma}{B}\right)^2$$

If σ is unknown, it can be estimated based on previous information or, for a population that is not too skewed, by using (range)/4.

If the desired confidence level is something other than 95%, 1.96 is replaced by the appropriate z critical value (e.g., 2.58 for 99% confidence).

■ **Example 9.11   Cost of Textbooks**

The financial aid office wishes to estimate the mean cost of textbooks per quarter for students at a particular university. For the estimate to be useful, it should be within $20 of the true population mean. How large a sample should be used to be 95% confident of achieving this level of accuracy?

To determine the required sample size, we must have a value for σ. The financial aid office is pretty sure that the amount spent on books varies widely, with most values between $50 and $450. A reasonable estimate of σ is then

$$\frac{\text{range}}{4} = \frac{450 - 50}{4} = \frac{400}{4} = 100$$

The required sample size is

$$n = \left(\frac{1.96\sigma}{B}\right)^2 = \left(\frac{(1.96)(100)}{20}\right)^2 = (9.8)^2 = 96.04$$

Rounding up, a sample size of 97 or larger is recommended.

▪ Exercises 9.28–9.48

9.28 Given a variable that has a t distribution with the specified degrees of freedom, what percentage of the time will its value fall in the indicated region?
a. 10 df, between -1.81 and 1.81
b. 10 df, between -2.23 and 2.23
c. 24 df, between -2.06 and 2.06
d. 24 df, between -2.80 and 2.80
e. 24 df, outside the interval from -2.80 to 2.80
f. 24 df, to the right of 2.80
g. 10 df, to the left of -1.81

9.29 The formula used to compute a confidence interval for the mean of a normal population when n is small is

$$\bar{x} \pm (t \text{ critical value})\frac{s}{\sqrt{n}}$$

What is the appropriate t critical value for each of the following confidence levels and sample sizes?
a. 95% confidence, $n = 17$
b. 90% confidence, $n = 12$
c. 99% confidence, $n = 24$
d. 90% confidence, $n = 25$
e. 90% confidence, $n = 13$
f. 95% confidence, $n = 10$

9.30 The two intervals (114.4, 115.6) and (114.1, 115.9) are confidence intervals for μ = true average resonance frequency (in hertz) for all tennis rackets of a certain type.
a. What is the value of the sample mean resonance frequency?
b. The confidence level for one of these intervals is 90% and for the other it is 99%. Which is which, and how can you tell?

9.31 Samples of two different types of automobiles were selected, and the actual speed for each car was determined when the speedometer registered 50 mph. The resulting 95% confidence intervals for true average actual speed were (51.3, 52.7) and (49.4, 50.6). Assuming that the two sample standard deviations are identical, which confidence interval is based on the larger sample size? Explain your reasoning.

9.32 Suppose that a random sample of 50 bottles of a particular brand of cough medicine is selected and

the alcohol content of each bottle is determined. Let μ denote the average alcohol content for the population of all bottles of the brand under study. Suppose that the sample of 50 results in a 95% confidence interval for μ of (7.8, 9.4).
a. Would a 90% confidence interval have been narrower or wider than the given interval? Explain your answer.
b. Consider the following statement: There is a 95% chance that μ is between 7.8 and 9.4. Is this statement correct? Why or why not?
c. Consider the following statement: If the process of selecting a sample of size 50 and then computing the corresponding 95% confidence interval is repeated 100 times, 95 of the resulting intervals will include μ. Is this statement correct? Why or why not?

9.33 Because of safety considerations, in May 2003 the Federal Aviation Administration (FAA) changed its guidelines for how small commuter airlines must estimate passenger weights. Under the old rule, airlines used 180 lb as a typical passenger weight (including carry-on luggage) in warm months and 185 lb as a typical weight in cold months. The *Alaska Journal of Commerce* (May 25, 2003) reported that Frontier Airlines conducted a study to estimate average passenger plus carry-on weights. They found an average summer weight of 183 lb and a winter average of 190 lb. Suppose that each of these estimates was based on a random sample of 100 passengers and that the sample standard deviations were 20 lb for the summer weights and 23 lb for the winter weights.
a. Construct and interpret a 95% confidence interval for the mean summer weight (including carry-on luggage) of Frontier Airlines passengers.
b. Construct and interpret a 95% confidence interval for the mean winter weight (including carry-on luggage) of Frontier Airlines passengers.
c. The new FAA recommendations are 190 lb for summer and 195 lb for winter. Comment on these recommendations in light of the confidence interval estimates from Parts (a) and (b).

9.34 "Heinz Plays Catch-up After Under-Filling Ketchup Containers" is the headline of an article that appeared on CNN.com (November 30, 2000). The article stated that Heinz had agreed to put an extra 1% of ketchup into each ketchup container sold in California for a 1-year period. Suppose that you want to make sure that Heinz is in fact fulfilling its end of the agreement. You plan to take a sample of 20-oz bottles shipped to California, measure the amount of ketchup in each bottle, and then use the resulting data to estimate the mean amount of ketchup in each bottle. A small pilot study showed that the amount of ketchup in 20-oz bottles varied from 19.9 to 20.3 oz. How many bottles should be included in the sample if you want to estimate the true mean amount of ketchup to within 0.1 oz with 95% confidence?

9.35 Example 9.3 gave the following airborne times for United Airlines flight 448 from Albuquerque to Denver on 10 randomly selected days:

57   54   55   51   56   48   52   51   59   59

a. Compute and interpret a 90% confidence interval for the mean airborne time for flight 448.
b. Give an interpretation of the 90% confidence level associated with the interval estimate in Part (a).
c. Based on your interval in Part (a), if flight 448 is scheduled to depart at 10 A.M., what would you recommend for the published arrival time? Explain.

9.36 The authors of the paper "Short-Term Health and Economic Benefits of Smoking Cessation: Low Birth Weight" (*Pediatrics* [1999]: 1312–1320) investigated the medical cost associated with babies born to mothers who smoke. The paper included estimates of mean medical cost for low-birth-weight babies for different ethnic groups. For a sample of 654 Hispanic low-birth-weight babies, the mean medical cost was

$55,007 and the standard error $\left(\dfrac{s}{\sqrt{n}} \right)$ was $3011.

For a sample of 13 Native American low-birth-weight babies, the mean and standard error were $73,418 and $29,577, respectively. Explain why the two standard errors are so different.

9.37 Seventy-seven students at the University of Virginia were asked to keep a diary of a conversation with their mothers, recording any lies they told during these conversations (*San Luis Obispo Telegram-Tribune*, August 16, 1995). It was reported that the mean number of lies per conversation was 0.5. Suppose that the standard deviation (which was not reported) was 0.4.

a. Suppose that this group of 77 is a random sample from the population of students at this university. Construct a 95% confidence interval for the mean number of lies per conversation for this population.

b. The interval in Part (a) does not include 0. Does this imply that all students lie to their mothers? Explain.

9.38 The article "Selected Characteristics of High-Risk Students and Their Enrollment Persistence" (*Journal of College Student Development* [1994]: 54–60) examined factors that affect whether students stay in college. The following summary statistics are based on data from a sample of 44 students who did not return to college after the first quarter (the nonpersisters) and a sample of 257 students who did return (the persisters):

	Number of Hours Worked per Week During the First Quarter	
	Mean	Standard Deviation
Nonpersisters	25.62	14.41
Persisters	18.10	15.31

a. Consider the 44 nonpersisters as a random sample from the population of all nonpersisters at the university where the data were collected. Compute a 98% confidence interval for the mean number of hours worked per week for nonpersisters.
b. Consider the 257 persisters as a random sample from the population of all persisters at the university where the data were collected. Compute a 98% confidence interval for the mean number of hours worked per week for persisters.
c. The 98% confidence interval for persisters is narrower than the corresponding interval for nonpersisters, even though the standard deviation for persisters is larger than that for nonpersisters. Explain why this happened.
d. Based on the interval in Part (a), do you think that the mean number of hours worked per week for nonpersisters is greater than 20? Explain.

9.39 A number of cities have implemented needle exchange programs in an attempt to slow the spread of AIDS among intravenous drug users. These programs are designed to remove infected needles from circulation. One such program in New Haven kept information on the length of time a needle remained in circulation. It was reported that the mean circulation time was 2.2 days ("What Happened to HIV Transmission Among Drug Injectors in New Haven?" *Chance* [Spring 1993]: 9–14). The sample size and standard deviation for circulation time were not given in the article, but suppose that they were $n = 25$ needles and $s = 1.2$ days. Compute a 90% confidence interval for μ, the mean circulation time for needles recovered by this exchange program.

9.40 The eating habits of 12 bats were examined in the article "Foraging Behavior of the Indian False Vampire Bat" (*Biotropica* [1991]: 63–67). These bats consume insects and frogs. For these 12 bats, the mean time to consume a frog was $\bar{x} = 21.9$ min. Suppose that the standard deviation was $s = 7.7$ min. Construct and interpret a 90% confidence interval for the mean suppertime of a vampire bat whose meal consists of a frog. What assumptions must be reasonable for the one-sample t interval to be appropriate?

9.41 Fat contents (in percentage) for 10 randomly selected hot dogs were given in the article "Sensory and Mechanical Assessment of the Quality of Frankfurters" (*Journal of Texture Studies* [1990]: 395–409). Use the following data to construct a 90% confidence interval for the true mean fat percentage of hot dogs:

25.2   21.3   22.8   17.0   29.8   21.0   25.5
16.0   20.9   19.5

9.42 A triathlon consisting of swimming, cycling, and running is one of the more strenuous amateur sporting events. The article "Cardiovascular and Thermal Response of Triathlon Performance" (*Medicine and Science in Sports and Exercise* [1988]: 385–389) reported on a research study involving nine randomly selected male triathletes. Maximum heart rate (in beats per minute) was recorded while the athlete performed each of the three events. The results were:

	$\bar{x}$	s
Swimming	188	7.2
Biking	186	8.5
Running	194	7.8

a. Assuming that the heart-rate distribution for each event is approximately normal, construct 95% confidence intervals for the true mean heart rate of triathletes for each event.
b. Do the intervals in Part (a) overlap? Based on the computed intervals, do you think there is evidence that the mean maximum heart rate is higher for running than for the other two events? Explain.

9.43 Five students visiting the student health center for a free dental examination during National Dental Hygiene Month were asked how many months had passed since their last visit to a dentist. Their responses were as follows:

6   17   11   22   29

Assuming that these five students can be considered a random sample of all students participating in the free checkup program, construct a 95% confidence

interval for the mean number of months elapsed since the last visit to a dentist for the population of students participating in the program.

9.44 The article "First Year Academic Success: A Prediction Combining Cognitive and Psychosocial Variables for Caucasian and African American Students" (*Journal of College Student Development* [1999]: 599–610) reported that the sample mean and standard deviation for high school grade point average (GPA) for students enrolled at a large research university were 3.73 and 0.45, respectively. Suppose that the mean and standard deviation were based on a random sample of 900 students at the university.
a. Construct a 95% confidence interval for the mean high school GPA for students at this university.
b. Suppose that you wanted to make a statement about the range of GPAs for students at this university. Is it reasonable to say that 95% of the students at the university have GPAs in the interval you computed in Part (a)? Explain.

9.45 The following data are the calories per half-cup serving for 16 popular chocolate ice cream brands reviewed by *Consumer Reports* (July 1999):

270   150   170   140   160   160   160   290
190   190   160   170   150   110   180   170

Is it reasonable to use the t confidence interval to compute a confidence interval for μ, the true mean calories per half-cup serving of chocolate ice cream? Explain why or why not.

9.46 The Bureau of Alcohol, Tobacco, and Firearms (BATF) has been concerned about lead levels in California wines. In a previous testing of wine specimens, lead levels ranging from 50 to 700 parts per billion were recorded (*San Luis Obispo Telegram-Tribune*, June 11, 1991). How many wine specimens should be tested if the BATF wishes to estimate the true mean lead level for California wines to within 10 parts per billion with 95% confidence?

9.47 The article "*National Geographic*, the Doomsday Machine," which appeared in the March 1976 issue of the *Journal of Irreproducible Results* (yes, there really is a journal by that name — it's a spoof of technical journals!) predicted dire consequences resulting from a nationwide buildup of *National Geographic* magazines. The author's predictions are based on the observation that the number of subscriptions for *National Geographic* is on the rise and that no one ever throws away a copy of *National Geographic*. A key to the analysis presented in the article is the weight of an issue of the magazine. Suppose that you were assigned the task of estimating the average weight of an issue of *National Geographic*.

How many issues should you sample to estimate the average weight to within 0.1 oz with 95% confidence? Assume that σ is known to be 1 oz.

9.48 The formula described in this section for determining sample size corresponds to a confidence level of 95%. What would be the appropriate formula for determining sample size when the desired confidence level is 90%? 98%?

▪ 9.4 Communicating and Interpreting the Results of Statistical Analyses

The purpose of most surveys and many research studies is to produce estimates of population characteristics. One way of providing such an estimate is to construct and report a confidence interval for the population characteristic of interest.

▪ Communicating the Results of a Statistical Analysis

When using sample data to estimate a population characteristic, a point estimate or a confidence interval estimate might be used. Confidence intervals are generally preferred because a point estimate by itself does not convey any information about the accuracy of the estimate. For this reason, whenever you report the value of a point estimate, it is a good idea to also include an estimate of the bound on the error of estimation.

Reporting and interpreting a confidence interval estimate requires a bit of care. First, always report both the confidence interval and the confidence level associated with the method used to produce the interval. Then, remember that both the confidence interval and the confidence level should be interpreted. A good strategy is to begin with an interpretation of the confidence interval in the context of the problem and then to follow that with an interpretation of the confidence level. For example, if a 90% confidence interval for π, the proportion of students at a particular university who own a computer, is (.36, .58), we might say

interpretation of interval $\rightarrow$ { We can be 90% confident that between 36% and 58% of the students at this university own computers.

explanation of "90% confidence"; interpretation of confidence level $\rightarrow$ { We have used a method to produce this estimate that is successful in capturing the actual population proportion 90% of the time.

When providing an interpretation of a confidence interval, remember that the interval is an estimate of a population characteristic and be careful not to say that the interval applies to individual values in the population or to the values of sample statistics. For example, if a 99% confidence interval for μ, the mean amount of ketchup in bottles labeled as 12 oz, is (11.94, 11.98), this does not tell us that 99% of 12-oz ketchup bottles contain between 11.94 and 11.98 oz of ketchup. Nor does it tell us that 99% of samples of the same size would have sample means in this particu-

lar range. The confidence interval is an estimate of the *mean* for all bottles in the *population* of interest.

■ Interpreting the Results of a Statistical Analysis

Unfortunately, there is no customary way of reporting the estimates of population characteristics in published sources. Possibilities include

confidence interval

estimate $\pm$ bound on error

estimate $\pm$ standard error

If the population characteristic is a population mean, then you may also see

sample mean $\pm$ sample standard deviation

If the interval reported is described as a confidence interval, a confidence level should accompany it. These intervals can be interpreted just as we have interpreted the confidence intervals in this chapter, and the confidence level specifies the long-run error rate associated with the method used to construct the interval (e.g., a 95% confidence level specifies a 5% long-run error rate).

A form particularly common in news articles is estimate $\pm$ bound on error, where the bound on error is also sometimes called the **margin of error**. The bound on error reported is usually 2 times the standard deviation of the estimate. This method of reporting is a little more informal than a confidence interval and, if the sample size is reasonably large, is roughly equivalent to reporting a 95% confidence interval. You can interpret these intervals as you would a confidence interval with approximate confidence level of 95%.

You must use care in interpreting intervals reported in the form of an estimate $\pm$ standard error. Recall from Section 9.2 that the general form of a confidence interval is

estimate $\pm$ (critical value)(standard deviation of the estimate)

In journal articles, the estimated standard deviation of the estimate is usually referred to as the *standard error*. The critical value in the confidence interval formula was determined by the form of the sampling distribution of the estimate and by the confidence level. Note that the reported form, estimate $\pm$ standard error, is equivalent to a confidence interval with the critical value set equal to 1. For a statistic whose sampling distribution is (approximately) normal (such as the mean of a large sample or a large-sample proportion), a critical value of 1 corresponds to an approximate confidence level of about 68%. Because a confidence level of 68% is rather low, you may want to use the given information and the confidence interval formula to convert to an interval with a higher confidence level.

When researchers are trying to estimate a population mean, they sometimes report sample mean $\pm$ sample standard deviation. Be particularly careful here. To convert this information into a useful interval estimate of the population mean, you must first convert the sample standard deviation to the standard error of the sample mean (by dividing by $\sqrt{n}$) and then use the standard error and an appropriate critical value to construct a confidence interval.

For example, suppose that a random sample of size 100 is used to estimate the population mean. If the sample resulted in a sample mean of 500 and a sample

standard deviation of 20, you might find the published results summarized in any of the following ways:

> 95% confidence interval for the population mean: (496.08, 503.92)
>
> mean ± bound on error: 500 ± 4
>
> mean ± standard error: 500 ± 2
>
> mean ± standard deviation: 500 ± 20

▪ What to Look For in Published Data

Here are some questions to ask when you encounter interval estimates in research reports.

- Is the reported interval a confidence interval, mean ± bound on error, mean ± standard error, or mean ± standard deviation? If the reported interval is not a confidence interval, you may want to construct a confidence interval from the given information.

- What confidence level is associated with the given interval? Is the choice of confidence level reasonable? What does the confidence level say about the long-run error rate of the method used to construct the interval?

- Is the reported interval relatively narrow or relatively wide? Has the population characteristic been estimated precisely?

For example, the article "Increased Vital and Total Lung Capacity in Tibetan Compared to Han Residents of Lhasa" (*American Journal of Physical Anthropology* [1991]: 341–351) compared various physical characteristics of people who live at high altitudes with those of people who live at sea level. The article includes the following statements: "We studied 38 Tibetan and 43 Han residents. . . . The Tibetan compared with the Han subjects had a larger total lung capacity [6.80 ± 0.19 (mean ± SEM) vs. 6.24 ± 0.18 liters]."

The reported intervals are of the form estimate ± standard error. We can use this information to construct a confidence interval for the mean total lung capacity for residents of each of the two locations. Because the sample sizes are both large, we can use the t confidence interval formula

> mean ± (t critical value)(standard deviation of the mean)

The Tibetan sample has df = 38 − 1 = 37 and the Han sample has df = 43 − 1 = 42. For both of these numbers of degrees of freedom, the closest df value in Appendix Table 3 is for df = 40, and the corresponding t critical value for a 95% confidence level is 2.02. The corresponding intervals are

> Tibetan residents: 6.80 ± 2.02(0.19) = 6.80 ± 0.38 = (6.42, 7.18)
>
> Han residents: 6.24 ± 2.02(0.18) = 6.24 ± 0.36 = (5.88, 6.60)

The chosen confidence level of 95% implies that the method used to construct each of the intervals has a 5% long-run error rate. Assuming that it is reasonable to view these samples as random samples, we can interpret these intervals as follows: Based on the information provided in the sample of Tibetan residents, we can be 95% confident that the mean total lung capacity of Tibetan residents is between 6.42 and 7.18, and based on the information provided in the sample of Han residents, we can

be 95% confident that the mean total lung capacity of Han residents is between 5.88 and 6.60. These intervals are not very narrow, indicating that the value of the population mean has not been estimated as precisely as we might like in either case. This is not surprising, given the reported sample sizes and the variability in each sample. Note that the two intervals overlap. This may cause us to be skeptical of the statement that Tibetans have a higher total lung capacity than Han residents. Formal methods for directly comparing two groups, covered in Chapter 11, could be used to further investigate this issue.

■ A Word to the Wise: Cautions and Limitations

When working with point and confidence interval estimates, here are a few things you need to keep in mind:

1. In order for an estimate to be useful, we must know something about accuracy. You should beware of point estimates that are not accompanied by a bound on error or some other measure of accuracy.

2. A confidence interval estimate that is wide indicates that we don't have precise information about the population characteristic being estimated. Don't be fooled by a high confidence level if the resulting interval is wide. High confidence, although desirable, is not the same thing as saying that we have precise information about the value of a population characteristic.

    The width of a confidence interval is affected by the confidence level, the sample size, and the standard deviation of the statistic (e.g., p or $\bar{x}$) used as the basis for constructing the interval. The best strategy for decreasing the width of a confidence interval is to take a larger sample. It is far better to think about this before collecting data and to use the required sample size formulas to determine a sample size that will result in a confidence interval estimate that is narrow enough to provide useful information.

3. The accuracy of estimates depends on the sample size, not on the population size. This may be counter to intuition, but as long as the sample size is small relative to the population size (n less than 10% of the population size), the bound on the error of estimation for estimating a population proportion with 95% confidence is approximately $2\sqrt{\dfrac{p(1 - p)}{n}}$ and the bound on error for estimating a population mean with 95% confidence is approximately $2\dfrac{s}{\sqrt{n}}$.

    Note that each of these expressions involves the sample size n, and both bounds decrease as the sample size increases. Neither approximate bound on error depends on the population size.

    The size of the population N does need to be considered if sampling is without replacement and the sample size is more than 10% of the population size. In this case, a **finite population correction factor** $\sqrt{\dfrac{N - n}{N - 1}}$ is used to adjust the bound on error (the given bound is multiplied by the correction factor). Because this correction factor is always less than 1, the adjusted bound on error is smaller.

4. Assumptions and "plausibility" conditions are important. The confidence interval procedures of this chapter require certain assumptions. If these

assumptions are met, the confidence intervals provide us with a method for using sample data to estimate population characteristics with confidence. When the assumptions associated with a confidence interval procedure are in fact true, the confidence level specifies a correct success rate for the method. However, assumptions (such as the assumption of a normal population distribution) are rarely exactly met in practice. Fortunately, in most cases, as long as the assumptions are approximately met, the confidence interval procedures still work well.

In general, we can determine only whether assumptions are "plausible" or approximately met and, consequently, whether or not we are in the situation where we expect the inferential procedure to work reasonably well. This is usually confirmed by knowledge of the data collection process and by using the sample data to check certain "plausibility conditions."

The formal assumptions for the z confidence interval for a population proportion are the following:

  a. The sample is a random sample from the population of interest.

  b. The sample size is large enough for the sampling distribution of p to be approximately normal.

  c. Sampling is without replacement.

Whether or not the random sample assumption is plausible depends on how the sample was selected and what the intended population is. Plausibility conditions for the other two assumptions are the following:

  $np \geq 10$ and $n(1 - p) \geq 10$ (so the sampling distribution of p is approximately normal), and

  n is less than 10% of the population size (so sampling with replacement approximates sampling without replacement)

The formal assumptions for the t confidence interval for a population mean are the following:

  a. The sample is a random sample from the population of interest.

  b. The population distribution is normal, so that the distribution of

  $$t = \frac{\bar{x} - \mu}{s/\sqrt{n}} \text{ has a } t \text{ distribution.}$$

The plausibility of the random sample assumption, as was the case for proportions, depends on how the sample was selected and on what the population of interest is. The plausibility conditions for the normal population distribution assumption are the following:

  A normal probability plot of the data is reasonably straight (indicating that the population distribution is approximately normal), or

  The data distribution is approximately symmetric and there are no outliers. This can be confirmed by looking at a dotplot, boxplot, stem-and-leaf display, or histogram of the data.

Alternatively, if n is large ($n \geq 30$), the sampling distribution of $\bar{x}$ will be approximately normal even for nonnormal population distributions. This implies that use of the t interval is appropriate even if population normality is not plausible.

In the end, you must decide that the assumptions are met or that they are plausible and that the chosen inferential method will provide reason-

able results. This is also true for the inferential methods introduced in Chapters 10–15.

5. Watch out for the plus or minus sign (±) when reading published reports. Don't fall into the trap of thinking "confidence interval" every time you see a plus or minus sign in an expression. As was discussed earlier in this section, published reports are not consistent, and in addition to confidence intervals, it is common to see estimate ± standard error and estimate ± sample standard deviation reported.

▪ Activity 9.1: Getting a Feel for Confidence Level

Technology Activity (Applet): Open the applet (on the text's CD) called ConfidenceIntervals. You should see a screen like the one shown:

Simulating Confidence Intervals

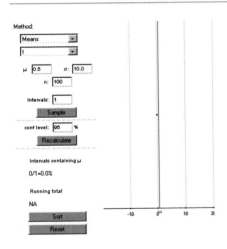

Getting Started: If the Method box in your applet window does not say "Means," use the drop-down menu to select Means. In the box just below, select "t" from the drop-down menu. This applet will select a random sample from a specified normal population distribution and then use the sample to construct a confidence interval for the population mean. The interval is then plotted on the display at the right, and we can see if the resulting interval contains the actual value of the population mean.

For the purposes of this activity, we sample from a normal population with mean 100 and standard deviation 5. We begin with a sample size of $n = 10$. In the applet window, set $\mu = 100$, $\sigma = 5$, and $n = 10$. Leave the conf-level box set at 95%. Click the Recalculate button to rescale the picture on the right. Now click on the sample button. You should see

a confidence interval appear on the display on the right-hand side. If the interval contains the actual mean of 100, the interval is drawn in green; if 100 is not in the confidence interval, the interval is shown in red. Your screen should look something like this:

Simulating Confidence Intervals

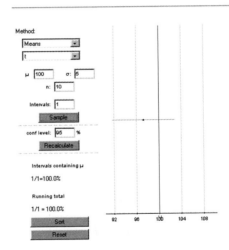

Part 1: Click on the Sample button several more times, and notice how the confidence interval estimate changes from sample to sample. Also notice that at the bottom of the left-hand side of the display, the applet is keeping track of the proportion of all the intervals calculated so far that include the actual value of μ. If we were to construct a large number of intervals, this proportion should closely approximate the capture rate for the confidence interval method.

To look at more than one interval at a time, change the Intervals box from 1 to 100 and then click the sample button. You should see a screen similar to the one below, with 100 intervals in the display on the right-hand side:

Simulating Confidence Intervals

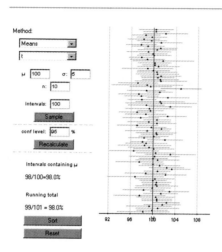

Again, intervals containing 100 (the value of μ in this case) are green and those that do not contain 100 are red. Also note that the capture proportion on the left-hand side has been updated to reflect what happened with the 100 newly generated intervals.

Continue generating intervals until you have seen at least 1000 intervals, and then answer the following questions:

a. How does the proportion of intervals constructed that contain $\mu = 100$ compare to the stated confidence level of 95%? On how many intervals was your

proportion based? (Note: If you followed the instructions, this number should be at least 1000.)

Experiment with three other confidence levels of your choice, and then answer the following question:

b. In general, is the proportion of computed t confidence intervals that contain $\mu = 100$ close to the stated confidence level?

Part 2: When the population is normal but σ is unknown, we construct a confidence interval for a population mean using a t critical value rather than a z critical value. How important is this distinction?

Let's investigate. Use the drop-down menu to change the box just below the method box that says "Means" from "t" to "z with s." The applet will now construct intervals using the sample standard deviation but will use a z critical value rather than the t critical value.

Use the applet to construct at least 1000 95% intervals, and then answer the following questions:

c. How does the proportion of the computed intervals that include the actual value of the population mean compare to the stated confidence level of 95%? Is this surprising? Explain why or why not.

Now experiment with some different sample sizes. What happens when $n = 20$? when $n = 50$? $n = 100$? Use what you have learned to write a paragraph explaining what these simulations tell you about the advisability of using a z critical value in the construction of a confidence interval for μ when σ is unknown.

▪ Activity 9.2: An Alternative Confidence Interval for a Population Proportion

Technology Activity (Applet): This activity presumes that you have already worked through Activity 9.1.

Background: In Section 9.2, it was suggested that a confidence interval of the form

$$p_{mod} \pm (z \text{ critical value})\sqrt{\frac{p_{mod}(1 - p_{mod})}{n}}$$

where $p_{mod} = \dfrac{\text{successes} + 2}{n + 4}$, can be used as an alternative to the usual large-sample z confidence interval. This alternative interval is preferred by many statisticians because, in repeated sampling, the proportion of intervals constructed that include the actual value of the population proportion, π, tends to be closer to the stated confidence level. In this activity, we explore how the capture rates for the two different interval estimation methods compare.

Open the applet (on the text's CD) called ConfidenceIntervals. You should see a screen like the one shown at right:

Simulating Confidence Intervals

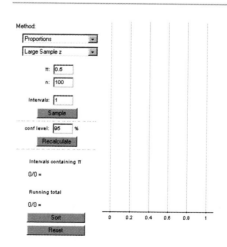

Select "Proportion" from the Method box drop-down menu, and then select "Large Sample z" from the drop-down menu of the second box. We consider sampling from a population with $\pi = .3$ using a sample of size 40. In the applet window, enter $\pi = .3$ and $n = 40$. Note that $n = 40$ is large enough to satisfy $n\pi \geq 10$ and $n(1 - \pi) \geq 10$. Set the Intervals box to 100, and then use the applet to construct a large number (at least 1000) of 95% confidence intervals.

1. How does the proportion of intervals constructed that include $\pi = .3$, the population proportion, compare to 95%? Does this surprise you? Explain.

Now use the drop-down menu to change "Large Sample z" to "Modified." Now the applet will construct the alternative confidence interval that is based on p_{mod}. Use the applet to construct a large number of 95% confidence intervals (at least 1000).

2. How does the proportion of intervals constructed that include $\pi = .3$, the population proportion, compare to 95%? Is this proportion closer to 95% than was the case for the large-sample z interval?

3. Now experiment with different combinations of values of sample size and population proportion π. Can you find a combination for which the large-sample z interval has a capture rate that is close to 95%? Can you find a combination for which it has a capture rate that is even farther from 95% than it was for $n = 40$ and $\pi = .3$? How does the modified interval perform in each of these cases?

■ Summary of Key Concepts and Formulas

Term or Formula	Comment
Point estimate	A single number, based on sample data, that represents a plausible value of a population characteristic.
Unbiased statistic	A statistic that has a sampling distribution with a mean equal to the value of the population characteristic to be estimated.
Confidence interval	An interval that is computed from sample data and provides a range of plausible values for a population characteristic.
Confidence level	A number that provides information on how much "confidence" we can have in the method used to construct a confidence interval estimate. The confidence level specifies the percentage of all possible samples that will produce an interval containing the true value of the population characteristic.
$p \pm (z \text{ critical value})\sqrt{\dfrac{p(1 - p)}{n}}$	A formula used to construct a confidence interval for π when the sample size is large.
$n = \pi(1 - \pi)\left(\dfrac{1.96}{B}\right)^2$	A formula used to compute the sample size necessary for estimating π to within an amount B with 95% confidence. (For other confidence levels, replace 1.96 with an appropriate z critical value.)
$\bar{x} \pm (z \text{ critical value})\dfrac{\sigma}{\sqrt{n}}$	A formula used to construct a confidence interval for μ when σ is known and either the sample size is large or the population distribution is normal.
$\bar{x} \pm (t \text{ critical value})\dfrac{s}{\sqrt{n}}$	A formula used to construct a confidence interval for μ when σ is unknown and either the sample size is large or the population distribution is normal.
$n = \left(\dfrac{1.96\sigma}{B}\right)^2$	A formula used to compute the sample size necessary for estimating μ to within an amount B with 95% confidence. (For other confidence levels, replace 1.96 with an appropriate z critical value.)

▪ Supplementary Exercises 9.49–9.64

9.49 Retailers report that the use of cents-off coupons is increasing. The Scripps Howard News Service (July 9, 1991) reported the proportion of all households that use coupons as .77. Suppose that this estimate was based on a random sample of 800 households (i.e., $n = 800$ and $p = .77$). Construct a 95% confidence interval for π, the true proportion of all households that use coupons.

9.50 A manufacturer of small appliances purchases plastic handles for coffeepots from an outside vendor. If a handle is cracked, it is considered defective and must be discarded. A large shipment of plastic handles is received. The proportion of defective handles π is of interest. How many handles from the shipment should be inspected to estimate π to within 0.1 with 95% confidence?

9.51 Scripps News Service (September 14, 1991) reported that 4% of the members of the American Bar Association (ABA) are African American. Suppose that this figure is based on a random sample of 400 ABA members.

a. Is the sample size large enough to justify the use of the large-sample confidence interval for a population proportion?

b. Construct and interpret a 90% confidence interval for π, the true proportion of all ABA members who are African American.

9.52 An article in the *Chicago Tribune* (August 29, 1999) reported that in a poll of residents of the Chicago suburbs, 43% felt that their financial situation had improved during the last year. The following statement is from the article: "The findings of this Tribune poll are based on interviews with 930 randomly selected suburban residents. The sample included suburban Cook County plus DuPage, Kane, Lake, McHenry, and Will Counties. In a sample of this size, one can say with 95% certainty that results will differ by no more than 3 percent from results obtained if all residents had been included in the poll."

Comment on this statement. Give a statistical argument to justify the claim that the estimate of 43% is within 3% of the true proportion of residents who feel that their financial situation has improved.

9.53 The McClatchy News Service (*San Luis Obispo Telegram-Tribune*, June 13, 1991) reported on a study of violence on television during prime-time hours. The following table summarizes the information reported for four networks:

Network	Mean Number of Violent Acts per Hour
ABC	15.6
CBS	11.9
FOX	11.7
NBC	11.0

Suppose that each of these sample means was computed on the basis of viewing $n = 50$ randomly selected prime-time hours and that the population standard deviation for each of the four networks is known to be $\sigma = 5$.

a. Compute a 95% confidence interval for the true mean number of violent acts per prime-time hour for ABC.

b. Compute 95% confidence intervals for the mean number of violent acts per prime-time hour for each of the other three networks.

c. The National Coalition on Television Violence claims that shows on ABC are more violent than those on the other networks. Based on the confidence intervals from Parts (a) and (b), do you agree with this conclusion? Explain.

9.54 The *Chronicle of Higher Education* (January 13, 1993) reported that 72.1% of those responding to a national survey of college freshmen were attending the college of their first choice. Suppose that $n = 500$ students responded to the survey (the actual sample size was much larger).

a. Using the sample size $n = 500$, calculate a 99% confidence interval for the proportion of college students who are attending their first choice of college.

b. Compute and interpret a 95% confidence interval for the proportion of students who are *not* attending their first choice of college.

c. The actual sample size for this survey was much larger than 500. Would a confidence interval based on the actual sample size have been narrower or wider than the one computed in Part (a)?

9.55 Increases in worker injuries and disability claims have prompted renewed interest in workplace design and regulation. As one particular aspect of this, employees required to do regular lifting should not have to handle unsafe loads. The article "Anthropometric, Muscle Strength, and Spinal Mobility Characteristics as Predictors of the Rating of Acceptable Loads in Parcel Sorting" (*Ergonomics* [1992]: 1033–1044) reported on a study involving a random sample of $n = 18$ male postal workers. The sample mean rating of acceptable load attained with a work-simulating test was found to be $\bar{x} = 9.7$ kg, and the sample standard deviation was $s = 4.3$ kg. Suppose that in the population of all male postal workers, the distribution of rating of acceptable load can be modeled approximately using a normal distribution with mean value μ. Construct and interpret a 95% confidence interval for μ.

9.56 The Gallup Organization conducted a telephone survey on attitudes toward AIDS (*Gallup Monthly*, 1991). A total of 1014 individuals were contacted. Each individual was asked whether they agreed with the following statement: "Landlords should have the right to evict a tenant from an apartment because that person has AIDS." One hundred one individuals in the sample agreed with this statement. Use these data to construct a 90% confidence interval for the proportion who are in agreement with this statement. Give an interpretation of your interval.

9.57 A manufacturer of college textbooks is interested in estimating the strength of the bindings produced by a particular binding machine. Strength can be measured by recording the force required to pull the pages from the binding. If this force is measured in pounds, how many books should be tested to estimate with 95% confidence to within 0.1 lb, the average force required to break the binding? Assume that σ is known to be 0.8 lb.

9.58 Recent high-profile legal cases have many people reevaluating the jury system. Many believe that juries in criminal trials should be able to convict on less than a unanimous vote. To assess support for this idea, investigators asked each individual in a random sample of Californians whether they favored allowing conviction by a 10–2 verdict in criminal cases not involving the death penalty. The Associated Press (*San Luis Obispo Telegram-Tribune*, September 13, 1995) reported that 71% supported the 10–2 verdict. Suppose that the sample size for this survey was $n = 900$. Compute and interpret a 99% confidence interval for the proportion of Californians who favor the 10–2 verdict.

9.59 The Center for Urban Transportation Research released a report stating that the average commuting distance in the United States is 10.9 mi (*USA Today*, August 13, 1991). Suppose that this average is actually the mean of a random sample of 300 commuters and that the sample standard deviation is 6.2 mi. Estimate the true mean commuting distance using a 99% confidence interval.

9.60 In 1991, California imposed a "snack tax" (a sales tax on snack food) in an attempt to help balance the state budget. A proposed alternative tax was a 12¢-per-pack increase in the cigarette tax. In a poll of 602 randomly selected California registered voters, 445 responded that they would have preferred the cigarette tax increase to the snack tax (*Reno Gazette-Journal*, August 26, 1991). Estimate the true proportion of California registered voters who preferred the cigarette tax increase; use a 95% confidence interval.

9.61 The confidence intervals presented in this chapter give both lower and upper bounds on plausible values for the population characteristic being estimated. In some instances, only an upper bound or only a lower bound is appropriate. Using the same reasoning that gave the large sample interval in Section 9.3, we can say that when n is large, 99% of all samples have

$$\mu < \bar{x} + 2.33\frac{s}{\sqrt{n}}$$

(because the area under the z curve to the left of 2.33 is .99). Thus, $\bar{x} + 2.33\frac{s}{\sqrt{n}}$ is a 99% upper confidence bound for μ. Use the data of Example 9.8 to calculate the 99% upper confidence bound for the true average salary for married men in two-income families.

9.62 The Associated Press (December 16, 1991) reported that in a random sample of 507 people, only 142 correctly described the Bill of Rights as the first 10 amendments to the U.S. Constitution. Calculate a 95% confidence interval for the proportion of the entire population that could give a correct description.

9.63 When n is large, the statistic s is approximately unbiased for estimating σ and has approximately a normal distribution. The standard deviation of this statistic when the population distribution is normal is $\sigma_s \approx \frac{\sigma}{\sqrt{2n}}$, which can be estimated by $\frac{s}{\sqrt{2n}}$. A large-sample confidence interval for the population standard deviation σ is then

$$s \pm (z \text{ critical value})\frac{s}{\sqrt{2n}}$$

Use the data of Exercise 9.59 to obtain a 95% confidence interval for the true standard deviation of commuting distance.

9.64 The interval from -2.33 to 1.75 captures an area of .95 under the z curve. This implies that another large-sample 95% confidence interval for μ has lower limit $\bar{x} - 2.33\frac{\sigma}{\sqrt{n}}$ and upper limit $\bar{x} + 1.75\frac{\sigma}{\sqrt{n}}$. Would you recommend using this 95% interval over the 95% interval $\bar{x} \pm 1.96\frac{\sigma}{\sqrt{n}}$ discussed in the text? Explain. (Hint: Look at the width of each interval.)

▪ References

Devore, Jay L. *Probability and Statistics for Engineering and the Sciences*, 6th ed. Pacific Grove, CA: Brooks/Cole, 2003. (This book gives a somewhat general introduction to confidence intervals.)

Freedman, David, Robert Pisani, and Roger Purves, *Statistics*, 3d ed. New York: W. W. Norton, 1997. (This book contains an informal discussion of confidence intervals.)

10 · Hypothesis Testing Using a Single Sample

In Chapter 9, we considered situations in which the primary goal was to estimate the unknown value of some population characteristic. Sample data can also be used to decide whether some claim or *hypothesis* about a population characteristic is plausible. For example, the lead story in the August 27, 1995, *San Francisco Examiner* focused on health care problems resulting from an increase in the number of Americans without health insurance. Serious difficulties arise for state and local agencies if the coverage rate for the entire population is less than 90%. The same day, the *Los Angeles Times* reported that in its survey of 3297 California adults, 2750, or 83.4%, had health insurance coverage. Let π denote the proportion of all California adults who have health insurance. The hypothesis testing methods presented in this chapter can be used to decide whether the sample data from this survey provide strong support for the hypothesis $\pi < .90$.

As another example, a report released by the National Association of Colleges and Employers stated that the average starting salary for students graduating in 2002 with a degree in marketing was $35,822 ("Business Majors Holding Their Own, But Starting Salary Offers Continue to Fall for Technical Disciplines," April 7, 2003, available at www.naceweb.org/press). Suppose that you are interested in determining whether the mean starting salary for students graduating with a marketing degree from your university this year is greater than the 2002 average of $35,822. You select a random sample of $n = 10$ marketing graduates from the current graduating class of your university and determine the starting salary of each one. If this sample produced a mean starting salary of $35,758 and a standard deviation of $1214, is it reasonable to conclude that μ, the current mean starting salary for marketing graduates at your university, is greater than $35,822 (i.e., $\mu > 35,822$)? We will see in this chapter how these sample data can be analyzed to decide whether $\mu > 35,822$ is a reasonable conclusion.

▪ 10.1 Hypotheses and Test Procedures

A hypothesis is a claim or statement about either the value of a single population characteristic or the values of several population characteristics. The following are examples of legitimate hypotheses:

$\mu = 1000$, where μ is the mean number of characters in an e-mail message

$\pi < .01$, where π is the proportion of e-mail messages that are undeliverable

In contrast, the statements $\overline{x} = 1000$ and $p = .01$ are *not* hypotheses, because $\overline{x}$ and p are *sample* characteristics.

A **test of hypotheses** or **test procedure** is a method for using sample data to decide between two competing claims (hypotheses) about a population characteristic. One hypothesis might be $\mu = 1000$ and the other $\mu \neq 1000$, or one hypothesis might be $\pi = .01$ and the other $\pi < .01$. If it were possible to carry out a census of the entire population, we would know which of the two hypotheses is correct, but usually we must decide between them using information from a sample.

A criminal trial is a familiar situation in which a choice between two contradictory claims must be made. The person accused of the crime must be judged either guilty or not guilty. Under the U.S. system of justice, the individual on trial is initially presumed not guilty. Only strong evidence to the contrary causes the not guilty claim to be rejected in favor of a guilty verdict. The burden is thus put on the prosecution to prove the guilty claim. The French perspective in criminal proceedings is the opposite of ours. There, once enough evidence has been presented to justify bringing an individual to trial, the initial assumption is that the accused is guilty. The burden of proof then falls on the accused to establish otherwise.

As in a judicial proceeding, we initially assume that a particular hypothesis, called the *null hypothesis*, is the correct one. We then consider the evidence (the sample data) and reject the null hypothesis in favor of the competing hypothesis, called the *alternative hypothesis*, only if there is *convincing* evidence against the null hypothesis.

▪ **Definition**

The **null hypothesis**, denoted by H_0, is a claim about a population characteristic that is initially assumed to be true.

The **alternative hypothesis**, denoted by H_a, is the competing claim.

In carrying out a test of H_0 versus H_a, the hypothesis H_0 is rejected in favor of H_a only if sample evidence strongly suggests that H_0 is false. If the sample does not contain such evidence, H_0 is not rejected. The two possible conclusions are then *reject H_0* or *fail to reject H_0*.

▪ Example 10.1   Tennis Ball Diameters

Because of variation in the manufacturing process, tennis balls produced by a particular machine do not have identical diameters. Let μ denote the true average diameter for tennis balls currently being produced. Suppose that the machine was initially calibrated to achieve the design specification $\mu = 3$ in. However, the manufacturer is now concerned that the diameters no longer conform to this specification. That is, $\mu \neq 3$ in. must now be considered a possibility. If sample evidence suggests that $\mu \neq 3$ in., the production process will have to be halted while the machine is recalibrated. Because this production break is costly, the manufacturer wants to be quite sure that $\mu \neq 3$ in. before undertaking recalibration. Under these circumstances, a sensible choice of hypotheses is

H_0: $\mu = 3$ (the specification is being met, so recalibration is unnecessary)

H_a: $\mu \neq 3$ (the specification is not being met, so recalibration is necessary)

Only compelling sample evidence would then result in H_0 being rejected in favor of H_a.

■ Example 10.2   Light Bulb Lifetimes

Kmart brand 60-W light bulbs state on the package "Avg. Life 1000 Hr." Let μ denote the true mean life of Kmart 60-W light bulbs. Then the advertised claim is $\mu = 1000$ hr. People who purchase this brand would be unhappy if μ is actually less than the advertised value. Suppose that a sample of Kmart light bulbs is selected and the lifetime for each bulb in the sample is recorded. The sample results can then be used to test the hypothesis $\mu = 1000$ hr against the hypothesis $\mu < 1000$ hr. The accusation that the company is overstating the mean lifetime is a serious one, and it is reasonable to require compelling evidence from the sample before concluding that $\mu < 1000$. This suggests that the claim $\mu = 1000$ should be selected as the null hypothesis and that $\mu < 1000$ should be selected as the alternative hypothesis. Then

H_0: $\mu = 1000$

would be rejected in favor of

H_a: $\mu < 1000$

only when sample evidence strongly suggests that the initial assumption, $\mu = 1000$ hr, is no longer tenable.

Because the alternative hypothesis in Example 10.2 asserted that $\mu < 1000$ (true average lifetime is less than the advertised value), it might have seemed sensible to state H_0 as the inequality $\mu \geq 1000$. The assertion $\mu \geq 1000$ is in fact the *implicit* null hypothesis, but we state H_0 explicitly as a claim of equality. There are several reasons for this. First of all, the development of a decision rule is most easily understood if there is only a single value of μ (or π or whatever other population characteristic is under consideration) when H_0 is true. Second, suppose that the sample data provided compelling evidence that $H_0: \mu = 1000$ should be rejected in favor of $H_a: \mu < 1000$. This means that we were convinced by the sample data that the true mean was smaller than 1000. It follows that we would have also been convinced that the true mean could not have been 1001 or 1010 or any other value that was larger than 1000. As a consequence, the conclusion when testing $H_0: \mu = 1000$ versus $H_a: \mu < 1000$ is always the same as the conclusion for a test where the null hypothesis is $H_0: \mu \geq 1000$. For the sake of simplicity, we state the null hypothesis H_0 as a claim of equality.

The form of a null hypothesis is

H_0:   population characteristic = hypothesized value

where the hypothesized value is a specific number determined by the problem context.

The alternative hypothesis has one of the following three forms:

H_a: population characteristic > hypothesized value

H_a: population characteristic < hypothesized value

H_a: population characteristic ≠ hypothesized value

Thus, we might test H_0: $\pi = .1$ versus H_a: $\pi < .1$; but we won't test H_0: $\mu = 50$ versus H_a: $\mu > 100$. The number appearing in the alternative hypothesis must be identical to the hypothesized value in H_0.

Example 10.3 illustrates how the selection of H_0 (the claim initially believed true) and H_a depend on the objectives of a study.

▪ Example 10.3   Evaluating a New Medical Treatment

A medical research team has been given the task of evaluating a new laser treatment for certain types of tumors. Consider the following two scenarios:

Scenario 1: The current standard treatment is considered reasonable and safe by the medical community, has no major side effects, and has a known success rate of 0.85 (85%).

Scenario 2: The current standard treatment sometimes has serious side effects, is costly, and has a known success rate of 0.30 (30%).

In the first scenario, research efforts would probably be directed toward determining whether the new treatment has a higher success rate than the standard treatment. Unless convincing evidence of this is presented, it is unlikely that current medical practice would be changed. With π representing the true proportion of successes for the laser treatment, the following hypotheses would be tested:

$$H_0:\ \pi = .85 \qquad \text{versus} \qquad H_a:\ \pi > .85$$

In this case, rejection of the null hypothesis is indicative of compelling evidence that the success rate is higher for the new treatment.

In the second scenario, the current standard treatment does not have much to recommend it. The new laser treatment may be considered preferable because of cost or because it has fewer or less serious side effects, as long as the success rate for the new procedure is no worse than that of the standard treatment. Here, researchers might decide to test the hypothesis

$$H_0:\ \pi = .30 \qquad \text{versus} \qquad H_a:\ \pi < .30$$

If the null hypothesis is rejected, the new treatment will not be put forward as an alternative to the standard treatment, because there is strong evidence that the laser method has a lower success rate.

If the null hypothesis is not rejected, we are able to conclude only that there is not convincing evidence that the success rate for the laser treatment is lower than that for the standard. This is *not* the same as saying that we have evidence that the laser treatment is as good as the standard treatment. If medical practice were to embrace the new procedure, it would not be because it has a higher success rate but rather because it costs less or has fewer side effects, and there is not strong evidence that it has a lower success rate than the standard treatment.

You should be careful in setting up the hypotheses for a test. Remember that a statistical hypothesis test is only capable of demonstrating strong support for the alternative hypothesis (by rejection of the null hypothesis). When the null hypothesis is not rejected, it does not mean strong support for H_0— only lack of strong evidence against it. In the light bulb scenario of Example 10.2, if H_0: $\mu = 1000$ is rejected in favor of H_a: $\mu < 1000$, it is because we have strong evidence for believing that true average lifetime is less than the advertised value. However, nonrejection of H_0 does not necessarily provide strong support for the advertised claim. If the objective is to demonstrate that the average lifetime is greater than 1000 hr, the hypotheses to be tested are H_0: $\mu = 1000$ versus H_a: $\mu > 1000$. Now rejection of H_0 indicates strong evidence that $\mu > 1000$. When deciding which alternative hypothesis to use, *keep the research objectives in mind*.

■ Exercises 10.1–10.10

10.1 Explain why the statement $\bar{x} = 50$ is not a legitimate hypothesis.

10.2 For the following pairs, indicate which do not comply with the rules for setting up hypotheses, and explain why:
a. H_0: $\mu = 15$, H_a: $\mu = 15$
b. H_0: $\pi = .4$, H_a: $\pi > .6$
c. H_0: $\mu = 123$, H_a: $\mu < 123$
d. H_0: $\mu = 123$, H_a: $\mu = 125$
e. H_0: $p = .1$, H_a: $p \neq .1$

10.3 To determine whether the pipe welds in a nuclear power plant meet specifications, a random sample of welds is selected and tests are conducted on each weld in the sample. Weld strength is measured as the force required to break the weld. Suppose that the specifications state that the mean strength of welds should exceed 100 lb/in.². The inspection team decides to test H_0: $\mu = 100$ versus H_a: $\mu > 100$. Explain why this alternative hypothesis was chosen rather than $\mu < 100$.

10.4 Do state laws that allow private citizens to carry concealed weapons result in a reduced crime rate? The author of a study carried out by the Brookings Institution is reported as saying, "The strongest thing I could say is that I don't see any strong evidence that they are reducing crime" (*San Luis Obispo Tribune*, January 23, 2003).

a. Is this conclusion consistent with testing

H_0: concealed weapons laws reduce crime
versus
H_a: concealed weapons laws do not reduce crime

or with testing

H_0: concealed weapons laws do not reduce crime
versus
H_a: concealed weapons laws reduce crime

Explain.

b. Does the stated conclusion indicate that the null hypothesis was rejected or not rejected? Explain.

10.5 The mean length of long-distance telephone calls placed with a particular phone company was known to be 7.3 min under an old rate structure. In an attempt to be more competitive with other long-distance carriers, the phone company lowered long-distance rates, thinking that its customers would be encouraged to make longer calls and thus that there would not be a big loss in revenue. Let μ denote the true mean length of long-distance calls after the rate reduction. What hypotheses should the phone company test to determine whether the mean length of long-distance calls increased with the lower rates?

10.6 A certain university has decided to introduce the use of plus and minus with letter grades, as long as there is evidence that more than 60% of the faculty favor the change. A random sample of faculty will be selected, and the resulting data will be used to test the relevant hypotheses. If π represents the true proportion of all faculty that favor a change to plus–minus grading, which of the following pair of hypotheses should the administration test:

H_0: $\pi = .6$ versus H_a: $\pi < .6$

or

H_0: $\pi = .6$ versus H_a: $\pi > .6$

Explain your choice.

10.7 A certain television station has been providing live coverage of a particularly sensational criminal trial. The station's program director wishes to know whether more than half the potential viewers prefer a return to regular daytime programming. A survey of randomly selected viewers is conducted. Let π represent the true proportion of viewers who prefer regular daytime programming. What hypotheses should the program director test to answer the question of interest?

10.8 Researchers have postulated that because of differences in diet, Japanese children have a lower mean blood cholesterol level than U.S. children do. Suppose that the mean level for U.S. children is known to be 170. Let μ represent the true mean blood cholesterol level for Japanese children. What hypotheses should the researchers test?

10.9 A county commissioner must vote on a resolution that would commit substantial resources to the construction of a sewer in an outlying residential area. Her fiscal decisions have been criticized in the past, so she decides to take a survey of constituents to find out whether they favor spending money for a sewer system. She will vote to appropriate funds only if she can be fairly certain that a majority of the people in her district favor the measure. What hypotheses should she test?

10.10 Many older homes have electrical systems that use fuses rather than circuit breakers. A manufacturer of 40-amp fuses wants to make sure that the mean amperage at which its fuses burn out is in fact 40. If the mean amperage is lower than 40, customers will complain because the fuses require replacement too often. If the mean amperage is higher than 40, the manufacturer might be liable for damage to an electrical system as a result of fuse malfunction. To verify the mean amperage of the fuses, a sample of fuses is selected and tested. If a hypothesis test is performed using the resulting data, what null and alternative hypotheses would be of interest to the manufacturer?

▪ 10.2 Errors in Hypothesis Testing

Once hypotheses have been formulated, we need a method for using sample data to determine whether H_0 should be rejected. The method that we use for this purpose is called a **test procedure**. Just as a jury may reach the wrong verdict in a trial, there is some chance that the use of a test procedure with sample data may lead us to the wrong conclusion about a population characteristic. In this section, we discuss the kinds of errors that can occur and consider how the choice of a test procedure influences the chances of these errors.

One erroneous conclusion in a criminal trial is for a jury to convict an innocent person, and another is for a guilty person to be set free. Similarly, there are two different types of errors that might be made when making a decision in a hypothesis-testing problem. One type of error involves rejecting H_0 even though the null hypothesis is true. The second type of error results from failing to reject H_0 when it is false. These errors are known as Type I and Type II errors, respectively.

▪ **Definition**

Type I error: the error of rejecting H_0 when H_0 is true

Type II error: the error of failing to reject H_0 when H_0 is false

The only way to guarantee that neither type of error occurs is to base the decision on a census of the entire population. The risk of error is the price researchers pay for basing an inference on a sample. With any reasonable procedure, there is some chance that a Type I error will be made and some chance that a Type II error will result.

▪ Example 10.4   On-Time Arrivals

The U.S. Department of Transportation reported that during a recent period, 77% of all domestic passenger flights arrived on time (meaning within 15 min of the scheduled arrival). Suppose that an airline with a poor on-time record decides to offer its employees a bonus if, in an upcoming month, the airline's proportion of on-time flights exceeds the overall industry rate of 0.77. Let π be the true proportion of the airline's flights that are on time during the month of interest. A random sample of flights might be selected and used as a basis for choosing between

$$H_0: \quad \pi = .77 \quad \text{and} \quad H_a: \quad \pi > .77$$

In this context, a Type I error (rejecting a true H_0) results in the airline rewarding its employees when in fact their true proportion of on-time flights did not exceed .77. A Type II error (not rejecting a false H_0) results in the airline employees *not* receiving a reward that in fact they deserved.

▪ Example 10.5   Slowing the Growth of Tumors

Researchers at the National Cancer Institute announced plans to begin studies of a cancer treatment thought to slow the growth of tumors (Associated Press, March 30, 1993). Having tested the treatment on only 13 patients, the researchers had very little information, but they stated that the treatment appears to be less toxic than standard chemotherapy treatments. An experiment to study the treatment more extensively was reported as being in the planning stages.

Let μ denote the true mean growth rate of tumors for patients receiving the new treatment. Data resulting from the planned experiments can be used to test

$$H_0: \quad \mu = \text{mean growth rate of tumors without treatment}$$

versus

$$H_a: \quad \mu < \text{mean growth rate of tumors without treatment}$$

The null hypothesis states that the new treatment is not effective — that the mean growth rate of tumors for patients receiving the new treatment is the same as for patients who are not treated. The alternative hypothesis states that the new treatment is effective in reducing the mean growth rate of tumors. In this context, a Type I error consists of incorrectly concluding that the new treatment is effective in slowing the growth rate of tumors. A Type II error consists of concluding that the new treatment is ineffective when, in fact, the mean growth rate of tumors is reduced.

Examples 10.4 and 10.5 illustrate the two different types of error that might occur when testing hypotheses. Type I and Type II errors — and the associated consequences of making such errors — are quite different. We would like a small probability of drawing an incorrect conclusion. The accompanying box introduces the terminology and notation used to describe error probabilities.

> **▪ Definition**
>
> The probability of a Type I error is denoted by α and is called the **level of significance** of the test. Thus, a test with $\alpha = .01$ is said to have a level of significance of .01 or to be a level .01 test.
>
> The probability of a Type II error is denoted by β.

▪ Example 10.6   Blood Test for Ovarian Cancer

Women with ovarian cancer are usually not diagnosed until the disease is in an advanced stage, when it is most difficult to treat. A new blood test has been developed that appears to be able to identify ovarian cancer at its earliest stages. In a report issued by the National Cancer Institute and the Food and Drug Administration (February 8, 2002), the following information from a preliminary evaluation of the blood test was given:

- The test was given to 50 women known to have ovarian cancer, and it correctly identified all of them as having cancer.
- The test was given to 66 women known not to have ovarian cancer, and it correctly identified 63 of these 66 as being cancer free.

We can think of using this blood test to choose between two hypotheses:

H_0: woman has ovarian cancer
H_a: woman does not have ovarian cancer

Note that although these are not "statistical hypotheses" (statements about a population characteristic), the possible decision errors are analogous to Type I and Type II errors.

In this situation, believing that a woman with ovarian cancer is cancer free would be a Type I error — rejecting the hypothesis of ovarian cancer when it is, in fact, true. Believing that a woman who is actually cancer free does have ovarian cancer is a Type II error — not rejecting the null hypothesis when it is, in fact, false. Based on the preliminary study results, we can estimate the error probabilities. The probability of a Type I error, α, is approximately $0/50 = 0$. The probability of a Type II error, β, is approximately $3/66 = .046$.

The ideal test procedure would result in both $\alpha = 0$ and $\beta = 0$. However, if we must base our decision on incomplete information — a sample rather than a census — it is impossible to achieve this ideal. The standard test procedures allow us to control α, but they provide no direct control over β. Because α represents the probability of rejecting a true null hypothesis, selecting a significance level $\alpha = .05$ results in a test procedure that, used over and over with different samples, rejects a *true* H_0 about 5 times in 100. Selecting $\alpha = .01$ results in a test procedure with a

Type I error rate of 1% in long-term repeated use. Choosing a small value for α implies that the user wants to use a procedure for which the risk of a Type I error is quite small.

One question arises naturally at this point: If we can select α, the probability of making a Type I error, why would we ever select $\alpha = .05$ rather than $\alpha = .01$? Why not always select a very small value for α? To achieve a small probability of making a Type I error, we would need the corresponding test procedure to require the evidence against H_0 to be very strong before the null hypothesis can be rejected. Although this makes a Type I error unlikely, it increases the risk of a Type II error (*not* rejecting H_0 when it should have been rejected). Frequently the investigator must balance the consequences of Type I and Type II errors. If a Type II error has more serious consequences, it may be a good idea to select a somewhat larger value for α.

In general, there is a compromise between small α and small β, leading to the following widely accepted principle for specifying a test procedure.

After assessing the consequences of Type I and Type II errors, identify the largest α that is tolerable for the problem. Then employ a test procedure that uses this maximum acceptable value — rather than anything smaller — as the level of significance (because using a smaller α increases β). In other words, don't make α smaller than it needs to be.

Thus, if you decide that $\alpha = .05$ is tolerable, you should not use a test with $\alpha = .01$, because the smaller α inevitably results in a larger β. The values of α most commonly used in practice are .05 and .01 (a 1 in 20 or 1 in 100 chance of rejecting H_0 when it is actually true), but the choice in any given problem depends on the seriousness of a Type I error relative to a Type II error in that context.

▪ Example 10.7   Lead in Tap Water

The Associated Press (May 13, 1993) reported that the Environmental Protection Agency (EPA) had warned 819 communities that their tap water contained too much lead. Drinking water is considered unsafe if the mean concentration of lead is 15 ppb (parts per billion) or greater. The EPA requires the cited communities to take corrective actions and to monitor lead levels. With μ denoting the mean concentration of lead, a cited community could test

$$H_0: \quad \mu = 15 \qquad \text{versus} \qquad H_a: \quad \mu < 15$$

The null hypothesis (which is equivalent to the assertion $\mu \geq 15$) states that the mean lead concentration is excessive by EPA standards. The alternative hypothesis states that the mean lead concentration is at an acceptable level and that the water system meets EPA standards for lead.

In this context, a Type I error leads to the conclusion that a water source meets EPA standards for lead when, in fact, it does not. Possible consequences of this type of error include health risks associated with excessive lead consumption (e.g., increased blood pressure, hearing loss, and, in severe cases, anemia and kidney damage). A Type II error is to conclude that the water does not meet EPA standards for lead when, in fact, it actually does. Possible consequences of a Type II error include elimination of a community water source. Because a Type I error might result in potentially serious public health risks, a small value of α (Type I error probability), such as $\alpha = .01$, could be selected. Of course, selecting a small value for α increases the risk of a Type II error. If the community has only one water source, a Type II error could also have very serious consequences for the community, and we might want to rethink our choice of α.

■ Exercises 10.11–10.21

10.11 Researchers at the University of Washington and Harvard University analyzed records of breast cancer screening and diagnostic evaluations ("Mammogram Cancer Scares More Frequent than Thought," *USA Today*, April 16, 1998). Discussing the benefits and downsides of the screening process, the article states that, although the rate of false-positives is higher than previously thought, if radiologists were less aggressive in following up on suspicious tests, the rate of false-positives would fall but the rate of missed cancers would rise. Suppose that such a screening test is used to decide between a null hypothesis of H_0: no cancer is present and an alternative hypothesis of H_a: cancer is present. (Although these are not hypotheses about a population characteristic, this exercise illustrates the definitions of Type I and Type II errors.)

a. Would a false-positive (thinking that cancer is present when in fact it is not) be a Type I error or a Type II error?

b. Describe a Type I error in the context of this problem, and discuss the consequences of making a Type I error.

c. Describe a Type II error in the context of this problem, and discuss the consequences of making a Type II error.

d. What aspect of the relationship between the probability of Type I and Type II errors is being described by the statement in the article that if radiologists were less aggressive in following up on suspicious tests, the rate of false-positives would fall but the rate of missed cancers would rise?

10.12 Medical personnel are required to report suspected cases of child abuse. Because some diseases have symptoms that mimic those of child abuse, doctors who see a child with these symptoms must decide between two competing hypotheses:

H_0:   symptoms are due to child abuse
H_a: symptoms are due to disease

(Although these are not hypotheses about a population characteristic, this exercise illustrates the definitions of Type I and Type II errors.) The article

"Blurred Line Between Illness, Abuse Creates Problem for Authorities" (*Macon Telegraph*, February 28, 2000) included the following quote from a doctor in Atlanta regarding the consequences of making an incorrect decision: "If it's disease, the worst you have is an angry family. If it is abuse, the other kids (in the family) are in deadly danger."

a. For the given hypotheses, describe Type I and Type II errors.

b. Based on the quote regarding consequences of the two kinds of error, which type of error does the doctor quoted consider more serious? Explain.

10.13 The National Cancer Institute conducted a 2-year study to determine whether cancer death rates for areas near nuclear power plants are higher than for areas without nuclear facilities (*San Luis Obispo Telegram-Tribune*, September 17, 1990). A spokesperson for the Cancer Institute said, "From the data at hand, there was no convincing evidence of any increased risk of death from any of the cancers surveyed due to living near nuclear facilities. However, no study can prove the absence of an effect."

a. Let π denote the true proportion of the population in areas near nuclear power plants who die of cancer during a given year. The researchers at the Cancer Institute might have considered the two rival hypotheses of the form

H_0: π = value for areas without nuclear facilities
H_a: π > value for areas without nuclear facilities

Did the researchers reject H_0 or fail to reject H_0?

b. If the Cancer Institute researchers were incorrect in their conclusion that there is no increased cancer risk associated with living near a nuclear power plant, are they making a Type I or a Type II error? Explain.

c. Comment on the spokesperson's last statement that no study can *prove* the absence of an effect. Do you agree with this statement?

10.14 Ann Landers, in her advice column of October 24, 1994 (*San Luis Obispo Telegram-Tribune*), described the reliability of DNA paternity testing as follows: "To get a completely accurate result, you would have to be tested, and so would (the man) and your mother. The test is 100 percent accurate if the man is *not* the father and 99.9 percent accurate if he is."

a. Consider using the results of DNA paternity testing to decide between the following two hypotheses:

H_0: a particular man is the father
H_a: a particular man is not the father

In the context of this problem, describe Type I and Type II errors. (Although these are not hypotheses about a population characteristic, this exercise illustrates the definitions of Type I and Type II errors.)

b. Based on the information given, what are the values of α, the probability of Type I error, and β, the probability of Type II error?

c. Ann Landers also stated, "If the mother is not tested, there is a 0.8 percent chance of a false positive." For the hypotheses given in Part (a), what are the values of α and β if the decision is based on DNA testing in which the mother is not tested?

10.15 Pizza Hut, after test-marketing a new product called the Bigfoot Pizza, concluded that introduction of the Bigfoot nationwide would increase its sales by more than 14% (*USA Today*, April 2, 1993). This conclusion was based on recording sales information for a random sample of Pizza Hut restaurants selected for the marketing trial. With μ denoting the mean percentage increase in sales for all Pizza Hut restaurants, consider using the sample data to decide between H_0: $\mu = 14$ and H_a: $\mu > 14$.

a. Is Pizza Hut's conclusion consistent with a decision to reject H_0 or to fail to reject H_0?

b. If Pizza Hut is incorrect in its conclusion, is the company making a Type I or a Type II error?

10.16 A television manufacturer claims that (at least) 90% of its TV sets will need no service during the first 3 years of operation. A consumer agency wishes to check this claim, so it obtains a random sample of $n = 100$ purchasers and asks each whether the set purchased needed repair during the first 3 years after purchase. Let p be the sample proportion of responses indicating no repair (so that no repair is identified with a success). Let π denote the true proportion of successes for all sets made by this manufacturer. The agency does not want to claim false advertising unless sample evidence strongly suggests that $\pi < .9$. The appropriate hypotheses are then H_0: $\pi = .9$ versus H_a: $\pi < .9$.

a. In the context of this problem, describe Type I and Type II errors, and discuss the possible consequences of each.

b. Would you recommend a test procedure that uses $\alpha = .10$ or one that uses $\alpha = .01$? Explain.

10.17 A manufacturer of hand-held calculators receives large shipments of printed circuits from a supplier. It is too costly and time-consuming to inspect all incoming circuits, so when each shipment arrives, a sample is selected for inspection. Information from the sample is then used to test H_0: $\pi = .05$ versus H_a: $\pi > .05$, where π is the true proportion of defective circuits in the shipment. If the null hypothesis is not rejected, the shipment is accepted, and the circuits are used in the production of calculators. If the null hypothesis is rejected, the entire shipment is returned to the supplier because of inferior quality. (A shipment is defined to be of inferior qualify if it contains more than 5% defective circuits.)

a. In this context, define Type I and Type II errors.

b. From the calculator manufacturer's point of view, which type of error is considered more serious?

c. From the printed circuit supplier's point of view, which type of error is considered more serious?

10.18 Water samples are taken from water used for cooling as it is being discharged from a power plant into a river. It has been determined that as long as the mean temperature of the discharged water is at most 150°F, there will be no negative effects on the river's ecosystem. To investigate whether the plant is in compliance with regulations that prohibit a mean discharge water temperature above 150°F, researchers will take 50 water samples at randomly selected times and record the temperature of each sample. The resulting data will be used to test the hypotheses H_0: $\mu = 150°F$ versus H_a: $\mu > 150°F$. In the context of this example, describe Type I and Type II errors. Which type of error would you consider more serious? Explain.

10.19 Occasionally, warning flares of the type contained in most automobile emergency kits fail to ignite. A consumer advocacy group wants to investigate a claim against a manufacturer of flares brought by a person who claims that the proportion of defective flares is much higher than the value of .1 claimed by the manufacturer. A large number of flares will be tested, and the results will be used to decide between H_0: $\pi = .1$ and H_a: $\pi > .1$, where π represents the true proportion of defective flares made by this manufacturer. If H_0 is rejected, charges of false advertising will be filed against the manufacturer.

a. Explain why the alternative hypothesis was chosen to be H_a: $\pi > .1$.

b. In this context, describe Type I and Type II errors, and discuss the consequences of each.

10.20 Suppose that you are an inspector for the Fish and Game Department and that you are given the task of determining whether to prohibit fishing along part of the Oregon coast. You will close an area to fishing if it is determined that fish in that region have an unacceptably high mercury content.

a. Assuming that a mercury concentration of 5 ppm is considered the maximum safe concentration, which of the following pairs of hypotheses would you test:

$$H_0: \quad \mu = 5 \qquad \text{versus} \qquad H_a: \quad \mu > 5$$

or

$$H_0: \quad \mu = 5 \qquad \text{versus} \qquad H_a: \quad \mu < 5$$

Give the reasons for your choice.

b. Would you prefer a significance level of .1 or .01 for your test? Explain.

10.21 An automobile manufacturer is considering using robots for part of its assembly process. Converting to robots is an expensive process, so it will be undertaken only if there is strong evidence that the

proportion of defective installations is lower for the robots than for human assemblers. Let π denote the true proportion of defective installations for the robots. It is known that human assemblers have a defect proportion of .02.

a. Which of the following pairs of hypotheses should the manufacturer test:

$$H_0: \quad \pi = .02 \qquad \text{versus} \qquad H_a: \quad \pi < .02$$

or

$$H_0: \quad \pi = .02 \qquad \text{versus} \qquad H_a: \quad \pi > .02$$

Explain your answer.

b. In the context of this exercise, describe Type I and Type II errors.

c. Would you prefer a test with $\alpha = .01$ or $\alpha = .1$? Explain your reasoning.

■ 10.3 Large-Sample Hypothesis Tests for a Population Proportion

Now that some general concepts of hypothesis testing have been introduced, we are ready to turn our attention to the development of procedures for using sample information to decide between the null and the alternative hypotheses. There are two possible decisions: We either reject H_0 or else fail to reject H_0. The fundamental idea behind hypothesis-testing procedures is this: *We reject the null hypothesis if the observed sample is very unlikely to have occurred when H_0 is true.* In this section, we consider testing hypotheses about a population proportion when the sample size n is large.

Let π denote the proportion of individuals or objects in a specified population that possess a certain property. A random sample of n individuals or objects is selected from the population. The sample proportion

$$p = \frac{\text{number in the sample that possess property}}{n}$$

is the natural statistic for making inferences about π.

The large-sample test procedure is based on the same properties of the sampling distribution of p that were used previously to obtain a confidence interval for π, namely:

1. $\mu_p = \pi$.

2. $\sigma_p = \sqrt{\dfrac{\pi(1-\pi)}{n}}$.

3. When n is large, the sampling distribution of p is approximately normal.

These three results imply that the standardized variable

$$z = \frac{p - \pi}{\sqrt{\dfrac{\pi(1-\pi)}{n}}}$$

has approximately a standard normal distribution when n is large. Let's look at Example 10.8 to see how this information allows us to make a decision.

▪ **Example 10.8**   **Perceived Risks of Smoking**

The authors of the article "Perceived Risks of Heart Disease and Cancer Among Cigarette Smokers" (*Journal of the American Medical Association* [1999]: 1019–1021) expressed the concern that a majority of smokers do not view themselves as being at increased risk of heart disease or cancer. Because of this, the authors called for a public health campaign to educate smokers about the associated risks. In support of this recommendation, the authors offered the results of a study of 737 current smokers selected at random from U.S. households with telephones. Of the 737 smokers surveyed, 295 indicated that they believed they have a higher than average risk of cancer. Do these data suggest that π, the true proportion of smokers who view themselves as being at increased risk of cancer is, in fact, less than .5, as claimed by the authors of the paper?

This question can be answered by testing hypotheses, where

π = true proportion of smokers who believe that they are at increased risk of cancer

H_0: $\pi = .5$

H_a: $\pi < .5$ (The proportion of smokers who believe that they are at increased risk of cancer is less than .5. That is, less than half of all smokers believe that they are at increased risk of cancer.)

Recall that in a hypothesis test, the null hypothesis is rejected only if there is convincing evidence against it — in this case, convincing evidence that $\pi < .5$. If H_0 is rejected, the authors would have strong support for the claim made in the article and for their recommended action.

For this sample,

$$p = \frac{295}{737} = .40$$

The observed sample proportion is certainly less than .5, but this could just be due to sampling variability. That is, when $\pi = .5$ (meaning H_0 is true), the sample proportion p usually differs somewhat from .5 simply because of chance variation from one sample to another. Is it plausible that a sample proportion of $p = .40$ occurred as a result of this chance variation, or is it unusual to observe a sample proportion this small when $\pi = .5$?

To answer this question, we form a *test statistic*, the quantity used as a basis for making a decision between H_0 and H_a. Creating a test statistic involves replacing π with the hypothesized value in the z variable $z = \dfrac{p - \pi}{\sqrt{\pi(1 - \pi)/n}}$ to obtain

$$z = \frac{p - .5}{\sqrt{\dfrac{(.5)(.5)}{n}}}$$

If the null hypothesis is true, this statistic should have approximately a standard normal distribution, because when the sample size is large and H_0 is true,

1. $\mu_p = .5$,

2. $\sigma_p = \sqrt{\dfrac{(.5)(.5)}{n}}$, and

3. p has approximately a normal distribution.

The calculated value of z expresses the distance between p and the hypothesized value as a number of standard deviations. If, for example, $z = -3$, then the value of p that came from the sample is 3 standard deviations (of p) less than what we would have expected if the null hypothesis were true. How likely is it that a z value at least this contradictory to H_0 would be observed if in fact H_0 is true? Because the test statistic z is constructed using the hypothesized value from the null hypothesis, when H_0 is true, the test statistic has (approximately) a standard normal distribution. Therefore

$$P(z \leq -3 \text{ when } H_0 \text{ is true}) = \text{area under the } z \text{ curve to the left of } -3.00$$
$$= .0013$$

That is, if H_0 is true, very few samples (much less than 1% of all samples) produce a value of z at least as contradictory to H_0 as $z = -3$. Because this z value is in the most extreme 1%, it is sensible to reject H_0.

For our data,

$$z = \frac{p - .5}{\sqrt{\dfrac{(.5)(.5)}{n}}} = \frac{.40 - .5}{\sqrt{\dfrac{(.5)(.5)}{737}}} = \frac{-.10}{.018} = -5.56$$

That is, $p = .40$ is more than 5.5 standard deviations less than what we would expect it to be if the null hypothesis $H_0: \pi = .5$ were true. The sample data appear to be much more consistent with the alternative hypothesis, $H_a: \pi < .5$. In particular,

$$P(\text{value of } z \text{ is at least as contradictory to } H_0 \text{ as } -5.56 \text{ when } H_0 \text{ is true})$$
$$= P(z \leq -5.56 \text{ when } H_0 \text{ is true})$$
$$= \text{area under the } z \text{ curve to the left of } -5.56$$
$$\approx 0$$

There is virtually no chance of seeing a sample proportion and corresponding z value this extreme as a result of chance variation alone when H_0 is true. If p is more than 5.5 standard deviations away from .5, how can we believe that $\pi = .5$? The evidence for rejecting H_0 in favor of H_a is very compelling.

The preceding example illustrates the rationale behind large-sample procedures for testing hypotheses about p (and other test procedures as well). We begin by assuming that the null hypothesis is correct. The sample is then examined in light of this assumption. If the observed sample proportion would not be unusual when H_0 is true, then chance variability from one sample to another is a plausible explanation for what has been observed, and H_0 should not be rejected. On the other hand, if the observed sample proportion would have been quite unlikely when H_0 is true, then we would take the sample as convincing evidence against the null hypothesis and we should reject H_0. We base a decision to reject or to fail to reject the null hypothesis on an assessment of how extreme or unlikely the observed sample is if H_0 is true.

The assessment of how contradictory the observed data are to H_0 is based on first computing the value of the test statistic

$$z = \frac{p - \text{hypothesized value}}{\sqrt{\dfrac{(\text{hypothesized value})(1 - \text{hypothesized value})}{n}}}$$

We then calculate the *P-value*, the probability, assuming that H_0 is true, of obtaining a z value at least as contradictory to H_0 as what was actually observed.

▪ **Definition**

A **test statistic** is the function of sample data on which a conclusion to reject or fail to reject H_0 is based.

The **P-value** (also sometimes called the **observed significance level**) is a measure of inconsistency between the hypothesized value for a population characteristic and the observed sample. It is the probability, assuming that H_0 is true, of obtaining a test statistic value at least as inconsistent with H_0 as what actually resulted.

▪ **Example 10.9   Support for Legalized Gambling**

A number of initiatives on the topic of legalized gambling have appeared on state ballots in recent years. Suppose that a political candidate has decided to support legalization of casino gambling if he is convinced that more than two-thirds of U.S. adults approve of casino gambling. *USA Today* (June 17, 1999) reported the results of a Gallup poll in which 1523 adults (selected at random from households with telephones) were asked whether they approved of casino gambling. The number in the sample who approved was 1035. Does the sample provide convincing evidence that more than two-thirds approve?

With

$$\pi = \text{true proportion of U.S. adults who approve of casino gambling}$$

the relevant hypotheses are

$$H_0: \quad \pi = 2/3 = .667$$
$$H_a: \quad \pi > .667$$

The sample proportion is

$$p = \frac{1035}{1523} = .680$$

Does the value of p exceed two-thirds by enough to cast substantial doubt on H_0? Because the sample size is large, the statistic

$$z = \frac{p - .667}{\sqrt{\dfrac{(.667)(1 - .667)}{n}}}$$

has approximately a standard normal distribution when H_0 is true. The calculated value of the test statistic is

$$z = \frac{.680 - .667}{\sqrt{\dfrac{(.667)(1 - .667)}{1523}}} = \frac{.013}{.012} = 1.08$$

The probability that a z value at least this inconsistent with H_0 would be observed if in fact H_0 is true is

$$P\text{-value} = P(z \geq 1.08 \text{ when } H_0 \text{ is true})$$

$$= \text{area under the } z \text{ curve to the right of } 1.08$$

$$= 1 - .8599$$

$$= .1401$$

This probability indicates that when $\pi = .667$, it would not be all that unusual to observe a sample proportion as large as .680. When H_0 is true, roughly 14% of all samples would have a sample proportion larger than .680, so a sample proportion of .680 is reasonably consistent with the null hypothesis. Although .680 is larger than the hypothesized value of $\pi = .667$, chance variation from sample to sample is a plausible explanation for what was observed. There is not strong evidence that the proportion who favor casino gambling is greater than two-thirds.

———

As illustrated by Examples 10.8 and 10.9, small P-values indicate that sample results are inconsistent with H_0, whereas larger P-values are interpreted as meaning that the data are consistent with H_0 and that sampling variability alone is a plausible explanation for what was observed in the sample. As you probably noticed, the two cases examined (P-value ≈ 0 and P-value $= .1401$) were such that a decision between rejecting or not rejecting H_0 was clear-cut. A decision in other cases might not be so obvious. For example, what if the sample had resulted in a P-value of .04? Is this unusual enough to warrant rejection of H_0? How small must the P-value be before H_0 should be rejected?

The answer depends on the significance level α (the probability of a Type I error) selected for the test. For example, suppose that we set $\alpha = .05$. This implies that the probability of rejecting a true null hypothesis is .05. To obtain a test procedure with this probability of Type I error, we would reject the null hypothesis if the sample result is among the most unusual 5% of all samples when H_0 is true. That is, H_0 is rejected if the computed P-value $\leq .05$. If we had selected $\alpha = .01$, H_0 would be rejected only if we observed a sample result so extreme that it would be among the most unusual 1% if H_0 is true (i.e., if P-value $\leq .01$).

A decision as to whether H_0 should be rejected results from comparing the P-value to the chosen α:

H_0 should be rejected if P-value $\leq \alpha$.

H_0 should not be rejected if P-value $> \alpha$.

Suppose, for example, that P-value $= .0352$ and that a significance level of .05 is chosen. Then, because

$$P\text{-value} = .0352 \leq .05 = \alpha$$

H_0 would be rejected. This would not be the case, though, for $\alpha = .01$, because then P-value $> \alpha$.

▪ Computing a *P-value* for a Large-Sample Test Concerning π

The computation of the P-value depends on the form of the inequality in H_a. Suppose, for example, that we wish to test

$$H_0: \quad \pi = .6 \qquad \text{versus} \qquad H_a: \quad \pi > .6$$

based on a large sample. The appropriate test statistic is

$$z = \frac{p - .6}{\sqrt{\dfrac{(.6)(1 - .6)}{n}}}$$

Values of p contradictory to H_0 and much more consistent with H_a are those much *larger* than .6 (because $\pi = .6$ when H_0 is true and $\pi > .6$ when H_0 is false and H_a is true). Such values of p correspond to z values considerably greater than 0. If $n = 400$ and $p = .679$, then

$$z = \frac{.679 - .6}{\sqrt{\dfrac{(.6)(1 - .6)}{400}}} = \frac{.079}{.025} = 3.16$$

The value $p = .679$ is a bit more than 3 standard deviations larger than what we would have expected if H_0 were true. Thus

$$
\begin{aligned}
P\text{-value} &= P(z \text{ at least as contradictory to } H_0 \text{ as } 3.16 \text{ when } H_0 \text{ is true}) \\
&= P(z \geq 3.16 \text{ when } H_0 \text{ is true}) \\
&= \text{area under the } z \text{ curve to the right of } 3.16 \\
&= 1 - .9992 \\
&= .0008
\end{aligned}
$$

This P-value is illustrated in Figure 10.1. If H_0 is true, in the long run only 8 out of 10,000 samples would result in a z value more extreme than what actually resulted; most of us would consider such a z quite unusual. Using a significance level of .01, we reject the null hypothesis because P-value $= .0008 \leq .01 = \alpha$.

Figure 10.1 Calculating a *P*-value.

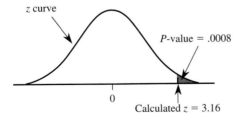

Now consider testing $H_0: \pi = .3$ versus $H_a: \pi \neq .3$. A value of p *either* much greater than .3 *or* much less than .3 is inconsistent with H_0 and provides support for H_a. Such a p corresponds to a z value far out in *either* tail of the z curve. If

$$z = \frac{p - .3}{\sqrt{\dfrac{(.3)(1 - .3)}{n}}} = 1.75$$

then (as shown in Figure 10.2)

$$P\text{-value} = P(z \text{ value at least as inconsistent with } H_0 \text{ as } 1.75 \text{ when } H_0 \text{ is true})$$
$$= P(z \geq 1.75 \text{ or } z \leq -1.75 \text{ when } H_0 \text{ is true})$$
$$= (z \text{ curve area to the right of } 1.75) + (z \text{ curve area to the left of } -1.75)$$
$$= (1 - .9599) + .0401$$
$$= .0802$$

The P-value in this situation is also .0802 if $z = -1.75$, because 1.75 and -1.75 are equally inconsistent with H_0.

Figure 10.2 *P*-value as the sum of two tail areas.

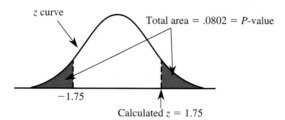

▪ **Determination of the *P*-Value When the Test Statistic Is *z***

1. **Upper-tailed test**:
   H_a: $\pi >$ hypothesized value.
   P-value computed as illustrated:

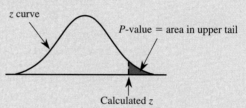

2. **Lower-tailed test**:
   H_a: $\pi <$ hypothesized value.
   P-value computed as illustrated:

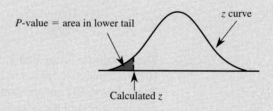

3. **Two-tailed test**:
   H_a: $\pi \neq$ hypothesized value.
   P-value computed as illustrated:

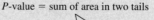
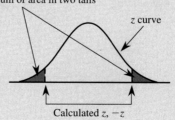

The symmetry of the z curve implies that when the test is two-tailed (the "not equal" alternative), it is not necessary to add two curve areas. Instead,

If z is positive, *P*-value = 2(area to the right of z).

If z is negative, *P*-value = 2(area to the left of z).

▪ Example 10.10   Water Conservation

In December 2003 a countywide water conservation campaign was conducted in a particular county. In January 2004 a random sample of 500 homes was selected, and water usage was recorded for each home in the sample. The county supervisors want to know whether the data support the claim that fewer than half the households in the county reduced water consumption. The relevant hypotheses are

$$H_0: \ \pi = .5 \quad \text{versus} \quad H_a: \ \pi < .5$$

where π is the true proportion of households in the county with reduced water usage.

Suppose that the sample results were $n = 500$ and $p = .440$. Because the sample size is large and this is a lower-tailed test, we can compute the P-value by first calculating the value of the z test statistic

$$z = \frac{p - .5}{\sqrt{\dfrac{(.5)(1 - .5)}{n}}}$$

and then finding the area under the z curve to the left of this z.

Based on the observed sample data,

$$z = \frac{.440 - .5}{\sqrt{\dfrac{(.5)(1 - .5)}{500}}} = \frac{-.060}{.0224} = -2.68$$

The P-value is then equal to the area under the z curve and to the left of -2.68. From the entry in the -2.6 row and $.08$ column of Appendix Table 2, we find that

$$P\text{-value} = .0037$$

Using a .01 significance level, we reject H_0 (because $.0037 \le .01$), suggesting that the proportion with reduced water usage was less than .5. Notice that rejection of H_0 would not be justified if a *very* small significance level, such as .001, had been selected.

Example 10.10 illustrates the calculation of a P-value for a lower-tailed test. The use of P-values in upper-tailed and two-tailed tests is illustrated in Examples 10.11 and 10.12. But first we summarize large-sample tests of hypotheses about a population proportion and introduce a step-by-step procedure for carrying out a hypothesis test.

▪ Summary of Large-Sample z Test for π

Null hypothesis: $H_0: \pi =$ hypothesized value

Test statistic: $z = \dfrac{p - \text{hypothesized value}}{\sqrt{\dfrac{(\text{hypothesized value})(1 - \text{hypothesized value})}{n}}}$

(continued)

Alternative Hypothesis:

H_a: $\pi >$ hypothesized value

H_a: $\pi <$ hypothesized value

H_a: $\pi \neq$ hypothesized value

P-Value:

Area under z curve to right of calculated z

Area under z curve to left of calculated z

(1) 2(area to right of z) if z is positive, or

(2) 2(area to left of z) if z is negative

Assumptions: 1. p is the sample proportion from a *random sample*.
2. The *sample size is large*. This test can be used if n satisfies both n(hypothesized value) ≥ 10 and $n(1 -$ hypothesized value) ≥ 10.
3. If sampling is without replacement, the sample size is no more than 10% of the population size.

We recommend that the following sequence of steps be used when carrying out a hypothesis test.

■ **Steps in a Hypothesis-Testing Analysis**

1. Describe the population characteristic about which hypotheses are to be tested.
2. State the null hypothesis H_0.
3. State the alternative hypothesis H_a.
4. Select the significance level α for the test.
5. Display the test statistic to be used, with substitution of the hypothesized value identified in Step 2 but without any computation at this point.
6. Check to make sure that any assumptions required for the test are reasonable.
7. Compute all quantities appearing in the test statistic and then the value of the test statistic itself.
8. Determine the *P*-value associated with the observed value of the test statistic.
9. State the conclusion (which is to reject H_0 if *P*-value $\leq \alpha$ and not to reject H_0 otherwise). The conclusion should then be stated in the context of the problem, and the level of significance should be included.

Steps 1–4 constitute a statement of the problem, Steps 5–8 give the analysis that leads to a decision, and Step 9 provides the conclusion.

■ **Example 10.11   Credit Card Debt**

The article "Credit Cards and College Students: Who Pays, Who Benefits?" (*Journal of College Student Development* [1998]: 50–56) described a study of credit card payment practices of college students. According to the authors of the article, the credit card industry asserts that at most 50% of college students carry a credit card balance from month to month. However, the authors of the article report that, in a random sample of 310 college students, 217 carried a balance each month. Does this sample provide sufficient evidence to reject the industry claim? We answer this question by carrying out a hypothesis test using a .05 significance level.

1. Population characteristic of interest:

π = true proportion of college students who carry a balance from month to month.

2. Null hypothesis: $H_0: \pi = .5$.

3. Alternative hypothesis: $H_a: \pi > .5$ (the proportion of students who carry a balance from month to month is greater than .5).

4. Significance level: $\alpha = .05$.

5. Test statistic:

$$z = \frac{p - \text{hypothesized value}}{\sqrt{\dfrac{(\text{hypothesized value})(1 - \text{hypothesized value})}{n}}} = \frac{p - .5}{\sqrt{\dfrac{(.5)(1 - .5)}{n}}}$$

6. Assumptions: This test requires a random sample and a large sample size. The given sample was a random sample with $n = 310$. Because $310(.5) \geq 10$ and $310(1 - .5) \geq 10$, the large-sample test is appropriate. The sample size is small compared to the population (college students) size.

7. Computations: $n = 310$ and $p = 217/310 = .700$, so

$$z = \frac{.700 - .5}{\sqrt{\dfrac{(.5)(1 - .5)}{310}}} = \frac{.200}{.028} = 7.14$$

8. *P*-value: This is an upper-tailed test (the inequality in H_a is "greater than"), so the *P*-value is the area to the right of the computed z value. Because $z = 7.14$ is so far out in the upper tail of the standard normal distribution, the area to its right is negligible. Thus *P*-value ≈ 0.

9. Conclusion: Because *P*-value $\leq \alpha$ $(0 \leq .05)$, H_0 is rejected at the .05 level of significance.

We conclude that the proportion of students who carry a credit card balance from month to month is greater than .5. That is, the sample provides convincing evidence that the industry claim is not correct.

▪ Example 10.12   Single-Family Homes

The Public Policy Institute of California reported that 71% of people nationwide prefer to live in a single-family home. To determine whether the preferences of Californians are consistent with this nationwide figure, a random sample of 2002 Californians were interviewed. Of those interviewed, 1682 said that they consider a single-family home the ideal (Associated Press, November 13, 2001). Can we reasonably conclude that the proportion of Californians who prefer a single-family home is different from the national figure? We answer this question by carrying out a hypothesis test with $\alpha = .01$.

1. π = proportion of all Californians who prefer a single-family home.

2. $H_0: \pi = .71$.

3. $H_a: \pi \neq .71$ (differs from the national proportion).

4. Significance level: $\alpha = .01$.

5. Test statistic:

$$z = \frac{p - \text{hypothesized value}}{\sqrt{\dfrac{(\text{hypothesized value})(1 - \text{hypothesized value})}{n}}} = \frac{p - .71}{\sqrt{\dfrac{(.71)(.29)}{n}}}$$

6. Assumptions: This test requires a random sample and a large sample size. The given sample was a random sample, the population size is much larger than the sample size, and the sample size was $n = 2002$. Because $2002(.71) \geq 10$ and $2002(.29) \geq 10$, the large-sample test is appropriate.

7. Computations: $p = 1682/2002 = .84$, from which

$$z = \frac{.84 - .71}{\sqrt{\dfrac{(.71)(.29)}{2002}}} = \frac{.13}{.0101} = 12.87$$

8. P-value: The area under the z curve to the right of 12.87 is approximately 0, so P-value $\approx 2(0) = 0$.

9. Conclusion: At significance level .01, we reject H_0 because P-value $\approx 0 < .01 = \alpha$. The data provide convincing evidence that the proportion in California who prefer a single-family home differs from the nationwide proportion.

Most statistical computer packages and graphing calculators can calculate and report P-values for a variety of hypothesis-testing situations, including the large-sample test for a proportion. MINITAB was used to carry out the test of Example 10.10, and the resulting computer output follows (MINITAB uses p instead of π to denote the population proportion):

Test and Confidence Interval for One Proportion

Test of p = 0.5 vs p < 0.5

Sample	X	N	Sample p	95.0 % CI	Z-Value	P-Value
1	220	500	0.440000	(0.396491, 0.483509)	−2.68	0.004

From the MINITAB output, $z = -2.68$, and the associated P-value is .004. The small difference in the P-value is the result of rounding.

It is also possible to compute the value of the z test statistic and then use a statistical computer package or graphing calculator to determine the corresponding P-value as an area under the standard normal curve. For example, the user can specify a value and MINITAB will determine the area to the left of this value for any particular normal distribution. Because of this, the computer can be used in place of Appendix Table 2. In Example 10.10 the computed z was -2.68. Using MINITAB gives the following output:

Normal with mean = 0 and standard deviation = 1.00000

x	P(X ≤ x)
−2.6800	0.0037

Thus we learn that the area to the left of $-2.68 = .0037$, which agrees with the value obtained by using the tables.

■ Exercises 10.22–10.40

10.22 Use the definition of the P-value to explain the following:
a. Why H_0 would certainly be rejected if P-value = .0003
b. Why H_0 would definitely not be rejected if P-value = .350

10.23 For which of the following P-values will the null hypothesis be rejected when performing a level .05 test:
a. .001   **b.** .021   **c.** .078   **d.** .047   **e.** .148

10.24 Pairs of P-values and significance levels, α, are given. For each pair, state whether the observed P-value leads to rejection of H_0 at the given significance level.
a. P-value = .084, α = .05
b. P-value = .003, α = .001
c. P-value = .498, α = .05
d. P-value = .084, α = .10
e. P-value = .039, α = .01
f. P-value = .218, α = .10

10.25 Let π denote the proportion of grocery store customers that use the store's club card. For a large sample z test of H_0: π = .5 versus H_a: π > .5, find the P-value associated with each of the given values of the test statistic:
a. 1.40   **b.** 0.93   **c.** 1.96   **d.** 2.45   **e.** −0.17

10.26 Assuming a random sample from a large population, for which of the following null hypotheses and sample sizes n is the large-sample z test appropriate:
a. H_0: π = .2, n = 25   **b.** H_0: π = .6, n = 210
c. H_0: π = .9, n = 100   **d.** H_0: π = .05, n = 75

10.27 In a national survey of 2013 adults, 1590 responded that lack of respect and courtesy in American society is a serious problem, and 1283 indicated that they believe that rudeness is a more serious problem than in past years (Associated Press, April 3, 2002). Is there convincing evidence that more than three-quarters of U.S. adults believe that rudeness is a worsening problem? Test the relevant hypotheses using a significance level of .05.

10.28 The success of the U.S. census depends on people filling out and returning census forms. Despite extensive advertising, many Americans are skeptical about claims that the Census Bureau will guard the information it collects from other government agencies. In a *USA Today* poll (March 13, 2000), only 432 of 1004 adults surveyed said that they believe the Census Bureau when it says the information you give about yourself is kept confidential. Is there convincing evidence that, despite the advertising campaign, fewer than half of U.S. adults believe the Census Bureau will keep information confidential? Use a significance level of .01.

10.29 The article "Americans Seek Spiritual Guidance on Web" (*San Luis Obispo Tribune*, October 12, 2002) reported that 68% of the general population belong to a religious community. In a survey on Internet use, 84% of "religion surfers" (defined as those who seek spiritual help online or who have used the web to search for prayer and devotional resources) belong to a religious community. Suppose that this result was based on a sample of 512 religion surfers. Is there convincing evidence that the proportion of religion surfers who belong to a religious community is different from .68, the proportion for the general population? Use α = .05.

10.30 Teenagers (age 15 to 20) make up 7% of the driving population. The article "More States Demand Teens Pass Rigorous Driving Tests" (*San Luis Obispo Tribune*, January 27, 2000) described a study of auto accidents conducted by the Insurance Institute for Highway Safety. The Institute found that 14% of the accidents studied involved teenage drivers. Suppose that this percentage was based on examining records from 500 randomly selected accidents. Does the study provide convincing evidence that the proportion of accidents involving teenage drivers differs from .07, the proportion of teens in the driving population?

10.31 Students at the Akademia Podlaka conducted an experiment to determine whether the Belgium-minted Euro coin was equally likely to land heads up or tails up. Coins were spun on a smooth surface, and in 250 spins, 140 landed with the heads side up (*New Scientist*, January 4, 2002). Should the students interpret this result as convincing evidence that the proportion of the time the coin would land heads up is not .5? Test the relevant hypotheses using α = .01. Would your conclusion be different if a significance level of .05 had been used? Explain.

10.32 Although arsenic is known to be a poison, it also has some beneficial medicinal uses. In one study of the use of arsenic to treat acute promyelocytic leukemia (APL), a rare type of blood cell cancer, APL patients were given an arsenic compound as part of their treatment. Of those receiving arsenic, 42% were in remission and showed no signs of leukemia in a subsequent examination (*Washington Post*, November 5, 1998). It is known that 15% of

APL patients go into remission after the conventional treatment. Suppose that the study had included 100 randomly selected patients (the actual number in the study was much smaller). Is there sufficient evidence to conclude that the proportion in remission for the arsenic treatment is greater than .15, the remission proportion for the conventional treatment? Test the relevant hypotheses using a .01 significance level.

10.33 According to the article "Which Adults Do Underage Youth Ask for Cigarettes?" (*American Journal of Public Health* [1999]: 1561–1564), 43.6% of the 149 18- to 19-year-olds in a random sample have been asked to buy cigarettes for an underage smoker.
a. Is there convincing evidence that fewer than half of 18- to 19-year-olds have been approached to buy cigarettes by an underage smoker?
b. The article went on to state that of the 110 non-smoking 18- to 19-year-olds, only 38.2% had been approached to buy cigarettes for an underage smoker. Is there evidence that less than half of nonsmoking 18- to 19-year-olds have been approached to buy cigarettes?

10.34 Many people have misconceptions about how profitable small, consistent investments can be. In a survey of 1010 randomly selected U.S. adults (Associated Press, October 29, 1999), only 374 responded that they thought that an investment of $25 per week over 40 years with a 7% annual return would result in a sum of over $100,000 (the correct amount is $286,640). Is there sufficient evidence to conclude that less than 40% of U.S. adults are aware that such an investment would result in a sum of over $100,000? Test the relevant hypotheses using $\alpha = .05$.

10.35 The same survey described in Exercise 10.34 also asked the individuals in the sample what they thought was their best chance to obtain more than $500,000 in their lifetime. Twenty-eight percent responded "win a lottery or sweepstakes." Does this provide convincing evidence that more than one-fourth of U.S. adults see a lottery or sweepstakes win as their best chance of accumulating $500,000? Carry out a test using a significance level of .01.

10.36 The Associated Press (February 27, 1995) reported that 71% of Americans, age 25 and older, are overweight, a substantial increase over the 58% figure from 1983. Although this information came from a Harris Poll (a survey) rather than a census of the population, let's assume for the purposes of this exercise that the nationwide population proportion is exactly .71. Suppose that an investigator wishes to know whether the proportion of such individuals in her state who are overweight differs from the na-

tional proportion. A random sample of size $n = 600$ results in 450 who are classified as overweight.
a. What can the investigator conclude? Answer this question by carrying out a hypothesis test with $\alpha = .01$.
b. For the hypotheses in Part (a), describe Type I and Type II errors.
c. If the conclusion to the test in Part (a) is in error, which type of error (Type I or Type II) are you making?

10.37 In an experiment, 25 of 37 panelists preferred the flavor of deep-fat fried chicken nuggets to baked chicken nuggets ("Effect of Preparation Methods on Total Fat Content, Moisture Content, and Sensory Characteristics of Breaded Chicken Nuggets and Beef Steak Fingers," *Family and Consumer Sciences Research Journal* [1999]: 18–27).
a. Is there evidence to show that more than two-thirds of consumers prefer deep-fat fried chicken nuggets? Assume that the 37 panelists can be viewed as a random sample of consumers.
b. Describe a Type I error.
c. Which sort of error, Type I or Type II, could not have been committed in this test.
d. In another part of this experiment, it was noted that 23 of 37 panelists preferred the flavor of baked beef steak fingers to hot-air fried beef steak fingers. Is there evidence that the majority of consumers prefer the flavor of baked beef steak fingers to hot-air fried beef steak fingers?

10.38 The state of Georgia's HOPE scholarship program guarantees fully paid tuition to Georgia public universities for Georgia high school seniors who have a B average in academic requirements as long as they maintain a B average in college. Of 137 randomly selected students enrolling in the Ivan Allen College at the Georgia Institute of Technology (social science and humanities majors) in 1996 who had a B average going into college, 53.2% had a GPA below 3.0 at the end of their first year ("Who Loses HOPE? Attrition from Georgia's College Scholarship Program," *Southern Economic Journal* [1999]: 379–390). Do these data provide convincing evidence that a majority of students at Ivan Allen College who enroll with a HOPE scholarship lose their scholarship?

10.39 Academic dishonesty continues to be a problem on college and university campuses. One study ("Cheating at Small Colleges: An Examination of Student and Faculty Behaviors," *Journal of College Student Development* [1994]: 255–260) reported that in a random sample of 480 students, 124 said that they had looked at notes during an exam. Assuming that the survey was designed and carried out in a manner resulting in truthful responses, does it appear that more than 20% of all students in the sampled population engaged in this particular form of cheating?

10.40 A random sample of 2700 California lawyers revealed only 1107 who felt that the ethical standards of most lawyers are high (Associated Press, November 12, 1994). Does this provide strong evidence for concluding that fewer than 50% of all California lawyers feel this way? Carry out a test of appropriate hypotheses.

▪ 10.4 Hypothesis Tests for a Population Mean

We now turn our attention to developing a method for testing hypotheses about a population mean. The test procedures in this case are based on the same two results that led to the z and t confidence intervals in Chapter 9. These results follow:

1. When either n is large or the population distribution is approximately normal, then

$$z = \frac{\bar{x} - \mu}{\dfrac{\sigma}{\sqrt{n}}}$$

   has approximately a standard normal distribution.

2. When either n is large or the population distribution is approximately normal, then

$$t = \frac{\bar{x} - \mu}{\dfrac{s}{\sqrt{n}}}$$

   has approximately a t distribution with df $= n - 1$.

A consequence of these two results is that if we are interested in testing a null hypothesis of the form

H_0: μ = hypothesized value

then, depending on whether σ is known or unknown, we can use either the z or the t test statistic as long as n is large or the population distribution is approximately normal.

Case 1: σ known

Test statistic: $z = \dfrac{\bar{x} - \text{hypothesized value}}{\dfrac{\sigma}{\sqrt{n}}}$

P-value: Computed as an area under the z curve

Case 2: σ unknown

Test statistic: $t = \dfrac{\bar{x} - \text{hypothesized value}}{\dfrac{s}{\sqrt{n}}}$

P-value: Computed as an area under the t curve with df $= n - 1$

Because it is rarely the case that σ, the population standard deviation, is known, we focus our attention on the test procedure for the case in which σ is unknown.

When testing a hypothesis about a population mean, the null hypothesis specifies a particular hypothesized value for μ, specifically, $H_0: \mu =$ hypothesized value. The alternative hypothesis has one of the following three forms, depending on the research question being addressed:

$$H_a: \quad \mu > \text{hypothesized value}$$
$$H_a: \quad \mu < \text{hypothesized value}$$
$$H_a: \quad \mu \neq \text{hypothesized value}$$

If n is large or if the population distribution is approximately normal, the test statistic

$$t = \frac{\bar{x} - \text{hypothesized value}}{\dfrac{s}{\sqrt{n}}}$$

can be used. For example, if the null hypothesis to be tested is $H_0: \mu = 100$, the test statistic becomes

$$t = \frac{\bar{x} - 100}{\dfrac{s}{\sqrt{n}}}$$

Consider the alternative hypothesis $H_a: \mu > 100$, and suppose that a sample of size $n = 24$ gives $\bar{x} = 104.20$ and $s = 8.23$. The resulting test statistic value is

$$t = \frac{104.20 - 100}{\dfrac{8.23}{\sqrt{24}}} = \frac{4.20}{1.6799} = 2.50$$

Because this is an upper-tailed test, if the test statistic had been z rather than t, the P-value would be the area under the z curve to the right of 2.50. With a t statistic, the P-value is the area under an appropriate t curve (here with df $= 24 - 1 = 23$) to the right of 2.50. Appendix Table 4 is a tabulation of t curve tail areas. Each column of the table is for a different number of degrees of freedom: 1, 2, 3, ..., 30, 35, 40, 60, 120, and a last column for df $= \infty$, which is the same as for the z curve. The table gives the area under each t curve to the right of values ranging from 0.0 to 4.0 in increments of 0.1. Part of this table appears in Figure 10.3. For example,

(area under the t curve with 23 df to the right of 2.5) $= .010$

$= P$-value for an upper-tailed t test

Suppose that $t = -2.7$ for a lower-tailed test based on 23 df. Then, because any t curve is symmetric about 0,

P-value $=$ area to the left of $-2.7 =$ area to the right of $2.7 = .006$

As is the case for z tests, we double the captured tail area to obtain the P-value for two-tailed t tests. Thus, if $t = 2.6$ or if $t = -2.6$ for a two-tailed test with 23 df, then

P-value $= 2(.008) = .016$

Once past 30 df, the tail areas change very little, so the last column (∞) in Appendix Table 4 provides a good approximation.

The following two boxes give a general description of the test procedure and show how the P-value is obtained as a t curve area.

Figure 10.3 Part of Appendix Table 4: *t* curve tail areas.

df	1	2	. . .	22	23	24	. . .	60	120
t									
0.0									
0.1									
⋮				⋮	⋮	⋮			
2.5			. . .	.010	.010	.010	. . .		
2.6			. . .	.008	.008	.008	. . .		
2.7			. . .	.007	.006	.006	. . .		
2.8			. . .	.005	.005	.005	. . .		
⋮				⋮	⋮	⋮			
4.0									

Area under 23-df t curve to right of 2.7

■ **Finding *P*-Values for a *t* Test**

1. **Upper-tailed test**:
   $H_a: \mu >$ hypothesized value.
   P-value computed as illustrated:

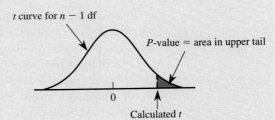

2. **Lower-tailed test**:
   $H_a: \mu <$ hypothesized value.
   P-value computed as illustrated:

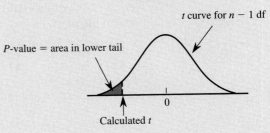

3. **Two-tailed test**:
   $H_a: \mu \neq$ hypothesized value.
   P-value computed as illustrated:

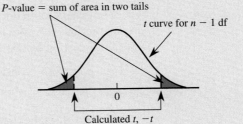

Appendix Table 4 gives upper-tail *t* curve areas to the right of values 0.0, 0.1, . . . , 4.0. These areas are *P*-values for upper-tailed tests and, by symmetry, also for lower-tailed tests. Doubling an area gives the *P*-value for a two-tailed test.

▪ **One-Sample *t* Test for a Population Mean**

Null hypothesis: $H_0: \mu =$ hypothesized value.

Test statistic: $t = \dfrac{\bar{x} - \text{hypothesized value}}{\dfrac{s}{\sqrt{n}}}$

Alternative Hypothesis:

H_a: $\mu >$ hypothesized value

H_a: $\mu <$ hypothesized value

H_a: $\mu \neq$ hypothesized value

P-Value:

Area to right of calculated t under t curve with df $= n - 1$

Area to the left of calculated t under t curve with df $= n - 1$

(1) 2(area to right of t) if t is positive, or
(2) 2(area to left of t) if t is negative

Assumptions: 1. $\bar{x}$ and s are the sample mean and sample standard deviation, respectively, from a *random sample*.
2. The *sample size is large* (generally $n \geq 30$) or *the population distribution is at least approximately normal*.

▪ Example 10.13   Time Stands Still (or So It Seems)

A study conducted by researchers at Pennsylvania State University investigated whether time perception, a simple indication of a person's ability to concentrate, is impaired during nicotine withdrawal. The study results were presented in the paper "Smoking Abstinence Impairs Time Estimation Accuracy in Cigarette Smokers" (*Psychopharmacology Bulletin* [2003]: 90–95). After a 24-hr smoking abstinence, 20 smokers were asked to estimate how much time had passed during a 45-sec period. Suppose the resulting data on perceived elapsed time (in seconds) were as shown (these data are artificial but are consistent with summary quantities given in the paper):

69	65	72	73	59	55	39	52	67	57
56	50	70	47	56	45	70	64	67	53

From these data, we obtain

$$n = 20 \qquad \bar{x} = 59.30 \qquad s = 9.84$$

The researchers wanted to determine whether smoking abstinence had a negative impact on time perception, causing elapsed time to be overestimated. We can answer this question by testing

H_0: $\mu = 45$ (no consistent tendency to overestimate the time elapsed)

versus

H_a: $\mu > 45$ (tendency for elapsed time to be overestimated)

The null hypothesis is rejected only if there is convincing evidence that $\mu > 45$. The observed $\bar{x}$ value, 59.30, is certainly larger than 45, but can a sample mean as large as this be plausibly explained by chance variation from one sample to another when

$\mu = 45$? To answer this question, we carry out a hypothesis test with a significance level of .05 using the step-by-step procedure described in Section 10.3.

1. Population characteristic of interest:

   μ = mean perceived elapsed time for smokers who have abstained from smoking for 24 hr.

2. Null hypothesis: $H_0: \mu = 45$.

3. Alternative hypothesis: $H_a: \mu > 45$.

4. Significance level: $\alpha = .05$.

5. Test statistic: $t = \dfrac{\overline{x} - \text{hypothesized value}}{\dfrac{s}{\sqrt{n}}} = \dfrac{\overline{x} - 45}{\dfrac{s}{\sqrt{n}}}$

6. Assumptions: This test requires a random sample and either a large sample size or a normal population distribution. The authors of the paper believed that it was reasonable to consider this sample as representative of smokers in general, so we regard it as if it were a random sample. Because the sample size is only 20, for the t test to be appropriate, we must be able to assume that the population distribution of perceived elapsed times is at least approximately normal. Is this reasonable? The following graph gives a boxplot of the data:

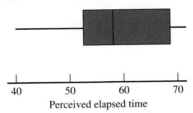

Perceived elapsed time

Although the boxplot is not perfectly symmetric, it is not too skewed and there are no outliers, so we judge the use of the t test to be reasonable.

7. Computations: $n = 20$, $\overline{x} = 59.30$, and $s = 9.84$. Therefore

$$t = \frac{59.30 - 45}{\dfrac{9.84}{\sqrt{20}}} = \frac{14.30}{2.20} = 6.50$$

8. *P*-value: This is an upper-tailed test (the inequality in H_a is "greater than"), so the *P*-value is the area to the right of the computed t value. Because df $= 20 - 1 = 19$, we can use the df $= 19$ column of Appendix Table 4 to find the *P*-value. With $t = 6.50$, we obtain *P*-value = area to the right of $6.50 \approx 0$ (because 6.50 is greater than 4.0, the largest tabulated value).

9. Conclusion: Because *P*-value $\leq \alpha$, we reject H_0 at the .05 level of significance. There is virtually no chance of seeing a sample mean (and hence a t value) this extreme as a result of just chance variation when H_0 is true. There is convincing evidence that the mean perceived time elapsed is greater than the actual time elapsed of 45 sec.

This paper also looked at perception of elapsed time for a sample of nonsmokers and for a sample of smokers who had not abstained from smoking. The investigators found that the null hypothesis of $\mu = 45$ could not be rejected for either of these groups.

▪ Example 10.14   Personal Use of Company Technology

One concern employers have about the use of technology is the amount of time that employees spend each day making personal use of company technology, such as personal phone calls, non-business-related e-mail, Internet use, and computer games. The Associated Press (September 7, 1999) reported that a management consultant believes that, on average, workers spend 75 min a day making personal use of company technology. Suppose that the CEO of a large corporation wants to determine whether the average amount of time spent on personal use of company technology for her employees is greater than the reported value of 75 min. Each person in a random sample of 10 employees was contacted and asked about daily personal use of company technology. (Participants would probably have to be guaranteed anonymity to obtain truthful responses.) The resulting data are the following:

$$66 \quad 70 \quad 75 \quad 88 \quad 69 \quad 89 \quad 71 \quad 71 \quad 63 \quad 86$$

Summary quantities are $n = 10$, $\bar{x} = 74.80$, and $s = 9.45$.

Do these data provide evidence that the mean for this company is greater than 75 min? To answer this question, let's carry out a hypothesis test with $\alpha = .05$.

1. μ = mean daily time spent in personal use of company technology for this company.

2. $H_0: \mu = 75$.

3. $H_a: \mu > 75$.

4. $\alpha = .05$.

5. Test statistic: $t = \dfrac{\bar{x} - \text{hypothesized value}}{\dfrac{s}{\sqrt{n}}} = \dfrac{\bar{x} - 75}{\dfrac{s}{\sqrt{n}}}$

6. This test requires a random sample and either a large sample or a normal population distribution. The given sample was a random sample of employees. Because the sample size is small, we must be willing to assume that the population distribution of times is at least approximately normal. The following normal probability plot appears to be reasonably straight, and although the normal probability plot and the boxplot reveal some skewness in the sample, there are no outliers:

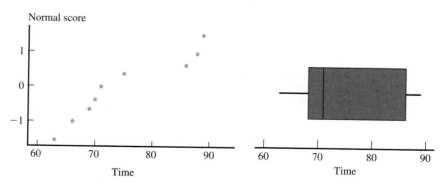

Correlations (Pearson)
Correlation of Time and Normal Score = 0.943

Also, the correlation between the expected normal scores and the observed data for this sample is .943, which is well above the critical r value for $n = 10$ of .880 (see Chapter 5 for critical r values). Based on these observations, it is plausible that the population distribution is approximately normal, so we proceed with the t test.

7. Test statistic: $t = \dfrac{74.80 - 75}{\dfrac{9.45}{\sqrt{10}}} = -0.07$

8. From the df = 9 column of Appendix Table 4 and by rounding the test statistic value to -0.1, we get

$$P\text{-value} = \text{area to the right of } -.1$$
$$= 1 - \text{area to the left of } -.1$$
$$= 1 - .461$$
$$= .539$$

as shown:

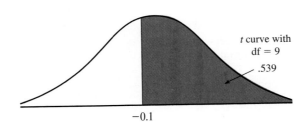

t curve with df = 9

.539

-0.1

9. Because the P-value $> \alpha$, we fail to reject H_0. There is not sufficient evidence to conclude that the mean time spent on personal use of company technology is greater than 75 min per day for this company.

MINITAB could also have been used to carry out the test, as shown in the accompanying output:

T-Test of the Mean

Test of mu = 75.00 vs mu > 75.00

Variable	n	Mean	StDev	SE Mean	t	P
Time	10	74.80	9.45	2.99	−0.07	0.53

Although we had to round the computed t value to -0.1 to use Appendix Table 4, MINITAB was able to compute the P-value corresponding to the actual value of the test statistic.

▪ Example 10.15   Cricket Love

The article "Well Fed Crickets Bowl Maidens Over" (*Nature Science Update*, February 11, 1999) reported that female field crickets are attracted to males that have high chirp rates and hypothesized that chirp rate is related to nutritional status.

The usual chirp rate for male field crickets was reported to vary around a mean of 60 chirps per second. To investigate whether chirp rate was related to nutritional status, investigators fed male crickets a high protein diet for 8 days, after which chirp rate was measured. The mean chirp rate for the crickets on the high protein diet was reported to be 109 chirps per second. Is this convincing evidence that the mean chirp rate for crickets on a high protein diet is greater than 60 (which would then imply an advantage in attracting the ladies)? Suppose that the sample size and sample standard deviation are $n = 32$ and $s = 40$. We test the relevant hypotheses with $\alpha = .01$.

1. μ = mean chirp rate for crickets on a high protein diet.

2. $H_0: \mu = 60$.

3. $H_a: \mu > 60$.

4. $\alpha = .01$.

5. Test statistic: $\quad t = \dfrac{\overline{x} - \text{hypothesized value}}{\dfrac{s}{\sqrt{n}}} = \dfrac{\overline{x} - 60}{\dfrac{s}{\sqrt{n}}}$

6. This test requires a random sample and either a large sample or a normal population distribution. Because the sample size is large ($n = 32$), it is reasonable to proceed with the t test as long as we are willing to consider the 32 male field crickets in this study as if they were a random sample from the population of male field crickets.

7. Computation: $\quad t = \dfrac{109 - 60}{\dfrac{40}{\sqrt{32}}} = \dfrac{49}{7.07} = 6.93$

8. This is an upper-tailed test, so the P-value is the area under the t curve with df = 31 and to the right of 6.93. From Appendix Table 4, P-value ≈ 0.

9. Because P-value ≈ 0, which is less than the significance level, α, we reject H_0. There is convincing evidence that the mean chirp rate is higher for male field crickets that eat a high protein diet.

▪ Statistical Versus Practical Significance

Carrying out a hypothesis test amounts to deciding whether the value obtained for the test statistic could plausibly have resulted when H_0 is true. When the value of the test statistic leads to rejection of H_0, it is customary to say that the result is **statistically significant** at the chosen level α. The finding of statistical significance means that, in the investigator's opinion, the observed deviation from what was expected under H_0 cannot plausibly be attributed only to chance variation. However, statistical significance cannot be equated with the conclusion that the true situation differs from what H_0 states in any practical sense. That is, even after H_0 has been rejected, the data may suggest that there is no *practical* difference between the true value of the population characteristic and what the null hypothesis states that value to be. This is illustrated in Example 10.16.

■ Example 10.16   "Significant" but Unimpressive Test Score Improvement

Let μ denote the true average score on a standardized test for children in a certain region of the United States. The average score for all children in the United States is 100. Regional education authorities are interested in testing $H_0: \mu = 100$ versus $H_a: \mu > 100$ using a significance level of .001. A sample of 2500 children resulted in the values $n = 2500$, $\bar{x} = 101.0$, and $s = 15.0$. Then

$$t = \frac{101.0 - 100}{\dfrac{15}{\sqrt{2500}}} = 3.33$$

This is an upper-tailed test, so (using the z column of Appendix Table 4 because df = 2499) P-value = area to the right of 3.33 ≈ .000. Because P-value < .001, we reject H_0. The true mean score for this region does appear to exceed 100.

However, with $n = 2500$, the point estimate $\bar{x} = 101.0$ is almost surely very close to the true value of μ. Therefore it looks as though H_0 was rejected because $\mu \approx 101$ rather than 100. And, from a practical point of view, a 1-point difference is most likely of no practical importance. A statistically significant result does not necessarily mean that there are any practical consequences.

■ Exercises 10.41–10.58

10.41 Newly purchased automobile tires of a certain type are supposed to be filled to a pressure of 30 psi. Let μ denote the true average pressure. Find the P-value associated with each of the following given z statistic values for testing $H_0: \mu = 30$ versus $H_a: \mu \neq 30$ when σ is known:
a. 2.10   b. −1.75   c. −0.58   d. 1.44   e. −5.00

10.42 The desired percentage of silicon dioxide in a certain type of cement is 5.0%. A random sample of $n = 36$ specimens gave a sample average percentage of $\bar{x} = 5.21$. Let μ be the true average percentage of silicon dioxide in this type of cement, and suppose that σ is known to be 0.38. Test $H_0: \mu = 5$ versus $H_a: \mu \neq 5$ using a significance level of .01.

10.43 Give as much information as you can about the P-value of a t test in each of the following situations:
a. Upper-tailed test, df = 8, $t = 2.0$
b. Upper-tailed test, $n = 14$, $t = 3.2$
c. Lower-tailed test, df = 10, $t = -2.4$
d. Lower-tailed test, $n = 22$, $t = -4.2$
e. Two-tailed test, df = 15, $t = -1.6$
f. Two-tailed test, $n = 16$, $t = 1.6$
g. Two-tailed test, $n = 16$, $t = 6.3$

10.44 Give as much information as you can about the P-value of a t test in each of the following situations:
a. Two-tailed test, df = 9, $t = 0.73$
b. Upper-tailed test, df = 10, $t = -0.5$
c. Lower-tailed test, $n = 20$, $t = -2.1$
d. Lower-tailed test, $n = 20$, $t = -5.1$
e. Two-tailed test, $n = 40$, $t = 1.7$

10.45 Paint used to paint lines on roads must reflect enough light to be clearly visible at night. Let μ denote the true average reflectometer reading for a new type of paint under consideration. A test of $H_0: \mu = 20$ versus $H_a: \mu > 20$ based on a sample of 15 observations gave $t = 3.2$. What conclusion is appropriate at each of the following significance levels?
a. $\alpha = .05$ b. $\alpha = .01$ c. $\alpha = .001$

10.46 A certain pen has been designed so that true average writing lifetime under controlled conditions (involving the use of a writing machine) is at least 10 hr. A random sample of 18 pens is selected, the writing lifetime of each is determined, and a normal probability plot of the resulting data supports the use of a one-sample t test. The relevant hypotheses are $H_0: \mu = 10$ versus $H_a: \mu < 10$.

a. If $t = -2.3$ and $\alpha = .05$ is selected, what conclusion is appropriate?

b. If $t = -1.83$ and $\alpha = .01$ is selected, what conclusion is appropriate?

c. If $t = 0.47$, what conclusion is appropriate?

10.47 The true average diameter of ball bearings of a certain type is supposed to be 0.5 in. What conclusion is appropriate when testing $H_0: \mu = 0.5$ versus $H_a: \mu \neq 0.5$ in each of the following situations:

a. $n = 13, t = 1.6, \alpha = .05$
b. $n = 13, t = -1.6, \alpha = .05$
c. $n = 25, t = -2.6, \alpha = .01$
d. $n = 25, t = -3.6$

10.48 There is now broad evidence that stressful work conditions and critical personal characteristics contribute to the development of coronary dysfunction and disease. The article "Job Stressors and Coping Characteristics in Work-Related Disease" (*Work and Stress* [1994]: 130–140) reported that, for a random sample of 42 blue-collar workers suffering from ischemic heart disease (IHD), the sample mean systolic blood pressure and sample standard deviation were 145.3 and 17.6, respectively. Suppose that for the population of all blue-collar non-IHD workers, the mean is 138 (a value consistent with information given in the article). Does it appear that the true average blue-collar IHD pressure exceeds the non-IHD value? State and test the relevant hypotheses at significance level .01 using the suggested sequence of steps.

10.49 Typically, only very brave students are willing to speak out in a college classroom. Student participation may be especially difficult if the individual is from a different culture or country. The article "An Assessment of Class Participation by International Graduate Students" (*Journal of College Student Development* [1995]: 132–140) considered a numerical "speaking-up" scale, with possible values from 3 to 15 (a low value means that a student rarely speaks). For a random sample of 64 males from Asian countries where English is not the official language, the sample mean and sample standard deviation were 8.75 and 2.57, respectively. Suppose that the mean for the population of all males having English as their native language is 10.0 (suggested by data in the article). Does it appear that the population mean for males from non-English-speaking Asian countries is smaller than 10.0?

10.50 Speed, size, and strength are thought to be important factors in football performance. The article "Physical and Performance Characteristics of NCAA Division I Football Players" (*Research Quarterly for Exercise and Sport* [1990]: 395–401) reported on physical characteristics of Division I starting football players in the 1988 football season. Information for teams ranked in the top 20 was easily obtained, and it was reported that the mean weight of starters on top-20 teams was 105 kg. A random sample of 33 starting players (various positions were represented) from Division I teams that were not ranked in the top 20 resulted in a sample mean weight of 103.3 kg and a sample standard deviation of 16.3 kg. Is there sufficient evidence to conclude that the mean weight for non-top-20 starters is less than 105, the known value for top-20 teams?

10.51 A well-designed and safe workplace can contribute greatly to increasing productivity. It is especially important that workers not be asked to perform tasks, such as lifting, that exceed their capabilities. The following data on maximum weight of lift (MWOL, in kilograms) for a frequency of 4 lifts per minute was reported in the article "The Effects of Speed, Frequency, and Load on Measured Hand Forces for a Floor-to-Knuckle Lifting Task" (*Ergonomics* [1992]: 833–843):

25.8   36.6   26.3   21.8   27.2

Suppose that it is reasonable to regard the sample as a random sample from the population of healthy males, age 18–30. Do the data suggest that the population mean MWOL exceeds 25? Carry out a test of the relevant hypotheses using a .05 significance level.

10.52 Are young women delaying marriage and marrying at a later age? This question was addressed in a report issued by the Census Bureau (Associated Press, June 8, 1991). The report stated that in 1970 (based on census results) the mean age of brides marrying for the first time was 20.8 years. In 1990 (based on a sample, because census results were not yet available), the mean was 23.9. Suppose that the 1990 sample mean had been based on a random sample of size 100 and that the sample standard deviation was 6.4. Is there sufficient evidence to support the claim that in 1990 women were marrying later in life than in 1970? Test the relevant hypotheses using $\alpha = .01$. (Note: It is probably not reasonable to think that the distribution of age at first marriage is normal in shape.)

10.53 In the article "Religious Outlook, Culture War Politics, and Antipathy Toward Christian Fundamentalists" (*Public Opinion Quarterly* [1999]: 29–61), it is stated that nonfundamentalists have negative feelings overall toward Christian fundamentalists. The information offered to support this claim is that the mean "temperature" rating given to Christian fundamentalists for a random sample of

960 nonfundamentalists is 47° (higher temperatures mean more approval). The sample standard deviation is 21°. Test the hypothesis that the mean rating given by all nonfundamentalists to Christian fundamentalists is below 57° (the average rating given across all religious, racial, and political groups).

10.54 In recent years, female athletes have been identified as a population particularly at risk for developing eating disorders. The article "Disordered Eating in Female Collegiate Gymnasts" (*Journal of Sport and Exercise Psychology* [1993]: 424–436) reported that for a sample of 17 gymnasts who were classified as binge eaters, the sample mean difference between current weight and ideal weight was 6.65 lb. The following boxplot is based on data consistent with this information; the sample standard deviation is 2.97 lb:

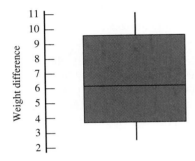

Suppose that the true average weight difference for female gymnasts with normal eating habits is 4.5 lb (a value suggested in the article).
a. Does the boxplot indicate that μ, the true average weight difference for bingers, exceeds the value for normal eaters? Explain.
b. Carry out a test to determine whether there is convincing evidence that the mean weight difference for bingers exceeds the value for normal eaters.

10.55 An article titled "Teen Boys Forget Whatever It Was" appeared in the Australian newspaper *The Mercury* (April 21, 1997). It described a study of academic performance and attention span and reported that the mean time to distraction for teenage boys working on an independent task was 4 min. Although the sample size was not given in the article, suppose that this mean was based on a random sample of 50 teenage Australian boys and that the sample standard deviation was 1.4 min. Is there convincing evidence that the average attention span for teenage boys is less than 5 min? Test the relevant hypotheses using $\alpha = .01$.

10.56 According to the article "Workaholism in Organizations: Gender Differences" (*Sex Roles* [1999]: 333–346), the following data were reported on 1996 income for random samples of male and female

MBA graduates from a certain Canadian business school:

	n	$\bar{x}$	s
Males	258	$133,442	$131,090
Females	233	$105,156	$98,525

Note: These salary figures are in Canadian dollars.

a. Test the hypothesis that the mean salary of male MBA graduates from this school was in excess of $100,000 in 1996.
b. Is there convincing evidence that the mean salary for all female MBA graduates is above $100,000? Test using $\alpha = .10$.
c. If a significance level of .05 or .01 were used instead of .10 in the test of Part (b), would you still reach the same conclusion? Explain.

10.57 Many consumers pay careful attention to stated nutritional contents on packaged foods when making purchases. It is therefore important that the information on packages be accurate. A random sample of $n = 12$ frozen dinners of a certain type was selected from production during a particular period, and the calorie content of each one was determined. (This determination entails destroying the product, so a census would certainly not be desirable!) Here are the resulting observations, along with a boxplot and normal probability plot:

255 244 239 242 265 245 259 248
225 226 251 233

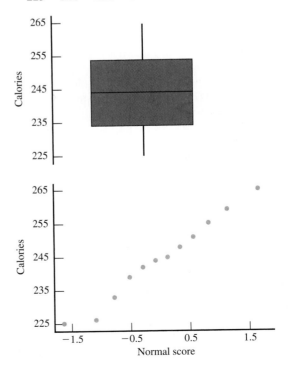

a. Is it reasonable to test hypotheses about true average calorie content μ by using a t test?
b. The stated calorie content is 240. Does the boxplot suggest that true average content differs from the stated value? Explain your reasoning.
c. Carry out a formal test of the hypotheses suggested in Part (b).

10.58 Much concern has been expressed in recent years regarding the practice of using nitrates as meat preservatives. In one study involving possible effects of these chemicals, bacteria cultures were grown in a medium containing nitrates. The rate of uptake of radio-labeled amino acid was then determined for each culture, yielding the following observations:

7251 6871 9632 6866 9094 5849 8957 7978
7064 7494 7883 8178 7523 8724 7468

Suppose that it is known that the true average uptake for cultures without nitrates is 8000. Do the data suggest that the addition of nitrates results in a decrease in the true average uptake? Test the appropriate hypotheses using a significance level of .10.

▪ 10.5 Power and Probability of Type II Error

In this chapter, we have introduced test procedures for testing hypotheses about population characteristics, such as μ and π. What characterizes a "good" test procedure? It makes sense to think that a good test procedure is one that has both a small probability of rejecting H_0 when it is true (a Type I error) and a high probability of rejecting H_0 when it is false. The test procedures presented in this chapter allow us to directly control the probability of rejecting a true H_0 by our choice of the significance level α. But what about the probability of rejecting H_0 when it is false? As we will see, several factors influence this probability. Let's begin by considering an example.

Suppose that the student body president at a university is interested in studying the amount of money that students spend on textbooks each semester. The director of the financial aid office believes that the average amount spent on books is $300 per semester and uses this figure to determine the amount of financial aid for which a student is eligible. The student body president plans to ask each individual in a random sample of students how much he or she spent on books this semester and has decided to use the resulting data to test

$$H_0: \quad \mu = 300 \qquad \text{versus} \qquad H_a: \quad \mu > 300$$

using a significance level of .05. If the true mean is 300 (or less than 300), the correct decision is to fail to reject the null hypothesis (incorrectly rejecting the null hypothesis is a Type I error). On the other hand, if the true mean is 325 or 310 or even 301, the correct decision is to reject the null hypothesis (not rejecting the null hypothesis is a Type II error). How likely is it that the null hypothesis will, in fact, be rejected?

If the true mean is 301, the probability that we reject $H_0: \mu = 300$ is not very great. This is because when we carry out the test, we are essentially looking at the sample mean and asking, Does this look like what we would expect to see if the population mean were 300? As illustrated in Figure 10.4, if the true mean is greater than but very close to 300, chances are that the sample mean will look pretty much like what we would expect to see if the population mean were 300, and we will be unconvinced that the null hypothesis should be rejected. If the true mean is 325, it is less likely that the sample will be mistaken for a sample from a population with mean 300; sample means will tend to cluster around 325, and so it is more likely that we will correctly reject H_0. If the true mean is 350, rejection of H_0 is even more likely.

Figure 10.4 Sampling distribution of $\bar{x}$ when $\mu = 300$, 305, and 325.

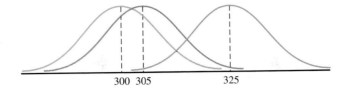

300 305   325

When we consider the probability of rejecting the null hypothesis, we are looking at what statisticians refer to as the **power** of the test.

The **power of a test** is the probability of rejecting the null hypothesis.

From the previous discussion, it should be apparent that the power of the test when a hypothesis about a population mean is being tested depends on the true value of the population mean, μ. Because the true value of μ is unknown (if we knew the value of μ, we wouldn't be doing the hypothesis test!), we cannot know what the power is for the actual true value of μ. It is possible, however, to gain some insight into the power of a test by looking at a number of "what if" scenarios. For example, we might ask, What is the power if the true mean is 325? or What is the power if the true mean is 310? and so on. That is, we can determine the power at $\mu = 325$, the power at $\mu = 310$, and the power at any other value of interest. Although it is technically possible to consider power when the null hypothesis is true, an investigator is usually concerned about the power only at values for which the null hypothesis is false.

In general, when testing a hypothesis about a population characteristic, there are three factors that influence the power of the test:

1. The size of the difference between the true value of the population characteristic and the hypothesized value (the value that appears in the null hypothesis)

2. The choice of significance level, α, for the test

3. The sample size

■ **Effect of Various Factors on the Power of a Test**

1. The larger the size of the discrepancy between the hypothesized value and the true value of the population characteristic, the higher the power.

2. The larger the significance level, α, the higher the power of the test.

3. The larger the sample size, the higher the power of the test.

Let's consider each of these three statements. The first statement has already been discussed in the context of the textbook example. Because power is the probability of rejecting the null hypothesis, it makes sense that the power will be higher when the true value of a population characteristic is quite different from the hypothesized value than when it is close to that value.

The effect of significance level on power is not quite as obvious. To understand the relationship between power and significance level, it helps to see the relationship between power and β, the probability of a Type II error.

When H_0 is false, power $= 1 - \beta$.

This relationship follows from the definitions of power and Type II error. A Type II error results from *not* rejecting a false H_0. Because power is the probability of rejecting H_0, it follows that *when H_0 is false*

power $=$ probability of rejecting a false H_0

$\quad\quad\quad = 1 -$ probability of not rejecting a false H_0

$\quad\quad\quad = 1 - \beta$

Recall from Section 10.2 that the choice of α, the Type I error probability, affects the value of β, the Type II error probability. Choosing a larger value for α results in a smaller value for β (and thus a larger value for $1 - \beta$). In terms of power, this means that choosing a larger value for α results in a larger value for the power of the test. That is, the larger the Type I error probability we are willing to tolerate, the more likely it is that the test will be able to detect any particular departure from H_0.

The third factor that affects the power of a test is the sample size. When H_0 is false, the power of a test is the probability that we will in fact "detect" that H_0 is false and, based on the observed sample, reject H_0. Intuition suggests that we will be more likely to detect a departure from H_0 with a large sample than with a small sample. This is in fact the case — the larger the sample size, the higher the power.

Consider testing the hypotheses presented previously:

$$H_0: \quad \mu = 300 \quad\quad \text{versus} \quad\quad H_a: \quad \mu > 300$$

The observations about power imply the following, for example:

1. For any value of μ exceeding 300, the power of a test based on a sample of size 100 is higher than the power of a test based on a sample of size 75 (assuming the same significance level).

2. For any value of μ exceeding 300, the power of a test using a significance level of .05 is higher than the power of a test using a significance level of .01 (assuming the same sample size).

3. For any value of μ exceeding 300, the power of the test is greater if the true mean is 350 than if the true mean is 325 (assuming the same sample size and significance level).

As was mentioned previously in this section, it is impossible to calculate the *actual* power of a test because in practice we do not know the true value of population characteristics. However, we can evaluate the power at a selected alternative value if we want to know whether the power would be high or low if this alternative value is the true value.

The following optional subsection shows how Type II error probabilities and power can be evaluated for selected tests.

▪ Calculating Power and Type II Error Probabilities for Selected Tests (Optional)

The test procedures presented in this chapter are designed to control the probability of a Type I error (rejecting H_0 when H_0 is true) at the desired level α. However, little has been said so far about calculating the value of β, the probability of a

Type II error (not rejecting H_0 when H_0 is false). Here we consider the determination of β and power for the hypothesis tests previously introduced.

When we carry out a hypothesis test, we specify the desired value of α, the probability of a Type I error. The probability of a Type II error, β, is the probability of not rejecting H_0 even though it is false. Suppose that we are testing

$$H_0: \quad \mu = 1.5 \quad \text{versus} \quad H_a: \quad \mu > 1.5$$

Because we do not know the true value of μ, we cannot calculate the actual value of β. However, the vulnerability of the test to Type II error can be investigated by calculating β for several different potential values of μ, such as $\mu = 1.55$, $\mu = 1.6$, and $\mu = 1.7$. Once the value of β has been determined, the power of the test at the corresponding alternative value is just $1 - \beta$.

▪ Example 10.17   Calculating Power

A cigarette manufacturer claims that the mean nicotine content of its cigarettes is 1.5 mg. We might investigate this claim by testing

$$H_0: \quad \mu = 1.5 \quad \text{versus} \quad H_a: \quad \mu > 1.5$$

where μ is the true mean nicotine content. A random sample of $n = 36$ cigarettes is to be selected, and the resulting data will be used to reach a conclusion. Suppose that the standard deviation of nicotine content (σ) is known to be 0.20 mg and that a significance level of .01 is to be used. Our test statistic is (because $\sigma = 0.20$)

$$z = \frac{\bar{x} - 1.5}{\dfrac{0.20}{\sqrt{n}}} = \frac{\bar{x} - 1.5}{\dfrac{0.20}{\sqrt{36}}} = \frac{\bar{x} - 1.5}{0.0333}$$

The inequality in H_a implies that

P-value = area under z curve to the right of calculated z

From Appendix Table 2, it is easily verified that the z critical value 2.33 captures an upper-tail z curve area of .01. Thus, P-value $\leq .01$ if and only if $z \geq 2.33$. This is equivalent to the decision rule

Reject H_0 if calculated $z \geq 2.33$

which becomes

Reject H_0 if $\dfrac{\bar{x} - 1.5}{0.0333} \geq 2.33$

Solving this inequality for $\bar{x}$, we get

$$\bar{x} \geq 1.5 + 2.33(0.0333)$$

or

$$\bar{x} \geq 1.578$$

So if $\bar{x} \geq 1.578$, we reject H_0, and if $\bar{x} < 1.578$, we fail to reject H_0. This decision rule corresponds to $\alpha = .01$.

Suppose now that $\mu = 1.6$ (so that H_0 is false). A Type II error will then occur if $\bar{x} < 1.578$. What is the probability that this occurs? If $\mu = 1.6$, the sampling distribution of $\bar{x}$ is approximately normal, centered at 1.6, and has a standard deviation of 0.0333. The probability of observing an $\bar{x}$ value less than 1.578 can then be determined by finding an area under a normal curve with mean 1.6 and standard deviation 0.0333, as illustrated in Figure 10.5.

Figure 10.5
β when $\mu = 1.6$ in Example 10.17.

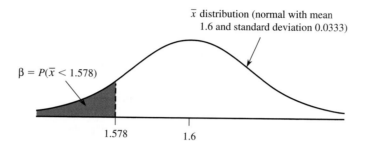

Because the curve in Figure 10.5 is not the standard normal (z) curve, we must first convert to a z score before using Appendix Table 2 to find the area. Here,

$$z \text{ score for } 1.578 = \frac{1.578 - \mu_x}{\sigma_x} = \frac{1.578 - 1.6}{0.0333} = -0.66$$

and

area under z curve to left of $-0.66 = .2546$

So, if $\mu = 1.6$, $\beta = .2546$. This means that if μ is 1.6, about 25% of all samples would still result in $\bar{x}$ values less than 1.578 and failure to reject H_0.

The power of the test at $\mu = 1.6$ is then

$$(\text{power at } \mu = 1.6) = 1 - (\beta \text{ when } \mu \text{ is } 1.6)$$
$$= 1 - .2546$$
$$= .7454$$

Thus, if the true mean is 1.6, the probability of rejecting H_0: $\mu = 1.5$ in favor of H_a: $\mu > 1.5$ is .7454. That is, if μ is 1.6 and the test is used repeatedly with random samples selected from the population, in the long run about 75% of the samples will result in the correct conclusion to reject H_0.

Now consider β and power when $\mu = 1.65$. The normal curve in Figure 10.5 would then be centered at 1.65. Because β is the area to the left of 1.578 and the curve has shifted to the right, β decreases. Converting 1.578 to a z score and using Appendix Table 2 gives $\beta = .0154$. Also,

$$(\text{power at } \mu = 1.65) = 1 - .0154 = .9846$$

As expected, the power at $\mu = 1.65$ is higher than the power at $\mu = 1.6$ because 1.65 is farther from the hypothesized value of 1.5.

MINITAB can calculate the power for specified values of σ, α, n, and the difference between the true and hypothesized values of μ. The following output shows power calculations corresponding to those in Example 10.17:

1-Sample Z Test

Testing mean = null (versus > null)
Alpha = 0.01   Sigma = 0.2   Sample Size = 36

Difference	Power
0.10	0.7497
0.15	0.9851

The slight differences between the power values computed by MINITAB and those previously obtained are due to rounding in Example 10.17.

The probability of a Type II error and the power for z tests concerning a population proportion are calculated in an analogous manner.

▪ Example 10.18   Power for Testing Hypotheses About Proportions

A package delivery service advertises that at least 90% of all packages brought to its office by 9 A.M. for delivery in the same city are delivered by noon that day. Let π denote the proportion of all such packages actually delivered by noon. The hypotheses of interest are

$$H_0: \quad \pi = .9 \quad \text{versus} \quad H_a: \quad \pi < .9$$

where the alternative hypothesis states that the company's claim is untrue. The value $\pi = .8$ represents a substantial departure from the company's claim. If the hypotheses are tested at level .01 using a sample of $n = 225$ packages, what is the probability that the departure from H_0 represented by this alternative value will go undetected?

At significance level .01, H_0 is rejected if P-value $\leq .01$. For the case of a lower-tailed test, this is the same as rejecting H_0 if

$$z = \frac{p - \mu_p}{\sigma_p} = \frac{p - .9}{\sqrt{\dfrac{(.9)(.1)}{225}}} = \frac{p - .9}{.02} \leq -2.33$$

(Because -2.33 captures a lower-tail z curve area of .01, the smallest 1% of all z values satisfy $z \leq -2.33$.) This inequality is equivalent to $p \leq .853$, so H_0 is *not* rejected if $p > .853$. When $\pi = .8$, p has approximately a normal distribution with

$$\mu_p = 0.8 \qquad \sigma_p = \sqrt{\frac{(.8)(.2)}{225}} = 0.0267$$

Then β is the probability of obtaining a sample proportion greater than .853, as illustrated in Figure 10.6.

Figure 10.6
β when $\pi = .8$ in Example 10.18.

Sampling distribution of p (normal with mean 0.8 and standard deviation 0.0267)

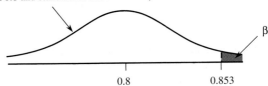

β

0.8   0.853

Converting to a z score results in

$$z = \frac{0.853 - 0.8}{0.0267} = 1.99$$

and Appendix Table 2 gives

$$\beta = 1 - .9767 = .0233$$

When $\pi = .8$ and a level .01 test is used, less than 3% of all samples of size $n = 225$ result in a Type II error. The power of the test at $\pi = .8$ is $1 - .0233 = .9767$. This means that the probability of rejecting H_0: $\pi = .9$ in favor of H_a: $\pi < .9$ when π is really .8 is .9767, which is quite high.

▪ β and Power for the t Test (Optional)

The power and β values for t tests can be determined by using a set of curves specially constructed for this purpose or by utilizing appropriate software. As with the z test, the value of β depends not only on the true value of μ but also on the selected significance level α; β increases as α is made smaller. In addition, β depends on the number of degrees of freedom, $n - 1$. For any fixed level α, it should be easier for the test to detect a specific departure from H_0 when n is large than when n is small. This is indeed the case; for a fixed alternative value, β decreases as $n - 1$ increases.

Unfortunately, there is one other quantity on which β depends: the population standard deviation σ. As σ increases, so does $\sigma_{\bar{x}}$. This in turn makes it more likely that an $\bar{x}$ value far from μ will be observed, resulting in an erroneous conclusion. Once α is specified and n is fixed, the determination of β at a particular alternative value of μ requires that a value of σ be chosen, because each different value of σ yields a different value of β. (This did not present a problem with the z test because when using a z test, the value of σ is known.) If the investigator can specify a range of plausible values for σ, then using the largest such value will give a pessimistic β (one on the high side) and a pessimistic value of power (one on the low side).

Figure 10.7 shows three different β curves for a one-tailed t test (appropriate for H_a: $\mu >$ hypothesized value or for H_a: $\mu <$ hypothesized value). A more complete

Figure 10.7 β curves for the one-tailed t test.

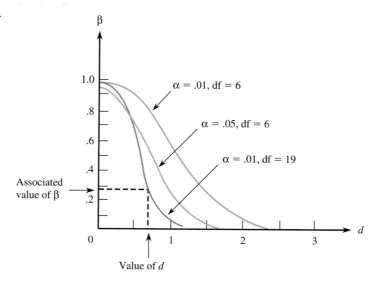

set of curves for both one- and two-tailed tests when $\alpha = .05$ and when $\alpha = .01$ appears in Appendix Table 5. To determine β, first compute the quantity

$$d = \frac{|\text{alternative value} - \text{hypothesized value}|}{\sigma}$$

Then locate d on the horizontal axis, move directly up to the curve for $n - 1$ df, and move over to the vertical axis to read β.

▪ Example 10.19 β and Power for t Tests

Consider testing

$$H_0: \quad \mu = 100 \qquad \text{versus} \qquad H_a: \quad \mu > 100$$

and focus on the alternative value $\mu = 110$. Suppose that $\sigma = 10$, the sample size is $n = 7$, and a significance level of .01 has been selected. For $\sigma = 10$,

$$d = \frac{|110 - 100|}{10} = \frac{10}{10} = 1$$

Figure 10.7 (using df $= 7 - 1 = 6$) gives $\beta \approx .6$. The interpretation is that if $\sigma = 10$ and a level .01 test based on $n = 7$ is used when $\mu = 110$ (and thus H_0 is false), roughly 60% of all samples result in erroneously not rejecting H_0! Equivalently, the power of the test at $\mu = 110$ is only $1 - .6 = .4$. The probability of rejecting H_0 when $\mu = 110$ is not very large. If a level .05 test is used instead, then $\beta \approx .3$, which is still rather large. Using a level .01 test with $n = 20$ (df $= 19$) yields, from Figure 10.7, $\beta \approx .05$. At the alternative value $\mu = 110$, for $\sigma = 10$ the level .01 test based on $n = 20$ has smaller β than the level .05 test with $n = 7$. Substantially increasing n counterbalances using the smaller α.

Now consider the alternative $\mu = 105$, again with $\sigma = 10$, so that

$$d = \frac{|105 - 100|}{10} = \frac{5}{10} = 0.5$$

Then, from Figure 10.7, $\beta = .95$ when $\alpha = .01, n = 7$; $\beta = .7$ when $\alpha = .05, n = 7$; and $\beta = .65$ when $\alpha = .01, n = 20$. These values of β are all quite large; with $\sigma = 10$, $\mu = 105$ is too close to the hypothesized value of 100 for any of these three tests to have a good chance of detecting such a departure from H_0. A substantial decrease in β necessitates using a much larger sample size. For example, from Appendix Table 5, $\beta = .08$ when $\alpha = .05$ and $n = 40$.

The curves in Figure 10.7 also give β when testing $H_0: \mu = 100$ versus $H_a: \mu < 100$. If the alternative value $\mu = 90$ is of interest and $\sigma = 10$,

$$d = \frac{|90 - 100|}{10} = \frac{10}{10} = 1$$

and values of β are the same as those given in the first paragraph of this example.

Because curves for only selected degrees of freedom appear in Appendix Table 5, other degrees of freedom require a visual approximation. For example, the 27-df curve (for $n = 28$) lies between the 19-df and 29-df curves, which do appear, and it is closer to the 29-df curve. This type of approximation is adequate because it is the general magnitude of β — large, small, or moderate — that is of primary concern.

MINITAB can also evaluate power for the t test. For example, the following output shows MINITAB calculations for power at $\mu = 110$ for samples of size 7 and 20 when $\alpha = .01$. The corresponding approximate values from Appendix Table 5 found in Example 10.19 are fairly close to the MINITAB values.

1-Sample t Test

Testing mean = null (versus > null)
Calculating power for mean = null + 10
Alpha = 0.01   Sigma = 10

Sample Size	Power
7	0.3968
20	0.9653

The β curves in Appendix Table 5 are those for t tests. When the alternative value in H_a corresponds to a value of d relatively close to 0, β for a t test may be rather large. One might ask whether there is another type of test that has the same level of significance α as does the t test and smaller values of β. The following result provides the answer to this question.

When the population distribution is normal, the t test for testing hypotheses about μ has smaller β than does any other test procedure that has the same level of significance α.

Stated another way, among all tests with level of significance α, the t test makes β as small as it can possibly be when the population distribution is normal. In this sense, the t test is a best test. Statisticians have also shown that when the population distribution is not too far from a normal distribution, no test procedure can improve on the t test by very much (i.e., no test procedure can have the same α and substantially smaller β). However, when the population distribution is believed to be strongly nonnormal (heavy-tailed, highly skewed, or multimodal), the t test should not be used. Then it's time to consult your friendly neighborhood statistician, who can provide you with alternative methods of analysis.

■ Exercises 10.59–10.65

10.59 The power of a test is influenced by the sample size and the choice of significance level.
a. Explain how increasing the sample size affects the power (when significance level is held fixed).
b. Explain how increasing the significance level affects the power (when sample size is held fixed).

10.60 Water samples are taken from water used for cooling as it is being discharged from a power plant into a river. It has been determined that as long as the mean temperature of the discharged water is at most 150°F, there will be no negative effects on the river ecosystem. To investigate whether the plant is in compliance with regulations that prohibit a mean discharge water temperature above 150°F, a scientist will take 50 water samples at randomly selected times and will record the water temperature of each sample. She will then use a z statistic

$$z = \frac{\bar{x} - 150}{\dfrac{\sigma}{\sqrt{n}}}$$

to decide between the hypotheses H_0: $\mu = 150$ and H_a: $\mu > 150$, where μ is the true mean temperature of discharged water. Assume that σ is known to be 10.

a. Explain why use of the z statistic is appropriate in this setting.

b. Describe Type I and Type II errors in this context.

c. The rejection of H_0 when $z \geq 1.8$ corresponds to what value of α? (That is, what is the area under the z curve to the right of 1.8?)

d. Suppose that the true value for μ is 153 and that H_0 is to be rejected if $z \geq 1.8$. Draw a sketch (similar to that of Figure 10.5) of the sampling distribution of $\bar{x}$, and shade the region that would represent β, the probability of making a Type II error.

e. For the hypotheses and test procedure described, compute the value of β when $\mu = 153$.

f. For the hypotheses and test procedure described, what is the value of β if $\mu = 160$?

g. If H_0 is rejected when $z \geq 1.8$ and $\bar{x} = 152.4$, what is the appropriate conclusion? What type of error might have been made in reaching this conclusion?

10.61 Let μ denote the true average lifetime for a certain type of pen under controlled laboratory conditions. A test of H_0: $\mu = 10$ versus H_a: $\mu < 10$ will be based on a sample of size 36. Suppose that σ is known to be 0.6, from which $\sigma_{\bar{x}} = 0.1$. The appropriate test statistic is then

$$z = \frac{\bar{x} - 10}{0.1}$$

a. What is α for the test procedure that rejects H_0 if $z \leq -1.28$?

b. If the test procedure of Part (a) is used, calculate β when $\mu = 9.8$, and interpret this error probability.

c. Without doing any calculation, explain how β when $\mu = 9.5$ compares to β when $\mu = 9.8$. Then check your assertion by computing β when $\mu = 9.5$.

d. What is the power of the test when $\mu = 9.8$? when $\mu = 9.5$?

10.62 The city council in a large city has become concerned about the trend toward exclusion of renters with children in apartments within the city. The housing coordinator has decided to select a random sample of 125 apartments and determine for each whether children are permitted. Let π be the true proportion of apartments that prohibit children. If π exceeds .75, the city council will consider appropriate legislation.

a. If 102 of the 125 sampled apartments exclude renters with children, would a level .05 test lead you

to the conclusion that more than 75% of all apartments exclude children?

b. What is the power of the test when $\pi = .8$ and $\alpha = .05$?

10.63 The amount of shaft wear after a fixed mileage was determined for each of 7 randomly selected internal combustion engines, resulting in a mean of 0.0372 in. and a standard deviation of 0.0125 in.

a. Assuming that the distribution of shaft wear is normal, test at level .05 the hypotheses H_0: $\mu = .035$ versus H_a: $\mu > .035$.

b. Using $\sigma = 0.0125$, $\alpha = .05$, and Appendix Table 5, what is the value of β, the probability of a Type II error, when $\mu = .04$?

c. What is the power of the test when $\mu = .04$ and $\alpha = .05$?

10.64 Optical fibers are used in telecommunications to transmit light. Current technology allows production of fibers that transmit light about 50 km (*Research at Rensselaer*, 1984). Researchers are trying to develop a new type of glass fiber that will increase this distance. In evaluating a new fiber, it is of interest to test H_0: $\mu = 50$ versus H_a: $\mu > 50$, with μ denoting the true average transmission distance for the new optical fiber.

a. Assuming $\sigma = 10$ and $n = 10$, use Appendix Table 5 to find β, the probability of a Type II error, for each of the given alternative values of μ when a level .05 test is employed:

i. 52
ii. 55
iii. 60
iv. 70

b. What happens to β in each of the cases in Part (a) if σ is actually larger than 10? Explain your reasoning.

10.65 Let μ denote the true average diameter for bearings of a certain type. A test of H_0: $\mu = 0.5$ versus H_a: $\mu \neq 0.5$ will be based on a sample of n bearings. The diameter distribution is believed to be normal. Determine the value of β in each of the following cases:

a. $n = 15$, $\alpha = .05$, $\sigma = 0.02$, $\mu = 0.52$
b. $n = 15$, $\alpha = .05$, $\sigma = 0.02$, $\mu = 0.48$
c. $n = 15$, $\alpha = .01$, $\sigma = 0.02$, $\mu = 0.52$
d. $n = 15$, $\alpha = .05$, $\sigma = 0.02$, $\mu = 0.54$
e. $n = 15$, $\alpha = .05$, $\sigma = 0.04$, $\mu = 0.54$
f. $n = 20$, $\alpha = .05$, $\sigma = 0.04$, $\mu = 0.54$

g. Is the way in which β changes as n, α, σ, and μ vary consistent with your intuition? Explain.

▪ 10.6 Communicating and Interpreting the Results of Statistical Analyses

The step-by-step procedure that we have proposed for testing hypotheses provides a systematic approach for carrying out a complete test. However, you rarely see the results of a hypothesis test reported in publications in such a complete way.

▪ Communicating the Results of Statistical Analyses

When summarizing the results of a hypothesis test, it is important that you include several things in the summary to have all the relevant information. These are:

1. *Hypotheses.* Whether specified in symbols or described in words, it is important that both the null and the alternative hypotheses be clearly stated. If you are using symbols to define the hypotheses, be sure to describe them in the context of the problem at hand (e.g., μ = population mean calorie intake).

2. *Test procedure.* You should be clear about what test procedure was used (e.g., large-sample z test for proportions) and why you think it was reasonable to use this procedure. The plausibility of any required assumptions should be satisfactorily addressed.

3. *Test statistic.* Be sure to include the value of the test statistic and the P-value. Including the P-value allows a reader who may have chosen a different significance level to see whether she would have reached the same or a different conclusion.

4. *Conclusion in context.* Never end the report of a hypothesis test with the statement "I rejected (or did not reject) H_0." Always provide a conclusion that is in the context of the problem and that answers the original research question which the hypothesis test was designed to answer. Be sure also to indicate the level of significance used as a basis for the decision.

▪ Interpreting the Results of Statistical Analyses

When the results of a hypothesis test are reported in a journal article or other published source, it is common to find only the value of the test statistic and the associated P-value accompanying the discussion of conclusions drawn from the data. Sometimes, even the exact P-value doesn't appear, but instead "coded" information is given. For example, * = significant (P-value < .05), ** = very significant (P-value < .01), and *** = highly significant (P-value < .001). Often, especially in newspaper articles, only sample summary statistics are given, with the conclusion immediately following. You may have to fill in some of the intermediate steps for yourself to see whether or not the conclusion is justified.

For example, the article "Clerk Tosses Out Two Recall Petitions: Random Check Fails to Confirm Enough Valid Signatures" (*San Luis Obispo Tribune*, September 9, 1999) stated that a random check of the petitions showed that there were not enough valid signatures to force a recall election. For one of the petitions, 11,800 signatures were submitted and 10,384 valid signatures were required to force a recall election. The county clerk selected a random sample of 592 signatures and verified the signature, address, and registration status against voter registration records. It was determined that 411 of the 592 were valid. Based on this information, was the county clerk's decision to refuse to place the recall on the ballot justified?

We have to fill in the details of the implied hypothesis test to be able to answer this question. Because only 10,384 valid signatures are required to force a recall election, there would have been enough valid signatures if 10,384/11,800 or 88% of the signatures were valid.

The z test for a population proportion would be appropriate for testing

$$H_0: \quad \pi = .88 \quad \text{versus} \quad H_a: \quad \pi < .88$$

where π represents the true proportion of valid signatures on the petition. This alternative hypothesis was selected because the clerk would want strong evidence that there were not enough valid signatures before refusing to place the recall on the ballot. Computer output for this test follows:

Test for One Proportion
Test of p = 0.88 vs p < 0.88

X	N	Sample p	Z-Value	P-Value
411	592	0.694257	−13.91	0.000

The stated P-value would lead to the rejection of H_0 for any reasonable choice of α, and so the data support the county clerk's decision to not place the recall on the ballot.

■ What to Look For in Published Data

Here are some questions to consider when you are reading a report that contains the results of a hypothesis test:

- What hypotheses are being tested? Are the hypotheses about a population mean, a population proportion, or some other population characteristic?

- Was the appropriate test used? Does the validity of the test depend on any assumptions about the population from which the sample was selected? If so, are the assumptions reasonable?

- What is the P-value associated with the test? Was a significance level selected for the test (as opposed to simply reporting the P-value)? Is the chosen significance level reasonable?

- Are the conclusions drawn consistent with the results of the hypothesis test?

For example, consider the following statement (based on information in the article "Serum Transferrin Receptor for the Detection of Iron Deficiency in Pregnancy," *American Journal of Clinical Nutrition* [1991]: 1077–1081): "In a total sample of 176 pregnant women, mean serum receptor concentration did not differ significantly from 5.63, the mean for women who are not pregnant ($P > 0.10$)." The statement does not indicate what test was performed or what the value of the test statistic was. It appears that the hypotheses of interest are $H_0: \mu = 5.63$ versus $H_a: \mu \neq 5.63$, where μ represents the true mean serum receptor concentration for pregnant women. Because the sample size is large, the one-sample t test would be appropriate if the sample can be considered a random sample. With the large sample size, no assumptions about the shape of the population distribution of serum receptor concentration values are necessary. Because the reported P-value is so large

(P-value $> .10$), there is no reason to reject H_0. We cannot conclude that the mean for pregnant women differs from the known mean of 5.63 for women who are not pregnant.

▪ A Word to the Wise: Cautions and Limitations

There are several things you should watch for when conducting a hypothesis test or when evaluating a written summary of such a test.

1. The result of a hypothesis test can never show strong support for the null hypothesis. Make sure that you don't confuse "There is no reason to believe the null hypothesis is not true" with the statement "There is convincing evidence that the null hypothesis is true." These are very different statements!

2. If you have complete information for the population, don't carry out a hypothesis test! It should be obvious that no test is needed to answer questions about a population if you have complete information and don't need to generalize from a sample, but people sometimes forget this fact. For example, in an article on growth in the number of prisoners by state, the *San Luis Obispo Tribune* (August 13, 2001) reported "California's numbers showed a statistically insignificant change, with 66 fewer prisoners at the end of 2000." The use of the term "statistically insignificant" implies some sort of statistical inference, which is not appropriate when a complete accounting of the entire prison population is known. Perhaps the author confused statistical and practical significance. Which brings us to . . .

3. Don't confuse statistical significance with practical significance. When statistical significance has been declared, be sure to step back and evaluate the result in light of its practical importance. For example, we may be convinced that the proportion who respond favorably to a proposed medical treatment is greater than .4, the known proportion who respond favorably for the currently recommended treatments. But if our estimate of this proportion for the proposed treatment is .405, is this of any practical interest? It might be if the proposed treatment is less costly or has fewer side effects, but in other cases it may not be of any real interest. Results must always be interpreted in context.

▪ Activity 10.1: Comparing the *t* and *z* Distributions

Technology Activity: Requires use of a computer or a graphing calculator.

The instructions that follow assume the use of MINITAB. If you are using a different software package or a graphing calculator, your instructor will provide alternative instructions.

Background: Suppose that a random sample is selected from a population that is known to have a normal distribution. Then the statistic

$$z = \frac{\bar{x} - \mu}{\dfrac{\sigma}{\sqrt{n}}}$$

has a standard normal (z) distribution. Because it is rarely the case that σ is known, inferences for

population means are usually based on the statistic $t = \dfrac{\bar{x} - \mu}{s/\sqrt{n}}$, which has a t distribution rather than a z distribution. The informal justification for this is that the use of s to estimate σ introduces additional variability, resulting in a statistic whose distribution is more spread out than is the z distribution.

In this activity, you will use simulation (in the steps listed in what follows) to sample from a known normal population and then investigate how the behavior of $t = \dfrac{\bar{x} - \mu}{s/\sqrt{n}}$ compares with $z = \dfrac{\bar{x} - \mu}{\sigma/\sqrt{n}}$.

1. Generate 200 random samples of size 5 from a normal population with mean 100 and standard deviation 10.

Using MINITAB, go to the Calc Menu. Then

Calc → Random Data → Normal
In the Generate box, enter 200
In the Store in Columns box, enter c1–c5
In the Mean box, enter 100
In the Standard Deviation box, enter 10
Click on OK

You should now see 200 rows of data in each of the first 5 columns of the MINITAB worksheet.

2. Each row contains five values that have been randomly selected from a normal population with mean 100 and standard deviation 10. Viewing each row as a sample of size 5 from this population, calculate the mean and standard deviation for each of the 200 samples (the 200 rows) by using MINITAB's row statistics functions, which can also be found under the Calc menu:

Calc → Row statistics
Choose the Mean button
In the Input Variables box, enter c1–c5
In the Store Result In box, enter c7
Click on OK

You should now see the 200 sample means in column 7 of the MINITAB worksheet. Name this column "x-bar" by typing the name in the gray box at the top of c7.

Now follow a similar process to compute the 200 sample standard deviations, and store them in c8. Name c8 "s."

3. Next, calculate the value of the *z* statistic for each of the 200 samples. We can calculate *z* in this example because we know that the samples were selected from a population for which $\sigma = 10$. Use the calculator function of MINITAB to compute

$$z = \frac{\bar{x} - \mu}{\dfrac{\sigma}{\sqrt{n}}} = \frac{\bar{x} - 100}{\dfrac{10}{\sqrt{5}}}$$

as follows:

Calc → Calculator
In the Store Results In box, enter c10
In the Expression box, type in the following:
(c7 − 100)/(10/sqrt(5))
Click on OK

You should now see the *z* values for the 200 samples in c10. Name c10 "z."

4. Now calculate the value of the *t* statistic for each of the 200 samples. Use the calculator function of MINITAB to compute

$$t = \frac{\bar{x} - \mu}{\dfrac{s}{\sqrt{n}}} = \frac{\bar{x} - 100}{\dfrac{s}{\sqrt{5}}}$$

as follows:

Calc → Calculator
In the Store Results In box, enter c11
In the Expression box, type in the following:
(c7 − 100)/(c8/sqrt(5))
Click on OK

You should now see the *t* values for the 200 samples in c10. Name c10 "t."

5. Graphs, at last! Now construct histograms of the 200 *z* values and the 200 *t* values. These two graphical displays will provide insight into how each of these two statistics behave in repeated sampling. Use the same scale for the two histograms so that it will be easier to compare the two distributions.

Graph → Histogram
In the Graph Variables box, enter c10 for Graph 1 and c11 for Graph 2.
Click the Frame drop-down menu, and select "Multiple Graphs."
Then under the scale choices, select "Same X and Same Y."

6. Now use the histograms from Step 5 to answer the following questions:

a. Write a brief description of the shape, center, and spread for the histogram of the *z* values. Is what you see in the histogram consistent with what you would expect to see? Explain. (Hint: In theory, what is the distribution of the *z* statistic?)

b. How does the histogram of the *t* values compare to the *z* histogram? Be sure to comment on center, shape, and spread.

c. Is your answer to Part (b) consistent with what you would expect for a statistic that has a *t* distribution? Explain.

d. The *z* and *t* histograms are based on only 200 samples, and they only approximate the corresponding sampling distributions. The 5th percentile for the standard normal distribution is −1.645, and the 95th percentile is +1.645. For a *t* distribution with df = 5 − 1 = 4, the 5th and 95th percentiles are −2.13 and +2.13, respectively. How do these percentiles compare to those of the distributions displayed in the histograms? (Hint: Sort the 200 *z* values. In MINITAB, choose "Sort" from the Manip menu. Once the values are sorted, percentiles from the histogram can be found by counting in 10 [which is 5% of 200] values from either end of the sorted list. Then repeat this with the *t* values.)

e. Are the results of your simulation and analysis consistent with the statement that the statistic $z = \dfrac{\bar{x} - \mu}{\sigma / \sqrt{n}}$ has a standard normal (*z*) distribution and the statistic $t = \dfrac{\bar{x} - \mu}{s / \sqrt{n}}$ has a *t* distribution? Explain.

▪ Summary of Key Concepts and Formulas

Term or Formula	Comment
Hypothesis	A claim about the value of a population characteristic.
Null hypothesis, H_0	The hypothesis initially assumed to be true. It has the form H_0: population characteristic = hypothesized value.
Alternative hypothesis, H_a	A hypothesis that specifies a claim that is contradictory to H_0 and that is judged the more plausible claim when H_0 is rejected.
Type I error	Rejection of H_0 when H_0 is true; the probability of a Type I error is denoted by α and is referred to as the significance level for the test.
Type II error	Nonrejection of H_0 when H_0 is false; the probability of a Type II error is denoted by β.
Test statistic	The quantity computed from sample data and used to make a decision between H_0 and H_a.
P-value	The probability, computed assuming that H_0 is true, of obtaining a value of the test statistic at least as contradictory to H_0 as what actually resulted. H_0 is rejected if P-value $\leq \alpha$ and not rejected if P-value $> \alpha$, where α is the chosen significance level.
$z = \dfrac{p - \text{hypothesized value}}{\sqrt{\dfrac{(\text{hyp. value})(1 - \text{hyp. value})}{n}}}$	A test statistic for testing H_0: π = hypothesized value when the sample size is large. The P-value is determined from the z curve.
$z = \dfrac{\bar{x} - \text{hypothesized value}}{\dfrac{\sigma}{\sqrt{n}}}$	A test statistic for testing H_0: μ = hypothesized value when σ is known and either the population distribution is normal or the sample size is large. The P-value is determined from the z curve.
$t = \dfrac{\bar{x} - \text{hypothesized value}}{\dfrac{s}{\sqrt{n}}}$	A test statistic for testing H_0: μ = hypothesized value when σ is unknown and either the population distribution is normal or the sample size is large. The P-value is determined from the t curve with df = $n - 1$.
Power	The power of a test is the probability of rejecting the null hypothesis. Power is affected by the size of the difference between the hypothesized value and the true value, the sample size, and the significance level.

■ Supplementary Exercises 10.66–10.79

10.66 Duck hunting in populated areas faces opposition on the basis of safety and environmental issues. The *San Luis Obispo Telegram-Tribune* (June 18, 1991) reported the results of a survey to assess public opinion regarding duck hunting on Morro Bay (located along the central coast of California). A random sample of 750 local residents included 560 who strongly opposed hunting on the bay. Does this sample provide sufficient evidence to conclude that the majority of local residents oppose hunting on Morro Bay? Test the relevant hypotheses using $\alpha = .01$.

10.67 Drug testing of job applicants is becoming increasingly common. The Associated Press (May 24, 1990) reported that 12.1% of those tested in California tested positive. Suppose that this figure had been based on a random sample of size 600, with 73 testing positive. Does this sample support a claim that more than 10% of job applicants in California test positive for drug use?

10.68 Seat belts help prevent injuries in automobile accidents, but they certainly don't offer complete protection in extreme situations. A random sample of 319 front-seat occupants involved in head-on collisions in a certain region resulted in 95 people who sustained no injuries ("Influencing Factors on the Injury Severity of Restrained Front Seat Occupants in Car-to-Car Head-on Collisions," *Accident Analysis and Prevention* [1995]: 143–150). Does this suggest that the true (population) proportion of uninjured occupants exceeds .25? State and test the relevant hypotheses using a significance level of .05.

10.69 White remains the most popular car color in the United States, but its popularity appears to be slipping. According to an annual survey by DuPont (*Los Angeles Times*, February 22, 1994), white was the color of 20% of the vehicles purchased during 1993, a decline of 4% from the previous year. (According to a DuPont spokesperson, white represents "innocence, purity, honesty, and cleanliness.") A random sample of 400 cars purchased during this period in a certain metropolitan area resulted in 100 cars that were white. Does the proportion of all cars purchased in this area that are white appear to differ from the national percentage? Test the relevant hypotheses using $\alpha = .05$. Does your conclusion change if $\alpha = .01$ is used?

10.70 When a published article reports the results of many hypothesis tests, the *P*-values are not usually given. Instead, the following type of coding scheme is frequently used: *$p < .05$, **$p < .01$, ***$p < .001$, ****$p < .0001$. Which of the symbols would be used to code for each of the following *P*-values?

a. .037   **b.** .0026   **c.** .072   **d.** .0003

10.71 A random sample of $n = 44$ individuals with a B.S. degree in accounting who started with a Big Eight accounting firm and subsequently changed jobs resulted in a sample mean time to change of 35.02 months and a sample standard deviation of 18.94 months ("The Debate over Post-Baccalaureate Education: One University's Experience," *Issues in Accounting Education* [1992]: 18–36). Can it be concluded that the true average time to change exceeds 2 years? Test the appropriate hypotheses using a significance level of .01.

10.72 What motivates companies to offer stock ownership plans to their employees? In a random sample of 87 companies having such plans, 54 said that the primary rationale was tax related ("The Advantages and Disadvantages of ESOPs: A Long-Range Analysis," *Journal of Small Business Management* [1991]: 15–21). Does this information provide strong support for concluding that more than half of all such firms feel this way?

10.73 The article "Caffeine Knowledge, Attitudes, and Consumption in Adult Women" (*Journal of Nutrition Education* [1992]: 179–184) reported the following summary statistics on daily caffeine consumption for a random sample of adult women: $n = 47$, $\bar{x} = 215$ mg, $s = 235$ mg, and range = 5 to 1176.

a. Does it appear plausible that the population distribution of daily caffeine consumption is normal? Is it necessary to assume a normal population distribution to test hypotheses about the value of the population mean consumption? Explain your reasoning.

b. Suppose that it had previously been believed that mean consumption was at most 200 mg. Does the given information contradict this prior belief? Test the appropriate hypotheses at significance level .10.

10.74 Past experience has indicated that the true response rate is 40% when individuals are approached with a request to fill out and return a particular questionnaire in a stamped and addressed envelope. An investigator believes that if the person distributing the questionnaire is stigmatized in some obvious way, potential respondents would feel sorry for the distributor and thus tend to respond at a rate higher than 40%. To investigate this theory, a distributor is fitted with an eye patch. Of the 200 questionnaires distributed by this individual, 109 were returned. Does this strongly suggest that the response rate in this situation exceeds the rate in the past? State and test the appropriate hypotheses at significance level .05.

10.75 An automobile manufacturer who wishes to advertise that one of its models achieves 30 mpg (miles per gallon) decides to carry out a fuel efficiency test. Six nonprofessional drivers are selected, and each one drives a car from Phoenix to Los Angeles. The resulting fuel efficiencies (in miles per gallon) are:

27.2   29.3   31.2   28.4   30.3   29.6

Assuming that fuel efficiency is normally distributed under these circumstances, do the data contradict the claim that true average fuel efficiency is (at least) 30 mpg?

10.76 A student organization uses the proceeds from a particular soft-drink dispensing machine to finance its activities. The price per can had been $0.50 for a long time, and the average daily revenue during that period had been $50.00. The price was recently increased to $0.60 per can. A random sample of $n = 20$ days after the price increase yielded a sample average revenue and sample standard deviation of $47.30 and $4.20, respectively. Does this information suggest that the true average daily revenue has decreased from its value before the price increase? Test the appropriate hypotheses using $\alpha = .05$.

10.77 A hot-tub manufacturer advertises that with its heating equipment, a temperature of 100°F can be achieved in at most 15 min. A random sample of 25 tubs is selected, and the time necessary to achieve a 100°F temperature is determined for each tub. The sample average time and sample standard deviation are 17.5 min and 2.2 min, respectively. Does this information cast doubt on the company's claim? Carry out a test of hypotheses using significance level .05.

10.78 Let π denote the proportion of voters in a certain state who favor a particular proposed constitutional amendment. Consider testing $H_0: \pi = .5$ versus $H_a: \pi > .5$ at significance level .05 based on a sample of size $n = 50$.

a. Suppose that H_0 is in fact true. Use Appendix Table 1 (our table of random numbers) to simulate selecting a sample, and use the resulting data to carry out the test.

b. If you repeated Part (a) a total of 100 times (a simulation consisting of 100 replications), how many times would you expect H_0 to be rejected?

c. Now suppose that $\pi = .6$, which implies that H_0 is false. Again, use Appendix Table 1 to simulate selecting a sample, and carry out the test. If you repeated this a total of 100 times, would you expect H_0 to be rejected more frequently than when H_0 is true?

10.79 A type of lie detector that measures brain waves was developed by a professor of neurobiology at Northwestern University (Associated Press, July 7, 1988). He said, "It would probably not falsely accuse any innocent people and it would probably pick up 70% to 90% of guilty people." Suppose that the result of this lie detector test is allowed as evidence in a criminal trial as the sole basis of a decision between two rival hypotheses: accused is innocent versus accused is guilty. Although these are not "statistical hypotheses" (statements about a population characteristic), the possible decision errors are analogous to Type I and Type II errors.

In this situation, a Type I error is finding an innocent person guilty — rejecting the null hypothesis of innocence when it is, in fact, true. A Type II error is finding a guilty person innocent — not rejecting the null hypothesis of innocence when it is, in fact, false. If the developer of the lie detector is correct in his statements, what is the probability of a Type I error, α? What can you say about the probability of a Type II error, β?

▪ References

The books by Freedman et al. and by Moore listed in previous chapter references are excellent sources. Their orientation is primarily conceptual, with a minimum of mathematical development, and both sources offer many valuable insights.

11 ▪ Comparing Two Populations or Treatments

Many investigations are carried out for the purpose of comparing two populations or treatments. For example, the article "Learn More, Earn More?" (*ETS Policy Notes* [1999]: 1–12) described how the job market treats high school graduates who do not go to college. By comparing data from a random sample of students who had high grades in high school with data from a random sample of students who had low grades, the authors of the article were able to investigate whether the proportion employed for those with high grades was higher than the proportion employed for those with low grades. The authors were also interested in whether the mean monthly salary was higher for those with good grades than for those who did not earn good grades. To answer these questions, hypothesis tests that compare the proportions or means for two different populations were used. In this chapter we introduce hypothesis tests and confidence intervals that can be used to compare two populations or treatments.

▪ 11.1 Inferences Concerning the Difference Between Two Population or Treatment Means Using Independent Samples

In this section, we consider using sample data to compare two population means or two treatment means. An investigator may wish to estimate the difference between two population means or to test hypotheses about this difference. For example, a university financial aid director may want to determine whether the mean cost of textbooks is different for students enrolled in the engineering college than for students enrolled in the liberal arts college. Here, two populations (one consisting of all students enrolled in the engineering college and the other consisting of all students enrolled in the liberal arts college) are to be compared on the basis of their respective means. Information from two random samples, one from each population, could be the basis for making such a comparison.

In other cases, an experiment might be carried out to compare two different treatments or to compare the effect of a treatment with the effect of no treatment (treatment versus control). For example, an agricultural experimenter might wish to compare weight gains for animals placed on two different diets (each diet is a treatment), or an educational researcher might wish to compare online instruction to traditional classroom instruction by studying the difference in mean scores on a common final exam (each type of instruction is a treatment).

In previous chapters, the symbol μ was used to denote the mean of a single population under study. When comparing two populations or treatments, we must use notation that distinguishes between the characteristics of the first and those of the second. This is accomplished by using subscripts on quantities such as μ and σ^2. Similarly, subscripts on sample statistics, such as $\bar{x}$, indicate to which sample these quantities refer.

■ Notation

	Mean	Variance	Standard Deviation
Population or Treatment 1	μ_1	σ_1^2	σ_1
Population or Treatment 2	μ_2	σ_2^2	σ_2

	Sample Size	Mean	Variance	Standard Deviation
Sample from Population or Treatment 1	n_1	$\bar{x}_1$	s_1^2	s_1
Sample from Population or Treatment 2	n_2	$\bar{x}_2$	s_2^2	s_2

Comparison of means focuses on the difference $\mu_1 - \mu_2$. When $\mu_1 - \mu_2 = 0$, the two population or treatment means are identical. That is,

$$\mu_1 - \mu_2 = 0 \text{ is equivalent to } \mu_1 = \mu_2$$

Similarly,

$$\mu_1 - \mu_2 > 0 \text{ is equivalent to } \mu_1 > \mu_2$$

and

$$\mu_1 - \mu_2 < 0 \text{ is equivalent to } \mu_1 < \mu_2$$

We first consider the problem of comparing two population means. Before developing inferential procedures concerning $\mu_1 - \mu_2$, we must consider how the two samples, one from each population, are selected. Two samples are said to be **independent** samples if the selection of the individuals or objects that make up one sample does not influence the selection of individuals or objects in the other sample. However, when observations from the first sample are paired in some meaningful way with observations in the second sample, the samples are said to be **paired**. For example, to study the effectiveness of a speed-reading course, the reading speed of subjects could be measured before they take the class and again after they complete the course. This gives rise to two related samples — one from the population of individuals who have not taken this particular course (the "before" measurements) and one from the population of individuals who have had such a course (the "after" measurements). These samples are paired. The two samples are not independently chosen, because the selection of individuals from the first (before) population completely determines which individuals make up the sample from the second (after) population. In this section, we consider procedures based on independent samples. Methods for analyzing data resulting from paired samples are presented in Section 11.2.

Because $\bar{x}_1$ provides an estimate of μ_1 and because $\bar{x}_2$ gives an estimate of μ_2, it is natural to use $\bar{x}_1 - \bar{x}_2$ as a point estimate of $\mu_1 - \mu_2$. The value of $\bar{x}_1$ varies from sample to sample (it is a *statistic*), as does the value of $\bar{x}_2$. Because the difference $\bar{x}_1 - \bar{x}_2$ is calculated from sample values, it is also a statistic and therefore has a sampling distribution. Inferential methods will be based on information about the sampling distribution of $\bar{x}_1 - \bar{x}_2$.

▪ Properties of the Sampling Distribution of $\bar{x}_1 - \bar{x}_2$

If the random samples on which $\bar{x}_1$ and $\bar{x}_2$ are based are selected independently of one another, then

1. $\mu_{\bar{x}_1 - \bar{x}_2} = $ (mean value of $\bar{x}_1 - \bar{x}_2$) $= \mu_{\bar{x}_1} - \mu_{\bar{x}_2} = \mu_1 - \mu_2$

   Thus, the sampling distribution of $\bar{x}_1 - \bar{x}_2$ is always centered at the value of $\mu_1 - \mu_2$, so $\bar{x}_1 - \bar{x}_2$ is an unbiased statistic for estimating $\mu_1 - \mu_2$.

2. $\sigma^2_{\bar{x}_1 - \bar{x}_2} = $ (variance of $\bar{x}_1 - \bar{x}_2$) $= \sigma^2_{\bar{x}_1} + \sigma^2_{\bar{x}_2} = \dfrac{\sigma^2_1}{n_1} + \dfrac{\sigma^2_2}{n_2}$

   and

   $\sigma_{\bar{x}_1 - \bar{x}_2} = $ (standard deviation of $\bar{x}_1 - \bar{x}_2$) $= \sqrt{\dfrac{\sigma^2_1}{n_1} + \dfrac{\sigma^2_2}{n_2}}$

3. If n_1 and n_2 are both large or if the population distributions are (at least approximately) normal, then both $\bar{x}_1$ and $\bar{x}_2$ have (at least approximately) a normal distribution. This implies that the sampling distribution of $\bar{x}_1 - \bar{x}_2$ is also normal or approximately normal.

Properties 1 and 2 follow from the following general results:

1. The mean value of a difference in means is the difference of the two individual mean values.

2. The variance of a difference of *independent* quantities is the *sum* of the two individual variances.

When the sample sizes are large or when the population distributions are approximately normal, the properties of the sampling distribution of $\bar{x}_1 - \bar{x}_2$ imply that $\bar{x}_1 - \bar{x}_2$ can be standardized to obtain a variable with a sampling distribution that is approximately the standard normal (z) distribution. This gives the following result.

When n_1 and n_2 are both large or when the population distributions are (at least approximately) normal, the distribution of

$$z = \frac{\bar{x}_1 - \bar{x}_2 - (\mu_1 - \mu_2)}{\sqrt{\dfrac{\sigma_1^2}{n_1} + \dfrac{\sigma_2^2}{n_2}}}$$

is described (at least approximately) by the standard normal (z) distribution.

Although it is possible to base a test procedure and confidence interval on this result, the values of σ_1^2 and σ_2^2 are rarely known. As a result, the applicability of z is limited. When σ_1^2 and σ_2^2 are unknown, we must estimate them by using the corresponding sample variances, s_1^2 and s_2^2. The result on which a test procedure and confidence interval are based is given in the accompanying box.

When two random samples are independently selected and when n_1 and n_2 are both large or if the population distributions are normal, the standardized variable

$$t = \frac{\bar{x}_1 - \bar{x}_2 - (\mu_1 - \mu_2)}{\sqrt{\dfrac{s_1^2}{n_1} + \dfrac{s_2^2}{n_2}}}$$

has approximately a t distribution with

$$df = \frac{(V_1 + V_2)^2}{\dfrac{V_1^2}{n_1 - 1} + \dfrac{V_2^2}{n_2 - 1}}$$

where

$$V_1 = \frac{s_1^2}{n_1} \quad \text{and} \quad V_2 = \frac{s_2^2}{n_2}$$

Truncate the number of degrees of freedom (round down) to obtain an integer value.

 If one or both sample sizes are small, we can use normal probability plots or boxplots to evaluate whether it is reasonable to assume that the population distributions are normal.

▪ Test Procedures

In a test designed to compare two population means, the null hypothesis is of the form

H_0: $\mu_1 - \mu_2$ = hypothesized value

Often the hypothesized value is 0, indicating that there is no difference between the population means. The alternative hypothesis involves the same hypothesized value but uses one of three inequalities (less than, greater than, or not equal to), depending on the research question of interest. As an example, let μ_1 and μ_2 denote the average fuel efficiencies (in miles per gallon, mpg) for two models of a certain type of car, equipped with 4-cylinder and 6-cylinder engines, respectively. The hypotheses under consideration might be

$$H_0: \quad \mu_1 - \mu_2 = 5 \quad \text{versus} \quad H_a: \quad \mu_1 - \mu_2 > 5$$

The null hypothesis is equivalent to the claim that average efficiency for the 4-cylinder engine exceeds the average efficiency for the 6-cylinder engine by 5 mpg. The alternative hypothesis states that the difference between the true average efficiencies is more than 5 mpg.

A test statistic is obtained by replacing $\mu_1 - \mu_2$ in the standardized t variable (given in the previous box) with the hypothesized value that appears in H_0. Thus, the t statistic for testing $H_0: \mu_1 - \mu_2 = 5$ is

$$t = \frac{\bar{x}_1 - \bar{x}_2 - 5}{\sqrt{\dfrac{s_1^2}{n_1} + \dfrac{s_2^2}{n_2}}}$$

When H_0 is true and the sample sizes are large or when the population distributions are normal, the sampling distribution of the test statistic is approximately a t distribution. The P-value for the test is obtained by first computing the appropriate number of degrees of freedom and then using Appendix Table 4. The following box gives a general description of the test procedure.

▪ **Summary of the Two-Sample t Test for Comparing Two Population Means**

Null hypothesis: $\quad H_0: \mu_1 - \mu_2 =$ hypothesized value

Test statistic: $\quad t = \dfrac{\bar{x}_1 - \bar{x}_2 - \text{hypothesized value}}{\sqrt{\dfrac{s_1^2}{n_1} + \dfrac{s_2^2}{n_2}}}$

The appropriate number of degrees of freedom for the two-sample t test is

$$df = \frac{(V_1 + V_2)^2}{\dfrac{V_1^2}{n_1 - 1} + \dfrac{V_2^2}{n_2 - 1}}$$

where

$$V_1 = \frac{s_1^2}{n_1} \quad \text{and} \quad V_2 = \frac{s_2^2}{n_2}$$

The number of degrees of freedom should be truncated (rounded down) to an integer.

Alternative Hypothesis	P-Value
$H_a: \mu_1 - \mu_2 >$ hypothesized value	Area under appropriate t curve to the right of the computed t
$H_a: \mu_1 - \mu_2 <$ hypothesized value	Area under appropriate t curve to the left of the computed t
$H_a: \mu_1 - \mu_2 \neq$ hypothesized value	(1) 2(area to the right of the computed t) if t is positive, or (2) 2(area to the left of the computed t) if t is negative

Assumptions: 1. The two samples are *independently selected random samples*.
2. The *sample sizes are large* (in general, 30 or larger), or *the population distributions are (at least approximately) normal*.

■ Example 11.1   Brain Size

Do children diagnosed with attention deficit/hyperactivity disorder (ADHD) have smaller brains than children without this condition? This question was the topic of a research study descried in the paper "Developmental Trajectories of Brain Volume Abnormalities in Children and Adolescents with Attention-Deficit/Hyperactivity Disorder" (*Journal of the American Medical Association* [2002]: 1740–1747). Brain scans were completed for 152 children with ADHD and 139 children of similar age without ADHD. Summary values for total cerebral volume (in milliliters) are given in the following table:

	n	$\bar{x}$	s
Children with ADHD	152	1059.4	117.5
Children without ADHD	139	1104.5	111.3

Do these data provide evidence that the mean brain volume of children with ADHD is smaller than the mean for children without ADHD? Let's test the relevant hypotheses using a .05 level of significance.

1. μ_1 = true mean brain volume for children with ADHD.

    μ_2 = true mean brain volume for children without ADHD.

    $\mu_1 - \mu_2$ = difference in mean brain volume.

2. $H_0: \mu_1 - \mu_2 = 0$.

3. $H_a: \mu_1 - \mu_2 < 0$.

4. Significance level: $\alpha = .05$.

5. Test statistic: $t = \dfrac{\bar{x}_1 - \bar{x}_2 - \text{hypothesized value}}{\sqrt{\dfrac{s_1^2}{n_1} + \dfrac{s_2^2}{n_2}}} = \dfrac{\bar{x}_1 - \bar{x}_2 - 0}{\sqrt{\dfrac{s_1^2}{n_1} + \dfrac{s_2^2}{n_2}}}$

6. Assumptions: The paper states that the study controlled for age and that the participants were "recruited from the local community." This is not equivalent to random sampling, but the authors of the paper (five of whom were doctors at well-known medical institutions) believed that it was reasonable to regard these samples as representative of the two groups under study. Both sample sizes are large, so it is reasonable to proceed with the two-sample t test.

7. Calculation:

$$t = \frac{(1059.4 - 1104.5) - 0}{\sqrt{\dfrac{(117.5)^2}{152} + \dfrac{(111.3)^2}{139}}} = \frac{-45.10}{\sqrt{90.831 + 89.120}} = \frac{-45.10}{13.415} = -3.36$$

8. *P*-value: We first compute the number of degrees of freedom for the two-sample t test:

$$V_1 = \frac{s_1^2}{n_1} = 90.831 \qquad V_2 = \frac{s_2^2}{n_2} = 89.120$$

$$df = \frac{(V_1 + V_2)^2}{\dfrac{V_1^2}{n_1 - 1} + \dfrac{V_2^2}{n_2 - 1}} = \frac{(90.831 + 89.120)^2}{\dfrac{(90.831)^2}{151} + \dfrac{(89.120)^2}{138}} = \frac{32{,}382.362}{112.191} = 288.636$$

We truncate the number of degrees of freedom to 288. Appendix Table 4 shows that the area under the t curve with 288 df (using the z critical value column because 288 is larger than 120 df) to the left of -3.36 is approximately 0. Therefore

$$P\text{-value} \approx 0$$

9. Conclusion: Because P-value $\approx 0 < .05$, we reject H_0. There is convincing evidence that the mean brain volume for children with ADHD is smaller than the mean for children without ADHD.

▪ Example 11.2   Oral Contraceptive Use and Bone Mineral Density

To assess the impact of oral contraceptive use on bone mineral density (BMD), researchers in Canada carried out a study comparing BMD for women who had used oral contraceptives for at least 3 months to BMD for women who had never used oral contraceptives ("Oral Contraceptive Use and Bone Mineral Density in Premenopausal Women," *Canadian Medical Association Journal* [2001]: 1023–1029). Data on BMD (in grams per centimeter) consistent with summary quantities given in the paper appear in the following table (the actual sample sizes for the study were much larger):

Never used oral
contraceptives   0.82   0.94   0.96   1.31   0.94   1.21   1.26   1.09   1.13   1.14

Used oral
contraceptives   0.94   1.09   0.97   0.98   1.14   0.85   1.30   0.89   0.87   1.01

The authors of the paper believed that it was reasonable to view the samples used in the study as representative of the two populations of interest—women who used oral contraceptives for at least 3 months and women who never used oral contraceptives. For the purposes of this example, we will assume that it is also justifiable to consider the two samples given here as representative of the populations. We use the given information and a significance level of .05 to determine whether there is evidence that women who use oral contraceptives have a lower mean BMD than women who have never used oral contraceptives.

1. μ_1 = true mean BMD for women who never used oral contraceptives.

   μ_2 = true mean BMD for women who used oral contraceptives.

   $\mu_1 - \mu_2$ = difference in mean BMD.

2. $H_0: \mu_1 - \mu_2 = 0$.

3. $H_a: \mu_1 - \mu_2 > 0$.

4. Significance level: $\alpha = .05$.

5. Test statistic: $t = \dfrac{\bar{x}_1 - \bar{x}_2 - \text{hypothesized value}}{\sqrt{\dfrac{s_1^2}{n_1} + \dfrac{s_2^2}{n_2}}} = \dfrac{\bar{x}_1 - \bar{x}_2 - 0}{\sqrt{\dfrac{s_1^2}{n_1} + \dfrac{s_2^2}{n_2}}}$

6. Assumptions: For the two-sample t test to be appropriate, we must be willing to assume that the two samples can be viewed as independently selected random samples from the two populations of interest. As previously noted, we assume that this is reasonable. Because both of the sample sizes are small, it is also necessary to assume that the BMD distribution is approximately

normal for each of these two populations. Boxplots constructed using the sample data are shown here:

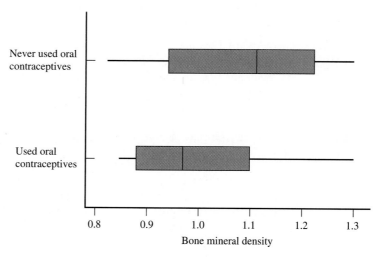

Because the boxplots are reasonably symmetric and because there are no outliers, the assumption of normality is plausible.

7. Calculation: For the given data, $\bar{x}_1 = 1.08$, $s_1 = 0.16$, $\bar{x}_2 = 1.00$, $s_2 = 0.14$, and

$$t = \frac{(1.08 - 1.00) - 0}{\sqrt{\dfrac{(0.16)^2}{10} + \dfrac{(0.14)^2}{10}}} = \frac{0.08}{\sqrt{0.003 + 0.002}} = \frac{0.08}{0.071} = 1.13$$

8. P-value: We first compute the number of degrees of freedom for the two-sample t test:

$$V_1 = \frac{s_1^2}{n_1} = 0.003 \qquad V_2 = \frac{s_2^2}{n_2} = 0.002$$

$$\text{df} = \frac{(V_1 + V_2)^2}{\dfrac{V_1^2}{n_1 - 1} + \dfrac{V_2^2}{n_2 - 1}} = \frac{(0.003 + 0.002)^2}{\dfrac{(0.003)^2}{9} + \dfrac{(0.002)^2}{9}} = \frac{0.000025}{0.0000014} = 17.86$$

We truncate the number of degrees of freedom to 17. Appendix Table 4 shows that the area under the t curve with 17 df to the right of 1.1 is .143, so P-value = .143.

9. Conclusion: Because P-value = .143 > .05, we fail to reject H_0. There is not convincing evidence to support the claim that mean bone mineral density is lower for women who used oral contraceptives.

Many statistical computer packages can perform the calculations for the two-sample t test. The accompanying MINITAB output shows the number of degrees of freedom, the test statistic value, and the P-value for the hypothesis test of Example 11.2 (in the output, OC stands for "oral contraceptive"). The P-value reported in the computer output differs from the value in Example 11.2 because Appendix Table 4 gives only tail areas for t values to one decimal place and so we rounded the test statistic to 1.1. As a consequence, the P-value given in the example is only approximate; the MINITAB P-value is more accurate.

Two-sample T for No OC vs OC

	N	Mean	StDev	SE Mean
No OC	10	1.080	0.160	0.051
OC	10	1.004	0.139	0.044

T-Test of difference = 0 (vs >): T-Value = 1.13 P-Value = 0.136 DF = 17

■ Comparing Treatments

When an experiment is carried out to compare two treatments (or to compare a single treatment with a control), the investigator is interested in the effect of the treatments on some response variable. The treatments are "applied" to individuals (as in an experiment to compare two different medications for decreasing blood pressure) or objects (as in an experiment to compare two different baking temperatures on the density of bread), and the value of some response variable (e.g., blood pressure, density) is recorded. Based on the resulting data, the investigator might wish to determine whether there is a difference in the mean response for the two treatments.

In many actual experimental situations, the individuals or objects to which the treatments will be applied are not selected at random from some larger population. A consequence of this is that it is not possible to generalize the results of the experiment to some larger population. However, *if the experimental design provides for random assignment of treatments to the individuals or objects used in the experiment (or for random assignment of the individuals or objects to treatments), it is possible to test hypotheses about treatment differences.*

It is common practice to use the two-sample t test statistic previously described if the experiment employs random assignment and if either the sample sizes are large or it is reasonable to think that the treatment response distributions (the distributions of response values that would result if the treatments were applied to a *very* large number of individuals or objects) are approximately normal.

■ Two-Sample t Test for Comparing Two Treatments

When

1. *treatments are randomly assigned* to individuals or objects (or vice versa; i.e., individuals or objects are randomly assigned to treatments), and

2. the *sample sizes are large* (in general, 30 or larger) or the *treatment response distributions are approximately normal*

the two-sample t test can be used to test $H_0: \mu_1 - \mu_2 =$ hypothesized value, where μ_1 and μ_2 represent the mean response for Treatments 1 and 2, respectively.

In this case, these two conditions replace the assumptions previously stated for comparing two population means. The validity of the assumption of normality of the treatment response distributions can be assessed by constructing a normal probability plot or a boxplot of the response values in each sample.

When the two-sample t test is used to compare two treatments when the individuals or objects used in the experiment are not randomly selected from some population, it is only an approximate test (the reported P-values are only approximate). However, this is still the most common way to analyze such data.

▪ **Example 11.3** Learning Styles

The article "Effects of Learning-Style Intervention on College Student's Achievement, Anxiety, Anger, and Curiosity" (*Journal of College Student Development* [1994]: 461–466) reported on an experiment designed to compare a treatment group with a control group. First-year nursing students enrolled in a science course were randomly assigned to one of two groups. Students in the control group were provided with conventional study-skill guidelines. Each student in the treatment group was evaluated to determine his or her preferred learning style. Students in the treatment group were then provided with homework prescriptions based on their identified learning style as well as with the conventional study-skill guidelines. Does the treatment (the homework prescriptions based on learning style) improve science grades? We will test the relevant hypotheses using a significance level of .01.

Group	Sample Size	Sample Mean	Sample SD
Control	100	2.75	0.85
Treatment	103	3.28	0.70

1. Let μ_1 denote the mean science score for the treatment *conventional study-skill guidelines*, and define μ_2 analogously for the treatment *homework prescriptions based on learning style in addition to conventional study-skill guidelines*. Then $\mu_1 - \mu_2$ is the difference between the mean science scores for the two treatments.

2. $H_0: \mu_1 - \mu_2 = 0$.

3. $H_a: \mu_1 - \mu_2 < 0$.

4. Significance level: $\alpha = .01$.

5. Test statistic: $t = \dfrac{\bar{x}_1 - \bar{x}_2 - \text{hypothesized value}}{\sqrt{\dfrac{s_1^2}{n_1} + \dfrac{s_2^2}{n_2}}} = \dfrac{\bar{x}_1 - \bar{x}_2 - 0}{\sqrt{\dfrac{s_1^2}{n_1} + \dfrac{s_2^2}{n_2}}}$

6. Assumptions: Subjects were randomly assigned to the treatment groups, and both sample sizes are large, so it is reasonable to use the two-sample t test.

7. Calculation: $t = \dfrac{(2.75 - 3.28) - 0}{\sqrt{\dfrac{(0.85)^2}{100} + \dfrac{(0.70)^2}{103}}} = \dfrac{-0.53}{0.1095} = -4.84$

8. *P*-value: We first compute the number of degrees of freedom for the two-sample t test:

$$V_1 = \frac{s_1^2}{n_1} = 0.0072$$

$$V_2 = \frac{s_2^2}{n_2} = 0.0048$$

$$\text{df} = \frac{(V_1 + V_2)^2}{\dfrac{V_1^2}{n_1 - 1} + \dfrac{V_2^2}{n_2 - 1}} = \frac{(0.0072 + 0.0048)^2}{\dfrac{(0.0072)^2}{99} + \dfrac{(0.0048)^2}{102}} = \frac{0.000144}{0.00000075} = 192.00$$

This is a lower-tailed test, so the P-value is the area under the t curve with df $= 192$ and to the left of -4.84. Because -4.84 is so far out in the lower tail of this t curve, P-value ≈ 0.

9. Conclusion: Because P-value $\leq \alpha$, H_0 is rejected. There is evidence that the mean science score for the treatment group is higher. The data support the conclusion that homework prescriptions based on learning style are effective in raising the mean science score.

You have probably noticed that evaluating the formula for number of degrees of freedom for the two-sample t test involves quite a bit of arithmetic. An alternative approach is to compute a conservative estimate of the P-value — one that is close to but larger than the actual P-value. If H_0 is rejected using this conservative estimate, then it will also be rejected if the actual P-value is used. *A conservative estimate of the P-value for the two-sample t test can be found by using the t curve with the number of degrees of freedom equal to the smaller of* $(n_1 - 1)$ *and* $(n_2 - 1)$.

▪ The Pooled t Test

The two-sample t test procedure just described is appropriate when it is reasonable to assume that the population distributions are approximately normal. If it is also known that the variances of the two populations are equal $(\sigma_1^2 = \sigma_2^2)$, an alternative procedure known as the *pooled t test* can be used. This test procedure combines information from both samples to obtain a "pooled" estimate of the common variance and then uses this pooled estimate of the variance in place of s_1^2 and s_2^2 in the t test statistic. This test procedure was widely used in the past, but it has fallen into some disfavor because it is quite sensitive to departures from the assumption of equal population variances. If the population variances are equal, the pooled t procedure has a slightly better chance of detecting departures from H_0 than does the two-sample t test of this section. However, P-values based on the pooled t procedure can be seriously in error if the population variances are not equal, so, in general, the two-sample t procedure is a better choice than the pooled t test.

▪ Comparisons and Causation

If the assignment of treatments to the individuals or objects used in a comparison of treatments is not made by the investigators, the study is observational. As an example, the article "Lead and Cadmium Absorption Among Children near a Nonferrous Metal Plant" (*Environmental Research* [1978]: 290–308) reported data on blood lead concentrations for two different samples of children. The first sample was drawn from a population residing within 1 km of a lead smelter, whereas those in the second sample were selected from a rural area much farther from the smelter. It was the parents of the children, rather than the investigators, who determined whether the children would be in the close-to-smelter group or the far-from-smelter group. As a second example, a letter in the *Journal of the American Medical Association* (May 19, 1978) reported on a comparison of doctors' longevity after medical school graduation for those with an academic affiliation and those in private practice. (The letter writer's stated objective was to see whether "publish or perish" really meant "publish *and* perish.") Here again, an investigator did not start out with a group of doctors, assigning some to academic and others to nonacademic careers. The doctors themselves selected their groups.

The difficulty with drawing conclusions based on an observational study is that a statistically significant difference may be due to some underlying factors that have not been controlled rather than to conditions that define the groups. Does the type of medical practice itself have an effect on longevity, or is the observed difference in lifetimes caused by other factors, which themselves led graduates to choose academic or nonacademic careers? Similarly, is the observed difference in blood lead concentration levels due to proximity to the smelter? Perhaps other physical and socioeconomic factors are related both to choice of living area and to concentration.

In general, rejection of $H_0: \mu_1 - \mu_2 = 0$ in favor of $H_a: \mu_1 - \mu_2 > 0$ suggests that, on average, higher values of the variable are *associated* with individuals in the first population or receiving the first treatment than with those in the second population or receiving the second treatment. But *association does not imply causation.* Strong statistical evidence for a causal relationship can be built up over time through many different comparative studies that point to the same conclusions (as in the many investigations linking smoking to lung cancer). A **randomized controlled experiment**, in which investigators assign subjects at random to the treatments or conditions being compared, is particularly effective in suggesting causality. With such random assignment, the investigator and other interested parties can have more confidence in the conclusion that an observed difference is caused by the difference in treatments or conditions. (Recall the discussion in Chapter 2 of the role of randomization in designing an experiment.)

▪ A Confidence Interval

A confidence interval for $\mu_1 - \mu_2$ is easily obtained from the basic t variable of this section. Both the derivation of and the formula for the interval are similar to those of the one-sample t interval discussed in Chapter 9.

▪ Two-Sample t Confidence Interval for the Difference Between Two Population or Treatment Means

The general formula for a confidence interval for $\mu_1 - \mu_2$ when

1. the two samples are *independently chosen random samples*, and
2. the *sample sizes are both large* (in general, $n_1 \geq 30$ and $n_2 \geq 30$) or the *population distributions are approximately normal* is

$$\bar{x}_1 - \bar{x}_2 \pm (t \text{ critical value})\sqrt{\frac{s_1^2}{n_1} + \frac{s_2^2}{n_2}}$$

The t critical value is based on

$$\text{df} = \frac{(V_1 + V_2)^2}{\frac{V_1^2}{n_1 - 1} + \frac{V_2^2}{n_2 - 1}}$$

where

$$V_1 = \frac{s_1^2}{n_1} \quad \text{and} \quad V_2 = \frac{s_2^2}{n_2}$$

The number of degrees of freedom should be truncated (rounded down) to an integer. The t critical values for the usual confidence levels are given in Appendix Table 3.

For a comparison of two treatments, when

1. *treatments are randomly assigned* to individuals or objects (or vice versa), and

2. the *sample sizes are large* (in general, 30 or larger) or the *treatment response distributions are approximately normal,*

the two-sample t confidence interval formula can be used to estimate $\mu_1 - \mu_2$.

■ **Example 11.4   Effect of Talking on Blood Pressure**

Does talking elevate blood pressure, contributing to the tendency for blood pressure to be higher when measured in a doctor's office than when measured in a less stressful environment (called the white coat effect)? The article "The Talking Effect and 'White Coat' Effect in Hypertensive Patients: Physical Effort or Emotional Content" (*Behavioral Medicine* [2001]: 149–157) described a study in which patients with high blood pressure were randomly assigned to one of two groups. Those in the first group (the talking group) were asked questions about their medical history and about the sources of stress in their lives in the minutes before their blood pressure was measured. Those in the second group (the counting group) were asked to count aloud from 1 to 100 four times before their blood pressure was measured. The following data values for diastolic blood pressure (in millimeters Hg) are consistent with summary quantities appearing in the paper:

Talking	104	110	107	112	108	103	108	118

$n_1 = 8$ $\bar{x}_1 = 108.75$ $s_1 = 4.74$

Counting	110	96	103	98	100	109	97	105

$n_1 = 8$ $\bar{x}_2 = 102.25$ $s_2 = 5.39$

Subjects were randomly assigned to the two treatments. Because both sample sizes are small, we must first investigate whether it is reasonable to assume that the diastolic blood pressure distributions are approximately normal for the two treatments. There are no outliers in either data set, and the boxplots are reasonably symmetric, suggesting that the assumption of approximate normality is reasonable.

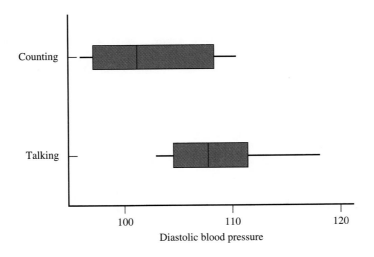

To estimate $\mu_1 - \mu_2$, the difference in mean diastolic blood pressure for the two treatments, we calculate a 95% confidence interval:

$$V_1 = \frac{s_1^2}{n_1} = \frac{(4.74)^2}{8} = 2.81 \qquad V_2 = \frac{s_2^2}{n_2} = \frac{(5.39)^2}{8} = 3.63$$

$$df = \frac{(V_1 + V_2)^2}{\dfrac{V_1^2}{n_1 - 1} + \dfrac{V_2^2}{n_2 - 1}} = \frac{(2.81 + 3.63)^2}{\dfrac{(2.81)^2}{7} + \dfrac{(3.63)^2}{7}} = \frac{41.47}{3.01} = 13.78$$

Truncating to an integer gives df = 13. In the 13-df row of Appendix Table 3, the t critical value for a 95% confidence level is 2.16. The interval is then

$$\bar{x}_1 - \bar{x}_2 \pm (t \text{ critical value})\sqrt{\frac{s_1^2}{n_1} + \frac{s_2^2}{n_2}}$$

$$= (108.75 - 102.25) \pm (2.16)\sqrt{\frac{(4.74)^2}{8} + \frac{(5.39)^2}{8}}$$

$$= 6.5 \pm (2.16)(2.54)$$

$$= 6.5 \pm 5.49$$

$$= (1.01, 11.99)$$

This interval is rather wide, because the two sample variances are large and the sample sizes are small. Notice that the interval does not include 0, so 0 is not one of the plausible values for $\mu_1 - \mu_2$. Based on this computed interval, we estimate that the mean diastolic blood pressure when talking is higher than the mean when counting by somewhere between 1.01 and 11.99 mm Hg. The 95% confidence level means that we used a method to produce this estimate that correctly captures the true value of $\mu_1 - \mu_2$ 95% of the time in repeated sampling.

Most statistical computer packages can compute the two-sample t confidence interval. MINITAB was used to construct a 95% confidence interval using the data of this example; the resulting output is shown here:

Two-Sample T-Test and CI: Talking, Counting

Two-sample T for Talking vs Counting

	N	Mean	StDev	SE Mean
Talking	8	108.75	4.74	1.7
Counting	8	102.25	5.39	1.9

95% CI for difference: (1.01, 11.99)

■ Exercises 11.1–11.28

11.1 Consider two populations for which $\mu_1 = 30$, $\sigma_1 = 2$, $\mu_2 = 25$, and $\sigma_2 = 3$. Suppose that two independent random samples of sizes $n_1 = 40$ and $n_2 = 50$ are selected. Describe the approximate sampling distribution of $\bar{x}_1 - \bar{x}_2$ (center, spread, and shape).

11.2 An individual can take either a scenic route to work or a nonscenic route. She decides that use of the nonscenic route can be justified only if it reduces true average travel time by more than 10 min.

a. If μ_1 refers to the scenic route and μ_2 to the nonscenic route, what hypotheses should be tested?

b. If μ_1 refers to the nonscenic route and μ_2 to the scenic route, what hypotheses should be tested?

11.3 The article "Affective Variables Related to Mathematics Achievement Among High-Risk College Freshmen" (*Psychological Reports* [1991]: 399–403) examined the relationship between attitudes toward mathematics and success at college-level mathematics. Twenty men and 38 women participated in the study. They were selected at random from those indentified as being at high risk of failure because they did not meet the usual admission requirements for the university. Each student was asked to respond to a series of questions, and the answers were combined to obtain a math anxiety score. For this particular scale, the *higher* the score, the *lower* the level of anxiety toward mathematics. Summary values appear in the following table:

	n	$\bar{x}$	s
Males	20	35.9	11.9
Females	38	36.6	12.3

Is there convincing evidence that, as many researchers have hypothesized, the mean anxiety score for women is different from that for men? Test the relevant hypotheses using a .05 level of significance.

11.4 Many people take ginkgo supplements advertised to improve memory. Are these over-the-counter supplements effective? In a study reported in the paper "Ginkgo for Memory Enhancement" (*Journal of the American Medical Association* [2002]: 835–840), elderly adults were assigned at random to either a treatment group or a control group. The 104 participants who were assigned to the treatment group took 40 mg of ginkgo 3 times a day for 6 weeks. The 115 participants assigned to the control group took a placebo pill 3 times a day for 6 weeks. At the end of 6 weeks, the Wechsler Memory Scale (a test of short-term memory) was administered. Higher scores indicate better memory function. Summary values are given in the following table:

	n	$\bar{x}$	s
Ginkgo	104	5.6	0.6
Placebo	115	5.5	0.6

Based on these results, is there evidence that taking 40 mg of ginkgo 3 times a day is effective in increasing mean performance on the Wechsler Memory Scale? Test the relevant hypotheses using $\alpha = .05$.

11.5 Fumonisins are environmental toxins produced by a type of mold; they are found in corn and in products made from raw corn. Based on a study of corn meal, the Center for Food Safety and Applied Nutrition provided recommendations on allowable fumonisin levels in human food and in animal feed. The study compared corn meal made from partially degermed corn (corn that has had the germ partially removed; the germ is the part of the kernel located at the bottom center of the kernel that is used to pro-

duce corn oil) with corn meal made from corn that has not been degermed. Specimens of corn meal were analyzed, and the total fumonisin level (in parts per million) was determined for each specimen. Summary statistics for total fumonisin level from the U.S. Food and Drug Administration web site are given here:

	$\bar{x}$	s
Partially Degermed	0.59	1.01
Not Degermed	1.21	1.71

a. If the given means and standard deviations had been based on a random sample of 10 partially degermed specimens and a random sample of 10 specimens made from corn that was not degermed, explain why it would not be appropriate to carry out a two-sample t test to determine whether there is a significant difference in the mean fumonisin level for the two types of corn meal.

b. Suppose instead that each of the samples had included 50 corn meal specimens. Explain why it would now be reasonable to carry out a two-sample t test.

c. Assuming that each sample size was 50, carry out a test to determine whether there is a significant difference in mean fumonisin level for the two types of corn meal. Use a significance level of .01.

11.6 When surgeons repair injuries, they use sutures (stitched knots) to hold together and stabilize the injured area. If these knots elongate and loosen through use, the injury may not heal properly because the tissues are not optimally positioned. Researchers at the University of California, San Francisco, tied a series of different types of knots with two types of suture material, Maxon and Ticron. Suppose that 112 tissue specimens were available and that for each specimen the type of knot and suture material were randomly assigned. The investigators tested the knots to see how much the loops elongated; the elongations (in millimeters) were measured and the resulting data are summarized here. For purposes of this exercise, assume that it is reasonable to regard the elongation distributions as approximately normal.

Knot Type	Maxon			Ticron		
	n	$\bar{x}$	sd	n	$\bar{x}$	sd
Square (control)	10	10.0	0.1	10	2.5	0.06
Duncan Loop	15	11.0	0.3	11	10.9	0.4
Overhand	15	11.0	0.9	11	8.1	1.0
Roeder	10	13.5	0.1	10	5.0	0.04
Snyder	10	13.5	2.0	10	8.1	0.06

a. Is there a significant difference in mean elongation between the square knot and the Duncan loop for Maxon thread?

b. Is there a significant difference in mean elongation between the square knot and the Duncan loop for Ticron thread?

c. For the Duncan loop data, is there a significant difference between the mean elongations of Maxon versus Ticron threads?

11.7 Snake experts believe that venomous snakes inject different amounts of venom when killing their prey. Researchers at the University of Wyoming tested this hypothesis to determine whether young prairie rattlesnakes use more venom to kill larger mice and less venom for smaller mice ("Venom Metering by Juvenile Prairie Rattlesnakes, *Crotalus* v. *Viridis*: Effects of Prey Size and Experience," *Animal Behavior* [1995]: 33–40). In the first trial, the researchers used three groups of seven randomly selected snakes that were inexperienced hunters. In the second trial, three different groups of seven randomly selected "experienced" snakes were used. The amount of venom (in milligrams) that each snake injected was recorded and categorized according to the size of the mouse:

	Small Mouse		Medium Mouse		Large Mouse	
	$\bar{x}$	s	$\bar{x}$	s	$\bar{x}$	s
Inexperienced	3.1	1.0	3.4	0.4	1.8	0.3
Experienced	2.6	0.3	2.9	0.6	4.7	0.3

For the small prey, is there a significant difference between the inexperienced and experienced snakes? for the medium prey? for the large prey?

11.8 The coloration of male guppies might affect the mating preference of the female guppy. To test this hypothesis, scientists first identified two types of guppies, Yarra and Paria, that display different colorations ("Evolutionary Mismatch of Mating Preferences and Male Colour Patterns in Guppies," *Animal Behaviour* [1997]: 343–351). The relative area of orange was calculated for fish of each type. A random sample of 30 Yarra guppies resulted in a mean relative area of 0.106 and a standard deviation of 0.055. A random sample of 30 Paria guppies resulted in a mean relative area of 0.178 and a standard deviation 0.058. Is there evidence of a difference in coloration? Test the relevant hypotheses to determine whether the mean area of orange is different for the two types of guppies.

11.9 The article "Trial Lawyers and Testosterone: Blue-Collar Talent in a White-Collar World" (*Jour-nal of Applied Social Psychology* [1998]: 84–94) compared trial lawyers and nontrial lawyers on the basis of mean testosterone level. Random samples of 35 male trial lawyers, 31 male nontrial lawyers, 13 female trial lawyers, and 18 female nontrial lawyers were selected for study. The article includes the following statement: "Trial lawyers had higher testosterone levels than did nontrial lawyers. This was true for men, $t(64) = 3.75$, $p < .001$, and for women, $t(29) = 2.26$, $p < .05$."

a. Based on the given information, is there a significant difference in the mean testosterone level for male trial and nontrial lawyers?

b. Based on the given information, is there a significant difference in the mean testosterone level for female trial and nontrial lawyers?

c. Do you have enough information to carry out a test to determine whether there is a significant difference in the mean testosterone levels of male and female trial lawyers? If so, carry out such a test. If not, what additional information would you need to be able to conduct the test?

11.10 The article "The Relevance of Sexual Orientation to Substance Abuse and Psychological Distress Among College Students" (*Journal of College Student Development* [1998]: 157–165) examined substance abuse patterns for individuals in a random sample of gay, lesbian, and bisexual students and in a random sample of heterosexual students at a particular university. One of the variables considered was frequency of marijuana use in the past year. It was reported that the mean and standard deviation for this variable were 15.68 and 43.69, respectively, for the sample of 37 gay, lesbian, and bisexual students and 12.28 and 44.20 for the sample of 156 heterosexual students.

a. Based on the given information, is it reasonable to assume that the population distributions of the variable *frequency of marijuana use* are approximately normal? Explain why or why not.

b. Taking your answer to Part (a) into account, do you think it is reasonable to use the two-sample *t* test to determine whether there is sufficient evidence to conclude that the mean frequency of marijuana use differs for the two groups? Explain why or why not.

c. If you concluded in Part (b) that the independent samples *t* test is appropriate, use it to carry out a test of the relevant hypotheses using $\alpha = .01$.

11.11 Do faculty and students have similar perceptions of what types of behavior are inappropriate in the classroom? This question was examined by the author of the article "Faculty and Student Perceptions of Classroom Etiquette" (*Journal of College Student Development* [1998]: 515–516). Each individual in a random sample of 173 students in general education classes at a large public university was asked to judge various behaviors on a scale from 1 (totally inappropriate) to 5 (totally appropriate). Individuals in a random sample of 98 faculty members also rated the same behaviors. The mean rating for three of the behaviors studied are shown here (the means are consistent with data provided by the author of the article). The sample standard deviations were not given, but for purposes of this exercise, assume that they are all equal to 1.0.

Student Behavior	Student Mean Rating	Faculty Mean Rating
Wearing hats in the classroom	2.80	3.63
Addressing instructor by first name	2.90	2.11
Talking on a cell phone	1.11	1.10

a. Is there sufficient evidence to conclude that the mean "appropriateness" score assigned to wearing a hat in class differs for students and faculty?

b. Is there sufficient evidence to conclude that the mean "appropriateness" score assigned to addressing an instructor by his or her first name is higher for students than for faculty?

c. Is there sufficient evidence to conclude that the mean "appropriateness" score assigned to talking on a cell phone differs for students and faculty? Does the result of your test imply that students and faculty consider it acceptable to talk on a cell phone during class? Explain.

11.12 The article "So Close, Yet So Far: Predictors of Attrition in College Seniors" (*Journal of College Student Development* [1999]: 343–354) attempted to describe differences between college seniors who disenroll before graduating and those who do graduate. Researchers randomly selected 42 nonreturning and 48 returning seniors, none of whom were transfer students. These 90 students rated themselves on personal contact and campus involvement. The resulting data are summarized here:

	Returning (*n* = 48)		Nonreturning (*n* = 42)	
	Mean	Standard Deviation	Mean	Standard Deviation
Personal Contact	3.22	0.93	2.41	1.03
Campus Involvement	3.21	1.01	3.31	1.03

a. Construct and interpret a 95% confidence interval for the difference in mean campus involvement rating for returning and nonreturning students. Does your interval support the statement that students who do not return are less involved, on average, than those who do? Explain.

b. Do students who don't return have a lower mean personal contact rating than those who do return? Test the relevant hypotheses using a significance level of .01.

11.13 Does clear-cutting of trees in an area cause local extinction of the tailed frog? We don't really know, but an article that begins to address that issue attempted to quantify aspects of the habitat and microhabitat of streams with and without tailed frogs. The following summary values are from "Distribution and Habitat of *Ascaphus truei* in Streams on Managed, Young Growth Forests in North Coastal California" (*Journal of Herpetology* [1999]: 71–79):

	Sites With Tailed Frogs		
	n	$\bar{x}$	*s*
Habitat Characteristics			
Stream gradient (%)	18	9.1	6.00
Water temperature (°C)	18	12.2	1.71
Microhabitat Characteristics			
Depth (cm)	82	5.32	2.27

	Sites Without Tailed Frogs		
	n	$\bar{x}$	*s*
Habitat Characteristics			
Stream gradient (%)	31	5.9	6.29
Water temperature (°C)	31	12.8	1.33
Microhabitat Characteristics			
Depth (cm)	267	8.46	5.95

Assume that the habitats and microhabitats examined were selected independently and at random from those that have tailed frogs and from those that do not have tailed frogs. Use a significance level of .01 for any hypothesis tests required to answer the following questions.

a. Is there evidence of a difference in mean stream gradient between streams with and streams without tailed frogs? (Note: Assume that it is reasonable to think that the distribution of stream gradients is approximately normal for both types of streams. It is possible for the gradient percentage to be either positive or negative, so the fact that the mean − 2 standard deviations is negative is not, by itself, an indication of nonnormality in this case.)

b. Is there evidence of a difference in the mean water temperature between streams with and streams without tailed frogs?

c. The article reported that the two depth distributions are both quite skewed. Does this imply that it would be unreasonable to use the independent samples *t* test to compare the mean depth for the two types of streams? If not, carry out a test to determine whether the sample data support the claim that the mean depth of streams without tailed frogs is greater than the mean depth of streams with tailed frogs.

11.14 Are girls less inclined to enroll in science courses than boys? One recent study ("Intentions of Young Students to Enroll in Science Courses in the Future: An Examination of Gender Differences" (*Science Education* [1999]: 55–76)) asked randomly selected fourth, fifth, and sixth graders how many science courses they intend to take. The resulting data were used to compute the following summary statistics:

	n	Mean	Standard Deviation
Males	203	3.42	1.49
Females	224	2.42	1.35

Calculate a 99% confidence interval for the difference between males and females in mean number of science courses planned. Interpret your interval. Based on your interval, how would you answer the question posed at the beginning of the exercise?

11.15 The article "Workaholism in Organizations: Gender Differences" (*Sex Roles* [1999]: 333–346) gave the following summary statistics on 1996 income (in Canadian dollars) for random samples of male and female MBA graduates from a particular Canadian business school:

	n	$\bar{x}$	*s*
Males	258	$133,442	$131,090
Females	233	$105,156	$98,525

a. For what significance levels would you conclude that the mean salary of female MBA graduates of this business school is above $100,000?

b. Is there convincing evidence that the mean salary for female MBA graduates of this business school is lower than the mean salary for the male graduates?

11.16 The article "Movement and Habitat Use by Lake Whitefish During Spawning in a Boreal Lake: Integrating Acoustic Telemetry and Geographic Information Systems" (*Transactions of the American Fisheries Society* [1999]: 939–952) included the fol-

lowing data on weights of 20 fish caught in 1995 and 1996:

1995	776 580 539 648 538 891 673 783 571 627
1996	571 627 727 727 867 1042 804 832 764 727

Is it appropriate to use the independent samples *t* test to compare the mean weight of fish for the two years? Explain why or why not.

11.17 The article "The Relationship of Task and Ego Orientation to Sportsmanship Attitudes and the Perceived Legitimacy of Injurious Acts" (*Research Quarterly for Exercise and Sport* [1991]: 79–87) examined the extent of approval of unsporting play and cheating. High school basketball players completed a questionnaire that was used to arrive at an approval score, with higher scores indicating greater approval. A random sample of 56 male players resulted in a mean approval rating for unsportsmanlike play of 2.76, whereas the mean for a random sample of 67 female players was 2.02. Suppose that the two sample standard deviations were 0.44 for males and 0.41 for females. Is it reasonable to conclude that the mean approval rating is higher for male players than for female players by more than 0.5? Use $\alpha = .05$.

11.18 The article "Conflict Resolution Styles in Gay, Lesbian, Heterosexual Nonparent, and Heterosexual Parent Couples" (*Journal of Marriage and Family* [1994]: 705–722) reported the results of administering the Ineffective Arguing Inventory (IAI) to both partners of 75 gay, 51 lesbian, 108 married nonparent, and 99 married parent couples. The IAI is a measure of dysfunction in couple conflict resolution. The mean and standard deviation of the IAI scores for each group are given in the following table:

	Type of Couple			
	Gay	Lesbian	Heterosexual Nonparent	Heterosexual Parent
Sample size	75	51	108	99
Mean	17.16	15.92	16.79	18.09
Standard deviation	6.19	5.93	5.81	6.55

a. Is there evidence to suggest that the mean IAI score is higher for heterosexual parent couples than for heterosexual nonparent couples? Test the relevant hypotheses using $\alpha = .01$.

b. Is there evidence to suggest that the mean IAI score is higher for gay couples than for lesbian couples? Test the relevant hypotheses using $\alpha = .05$.

c. Construct a 95% confidence interval for the difference in mean IAI score for gay couples and heterosexual nonparent couples. Interpret this interval.

d. Construct a 95% confidence interval for the difference in mean IAI score for heterosexual parent couples and heterosexual nonparent couples. Interpret this interval.

11.19 Techniques for processing poultry were examined in the article "Texture Profiles of Canned Boned Chicken as Affected by Chilling-Aging Times" (*Poultry Science* [1994]: 1475–1478). Whole chickens were chilled 0, 2, 8, or 24 hr before being cooked and canned. To determine whether the chilling time affected the texture of the canned chicken, trained tasters evaluated samples. One characteristic of interest was hardness. The following summary quantities were obtained; each mean is based on 36 ratings:

	Chilling Time			
	0 hr	2 hr	8 hr	24 hr
Mean hardness	7.52	6.55	5.70	5.65
Standard deviation	0.96	1.74	1.32	1.50

a. Do the data suggest that there is a difference in mean hardness for chicken chilled 0 hr before cooking and chicken chilled 2 hr before cooking? Use $\alpha = .05$.

b. Do the data suggest that there is a difference in mean hardness for chicken chilled 8 hr before cooking and chicken chilled 24 hr before cooking? Use $\alpha = .05$.

c. Use a 90% confidence interval to estimate the difference in mean hardness for chicken chilled 2 hr before cooking and chicken chilled 8 hr before cooking.

11.20 The Rape Myth Acceptance Scale (RMAS) was administered to 333 male students at a public university. Of these students, 155 were randomly selected from students who did not belong to a fraternity and 178 were randomly selected from students who belonged to a fraternity. The higher the score on the RMAS, the greater the acceptance of rape myths, so lower scores are considered desirable. The following summary statistics appeared in the article "Rape Supportive Attitudes Among Greek Students Before and After a Date Rape Prevention Program" (*Journal of College Student Development* [1994]: 450–455):

Sample	n	Mean RMAS Score	sd
Fraternity members	178	25.63	6.16
Not fraternity members	155	27.40	5.51

Do the data support the researchers' claim that the mean RMAS score is lower for fraternity members? Use $\alpha = .05$.

11.21 The discharge of industrial wastewater into rivers affects water quality. To assess the effect of a particular power plant on water quality, investigators collected 24 water specimens 16 km upstream and 4 km downstream from the plant. Alkalinity (in milligrams per liter) was determined for each specimen, resulting in the summary quantities in the following table.

Location	n	Mean	Standard Deviation
Upstream	24	75.9	1.83
Downstream	24	183.6	1.70

Do the data suggest that the true mean alkalinity is higher downstream than upstream by more than 50 mg/l? Use a .05 significance level.

11.22 According to the Associated Press (*San Luis Obispo Telegram-Tribune*, June 23, 1995), a study by Italian researchers indicated that low cholesterol and depression were linked. The researchers found that among 331 randomly selected patients hospitalized because they had attempted suicide, the mean cholesterol level was 198. The mean cholesterol level of 331 randomly selected patients admitted to the hospital for other reasons was 217. The sample standard deviations were not reported, but suppose that they were 20 for the group who had attempted suicide and 24 for the other group. Is there sufficient evidence to conclude that the mean cholesterol level is lower for those who have attempted suicide? Test the relevant hypotheses using $\alpha = .05$.

11.23 The article "The Sorority Rush Process: Self-Selection, Acceptance Criteria, and the Effect of Rejection" (*Journal of College Student Development* [1994]: 346–353) reported on a study of factors associated with the decision to rush a sorority. Fifty-four women who rushed a sorority and 51 women who did not were asked how often they drank alcoholic beverages. For the sorority rush group, the mean was 2.72 drinks per week and the standard deviation was 0.86. For the group who did not rush, the mean was 2.11 and the standard deviation was 1.02. Is there evidence to support the claim that those who rush a sorority drink more than those who do not rush? Test the relevant hypotheses using $\alpha = .01$. What assumptions are required for the two-sample t test to be appropriate?

11.24 Tennis elbow is thought to be aggravated by the impact experienced when hitting the ball. The article "Forces on the Hand in the Tennis One-Handed Backhand" (*International Journal of Sport Biomechanics* [1991]: 282–292) reported the force (in newtons) on the hand just after impact on a one-handed

Table for Exercise 11.24

	Data							n	$\bar{x}$	s
Advanced	44.70	26.31	55.75	28.54	46.99	39.46		6	40.3	11.3
Intermediate	15.58	19.16	24.13	10.56	32.88	21.47	14.32	33.09	21.4	8.3

backhand drive for six advanced players and eight intermediate players. Summary statistics from the article as well as data consistent with these summary quantities appear in the table above.

The authors of the article assumed in their analysis of the data that both force distributions (advanced and intermediate) were normal. Using the given information, determine whether the data support the hypothesis that the mean force after impact is greater for advanced tennis players than it is for intermediate players.

11.25 British health officials have expressed concern about problems associated with vitamin D deficiency among certain immigrants. Doctors have conjectured that such a deficiency is related to the amount of fiber in a person's diet. An experiment was designed to compare the vitamin D plasma half-life for two groups of healthy individuals. One group was placed on a normal diet, whereas the second group was placed on a high-fiber diet. The following table gives the resulting data (from "Reduced Plasma Half-Lives of Radio-Labeled 25(OH)D3 in Subjects Receiving a High-Fibre Diet," *British Journal of Nutrition* [1993]: 213–216):

Normal diet   19.1 24.0 28.6 29.7 30.0 34.8
High-fiber diet   12.0 13.0 13.6 20.5 22.7 23.7 24.8

Use the following MINITAB output to determine whether the data indicate that the mean half-life is higher for those on a normal diet than for those on a high-fiber diet. Assume that the treatments were assigned at random and that the two plasma half-life distributions are approximately normal. Test the appropriate hypotheses using $\alpha = .01$.

Two-sample T for normal vs high

	N	Mean	StDev	SE Mean
Normal	6	27.70	5.44	2.2
High	7	18.61	5.55	2.1

95% C.I. for mu normal − mu high: (2.3, 15.9)
T-Test mu normal = mu high (vs >): T = 2.97 P = 0.0070 DF = 10

11.26 A researcher at the Medical College of Virginia conducted a study of 60 randomly selected male soccer players and concluded that frequently heading the ball in soccer lowers players' IQs (*USA Today*, August 14, 1995). The soccer players were divided into two groups, based on whether they averaged 10 or more headers per game. Mean IQs were reported in the article, but the sample sizes and standard deviations were not given. Suppose that these values were as given in the following table:

	n	Sample Mean	Sample sd
Fewer Than 10 Headers	35	112	10
10 or More Headers	25	103	8

Do these data support the researcher's conclusion? Test the relevant hypotheses using $\alpha = .05$. Can you conclude that heading the ball *causes* lower IQ? Explain.

11.27 The effect of loneliness among college students was examined in the article "The Importance of Perceived Duration: Loneliness and Its Relationship to Self-Esteem and Academic Performance" (*Journal of College Student Development* [1994]: 456–460). Based on reported frequency and duration of loneliness, subjects were divided into two groups. The first group ($n_1 = 72$) was the short-duration loneliness group, and the second ($n_2 = 17$) was the long-duration loneliness group. A self-esteem inventory was administered to students in both groups. For the short-duration group, the reported mean self-esteem score was 76.78 and the standard deviation was 17.80. For the long-duration group, the mean and standard deviation were 64.00 and 15.68, respectively. Do the data support the researcher's claim that mean self-esteem is lower for students classified as having long-duration loneliness? Test the relevant hypotheses using $\alpha = .01$. Be sure to state any assumptions that are necessary for your test to be valid.

11.28 Do certain behaviors result in a severe drain on energy resources because a great deal of energy is expended in comparison to energy intake? The article "The Energetic Cost of Courtship and Aggression in a Plethodontid Salamander" (*Ecology* [1983]: 979–983) reported on one of the few studies concerned with behavior and energy expenditure. The following table gives summary statistics for oxygen consumption (in milliliters per gram per hour) for male–female salamander pairs (the determination of consumption values is rather complicated; it is partly for this reason that so few studies of this type have been carried out):

Behavior	Sample Size	Sample Mean	Sample sd
Noncourting	11	0.072	0.0066
Courting	15	0.099	0.0071

a. The pooled t test is a test procedure for testing $H_0: \mu_1 - \mu_2 =$ hypothesized value when it is reasonable to assume that the two population distributions are normal with equal standard deviations ($\sigma_1 = \sigma_2$). The test statistic for the pooled t test is obtained by replacing both s_1 and s_2 in the two-sample t test statistic with s_p, where

$$s_p = \sqrt{\frac{(n_1 - 1)s_1^2 + (n_2 - 1)s_2^2}{n_1 + n_2 - 2}}$$

When the population distributions are normal with equal standard deviations and H_0 is true, the resulting pooled t statistic has a t distribution with df = $n_1 + n_2 - 2$. For the reported data, the two sample standard deviations are similar. Use the pooled t test with $\alpha = .05$ to determine whether the mean oxygen consumption for courting pairs is higher than the mean oxygen consumption for noncourting pairs.

b. Would the conclusion in Part (a) have been different if the two-sample t test had been used rather than the pooled t test?

▪ 11.2 Inferences Concerning the Difference Between Two Population or Treatment Means Using Paired Samples

Two samples are said to be *independent* if the selection of the individuals or objects that make up one of the samples has no bearing on the selection of individuals or objects in the other sample. In some situations, an experiment with independent samples is not the best way to obtain information about a possible difference between the populations. For example, suppose that an investigator wants to determine whether regular aerobic exercise affects blood pressure. A random sample of people who jog regularly and a second random sample of people who do not exercise regularly are selected independently of one another. The researcher then uses the two-sample t test to conclude that a significant difference exists between the average blood pressures for joggers and nonjoggers. Is it reasonable to think that the difference in mean blood pressure is attributable to jogging? It is known that blood pressure is related to both diet and body weight. Might it not be the case that joggers in the sample tend to be leaner and adhere to a healthier diet than the nonjoggers and that *this* might account for the observed difference? On the basis of this study, the researcher would not be able to rule out the possibility that the observed difference in blood pressure is explained by weight differences between the people in the two samples and that aerobic exercise itself has no effect.

One way to avoid this difficulty is to match subjects by weight. The researcher would find pairs of subjects so that the jogger and nonjogger in each pair were similar in weight (although weights for different pairs might vary widely). The factor *weight* could then be ruled out as a possible explanation for an observed difference in average blood pressure between the two groups. Matching the subjects by weight results in two samples for which each observation in the first sample is coupled in a meaningful way with a particular observation in the second sample. Such samples are said to be **paired**.

Experiments can be designed to yield paired data in a number of different ways. Some studies involve using the same group of individuals with measurements recorded both before and after some intervening treatment. Other experiments use naturally occurring pairs, such as twins or husbands and wives, and some investigations construct pairs by matching on factors with effects that might otherwise obscure differences (or the lack of them) between the two populations of interest (as might weight in the jogging example). Paired samples often provide more information than independent samples because extraneous effects are screened out.

▪ Example 11.5   Alcohol Content of Wine

Are wine drinkers being misled by bottle labels? The *London Times* (August 5, 2001) reported the results of a study that compared the actual alcohol content of wine to the alcohol content printed on the label. Laboratory analysis was used to determine the alcohol content (percentage) of different brands and varieties selected at random from wines produced in 1999 and 2000. A representative subset of the resulting data is given in the following table:

	Wine					
	1	2	3	4	5	6
Actual Alcohol Content	14.2	14.5	14.0	14.9	13.6	12.6
Label Alcohol Content	14.0	14.0	13.5	15.0	13.0	12.5

We can regard the data as consisting of two samples — a sample of actual alcohol contents for wines produced in 1999 and 2000 and a sample of label alcohol contents for wines produced in 1999 and 2000. The samples are paired rather than independent because both samples are composed of observations on the same six bottles of wine.

Is there evidence that the label tends to understate the actual alcohol content? Let μ_1 denote the mean actual alcohol content for the population of all wines produced in 1999 and 2000. Similarly, let μ_2 denote the mean stated alcohol content for the population of all labels on wines produced in 1999 and 2000. Hypotheses of interest might be

$$H_0:\ \mu_1 - \mu_2 = 0 \qquad \text{versus} \qquad H_a:\ \mu_1 - \mu_2 > 0$$

with the null hypothesis indicating that the mean actual alcohol content and the mean label alcohol content are equal and the alternative hypothesis stating that the mean actual alcohol content exceeds the mean label alcohol content. Notice that in five of the six data pairs, the actual alcohol content is higher than the corresponding label value. Intuitively, this suggests that the population means may not be equal.

Disregarding the paired nature of the samples results in a loss of information. Both the actual and label alcohol contents vary from one type of wine to another. It is this variability that may obscure the difference when the two-sample t test is used. If we were to (incorrectly) use the two-sample t test for independent samples on the given data, the resulting t test statistic value would be .62. This value would not allow for rejection of the hypothesis even at level of significance .10. This result might surprise you at first, but remember that this test procedure ignores the information about how the samples are paired. Two plots of the data are given in Figure 11.1. The first plot (Figure 11.1(a)) ignores the pairing, and the two samples look quite similar. The plot in which pairs are identified (Figure 11.1(b)) does suggest a difference, because for five of the six pairs the actual alcohol content observation exceeds the label alcohol content observation.

Figure 11.1 Two plots of the paired data from Example 11.5: (a) pairing ignored; (b) pairs identified.

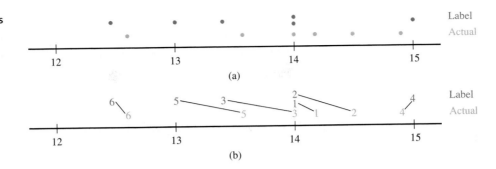

Example 11.5 suggests that the methods of inference developed for independent samples are not adequate for dealing with paired samples. When sample observations from the first population are paired in some meaningful way with sample observations from the second population, inferences can be based on the differences between the two observations within each sample pair. The n sample differences can then be regarded as having been selected from a large population of differences. Thus, in Example 11.5, we can think of the six (actual − label) differences as having been selected from an entire population of differences.

Let

μ_d = the mean value of the difference population

and let

σ_d = the standard deviation of the difference population

The relationship between μ_d and the two individual population means is

$\mu_d = \mu_1 - \mu_2$

Therefore, when the samples are paired, inferences about $\mu_1 - \mu_2$ are equivalent to inferences about μ_d. Because inferences about μ_d can be based on the n observed sample differences, the original two-sample problem becomes a familiar one-sample problem.

▪ Paired t Test

To compare two population or treatment means when the samples are paired, we first translate the hypothesis of interest from one about the value of $\mu_1 - \mu_2$ to an equivalent one involving μ_d:

Hypothesis	Equivalent Hypothesis When Samples Are Paired
H_0: $\mu_1 - \mu_2$ = hypothesized value	H_0: μ_d = hypothesized value
H_a: $\mu_1 - \mu_2$ > hypothesized value	H_a: μ_d > hypothesized value
H_a: $\mu_1 - \mu_2$ < hypothesized value	H_a: μ_d < hypothesized value
H_a: $\mu_1 - \mu_2 \neq$ hypothesized value	H_a: $\mu_d \neq$ hypothesized value

Sample differences (Sample 1 value − Sample 2 value) are then computed and used as the basis for testing hypotheses about μ_d. When the number of differences is large or when it is reasonable to assume that the population of differences is approximately normal, the one-sample t test based on the differences is the recommended test procedure. In general, the population of differences is normal if each of the two individual populations is normal. A normal probability plot or boxplot of the differences can be used to support this assumption.

■ **Summary of the Paired t Test for Comparing Two Population or Treatment Means**

Null hypothesis: $H_0: \mu_d =$ hypothesized value

Test statistic: $t = \dfrac{\bar{x}_d - \text{hypothesized value}}{\dfrac{s_d}{\sqrt{n}}}$

where n is the number of sample differences and $\bar{x}_d$ and s_d are the sample mean and standard deviation of the differences, respectively. This test is based on df $= n - 1$.

Alternative Hypothesis	P-Value
$H_a: \mu_d >$ hypothesized value	Area under the appropriate t curve to the right of the calculated t
$H_a: \mu_d <$ hypothesized value	Area under the appropriate t curve to the left of the calculated t
$H_a: \mu_d \neq$ hypothesized value	(1) 2(area to the right of t) if t is positive, or (2) 2(area to the left of t) if t is negative

Assumptions:
1. The samples are *paired*.
2. The n sample differences can be viewed as a *random sample* from a population of differences.
3. The *number of sample differences is large* (in general, at least 30) or the *population distribution of differences is approximately normal*.

■ Example 11.6   Improve Memory by Playing Chess?

Can taking chess lessons and playing chess daily improve memory? The online article "The USA Junior Chess Olympics Research: Developing Memory and Verbal Reasoning" (*New Horizons for Learning*, April 2001; available at www.newhorizons .org) described a study in which sixth-grade students who had not previously played chess participated in a program where they took chess lessons and played chess daily for 9 months. Each student took a memory test (the Test of Cognitive Skills) before starting the chess program and again at the end of the 9-month period. Data (read from a graph in the article) and computed differences are given in the table on the next page.

The author of the article proposed using these data to test the theory that students who participated in the chess program tend to achieve higher memory scores after completion of the program. We can consider the pretest scores as a sample of scores from the population of sixth-grade students who have not participated in the chess program and the posttest scores as a sample of scores from the population of sixth-grade students who have completed the chess training program. The samples were not independently chosen, because each sample is composed of the same 12 students.

Student	Memory Test Score		
	Pretest	Posttest	Difference
1	510	850	−340
2	610	790	−180
3	640	850	−210
4	675	775	−100
5	600	700	−100
6	550	775	−225
7	610	700	−90
8	625	850	−225
9	450	690	−240
10	720	775	−55
11	575	540	35
12	675	680	−5

Let

μ_1 = mean memory score for sixth-graders with no chess training

μ_2 = mean memory score for sixth-graders after chess training

and

$\mu_d = \mu_1 - \mu_2$ = mean memory score difference between students with no chess training and students who have completed chess training

The question of interest can be answered by testing the hypothesis

$$H_0: \ \mu_d = 0 \quad \text{versus} \quad H_a: \ \mu_d < 0$$

Using the 12 differences, we compute

$$\sum(\text{differences}) = -1735$$

$$\sum(\text{differences})^2 = 383{,}325$$

$$\bar{x}_d = \frac{\sum(\text{differences})}{n} = \frac{-1735}{12} = -144.58$$

$$s_d^2 = \frac{\sum(\text{differences})^2 - \dfrac{[\sum(\text{differences})]^2}{n}}{n-1}$$

$$= \frac{383{,}325 - \dfrac{(-1735)^2}{12}}{11}$$

$$= 12{,}042.99$$

$$s_d = \sqrt{s_d^2} = 109.74$$

We now use the paired t test with a significance level of .05 to carry out the hypothesis test.

1. μ_d = mean memory score difference between students with no chess training and students who have completed chess training.

2. H_0: $\mu_d = 0$.

3. H_a: $\mu_d < 0$.

4. Significance level: $\alpha = .05$.

5. Test statistic: $t = \dfrac{\bar{x}_d - \text{hypothesized value}}{\dfrac{s_d}{\sqrt{n}}}$

6. Assumptions: Although the sample of 12 sixth-graders was not a random sample, the author believed that it was reasonable to view the 12 sample differences as a random sample of all such differences. A boxplot of the differences is approximately symmetric and does not show any outliers, so the assumption of normality is not unreasonable. Therefore we proceed with the paired t test.

7. Calculation: $t = \dfrac{-144.6 - 0}{\dfrac{109.74}{\sqrt{12}}} = -4.56$

8. *P*-value: This is a lower-tailed test, so the *P*-value is the area to the left of the computed t value. The appropriate number of degrees of freedom for this test is df = $12 - 1 = 11$. From the 11-df column of Appendix Table 4, we find that *P*-value $< .001$ because the area to the left of -4.0 is .001 and the test statistic (-4.56) is even farther out in the lower tail.

9. Conclusion: Because *P*-value $\leq \alpha$, we reject H_0. The data support the theory that the mean memory score is higher for sixth-graders who have completed the chess training than the mean score before training.

Using the two-sample t test (for independent samples) for the data in Example 11.6 would have been incorrect, because the samples are not independent. Inappropriate use of the two-sample t test would have resulted in a computed test statistic value of -4.25. The conclusion would still be to reject the hypothesis of equal mean memory scores in this particular example, but this is not always the case.

▪ Example 11.7 Spatial Ability

It has been estimated that between 1945 and 1971, as many as 2 million children were born to mothers treated with diethylstilbestrol (DES), a nonsteroidal estrogen. The Food and Drug Administration banned this drug in 1971 because research indicated a link with the incidence of cervical cancer. The article "Effects of Prenatal Exposure to Diethylstilbestrol (DES) on Hemispheric Laterality and Spatial Ability in Human Males" (*Hormones and Behavior* [1992]: 62–75) discussed a study in which 10 males exposed to DES and their unexposed brothers underwent various tests. Here are the summary statistics on the results of a spatial ability test:

$$n = 10$$
$$\overline{x}_1 = \text{unexposed mean} = 13.8$$
$$\overline{x}_2 = \text{exposed mean} = 12.6$$
$$\overline{x}_d = 1.2$$
$$\text{standard error of the differences} = \frac{s_d}{\sqrt{n}} = 0.5$$

The investigators used a one-tailed test to see whether DES exposure was associated with reduced spatial ability.

1. μ_d = difference between true average score for unexposed males and true average score for exposed males.

2. $H_0: \mu_d = 0$.

3. $H_a: \mu_d > 0$.

4. Significance level: $\alpha = .05$.

5. Test statistic: $t = \dfrac{\overline{x}_d - \text{hypothesized value}}{\dfrac{s_d}{\sqrt{n}}}$

6. Assumptions: Without the raw data, it is difficult to assess the reasonableness of the assumptions. The author of the article judged the assumptions of the paired t test to be reasonable, so we will proceed.

7. Calculation: $t = \dfrac{1.2 - 0}{0.5} = 2.4$

8. *P*-value: This is an upper-tailed test. There are 10 sample differences, so df = 10 − 1 = 9. From the 9-df column and the 2.4 row of Appendix Table 4, we find that the area to the right of 2.4 is .020, so *P*-value = .020.

9. Conclusion: Because *P*-value $\leq \alpha$, H_0 is rejected. There is evidence that the mean spatial ability score is lower for those exposed to DES than for those who were unexposed.

Notice that the numerators $\overline{x}_d$ and $\overline{x}_1 - \overline{x}_2$ of the paired t and the two-sample t test statistics are always equal. The difference lies in the denominator. The variability in differences is usually much smaller than the variability in each sample separately (because measurements in a pair tend to be similar). As a result, the value of the paired t statistic is usually larger in magnitude than the value of the two-sample t statistic. Pairing typically reduces variability that might otherwise obscure small but nevertheless significant differences.

▪ A Confidence Interval

The one-sample t confidence interval for μ given in Chapter 9 is easily adapted to obtain an interval estimate for μ_d.

▪ **Paired t Confidence Interval for μ_d**

When

1. the samples are *paired*,
2. the n sample differences can be viewed as a *random sample* from a population of differences, and
3. the *number of sample differences is large* (in general, at least 30) or the *population distribution of differences is approximately normal*,

the paired t confidence interval for μ_d is

$$\bar{x}_d \pm (t \text{ critical value})\left(\frac{s_d}{\sqrt{n}}\right)$$

For a specified confidence level, the $(n-1)$ df row of Appendix Table 3 gives the appropriate t critical value.

▪ **Example 11.8   Lactic Acid in the Blood After Exercise**

The effect of exercise on the amount of lactic acid in the blood was examined in the article "A Descriptive Analysis of Elite-Level Racquetball" (*Research Quarterly for Exercise and Sport* [1991]: 109–114). Eight males were selected at random from those attending a week-long training camp. Blood lactate levels were measured before and after playing three games of racquetball, as shown in the following table:

Player	Before	After	Difference
1	13	18	−5
2	20	37	−17
3	17	40	−23
4	13	35	−22
5	13	30	−17
6	16	20	−4
7	15	33	−18
8	16	19	−3

We will use these data to estimate the mean change in blood lactate level using a 95% confidence interval. The eight men were selected at random from training camp participants. The following boxplot of the eight sample differences is not inconsistent with a difference population that is approximately normal, so the paired t confidence interval is appropriate:

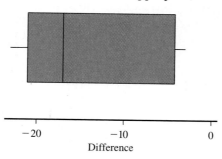

Difference

The t critical value for df $= 7$ and a 95% confidence level is 2.37, and therefore the confidence interval is

$$\bar{x}_d \pm (t \text{ critical value})\left(\frac{s_d}{\sqrt{n}}\right) = -13.63 \pm (2.37)\left(\frac{8.28}{\sqrt{8}}\right)$$

$$= -13.63 \pm 6.938$$

$$= (-20.568, -6.692)$$

Based on the sample data, we can be 95% confident that the difference in mean blood lactate level is between -20.568 and -6.692. That is, we are 95% confident that the mean increase in blood lactate level is somewhere between 6.692 and 20.568 after three games of racquetball.

When two populations must be compared to draw a conclusion on the basis of sample data, a researcher might choose to use independent samples or paired samples. In many situations, paired data provide a more effective comparison by screening out the effects of extraneous variables that might obscure differences between the two populations or that might suggest a difference when none exists.

▪ Exercises 11.29–11.40

11.29 Suppose that you are interested in investigating the effect of a drug that is to be used in the treatment of patients who have glaucoma in both eyes. A comparison between the mean reduction in eye pressure for this drug and for a standard treatment is desired. Both treatments are applied directly to the eye.
a. Describe how you would go about collecting data for your investigation.
b. Does your method result in paired data?
c. Can you think of a reasonable method of collecting data that would result in independent samples? Would such an experiment be as informative as a paired experiment? Comment.

11.30 Two different underground pipe coatings for preventing corrosion are to be compared. The effect of a coating (as measured by maximum depth of corrosion penetration on a piece of pipe) can vary with depth, orientation, soil type, pipe composition, and so on. Describe how an experiment that filters out the effects of these extraneous factors could be carried out.

11.31 The article "More Students Taking AP Tests" (*San Luis Obispo Tribune*, January 10, 2003) provided the following information on the percentage of students in grades 11 and 12 taking one or more AP exams and the percentage of exams that earned credit in 1997 and 2002 for seven high schools on the central coast of California:

School	Percentage of Students Taking One or More AP Exams		Percentage of Exams That Earned College Credit	
	1997	2002	1997	2002
1	13.6	18.4	61.4	52.8
2	20.7	25.9	65.3	74.5
3	8.9	13.7	65.1	72.4
4	17.2	22.4	65.9	61.9
5	18.3	43.5	42.3	62.7
6	9.8	11.4	60.4	53.5
7	15.7	17.2	42.9	62.2

a. Assuming that it is reasonable to regard these seven schools as a random sample of high schools located on the central coast of California, carry out an appropriate test to determine whether there is convincing evidence that the mean percentage of exams earning college credit at central coast high schools declined between 1997 and 2002.
b. Do you think it is reasonable to generalize the conclusion of the test in Part (a) to all California high schools? Explain.
c. Would it be reasonable to use the paired t test with the data on percentage of students taking one or more AP classes? Explain.

11.32 The Oregon Department of Health web site provides information on the cost-to-charge ratio (the percentage of billed charges that are actual costs to the hospital). The cost-to-charge ratios for both inpatient and outpatient care in 2002 for a random sample of six hospitals in Oregon are:

Hospital	2002 Inpatient Ratio	2002 Outpatient Ratio
1	68	54
2	100	75
3	71	53
4	74	56
5	100	74
6	83	71

Is there evidence that the mean cost-to-charge ratio for Oregon hospitals is lower for outpatient care than for inpatient care? Use a significance level of .05.

11.33 Babies born extremely prematurely run the risk of various neurological problems and tend to have lower IQ and verbal ability scores than babies who are not premature. The article "Premature Babies May Recover Intelligence, Study Says" (*San Luis Obispo Tribune*, February 12, 2003) summarized the results of medical research that suggests that the deficit observed at an early age may decrease as children age. Children who were born prematurely were given a test of verbal ability at age 3 years and again at age 8 years. The test is scaled so that a score of 100 is average for a normal-birth-weight child. Data that are consistent with summary quantities given in the paper for 50 children who were born prematurely were used to generate the following MINITAB output, where Age3 represents the verbal ability score at age 3 and Age8 represents the verbal ability score at age 8:

Paired T-Test and CI: Age8, Age3

Paired T for Age8 − Age3

	N	Mean	StDev	SE Mean
Age8	50	97.21	16.97	2.40
Age3	50	87.30	13.84	1.96
Difference	50	9.91	22.11	3.13

T-Test of mean difference = 0 (vs > 0): T-Value = 3.17
P-Value = 0.001

Use the MINITAB output to determine whether there is evidence that the mean verbal ability score for children born prematurely increases between age 3 and age 8. You may assume that it is reasonable to regard the sample of 50 children as a random sample from the population of all children born prematurely.

11.34 The article "Report: Mixed Progress in Math" (*USA Today*, August 3, 2001) gave data for all 50 states on the percentage of public school students who were at or above the proficient level in mathematics testing in 1996 and 2000. Explain why it is not necessary to use an inference procedure such as the paired t test if you want to know whether the mean percentage proficient for the 50 states increased from 1996 to 2002.

11.35 Do girls think they don't need to take as many science classes as boys? The article "Intentions of Young Students to Enroll in Science Courses in the Future: An Examination of Gender Differences" (*Science Education* [1999]: 55–76) gave information from a survey of children in grades 4, 5, and 6. The 224 girls participating in the survey each indicated the number of science courses they intended to take in the future, and they also indicated the number of science courses they thought boys their age should take in the future. For each girl, the investigators calculated the difference between the number of science classes she intends to take and the number she thinks boys should take.

a. Explain why these data are paired.

b. The mean of the differences was −0.83 (indicating that these 224 girls intended, on average, to take fewer classes than they thought boys should take), and the standard deviation was 1.51. Construct and interpret a 95% confidence interval for the mean difference.

11.36 In a study of memory recall, 8 students from a large psychology class were selected at random and given 10 min to memorize a list of 20 nonsense words. Each was asked to list as many of the words as he or she could remember both 1 hr and 24 hr later, as shown in the following table:

Subject	1	2	3	4	5	6	7	8
1 hr later	14	12	18	7	11	9	16	15
24 hr later	10	4	14	6	9	6	12	12

Is there evidence to suggest that the mean number of words recalled after 1 hr exceeds the mean recall after 24 hr by more than 3? Use a level .01 test.

11.37 As part of a study to determine the effects of allowing the use of credit cards for alcohol purchases in Canada (see "Changes in Alcohol Consumption Patterns Following the Introduction of Credit Cards in Ontario Liquor Stores," *Journal of Studies on Alcohol* [1999]: 378–382), randomly selected individuals were given a questionnaire asking them (among other things) how many drinks they had consumed during the previous week. A year later (after liquor stores started accepting credit cards for purchases), these same individuals were again asked how many drinks they had consumed in the previous week. The following summary statistics are consistent with those presented in the article:

	n	1994 Mean	1995 Mean	$\bar{d}$	s_d
Credit-Card Shoppers	96	6.72	6.34	0.38	5.52
Non–Credit Card Shoppers	850	4.09	3.97	0.12	4.58

a. The standard deviations of the differences are quite large. Explain how this could be the case.

b. Calculate a 95% confidence interval for the mean difference in drink consumption between 1994 and 1995 for credit card shoppers. Is there evidence that the mean number of drinks decreased?

c. Test the hypothesis that there was no change in the mean number of drinks between 1994 and 1995 for the non–credit card shoppers.

11.38 The effect of exercise on the amount of lactic acid in the blood was examined in the article "A Descriptive Analysis of Elite-Level Racquetball" (*Research Quarterly for Exercise and Sport* [1991]: 109–114). Eight men and seven women who were attending a week-long training camp participated in the experiment. Blood lactate levels were measured before and after playing three games of racquetball, as shown in the following table:

	Men			Women	
Player	Before	After	Player	Before	After
1	13	18	1	11	21
2	20	37	2	16	26
3	17	40	3	13	19
4	13	35	4	18	21
5	13	30	5	14	14
6	16	20	6	11	31
7	15	33	7	13	20
8	16	19			

a. Estimate the mean change in blood lactate level for male racquetball players using a 90% confidence interval.

b. Estimate the mean change for female players using a 90% confidence interval.

c. Based on the intervals from Parts (a) and (b), do you think that the mean change in blood lactate level is the same for men as it is for women? Explain.

11.39 Several methods of estimating the number of seeds in soil samples have been developed by ecologists. An article in the *Journal of Ecology* ("A Comparison of Methods for Estimating Seed Numbers in the Soil" [1990]: 1079–1093) considered three such methods. The following data give number of seeds detected by the direct method and by the stratified method for 27 soil specimens:

Specimen	Direct	Stratified
1	24	8
2	32	36
3	0	8
4	60	56
5	20	52
6	64	64
7	40	28
8	8	8
9	12	8
10	92	100
11	4	0
12	68	56
13	76	68
14	24	52
15	32	28
16	0	0
17	36	36
18	16	12
19	92	92
20	4	12
21	40	48
22	24	24
23	0	0
24	8	12
25	12	40
26	16	12
27	40	76

Do the data provide sufficient evidence to conclude that the mean number of seeds detected differs for the two methods? Test the relevant hypotheses using $\alpha = .05$.

11.40 Many people who quit smoking complain of weight gain. The results of an investigation of the relationship between smoking cessation and weight gain are given in the article "Does Smoking Cessation Lead to Weight Gain?" (*American Journal of Public Health* [1983]: 1303–1305). Three hundred twenty-two subjects, selected at random from those who successfully participated in a program to quit smoking, were weighed at the beginning of the program and again 1 year later. The mean change in weight was 5.15 lb, and the standard deviation of the weight changes was 11.45 lb. Is there sufficient evidence to conclude that the true mean change in weight is positive? Use $\alpha = .05$.

▪ 11.3 Large-Sample Inferences Concerning a Difference Between Two Population or Treatment Proportions

Large-sample methods for estimating and testing hypotheses about a single population proportion were presented in Chapters 9 and 10. The symbol π was used to represent the true proportion of individuals in the population who possess some characteristic (the successes). Inferences about the value of π were based on p, the corresponding sample proportion of successes.

Many investigations are carried out to compare the proportion of successes in one population (or resulting from one treatment) to the proportion of successes in a second population (or from a second treatment). As was the case for means, we use the subscripts 1 and 2 to distinguish between the two population proportions, sample sizes, and sample proportions.

▪ Notation

Population or Treatment 1: Proportion of "successes" = π_1.

Population or Treatment 2: Proportion of "successes" = π_2.

	Sample Size	Proportion of Successes
Sample from Population or Treatment 1	n_1	p_1
Sample from Population or Treatment 2	n_2	p_2

When comparing two populations or treatments on the basis of "success" proportions, it is common to focus on the quantity $\pi_1 - \pi_2$, the difference between the two proportions. Because p_1 provides an estimate of π_1 and p_2 provides an estimate of π_2, the obvious choice for an estimate of $\pi_1 - \pi_2$ is $p_1 - p_2$.

Because p_1 and p_2 each vary in value from sample to sample, so will the difference $p_1 - p_2$. For example, a first sample from each of two populations might yield

$$p_1 = .69 \qquad p_2 = .70 \qquad p_1 - p_2 = -.01$$

A second sample from each might result in

$$p_1 = .79 \qquad p_2 = .67 \qquad p_1 - p_2 = .12$$

and so on. Because the statistic $p_1 - p_2$ is the basis for drawing inferences about $\pi_1 - \pi_2$, we need to know something about its behavior.

▪ Properties of the Sampling Distribution of $p_1 - p_2$

If two random samples are selected independently of one another, the following properties hold:

1. $\mu_{p_1 - p_2} = \pi_1 - \pi_2$.

   This says that the sampling distribution of $p_1 - p_2$ is centered at $\pi_1 - \pi_2$, so $p_1 - p_2$ is an unbiased statistic for estimating $\pi_1 - \pi_2$.

2. $\sigma^2_{p_1-p_2} = \sigma^2_{p_1} + \sigma^2_{p_2} = \dfrac{\pi_1(1-\pi_1)}{n_1} + \dfrac{\pi_2(1-\pi_2)}{n_2}$

and

$\sigma_{p_1-p_2} = \sqrt{\dfrac{\pi_1(1-\pi_1)}{n_1} + \dfrac{\pi_2(1-\pi_2)}{n_2}}$

3. If both n_1 and n_2 are large (i.e., if $n_1\pi_1 \geq 10$, $n_1(1-\pi_1) \geq 10$, $n_2\pi_2 \geq 10$, and $n_2(1-\pi_2) \geq 10$), then p_1 and p_2 each have a sampling distribution that is approximately normal, and their difference $p_1 - p_2$ also has a sampling distribution that is approximately normal.

The properties in the box imply that when the samples are independently selected and when both sample sizes are large, the distribution of the standardized variable

$$z = \frac{p_1 - p_2 - (\pi_1 - \pi_2)}{\sqrt{\dfrac{\pi_1(1-\pi_1)}{n_1} + \dfrac{\pi_2(1-\pi_2)}{n_2}}}$$

is described approximately by the standard normal (z) curve.

■ A Large-Sample Test Procedure

Comparisons of π_1 and π_2 are often based on large, independently selected samples, and we restrict ourselves to this case. The most general null hypothesis of interest has the form

H_0: $\pi_1 - \pi_2 =$ hypothesized value

However, when the hypothesized value is something other than 0, the appropriate test statistic differs somewhat from the test statistic used for H_0: $\pi_1 - \pi_2 = 0$. Because this H_0 is almost always the relevant one in applied problems, we focus exclusively on it.

Our basic testing principle has been to use a procedure that controls the probability of a Type I error at the desired level α. This requires using a test statistic with a sampling distribution that is known when H_0 is true. That is, the test statistic should be developed under the assumption that $\pi_1 = \pi_2$ (as specified by the null hypothesis $\pi_1 - \pi_2 = 0$). In this case, π can be used to denote the common value of the two population proportions. The z variable obtained by standardizing $p_1 - p_2$ then simplifies to

$$z = \frac{p_1 - p_2}{\sqrt{\dfrac{\pi(1-\pi)}{n_1} + \dfrac{\pi(1-\pi)}{n_2}}}$$

Unfortunately, this cannot serve as a test statistic, because the denominator cannot be computed: H_0 says that there is a common value π, but it does not specify the value. A test statistic can be obtained, though, by first *estimating* π from the sample data and then using this estimate in the denominator of z.

When $\pi_1 = \pi_2$, either p_1 or p_2 is an estimate of the common proportion π. However, a better estimate than either p_1 or p_2 is a weighted average of the two, in which more weight is given to the sample proportion based on the larger sample.

■ **Definition**

The **combined estimate of the common population proportion** is

$$p_c = \frac{n_1 p_1 + n_2 p_2}{n_1 + n_2} = \frac{\text{total number of successes in the two samples}}{\text{total of the two sample sizes}}$$

The test statistic for testing $H_0: \pi_1 - \pi_2 = 0$ results from using p_c in place of π in the standardized variable z given previously. This z statistic has approximately a standard normal distribution when H_0 is true, so a test that has the desired significance level α can be obtained by calculating a P-value using the z table.

■ **Summary of Large-Sample z Tests for $H_0: \pi_1 - \pi_2 = 0$**

Null hypothesis: $H_0: \pi_1 - \pi_2 = 0$

Test statistic: $z = \dfrac{p_1 - p_2}{\sqrt{\dfrac{p_c(1 - p_c)}{n_1} + \dfrac{p_c(1 - p_c)}{n_2}}}$

Alternative Hypothesis	P-Value
$H_a: \pi_1 - \pi_2 > 0$	Area under the z curve to the right of the computed z
$H_a: \pi_1 - \pi_2 < 0$	Area under the z curve to the left of the computed z
$H_a: \pi_1 - \pi_2 \neq 0$	(1) 2(area to the right of z) if z is positive, or (2) 2(area to the left of z) if z is negative

Assumptions: 1. The samples are *independently chosen random samples*, or *treatments were assigned at random to individuals or objects* (or subjects were assigned at random to treatments).
2. Both *sample sizes are large*:

$$n_1 p_1 \geq 10, \quad n_1(1 - p_1) \geq 10, \quad n_2 p_2 \geq 10, \quad n_2(1 - p_2) \geq 10$$

■ **Example 11.9   Duct Tape to Remove Warts?**

Some people seem to believe that you can fix anything with duct tape. Even so, many were skeptical when researchers announced that duct tape may be a more effective and less painful alternative to liquid nitrogen, which doctors routinely use to freeze warts. The article "What a Fix-It: Duct Tape Can Remove Warts" (*San Luis Obispo Tribune*, October 15, 2002) described a study conducted at Madigan Army Medical Center. Patients with warts were randomly assigned to either the duct tape treatment or the more traditional freezing treatment. Those in the duct tape group wore duct tape over the wart for 6 days, then removed the tape, soaked the area in water, and used an emery board to scrape the area. This process was repeated for a maximum of 2 months or until the wart was gone. Data consistent with values in the article are summarized in the following table:

Treatment	n	Number with Wart Successfully Removed
Liquid nitrogen freezing	100	60
Duct tape	104	88

Do the data suggest that freezing is less successful than duct tape in removing warts? Let π_1 represent the true proportion of warts that would be successfully removed by freezing, and let π_2 represent the true proportion of warts that would be successfully removed with the duct tape treatment. We test the relevant hypotheses

$$H_0: \quad \pi_1 - \pi_2 = 0 \qquad \text{versus} \qquad H_a: \quad \pi_1 - \pi_2 < 0$$

using $\alpha = .01$. For these data,

$$p_1 = \frac{60}{100} = .60$$

$$p_2 = \frac{88}{104} = .85$$

Suppose that $\pi_1 = \pi_2$; let π denote the common value. Then the combined estimate of π is

$$p_c = \frac{n_1 p_1 + n_2 p_2}{n_1 + n_2} = \frac{100(.60) + 104(.85)}{100 + 104} = .73$$

The nine-step procedure can now be used to perform the hypothesis test:

1. $\pi_1 - \pi_2$ is the difference between the true proportions of warts removed for freezing and the duct tape treatment.
2. $H_0: \pi_1 - \pi_2 = 0 \quad (\pi_1 = \pi_2)$.
3. $H_a: \pi_1 - \pi_2 < 0 \quad (\pi_1 < \pi_2$, in which case the proportion of warts removed for freezing is lower than the proportion for duct tape).
4. Significance level: $\alpha = .01$.
5. Test statistic: $z = \dfrac{p_1 - p_2}{\sqrt{\dfrac{p_c(1 - p_c)}{n_1} + \dfrac{p_c(1 - p_c)}{n_2}}}$
6. Assumptions: The subjects were assigned randomly to the two treatments. Checking to make sure that the sample sizes are large enough, we have

$$n_1 p_1 = 100(.60) = 60 \geq 10$$
$$n_1(1 - p_1) = 100(.40) = 40 \geq 10$$
$$n_2 p_2 = 104(.85) = 88.4 \geq 10$$
$$n_2(1 - p_2) = 104(.15) = 15.6 \geq 10$$

7. Calculations:

$$n_1 = 100 \qquad n_2 = 104 \qquad p_1 = .60 \qquad p_2 = .85 \qquad p_c = .73$$

and so

$$z = \frac{.60 - .85}{\sqrt{\dfrac{(.73)(.27)}{100} + \dfrac{(.73)(.27)}{104}}} = \frac{-.25}{.062} = -4.03$$

8. P-value: This is a lower-tailed test, so the P-value is the area under the z curve and to the left of the computed $z = -4.03$. From Appendix Table 2, we find P-value ≈ 0.
9. Conclusion: Because P-value $\leq \alpha$, the null hypothesis is rejected at level .01. There is convincing evidence that the proportion of warts successfully removed is lower for freezing than for the duct tape treatment.

▪ **Example 11.10** AIDS and Housing Availability

The authors of the article "Accommodating Persons with AIDS: Acceptance and Rejection in Rental Situations" (*Journal of Applied Social Psychology* [1999]: 261– 270) stated that, even though landlords participating in a telephone survey indicated that they would generally be willing to rent to persons with AIDS, they wondered whether this was true in actual practice. To investigate, the researchers independently selected two random samples of 80 advertisements for rooms for rent from newspaper advertisements in three large cities. An adult male caller responded to each ad in the first sample of 80 and inquired about the availability of the room and was told that the room was still available in 61 of these calls. The same caller also responded to each ad in the second sample. In these calls, the caller indicated that he was currently receiving some treatment for AIDS and was about to be released from the hospital and would require a place to live. The caller was told that a room was available in 32 of these calls. Based on this information, the authors concluded that "reference to AIDS substantially decreased the likelihood of a room being described as available." Do the data support this conclusion? Let's carry out a hypothesis test with $\alpha = .01$.

No AIDS reference: $n_1 = 80$ $p_1 = 61/80 = .763$
AIDS reference: $n_2 = 80$ $p_2 = 32/80 = .400$

1. π_1 = proportion of rooms available when there is no AIDS reference

   π_2 = proportion of rooms available with AIDS reference

2. $H_0: \pi_1 - \pi_2 = 0$.

3. $H_a: \pi_1 - \pi_2 > 0$.

4. Significance level: $\alpha = .01$.

5. Test statistic: $z = \dfrac{p_1 - p_2}{\sqrt{\dfrac{p_c(1 - p_c)}{n_1} + \dfrac{p_c(1 - p_c)}{n_2}}}$

6. Assumptions: The two samples are independently chosen random samples. Checking to make sure that the sample sizes are large enough by using $n_1 = 80, p_1 = .763, n_2 = 80$, and $p_2 = .400$, we have

   $n_1 p_1 = 61.04 \geq 10$
   $n_1(1 - p_1) = 18.96 \geq 10$
   $n_2 p_2 = 32.00 \geq 10$
   $n_2(1 - p_2) = 48.00 \geq 10$

7. Calculations:

   $$p_c = \frac{n_1 p_1 + n_2 p_2}{n_1 + n_2} = \frac{80(.763) + 80(.400)}{80 + 80} = .582$$

   $$z = \frac{.763 - .400}{\sqrt{\dfrac{(.582)(.418)}{80} + \dfrac{(.582)(.418)}{80}}} = \frac{.363}{.078} = 4.65$$

8. *P*-value: The *P*-value for this test is the area under the *z* curve and to the right of the computed $z = 4.65$. Because 4.65 is so far out in the upper tail of the *z* curve, *P*-value ≈ 0

9. **Conclusion:** Because P-value $\leq \alpha$, the null hypothesis is rejected at level .01. There is strong evidence that the proportion of rooms reported as available is smaller with the AIDS reference than without it. This supports the claim made by the authors of the article.

MINITAB can also be used to carry out a two-sample z test to compare two population proportions, as shown in the following output:

Test and Confidence Interval for Two Proportions

Sample	X	N	Sample p
1	61	80	0.762500
2	32	80	0.400000

Estimate for p(1) − p(2): 0.3625
95% CI for p(1) − p(2): (0.220302, 0.504698)
Test for p(1) − p(2) = 0 (vs > 0): z = 4.65 P-Value = 0.000

▪ A Confidence Interval

A large-sample confidence interval for $\pi_1 - \pi_2$ is a special case of the general z interval formula

point estimate $\pm$ (z critical value)(estimated standard deviation)

The statistic $p_1 - p_2$ gives a point estimate of $\pi_1 - \pi_2$, and the standard deviation of this statistic is

$$\sigma_{p_1 - p_2} = \sqrt{\frac{\pi_1(1 - \pi_1)}{n_1} + \frac{\pi_2(1 - \pi_2)}{n_2}}$$

An estimated standard deviation is obtained by using the sample proportions p_1 and p_2 in place of π_1 and π_2, respectively, under the square-root symbol. Notice that this estimated standard deviation differs from the one used previously in the test statistic. Here, there isn't a null hypothesis that claims $\pi_1 = \pi_2$, so there is no assumed common value of π to estimate.

▪ **A Large-Sample Confidence Interval for $\pi_1 - \pi_2$**

When

1. the samples are *independently selected random samples* or *treatments were assigned at random to individuals or objects* (or vice versa), and

2. both *sample sizes are large*, that is,

$$n_1 p_1 \geq 10 \qquad n_1(1 - p_1) \geq 10 \qquad n_2 p_2 \geq 10 \qquad n_2(1 - p_2) \geq 10$$

a large-sample confidence interval for $\pi_1 - \pi_2$ is

$$(p_1 - p_2) \pm (z \text{ critical value})\sqrt{\frac{p_1(1 - p_1)}{n_1} + \frac{p_2(1 - p_2)}{n_2}}$$

■ **Example 11.11** Pets and Pesticides

Researchers at the National Cancer Institute released the results of a study that examined the effect of weed-killing herbicides on house pets (Associated Press, September 4, 1991). Dogs, some of whom were from homes where the herbicide was used on a regular basis, were examined for the presence of malignant lymphoma. The following data are compatible with summary values given in the report:

Group	Sample Size	Number with Lymphoma	p
Exposed	827	473	.572
Unexposed	130	19	.146

We use the given data to estimate the difference between the proportion of exposed dogs that develop lymphoma and the proportion of unexposed dogs that develop lymphoma.

Let π_1 denote the proportion of exposed dogs that develop lymphoma, and define π_2 similarly for unexposed dogs. The sample sizes are large enough for the large sample interval to be valid ($n_1 p_1 = 827(.572) \geq 10, n_1(1 - p_1) = 827(.428) \geq 10$, etc.). A 90% confidence interval for $\pi_1 - \pi_2$ is

$$(p_1 - p_2) \pm (z \text{ critical value})\sqrt{\frac{p_1(1 - p_1)}{n_1} + \frac{p_2(1 - p_2)}{n_2}}$$

$$= (.572 - .146) \pm (1.645)\sqrt{\frac{(.572)(.428)}{827} + \frac{(.146)(.854)}{130}}$$

$$= .426 \pm (1.645)(.0354)$$

$$= .426 \pm .058$$

$$= (.368, .484)$$

Assuming that it is reasonable to regard these two samples as being independently selected and as representative of the two populations of interest, we can say that we believe that the proportion of exposed dogs that develop lymphoma exceeds that for unexposed dogs by somewhere between .36 and .48. We used a method to construct this estimate that captures the true difference in proportions 90% of the time in repeated sampling.

■ **Exercises 11.41–11.56**

11.41 A university is interested in evaluating registration processes. Students can register for classes by using either a telephone registration system or an online system that is accessed through the university's web site. Independent random samples of 80 students who registered by phone and 60 students who registered online were selected. Of those who registered by phone, 57 reported that they were satisfied with the registration process. Of those who registered online, 50 reported that they were satisfied. Based on these data, is it reasonable to conclude that the proportion who are satisfied is higher for those who register online? Test the appropriate hypotheses using $\alpha = .05$.

11.42 Some commercial airplanes recirculate approximately 50% of the cabin air to increase fuel efficiency. The authors of the paper "Aircraft Cabin Air Recirculation and Symptoms of the Common Cold" (*Journal of the American Medical Association* [2002]: 483–486) studied 1100 airline passengers who flew from San Francisco to Denver between January and

April 1999. Some passengers traveled on airplanes that recirculated air, and others traveled on planes that did not recirculate air. Of the 517 passengers who flew on planes that did not recirculate air, 108 reported postflight respiratory symptoms, whereas 111 of the 583 passengers on planes that did recirculate air reported such symptoms. Is there sufficient evidence to conclude that the proportion of passengers with postflight respiratory symptoms differs for planes that do and do not recirculate air? Test the appropriate hypotheses using $\alpha = .05$. You may assume that it is reasonable to regard these two samples as being independently selected and as representative of the two populations of interest.

11.43 The article "A 'White' Name Found to Help in Job Search" (Associated Press, January 15, 2003) described an experiment to investigate whether it helps to have a white-sounding first name when looking for a job. Researchers sent 5000 résumés in response to ads that appeared in the *Boston Globe* and the *Chicago Tribune*. The résumés were identical except that 2500 of them had "white-sounding" first names, such as Brett and Emily, whereas the other 2500 had "black-sounding" names, such as Tamika and Rasheed. Résumés of the first type elicited 250 responses, and résumés of the second type elicited only 167 responses. Do these data support the theory that the proportion receiving positive responses is higher for résumés with white-sounding first names?

11.44 "Mountain Biking May Reduce Fertility in Men, Study Says" was the headline of an article appearing in the *San Luis Obispo Tribune* (December 3, 2002). This conclusion was based on an Austrian study that compared sperm counts of avid mountain bikers (those who ride at least 12 hr/wk) and nonbikers. Ninety percent of the avid mountain bikers studied had low sperm counts, compared to 26% of the nonbikers. Suppose that these percentages were based on independent samples of 100 avid mountain bikers and 100 nonbikers and that it is reasonable to view these samples as representative of Austrian avid mountain bikers and nonbikers.
a. Do these data provide convincing evidence that the proportion of Austrian avid mountain bikers with low sperm count is higher than the proportion of Austrian nonbikers?
b. Based on the outcome of the test in Part (a), is it reasonable to conclude that mountain biking for 12 hr/wk or more causes low sperm count? Explain.

11.45 In a study of a proposed approach to diabetes prevention, 339 people under the age of 20 who were thought to be at high risk of developing type I diabetes were assigned at random to two groups. One group received twice-daily injections of a low dose of insulin. The other group (the control group) did not receive any insulin but was closely monitored. Summary data (from the article "Diabetes Theory Fails Test," *USA Today*, June 25, 2001) follow:

Group	n	Number Developing Diabetes
Insulin	169	25
Control	170	24

a. Use the given data to construct a 90% confidence interval for the difference in the proportion that develop diabetes for the control group and the insulin group.
b. Give an interpretation of the confidence interval and the associated confidence level.
c. Based on your interval from Part (a), write a few sentences commenting on the effectiveness of the proposed prevention treatment.

11.46 Women diagnosed with breast cancer whose tumors have not spread may be faced with a decision between two surgical treatments — mastectomy (removal of the breast) or lumpectomy (removal of only the tumor). In a long-term study of the effectiveness of these two treatments, 701 women with breast cancer were randomly assigned to one of two treatment groups. One group received mastectomies, and the other group received lumpectomies and radiation. Both groups were followed for 20 years after surgery. It was reported that there was no statistically significant difference in the proportion surviving for 20 years for the two treatments (Associated Press, October 17, 2002). What hypotheses do you think the researchers tested to reach the given conclusion? Did the researchers reject or fail to reject the null hypothesis?

11.47 Do teachers find their work rewarding and satisfying? The article "Work-Related Attitudes" (*Psychological Reports* [1991]: 443–450) reported the results of a survey of random samples of 395 elementary school teachers and 266 high school teachers. Of the elementary school teachers, 224 said that they were very satisfied with their jobs, whereas 126 of the high school teachers were very satisfied with their work. Based on these data, is it reasonable to conclude that the proportion of very satisfied teachers is different for elementary school teachers than it is for high school teachers? Test the appropriate hypotheses using a .05 significance level.

11.48 The article "Foraging Behavior of the Indian False Vampire Bat" (*Biotropica* [1991]: 63–67) reported that 36 of 193 female bats in flight spent more than 5 min in the air before locating food. For male bats, 64 of 168 spent more than 5 min in the air. Is there sufficient evidence to conclude that the proportion of flights longer than 5 min differs for males and females? Test the relevant hypotheses using $\alpha = .01$.

11.49 In December 2001 the Department of Veterans Affairs announced that it would begin paying benefits to soldiers suffering from Lou Gehrig's disease who had served in the first Gulf War (*New York Times*, December 11, 2001). This decision was based on an analysis in which the disease's incidence rate (the proportion developing the disease) for all of the approximately 700,000 soldiers sent to the Persian Gulf between August 1990 and July 1991 was compared to the incidence rate for all of the approximately 1.8 million other soldiers who were not in the Persian Gulf during this time period. Based on these data, explain why it is not appropriate to perform a formal inference procedure (such as the two-sample z test) and yet it is still reasonable to conclude that the incidence rate is higher for Gulf War veterans than for those who did not serve in the Gulf War.

11.50 The article "Regrets on Early Sex" appeared in the Australian newspaper *The Herald* (January 2, 1998). The article stated that "while 54% of women thought that they should have waited longer before having sex, only 16% of the men felt that way." Suppose that this statement had been based on interviews with independently chosen random samples of 50 women and 50 men from a particular city in New Zealand. (The article summarizes a survey conducted in New Zealand.) Construct and interpret a 90% confidence interval for the difference in the true proportions of men and women who think they should have waited longer before having sex.

11.51 Gender differences in student needs and fears were examined in the article "A Survey of Counseling Needs of Male and Female College Students" (*Journal of College Student Development* [1998]: 205–208). Random samples of male and female students were selected from those attending a particular university. Of 234 males surveyed, 27.5% said that they were concerned about the possibility of getting AIDS. Of 568 female students surveyed, 42.7% reported being concerned about the possibility of getting AIDS. Is there sufficient evidence to conclude that the proportion of female students concerned about the possibility of getting AIDS is greater than the corresponding proportion for males?

11.52 In a study that was described by a writer for the *Washington Post* (*San Luis Obispo Tribune*, February 2, 2000), 500 patients undergoing abdominal surgery were randomly assigned to breathe one of two oxygen mixtures during surgery and for 2 hr afterward. One group received a mixture containing 30% oxygen, a standard generally used in surgery. The other group was given 80% oxygen. Wound infections developed in 28 of the 250 patients who received 30% oxygen and in 13 of the 250 patients who received 80% oxygen. Is there sufficient evidence to conclude that the proportion of patients who develop wound infections is lower for the 80% oxygen treatment than for the 30% oxygen treatment? Test the relevant hypotheses using a significance level of .05.

11.53 The state of Georgia's HOPE scholarship program guarantees fully paid tuition to Georgia public universities for Georgia high school seniors who have a B average in academic requirements as long as they maintain a B average in college (see "Who Loses HOPE? Attrition from Georgia's College Scholarship Program," *Southern Economic Journal* [1999]: 379–390). It was reported that 53.2% of a random sample of 137 students entering Ivan Allen College (social science and humanities) at the Georgia Institute of Technology with a HOPE scholarship lost the scholarship at the end of the first year because they had a GPA of less than 3.0. It was also reported that 72 of a random sample of 111 students entering the College of Computing with a B average had lost their HOPE scholarship by the end of the first year. Is there evidence that the proportion who lose HOPE scholarships is different for the College of Computing and the Ivan Allen College?

11.54 An Associated Press article (*San Luis Obispo Telegram-Tribune*, September 23, 1995) examined the changing attitudes of Catholic priests. National surveys of priests, age 26 to 35, were conducted in 1985 and again in 1993. The surveyed priests were asked whether they agreed with the following statement: Celibacy should be a matter of personal choice for priests. In 1985, 69% of those surveyed agreed; in 1993, 38% agreed. Suppose that the samples were randomly selected and that the sample sizes were both 200. Is there evidence that the proportion of priests who agreed that celibacy should be a matter of personal choice declined from 1985 to 1993? Use $\alpha = .05$.

11.55 Are college students who take a freshman orientation course more or less likely to stay in college than those who do not take such a course? The article "A Longitudinal Study of the Retention and Academic Performance of Participants in Freshmen Orientation Courses" (*Journal of College Student Development* [1994]: 444–449) reported that 50 of 94 randomly selected students who did not participate in an orientation course returned for a second year. Of 94 randomly selected students who did take the orientation course, 56 returned for a second year. Construct a 95% confidence interval for $\pi_1 - \pi_2$, the difference in the proportion returning for students who do not take an orientation course and those who do. Give an interpretation of this interval.

11.56 The article "Truth and DARE: Tracking Drug Education to Graduation" (*Social Problems* [1994]: 448–456) compared the drug use of 288 randomly selected high school seniors exposed to a drug education program (DARE) and 335 randomly selected high school seniors who were not exposed to such a program. Data for marijuana use are given in the following table:

	n	Number Who Use Marijuana
Exposed to DARE	288	141
Not Exposed to DARE	335	181

Is there evidence that the proportion using marijuana is lower for students exposed to the DARE program? Use $\alpha = .05$.

▪ 11.4 Distribution-Free Procedures for Inferences About a Difference Between Two Population or Treatment Means Using Independent Samples (Optional)

One approach to making inferences about $\mu_1 - \mu_2$ when n_1 and n_2 are small is to assume that the two population or treatment response distributions are normal and then to use the two-sample t test or confidence interval presented in Section 11.1. In some situations, however, the normality assumption may not be reasonable. The validity of the procedures developed in this section does not depend on the normality of the population distributions. The procedures can be used to compare two populations or treatments when it is reasonable to assume that the population or treatment response distributions have the same shape and spread.

▪ **Basic Assumptions in This Section**

The two population or treatment response distributions have the same shape and spread. The only possible difference between the distributions is that one may be shifted to one side or the other.

Distributions consistent with these assumptions are shown in Figure 11.2(a). The distributions shown in Figure 11.2(b) have different shapes and spreads, therefore the methods of this section would not be appropriate in this case. Inferences that involve comparing distributions with different shapes and spreads can be quite complicated; if you encounter that situation, good advice from a statistician is particularly important.

Procedures that do not require any overly specific assumptions about the population distributions are said to be **distribution free** (some texts use the term *nonparametric* instead of *distribution free*). The two-sample t test of Section 11.1 is not

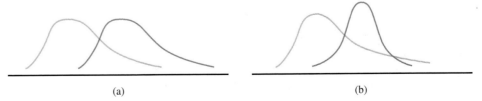

(a)   (b)

Figure 11.2 Two possible population distribution pairs:
(a) same shape and spread, differing only in location;
(b) very different shape and spread.

distribution free because its use is predicated on the specific assumption of (at least approximate) normality.

Inferences about $\mu_1 - \mu_2$ are made using information from two independent random samples, one consisting of n_1 observations from the first population and the other consisting of n_2 observations from the second population. Suppose that the two population distributions are in fact identical (so that $\mu_1 = \mu_2$). In this case, each of the $n_1 + n_2$ observations is actually drawn from the same population distribution. The distribution-free procedure presented here is based on regarding the $n_1 + n_2$ observations as a single data set and assigning ranks to the ordered values. The assignment is easiest when there are no ties among the $n_1 + n_2$ values (each observation is different from every one of the other observations), so assume for the moment that this is the case. Then the smallest among the $n_1 + n_2$ values receives rank 1, the second smallest rank 2, and so on, until finally the largest value is assigned rank $n_1 + n_2$. This procedure is illustrated in Example 11.12.

▪ Example 11.12   Comparing Fuel Efficiency

An experiment to compare fuel efficiencies for two types of subcompact automobile was carried out by first randomly selecting $n_1 = 5$ cars of Type 1 and $n_2 = 5$ cars of Type 2. Each car was then driven from Phoenix to Los Angeles by a nonprofessional driver, after which the fuel efficiency (in miles per gallon) was determined. The resulting data, with observations in each sample ordered from smallest to largest, are given here:

Type 1	39.3	41.1	42.4	43.0	44.4
Type 2	37.8	39.0	39.8	40.7	42.1

The data and the associated ranks are shown in the following dotplot:

● Sample 1

● Sample 2

The ranks of the five observations in the first sample are 3, 6, 8, 9, and 10. If these five observations had all been larger than every value in the second sample, the corresponding ranks would have been 6, 7, 8, 9, and 10. On the other hand, if all five Sample 1 observations had been less than each value in the second sample, the ranks would have been 1, 2, 3, 4, and 5. The ranks of the five observations in the first sample might be any set of five numbers from among 1, 2, 3, . . . , 9, 10 — there are actually 252 possibilities.

▪ Testing Hypotheses

Let's first consider testing

$$H_0: \mu_1 - \mu_2 = 0 \ (\mu_1 = \mu_2) \qquad \text{versus} \qquad H_a: \mu_1 - \mu_2 > 0 \ (\mu_1 > \mu_2)$$

If H_0 is true, all $n_1 + n_2$ observations in the two samples are actually drawn from identical population distributions. We would then expect that the observations in the first sample would be intermingled with those of the second sample when plotted along the number line. In this case, the ranks of the observations should also be intermingled. For example, with $n_1 = 5$ and $n_2 = 5$, the set of Sample 1 ranks 2, 3, 5, 8, 10 would be consistent with $\mu_1 = \mu_2$, as would the set 1, 4, 7, 8, 9. However, when $\mu_1 = \mu_2$, it would be quite unusual for all five values from Sample 1 to be larger than every value in Sample 2, resulting in the set 6, 7, 8, 9, 10 of Sample 1 ranks.

A convenient measure of the extent to which the ranks are intermingled is the sum of the Sample 1 ranks. These ranks in Example 11.12 were 3, 6, 8, 9, and 10, so

rank sum $= 3 + 6 + 8 + 9 + 10 = 36$

The largest possible rank sum when $n_1 = n_2 = 5$ is $6 + 7 + 8 + 9 + 10 = 40$. If μ_1 is much larger than μ_2, we would expect the rank sum to be near its largest possible value. This suggests that we reject H_0 for unusually large values of the rank sum.

Developing a test procedure requires information about the sampling distribution of the rank-sum statistic when H_0 is true. To illustrate this, consider again the case $n_1 = n_2 = 5$. There are 252 different sets of 5 from among the 10 ranks 1, 2, 3, ..., 9, 10. The key point is that, when H_0 is true, any 1 of these 252 sets has the same chance of being the Sample 1 ranks as does any other set, because all 10 observations come from the same population distribution. The chance under H_0 that any particular set occurs is 1/252 (because the possibilities are equally likely).

Table 11.1 displays the 12 sets of Sample 1 ranks that yield the largest rank-sum values. Each of the other 240 possible rank sets has a rank-sum value less than 36. If we observe rank sum $= 36$, we can compute

$$P(\text{rank sum} \geq 36 \text{ when } H_0 \text{ is true}) = \frac{12}{252} = .0476$$

That is, when H_0 is true, a rank sum at least as large as 36 would be observed only about 4.76% of the time. Thus, a test of $H_0: \mu_1 - \mu_2 = 0$ versus $H_a: \mu_1 - \mu_2 > 0$, based on $n_1 = 5$ and $n_2 = 5$ and a rank sum of 36, would have an associated P-value of .0476. It is this type of reasoning that allows us to reach a conclusion about whether or not to reject H_0.

Table 11.1 ■ The 12 Rank Sets That Have the Largest Rank Sums When $n_1 = 5$ and $n_2 = 5$

Sample 1 Ranks	Rank Sum	Sample 1 Ranks	Rank Sum
6 7 8 9 10	40	5 6 7 9 10	37
5 7 8 9 10	39	2 7 8 9 10	36
4 7 8 9 10	38	3 6 8 9 10	36
5 6 8 9 10	38	4 5 8 9 10	36
3 7 8 9 10	37	4 6 7 9 10	36
4 6 8 9 10	37	5 6 7 8 10	36

The process of looking at different rank-sum sets to carry out a test can be quite tedious. Fortunately, information about both one- and two-tailed P-values associated with values of the rank-sum statistic has been tabulated. Appendix Table 6 provides information on P-values for selected values of n_1 and n_2. For example, when n_1 and n_2 are both 5, Appendix Table 6 tells us that, for an upper-tailed test with a rank-sum statistic value of 36, P-value $< .05$. This is consistent with the value of

.0476 computed previously. Had the rank sum been 40, we would have concluded that the *P*-value was less than .01 (using the table). The use of this appendix table is further illustrated in the examples that follow.

■ **Summary of the Rank-Sum Test***

Null hypothesis: $H_0: \mu_1 - \mu_2 = 0$

Test statistic: rank sum = sum of ranks assigned to the observations in the first sample.

Alternative Hypothesis	Type of Test
H_a: $\mu_1 - \mu_2 > 0$	Upper-tailed
H_a: $\mu_1 - \mu_2 < 0$	Lower-tailed
H_a: $\mu_1 - \mu_2 \neq 0$	Two-tailed

Information about the *P*-value associated with this test is given in Appendix Table 6.

Assumptions: 1. *The samples are independent random samples*, or *treatments are randomly assigned to individuals or objects* (or subjects are randomly assigned to treatments).

   2. The two *population or treatment response distributions have the same shape and spread.*

■ **Example 11.13**   Parental Smoking and Infant Health

The extent to which an infant's health is affected by parents' smoking is an important public health concern. The article "Measuring the Exposure of Infants to Tobacco Smoke" (*New England Journal of Medicine* [1984]: 1075–1078) reported on a study in which various measurements were taken both from a random sample of infants who had been exposed to household smoke and from a sample of unexposed infants. The following data consist of observations on urinary concentration of cotanine, a major metabolite of nicotine (the values constitute a subset of the original data and were read from a plot that appeared in the article):

Unexposed ($n_1 = 7$)	8	11	12	14	20	43	111	
Rank	1	2	3	4	5	7	11	
Exposed ($n_2 = 8$)	35	56	83	92	128	150	176	208
Rank	6	8	9	10	12	13	14	15

Do the data suggest that the true average cotanine level is higher for exposed than for unexposed infants? The investigators used the rank-sum test to analyze the data.

1. $\mu_1 - \mu_2$ is the difference between true average cotanine concentration for unexposed and exposed infants.

2. $H_0: \mu_1 - \mu_2 = 0$.

3. $H_a: \mu_1 - \mu_2 < 0$ (unexposed average is less than exposed average).

---

*This procedure is often called the *Wilcoxon rank-sum test* or the *Mann–Whitney test*, after the statisticians who developed it. Some sources use a slightly different (but equivalent) test statistic formula.

4. Significance level: $\alpha = .01$.

5. Test statistic: rank sum = sum of Sample 1 ranks.

6. Assumptions: The authors of the article indicated that they believe the assumptions of the rank-sum test were reasonable.

7. Calculation: rank sum = $1 + 2 + 3 + 4 + 5 + 7 + 11 = 33$.

8. *P*-value: This is a lower-tailed test. With $n_1 = 7$ and $n_2 = 8$, Appendix Table 6 tells us that *P*-value $< .05$ if the rank sum ≤ 41 and that *P*-value $< .01$ if the rank sum ≤ 36. Because the rank sum = 33, we conclude that *P*-value $< .01$.

9. Conclusion: Because the *P*-value is less than α (.01), we reject H_0 and conclude that infants exposed to cigarette smoke do have a higher average cotanine level than do unexposed infants.

Many statistical computer packages can perform the rank-sum test and give exact *P*-values. Partial MINITAB output for the data of Example 11.13 follows (MINITAB uses the symbol W to denote the rank-sum statistic and the terms *ETA1* and *ETA2* in place of μ_1 and μ_2, but the test statistic value and the associated *P*-values are the same as for the test presented here):

```
Unexposed     N = 7     Median = 14.00
Exposed       N = 8     Median = 110.00
Point estimate for ETA1 − ETA2 is −79.00
< 95.7 Percent CI for ETA1 − ETA2 is (−156.00, −23.99)
W = 33.0
Test of ETA1 = ETA2 vs ETA1 < ETA2 is signfi cant at 0.0046
```

The output indicates that W = 33 and that the test is significant at .0046. These two values, 33 and .0046, are the rank-sum statistic and the *P*-value, respectively. In Example 11.13, we used Appendix Table 6 to determine that *P*-value $< .01$. This statement is consistent with the actual *P*-value given in the MINITAB output.

The test procedure just described can be easily modified to handle a hypothesized value other than 0. Consider as an example testing $H_0: \mu_1 - \mu_2 = 5$. This hypothesis is equivalent to $(\mu_1 - 5) - \mu_2 = 0$. That is, if 5 is subtracted from each Population 1 value, then according to H_0, the distribution of the resulting values coincides with the Population 2 distribution. This suggests that, if the hypothesized value of 5 is first subtracted from each Sample 1 observation, the test can then be carried out as before.

To test $H_0: \mu_1 - \mu_2 =$ hypothesized value, subtract the hypothesized value from each observation in the first sample and then determine the ranks of these observations when combined with the n_2 observations from the second sample.

■ **Example 11.14   Parental Smoking and Infant Health Revisited**

Reconsider the cotanine concentration data introduced in Example 11.13. Suppose that a researcher wished to know whether average concentration for exposed children exceeds that for unexposed children by more than 25. Recall that μ_1 denotes the true average concentration for unexposed children. The exposed average exceeds the unexposed average by 25 when $\mu_1 - \mu_2 = -25$ and by more than 25 when $\mu_1 - \mu_2 < -25$. The hypotheses of interest are therefore

$$H_0: \quad \mu_1 - \mu_2 = -25 \quad \text{versus} \quad H_a: \quad \mu_1 - \mu_2 < -25$$

These hypotheses can be tested by first subtracting -25 (or, equivalently, adding 25) to each Sample 1 observation.

Sample 1

Unexposed		8	11	12	14	20	43	111
Unexposed $- (-25)$		33	36	37	39	45	68	136
Rank		1	3	4	5	6	8	12

Sample 2

Exposed	35	56	83	92	128	150	176	208
Rank	2	7	9	10	11	13	14	15

1. $\mu_1 - \mu_2 =$ difference in true mean cotanine concentration for unexposed and exposed infants.

2. $H_0: \mu_1 - \mu_2 = -25$.

3. $H_a: \mu_1 - \mu_2 < -25$.

4. Significance level: $\alpha = .01$.

5. Test statistic: rank sum = sum of Sample 1 ranks.

6. Assumptions: Subtracting -25 does not change the shape or spread of a distribution, so if the assumptions were reasonable in Example 11.13, they are also reasonable here.

7. Calculation: rank sum $= 1 + 3 + 4 + 5 + 6 + 8 + 12 = 39$.

8. *P*-value: This is a lower-tailed test. With $n_1 = 7$ and $n_2 = 8$, Appendix Table 6 tells us that *P*-value $< .05$ if the rank sum ≤ 41 and that *P*-value $< .01$ if the rank sum ≤ 36. Because the rank sum $= 39$, we conclude that $.01 < P$-value $< .05$.

9. Conclusion: Because *P*-value $> .01$, we fail to reject H_0. Sample evidence does not suggest that the mean concentration level for exposed infants is more than 25 higher than the mean for unexposed infants.

Frequently the $n_1 + n_2$ observations in the two samples are not all different from one another. When this occurs, the rank assigned to each observation in a tied group is the average of the ranks that would be assigned if the values in the group all differed slightly from one another. Consider, for example, the 10 ordered values

5.6	6.0	6.0	6.3	6.8	7.1	7.1	7.1	7.9	8.2

If the two 6.0 values differed slightly from each other, they would be assigned ranks 2 and 3. Therefore, each one is assigned rank $\dfrac{(2 + 3)}{2} = 2.5$. If the three 7.1 observations were all slightly different, they would receive ranks 6, 7, and 8, so each of the three is assigned rank $\dfrac{(6 + 7 + 8)}{3} = 7$. The ranks for the above 10 observations are then

1	2.5	2.5	4	5	7	7	7	9	10

If the proportion of tied values is quite large, it is recommended that the rank-sum statistic be multiplied by a *correction factor*. Several of the references at the end of this chapter contain additional information.

Appendix Table 6 contains information about P-values for the rank-sum test when $n_1 \leq 8$ and $n_2 \leq 8$. More extensive tables exist for other combinations of sample size. There is also a test procedure based on using a normal distribution to approximate the sampling distribution of the rank-sum statistic. This alternative procedure is often used when the two sample sizes are larger than 8. One of the references at the end of this chapter can be consulted for details.

■ A Confidence Interval for $\mu_1 - \mu_2$

A confidence interval based on the rank-sum statistic is not as familiar to users of statistical methods as is the hypothesis-testing procedure. This is unfortunate, because the confidence interval has the same virtues to recommend it as does the rank-sum test. It does not require the assumption that the population distributions are normal, as does the two-sample t interval.

The actual derivation of the rank-sum confidence interval is quite involved, and computing these intervals by hand can be tedious, so we rely on computer software.

The **rank-sum confidence interval for $\mu_1 - \mu_2$** is the interval consisting of all hypothesized values for which

$$H_0: \quad \mu_1 - \mu_2 = \text{hypothesized value}$$

cannot be rejected when using a two-tailed test.

A 95% confidence interval consists of those hypothesized values for which the previous null hypothesis is not rejected by a test with significance level $\alpha = .05$. A 99% confidence interval is associated with a level $\alpha = .01$ test, and a 90% confidence interval is associated with a level $\alpha = .10$ test.

Thus, if $H_0: \mu_1 - \mu_2 = 100$ cannot be rejected at level .05, then 100 is included in the 95% confidence interval for $\mu_1 - \mu_2$.

Example 11.15 Strength of Bark Board

The article "Some Mechanical Properties of Impregnated Bark Board" (*Forest Products Journal* [1977]: 31–38) reported the following observations on crushing strength for epoxy-impregnated bark board (Sample 1) and bark board impregnated with another polymer (Sample 2):

Sample 1	10,860	11,120	11,340	12,130	13,070	14,380
Sample 2	4590	4850	5640	6390	6510	

A 95% confidence interval for $\mu_1 - \mu_2$ was requested using MINITAB. The resulting output follows:

```
Sample 1    N = 6    Median = 11735
Sample 2    N = 5    Median = 5640
Point estimate for ETA1 − ETA2 is 6490
96.4 percent CI for ETA1 − ETA2 is (4730, 8480)
```

cause it makes no assumption of normality. They stated that "because of the wide range of some measures, such as frequency of binges, the rank sum is more appropriate and somewhat more conservative." Data on number of binges during 1 week that are consistent with the findings of the article are given in the following table:

Placebo	8	3	15	3	4	10	6	4
Imipramine	2	1	2	7	3	12	1	5

Do these data strongly suggest that imipramine is effective in reducing the mean number of binges per week? Use a level .05 rank-sum test.

11.62 In an experiment to compare the bond strength of two different adhesives, each adhesive was used in five bondings of two surfaces, and the force necessary to separate the surfaces was determined for each bonding. For Adhesive 1, the resulting values were 229, 286, 245, 299, and 259, whereas the Adhesive 2 observations were 213, 179, 163, 247, and 225. Let μ_1 and μ_2 denote the true average bond strengths of Adhesives 1 and 2, respectively. Interpret the 90% distribution-free confidence interval estimate of $\mu_1 - \mu_2$ given in the MINITAB output shown here:

```
adhes. 1    N = 5      Median = 259.00
adhes. 2    N = 5      Median = 213.00
Point estimate for ETA1 − ETA2 is 61.00
90.5 Percent C.I. for ETA1 − ETA2 is (16.00, 95.98)
```

11.63 The article "A Study of Wood Stove Particulate Emissions" (*Journal of the Air Pollution Control Association* [1979]: 724–728) reported the following data on burn time (in hours) for samples of oak and pine:

Oak	1.72	0.67	1.55	1.56	1.42	1.23	1.77	0.48
Pine	0.98	1.40	1.33	1.52	0.73	1.20		

An estimate of the difference between mean burn time for oak and mean burn time for pine is desired. Interpret the interval given in the following MINITAB output:

```
Oak      N = 8      Median = 1.4850
Pine     N = 6      Median = 1.2650
Point estimate for ETA1 − ETA2 is 0.2100
95.5 Percent C.I. for ETA1 − ETA2 is (−0.4998, 0.5699)
```

■ 11.5 Communicating and Interpreting the Results of Statistical Analyses

Many different types of research involve comparing two populations or treatments. It is easy to find examples of the two-sample hypothesis tests introduced in this chapter in published sources in a wide variety of disciplines.

■ Communicating the Results of Statistical Analyses

As was the case with one-sample hypothesis tests, it is important to include a description of the hypotheses, the test procedure used, the value of the test statistic and the *P*-value, and a conclusion in context when summarizing the results of a two-sample test.

Correctly interpreting confidence intervals in the two-sample case is more difficult than in the one-sample case, so take particular care when providing a two-sample confidence interval interpretation. Because the two-sample confidence intervals of this chapter estimate a difference ($\mu_1 - \mu_2$ or $\pi_1 - \pi_2$), the most important thing to note is whether or not the interval includes 0. If both endpoints of the interval are positive, then it is correct to say that, based on the interval, you believe that μ_1 is greater than μ_2 (or that π_1 is greater than π_2 if you are working with proportions) and then the interval provides an estimate of how much greater. Similarly, if both interval endpoints are negative, you would say that μ_1 is less than μ_2 (or that π_1 is less than π_2), with the interval providing an estimate of the size of the difference. If 0 is included in the interval, it is plausible that μ_1 and μ_2 (or π_1 and π_2) are equal.

■ Interpreting the Results of Statistical Analyses

As with one-sample tests, it is common to find only the value of the test statistic and the associated *P*-value (or sometimes only the *P*-value) in published reports. You may have to think carefully about the missing steps to determine whether or not the conclusions are justified.

■ What to Look For in Published Data

Here are some questions to consider when you are reading a report that contains the result of a two-sample hypothesis test:

- Are only two groups being compared? If more than two groups are being compared two at a time, then a different type of analysis is preferable (see Chapter 15).
- Were the samples selected independently, or were the samples paired? If the samples were paired, was the performed analysis appropriate for paired samples?
- What hypotheses are being tested? Is the test one- or two-tailed?
- Does the validity of the test performed depend on any assumptions about the sampled populations (such as normality)? If so, do the assumptions appear to be reasonable?
- What is the *P*-value associated with the test? Does the *P*-value lead to rejection of the null hypothesis?
- Are the conclusions consistent with the results of the hypothesis test? In particular, if H_0 was rejected, does this indicate practical significance or only statistical significance?

As an example, we consider a study reported in the article "The Relationship Between Distress and Delight in Males' and Females' Reactions to Frightening Films" (*Human Communication Research* [1991]: 625–637). The investigators measured emotional responses of 50 males and 60 females after the subjects viewed a segment from a horror film. The article included the following statement: "Females were much more likely to express distress than were males. While males did express higher levels of delight than females, the difference was not statistically significant." The following summary information was also contained in the article:

Gender	Distress Index Mean	SD	Delight Index Mean	SD
Males	31.2	10.0	12.02	3.65
Females	40.4	9.1	9.09	5.55
	P-value < .001,		not significant (*P*-value > .05)	

The *P*-values are the only evidence of the hypothesis tests that support the given conclusions. The *P*-value < .001 for the distress index means that the hypothesis $H_0: \mu_F - \mu_M = 0$ was rejected in favor of $H_a: \mu_F - \mu_M > 0$, where μ_F and μ_M are the true mean distress indexes for females and males, respectively.

The nonsignificant *P*-value (*P*-value > .05) reported for the delight index means that the hypothesis $H_0: \mu_F - \mu_M = 0$ (where μ_F and μ_M now refer to mean delight index for females and males, respectively) could not be rejected. Chance sample-to-sample variability is a plausible explanation for the observed difference in sample means (12.02 − 9.09). Thus we would not want to put much emphasis on the author's statement that males express higher levels of delight than females, because it is based only on the fact that 12.02 > 9.09, which could plausibly be due entirely to chance.

The article describes the samples as consisting of undergraduates selected from the student body of a large Midwestern university. The authors extrapolate their results to American men and women in general. If this type of generalization is considered unreasonable, we could be more conservative and view the sampled populations as male and female university students or male and female Midwestern university students or even male and female students at this particular university.

The comparison of males and females was based on two independently selected groups (not paired). Because the sample sizes were large, the two-sample t test for means could reasonably have been used, and this would have required no specific assumptions about the two underlying populations.

▪ A Word to the Wise: Cautions and Limitations

The three cautions that appeared at the end of Chapter 10 apply here as well. They were (see Chapter 10 for more detail):

1. Remember that the result of a hypothesis test can never show strong support for the null hypothesis. In two-sample situations, this means that we shouldn't be *convinced* that there is no difference between population means or proportions based on the outcome of a hypothesis test.

2. If you have complete information (a census) of both populations, there is no need to carry out a hypothesis test or to construct a confidence interval — in fact, it would be inappropriate to do so.

3. Don't confuse statistical significance and practical significance. In the two-sample setting, it is possible to be convinced that two population means or proportions are not equal even in situations where the actual difference between them is small enough that it is of no practical interest. After rejecting a null hypothesis of no difference (statistical significance), it is useful to look at a confidence interval estimate of the difference to get a sense of practical significance.

And here's one new caution to keep in mind for two-sample tests:

4. Be sure to think carefully about how the data were collected, and make sure that an appropriate test procedure or confidence interval is used. A common mistake is to overlook pairing and to analyze paired samples as if they were independent. The question, Are the samples paired? is usually easy to answer — you just have to remember to ask!

▪ Activity 11.1: Helium-Filled Footballs? _____

Technology activity: Requires Internet access.

Background: Do you think that a football filled with helium will travel farther than a football filled with air? Two researchers at the Ohio State University investigated this question by performing an experiment in which 39 people each kicked a helium-filled football and an air-filled football. Half were assigned to kick the air-filled football first and then the helium-filled ball, whereas the other half kicked the helium-filled ball first followed by the air-filled ball. Distance (in yards) was measured for each kick.

In this activity, you will use the Internet to obtain the data from this experiment and then carry out a hypothesis test to determine whether the mean distance is greater for helium-filled footballs than for air-filled footballs.

1. Do you think that helium-filled balls will tend to travel farther than air-filled balls when kicked? Before looking at the data, write a few sentences indicating what you think the outcome of this experiment was and describing the reasoning that supports your prediction.

2. The data from this experiment can be found in the Data and Story Library at the following web site:

   http://lib.stat.cme.edu/DASL/Datafiles/
   heliumfootball.html

Go to this web site and print out the data for the 39 trials.

3. There are two samples in this data set. One consists of distances traveled for the 39 kicks of the air-filled football, and the other consists of the 39 distances for the helium-filled football. Are these samples independent or paired? Explain.

4. Carry out an appropriate hypothesis test to determine whether there is convincing evidence that the mean distance traveled is greater for a helium-filled football than for an air-filled football.

5. Is the conclusion in the test of Step 4 consistent with your initial prediction of the outcome of this experiment? Explain.

6. Write a paragraph for the sports section of your school newspaper describing this experiment and the conclusions that can be drawn from it.

▪ Activity 11.2: Thinking About Data Collection

Background: In this activity you will design two experiments that would allow you to investigate whether people tend to have quicker reflexes when reacting with their dominant hand than with their nondominant hand.

1. Working in a group, design an experiment to investigate the given research question that would result in independent samples. Be sure to describe how you plan to measure quickness of reflexes, what extraneous variables will be directly controlled, and the role that randomization plays in your design.

2. How would you modify the design from Step 1 so that the resulting data are paired? Is the way in which randomization is incorporated into the new design different from the way it is incorporated in the design from Step 1? Explain.

3. Which of the two proposed designs would you recommend, and why?

4. If assigned to do so by your instructor, carry out one of your experiments and analyze the resulting data. Write a brief report that describes the experimental design, includes both graphical and numerical summaries of the resulting data, and communicates the conclusions that follow from your data analysis.

▪ Summary of Key Concepts and Formulas

Term or Formula	Comment
Independent samples	Two samples for which the individuals or objects in the first sample are selected independently from those in the second sample.
Paired samples	Two samples for which each observation in one sample is paired in a meaningful way with a particular observation in the second sample.
$t = \dfrac{(\bar{x}_1 - \bar{x}_2) - \text{hypothesized value}}{\sqrt{\dfrac{s_1^2}{n_1} + \dfrac{s_2^2}{n_2}}}$	The test statistic for testing H_0: $\mu_1 - \mu_2 = $ hypothesized value when the samples are independently selected and when the sample sizes are large or it is reasonable to assume that both population distributions are normal.

Term or Formula	**Comment**

$$(\bar{x}_1 - \bar{x}_2) \pm (t \text{ critical value})\sqrt{\frac{s_1^2}{n_1} + \frac{s_2^2}{n_2}}$$

A formula for constructing a confidence interval for $\mu_1 - \mu_2$ when the samples are independently selected and when the sample sizes are large or it is reasonable to assume that the population distributions are normal.

$$df = \frac{(V_1 + V_2)^2}{\frac{V_1^2}{n_1 - 1} + \frac{V_2^2}{n_2 - 1}}$$

The formula for determining the number of degrees of freedom for the two-sample t test and confidence interval.

where $V_1 = \frac{s_1^2}{n_1}$ and $V_2 = \frac{s_2^2}{n_2}$

$\bar{x}_d$

The sample mean difference.

s_d

The standard deviation of the sample differences.

μ_d

The mean value for the population of differences.

σ_d

The standard deviation for the population of differences.

$$t = \frac{\bar{x}_d - \text{hypothesized value}}{\frac{s_d}{\sqrt{n}}}$$

The paired t test statistic for testing

$$H_0: \quad \mu_d = \text{hypothesized value}$$

$$\bar{x}_d \pm (t \text{ critical value})\frac{s_d}{n}$$

The paired t confidence interval formula.

$$p_c = \frac{n_1 p_1 + n_2 p_2}{n_1 + n_2}$$

The statistic for estimating the common population proportion when $\pi_1 = \pi_2$.

$$z = \frac{p_1 - p_2}{\sqrt{\frac{p_c(1 - p_c)}{n_1} + \frac{p_c(1 - p_c)}{n_2}}}$$

The test statistic for testing

$$H_0: \quad \pi_1 - \pi_2 = 0$$

when both sample sizes are large.

$$(p_1 - p_2) \pm (z \text{ critical value})\sqrt{\frac{p_1(1 - p_1)}{n_1} + \frac{p_2(1 - p_2)}{n_2}}$$

A formula for constructing a confidence interval for $\pi_1 - \pi_2$ when both sample sizes are large.

Rank sum = sum of Sample 1 ranks

The distribution-free test statistic for testing

$$H_0: \quad \mu_1 - \mu_2 = \text{hypothesized value}$$

when it is reasonable to assume that the two populations have the same shape and spread.

▪ Supplementary Exercises 11.64–11.80

11.64 The article "Softball Sliding Injuries" (*American Journal of Diseases of Children* [1988]: 715–716) provided a comparison of breakaway bases (designed to reduce injuries) and stationary bases. Consider the following data (which agree with summary values given in the paper):

	Number of Games Played	Number of Games Where a Player Suffered a Sliding Injury
Stationary Bases	1250	90
Breakaway Bases	1250	20

Does the use of breakaway bases reduce the proportion of games in which a player suffers a sliding injury? Answer by performing a level .01 test. What assumptions are necessary for your test to be valid? Do you think they are reasonable in this case?

11.65 The positive effect of water fluoridation on dental health is well documented. One study that validates this is described in the article "Impact of Water Fluoridation on Children's Dental Health: A Controlled Study of Two Pennsylvania Communities" (*1981 Proceedings of the Section on Social Statistics* [Alexandria, VA: American Statistical Association, 1981]: 262–265). Two communities were compared. One had adopted fluoridation in 1966, whereas the other had no such program. Of 143 randomly selected children from the town without fluoridated water, 106 had decayed teeth, and 67 of 119 randomly selected children from the town with fluoridated water had decayed teeth. Let π_1 denote the true proportion of children drinking fluoridated water who have decayed teeth, and let π_2 denote the analogous proportion for children drinking unfluoridated water. Estimate $\pi_1 - \pi_2$ using a 90% confidence interval. Does the interval contain 0? Interpret the interval.

11.66 Wayne Gretzky was one of ice hockey's most prolific scorers when he played for the Edmonton Oilers. During his last season with the Oilers, Gretzky played in 41 games and missed 17 games because of injury. The article "The Great Gretzky" (*Chance* [1991]: 16–21) looked at the number of goals scored by the Oilers in games with and without Gretzky, as shown in the following table:

	n	Sample Mean	Sample sd
Games with Gretzky	41	4.73	1.29
Games without Gretzky	17	3.88	1.18

If we view the 41 games with Gretzky as a random sample of all Oiler games in which Gretzky played and the 17 games without Gretzky as a random sample of all Oiler games in which Gretzky did not play, is there evidence that the mean number of goals scored by the Oilers is higher for games in which Gretzky played? Use $\alpha = .01$.

11.67 Here's one to sink your teeth into: The authors of the article "Analysis of Food Crushing Sounds During Mastication: Total Sound Level Studies" (*Journal of Texture Studies* [1990]: 165–178) studied the nature of sounds generated during eating. Peak loudness (in decibels at 20 cm away) was measured for both open-mouth and closed-mouth chewing of potato chips and of tortilla chips. Forty subjects participated, with 10 assigned at random to each combination of conditions (such as closed-mouth, potato chip, and so on). We are not making this up! Summary values taken from plots given in the article appear in the following table:

	n	$\bar{x}$	s
Potato Chip			
Open mouth	10	63	13
Closed mouth	10	54	16
Tortilla Chip			
Open mouth	10	60	15
Closed mouth	10	53	16

a. Construct a 95% confidence interval for the difference in mean peak loudness between open-mouth and closed-mouth chewing of potato chips. Interpret the resulting interval.
b. For closed-mouth chewing (the recommended method!), is there sufficient evidence to indicate that there is a difference between potato chips and tortilla chips with respect to mean peak loudness? Test the relevant hypotheses using $\alpha = .01$.
c. The means and standard deviations given here were actually for stale chips. When 10 measurements of peak loudness were recorded for closed-mouth chewing of fresh tortilla chips, the resulting mean and standard deviation were 56 and 14, respectively.

Is there sufficient evidence to conclude that chewing fresh tortilla chips is louder than chewing stale chips? Use $\alpha = .05$.

11.68 Thirteen males and 14 females participated in a study of grip and leg strength ("Sex Differences in Static Strength and Fatigability in Three Different Muscle Groups," *Research Quarterly for Exercise and Sport* [1990]: 238–242). Right-leg strength (in newtons) was recorded for each participant, resulting in the summary statistics given in the following table:

	n	$\bar{x}$	s
Males	13	2127	513
Females	14	1843	446

Estimate the difference in mean right-leg strength between males and females using a 95% confidence interval. Interpret the resulting interval.

11.69 Are young infants more likely to imitate actions that are modeled by a person or simulated by an object? This question was the basis of a research study summarized in the article "The Role of Person and Object in Eliciting Early Imitation" (*Journal of Experimental Child Psychology* [1991]: 423–433). One action examined was mouth opening. This action was modeled repeatedly by either a person or a doll, and the number of times that the infant imitated the behavior was recorded. Twenty-seven infants participated, with 12 exposed to a human model and 15 exposed to the doll. Summary values are given here:

	Person Model	Doll Model
$\bar{x}$	5.14	3.46
s	1.60	1.30

Is there sufficient evidence to conclude that the mean number of imitations is higher for infants who watch a human model than for infants who watch a doll? Test the relevant hypotheses using a .01 significance level.

11.70 Dentists make many people nervous (even more so than statisticians!). To see whether such nervousness elevates blood pressure, the blood pressure and pulse rates of 60 subjects were measured in a dental setting and in a medical setting ("The Effect of the Dental Setting on Blood Pressure Measurement," *American Journal of Public Health* [1983]: 1210–1214). For each subject, the difference (dental-setting blood pressure minus medical-setting blood pressure) was calculated. The analogous differences were also calculated for pulse rates. Summary statistics follow:

	Mean Difference	Standard Deviation of Differences
Systolic Blood Pressure	4.47	8.77
Pulse (beats/min)	−1.33	8.84

a. Do the data strongly suggest that true mean blood pressure is higher in a dental setting than in a medical setting? Use a level .01 test.
b. Is there sufficient evidence to indicate that true mean pulse rate in a dental setting differs from the true mean pulse rate in a medical setting? Use a significance level of .05.

11.71 An eating attitudes test (EAT) was administered to both a sample of female models and a control group of females, resulting in the following summary statistics ("Gender Differences in Eating Attitudes, Body Concept, and Self-Esteem Among Models," *Sex Roles* [1992]: 413–437):

Group	Sample Size	Sample Mean	Sample sd
Models	30	8.63	7.1
Controls	30	10.97	8.9

Let μ_1 denote the true average EAT score for models, and let μ_2 denote the true average EAT score for controls.
a. Verify that the value of the test statistic for testing $H_0: \mu_1 - \mu_2 = 0$ versus $H_a: \mu_1 - \mu_2 < 0$ is $t = -1.13$.
b. Calculate the P-value, and then use it to reach a conclusion at significance level .05.
c. If μ_1 had been the true average score for controls and if μ_2 had been the true average score for models, how would the hypotheses, value of t, and P-value have changed?

11.72 Do teenage boys worry more than teenage girls? This is one of the questions addressed by the authors of the article "The Relationship of Self-Esteem and Attributional Style to Young People's Worries" (*Journal of Psychology* [1987]: 207–215). A scale called the Worries Scale was administered to a group of teenagers, and the results are summarized in the following table:

Sex	n	Sample Mean Score	Sample sd
Girls	108	62.05	9.5
Boys	78	67.59	9.7

Is there sufficient evidence to conclude that teenage boys score higher on the Worries Scale than teenage girls? Use a significance level of $\alpha = .05$.

11.73 Data on self-esteem, leadership ability, and grade point average (GPA) were used to compare college students who were hired as resident assistants at Mississippi State University and those who were not hired ("Wellness as a Factor in University Housing," *Journal of College Student Development* [1994]: 248–254), as shown in the following table:

	Self-Esteem		Leadership		GPA	
	$\bar{x}$	s	$\bar{x}$	s	$\bar{x}$	s
Hired ($n = 69$)	83.28	12.21	62.51	3.05	2.94	0.61
Not Hired ($n = 47$)	81.96	12.78	62.43	3.36	2.60	0.79

a. Do the data suggest that the true mean self-esteem score for students hired by the university as resident assistants differs from the mean for students who are not hired? Test the relevant hypotheses using $\alpha = .05$.

b. Do the data suggest that the true mean leadership score for students hired by the university as resident assistants differs from the mean for students who are not hired? Test the relevant hypotheses using $\alpha = .05$.

c. Do the data suggest that the true mean GPA for students hired by the university as resident assistants differs from the mean for students who are not hired? Test the relevant hypotheses using $\alpha = .05$.

11.74 The article "Religion and Well-Being Among Canadian University Students: The Role of Faith Groups on Campus" (*Journal for the Scientific Study of Religion* [1994]: 62–73) compared the self-esteem of students who belonged to Christian clubs and students who did not belong to such groups. Each student in a random sample of $n = 169$ members of Christian groups (the affiliated group) completed a questionnaire designed to measure self-esteem. The same questionnaire was also completed by each student in a random sample of $n = 124$ students who did not belong to a religious club (the unaffiliated group). The mean self-esteem score for the affiliated group was 25.08, and the mean for the unaffiliated group was 24.55. The sample standard deviations weren't given in the article, but suppose that they were 10 for the affiliated group and 8 for the unaffiliated group. Is there evidence that the true mean self-esteem score differs for affiliated and unaffiliated students? Test the relevant hypotheses using a significance level of .01.

11.75 The article "Social Science Research on Lay Definitions of Sexual Harassment" (*Journal of Social Issues* [1995]: 21–37) reported on a survey of randomly selected undergraduate and graduate students. Each student was asked whether he or she thought that unwanted pressure for a date from a faculty member was an example of sexual harassment. The data are summarized in the following table:

	n	Number Who Perceived Pressure as Sexual Harassment
Undergraduates		
Males	1336	882
Females	1346	1104
Graduate Students		
Males	565	469
Females	575	529

a. For undergraduate students, is there evidence that the proportion who perceive pressure for a date as sexual harassment is lower for males than for females? Test using $\alpha = .05$.

b. For graduate students, is there evidence that the proportion who perceive pressure for a date as sexual harassment is lower for males than for females? Test using $\alpha = .05$.

c. Estimate the difference between the proportion of male graduate students who perceive pressure for a date as sexual harassment and the corresponding proportion for male undergraduate students using a 90% confidence interval. Interpret the resulting interval.

11.76 The following summary data on the ratio of strength to cross-sectional area for knee extensors are taken from the article "Knee Extensor and Knee Flexor Strength: Cross-Sectional Area Ratios in Young and Elderly Men" (*Journal of Gerontology* [1992]: M204–M210):

Group	n	$\bar{x}$	Standard Error, $s/\sqrt{n}$
Young men	13	7.47	0.22
Elderly men	12	6.71	0.28

Do these data suggest that the true average ratio for young men exceeds that for elderly men? Carry out a test of appropriate hypotheses using $\alpha = .05$.

11.77 The authors of the article "The Color-a-Person Body Dissatisfaction Test" (*Journal of Personality Assessment* [1991]: 395–413) used a nonverbal measure of body dissatisfaction to compare students at two different universities. A random sample of 47 women at the University of Cincinnati and a random sample of 24 women at the University of New Mexico participated in the study. Mean and standard deviations for a measure of self-esteem are given in the following table:

	Self-Esteem Score	
	$\bar{x}$	s
University of Cincinnati	32.55	4.41
University of New Mexico	31.25	4.92

Is it reasonable to conclude that women at the two universities differ with respect to mean self-esteem score? Use $\alpha = .05$.

11.78 Key terms in survey questions too often are not well understood, and such ambiguity can affect responses. As an example, the article "How Unclear Terms Affect Survey Data" (*Public Opinion Quarterly* [1992]: 218–231) described a survey in which each individual in a sample was asked, "Do you exercise or play sports regularly?" But what constitutes exercise? The following revised question was then asked of each individual in the same sample: "Do you do any sports or hobbies involving physical activities, or any exercise, including walking, on a regular basis?" The resulting data are shown in the following table:

	Yes	No
Initial Question	48	52
Revised Question	60	40

Is there any difference between the true proportions of yes responses to these questions? Can a procedure from this chapter be used to answer the question posed? If yes, use it; if not, explain why not.

11.79 An electronic implant that stimulates the auditory nerve has been used to restore partial hearing to a number of deaf people. In a study of implant acceptability (*Los Angeles Times*, January 29, 1985), 250 adults who were born deaf and 250 adults who went deaf after learning to speak were followed for a period of time after receiving an implant. Of those who were deaf from birth, 75 had removed the implant, whereas only 25 of those who went deaf after learning to speak had done so. Does this suggest that the true proportion who remove the implants differs for those who were born deaf and those who went deaf after learning to speak? Test the relevant hypotheses using a .01 significance level.

11.80 Samples of both surface soil and subsoil were taken from eight randomly selected agricultural locations in a particular county. The soil samples were analyzed to determine both surface pH and subsoil pH, with the results shown in the following table:

	Location			
	1	2	3	4
Surface pH	6.55	5.98	5.59	6.17
Subsoil pH	6.78	6.14	5.80	5.91

	Location			
	5	6	7	8
Surface pH	5.92	6.18	6.43	5.68
Subsoil pH	6.10	6.01	6.18	5.88

a. Compute a 90% confidence interval for the true average difference between surface and subsoil pH for agricultural land in this county.
b. What assumptions are necessary to validate the interval in Part (a)?

■ References

Conover, W. J., *Practical Nonparametric Statistics*, 3d ed. New York: Wiley, 1999. (An accessible presentation of distribution-free methods.)

Daniel, Wayne. *Applied Nonparametric Statistics*, 2d ed. Boston: PWS-Kent, 1990. (An elementary presentation of distribution-free methods, including the rank-sum test discussed in Section 11.4.)

Devore, Jay. *Probability and Statistics for Engineering and the Sciences*, 6th ed. Belmont, Calif.: Duxbury Press, 2004. (Contains a somewhat more comprehensive treatment of the inferential material presented in this and the previous two chapters, although the notation is a bit more mathematical than that of the present textbook.)

Mosteller, Frederick, and Richard Rourke. *Sturdy Statistics*. Reading, Mass.: Addison-Wesley, 1973. (A readable, intuitive development of distribution-free methods, including those based on ranks.)

12 · The Analysis of Categorical Data and Goodness-of-Fit Tests

Most of the techniques presented in previous chapters are designed for numerical data. It is often the case, however, that information is collected on categorical variables, such as political affiliation, gender, or college major. As with numerical data, categorical data sets can be univariate (consisting of observations on a single categorical variable), bivariate (observations on two categorical variables), or even multivariate. In this chapter, we first consider inferential methods for analyzing univariate categorical data sets and then turn to techniques appropriate for use with bivariate categorical data.

▪ 12.1 Chi-Square Tests for Univariate Categorical Data

Univariate categorical data sets arise in a variety of settings. If each student in a sample of 100 is classified according to whether he or she is enrolled full-time or part-time, data on a categorical variable with two categories result. Each airline passenger in a sample of 50 might be classified into one of three categories based on type of ticket — coach, business class, or first class. Each registered voter in a sample of 100 selected from those registered in a particular city might be asked which of the five city council members he or she favors for mayor. This would yield observations on a categorical variable with five categories.

Univariate categorical data are most conveniently summarized in a **one-way frequency table**. Suppose, for example, that each item returned to a department store is classified according to how it was resolved — cash refund, credit to charge account, merchandise exchange, or return refused. Records of 100 randomly selected returns are examined, and each outcome is recorded. The first few observations might be

Cash refund	Return refused	Exchange
Cash refund	Exchange	Credit to account

515

Counting the number of observations of each type might then result in the following one-way table:

	\multicolumn{4}{c}{Outcome}
	Cash	Credit	Exchange	Refused
Frequency	34	18	31	17

For a categorical variable with k possible values (k different levels or categories), sample data are summarized in a one-way frequency table consisting of k cells, displayed either horizontally or vertically.

In this section, we consider testing hypotheses about the proportion of the population that falls into each of the possible categories. For example, the customer relations manager for a department store might be interested in determining whether the four possible dispositions for a return request occur equally often. If this is indeed the case, the long-run proportion of returns falling into each of the four categories is 1/4, or .25. The test procedure to be presented shortly would allow the manager to decide whether the hypothesis that all four category proportions are equal to .25 is plausible.

▪ **Notation**

k = number of categories of a categorical variable.

π_1 = true proportion for Category 1

π_2 = true proportion for Category 2

⋮

π_k = true proportion for Category k

(Note: $\pi_1 + \pi_2 + \cdots + \pi_k = 1$.)

The hypotheses to be tested have the form

H_0: π_1 = hypothesized proportion for Category 1

π_2 = hypothesized proportion for Category 2

⋮

π_k = hypothesized proportion for Category k

H_a: H_0 is not true, so at least one of the true category proportions differs from the corresponding hypothesized value.

For the example involving department store returns, let

π_1 = the proportion of all returns that result in a cash refund

π_2 = the proportion of all returns that result in an account credit

π_3 = the proportion of all returns that result in an exchange

and

π_4 = the proportion of all returns that are refused

The null hypothesis of interest is then

$$H_0: \quad \pi_1 = .25, \pi_2 = .25, \pi_3 = .25, \pi_4 = .25$$

A null hypothesis of the type just described can be tested by first selecting a random sample of size n and then classifying each sample response into one of the k possible categories. To decide whether the sample data are compatible with the null hypothesis, we compare the observed cell counts (frequencies) to the cell counts that would have been expected when the null hypothesis is true. The expected cell counts are

Expected cell count for Category 1 $= n\pi_1$

Expected cell count for Category 2 $= n\pi_2$

and so on. The expected cell counts when H_0 is true result from substituting the corresponding hypothesized proportion for each π_i.

▪ **Example 12.1** **Birthdays and Psychiatric Admissions**

A number of psychological studies have considered the relationship between various deviant behaviors and other variables, such as the lunar phase. The article "Psychiatric and Alcoholic Admissions Do Not Occur Disproportionately Close to Patients' Birthdays" (*Psychological Reports* [1992]: 944–946) focused on the existence of any relationship between date of patient admission for specified treatment and patient's birthday. Admission date was partitioned into four categories according to how close it was to the patient's birthday:

1. Within 7 days of birthday
2. Between 8 and 30 days, inclusive, from the birthday
3. Between 31 and 90 days, inclusive, from the birthday
4. More than 90 days from the birthday

Let π_1, π_2, π_3, and π_4 denote the true proportions in Categories 1, 2, 3, and 4, respectively. If there is no relationship between admission date and birthday, then, because there are 15 days included in the first category (from 7 days before the patient's birthday to 7 days after, including, of course, the birthday itself),

$$\pi_1 = \frac{15}{365} = .041$$

(For simplicity, we are ignoring February 29 of leap years.) Similarly, in the absence of any relationship,

$$\pi_2 = \frac{46}{365} = .126$$

$$\pi_3 = .329$$

$$\pi_4 = .504$$

The hypotheses of interest are then

$$H_0: \quad \pi_1 = .041, \pi_2 = .126, \pi_3 = .329, \pi_4 = .504$$

$$H_a: \quad H_0 \text{ is not true}$$

The cited article gave data for $n = 200$ patients admitted for alcoholism treatment. If H_0 is true, the expected counts are

$$\left(\begin{array}{c}\text{expected count}\\\text{for Category 1}\end{array}\right) = n\left(\begin{array}{c}\text{hypothesized proportion}\\\text{for Category 1}\end{array}\right) = 200(.041) = 8.2$$

$$\left(\begin{array}{c}\text{expected count}\\\text{for Category 2}\end{array}\right) = n\left(\begin{array}{c}\text{hypothesized proportion}\\\text{for Category 2}\end{array}\right) = 200(.126) = 25.2$$

$$\left(\begin{array}{c}\text{expected count}\\\text{for Category 3}\end{array}\right) = n\left(\begin{array}{c}\text{hypothesized proportion}\\\text{for Category 3}\end{array}\right) = 200(.329) = 65.8$$

$$\left(\begin{array}{c}\text{expected count}\\\text{for Category 4}\end{array}\right) = n\left(\begin{array}{c}\text{hypothesized proportion}\\\text{for Category 4}\end{array}\right) = 200(.504) = 100.8$$

Observed and expected cell counts are given in the following table:

	Category			
	1	2	3	4
Observed Count	11	24	69	96
Expected Count	8.2	25.2	65.8	100.8

Because the observed counts are based on a *sample* of admissions, it would be somewhat surprising to see *exactly* 4.1% of the sample falling in the first category, exactly 12.6% in the second, and so on, even when H_0 is true. If the differences between the observed and expected cell counts can reasonably be attributed to sampling variation, the data are considered compatible with H_0. On the other hand, if the discrepancy between the observed and the expected cell counts is too large to be attributed solely to chance differences from one sample to another, H_0 should be rejected in favor of H_a. Thus, we need an assessment of how different the observed and expected counts are.

The goodness-of-fit statistic, denoted by X^2, is a quantitative measure of the extent to which the observed counts differ from those expected when H_0 is true. (The Greek letter χ is often used in place of X. The symbol X^2 is referred to as the chi-square [χ^2] statistic. In using X^2 rather than χ^2, we are adhering to the convention of denoting sample quantities by Roman letters.)

The **goodness-of-fit statistic**, X^2, results from first computing the quantity

$$\frac{(\text{observed cell count} - \text{expected cell count})^2}{\text{expected cell count}}$$

for each cell, where, for a sample of size n,

$$\left(\begin{array}{c}\text{expected cell}\\\text{count}\end{array}\right) = n\left(\begin{array}{c}\text{hypothesized value of corresponding}\\\text{population proportion}\end{array}\right)$$

The X^2 statistic is the sum of these quantities for all k cells:

$$X^2 = \sum_{\text{all cells}} \frac{(\text{observed cell count} - \text{expected cell count})^2}{\text{expected cell count}}$$

The value of the X^2 statistic reflects the magnitude of the discrepancies between observed and expected cell counts. When the differences are sizable, the value of X^2 tends to be large. Therefore, large values of X^2 suggest rejection of H_0. A small value of X^2 (it can never be negative) occurs when the observed cell counts are quite similar to those expected when H_0 is true and so would be consistent with H_0.

As with previous test procedures, a conclusion is reached by comparing a P-value to the significance level for the test. The P-value is computed as the probability of observing a value of X^2 at least as large as the observed value when H_0 is true. This requires information about the sampling distribution of X^2 when H_0 is true. When the null hypothesis is correct and the sample size is sufficiently large, the behavior of X^2 is described approximately by a **chi-square distribution**. A chi-square curve has no area associated with negative values and is asymmetric, with a longer tail on the right. There are actually many chi-square distributions, each one identified with a different number of degrees of freedom. Curves corresponding to several chi-square distributions are shown in Figure 12.1.

Figure 12.1 Chi-square curves.

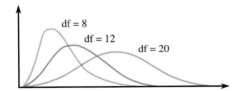

For a test procedure based on the X^2 statistic, the associated P-value is the area under the appropriate chi-square curve and to the right of the computed X^2 value. Appendix Table 9 gives upper-tail areas for chi-square distributions with up to 20 df. Our chi-square table has a different appearance from the t table used in previous chapters. In the t table, there is a single "value" column on the far left and then a column of P-values (tail areas) for each different number of degrees of freedom. A single column of t values works for the t table because all t curves are centered at 0, and the t curves approach the z curve as the number of degrees of freedom increases. However, because the chi-square curves move farther and farther to the right and spread out more as the number of degrees of freedom increases, a single "value" column is impractical in this situation.

To find the area to the right for a particular X^2 value, locate the appropriate df column in Appendix Table 9. Determine which listed value is closest to the X^2 value of interest, and read the right-tail area corresponding to this value from the left-hand column of the table. For example, for a chi-square distribution with df = 2, the area to the right of $X^2 = 4.93$ is .085, as shown in Figure 12.2. For this same chi-square distribution (df = 2), the area to the right of 6.25 is approximately .045 (the area to the right of 6.20, the closest entry in the table for df = 2).

Figure 12.2 A chi-square upper-tail area.

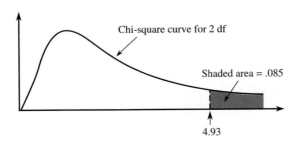

■ Goodness-of-Fit Tests

When H_0 is true, the X^2 goodness-of-fit statistic has approximately a chi-square distribution with df $= (k - 1)$, as long as none of the expected cell counts are too small. *It is generally agreed that use of the chi-square distribution is appropriate when the sample size is large enough for every expected cell count to be at least 5.* If any of the expected cell frequencies are less than 5, categories can be combined in a sensible way to create acceptable expected cell counts. Just remember to compute the number of degrees of freedom based on the reduced number of categories.

■ Goodness-of-Fit Test Procedure

Hypotheses: H_0: $\pi_1 =$ hypothesized proportion for Category 1
$\vdots$

$\pi_k =$ hypothesized proportion for Category k

H_a: H_0 is not true

Test statistic: $X^2 = \sum_{\text{all cells}} \dfrac{(\text{observed cell count} - \text{expected cell count})^2}{\text{expected cell count}}$

P-values: When H_0 is true and all expected counts are at least 5, X^2 has approximately a chi-square distribution with df $= k - 1$. Therefore, the P-value associated with the computed test statistic value is the area to the right of X^2 under the df $= k - 1$ chi-square curve. Upper-tail areas for chi-square distributions are found in Appendix Table 9.

Assumptions:
1. Observed cell counts are based on a *random sample*.
2. The *sample size is large*. The sample size is large enough for the chi-square test to be appropriate as long as every expected cell count is at least 5.

■ Example 12.2 Birthdays and Psychiatric Admissions Revisited

We use the admissions data of Example 12.1 to test the hypothesis that admission date is unrelated to birthday. Let's use a .05 level of significance and the nine-step hypothesis-testing procedure illustrated in previous chapters.

1. Let π_1, π_2, π_3, and π_4 denote the proportions of all admissions for treatment of alcoholism falling in the four categories.
2. H_0: $\pi_1 = .041$, $\pi_2 = .126$, $\pi_3 = .329$, $\pi_4 = .504$.
3. H_a: H_0 is not true.
4. Significance level: $\alpha = .05$.
5. Test statistic: $X^2 = \sum_{\text{all cells}} \dfrac{(\text{observed cell count} - \text{expected cell count})^2}{\text{expected cell count}}$
6. Assumptions: The expected cell counts (from Example 12.1) are 8.2, 25.2, 65.8, and 100.8, all of which are greater than 5. The article did not indicate how the patients were selected. We can proceed with the chi-square test if it is reasonable to assume that the 200 patients in the sample can be regarded as a random sample of patients admitted for treatment of alcoholism.

7. Calculation:

$$X^2 = \frac{(11-8.2)^2}{8.2} + \frac{(24-25.2)^2}{25.2} + \frac{(69-65.8)^2}{65.8} + \frac{(96-100.8)^2}{100.8}$$

$$= 0.96 + 0.06 + 0.16 + 0.23$$

$$= 1.41$$

8. *P*-value: The *P*-value is based on a chi-square distribution with df $= 4-1 = 3$. The computed value of X^2 is smaller than 6.25 (the smallest entry in the df $= 3$ column of Appendix Table 9), so *P*-value $> .10$.

9. Conclusion: Because *P*-value $> \alpha$, H_0 cannot be rejected. There is not sufficient evidence to conclude that date admitted for treatment and birthday are related.

▪ Example 12.3   Color of Stolen Cars

Does the color of a car influence the chance that it will be stolen? The Associated Press (*San Luis Obispo Telegram-Tribune*, September 2, 1995) reported the following information for a random sample of 830 stolen vehicles: 140 were white, 100 were blue, 270 were red, 230 were black, and 90 were other colors. We use the X^2 goodness-of-fit test and a significance level of $\alpha = .01$ to test the hypothesis that proportions stolen are identical to population color proportions. Suppose that it is known that 15% of all cars are white, 15% are blue, 35% are red, 30% are black, and 5% are other colors. If these same population color proportions hold for stolen cars, the expected counts are

Expected count for white $= 830(0.15) = 124.5$
Expected count for blue $= 830(0.15) = 124.5$
Expected count for red $= 830(0.35) = 290.5$
Expected count for black $= 830(0.30) = 249.0$
Expected count for other colors $= 830(0.05) = 41.5$

These expected counts have been entered in Table 12.1.

Table 12.1 ▪ Observed and Expected Counts for Example 12.3

Category	Color	Observed Count	Expected Count
1	White	140	124.5
2	Blue	100	124.5
3	Red	270	290.5
4	Black	230	249.0
5	Other colors	90	41.5

1. Let $\pi_1, \pi_2, \ldots, \pi_5$ denote the true proportions of stolen cars that fall into the five color categories.
2. H_0: $\pi_1 = .15$, $\pi_2 = .15$, $\pi_3 = .35$, $\pi_4 = .30$, $\pi_5 = .05$
3. H_a: H_0 is not true.
4. Significance level: $\alpha = .01$.

5. Test statistic: $X^2 = \sum_{\text{all cells}} \dfrac{(\text{observed cell count} - \text{expected cell count})^2}{\text{expected cell count}}$

6. Assumptions: The sample was a random sample of stolen vehicles. All expected counts are greater than 5, so the sample size is large enough to use the chi-square test.

7. Calculation:

$$X^2 = \frac{(140 - 124.5)^2}{124.5} + \frac{(100 - 124.5)^2}{124.5} + \frac{(270 - 290.5)^2}{290.5} + \frac{(230 - 249.0)^2}{249.0}$$

$$+ \frac{(90 - 41.5)^2}{41.5}$$

$$= 1.93 + 4.82 + 1.45 + 1.45 + 56.68$$

$$= 66.33$$

8. *P*-value: All expected counts exceed 5, so the *P*-value can be based on a chi-square distribution with df = 5 − 1 = 4. The computed value is larger than 18.46, the largest value in the df = 4 column of Appendix Table 9, so *P*-value < .001.

9. Conclusion: Because *P*-value ≤ α, H_0 is rejected. There is convincing evidence that at least one of the color proportions for stolen cars differs from the corresponding proportion for all cars.

▪ Exercises 12.1–12.13

12.1 From the given information in each of the following cases (a)–(e), state what you know about the *P*-value for a chi-square test and give the conclusion for a significance level of $\alpha = .01$:

a. $X^2 = 7.5$, df = 2   b. $X^2 = 13.0$, df = 6
c. $X^2 = 18.0$, df = 9   d. $X^2 = 21.3$, df = 4
e. $X^2 = 5.0$, df = 3

12.2 A particular paperback book is published in a choice of four different covers. A certain bookstore keeps copies of each cover on its racks. To test the hypothesis that sales are equally divided among the four choices, a random sample of 100 purchases is identified.

a. If the resulting X^2 value is 6.4, what conclusion would you reach when using a test with significance level .05?
b. What conclusion would be appropriate at significance level .01 if $X^2 = 15.3$?
c. If there were six different covers rather than just four, what would you conclude if $X^2 = 13.7$ and a test with $\alpha = .05$ was used?

12.3 Packages of mixed nuts made by a certain company contain four types of nuts. The percentages of nuts of Types 1, 2, 3, and 4 are supposed to be 40%, 30%, 20%, and 10%, respectively. A random

sample of nuts is selected, and each one is categorized by type.
a. If the sample size is 200 and the resulting test statistic value is $X^2 = 19.0$, what conclusion would be appropriate for a significance level of .001?
b. If the random sample had consisted of only 40 nuts, would you use the chi-square test here? Explain your reasoning.

12.4 The article "Family Planning: Football Style — The Relative Age Effect in Football" (*International Review for Sociology of Sport* [1992]: 77–88) investigated the relationship between month of birth and achievement in sports. Birth dates were collected for players on teams competing in the 1990 World Cup soccer games, and they are summarized in the following table:

Birthday	Frequency
Quarter 1 (August–October)	150
Quarter 2 (November–January)	138
Quarter 3 (February–April)	140
Quarter 4 (May–July)	100

The authors of the article used a chi-square goodness-of-fit test to determine whether the data are consistent with the hypothesis that the birthdays of world-class soccer players are evenly distributed across the four quarters.

a. Perform a test of this hypothesis, using a .05 significance level.

b. The article stated that this analysis demonstrates that the distribution of players' birthdays is not random and that the number of players is related to the "quarters of the football year." Does your analysis support this conclusion? Explain.

12.5 Each observation in a random sample of 100 bicycle accidents resulting in death was classified according to the day of the week on which the accident occurred. Data consistent with information given on the web site **www.highwaysafety.com** are given in the following table:

Day of Week	Frequency
Sunday	14
Monday	13
Tuesday	12
Wednesday	15
Thursday	14
Friday	17
Saturday	15

Based on these data, is it reasonable to conclude that the proportion of accidents is not the same for all days of the week? Use $\alpha = .05$.

12.6 The color vision of birds plays a role in their foraging behavior: Birds use color to select and avoid certain types of food. The authors of the article "Colour Avoidance in Northern Bobwhites: Effects of Age, Sex, and Previous Experience" (*Animal Behaviour* [1995]: 519–526) studied the pecking behavior of 1-day-old bobwhites. In an area painted white, they inserted four pins with different colored heads. The color of the pin chosen on the bird's first peck was noted for each of 33 bobwhites, resulting in the following data:

Color	First Peck Frequency
Blue	16
Green	8
Yellow	6
Red	3

Do the data provide evidence of a color preference? Test using $\alpha = .01$.

12.7 An article about the California lottery that appeared in the *San Luis Obispo Tribune* (December 15, 1999) gave the following information on the age distribution of adults in California: 35% are between 18 and 34 years old, 51% are between 35 and 64 years old, and 14% are 65 years old or older. The article also gave information on the age distribution of those who purchase lottery tickets. The following table is consistent with the values given in the article:

Age of Purchaser	Frequency
18–34	36
35–64	130
65 and over	34

Suppose that the data resulted from a random sample of 200 lottery ticket purchasers. Based on these sample data, is it reasonable to conclude that one or more of these three age groups buy a disproportionate share of lottery tickets? Use a chi-square goodness-of-fit test with $\alpha = .05$.

12.8 According to Census Bureau data, in 1998 the California population consisted of 50.7% whites, 6.6% blacks, 30.6% Hispanics, 10.8% Asians, and 1.3% other ethnic groups. Suppose that a random sample of 1000 students graduating from California colleges and universities in 1998 resulted in the following data on ethnic group; these data are consistent with summary statistics contained in the article "Crumbling Public School System a Threat to California's Future" (*Investor's Business Daily*, November 12, 1999):

Ethnic Group	Number in Sample
White	679
Black	51
Hispanic	77
Asian	190
Other	3

Do these data provide evidence that the proportion of students graduating from colleges and universities in California for these ethnic group categories differs from the respective proportions in the population for California? Test the appropriate hypotheses using $\alpha = .01$.

12.9 Criminologists have long debated whether there is a relationship between weather and violent crime. The author of the article "Is There a Season for Homicide?" (*Criminology* [1988]: 287–296) classified 1361 homicides according to season, resulting in the following data:

Winter	Spring	Summer	Fall
328	334	372	327

Do these data support the theory that the homicide rate is not the same over the four seasons? Test the relevant hypotheses using a significance level of .05.

12.10 When public opinion surveys are conducted by mail, a cover letter explaining the purpose of the survey is usually included. To determine whether the wording of the cover letter influences the response rate, investigators used three different cover letters in a survey of students at a Midwestern university ("The Effectiveness of Cover-Letter Appeals," *Journal of Social Psychology* [1984]: 85–91). Suppose that each of the three cover letters accompanied questionnaires sent to an equal number of randomly selected students. Returned questionnaires were then classified according to the type of cover letter (1, 2, or 3). Use the accompanying data to test the hypothesis that $\pi_1 = 1/3$, $\pi_2 = 1/3$, and $\pi_3 = 1/3$, where π_1, π_2, and π_3 are the true proportions that would be returned when accompanied by cover letters 1, 2, and 3, respectively. Use a .05 significance level.

	Cover-Letter Type		
	1	2	3
Returned	48	44	39

12.11 A certain genetic characteristic of a particular plant can appear in one of three forms (phenotypes). A researcher has developed a theory, according to which the hypothesized proportions are $\pi_1 = .25$, $\pi_2 = .50$, and $\pi_3 = .25$. A random sample of 200 plants yields $X^2 = 4.63$.
a. Carry out a test of the null hypothesis that the theory is correct, using level of significance $\alpha = .05$.
b. Suppose that a random sample of 300 plants had resulted in the same value of X^2. How would your analysis and conclusion differ from those in Part (a)?

12.12 The article "Linkage Studies of the Tomato" (*Transactions of the Royal Canadian Institute* [1931]:

1–19) reported data on phenotypes resulting from crossing tall cut-leaf tomatoes with dwarf potato-leaf tomatoes. There are four possible phenotypes: (1) tall cut-leaf, (2) tall potato-leaf, (3) dwarf cut-leaf, and (4) dwarf potato-leaf. The data are:

	Phenotype			
	1	2	3	4
Frequency	926	288	293	104

Mendel's laws of inheritance imply that $\pi_1 = 9/16$, $\pi_2 = 3/16$, $\pi_3 = 3/16$, and $\pi_4 = 1/16$. Are the data from this experiment consistent with Mendel's laws? Use a .01 significance level.

12.13 It is hypothesized that when homing pigeons are disoriented in a certain manner, they exhibit no preference for direction of flight after takeoff. To test this, investigators disoriented and let loose 120 pigeons and recorded the direction of flight of each pigeon. The resulting data are given in the following table:

Direction	Frequency
0° to <45°	12
45° to <90°	16
90° to <135°	17
135° to <180°	15
180° to <225°	13
225° to <270°	20
270° to <315°	17
315° to <360°	10

Use the goodness-of-fit test with significance level .10 to determine whether the data are consistent with this hypothesis.

▪ 12.2 Tests for Homogeneity and Independence in a Two-Way Table

Data resulting from observations made on two different categorical variables can also be summarized using a tabular format. As an example, suppose that residents of a particular city can watch national news on affiliate stations of ABC, CBS, NBC, or PBS. A researcher wishes to know whether there is any relationship between political philosophy (liberal, moderate, or conservative) and preferred news program among those residents who regularly watch the national news. Let x denote the variable *political philosophy* and y the variable *preferred network*. A random sample of 300 regular watchers is selected, and each individual is asked for his or her x and y values. The data set is bivariate and might initially be displayed as follows:

Observation	x Value	y Value
1	Liberal	CBS
2	Conservative	ABC
3	Conservative	PBS
⋮	⋮	⋮
299	Moderate	NBC
300	Liberal	PBS

Bivariate categorical data of this sort can most easily be summarized by constructing a **two-way frequency table**, or **contingency table**. This is a rectangular table that consists of a row for each possible value of x (each category specified by this variable) and a column for each possible value of y. There is then a cell in the table for each possible (x, y) combination. Once such a table has been constructed, the number of times each particular (x, y) combination occurs in the data set is determined, and these numbers (frequencies) are entered in the corresponding cells of the table. The resulting numbers are called **observed cell counts**. The table for the example relating political philosophy to preferred network contains 3 rows and 4 columns (because x and y have 3 and 4 possible values, respectively). Table 12.2 is one possible table, with the rows and columns labeled with the possible x and y values. These are often referred to as the *row* and *column categories*.

Table 12.2 ■ An Example of a 3 × 4 Frequency Table

	ABC	CBS	NBC	PBS	Row Marginal Total
Liberal	20	20	25	15	**80**
Moderate	45	35	50	20	**150**
Conservative	15	40	10	5	**70**
Column Marginal Total	**80**	**95**	**85**	**40**	**300**

Marginal totals are obtained by adding the observed cell counts in each row and also in each column of the table. The row and column marginal totals, along with the total of all observed cell counts in the table — the **grand total** — have been included in Table 12.2. The marginal totals provide information on the distribution of observed values for each variable separately. In this example, the row marginal totals reveal that the sample consisted of 80 liberals, 150 moderates, and 70 conservatives. Similarly, column marginal totals indicate how often each of the preferred program categories occurred: 80 preferred ABC news, 95 preferred CBS, and so on. The grand total, 300, is the number of observations in the bivariate data set — in this case, the sample size (although occasionally such a table results from a census of the entire population).

Two-way frequency tables are often characterized by the number of rows and columns in the table (specified in that order: rows first, then columns). Table 12.2 is called a 3 × 4 table. The smallest two-way frequency table is a 2 × 2 table, which has only two rows and two columns and thus four cells.

Two-way tables arise naturally in two different types of investigations. A researcher may be interested in comparing two or more groups on the basis of a single categorical variable and so may obtain a sample separately from each group. For example, data could be collected at a university to compare students, faculty, and staff on the basis of primary mode of transportation to campus (car, bicycle, motorcycle, bus, or on foot). One random sample of 200 students, another of 100 faculty members, and a third of 150 staff members might be chosen, and the selected individuals could be interviewed to obtain the necessary transportation information. Data from such a study could easily be summarized in a 3×5 two-way frequency table with row categories of student, faculty, and staff and column categories corresponding to the five possible modes of transportation. The observed cell counts could then be used to gain insight into differences and similarities among the three groups with respect to the means of transportation. This type of bivariate categorical data is characterized by having one set of marginal totals fixed (the sample sizes from the different groups). In the 3×5 situation just discussed, the row totals would be fixed at 200, 100, and 150.

A two-way table also arises when the values of two different categorical variables are observed for all individuals or items in a single sample. For example, a sample of 500 registered voters might be selected. Each voter could then be asked both if he or she favored a particular property tax initiative and if he or she was a registered Democrat, Republican, or Independent. This would result in a bivariate data set with x representing the variable *political affiliation* (with categories Democrat, Republican, and Independent) and y representing the variable *response* (favors initiative or opposes initiative). The corresponding 3×2 frequency table could then be used to investigate any association between position on the tax initiative and political affiliation. This type of bivariate categorical data is characterized by having only the grand total fixed.

▪ Comparing Two or More Populations: A Test of Homogeneity

When the value of a categorical variable is recorded for members of independent random samples obtained from each population under study, the central issue is whether the category proportions are the same for all the populations. As in Section 12.1, the test procedure uses a chi-square statistic that compares the observed counts to those that would be expected if there were no differences between the populations.

▪ Example 12.4   Smoking During Pregnancy

Numerous studies have linked maternal smoking during pregnancy with increased health risks for newborn babies. The paper "Short-Term Health and Economic Benefits of Smoking Cessation: Low Birth Weight" (*Pediatrics* [1999]: 1312–1320) compared the smoking habits of African American, Asian American, Hispanic, and non-Hispanic white mothers who smoked while pregnant. Independent samples were selected from each of the four populations (African American women who smoked during pregnancy, Asian American women who smoked during pregnancy, etc.). Data that are consistent with summary quantities that appear in the paper are given in Table 12.3, a 4×3 frequency table. The authors were interested in determining whether the four groups differed with respect to amount smoked per day.

Table 12.3 ■ Observed Counts for Example 12.4

	Amount Smoked per Day			
	1–10 Cigarettes	11–19 Cigarettes	≥20 Cigarettes	Row Marginal Total
African American	160	36	4	**200**
Asian American	162	34	4	**200**
Hispanic	243	51	6	**300**
Non-Hispanic White	248	132	20	**400**
Column Marginal Total	**813**	**253**	**34**	**1100**
Column Proportion of Total	.739	.230	.031	

Estimates of expected cell counts can be thought of in the following manner: There were 1100 responses on amount smoked per day, of which 813 were "1–10 cigarettes." The proportion of the total responding "1–10 cigarettes" is then

$$\frac{813}{1100} = .739$$

If there were no difference in response for the different groups, we would then expect about 73.9% of the African American mothers to have responded "1–10 cigarettes," 73.9% of the Asian American mothers to have responded "1–10 cigarettes," and so on. Therefore the estimated expected cell counts for the four cells in the "1–10 cigarettes" column are

$$\left(\begin{array}{l}\text{Expected count for African American}\\ \textit{and}\ \text{1–10 cigarettes cell}\end{array}\right) = (.739)(200) = 147.8$$

$$\left(\begin{array}{l}\text{Expected count for Asian American}\\ \textit{and}\ \text{1–10 cigarettes cell}\end{array}\right) = (.739)(200) = 147.8$$

$$\left(\begin{array}{l}\text{Expected count for Hispanic}\\ \textit{and}\ \text{1–10 cigarettes cell}\end{array}\right) = (.739)(300) = 221.7$$

$$\left(\begin{array}{l}\text{Expected count for non-Hispanic}\\ \text{white}\ \textit{and}\ \text{1–10 cigarettes cell}\end{array}\right) = (.739)(400) = 295.6$$

Note that the expected cell counts need not be whole numbers. The expected cell counts for the remaining cells can be computed in a similar manner. For example,

$$\frac{253}{1100} = .230$$

of all responses were in the "11–19 cigarettes" category. Therefore

$$\left(\begin{array}{l}\text{Expected count for African American}\\ \textit{and}\ \text{11–19 cigarettes cell}\end{array}\right) = (.230)(200) = 46.0$$

$$\left(\begin{array}{l}\text{Expected count for Asian American}\\ \textit{and}\ \text{11–19 cigarettes cell}\end{array}\right) = (.230)(200) = 46.0$$

$$\left(\begin{array}{l}\text{Expected count for Hispanic}\\ \textit{and}\ \text{11–19 cigarettes cell}\end{array}\right) = (.230)(300) = 69.0$$

$$\left(\begin{array}{l}\text{Expected count for non-Hispanic}\\ \text{white}\ \textit{and}\ \text{11–19 cigarettes cell}\end{array}\right) = (.230)(400) = 92.0$$

It is common practice to display the observed cell counts and the corresponding expected cell counts in the same table, with the expected cell counts enclosed in parentheses. Expected cell counts for the remaining cells have been computed and entered into Table 12.4. Except for small differences resulting from rounding, each marginal total for the expected cell counts is identical to that of the corresponding observed counts.

Table 12.4   ■   Observed and Expected Counts for Example 12.4

	Observed and Expected Counts			
	1–10 Cigarettes	11–19 Cigarettes	≥20 Cigarettes	Row Marginal Total
African American	160 (147.8)	36 (46.0)	4 (6.2)	200
Asian American	162 (147.8)	34 (46.0)	4 (6.2)	200
Hispanic	243 (221.7)	51 (69.0)	6 (9.3)	300
Non-Hispanic White	248 (295.6)	132 (92.0)	20 (12.4)	400
Column Marginal Total	813	253	34	1100

A quick comparison of the observed and expected cell counts in Table 12.4 reveals some large discrepancies, suggesting that the proportions falling into the amount smoked per day categories may not be the same for all four groups.

In Example 12.4, the expected count for a cell corresponding to a particular group–response combination was computed in two steps. First, the response *marginal proportion* was computed (e.g., 813/1100 for the "1–10 cigarette" response). Then this proportion was multiplied by a marginal group total (e.g., 200(813/1100) for the African American group). This is equivalent to first multiplying the row and column marginal totals and then dividing by the grand total:

$$\frac{(200)(813)}{1100}$$

To compare two or more groups on the basis of a categorical variable, calculate an **expected cell count** for each cell by selecting the corresponding row and column marginal totals and then computing

$$\text{expected cell count} = \frac{(\text{row marginal total})(\text{column marginal total})}{\text{grand total}}$$

These quantities represent what would be expected when there is no difference between the groups under study.

In this section, all expected cell counts are estimated from sample data. All references to "expected counts" should be interpreted as *estimated* expected counts.

The X^2 statistic, introduced in Section 12.1, can now be used to compare the observed cell counts to the expected cell counts. A large value of X^2 results when there are substantial discrepancies between the observed and expected counts and suggests that the hypothesis of no differences between the populations should be rejected. A formal test procedure is described in the accompanying box.

■ **Comparing Two or More Populations Using the X^2 Statistic**

Null hypothesis: H_0: The true category proportions are the same for all the populations (homogeneity of populations).

Alternative hypothesis: H_a: The true category proportions are not all the same for all of the populations.

Test statistic: $X^2 = \sum_{\text{all cells}} \dfrac{(\text{observed cell count} - \text{expected cell count})^2}{\text{expected cell count}}$

The expected cell counts are estimated from the sample data (assuming that H_0 is true) using the formula

$$\text{expected cell count} = \frac{(\text{row marginal total})(\text{column marginal total})}{\text{grand total}}$$

P-values: When H_0 is true, X^2 has approximately a chi-square distribution with df = (number of rows − 1)(number of columns − 1). The P-value associated with the computed test statistic value is the area to the right of X^2 under the chi-square curve with the appropriate number of degrees of freedom. Upper-tail areas for chi-square distributions are found in Appendix Table 9.

Assumptions: 1. The data consist of *independently chosen random samples*.
2. *The sample size is large*: All expected counts are at least 5. If some expected counts are less than 5, rows or columns of the table may be combined to achieve a table with satisfactory expected counts.

■ Example 12.5   Smoking During Pregnancy Revisited

The following table of observed and expected cell counts appeared in Example 12.4:

	1–10 Cigarettes	11–19 Cigarettes	≥20 Cigarettes
African American	160	36	4
	(147.8)	(46.0)	(6.2)
Asian American	162	34	4
	(147.8)	(46.0)	(6.2)
Hispanic	243	51	6
	(221.7)	(69.0)	(9.3)
Non-Hispanic White	248	132	20
	(295.6)	(92.0)	(12.4)

Hypotheses:

H_0: Proportions in each response (amount smoked per day) category are the same for all four groups.

H_a: H_0 is not true.

Significance level: A significance level of $\alpha = .05$ will be used.

Test statistic: $X^2 = \sum_{\text{all cells}} \dfrac{(\text{observed cell count} - \text{expected cell count})^2}{\text{expected cell count}}$

Assumptions: All expected cell counts are at least 5, and the random samples were independently chosen, so use of the X^2 test is appropriate.

Calculation: $X^2 = \dfrac{(160 - 147.8)^2}{147.8} + \cdots + \dfrac{(20 - 12.4)^2}{12.4} = 46.9$

P-value: The two-way table for this example has 4 rows and 3 columns, so the appropriate number of degrees of freedom is $(4 - 1)(3 - 1) = 6$. Because 46.9 is greater than 22.45, the largest entry in the 6-df column of Appendix Table 9, P-value $< .001$.

Conclusion: P-value $\leq \alpha$, so H_0 is rejected. There is strong evidence to support the claim that the proportions in the amount smoked per day response categories are not the same for the four groups. In particular, the response proportions for non-Hispanic white mothers who smoked during pregnancy seem to be quite different from those for the other three groups.

▪ Example 12.6 Drinking Habits of College Students

The data on drinking behavior for independently chosen random samples of male and female students given in Table 12.5 are similar to data that appeared in the article "Relationship of Health Behaviors to Alcohol and Cigarette Use by College Students" (*Journal of College Student Development* [1992]: 163–170). Does there appear to be a gender difference with respect to drinking behavior? (Note: Low = 1–7 drinks/week, moderate = 8–24 drinks/week, high = 25 or more drinks/week.)

Table 12.5 ▪ Observed and Expected Counts for Example 12.6

Drinking Level	Gender Male	Gender Female	Row Marginal Total
None	140 (158.6)	186 (167.4)	326
Low	478 (554.0)	661 (585.0)	1139
Moderate	300 (230.1)	173 (242.9)	473
High	63 (38.4)	16 (40.6)	79
Column Marginal Total	981	1036	2017

The relevant hypotheses are

H_0: True proportions for the four drinking levels are the same for males and females.

H_a: H_0 is not true.

Significance level: $\alpha = .01$.

Test statistic: $X^2 = \sum_{\text{all cells}} \dfrac{(\text{observed cell count} - \text{expected cell count})^2}{\text{expected cell count}}$

Assumptions: Table 12.5 contains the computed expected counts, all of which are greater than 5. The data consist of independently chosen random samples.

Calculation: $X^2 = \dfrac{(140 - 158.6)^2}{158.6} + \cdots + \dfrac{(16 - 40.6)^2}{40.6} = 96.6$

P-value: The table has 4 rows and 2 columns, so df $= (4-1)(2-1) = 3$. The computed value of X^2 is greater than the largest entry in the 3-df column of Appendix Table 9, so P-value $< .001$.

Conclusion: Because P-value $\le \alpha$, H_0 is rejected. The data indicate that males and females differ with respect to drinking level.

■ Testing for Independence of Two Categorical Variables

The X^2 test statistic and test procedure can also be used to investigate association between two categorical variables in a single population. Suppose that each population member has a value of a first categorical variable and also of a second such variable. As an example, television viewers in a particular city might be categorized with respect to both preferred network (ABC, CBS, NBC, or PBS) and favorite type of programming (comedy, drama, or information and news). The question of interest is often whether knowledge of one variable's value provides any information about the value of the other variable — that is, are the two variables independent?

Continuing the example, suppose that those who favor ABC prefer the three types of programming in proportions .4, .5, and .1 and that these proportions are also correct for individuals favoring any of the other three networks. Then, learning an individual's preferred network provides no added information about that individual's favorite type of programming. The categorical variables *preferred network* and *favorite program type* would be independent.

To see how expected counts are obtained in this situation, recall from Chapter 6 that if two outcomes A and B are independent, then

$$P(A \text{ and } B) = P(A)P(B)$$

so the proportion of time that the two outcomes occur together in the long run is the product of the two individual long-run relative frequencies. Similarly, two categorical variables are independent in a population if, for each particular category of the first variable and each particular category of the second variable,

$$\begin{pmatrix} \text{proportion of individuals} \\ \text{in a particular category} \\ \text{combination} \end{pmatrix} = \begin{pmatrix} \text{proportion in} \\ \text{specified category} \\ \text{of first variable} \end{pmatrix} \begin{pmatrix} \text{proportion in} \\ \text{specified category} \\ \text{of second variable} \end{pmatrix}$$

Thus, if 30% of all viewers prefer ABC and the proportions of program type preferences are as previously given, then, assuming that the two variables are independent, the proportion of individuals who both favor ABC and prefer comedy is $(.3)(.4) = .12$ (or 12%).

Multiplying the right-hand side of this expression by the sample size gives us the expected number of individuals in the sample who are in both specified categories of the two variables when the variables are independent. However, these expected counts cannot be calculated, because the individual population proportions are not known. The solution to this dilemma is to estimate each population proportion using the corresponding sample proportion:

$$\begin{pmatrix} \text{estimated expected number} \\ \text{in specified categories} \\ \text{of the two variables} \end{pmatrix} = (\text{sample size}) \left[\begin{pmatrix} \text{observed number} \\ \text{in category of} \\ \text{first variable} \\ \hline \text{sample size} \end{pmatrix} \right] \left[\begin{pmatrix} \text{observed number} \\ \text{in category of} \\ \text{second variable} \\ \hline \text{sample size} \end{pmatrix} \right]$$

$$= \frac{\begin{pmatrix} \text{observed number in} \\ \text{category of first variable} \end{pmatrix} \begin{pmatrix} \text{observed number in} \\ \text{category of second variable} \end{pmatrix}}{\text{sample size}}$$

Suppose that the observed counts are displayed in a rectangular table in which rows correspond to the categories of the first variable and columns to the categories of the second variable. Then the numerator in the preceding expression for expected counts is just the product of the row and column marginal totals. This is exactly how expected counts were computed in the test for homogeneity of several populations, even though the reasoning used to arrive at the formula is different.

▪ **X^2 Test for Independence**

Null hypothesis: H_0: The two variables are independent.

Alternative hypothesis: H_a: The two variables are not independent.

Test statistic: $X^2 = \displaystyle\sum_{\text{all cells}} \frac{(\text{observed cell count} - \text{expected cell count})^2}{\text{expected cell count}}$

The expected cell counts are estimated (assuming H_0 is true) by using the formula

$$\text{expected cell count} = \frac{(\text{row marginal total})(\text{column marginal total})}{\text{grand total}}$$

P-values: When H_0 is true and the assumptions of the X^2 test are satisfied, X^2 has approximately a chi-square distribution with

df = (number of rows − 1)(number of columns − 1)

The P-value associated with the computed test statistic value is the area to the right of X^2 under the chi-square curve with the appropriate number of degrees of freedom. Upper-tail areas for chi-square distributions are found in Appendix Table 9.

Assumptions: 1. The observed counts are from a *random sample*.
2. The *sample size is large*; that is, all expected counts are at least 5. If some expected counts are less than 5, rows or columns of the table should be combined to achieve a table with satisfactory expected counts.

▪ **Example 12.7** Risky Behavior

The article "Factors Associated with Sexual Risk-Taking Behaviors Among Adolescents" (*Journal of Marriage and Family* [1994]: 622–632) examined the relationship between gender and contraceptive use by sexually active teens. Each person in a random sample of sexually active teens was classified according to gender and contraceptive use (with three categories: rarely or never use, use sometimes or most of the time, and always use), resulting in a 3×2 table. Data consistent with percentages given in the article are summarized in Table 12.6.

Table 12.6 ▪ Observed Counts for Example 12.7

Contraceptive Use	Gender		Row Marginal Total
	Female	Male	
Rarely/Never	210	350	560
Sometimes/Most Times	190	320	510
Always	400	530	930
Column Marginal Total	800	1200	2000

The authors were interested in determining whether there is an association between gender and contraceptive use. Using a .05 significance level, we will test

H_0:   Gender and contraceptive use are independent.

H_a: Gender and contraceptive use are not independent.

Significance level: $\alpha = .05$.

Test statistic: $X^2 = \sum_{\text{all cells}} \dfrac{(\text{observed cell count} - \text{expected cell count})^2}{\text{expected cell count}}$

Assumptions: Before we can check the assumptions, we must first compute the expected cell counts:

Cell Row	Cell Column	Expected Cell Count
1	1	$\dfrac{(560)(800)}{2000} = 224.0$
1	2	$\dfrac{(560)(1200)}{2000} = 336.0$
2	1	$\dfrac{(510)(800)}{2000} = 204.0$
2	2	$\dfrac{(510)(1200)}{2000} = 306.0$
3	1	$\dfrac{(930)(800)}{2000} = 372.0$
3	2	$\dfrac{(930)(1200)}{2000} = 558.0$

All expected cell counts are greater than 5, and the observed cell counts are based on a random sample, so the X^2 test can be used. The observed and expected counts are given together in Table 12.7.

Table 12.7 ■ **Observed and Expected Counts for Example 12.7**

Contraceptive Use	Gender		Row Marginal Total
	Female	Male	
Rarely/Never	210 (224)	350 (336)	560
Sometimes/Most Times	190 (204)	320 (306)	510
Always	400 (372)	530 (558)	930
Column Marginal Total	800	1200	2000

Calculation: $X^2 = \dfrac{(210 - 224)^2}{224} + \cdots + \dfrac{(530 - 558)^2}{558} = 6.572$

P-value: The table has 3 rows and 2 columns, so df = $(3 - 1)(2 - 1) = 2$. The entry closest to 6.572 in the 2-df column of Appendix Table 9 is 6.70, so the approximate *P*-value for this test is *P*-value $\approx .035$.

Conclusion: Because *P*-value $\leq \alpha$, we reject H_0 and conclude that there is an association between gender and contraceptive use.

■ Example 12.8   Television Viewing and Fitness Level

Table 12.8 was constructed using data from the article "Television Viewing and Physical Fitness in Adults" (*Research Quarterly for Exercise and Sport* [1990]: 315–320). The author hoped to determine whether time spent watching television is associated with cardiovascular fitness. Subjects were asked about their television-viewing time (per day, rounded to the nearest hour) and were classified as physically fit if they scored in the excellent or very good category on a step test. Expected cell counts (computed under the assumption of no association between television viewing and fitness) appear in parentheses in the table.

Table 12.8 ■ **Observed and Expected Counts for Example 12.8**

TV Viewing Time (hours)	Fitness		Row Marginal Total
	Physically Fit	Not Physically Fit	
0	35 (25.5)	147 (156.5)	182
1–2	101 (102.2)	629 (627.8)	730
3–4	28 (35.0)	222 (215.0)	250
5 or More	4 (5.3)	34 (32.7)	38
Column Marginal Total	168	1032	1200

The X^2 test with a significance level of .01 will be used to test the relevant hypotheses:

H_0: Fitness and television viewing are independent.

H_a: Fitness and television viewing are not independent.

Significance level: $\alpha = .01$.

Test statistic: $X^2 = \sum_{\text{all cells}} \dfrac{(\text{observed cell count} - \text{expected cell count})^2}{\text{expected cell count}}$

Assumptions: All expected cell counts are at least 5. Assuming that the data can be viewed as a random sample of adults, the X^2 test can be used.

Calculation: $X^2 = \dfrac{(35 - 25.5)^2}{25.5} + \cdots + \dfrac{(34 - 32.7)^2}{32.7} = 6.13$

P-value: The table has 4 rows and 2 columns, so df = $(4 - 1)(2 - 1) = 3$. The computed value of $X^2 = 6.13$ is smaller than 6.25, the smallest entry in the 3-df column of Appendix Table 9. Therefore, P-value > .10.

Conclusion: Because P-value > α, H_0 is not rejected. There is not sufficient evidence to conclude that an association exists between television viewing and fitness.

Most statistical computer packages can calculate expected cell counts, the value of the X^2 statistic, and the associated P-value. MINITAB output for the data of Example 12.8 is given in Figure 12.3. The discrepancy between $X^2 = 6.13$ and $X^2 = 6.161$ is due to rounding in the hand calculations. Note also that MINITAB gives P-value = .104. This is consistent with the statement P-value > .10 from Example 12.8.

Figure 12.3 MINITAB output for Example 12.8.

Chi-Square Test: Fit, Not Fit

Expected counts are printed below observed counts

	Fit	Not Fit	Total
1	35	147	182
	25.48	156.52	
2	101	629	730
	102.20	627.80	
3	28	222	250
	35.00	215.00	
4	4	34	38
	5.32	32.68	
Total	168	1032	1200

Chi-Sq = 3.557 + 0.579 +
        0.014 + 0.002 +
        1.400 + 0.228 +
        0.328 + 0.053 = 6.161

DF = 3, P-Value = 0.104

In some investigations, values of more than two categorical variables are recorded for each individual in the sample. For example, in addition to the variables *television-viewing time* and *fitness level*, the researchers in the study referenced in Example 12.8 might also have recorded age group (with categories under 21, 21–30, 31–40, etc.) for each subject. A number of interesting questions could then be explored: Are all three variables independent of one another? Is it possible that age and fitness level are dependent but that the relationship between them does not depend on television-viewing time? For a particular age group, are television-viewing time and fitness independent?

The X^2 test procedure described in this section for analysis of bivariate categorical data can be extended for use with *multivariate categorical data*. Appropriate hypothesis tests can then be used to provide insight into the relationships between variables. However, the computations required to calculate expected cell counts and to compute the value of X^2 are quite tedious, so they are seldom done without the aid of a computer. Most statistical computer packages can perform this type of analysis. Consult the references at the end of this chapter for further information on the analysis of categorical data.

▪ Exercises 12.14–12.34

12.14 A particular state university system has six campuses. On each campus, a random sample of students will be selected, and each student will be categorized with respect to political philosophy as liberal, moderate, or conservative. The null hypothesis of interest is that the proportion of students falling in these three categories is the same at all six campuses.
a. On how many degrees of freedom will the resulting X^2 test be based?
b. How does your answer in Part (a) change if there are seven campuses rather than six?
c. How does your answer in Part (a) change if there are four rather than three categories for political philosophy?

12.15 A random sample of 1000 registered voters in a certain county is selected, and each voter is categorized with respect to both educational level (four categories) and preferred candidate in an upcoming supervisorial election (five possibilities). The hypothesis of interest is that educational level and preferred candidate are independent.
a. If $X^2 = 7.2$, what would you conclude at significance level .10?
b. If there were only four candidates vying for election, what would you conclude if $X^2 = 14.5$ and $\alpha = .05$?

12.16 The paper "No Evidence of Impaired Neurocognitive Performance in Collegiate Soccer Players" (*American Journal of Sports Medicine* [2002]: 157–162) included the accompanying table, which resulted from classifying individuals in three independent random samples according to whether or not they had suffered one or more concussions. The three samples were college soccer players, college athletes participating in a sport other than soccer, and college students who were not participating in

any sport; soccer players were singled out for study because of the practice of "heading" the ball (hitting the soccer ball with the head).

	No Concussion	At Least One Concussion	Row Marginal Total
Soccer Athletes	46	45	**91**
Nonsoccer Athletes	68	28	**96**
Nonathletes	8	45	**53**

Is there evidence that the proportion who suffer at least one concussion is not the same for all three groups compared? Test the relevant hypotheses using $\alpha = .01$.

12.17 A survey was conducted in the San Francisco Bay area in which each participating individual was classified according to the type of vehicle used most often and city of residence. A subset of the resulting data is given in the following table (*The Relationship of Vehicle Type Choice to Personality, Lifestyle, Attitudinal, and Demographic Variables*, Technical Report UCD-ITS-RR02-06, DaimlerChrysler Corp., 2002):

Vehicle Type	City		
	Concord	Pleasant Hills	North San Francisco
Small	68	83	221
Compact	63	68	106
Midsize	88	123	142
Large	24	18	11

Do the data provide convincing evidence of an association between city of residence and vehicle type? Use a significance level of .05. You may assume that it is reasonable to regard the sample as a random sample of Bay Area residents.

12.18 Do women have different patterns of work behavior than men? The article "Workaholism in Organizations: Gender Differences" (*Sex Roles: A Journal of Research* [1999]: 333–346) attempted to answer this question. Each person in a random sample of 423 graduates of a business school in Canada were polled and classified by gender and workaholism type, resulting in the following table:

Workaholism Types	Gender	
	Female	Male
Work Enthusiasts	20	41
Workaholics	32	37
Enthusiastic Workaholics	34	46
Unengaged Workers	43	52
Relaxed Workers	24	27
Disenchanted Workers	37	30

a. Test the hypothesis that gender and workaholism type are independent.
b. The author writes that "women and men fell into each of the six workaholism types to a similar degree." Does the outcome of the test you performed in Part (a) support this conclusion? Explain.

12.19 The following table is based on data from 21,870 male physicians, age 40–84, who are participating in the Physicians Health Study ("Light-to-Moderate Alcohol Consumption and Risk of Stroke Among U.S. Male Physicians," *New England Journal of Medicine* [1999]: 1557–1564):

	Alcohol Consumption (number of drinks)				
	<1/ wk	1/ wk	2–4/ wk	5–6/ wk	1/ day
Never Smoked	3577	1711	2430	1211	1910
Smoked in the Past	1595	1068	1999	1264	2683
Currently Smokes	524	289	470	296	849

a. Calculate row proportions by dividing each observed count by the corresponding row total. Are the proportions falling in each of the alcohol consumption categories similar for the three smoking categories?

b. Test the hypothesis that smoking status and alcohol consumption are independent.
c. Is the result of your test in Part (b) consistent with what you expected based on your answer to Part (a)? If not, explain why your initial impression based on the proportions may not have been accurate? What aspect of the data set factors into your explanation?

12.20 The article "Motion Sickness in Public Road Transport: The Effect of Driver, Route, and Vehicle" (*Ergonomics* [1999]: 1646–1664) reported that seat position within a bus may have some effect on whether one experiences motion sickness. The following table classifies each person in a random sample of bus riders by the location of his or her seat and whether nausea was reported:

	Nausea	No Nausea
Front	58	870
Middle	166	1163
Rear	193	806

Based on these data, can you conclude that there is an association between seat location and nausea? Test the relevant hypotheses using $\alpha = .05$.

12.21 The article "Mortality Among Recent Purchasers of Handguns" (*New England Journal of Medicine* [1999]: 1583–1589) examined the relationship between handgun purchases and cause of death. Suppose that a random sample of 4000 death records was examined and that cause of death was noted for each individual in the sample. Gun registration records were also examined to determine which of these 4000 individuals had purchased a handgun within the previous year. Data consistent with summary values given in the article are shown in the following table:

	Suicide	Not Suicide
Handgun Purchased	4	12
No Handgun Purchased	63	3921

Is it reasonable to use the chi-square test to determine whether there is an association between handgun purchase within the year before death and whether the death was a suicide? Justify your answer.

12.22 Each boy in a sample of Mexican American males, age 10–18, was classified according to smoking status and response to a question asking whether he liked to do risky things. The following table is based on data given in the article "The Association Between Smoking and Unhealthy Behaviors Among a National Sample of Mexican-American Adolescents" (*Journal of School Health* [1998]: 376–379):

	Smoking Status	
	Smoker	Nonsmoker
Likes Risky Things	45	46
Doesn't Like Risky Things	36	153

Assume that it is reasonable to regard the sample as a random sample of Mexican American male adolescents.

a. Is there sufficient evidence to conclude that there is an association between smoking status and desire to do risky things? Test the relevant hypotheses using $\alpha = .05$.

b. Based on your conclusion in Part (a), is it reasonable to conclude that smoking *causes* an increase in the desire to do risky things? Explain.

12.23 The article "Cooperative Hunting in Lions: The Role of the Individual" (*Behavioral Ecology and Sociobiology* [1992]: 445–454) discussed the different roles taken by lionesses as they stalk and capture prey. The authors were interested in the effect of position in the line as stalking occurs; an individual lioness may be in the center of the line or on the wing (end of the line) as the group advances toward their prey. In addition to position, the role of the lioness was also considered. A lioness could initiate a chase (be the first one to charge the prey), or she could participate and join the chase after it has been initiated. Data from the article are summarized in the following table:

	Role	
Position	Initiates Chase	Participates in Chase
Center	28	48
Wing	66	41

Is there evidence of an association between position and role? Test the relevant hypotheses using $\alpha = .01$. What assumptions about how the data were collected must be true for the chi-square test to be an appropriate way to analyze these data?

12.24 The authors of the article "A Survey of Parent Attitudes and Practices Regarding Under Age Drinking" (*Journal of Youth and Adolescence* [1995]: 315–334) conducted a telephone survey of parents with preteen and teenage children. One of the questions asked was "How effective do you think you are in talking to your children about drinking?" Responses are summarized in the accompanying 3 × 2 table:

Response	Age of Children	
	Preteen	Teen
Very Effective	126	149
Somewhat Effective	44	41
Not at All Effective or Don't Know	51	26

Using a significance level of .05, carry out a test to determine whether there is an association between age of children and parental response.

12.25 The article "Regional Differences in Attitudes Toward Corporal Punishment" (*Journal of Marriage and Family* [1994]: 314–324) presented data resulting from a random sample of 978 adults. Each individual in the sample was asked whether he or she agreed with the following statement: "Sometimes it is necessary to discipline a child with a good, hard spanking." Respondents were also classified according to the region of the United States in which they lived. The resulting data are summarized in the following table:

Region	Response	
	Agree	Disagree
Northeast	130	59
West	146	42
Midwest	211	52
South	291	47

Is there an association between response (agree, disagree) and region of residence? Use $\alpha = .01$.

12.26 A story describing a date rape was read by 352 high school students. To investigate the effect of the victim's clothing on subject's judgment of the situation described, the investigators included with the story either a photograph of the victim dressed provocatively or a photo of the victim dressed conservatively, or no picture. Each student was asked whether the situation described in the story was one of rape. Data from the article "The Influence of Victim's Attire on Adolescent Judgments of Date Rape" (*Adolescence* [1995]: 319–323) are given in the following table:

	Provocative Picture	Conservative Picture	No Picture
Rape	80	104	92
Not Rape	47	12	17

Is there evidence that the proportion who believe that the story described a rape differs for the three different photo groups? Test the relevant hypotheses using $\alpha = .01$.

12.27 Can people tell the difference between a female nose and a male nose? This important (?) research question was examined in the article "You Can Tell by the Nose: Judging Sex from an Isolated Facial Feature" (*Perception* [1995]: 969–973). Eight white males and eight white females posed for nose photos. The article states that none of the volunteers wore nose studs or had prominent nasal hair. Each person placed a black Lycra tube over his or her head in such a way that only the nose protruded through a hole in the material. Photos were then taken from three different angles: front view, three-quarters view, and profile. These photos were shown to a sample of undergraduate students. Each student in the sample was shown one of the nose photos and asked whether it was a photo of a male or a female, and the response was classified as either correct or incorrect. The following table was constructed using summary values reported in the article:

| | View | | |
Sex ID	Front	Profile	Three-Quarters
Correct	23	26	29
Incorrect	17	14	11

Is there evidence that the proportion of correct sex identifications differs for the three different nose views?

12.28 The following table appeared in the article "Europe's Receptivity to New Religious Movements: Round Two" (*Journal for the Scientific Study of Religion* [1993]: 389–397):

Country	Percentage Who Believe in Fortune-Tellers
Great Britain	42
West Germany	32
East Germany	22
Slovenia	55
Ireland	27
Northern Ireland	33

Suppose that the percentages in the table were based on random samples of size 200 from each of the 6 countries.
a. Use the given percentages to construct a two-way table with rows corresponding to the six countries and columns corresponding to the categories "believe in fortune-tellers" and "don't believe in fortune-tellers." Enter the observed counts in the table. (Hint: The observed counts for Great Britain are (.42)(200) = 84 for the "believe" column and (.58)(200) = 116 for the "don't believe" column.)
b. Is there evidence that the proportion who believe in fortune-tellers is not the same for all six countries? Test the relevant hypotheses using $\alpha = .01$.

12.29 What factors influence people to respond to mail surveys? Researchers have found that the appearance of the survey and the nature of the first question have an influence on the overall response rate, but do they affect the speed of response? The article "The Impact of Cover Design and First Questions on Responses for a Mail Survey of Skydivers" (*Leisure Science* [1991]: 67–76) presented the results of an experiment to investigate rate of response. Surveys with a plain cover were mailed to a random sample of 427 skydivers. Surveys with a cover graphic of a skydiver were mailed to a separate random sample of 414 skydivers. The postmark on the returned survey was used to classify each person who received a survey in one of the following response categories: returned in 1–7 days, returned in 8–14 days, returned in 15–31 days, returned in 32–60 days, and not returned. The resulting data are summarized in the two-way table below.

Do the data support the theory that the proportions falling into the various response categories are not the same for the two cover designs? Use a significance level of .05 to test the relevant hypotheses.

12.30 The article "Heavy Drinking and Problems Among Wine Drinkers" (*Journal of Studies on Alcohol* [1999]: 467–471) analyzed drinking problems among Canadians. The article gave the percentage of drinkers who never drank to intoxication for seven different types of drinkers. From the

Table for Exercise 12.29

| | Response Category | | | | | |
Cover Design	1–7 Days	8–14 Days	15–31 Days	32–60 Days	Not Returned	Total
Graphic	70	76	51	19	198	**414**
Plain	84	53	51	32	207	**427**
Total	**154**	**129**	**102**	**51**	**405**	**841**

information given in the article, we can obtain the following table:

	Never Intoxicated	Sometimes Intoxicated
Beer Only	515	741
Wine Only	875	232
Spirits Only	387	372
Beer and Wine	774	560
Beer and Spirits	312	727
Wine and Spirits	708	349
Beer, Wine, and Spirits	1032	1119

Assume that each of the seven samples can be viewed as a random sample from the respective group.
a. Is there sufficient evidence to conclude that the proportion who never become intoxicated is not the same for the seven groups?
b. Is the test that you performed in Part (a) a test of independence or a test of homogeneity? Explain.

12.31 When charities solicit donations, they must know what approach will encourage people to donate. Would a picture of a needy child tug at the heart-strings and convince someone to contribute to the charity? As reported in the article "The Effect of Photographs of the Handicapped on Donation" (*Journal of Applied Social Psychology* [1979]: 426–431), experimenters went door to door with four types of displays: a picture of a smiling child, a picture of an unsmiling child, a verbal message, and an identification of the charity only. At each door, they would randomly select which display to present. Suppose that 18 of 30 subjects contributed when shown the picture of the smiling child, 14 of 30 contributed when shown the picture of an unsmiling child, 16 of 26 contributed when shown the verbal message, and 18 of 22 contributed when shown only the identification of the charity. What conclusions can be drawn from these data? Explain your procedure and reasoning.

12.32 Jail inmates can be classified into one of the following four categories according to the type of crime committed: violent crime, crime against property, drug offenses, and public-order offenses. Suppose that random samples of 500 male inmates and 500 female inmates are selected and that each inmate is classified according to type of offense. The data in the following table are based on summary values given in the article "Profile of Jail Inmates" (*USA Today*, April 25, 1991):

Type of Crime	Gender	
	Male	Female
Violent	117	66
Property	150	160
Drug	109	168
Public Order	124	106

We would like to know whether male and female inmates differ with respect to type of offense.
a. Is this a test of homogeneity or a test of independence?
b. Test the relevant hypotheses using a significance level of .05.

12.33 Job satisfaction of professionals was examined in the article "Psychology of the Scientist: Work-Related Attitudes of U.S. Scientists" (*Psychological Reports* [1991]: 443–450). Each person in a random sample of 778 teachers was classified according to a job satisfaction variable and also by teaching level, resulting in the following two-way table:

Teaching Level	Job Satisfaction	
	Satisfied	Unsatisfied
College	74	43
High School	224	171
Elementary	126	140

Can we conclude that there is an association between job satisfaction and teaching level? Test the relevant hypotheses using $\alpha = .05$.

12.34 The article "Victim's Race Affects Killer's Sentence, Study Finds" (*New York Times*, April 20, 2001) included the following information:

Defendant's Race	Victim's Race	Death Penalty	No Death Penalty
Not white	White	33	251
White	White	33	508
Not white	Not white	29	587
White	Not white	4	76

The table was based on a study of *all* homicide cases in North Carolina for the period 1993–1997 where it was possible that the person convicted of murder could receive the death penalty. Explain why it would not be necessary (or appropriate) to use a chi-square test to determine whether there was an association between the defendant–victim race combination and whether or not the convicted murderer received a death sentence for convicted murders in North Carolina during 1993 to 1997.

■ 12.3 Communicating and Interpreting the Results of Statistical Analyses

Many studies, particularly those in the social sciences, result in categorical data. The questions of interest in such studies often lead to an analysis that involves using a chi-square test.

■ Communicating the Results of Statistical Analyses

Three different chi-square tests were introduced in this chapter — the goodness-of-fit test, the test for homogeneity, and the test for independence. They are used in different settings and to answer different questions. When summarizing the results of a chi-square test, be sure to indicate which chi-square test was performed. One way to do this is to be clear about the nature of the hypotheses being tested.

It is also a good idea to include a table of observed and expected counts in addition to reporting the computed value of the test statistic and the P-value. And finally, make sure to give a conclusion in context, and make sure that the conclusion is worded appropriately for the type of test conducted. For example, don't use terms such as *independence* and *association* to describe the conclusion if the test performed was a test for homogeneity.

■ Interpreting the Results of Statistical Analyses

As with the other hypothesis tests considered, it is common to find the result of a chi-square test summarized by giving the value of the chi-square test statistic and an associated P-value. Because categorical data can be summarized compactly in frequency tables, the data often are given in the article (unlike data for numerical variables, which are rarely given).

■ What to Look For in Published Data

Here are some questions to consider when you are reading an article that contains the results of a chi-square test:

- Are the variables of interest categorical rather than numerical?
- Are the data given in the article in the form of a frequency table?
- If a two-way frequency table is involved, is the question of interest one of homogeneity or one of independence?
- What null hypothesis is being tested? Are the results of the analysis reported in the correct context (homogeneity, etc.)?
- Is the sample size large enough to make use of a chi-square test reasonable? (Are all expected counts at least 5?)
- What is the value of the test statistic? Is the associated P-value given? Should the null hypothesis be rejected?

- Are the conclusions drawn by the authors consistent with the results of the test?

- How different are the observed and expected counts? Does the result have practical significance as well as statistical significance?

The authors of the article "Predicting Professional Sports Game Outcomes from Intermediate Game Scores" (*Chance* [1992]: 18–22) used a chi-square test to determine whether there was any merit to the idea that basketball games are not settled until the last quarter, whereas baseball games are over by the seventh inning. They also considered football and hockey. Data were collected for 189 basketball games, 92 baseball games, 80 hockey games, and 93 football games. The analyzed games were sampled randomly from all games played during the 1990 season for baseball and football and for the 1990–1991 season for basketball and hockey. For each game, the late-game leader was determined, and then it was noted whether the late-game leader actually ended up winning the game. The resulting data are summarized in the following table:

Sport	Late-Game Leader Wins	Late-Game Leader Loses
Basketball	150	39
Baseball	86	6
Hockey	65	15
Football	72	21

The authors stated that the "*late-game leader* is defined as the team that is ahead after three quarters in basketball and football, two periods in hockey, and seven innings in baseball. The chi-square value (with three degrees of freedom) is 10.52 ($P < .015$)." They also concluded that "the sports of basketball, hockey, and football have remarkably similar percentages of late-game reversals, ranging from 18.8% to 22.6%. The sport that is an anomaly is baseball. Only 6.5% of baseball games resulted in late reversals. . . . [The chi-square test] is statistically significant due almost entirely to baseball."

In this particular analysis, the authors are comparing four populations (games from each of the four sports) on the basis of a categorical variable with two categories (late-game leader wins and late-game leader loses). The appropriate null hypothesis is then

H_0: The true proportion in each category (leader wins, leader loses) is the same for all four sports.

Based on the reported value of the chi-square statistic and the associated P-value, this null hypothesis is rejected, leading to the conclusion that the category proportions are not the same for all four sports.

The validity of the chi-square test requires that the sample sizes be large enough so that no expected counts are less than 5. Is this reasonable here? The following MINITAB output shows the expected cell counts and the computation of the X^2 statistic:

Chi-Square Test

Expected counts are printed below observed counts

	Leader W	Leader L	Total
1	150	39	189
	155.28	33.72	
2	86	6	92
	75.59	16.41	
3	65	15	80
	65.73	14.27	
4	72	21	93
	76.41	16.59	
Total	373	81	454

Chi-Sq = 0.180 + 0.827 +
1.435 + 6.607 +
0.008 + 0.037 +
0.254 + 1.171 = 10.518
DF = 3, P-Value = 0.015

The smallest expected count is 14.27, so the sample sizes are large enough to justify the use of the X^2 test. Note also that the two cells in the table that correspond to baseball contribute a total of $1.435 + 6.607 = 8.042$ to the value of the X^2 statistic of 10.518. This is due to the large discrepancies between the observed and expected counts for these two cells. There is reasonable agreement between the observed and the expected counts in the other cells. This is probably the basis for the authors' conclusion that baseball is the only anomaly and that the other sports were similar.

▪ A Word to the Wise: Cautions and Limitations

Be sure to keep the following in mind when analyzing categorical data using one of the chi-square tests presented in this chapter:

1. Don't confuse tests for homogeneity with tests for independence. The hypotheses and conclusions are different for the two types of test. Tests for homogeneity are used when the individuals in each of two or more independent samples are classified according to a single categorical variable. Tests for independence are used when individuals in a *single* sample are classified according to two categorical variables.

2. As was the case for the hypothesis tests of earlier chapters, remember that we can never say we have strong support for the null hypothesis. For example, if we do not reject the null hypothesis in a chi-square test for independence, we cannot conclude that there is convincing evidence that the variables are independent. We can only say that we were not convinced that there is an association between the variables.

3. Be sure that the assumptions for the chi-square test are reasonable. *P*-values based on the chi-square distribution are only approximate, and if the large sample conditions are not met, the true *P*-value may be quite different from the approximate one based on the chi-square distribution. This can sometimes

lead to erroneous conclusions. Also, for the chi-square test of homogeneity, the assumption of *independent* samples is particularly important.

4. Don't jump to conclusions about causation. Just as a strong correlation between two numerical variables does not mean that there is a cause-and-effect relationship between them, an association between two categorical variables does not imply a causal relationship.

▪ Activity 12.1: Pick a Number, Any Number . . .

Background: There is evidence to suggest that human beings are not very good random number generators. In this activity, you will investigate this phenomenon by collecting and analyzing a set of human-generated "random" digits.

For this activity, work in a group with two or three other students.

1. Each member of the group should complete this step individually. Ask 25 different people to pick a digit from 0 to 9 at random. Record the responses.

2. Combine the responses you collected with those of the other members of your group to form a single sample. Summarize the resulting data in a one-way frequency table.

3. If people are adept at picking digits at random, what would you expect for the proportion of the responses in the sample that were 0? that were 1?

4. State a null hypothesis and an alternative hypothesis that could be tested to determine whether there is evidence that the 10 digits from 0 to 9 are not selected an equal proportion of the time when people are asked to pick a digit at random.

5. Carry out the appropriate hypothesis test, and write a few sentences indicating whether or not the data support the theory that people are not good random number generators.

▪ Activity 12.2: Color and Perceived Taste

Background: Does the color of a food or beverage affect the way people perceive its taste? In this activity you will conduct an experiment to investigate this question and analyze the resulting data using a chi-square test.

You will need to recruit at least 30 subjects for this experiment, so it is advisable to work in a large group (perhaps even the entire class) to complete this activity.

Subjects for the experiment will be assigned at random to one of two groups. Each subject will be asked to taste a sample of gelatin (e.g., Jell-O) and rate the taste as not very good, acceptable, or very good. Subjects assigned to the first group will be asked to taste and rate a cube of lemon-flavored gelatin. Subjects in the second group will be asked to taste and rate a cube of lemon-flavored gelatin that has been colored an unappealing color by adding food coloring to the gelatin mix before the gelatin sets.

Note: You may choose to use something other than gelatin, such as lemonade. Any food or beverage whose color can be altered using food coloring can be used. You can experiment with the food colors to obtain a color that you think is particularly unappealing!

1. As a class, develop a plan for collecting the data. How will subjects be recruited? How will they be as-

signed to one of the two treatment groups (unaltered color, altered color)? What extraneous variables will be directly controlled, and how will you control them?

2. After the class is satisfied with the data collection plan, assign members of the class to prepare the gelatin to be used in the experiment.

3. Carry out the experiment, and summarize the resulting data in a two-way table like the one shown:

Treatment	Taste Rating		
	Not Very Good	Acceptable	Very Good
Unaltered Color			
Altered Color			

4. The two-way table summarizes data from two independent samples (as long as subjects were assigned *at random* to the two treatments, the samples are independent). Carry out an appropriate test to determine whether the proportion for each of the three taste rating categories is the same when the color is altered as for when the color is not altered.

■ Summary of Key Concepts and Formulas

Term or Formula	Comment
One-way frequency table	A compact way of summarizing data on a categorical variable; it gives the number of times each of the possible categories in the data set occurs.
Goodness-of-fit statistic, $$X^2 = \sum_{\text{all cells}} \frac{(\text{observed cell count} - \text{expected cell count})^2}{\text{expected cell count}}$$	A statistic used to provide a comparison between observed counts and those expected when a given hypothesis is true. When none of the expected counts are too small, X^2 has approximately a chi-square distribution when the null hypothesis is true.
X^2 goodness-of-fit test	A hypothesis test performed to determine whether the true category proportions are different from those specified by the given null hypothesis.
Two-way frequency table (contingency table)	A rectangular table used to summarize a bivariate categorical data set; two-way tables are used to compare several populations on the basis of a categorical variable or to identify whether an association exists between two categorical variables.
X^2 test for homogeneity	The hypothesis test performed to determine whether true category proportions are the same for two or more populations.
X^2 test for independence	The hypothesis test performed to determine whether an association exists between two categorical variables.

■ Supplementary Exercises 12.35–12.44

12.35 Drivers born under the astrological sign of Capricorn are the worst drivers in Australia, according to an article that appeared in the Australian newspaper *The Mercury* (October 26, 1998). This statement was based on a study of insurance claims that resulted in the following data for male policyholders of a large insurance company:

Astrological Sign	Number of Policyholders
Aries	37,926
Taurus	37,179
Gemini	37,179
Cancer	38,126
Leo	37,354
Virgo	37,718
Libra	37,910
Scorpio	35,352
Sagittarius	34,175
Capricorn	54,906
Aquarius	35,666
Pisces	36,677

a. Assuming that it is reasonable to treat the policyholders of this particular insurance company as a random sample of insured drivers in Australia, are the observed data consistent with the hypothesis that the proportion of policyholders is the same for each of the 12 astrological signs?

b. Why do you think that the proportion of Capricorn policyholders is so much higher than would be expected if the proportions are the same for all astrological signs?

c. Suppose that a random sample of 1000 accident claims submitted to this insurance company is selected and that each claim is classified according to the astrological sign of the driver. The following table is consistent with accident rates given in the article:

Astrological Sign	Observed Number in Sample
Aries	83
Taurus	84
Gemini	83
Cancer	82
Leo	83
Virgo	81
Libra	83
Scorpio	85
Sagittarius	81
Capricorn	88
Aquarius	85
Pisces	82

Test the null hypothesis that the proportion of accident claims submitted by drivers of each astrological sign is consistent with the proportion of policyholders of each sign. Use the information on the distribution of policyholders from Part (a) to compute expected frequencies, and then carry out an appropriate test.

12.36 In 1989, each person in a random sample of prison inmates was classified according to the type of offense committed, resulting in the one-way frequency table below (these values are based on information in the article "Profile of Jail Inmates," *USA Today*, April 25, 1991).

In 1983, it was reported that for a random sample of inmates, 30.7% had been convicted of violent crimes, 38.6% of crimes against property, 9.3% of drug-related crimes, 20.6% of public-order offenses, and 0.8% of other types of crimes. Do these data provide sufficient evidence to conclude that the true 1989 proportions falling in the various offense categories are not all the same as in 1983? Test the relevant hypotheses using $\alpha = .05$.

12.37 In a study of 2989 cancer deaths, the location of death (home, acute-care hospital, or chronic-care facility) and age at death were recorded, resulting in the following two-way frequency table ("Where Cancer Patients Die," *Public Health Reports* [1983]: 173):

Age	Location		
	Home	Acute-Care	Chronic-Care
15–54	94	418	23
55–64	116	524	34
65–74	156	581	109
Over 74	138	558	238

Using a .01 significance level, test the null hypothesis that age at death and location of death are independent.

12.38 The *Los Angeles Times* (July 29, 1983) conducted a survey to find out why some Californians don't register to vote. Random samples of 100 Latinos, 100 non-Hispanic whites, and 100 African Americans who were not registered to vote were selected. The resulting data are summarized in the following table:

Reason for Not Registering	Ethnic Group		
	Latino	White	African American
Not a Citizen	45	8	0
Not Interested	19	33	19
Can't Meet Residency Requirements	9	35	23
Distrust of Politics	5	10	8
Too Difficult to Register	10	10	27
Other Reason	12	4	23

Do the data suggest that the true proportion falling into each response category is not the same for Latinos, whites, and African Americans? Use a .05 significance level.

12.39 The relative importance attached to work and home life by high school students was examined in "Work Role Salience as a Determinant of Career Maturity in High School Students" (*Journal of Voca-*

Table for Exercise 12.36

	Type of Offense				
	Violent	Property	Drug-Related	Public-Order	Other
Frequency	225	300	230	228	16

tional Behavior [1984]: 30–44). Do the data summarized in the accompanying two-way frequency table suggest that gender and relative importance assigned to work and home are not independent? Test using a .05 level of significance.

	Relative Importance		
Gender	Work > Home	Work = Home	Work < Home
Female	68	26	94
Male	75	19	57

12.40 The accompanying 2 × 2 frequency table is the result of classifying random samples of 112 librarians and 108 faculty members of the California State University system with respect to gender ("Job Satisfaction Among Faculty and Librarians: A Study of Gender, Autonomy, and Decision Making Opportunities," *Journal of Library Administration* [1984]: 43–56):

	Faculty	Librarians
Male	56	59
Female	52	53

Do the data strongly suggest that librarians and faculty members differ with respect to the proportion of males and females? Test using a .05 level of significance.

12.41 Is there any relationship between the age of an investor and the rate of return that the investor expects from an investment? A random sample of 972 common stock investors was selected, and each investor was placed in one of four age categories and in one of four categories according to rate believed attainable ("Patterns of Investment Strategy and Behavior Among Individual Investors," *Journal of Business* [1977]: 296–333). The resulting data are given in the following table:

	Rate Believed Attainable			
Investor Age	0–5%	6–10%	11–15%	Over 15%
Under 45	15	51	51	29
45–54	31	133	70	48
55–64	59	139	35	20
65 and Over	84	157	32	18

Does there appear to be an association between age and rate of return believed attainable? Test the appropriate hypotheses using a .01 significance level.

12.42 The article "Participation of Senior Citizens in the Swine Flu Inoculation Program" (*Journal of*

Gerontology [1979]: 201–208) described a study of the factors thought to influence a person's decision to obtain a flu vaccination. Each member of a sample of 122 senior citizens was classified according to belief about the likelihood of getting the flu and vaccine status to obtain the two-way frequency table shown here:

	Vaccine Status	
Belief	Received Vaccine	Didn't Receive Vaccine
Very Unlikely	25	24
Unlikely	30	11
Likely	6	8
Don't Know	5	13

Using a .05 significance level, determine whether there is an association between belief and vaccine status.

12.43 The article "An Instant Shot of 'Aah': Cocaine Use Among Methadone Clients" (*Journal of Psychoactive Drugs* [1984]: 217–227) reported the following data on frequency of cocaine use for individuals in three different treatment groups:

Cocaine Use per Week	Treatment		
	A	B	C
None	149	75	8
1–2 Times	26	27	15
3–6 Times	6	20	11
At Least 7 Times	4	10	10

Do the data suggest that the true proportion of individuals in each of the different cocaine use categories differs for the three treatments? Carry out an appropriate test at level .05.

12.44 The article "Identification of Cola Beverages" (*Journal of Applied Psychology* [1962]: 356–360) reported on an experiment in which each of 79 subjects was presented with glasses of cola in pairs and asked to identify which glass contained a specific brand of cola. The following data appeared in the article:

	Number of Correct Identifications			
Cola	0	1	2	3 or 4
Coca-Cola	13	23	24	19
Pepsi Cola	12	20	26	21
Royal Crown	18	28	19	14

Do these data suggest that individuals' abilities to make correct identification differ for the different brands of cola?

▪ References

Agresti, Alan, and B. Finlay. *Statistical Methods for the Social Sciences*, 3d ed. Englewood Cliffs, NJ: Prentice-Hall, 1997. (This book includes a good discussion of measures of association for two-way frequency tables.)

Everitt, B. S. *The Analysis of Contingency Tables*. New York: Halsted Press, 1977. (A compact but informative survey of methods for analyzing categorical data.)

Mosteller, Frederick, and Robert Rourke. *Sturdy Statistics*. Reading, Mass.: Addison-Wesley, 1973. (Contains several readable chapters on the varied uses of the chi-square statistic.)

13 · Simple Linear Regression and Correlation: Inferential Methods

Regression and correlation were introduced in Chapter 5 as techniques for describing and summarizing bivariate data consisting of (x, y) pairs. For example, consider a regression of the dependent variable y = average SAT score on the independent variable x = spending per pupil (thousands of dollars) based on a sample of $n = 44$ New Jersey school districts ("Cost-Effectiveness in Public Education," *Chance* [1995]: 38–41). A scatterplot of the data shows a linear pattern; the sample correlation coefficient is $r = .4$; and the equation of the least-squares line is $\hat{y} = 766 + 15.0x$. When $x = 10$ is substituted into this equation, $\hat{y}$ is calculated to be 916. This number can be interpreted either as a *point estimate* of the average score for all districts that spend \$10,000 per pupil or as a *point prediction* for the score in a single district that spends this amount of money per pupil. In this chapter, we develop inferential methods for bivariate numerical data, including a confidence interval (interval estimate) for a mean y value, a prediction interval for a single y value, and a test of hypotheses regarding the extent of correlation in the entire population of (x, y) pairs.

▪ 13.1 Simple Linear Regression Model

A *deterministic relationship* is one in which the value of y is completely determined by the value of an independent variable x. Such a relationship can be described using traditional mathematical notation, such as $y = f(x)$, where $f(x)$ is a specified function of x. For example, we might have

$$y = f(x) = 10 + 2x$$

or

$$y = f(x) = 4 - (10)^{2x}$$

However, in most situations, the variables of interest are not deterministically related. For example, the value of y = first-year college grade point average is certainly not determined solely by x = high school grade point average, and y = crop yield is determined partly by factors other than x = amount of fertilizer used.

A description of the relation between two variables x and y that are not deterministically related can be given by specifying a **probabilistic model**. The general form of an **additive probabilistic model** allows y to be larger or smaller than $f(x)$ by a random amount e. The **model equation** is of the form

y = deterministic function of x + random deviation

$= f(x) + e$

Let x^* denote some particular value of x, and suppose that an observation on y is made when $x = x^*$. Then

$$y > f(x^*) \quad \text{if } e > 0$$
$$y < f(x^*) \quad \text{if } e < 0$$
$$y = f(x^*) \quad \text{if } e = 0$$

Thinking geometrically, if $e > 0$, the observed point (x^*, y) will lie above the graph of $y = f(x)$; $e < 0$ implies that this point will fall below the graph. This is illustrated in Figure 13.1. If $f(x)$ is a function used in a probabilistic model relating y to x and if observations on y are made for various values of x, the resulting (x, y) points will be distributed about the graph of $f(x)$, some falling above it and some falling below it.

Figure 13.1 A deviation from the deterministic part of a probabilistic model.

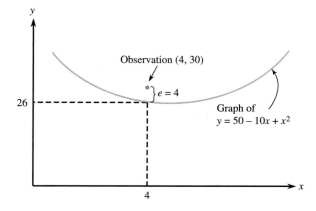

▪ Simple Linear Regression

The simple linear regression model is a special case of the general probabilistic model in which the deterministic function $f(x)$ is linear (so its graph is a straight line).

▪ **Definition**

The **simple linear regression model** assumes that there is a line with vertical or y intercept α and slope β, called the **true** or **population regression line**. When a value of the independent variable x is fixed and an observation on the dependent variable y is made,

$y = \alpha + \beta x + e$

Without the random deviation e, all observed (x, y) points would fall exactly on the population regression line. The inclusion of e in the model equation recognizes that points will deviate from the line.

Figure 13.2 shows several observations in relation to the population regression line.

Figure 13.2 Several observations resulting from the simple linear regression model.

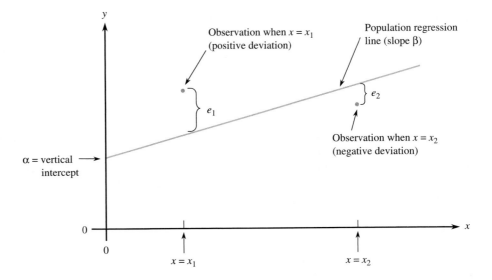

Before we make an observation on y for any particular value of x, we are uncertain about the value of e. It could be negative, positive, or even 0. Also, it might be quite large in magnitude (a point far from the population regression line) or quite small (a point very close to the line). In this chapter, we make some assumptions about the distribution of e in repeated sampling at any particular x value.

▪ **Basic Assumptions of the Simple Linear Regression Model**

1. The distribution of e at any particular x value has mean value 0. That is, $\mu_e = 0$.
2. The standard deviation of e (which describes the spread of its distribution) is the same for any particular value of x. This standard deviation is denoted by σ.
3. The distribution of e at any particular x value is normal.
4. The random deviations $e_1, e_2, \ldots, e_n$ associated with different observations are independent of one another.

Randomness in e implies that y itself is subject to uncertainty. The foregoing assumptions about the distribution of e imply that the distribution of y values in repeated sampling satisfies certain properties. Consider y when x has some fixed value x^*, so that

$$y = \alpha + \beta x^* + e$$

Because α and β are fixed numbers, $\alpha + \beta x^*$ is also a fixed number. The sum of a fixed number and a normally distributed variable is again a normally distributed variable (the bell-shaped curve is simply relocated), so y itself has a normal distribution. Furthermore, $\mu_e = 0$ implies that the mean value of y is just $\alpha + \beta x^*$, the height of the population regression line above the value x^*. Finally, because there is no variability in the fixed number $\alpha + \beta x^*$, the standard deviation of y is the same as that of e. These properties are summarized in the following box.

For any fixed x value, y itself has a normal distribution, with

$$\left(\begin{array}{c}\text{mean } y \text{ value} \\ \text{for fixed } x\end{array}\right) = \left(\begin{array}{c}\text{height of the population} \\ \text{regression line above } x\end{array}\right) = \alpha + \beta x$$

and

$$(\text{standard deviation of } y \text{ for a fixed } x) = \sigma$$

The slope β of the population regression line is the *average* change in y associated with a 1-unit increase in x. The y intercept α is the height of the population line when $x = 0$. The value of σ determines the extent to which (x, y) observations deviate from the population line; when σ is small, most observations are quite close to the line, but with large σ, there are likely to be some substantial deviations.

The key features of the model are illustrated in Figures 13.3 and 13.4. Notice that the three normal curves in Figure 13.3 have identical spreads. This is a consequence of $\sigma_e = \sigma$, which does not depend on x.

Figure 13.3 Illustration of the simple linear regression model.

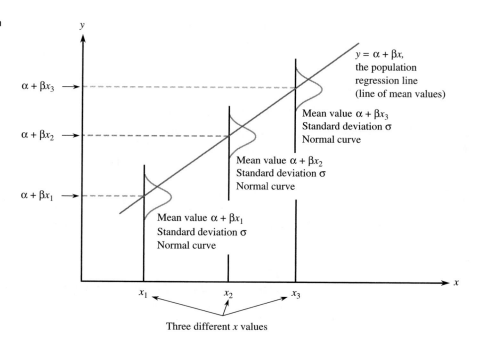

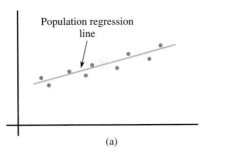
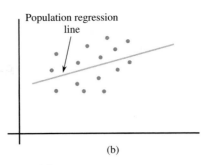

Figure 13.4 Data from the simple linear regression model:
(a) small σ; (b) large σ.

■ Example 13.1   Stand on Your Head to Lose Weight?

The authors of the article "On Weight Loss by Wrestlers Who Have Been Standing on Their Heads" (paper presented at the Sixth International Conference on Statistics, Combinatorics, and Related Areas, Forum for Interdisciplinary Mathematics, 1999) stated that "amateur wrestlers who are overweight near the end of the weight certification period, but just barely so, have been known to stand on their heads for a minute or two, get on their feet, step back on the scale, and establish that they are in the desired weight class. Using a headstand as the method of last resort has become a fairly common practice in amateur wrestling."

Does this really work? Data were collected in an experiment where weight loss was recorded for each wrestler after exercising for 15 min and then doing a headstand for 1 min 45 sec. Based on these data, the authors of the article concluded that there was in fact a demonstrable weight loss that was greater than that for a control group that exercised for 15 min but did not do the headstand. (The authors give a plausible explanation for why this might be the case based on the way blood and other body fluids collect in the head during the headstand and the effect of weighing while these fluids are draining immediately after standing.) The authors also concluded that a simple linear regression model was a reasonable way to relate the variable

y = weight loss (in pounds)

to

x = body weight before exercise and headstand (in pounds)

Suppose that the actual model equation has $\alpha = 0$, $\beta = 0.001$, and $\sigma = 0.09$ (these values are consistent with the findings in the article). The population regression line is shown in Figure 13.5.

If the distribution of the random errors at any fixed weight (x value) is normal, then the variable y = weight loss is normally distributed with

$\mu_y = 0 + 0.001x$

$\sigma_y = 0.09$

For example, when $x = 190$ (corresponding to a 190-lb wrestler), weight loss has mean value

$\mu_y = 0 + 0.001(190) = 0.19$

Figure 13.5 The population regression line for Example 13.1.

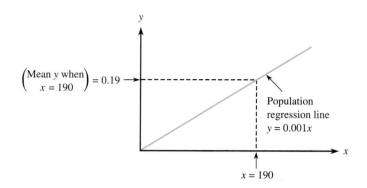

Because the standard deviation of y is $\sigma = 0.09$, the interval $0.19 \pm 2(0.09) = (0.01, 0.37)$ includes y values that are within 2 standard deviations of the mean value for y when $x = 190$. Roughly 95% of the weight loss observations made for 190-lb wrestlers will be in this range.

The slope $\beta = 0.001$ is the change in average weight associated with each additional pound of body weight.

More insight into model properties can be gained by thinking of the population of all (x, y) pairs as consisting of many smaller populations. Each one of these smaller populations contains pairs for which x has a fixed value. Suppose, for example, that the variables

x = grade point average in major courses

and

y = starting salary after graduation

are related according to the simple linear regression model. Then there is the population of all pairs with $x = 3.20$, the population of all pairs having $x = 2.75$, and so on. The model assumes that for each such population, y is normally distributed with the same standard deviation, and the *mean y value* (rather than y itself) is linearly related to x.

In practice, the judgment of whether the simple linear regression model is appropriate must be based on how the data were collected and on a scatterplot of the data. The sample observations should be independent of one another. In addition, the scatterplot should show a linear rather than a curved pattern, and the vertical spread of points should be relatively homogeneous throughout the range of x values. Figure 13.6 shows plots with three different patterns; only the first is consistent with the model assumptions.

■ Estimating the Population Regression Line

The values of α and β (y intercept and slope of the population regression line) will almost never be known to an investigator. Instead, these values must first be estimated from the sample data $(x_1, y_1), \ldots, (x_n, y_n)$. We now assume that these n (x, y) pairs were obtained independently of one another and that each observed y is related to the corresponding x by means of the model equation for simple linear regression.

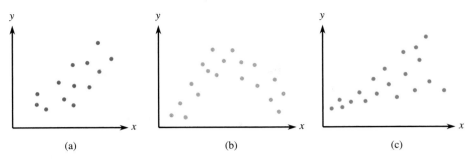

Figure 13.6 Some commonly encountered patterns in scatterplots: (a) pattern consistent with the simple linear regression model; (b) pattern suggesting a nonlinear probabilistic model; (c) pattern suggesting that variability in y changes with x.

Let a and b denote point estimates of α and β, respectively. These estimates come from applying the method of least squares introduced in Chapter 5; the sum of squared vertical deviations of points in the scatterplot from the least-squares line is smaller than for any other line.

The point estimates of β (the slope) and α (the y intercept of the population regression line) are the slope and y intercept, respectively, of the least-squares line. That is,

$$b = \text{point estimate of } \beta = \frac{S_{xy}}{S_{xx}}$$

$$a = \text{point estimate of } \alpha = \bar{y} - b\bar{x}$$

where

$$S_{xy} = \sum xy - \frac{(\sum x)(\sum y)}{n}$$

$$S_{xx} = \sum x^2 - \frac{(\sum x)^2}{n}$$

The estimated regression line is then just the least-squares line

$$\hat{y} = a + bx$$

Let x^* denote a specified value of the predictor variable x. Then $a + bx^*$ has two different interpretations:

1. It is a point estimate of the mean y value when $x = x^*$, and
2. It is a point prediction of an individual y value that would be observed when $x = x^*$.

▪ Example 13.2   Mother's Age and Baby's Birth Weight

Medical researchers have noted that adolescent females are much more likely to deliver low-birth-weight babies than are adult females. Because low-birth-weight babies have higher mortality rates, a number of studies have examined the relationship between birth weight and mother's age for babies born to young mothers.

One such study is described in the article "The Risk of Teen Mothers Having Low Birth Weight Babies: Implications of Recent Medical Research for School Health Personnel" (*Journal of School Health* [1998]: 271–274). The following data on

x = maternal age (in years)

and

y = birth weight of baby (in grams)

are consistent with summary values given in the referenced article and also with data published by the National Center for Health Statistics.

					Observation					
	1	**2**	**3**	**4**	**5**	**6**	**7**	**8**	**9**	**10**
x	15	17	18	15	16	19	17	16	18	19
y	2289	3393	3271	2648	2897	3327	2970	2535	3138	3573

A scatterplot (Figure 13.7) supports the appropriateness of the simple linear regression model.

Figure 13.7 Scatterplot of the data from Example 13.2.

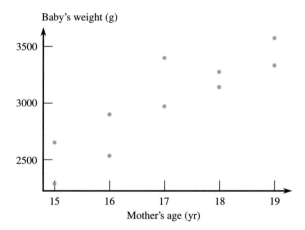

The summary statistics are

$$n = 10 \qquad \sum x = 170 \qquad \sum y = 30{,}041$$

$$\sum x^2 = 2910 \qquad \sum xy = 515{,}600 \qquad \sum y^2 = 91{,}785{,}351$$

from which

$$S_{xy} = \sum xy - \frac{(\sum x)(\sum y)}{n} = 515{,}600 - \frac{(170)(30{,}041)}{10} = 4903.0$$

$$S_{xx} = \sum x^2 - \frac{(\sum x)^2}{n} = 2910 - \frac{(170)^2}{10} = 20.0$$

$$\bar{x} = \frac{170}{10} = 17.0$$

$$\bar{y} = \frac{30{,}041}{10} = 3004.1$$

This gives

$$b = \frac{S_{xy}}{S_{xx}} = \frac{4903.0}{20.0} = 245.15$$

$$a = \bar{y} - b\bar{x} = 3004.1 - (245.15)(17.0) = -1163.45$$

The equation of the estimated regression line is then

$$\hat{y} = a + bx = -1163.45 + 245.15x$$

A point estimate of the average birth weight of babies born to 18-year-old mothers results from substituting $x = 18$ into the estimated equation:

$$\text{(estimated average } y \text{ when } x = 18) = a + bx$$
$$= -1163.45 + 245.15(18)$$
$$= 3249.25 \text{ g}$$

Similarly, we would predict the birth weight of a baby to be born to a particular 18-year-old mother to be

$$\text{(predicted } y \text{ value when } x = 18) = a + b(18) = 3249.25 \text{ g}$$

The point estimate and the point prediction are identical, because the same x value was used in each calculation. However, the interpretation of each is different. One represents our prediction of the weight of a single baby whose mother is 18, whereas the other represents our estimate of the average weight of *all* babies born to 18-year-old mothers. This distinction will become important in Section 13.4, when we consider interval estimates and predictions.

In Example 13.2, the x values in the sample ranged from 15 to 19. An estimate or prediction should not be attempted for any x value much outside this range. Without sample data for such values, there is no evidence that the estimated linear relationship can be extrapolated very far. Statisticians refer to this potential pitfall as the **danger of extrapolation**.

■ Estimating σ^2 and σ

The value of σ determines the extent to which observed points (x, y) tend to fall close to or far away from the population regression line. A point estimate of σ is based on

$$\text{SSResid} = \sum(y - \hat{y})^2$$

where $\hat{y}_1 = a + bx_1, \ldots, \hat{y}_n = a + bx_n$ are the fitted or predicted y values and the residuals are $y_1 - \hat{y}_1, \ldots, y_n - \hat{y}_n$. SSResid is a measure of the extent to which the sample data spread out about the estimated regression line.

■ **Definition**

The statistic for estimating the variance σ^2 is

$$s_e = \frac{\text{SSResid}}{n - 2}$$

(continued)

where

$$\text{SSResid} = \sum (y - \hat{y})^2 = \sum y^2 - a \sum y - b \sum xy$$

The subscript e in s_e^2 reminds us that we are estimating the variance of the "errors" or residuals.

The estimate of σ is the **estimated standard deviation**

$$s_e = \sqrt{s_e^2}$$

The number of degrees of freedom associated with estimating σ^2 or σ in simple linear regression is $n - 2$.

The estimates and number of degrees of freedom here have analogs in our previous work involving a single sample $x_1, x_2, \ldots, x_n$. The sample variance s^2 had numerator $\sum (x - \overline{x})^2$, a sum of squared deviations (residuals), and denominator $n - 1$, the number of degrees of freedom associated with s^2 and s. The use of $\overline{x}$ as an estimate of μ in the formula for s^2 reduces the number of degrees of freedom by 1, from n to $n - 1$. In simple linear regression, estimation of α and β results in a loss of 2 degrees of freedom, leaving $n - 2$ as the number of degrees of freedom for SSResid, s_e^2, and s_e.

The coefficient of determination was defined previously (see Chapter 5) as

$$r^2 = 1 - \frac{\text{SSResid}}{\text{SSTo}}$$

where

$$\text{SSTo} = \sum (y - \overline{y})^2 = \sum y^2 - \frac{(\sum y)^2}{n} = S_{yy}$$

The value of r^2 can now be interpreted as the proportion of observed y variation that can be explained by (or attributed to) the model relationship. The estimate s_e also gives another assessment of model performance. Roughly speaking, the value of σ represents the magnitude of a typical deviation of a point (x, y) in the population from the population regression line. Similarly, in a rough sense, s_e is the magnitude of a typical sample deviation (residual) from the least-squares line. The smaller the value of s_e, the closer the points in the sample fall to the line and the better the line does in predicting y from x.

▪ Example 13.3   Woodpecker Hole Depth

Forest managers are increasingly concerned about the damage done to animal populations when forests are clear-cut. Woodpeckers are a valuable forest asset, both because they provide nest and roost holes for other animals and birds and because they prey on many forest insect pests. The article "Artificial Trees as a Cavity Substrate for Woodpeckers" (*Journal of Wildlife Management* [1983]: 790–798) reported on a study of how woodpeckers behaved when provided with polystyrene cylinders as an alternative roost and nest cavity substrate. We give selected values of

x = ambient temperature (°C)

and

y = cavity depth (in centimeters)

These values were read from a scatterplot that appeared in the article:

Observation	Temperature	Depth	Predicted y Value	Residual
1	−6	21.1	22.195	−1.095
2	−3	26.0	21.160	4.840
3	−2	18.0	20.815	−2.815
4	1	19.2	19.780	−0.580
5	6	16.9	18.055	−1.155
6	10	18.1	16.675	1.425
7	11	16.8	16.330	0.470
8	19	11.8	13.569	−1.769
9	21	11.0	12.879	−1.879
10	23	12.1	12.189	−0.089
11	25	14.8	11.499	3.301
12	26	10.5	11.154	−0.654

The scatterplot (Figure 13.8) gives evidence of a negative linear relationship between x and y. The summary statistics are

$$n = 12 \qquad \sum x = 131 \qquad \sum y = 196.3$$

$$\sum x^2 = 2939 \qquad \sum xy = 1622.3 \qquad \sum y^2 = 3445.25$$

from which we calculate

$$b = -0.345043$$
$$a = 20.125053$$
$$\text{SSResid} = 54.4655$$
$$\text{SSTo} = 234.1093$$

Thus,

$$r^2 = 1 - \frac{\text{SSResid}}{\text{SSTo}} = 1 - \frac{54.4655}{234.1093} = 1 - 0.233 = .767$$

$$s_e^2 = \frac{\text{SSResid}}{n-2} = \frac{54.4655}{10} = 5.447$$

$$s_e = \sqrt{5.447} = 2.33$$

Approximately 76.7% of the observed variation in cavity depth y can be attributed to the probabilistic linear relationship with ambient temperature. The magnitude of a typical sample deviation from the least-squares line is about 2.3, which is reasonably small in comparison to the y values themselves. The model appears to

Figure 13.8 MINITAB scatterplot for Example 13.3.

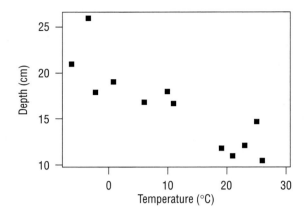

be useful for estimation and prediction; in Section 13.2, we show how a model utility test can be used to judge whether this is indeed the case.

A key assumption of the simple linear regression model is that the random deviation e in the model equation is normally distributed. In Section 13.3, we will indicate how the residuals can be used to determine whether this is plausible.

▪ Exercises 13.1–13.11

13.1 Let x be the size of a house (in square feet), and let y be the amount of natural gas used (in therms) during a specified period. Suppose that for a particular community, x and y are related according to the simple linear regression model with

> β = slope of population regression line
> = 0.017
>
> α = y intercept of population regression line
> = −5.0

a. What is the equation of the population regression line?
b. Graph the population regression line by first finding the point on the line corresponding to $x = 1000$ and then the point corresponding to $x = 2000$, and drawing a line through these points.
c. What is the mean value of gas usage for houses with 2100 ft^2 of space?
d. What is the average change in usage associated with a 1-ft^2 increase in size?
e. What is the average change in usage associated with a 100-ft^2 increase in size?
f. Would you use the model to predict mean usage for a 500-ft^2 house? Why or why not? (Note: There are no small houses in the community in which this model is valid.)

13.2 The flow rate in a device used for air quality measurement depends on the pressure drop x (in

inches of water) across the device's filter. Suppose that for x values between 5 and 20, these two variables are related according to the simple linear regression model with true regression line $y = −0.12 + 0.095x$.
a. What is the true average flow rate for a pressure drop of 10 in.? a drop of 15 in.?
b. What is the true average change in flow rate associated with a 1-in. increase in pressure drop? Explain.
c. What is the average change in flow rate when pressure drop decreases by 5 in.?

13.3 Data presented in the article "Manganese Intake and Serum Manganese Concentration of Human Milk-Fed and Formula-Fed Infants" (*American Journal of Clinical Nutrition* [1984]: 872–878) suggest that a simple linear regression model is reasonable for describing the relationship between y = serum manganese (Mn) and x = Mn intake (in milligrams per kilogram per day). Suppose that the true regression line is $y = −2 + 1.4x$ and that $\sigma = 1.2$. Then for a fixed x value, y has a normal distribution with mean $−2 + 1.4x$ and standard deviation 1.2.
a. What is the mean value of serum Mn when Mn intake is 4.0? when Mn intake is 4.5?
b. What is the probability that an infant whose Mn intake is 4.0 will have a serum Mn level greater than 5?
c. Approximately what proportion of infants whose Mn intake is 5 will have a serum Mn level greater than 5? less than 3.8?

13.4 A sample of small cars was selected, and the values of x = horsepower and y = fuel efficiency (in miles per gallon) were determined for each car. Fitting the simple linear regression model gave the estimated regression equation $\hat{y} = 44.0 - 0.150x$.

a. How would you interpret $b = -0.150$?

b. Substituting $x = 100$ gives $\hat{y} = 29.0$. Give two different interpretations of this number.

c. What happens if you predict efficiency for a car with a 300-hp engine? Why do you think this has occurred?

d. Interpret $r^2 = .680$ in the context of this problem.

e. Interpret $s_e = 3.0$ in the context of this problem.

13.5 Suppose that a simple linear regression model is appropriate for describing the relationship between y = house price and x = house size (in square feet) for houses in a large city. The true regression line is $y = 23,000 + 47x$ and $\sigma = 5000$.

a. What is the average change in price associated with 1 extra square foot of space? With an additional 100 ft^2 of space?

b. What proportion of 1800-ft^2 homes would be priced over $110,000? under $100,000?

13.6 In the context of simple linear regression, explain the following differences:

a. The difference between the line $y = \alpha + \beta x$ and the line $\hat{y} = a + bx$.

b. The difference between β and b.

c. The difference between $\alpha + \beta x^*$ and $a + bx^*$, where x^* denotes a particular value of the independent variable.

d. The difference between σ and s_e.

13.7 Legumes, such as peas and beans, are important crops whose production is greatly affected by pests. The article "Influence of Wind Speed on Residence Time of *Uroleucon ambrosiae alatae* on Bean Plants" (*Environmental Entomology* [1991]: 1375–1380) reported on a study in which aphids were placed on a bean plant and the elapsed time until half the aphids had departed was observed. Data on x = wind speed (in meters per second) and y = residence half-time was given and used to produce the following information: $a = 0.0119$, $b = 3.4307$, $n = 13$, SSTo $= 73.937$, and SSResid $= 27.890$.

a. What percentage of observed variation in residence half-time can be attributed to the simple linear regression model?

b. Give a point estimate of σ, and interpret the estimate.

c. Estimate the mean change in residence half-time associated with a 1-m/sec increase in wind speed.

d. Calculate a point estimate of true average residence half-time when wind speed is 1 m/sec.

13.8 The following data on x = treadmill run time to exhaustion (in minutes) and y = 20-km ski time (in minutes) was taken from the article "Physiological Characteristics and Performance of Top U.S. Biathletes" (*Medicine and Science in Sports and Exercise* [1995]: 1302–1310):

x	7.7	8.4	8.7	9.0	9.6	9.6	10.0
y	71.0	71.4	65.0	68.7	64.4	69.4	63.0

x	10.2	10.4	11.0	11.7
y	64.6	66.9	62.6	61.7

$$\sum x = 106.3 \quad \sum x^2 = 1040.95 \quad \sum y = 728.70$$

$$\sum xy = 7009.91 \quad \sum y^2 = 48,390.79$$

a. Draw the scatterplot for these data. Does the scatterplot suggest that the simple linear regression model is appropriate?

b. Determine the equation of the estimated regression line, and draw the line on your scatterplot.

c. What is your estimate of the average change in ski time associated with a 1-min increase in treadmill time?

d. What would you predict ski time to be for an individual whose treadmill time is 10 min?

e. Should the model be used as a basis for predicting ski time when treadmill time is 15 min? Explain.

f. Calculate and interpret the value of r^2.

g. Calculate and interpret the value of s_e.

13.9 The following summary quantities resulted from a study in which x was the number of photocopy machines serviced during a routine service call and y was the total service time (in minutes):

$$n = 16$$
$$\sum(y - \bar{y})^2 = 22,398.05$$
$$\sum(y - \hat{y})^2 = 2620.57$$

a. What proportion of observed variation in total service time can be explained by a linear probabilistic relationship between total service time and the number of machines serviced?

b. Calculate the value of the estimated standard deviation s_e. What is the number of degrees of freedom associated with this estimate?

13.10 Exercise 5.43 described a regression situation in which y = hardness of molded plastic and x = amount of time elapsed since termination of the molding process. Summary quantities included $n = 15$, SSResid $= 1235.470$, and SSTo $= 25,321.368$.

a. Calculate a point estimate of σ. On how many degrees of freedom is the estimate based?

b. What percentage of observed variation in hardness can be explained by the simple linear regression model relationship between hardness and elapsed time?

13.11 The following data on x = advertising share and y = market share for a particular brand of cigarettes during 10 randomly selected years is from the article "Testing Alternative Econometric Models on the Existence of Advertising Threshold Effect" (*Journal of Marketing Research* [1984]: 298–308):

x	.103	.072	.071	.077	.086	.047
y	.135	.125	.120	.086	.079	.076

x	.060	.050	.070	.052
y	.065	.059	.051	.039

a. Construct a scatterplot for these data. Do you think the simple linear regression model is appropriate for describing the relationship between x and y?

b. Calculate the equation of the estimated regression line and use it to obtain the predicted market share when the advertising share is .09.

c. Compute r^2. How would you interpret this value?

d. Calculate a point estimate of σ. On how many degrees of freedom is your estimate based?

■ 13.2 Inferences About the Slope of the Population Regression Line

The slope coefficient β in the simple linear regression model is the average or expected change in the dependent variable y associated with a 1-unit increase in the value of the independent variable x. For example, consider x = the size of a house (in square feet) and y = selling price of the house. If we assume that the simple linear regression model is appropriate for the population of houses in a particular city, β would be the average increase in selling price associated with a 1-ft^2 increase in size. As another example, if x = amount of time per week a computer system is used and y = the resulting annual maintenance expense, then β would be the expected change in expense associated with using the computer system one additional hour per week.

Because the value of β is almost always unknown, it has to be estimated from the n independently selected observations $(x_1, y_1), \ldots, (x_n, y_n)$. The slope b of the least-squares line gives a point estimate. As with any point estimate, though, it is desirable to have some indication of how accurately b estimates β. In some situations, the value of the statistic b may vary greatly from sample to sample, so b computed from a single sample may well be rather different from the true slope β. In other situations, almost all possible samples yield b values that are quite close to β, so the error of estimation is almost sure to be small. To proceed further, we need some facts about the sampling distribution of b: information about the shape of the sampling distribution curve, where the curve is centered relative to β, and how much the curve spreads out about its center.

■ Properties of the Sampling Distribution of b

When the four basic assumptions of the simple linear regression model are satisfied, the following conditions are met:

1. The mean value of b is β. That is, $\mu_b = \beta$, so the sampling distribution of b is always centered at the value of β. Thus, b is an unbiased statistic for estimating β.

2. The standard deviation of the statistic b is

$$\sigma_b = \frac{\sigma}{\sqrt{S_{xx}}}$$

3. The statistic b has a normal distribution (a consequence of the model assumption that the random deviation e is normally distributed).

The fact that b is unbiased means only that the sampling distribution is centered at the right place; it gives no information about dispersion. If σ_b is large, then the sampling distribution of b will be quite spread out around β and an estimate far from β may well result. For σ_b to be small, the numerator σ should be small (little variability about the population line) and/or the denominator $\sqrt{S_{xx}}$ or, equivalently, $S_{xx} = \Sigma(x - \bar{x})^2$ itself should be large. Because $\Sigma(x - \bar{x})^2$ is a measure of how much the observed x values spread out, β tends to be more precisely estimated when the x values in the sample are spread out rather than when they are close together.

The normality of b implies that the standardized variable

$$z = \frac{b - \beta}{\sigma_b}$$

has a standard normal distribution. However, inferential methods cannot be based on this variable, because the value of σ_b is not available (since the unknown σ appears in the numerator of σ_b). The obvious way out of this dilemma is to estimate σ with s_e, yielding an estimated standard deviation.

The **estimated standard deviation of the statistic b** is

$$s_b = \frac{s_e}{\sqrt{S_{xx}}}$$

When the four basic assumptions of the simple linear regression model are satisfied, the probability distribution of the standardized variable

$$t = \frac{b - \beta}{s_b}$$

is the t distribution with df $= (n - 2)$.

In the same way that $t = \dfrac{\bar{x} - \mu}{s/\sqrt{n}}$ was used in Chapter 9 to develop a confidence interval for μ, the t variable in the preceding box can be used to obtain a confidence interval (interval estimate) for β.

■ **Confidence Interval for β**

When the four basic assumptions of the simple linear regression model are satisfied, a **confidence interval for β**, the slope of the population regression line, has the form

$$b \pm (t \text{ critical value})(s_b)$$

where the t critical value is based on df $= n - 2$. Appendix Table 3 gives critical values corresponding to the most frequently used confidence levels.

The interval estimate of β is centered at b and extends out from the center by an amount that depends on the sampling variability of b. When s_b is small, the interval is narrow, implying that the investigator has relatively precise knowledge of β.

▪ Example 13.4   Athletic Performance and Cardiovascular Fitness

Is cardiovascular fitness (as measured by time to exhaustion from running on a treadmill) related to an athlete's performance in a 20-km ski race? The following data on

x = treadmill time to exhaustion (in minutes)

and

y = 20-km ski time (in minutes)

were taken from the article "Physiological Characteristics and Performance of Top U.S. Biathletes" (*Medicine and Science in Sports and Exercise* [1995]: 1302–1310):

x	7.7	8.4	8.7	9.0	9.6	9.6	10.0	10.2	10.4	11.0	11.7
y	71.0	71.4	65.0	68.7	64.4	69.4	63.0	64.6	66.9	62.6	61.7

A scatterplot of the data appears in the following figure:

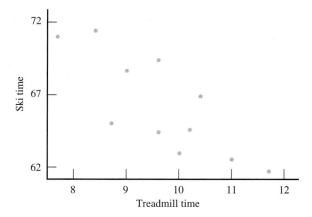

The plot shows a linear pattern, and the vertical spread of points does not appear to be changing over the range of x values in the sample. If we assume that the distribution of errors at any given x value is approximately normal, then the simple linear regression model seems appropriate.

The slope β in this context is the average change in ski time associated with a 1-min increase in treadmill time. The scatterplot shows a negative linear relationship, so the point estimate of β is negative.

Straightforward calculation gives

$$n = 11 \qquad \sum x = 106.3 \qquad \sum y = 728.70$$

$$\sum x^2 = 1040.95 \qquad \sum xy = 7009.91 \qquad \sum y^2 = 48{,}390.79$$

from which

$$b = -2.3335$$
$$a = 88.796$$
$$\text{SSResid} = 43.097$$
$$\text{SSTo} = 117.727$$

$r^2 = .634$ (63.4% of the observed variation in ski time can be explained by the simple linear regression model)

$s_e^2 = 4.789$

$s_e = 2.188$

$$s_b = \frac{s_e}{\sqrt{S_{xx}}} = \frac{2.188}{3.702} = 0.591$$

Calculation of the 95% confidence interval for β requires a t critical value based on df $= n - 2 = 11 - 2 = 9$, which (from Appendix Table 3) is 2.26. The resulting interval is then

$$b \pm (t \text{ critical value})(s_b) = -2.3335 \pm (2.26)(0.591)$$
$$= -2.3335 \pm 1.336$$
$$= (-3.671, -0.999)$$

We interpret this interval as follows: Based on the sample data, we are 95% confident that the true average decrease in ski time associated with a 1-min increase in treadmill time is between 1 and 3.7 min.

Output from any of the standard statistical computer packages routinely includes the computed values of a, b, SSResid, SSTo, and s_b. Figure 13.9 displays partial MINITAB output for the data of Example 13.4. The format from other packages is similar. Rounding occasionally leads to small discrepancies between hand-calculated and computer-calculated values, but there are no such discrepancies in this example.

Figure 13.9 Partial MINITAB output for the data of Example 13.4.

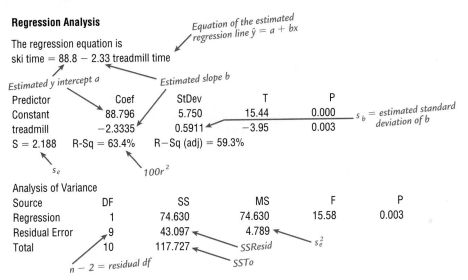

■ Hypothesis Tests Concerning β

Hypotheses about β can be tested using a t test similar to the t tests discussed in Chapters 10 and 11. The null hypothesis states that β has a specified hypothesized value. The t statistic results from standardizing b, the point estimate of β, under the assumption that H_0 is true. When H_0 is true, the sampling distribution of this statistic is the t distribution with df $= n - 2$.

▪ **Summary of Hypothesis Tests Concerning β**

Null hypothesis: $H_0: \beta =$ hypothesized value

Test statistic: $t = \dfrac{b - \text{hypothesized value}}{s_b}$

The test is based on df $= n - 2$.

Alternative Hypothesis	*P*-Value
$H_a:$ $\beta >$ hypothesized value	Area to the right of the computed t under the appropriate t curve
$H_a:$ $\beta <$ hypothesized value	Area to the left of the computed t under the appropriate t curve
$H_a:$ $\beta \neq$ hypothesized value	(1) 2(area to the right of t) if t is positive, or (2) 2(area to the left of the t) if t is negative

Assumptions: For this test to be appropriate, the four basic assumptions of the simple linear regression model must be met:
1. The distribution of e at any particular x value has mean value 0 (i.e., $\mu_e = 0$).
2. The standard deviation of e is σ, which does not depend on x.
3. The distribution of e at any particular x value is normal.
4. The random deviations $e_1, e_2, \ldots, e_n$, associated with different observations are independent of one another.

Frequently, the null hypothesis of interest is $\beta = 0$. When this is the case, the population regression line is a horizontal line, and the value of y in the simple linear regression model does not depend on x. That is,

$$y = \alpha + 0x + e$$

or equivalently,

$$y = \alpha + e$$

In this situation, knowledge of x is of no use in predicting y. On the other hand, if $\beta \neq 0$, there is a useful linear relationship between x and y, and knowledge of x is useful for predicting y. This is illustrated by the scatterplots in Figure 13.10.

The test of $H_0: \beta = 0$ versus $H_a: \beta \neq 0$ is called the *model utility test for simple linear regression*.

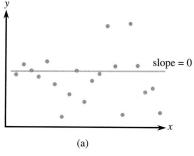
(a)

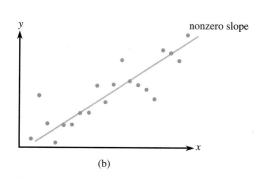
(b)

Figure 13.10 (a) $\beta = 0$; (b) $\beta \neq 0$.

■ **Model Utility Test for Simple Linear Regression**

The **model utility test for simple linear regression** is the test of

$$H_0: \quad \beta = 0$$

versus

$$H_a: \quad \beta \neq 0$$

The null hypothesis specifies that there is *no* useful linear relationship between x and y, whereas the alternative hypothesis specifies that there *is* a useful linear relationship between x and y. If H_0 is rejected, we conclude that the simple linear regression model is useful for predicting y. The test procedure in the previous box (with hypothesized value = 0) is used to carry out the model utility test; in particular, the test statistic is the ***t* ratio** $\dfrac{b}{s_b}$.

If a scatterplot and the r^2 value do not provide convincing evidence for a useful linear relationship, we recommend that the model utility test be carried out before the estimated model is used to make other inferences.

■ **Example 13.5**   **Image Quality and Reading Speed**

Computers seem to have invaded almost every aspect of our lives. In recent years, this has led to an explosion in the number of investigations in which the objective is to study and document the varied effects of the computer revolution. Many of these effects have dealt with the work environment of people who spend a substantial amount of time staring at computer monitors. Image quality of monitors is an important characteristic, affecting, among other things, extent of eye strain and work efficiency. The article "Image Quality Determines Differences in Reading Performance and Perceived Image Quality with CRT and Hard Copy Displays" (*Human Factors* [1991]: 459–469) reported on an experiment in which image quality (x) and average time for a group of subjects to read certain passages (y, in seconds) were determined. The following data were read from a graph in the article:

x	4.30	4.55	5.55	5.65	5.95	6.30	6.45	6.45
y	8.00	8.30	7.80	7.25	7.70	7.50	7.60	7.20

The summary statistics necessary for a simple linear regression analysis are as follows:

$$n = 8 \qquad \sum x = 45.20 \qquad \sum y = 61.35$$

$$\sum x^2 = 260.215 \qquad \sum xy = 344.9425 \qquad \sum y^2 = 471.4325$$

from which

$$b = -0.348501$$
$$a = 9.637778$$
$$\text{SSResid} = 0.367626$$
$$s_e = 0.247530$$
$$r^2 = .615$$
$$S_{xx} = 4.835$$

Because $r^2 = .615$, about 61.5% of observed variation in reading time can be explained by the simple linear regression model. It appears from this that there is a useful linear relation between the two variables, but a confirmation requires a formal model utility test. We will use a significance level of .05 to carry out this test.

1. β = the true average change in reading time associated with a 1-unit increase in image quality.

2. $H_0: \beta = 0$.

3. $H_a: \beta \neq 0$.

4. $\alpha = .05$.

5. Test statistic: $t = \dfrac{b - \text{hypothesized value}}{s_b} = \dfrac{b - 0}{s_b} = \dfrac{b}{s_b}$

6. Assumptions: The following scatterplot of the data shows a linear pattern, and the variability of points does not appear to be changing with x:

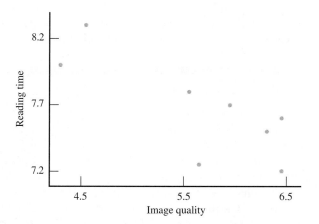

Assuming that the distribution of errors at any given x value is approximately normal, the assumptions of the simple linear regression model are appropriate.

7. Calculation: The calculation of t requires

$$s_b = \frac{s_e}{\sqrt{S_{xx}}} = \frac{0.247530}{2.1989} = 0.11257$$

yielding

$$t = \frac{-0.348501 - 0}{0.11257} = -3.10$$

8. *P*-value: Appendix Table 4 shows that for a t test based on 6 df, $P(t > 3.10) = P(t < -3.10) = .011$. The inequality in H_a requires a two-tailed test, so

$$P\text{-value} = 2(.011) = .022$$

9. Conclusion: Because .022 is smaller than the significance level .05, H_0 is rejected. We conclude, as did the investigators, that there is a useful linear relationship between image quality and reading time.

Figure 13.11 shows partial MINITAB output from a simple linear regression analysis. The Coef column gives $b = -0.3485$; $s_b = 0.1125$ is in the StDev column;

Figure 13.11 MINITAB output for the data of Example 13.5.

Regression Analysis

The regression equation is
reading time = 9.64 − 0.349 image quality

Predictor	Coef	StDev	T	P
Constant	9.6378	0.6419	15.01	0.000
image qu	−0.3485	0.1125	−3.10	0.021

S = 0.2475   R-Sq = 61.5%   R-Sq(adj) = 55.1%

Analysis of Variance

Source	DF	SS	MS	F	P
Regression	1	0.58722	0.58722	9.59	0.021
Residual Error	6	0.36746	0.06124		
Total	7	0.95469			

the T column (for t ratio) contains the value of the test statistic for testing H_0: $\beta = 0$; and the P-value for the model utility test is given in the last column as 0.021 (slightly different from the one given in Step 8 because of rounding). Other commonly used statistical packages also include this information in their output.

When H_0: $\beta = 0$ cannot be rejected by the model utility test at a reasonably small significance level, the search for a useful model must continue. One possibility is to relate y to x using a nonlinear model — an appropriate strategy if the scatterplot shows curvature. Alternatively, a multiple regression model using more than one predictor variable can be employed. We introduce such models in Chapter 14.

▪ Exercises 13.12–13.26

13.12 What is the difference between σ and σ_b? What is the difference between σ_b and s_b?

13.13 Suppose that a single y observation is made at each of the x values 5, 10, 15, 20, and 25.
a. If $\sigma = 4$, what is the standard deviation of the statistic b?
b. Now suppose that a second observation is made at every x value (for a total of 10 observations). Is the resulting value of σ_b half of what it was in Part (a)?
c. How many observations at each x value in Part (a) are required to yield a σ_b value that is half the value calculated in Part (a)? Verify your conjecture.

13.14 Refer back to Example 13.3 in which the simple linear regression model was fit to data on x = ambient temperature and y = cavity depth. For the purpose of estimating β as accurately as possible, would it have been preferable to make observations at the x values −6, −5, −4, −3, −2, −1, 21, 22, 23, 24, 25, and 26? Explain your reasoning.

13.15 Exercise 13.10 presented information from a study in which y was the hardness of molded plastic and x was the time elapsed since termination of the molding process. Summary quantities included

$$n = 15$$
$$b = 2.50$$
$$\text{SSResid} = 1235.470$$
$$\sum(x - \bar{x})^2 = 4024.20$$

a. Calculate the estimated standard deviation of the statistic b.
b. Obtain a 95% confidence interval for β, the slope of the true regression line.
c. Does the interval in Part (b) suggest that β has been precisely estimated? Explain.

13.16 A study was carried out to relate sales revenue y (in thousands of dollars) to advertising expenditure x (also in thousands of dollars) for fast-food outlets during a 3-month period. A sample of 15 outlets yielded the following summary quantities:

$$\sum x = 14.10 \quad \sum y = 1438.50 \quad \sum x^2 = 13.92$$

$$\sum y^2 = 140,354 \quad \sum xy = 1387.20$$

$$\sum (y - \bar{y})^2 = 2401.85 \quad \sum (y - \hat{y})^2 = 561.46$$

a. What proportion of observed variation in sales revenue can be attributed to the linear relationship between revenue and advertising expenditure?
b. Calculate s_e and s_b.
c. Obtain a 90% confidence interval for β, the average change in revenue associated with a $1000 (i.e., a 1-unit) increase in advertising expenditure.

13.17 An experiment to study the relationship between x = time spent exercising (in minutes) and y = amount of oxygen consumed during the exercise period resulted in the following summary statistics:

$$n = 20 \quad \sum x = 50 \quad \sum y = 16,705 \quad \sum x^2 = 150$$

$$\sum y^2 = 14,194,231 \quad \sum xy = 44,194$$

a. Estimate the slope and y intercept of the population regression line.
b. One sample observation on oxygen usage was 757 for a 2-min exercise period. What amount of oxygen consumption would you predict for this exercise period, and what is the corresponding residual?
c. Compute a 99% confidence interval for the true average change in oxygen consumption associated with a 1-min increase in exercise time.

13.18 Exercise 5.66 presented data on x = average hourly wage and y = quit rate for a sample of industries. Here is the accompanying MINITAB output:

The regression equation is
quit rate = 4.86 − 0.347 wage

Predictor	Coef	Stdev	t-ratio	p
Constant	4.8615	0.5201	9.35	0.000
wage	−0.34655	0.05866	−5.91	0.000

s = 0.4862   R-sq = 72.9%   R-sq(adj) = 70.8%
Analysis of Variance

Source	DF	SS	MS	F	p
Regression	1	8.2507	8.2507	34.90	0.000
Error	13	3.0733	0.2364		
Total	14	11.3240			

a. Based on the given *P*-value, does there appear to be a useful linear relationship between average wage and quit rate? Explain your reasoning.
b. Calculate an estimate of the average change in quit rate associated with a $1 increase in average hourly wage, and do so in a way that conveys information about the precision and reliability of the estimate.

13.19 The article "Cost-Effectiveness in Public Education," cited in the chapter introduction, reported that, for a sample of n = 44 school districts, a regression of y = average SAT score on x = expenditure per pupil (in thousands of dollars) gave b = 15.0 and s_b = 5.3.
a. Does the simple linear regression model specify a useful relationship between x and y?
b. Calculate and interpret a confidence interval for β based on a 95% confidence level.

13.20 The article "Root Dentine Transparency: Age Determination of Human Teeth Using Computerized Densitometric Analysis" (*American Journal of Physical Anthropology* [1991]: 25–30) described a study in which the objective was to predict age (y) from percentage of a tooth's root with transparent dentine. The following data are for anterior teeth:

x	15	19	31	39	41	44	47	48	55	65
y	23	52	65	55	32	60	78	59	61	60

Use the following MINITAB output to decide whether the simple linear regression model is useful:

The regression equation is
age = 32.1 + 0.555 percent

Predictor	Coef	Stdev	t-ratio	p
Constant	32.08	13.32	2.41	0.043
percent	0.5549	0.3101	1.79	0.111

s = 14.30   R-sq = 28.6%   R-sq(adj) = 19.7%
Analysis of Variance

Source	DF	SS	MS	F	p
Regression	1	654.8	654.8	3.20	0.111
Error	8	1635.7	204.5		
Total	9	2290.5			

13.21 Data on mean response time for a group of individuals with closed-head injury (CHI) and a matched control group without head injury on 10 different tasks were given in the article "Cognitive Slowing in Closed-Head Injury" (*Brain and Cognition* [1996]: 429–440). Each observation was based on a different study and used different subjects, so it is reasonable to assume that the observations are independent. The following data were read from a plot (and are a subset of the complete data set) given in the article:

Study	Mean Response Time Control	Mean Response Time CHI	Study	Mean Response Time Control	Mean Response Time CHI
1	250	303	6	740	1044
2	360	491	7	880	1421
3	475	659	8	920	1329
4	525	683	9	1010	1481
5	610	922	10	1200	1815

a. Fit a linear regression model that would allow you to predict the mean response time for those suffering a CHI from the mean response time on the same task for individuals with no head injury.

b. Do the sample data support the hypothesis that there is a useful linear relationship between the mean response time for individuals with no head injury and the mean response time for individuals with CHI? Test the appropriate hypotheses using $\alpha = .05$.

c. It is also possible to test hypotheses about the y intercept in a linear regression model. For these data, the null hypothesis $H_0: \alpha = 0$ cannot be rejected at the .05 significance level, suggesting that a model with a y intercept of 0 might be an appropriate model. Fitting such a model results in an estimated regression equation of

CHI = 1.48(Control)

Interpret the estimated slope of 1.48.

13.22 The article "Effects of Enhanced UV-B Radiation on Ribulose-1,5-Biphosphate, Carboxylase in Pea and Soybean" (*Environmental and Experimental Botany* [1984]: 131–143) included the following data on pea plants, with y = sunburn index and x = distance (in centimeters) from an ultraviolet light source:

x	0.18	0.21	0.25	0.26	0.30	0.32	0.36	0.40
y	4.0	3.7	3.0	2.9	2.6	2.5	2.2	2.0

x	0.40	0.50	0.51	0.54	0.61	0.62	0.63
y	2.1	1.5	1.5	1.5	1.3	1.2	1.1

$$\Sigma x = 609 \quad \Sigma y = 33.1 \quad \Sigma x^2 = 28{,}037$$

$$\Sigma y^2 = 84.45 \quad \Sigma xy = 1156.8$$

Estimate the mean change in the sunburn index associated with an increase of 1 cm in distance in a way that includes information about the precision of estimation.

13.23 Exercise 13.16 described a regression analysis in which y = sales revenue and x = advertising expenditure. Summary quantities given there yielded $n = 15$, $b = 52.27$, and $s_b = 8.05$.

a. Test the hypothesis $H_0: \beta = 0$ versus $H_a: \beta \neq 0$ using a significance level of .05. What does your conclusion say about the nature of the relationship between x and y?

b. Consider the hypothesis $H_0: \beta = 40$ versus $H_a: \beta > 40$. The null hypothesis states that the average change in sales revenue associated with a 1-unit increase in advertising expenditure is (at most) $40,000. Carry out a test using significance level .01.

13.24 The article "Technology, Productivity, and Industry Structure" (*Technological Forecasting and Social Change* [1983]: 1–13) included the following data on x = research and development expenditure and y = growth rate for eight different industries:

x	2024	5038	905	3572	1157	327	378	191
y	1.90	3.96	2.44	0.88	0.37	−0.90	0.49	1.01

a. Would a simple linear regression model provide useful information for predicting growth rate from research and development expenditure? Use a .05 level of significance.

b. Use a 90% confidence interval to estimate the average change in growth rate associated with a 1-unit increase in expenditure. Interpret the resulting interval.

13.25 The article "Effect of Temperature on the pH of Skim Milk" (*Journal of Dairy Research* [1988]: 277–280) reported on a study involving x = temperature (°C) under specified experimental conditions and y = milk pH. The following data (read from a graph) are a representative subset of the data that appeared in the article:

x	4	4	24	24	25	38	38	40
y	6.85	6.79	6.63	6.65	6.72	6.62	6.57	6.52

x	45	50	55	56	60	67	70	78
y	6.50	6.48	6.42	6.41	6.38	6.34	6.32	6.34

$$\Sigma x = 678 \quad \Sigma y = 104.54 \quad \Sigma x^2 = 36{,}056$$

$$\Sigma y^2 = 683.4470 \quad \Sigma xy = 4376.36$$

Do these data strongly suggest that there is a negative linear relationship between temperature and pH? State and test the relevant hypotheses using a significance level of .01.

13.26 In anthropological studies, an important characteristic of fossils is cranial capacity. Frequently, skulls are at least partially decomposed, so it is necessary to use other characteristics to obtain information about capacity. One such measure that has been used is the length of the lambda-opisthion chord. The article "Vertesszollos and the Presapiens Theory" (*American Journal of Physical Anthropology* [1971]) reported the following data for $n = 7$ *Homo erectus* fossils:

x (chord length, mm)	78	75	78	81
y (capacity, cm³)	850	775	750	975

x (chord length, mm)	84	86	87
y (capacity, cm³)	915	1015	1030

Suppose that from previous evidence, anthropologists had believed that for each 1-mm increase in chord length, cranial capacity would be expected to increase by 20 cm³. Do these new experimental data strongly contradict the prior belief?

▪ 13.3 Checking Model Adequacy

The simple linear regression model equation is

$$y = \alpha + \beta x + e$$

where e represents the random deviation of an observed y value from the population regression line $\alpha + \beta x$. The inferential methods presented in Section 13.2 required some assumptions about e. These assumptions include:

1. At any particular x value, the distribution of e is a normal distribution.
2. At any particular x value, the standard deviation of e is σ, which is constant over all values of x (i.e., σ does not depend on x).

Inferences based on the simple linear regression model continue to be reliable when model assumptions are slightly violated (e.g., mild nonnormality of the random deviation distribution). However, using an estimated model in the face of grossly violated assumptions can result in misleading conclusions. Therefore, it is desirable to have easily applied methods available for identifying such serious violations and for suggesting how a satisfactory model can be obtained.

▪ Residual Analysis

If the deviations $e_1, e_2, \ldots, e_n$ from the population line were available, they could be examined for any inconsistencies with model assumptions. For example, a normal probability plot would suggest whether or not the normality assumption was tenable. But, because

$$e_1 = y_1 - (\alpha + \beta x_1)$$
$$\vdots$$
$$e_n = y_n - (\alpha + \beta x_n)$$

these deviations can be calculated only if the equation of the population line is known. In practice, this will never be the case. Instead, diagnostic checks must be based on the residuals

$$y_1 - \hat{y}_1 = y_1 - (a + bx_1)$$
$$\vdots$$
$$y_n - \hat{y}_n = y_n - (a + bx_n)$$

which are the deviations from the *estimated* line.

The values of the residuals will vary from sample to sample. When all model assumptions are met, the mean value of the residuals at any particular x value is 0. Any observation that gives a large positive or negative residual should be examined carefully for any anomalous circumstances, such as a recording error or exceptional experimental conditions. Identifying residuals with unusually large magnitudes is made easier by inspecting **standardized residuals**.

Recall that a quantity is standardized by subtracting its mean value (0 in this case) and dividing by its true or estimated standard deviation. Thus

$$\text{standardized residual} = \frac{\text{residual}}{\text{estimated standard deviation of residual}}$$

The value of a standardized residual tells how many standard deviations the corresponding residual lies from its expected value, 0.

Because residuals at different x values have different standard deviations (depending on the value of x for that observation),* computing the standardized residuals can be tedious. Fortunately, many computer regression programs provide standardized residuals as part of the output.

In Chapter 7, a normal probability plot was introduced as a technique for deciding whether the n observations in a random sample could plausibly have come from a normal population distribution. To assess whether the assumption that $e_1, e_2, \ldots, e_n$ all come from the same normal distribution is reasonable, we recommend a normal probability plot of the standardized residuals.

▪ Example 13.6   Landslides and Timber Growth

Landslides are common events in tree-growing regions of the Pacific Northwest, so their effect on timber growth is of special concern to foresters. The article "Effects of Landslide Erosion on Subsequent Douglas Fir Growth and Stocking Levels in the Western Cascades, Oregon" (*Soil Science Society of America Journal* [1984]: 667–671) reported on the results of a study in which growth in a landslide area was compared with growth in a previously clear-cut area. We present data on clear-cut growth, with x = tree age (in years) and y = 5-year height growth (in centimeters). The scatterplot in Figure 13.12 is consistent with the assumptions of the simple linear regression model.

Figure 13.12 MINITAB scatterplot for the data of Example 13.6.

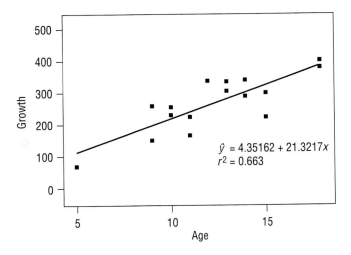

The residuals, their standard deviations, and the standardized residuals are given in Table 13.1. Except for $x = 5$ and $x = 18$, the two most extreme values, the residuals have roughly equal standard deviations. The residual with the largest magnitude, -99.2, initially seems quite extreme, but the corresponding standardized residual is only -2.04. That is, the residual is approximately 2 standard deviations below its expected value of 0, which is not terribly unusual in a sample of this size.

*The estimated standard deviation of the ith residual, $y_i - \hat{y}_i$, is

$$s_e\sqrt{1 - \frac{1}{n} - \frac{(x_i - \bar{x})^2}{S_{xx}}}$$

Table 13.1 ▪ Data, Residuals, and Standardized Residuals for Example 13.6

Observation	x	y	$\hat{y}$	Residual	Estimated Standard Deviation of Residual	Standardized Residual
1	5	70	111.0	−41.0	40.9	−1.00
2	9	150	196.2	−46.2	48.1	−0.96
3	9	260	196.2	63.8	48.1	1.33
4	10	230	217.6	12.4	49.0	0.25
5	10	255	217.6	37.4	49.0	0.76
6	11	165	238.9	−73.9	49.5	−1.49
7	11	225	238.9	−13.9	49.5	−0.28
8	12	340	260.2	79.8	49.8	1.60
9	13	305	281.5	23.5	49.7	0.47
10	13	335	281.5	53.5	49.7	1.08
11	14	290	302.9	−12.9	49.4	−0.26
12	14	340	302.9	37.1	49.4	0.75
13	15	225	324.2	−99.2	48.7	−2.04
14	15	300	324.2	−24.2	48.7	−0.50
15	18	380	388.1	−8.1	44.6	−0.18
16	18	400	388.1	11.9	44.6	0.27

On the standardized scale, no residual here is surprisingly large. Before standardization, there are some large residuals simply because there appears to be a substantial amount of variability about the true regression line ($s_e = 51.43$, $r^2 = .683$).

Figure 13.13 displays a normal probability plot of the standardized residuals and also one of the residuals. Notice that in this case the plots are nearly identical; it is usually the case that the two plots are similar. Although it is preferable to work with the standardized residuals, if you do not have access to a computer package or calculator that will produce standardized residuals, a plot of the unstandardized

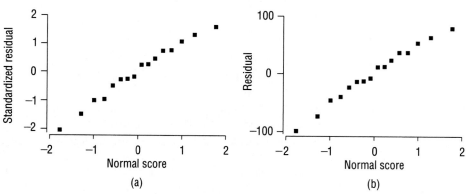

Figure 13.13 Normal probability plots for Example 13.6 (from MINITAB): (a) standardized residuals; (b) residuals.

residuals should suffice. Few plots are straighter than these! The plots would not cause us to question the assumption of normality.

■ Plotting the Residuals

A plot of the $(x, \text{residual})$ pairs is called a **residual plot**, and a plot of the $(x, \text{standardized residual})$ pairs is a **standardized residual plot**. Residual and standardized residual plots typically exhibit the same general shapes. If you are using a computer package or graphing calculator that calculates standardized residuals, we recommend using the standardized residual plot. If not, it is acceptable to use the residual plot instead.

A standardized residual or residual plot is often helpful in identifying unusual or highly influential observations and in checking for violations of model assumptions. A desirable plot is one that exhibits no particular pattern (such as curvature or a much greater spread in one part of the plot than in another) or one that has no point that is far removed from all the others. A point falling far above or far below the horizontal line at height 0 corresponds to a large standardized residual, which can indicate some kind of unusual behavior, such as a recording error, a nonstandard experimental condition, or an atypical experimental subject. A point that has an x value that differs greatly from others in the data set could have exerted excessive influence in determining the fitted line.

A standardized residual plot, such as the one pictured in Figure 13.14(a) (see next page), is desirable, because no point lies much outside the horizontal band between -2 and 2 (so there is no unusually large residual corresponding to an outlying observation); there is no point far to the left or right of the others (thus no observation that might greatly influence the fit), and there is no pattern to indicate that the model should somehow be modified. When the plot has the appearance of Figure 13.14(b), the fitted model should be changed to incorporate curvature (a nonlinear model).

The increasing spread from left to right in Figure 13.14(c) suggests that the variance of y is not the same at each x value but rather increases with x. A straight-line model may still be appropriate, but the best-fit line should be selected by using *weighted least squares* rather than ordinary least squares. This involves giving more weight to observations in the region exhibiting low variability and less weight to observations in the region exhibiting high variability. A specialized regression analysis textbook or a statistician should be consulted for details.

The standardized residual plots of Figures 13.14(d) and 13.14(e) show an extreme outlier and a potentially influential observation, respectively. Consider deleting the observation corresponding to such a point from the data set and refitting the same model. Substantial changes in estimates and various other quantities warn of instability in the data. The investigator should certainly carry out a more careful analysis and perhaps collect more data before drawing any firm conclusions. Improved computing power has allowed statisticians to develop and implement a variety of diagnostic tests for identifying unusual observations in a regression data set.

■ Example 13.7   Tree Age and 5-Year Growth

Figure 13.15 displays a standardized residual plot and a residual plot for the data of Example 13.6 regarding tree age and 5-year growth. The first observation was at $x_1 = 5$, and the corresponding standardized residual was -1.00, so the first plotted

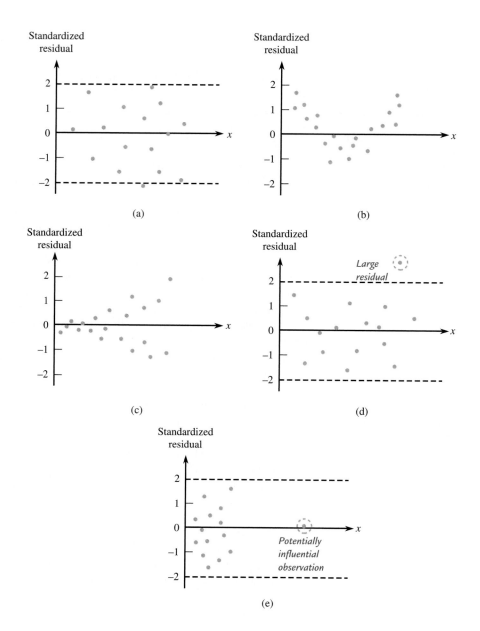

Figure 13.14 Examples of residual plots: (a) satisfactory plot; (b) plot suggesting that a curvilinear regression model is needed; (c) plot indicating nonconstant variance; (d) plot showing a large residual; (e) plot showing a potentially influential observation.

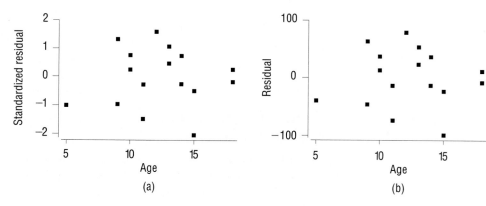

Figure 13.15 Plots for the data of Example 13.6 (from MINITAB): (a) standardized residual plot; (b) residual plot.

point in the standardized residual plot is (5, −1.00). Other points are similarly obtained and plotted. The standardized residual plot shows no unusual behavior that might call for model modifications or further analysis. Note that the general pattern in the residual plot is similar to that of the standardized residual plot.

▪ Example 13.8 Snow Cover and Temperature

The article "Snow Cover and Temperature Relationships in North America and Eurasia" (*Journal of Climate and Applied Meteorology* [1983]: 460–469) explored the relationship between October–November continental snow cover (*x*, in millions of square kilometers) and December–February temperature (*y*, in °C). The following data refer to Eurasia during the $n = 13$ time periods (1969–1970, 1970–1971, . . . , 1981–1982):

x	y	Standardized Residual	x	y	Standardized Residual
13.00	−13.5	−0.11	22.40	−18.9	−1.54
12.75	−15.7	−2.19	16.20	−14.8	0.04
16.70	−15.5	−0.36	16.70	−13.6	1.25
18.85	−14.7	1.23	13.65	−14.0	−0.28
16.60	−16.1	−0.91	13.90	−12.0	−1.54
15.35	−14.6	−0.12	14.75	−13.5	0.58
13.90	−13.4	0.34			

A simple linear regression analysis done by the authors yielded $r^2 = .52, r = .72$, suggesting a significant linear relationship. This is confirmed by a model utility test. The scatterplot and standardized residual plot are displayed in Figure 13.16 (see next page). There are no unusual patterns, although one standardized residual, −2.19, is a bit on the large side. The most interesting feature is the observation (22.40, −18.9), corresponding to a point far to the right of the others in these plots. This observation may have had a substantial influence on all aspects of the fit. The estimated slope when all 13 observations are included is $b = −0.459$, and $s_b = 0.133$. When the potentially influential observation is deleted, the estimate of β based on the remaining 12 observations is $b = −0.228$. Thus

$$\text{change in slope} = \text{original } b - \text{new } b$$
$$= −0.459 − (−0.288)$$
$$= −0.231$$

The change expressed in standard deviations is $\dfrac{−0.231}{0.133} = −1.74$. Because b has changed by substantially more than 1 standard deviation, the observation under consideration appears to be highly influential.

In addition, r^2 based just on the 12 observations is only .13, and the t ratio for testing $\beta = 0$ is not significant. Evidence for a linear relationship is much less conclusive in light of this analysis. The investigators should seek a climatological explanation for the influential observation and collect more data, which can be used to find an effective relationship.

Figure 13.16 Plots for the data of Example 13.8 (from MINITAB): (a) scatterplot; (b) standardized residual plot.

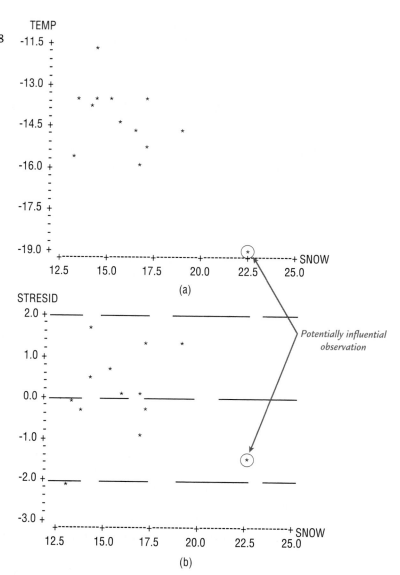

(a)

(b)

Potentially influential observation

■ **Example 13.9** Treadmill Time and Ski Time Revisited

Example 13.4 presented data on x = treadmill time and y = ski time. A simple linear regression model was fit to the data, and a confidence interval for β, the average change in ski time associated with a 1-min increase in treadmill time, was constructed. The validity of the confidence interval depends on the assumptions that the distribution of the residuals from the population regression line at any fixed x is approximately normal and that the variance of this distribution does not depend on x. Constructing a normal probability plot of the standardized residuals and a standardized residual plot will provide insight into whether these assumptions are in fact reasonable.

MINITAB was used to fit the simple linear regression model and compute the standardized residuals, resulting in the values shown in Table 13.2.

Table 13.2 ▪ Data, Residuals, and Standardized
Residuals for Example 13.9

Observation	Treadmill	Ski Time	Residual	Standardized Residual
1	7.7	71.0	0.172	0.10
2	8.4	71.4	2.206	1.13
3	8.7	65.0	−3.494	−1.74
4	9.0	68.7	0.906	0.44
5	9.6	64.4	−1.994	−0.96
6	9.6	69.4	3.006	1.44
7	10.0	63.0	−2.461	−1.18
8	10.2	64.6	−0.394	−0.19
9	10.4	66.9	2.373	1.16
10	11.0	62.6	−0.527	−0.27
11	11.7	61.7	0.206	0.12

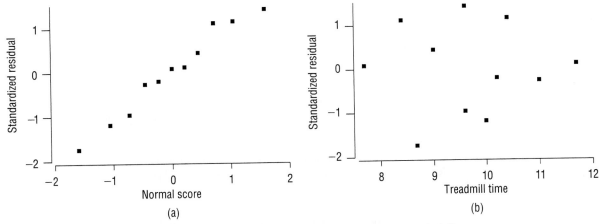

Figure 13.17 Plots for Example 13.9: (a) normal probability plot of standardized residuals; (b) standardized residual plot.

Figure 13.17 shows a normal probability plot of the standardized residuals and a standardized residual plot. The normal probability plot is quite straight, and the standardized residual plot does not show evidence of any patterns or of increasing spread. These observations support the use of the confidence interval in Example 13.4.

▪ Example 13.10 More on Image Quality and Reading Time

Use of the model utility test in Example 13.5 led to the conclusion that there was a useful linear relationship between y = average reading time and x = image quality. The validity of this test requires that the random error distribution be normal and

Table 13.3 ■ Data, Residuals, and Standardized Residuals for Example 13.10

Observation	Image Quality	Reading Time	Residual	Standardized Residual
1	4.30	8.00	−0.1392	−0.80
2	4.55	8.30	0.2479	1.27
3	5.55	7.80	0.0964	0.42
4	5.65	7.25	−0.4188	−1.81
5	5.95	7.70	0.1358	0.59
6	6.30	7.50	0.0578	0.26
7	6.45	7.60	0.2101	0.98
8	6.45	7.20	−0.1899	−0.89

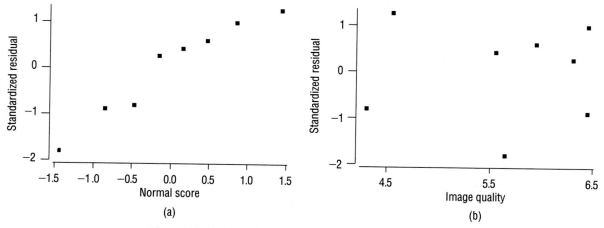

Figure 13.18 Plots for Example 13.10: (a) normal probability plot of standardized residuals; (b) standardized residual plot.

that the variability of the random error not change with x. Let's construct a normal probability plot of the standardized residuals and a standardized residual plot to see whether these assumptions are reasonable. Table 13.3 gives the residuals and standardized residuals that are needed to construct the plots shown in Figure 13.18.

The normal probability plot is quite straight, and the standardized residual plot does not show evidence of any pattern or increasing spread. This supports the use of the model utility test in Example 13.5.

Occasionally, you will see a residual plot or a standardized residual plot with $\hat{y}$ plotted on the horizontal axis rather than x. Because $\hat{y}$ is just a linear function of x, using $\hat{y}$ rather than x changes the scale of the horizontal axis but does not change the pattern of the points in the plot. As a consequence, plots that use $\hat{y}$ on the horizontal axis can be interpreted in the same manner as plots that use x.

When the distribution of the random deviation e has heavier tails than does the normal distribution, observations with large standardized residuals are not that unusual. Such observations can have great effects on the estimated regression line when the least-squares approach is used. Recently, statisticians have proposed a number of alternative methods — called **robust**, or **resistant**, methods — for fitting a line. Such methods give less weight to outlying observations than does the least-squares method without deleting the outliers from the data set. The most widely used robust procedures require a substantial amount of computation, so a good computer program is necessary. Associated confidence interval and hypothesis-testing procedures are still in the developmental stage.

▪ Exercises 13.27–13.33

13.27 Exercise 13.8 gave data on x = treadmill run time to exhaustion and y = 20-km ski time. The x values and corresponding standardized residuals from a simple linear regression are as follows:

x	7.7	8.4	8.7	9.0	9.6	9.6
St. resid.	0.10	1.13	−1.74	0.44	−0.96	1.44

x	10.0	10.2	10.4	11.0	11.7
St. resid.	−1.18	−0.19	1.16	−0.27	0.12

Construct a standardized residual plot. Does the plot exhibit any unusual features?

13.28 The article "Vital Dimensions in Volume Perception: Can the Eye Fool the Stomach?" (*Journal of Marketing Research* [1999]: 313–326) gave the following data on the dimensions of 27 representative food products (Gerber's baby food, Cheez Whiz, Skippy peanut butter, and Ahmed's tandoori paste, to name a few):

Product	Maximum Width	Minimum Width	Product	Maximum Width	Minimum Width
1	2.50	1.80	15	2.70	2.60
2	2.90	2.70	16	3.10	2.90
3	2.15	2.00	17	5.10	5.10
4	2.90	2.60	18	10.20	10.20
5	3.20	3.15	19	3.50	3.50
6	2.00	1.80	20	2.70	1.20
7	1.60	1.50	21	3.00	1.70
8	4.80	3.80	22	2.70	1.75
9	5.90	5.00	23	2.50	1.70
10	5.80	4.75	24	2.40	1.20
11	2.90	2.80	25	4.40	1.20
12	2.45	2.10	26	7.50	7.50
13	2.60	2.20	27	4.25	4.25
14	2.60	2.60			

a. Fit the simple linear regression model that would allow prediction of the maximum width (in centimeters) of a food container based on its minimum width (in centimeters).

b. Calculate the standardized residuals (or just the residuals if you don't have access to a computer program that gives standardized residuals), and plot them to determine whether there are any outliers. Would eliminating these outliers increase the accuracy of predictions using this regression? If so, eliminate any outlying data points and refit the regression.

c. Interpret the estimated slope and, if appropriate, the intercept.

d. Do you think that the assumptions of the simple linear regression model are reasonable? Give statistical evidence for your answer.

13.29 The authors of the article "Age, Spacing, and Growth Rate of *Tamarix* as an Indication of Lake Boundary Fluctuations at Sebkhet Kelbia, Tunisia" (*Journal of Arid Environments* [1982]: 43–51) used a simple linear regression model to describe the relationship between y = vigor (average width in centimeters of the last two annual rings in the tamarisk tree) and x = stem density (stems per square meter). The estimated model was based on the following data (also given are the standardized residuals):

x	4	5	6	9	14
y	0.75	1.20	0.55	0.60	0.65
St. resid.	−0.28	1.92	−0.90	−0.28	0.54

x	15	15	19	21	22
y	0.55	0.00	0.35	0.45	0.40
St. resid.	0.24	−2.05	−0.12	0.60	0.52

a. What assumptions are required for the simple linear regression model to be appropriate?

b. Construct a normal probability plot of the standardized residuals. Does the assumption that the random deviation distribution is normal appear to be reasonable? Explain.

c. Construct a standardized residual plot. Are there any unusually large residuals?

d. Is there anything about the standardized residual plot that would cause you to question the use of the simple linear regression model to describe the relationship between x and y?

13.30 The article "Effects of Gamma Radiation on Juvenile and Mature Cuttings of Quaking Aspen" (*Forest Science* [1967]: 240–245) reported the following data on x = exposure time to radiation (kiloroentgens per 16 hr) and y = dry weight of roots (in milligrams):

x	0	2	4	6	8
y	110	123	119	86	62

a. Construct a scatterplot for these data. Does the plot suggest that the simple linear regression model is appropriate?
b. The estimated regression line for these data is $\hat{y} = 127 - 6.65x$, and the standardized residuals are as given:

x	0	2	4	6	8
St. resid.	−1.55	0.68	1.25	−0.05	−1.06

Construct a standardized residual plot. What does the plot suggest about the adequacy of the simple linear regression model?

13.31 Carbon aerosols have been identified as a contributing factor in a number of air quality problems. In a chemical analysis of diesel engine exhaust, x = mass (micrograms per square centimeter) and y = elemental carbon (micrograms per square centimeter) were recorded ("Comparison of Solvent Extraction and Thermal Optical Carbon Analysis Methods: Application to Diesel Vehicle Exhaust Aerosol," *Environmental Science Technology* [1984]: 231–234). The estimated regression line for this data set is $\hat{y} = 31 + 0.737x$. The observed x and y values and the corresponding standardized residuals are as follows:

x	y	Standardized Residual
164.2	181	2.52
156.9	156	0.82
109.8	115	0.27
111.4	132	1.64
87.0	96	0.08
161.8	170	1.72
230.9	193	−0.73
106.5	110	0.05
97.6	94	−0.77
79.7	77	−1.11
118.7	106	−1.07
248.8	204	−0.95
102.4	98	−0.73
64.2	76	−0.20
89.4	89	−0.68
108.1	102	−0.75
89.4	91	−0.51

x	y	Standardized Residual
76.4	97	0.85
131.7	128	0.00
100.8	88	−1.49
78.9	86	−0.27
387.8	310	−0.89
135.0	141	0.91
82.9	90	−0.18
117.9	130	1.05

a. Construct a standardized residual plot. Are there any unusually large residuals? Do you think that there are any influential observations?
b. Is there any pattern in the standardized residual plot that would indicate that the simple linear regression model is not appropriate?
c. Based on your plot in Part (a), do you think that it is reasonable to assume that the variance of y is the same at each x value? Explain.

13.32 An investigation of the relationship between traffic flow x (thousands of cars per 24 hr) and lead content y of bark on trees near the highway (milligrams per gram, dry weight) yielded the following data (both residuals and standardized residuals from the least-squares line $\hat{y} = 28.7 + 33.3x$ are also given):

x	8.3	8.3	12.1	12.1	17.0
y	227	312	362	521	640
Residual	−78.1	6.9	−69.6	89.4	45.3
St. resid.	−0.99	0.09	−0.81	1.04	0.51

x	17.0	17.0	24.3	24.3	24.3
y	539	728	945	738	759
Residual	−55.7	133.3	107.2	−99.8	−78.8
St. resid.	−0.63	1.51	1.35	−1.25	−0.99

a. Plot the $(x, \text{residual})$ pairs. Does the resulting plot suggest that a simple linear regression model is an appropriate choice? Explain your reasoning.
b. Construct a standardized residual plot. Does the plot differ significantly in general appearance from the plot in Part (a)?

13.33 The following data on x = U.S. population (in millions) and y = crime index (in millions) appeared in the article "The Normal Distribution of Crime" (*Journal of Police Science and Administration* [1975]: 312–318):

Year	1963	1964	1965	1966	1967	1968
x	188.5	191.3	193.8	195.9	197.9	199.9
y	2.26	2.60	2.78	3.24	3.80	4.47

Year	1969	1970	1971	1972	1973
x	201.9	203.2	206.3	208.2	209.9
y	4.99	5.57	6.00	5.89	8.64

The author commented that "the simple linear regression analysis remains one of the most useful tools for crime prediction." When observations are made sequentially in time, the residuals or standardized residuals should be plotted in time order (i.e., first the one for time $t = 1$ (1963 here), then the one for time $t = 2$, etc.). Notice that here x increases with time, so an equivalent plot is of residuals or standardized residuals versus x. Using $\hat{y} = 47.26 + 0.260x$, calculate the residuals and plot the $(x, \text{residual})$ pairs. Does the plot exhibit a pattern that casts doubt on the appropriateness of the simple linear regression model? Explain.

▪ 13.4 Inferences Based on the Estimated Regression Line (Optional)

The number obtained by substituting a particular x value, x^*, into the equation of the estimated regression line has two different interpretations: It is a point estimate of the average y value when $x = x^*$, and it is also a point prediction of a single y value to be observed when $x = x^*$. How precise is this estimate or prediction? That is, how close is $a + bx^*$ to the actual mean value $\alpha + \beta x^*$ or to a particular y observation? Because both a and b vary in value from sample to sample (each one is a statistic), the statistic $a + bx^*$ also has different values for different samples. The way in which this statistic varies in value with different samples is summarized by its sampling distribution. Properties of the sampling distribution are used to obtain both a confidence interval formula for $\alpha + \beta x^*$ and a prediction interval formula for a particular y observation. The width of the corresponding interval conveys information about the precision of the estimate or prediction.

▪ Properties of the Sampling Distribution of $a + bx$ for a Fixed x Value

Let x^* denote a particular value of the independent variable x. When the four basic assumptions of the simple linear regression model are satisfied, the sampling distribution of the statistic $a + bx^*$ has the following properties:

1. The mean value of $a + bx^*$ is $\alpha + \beta x^*$, so $a + bx^*$ is an unbiased statistic for estimating the average y value when $x = x^*$.

2. The standard deviation of the statistic $a + bx^*$, denoted by σ_{a+bx^*}, is given by

$$\sigma_{a+bx^*} = \sigma\sqrt{\frac{1}{n} + \frac{(x^* - \overline{x})^2}{S_{xx}}}$$

3. The distribution of $a + bx^*$ is normal.

As you can see from the formula for σ_{a+bx^*}, the standard deviation of $a + bx^*$ is larger when $(x^* - \overline{x})^2$ is large than when it is small; that is, $a + bx^*$ tends to be a more precise estimate of $\alpha + \beta x^*$ when x^* is close to the center of the x values at which observations were made than when x^* is far from the center.

The standard deviation σ_{a+bx^*} cannot be calculated from the sample data, because the value of σ is unknown. However, σ_{a+bx^*} can be estimated by using s_e in place of σ. Using the mean and estimated standard deviation to standardize $a + bx^*$ gives a variable with a t distribution.

The **estimated standard deviation of the statistic $a + bx^*$**, denoted by s_{a+bx^*}, is given by

$$s_{a+bx^*} = s_e \sqrt{\frac{1}{n} + \frac{(x^* - \bar{x})^2}{S_{xx}}}$$

When the four basic assumptions of the simple linear regression model are satisfied, the probability distribution of the standardized variable

$$t = \frac{a + bx^* - (\alpha + \beta x^*)}{s_{a+bx^*}}$$

is the t distribution with df $= n - 2$.

▪ Inferences About the Mean y Value $\alpha + \beta x^*$

In previous chapters, standardized variables were manipulated algebraically to give confidence intervals of the form

(point estimate) $\pm$ (critical value)(estimated standard deviation)

A parallel argument leads to the following interval.

▪ Confidence Interval for a Mean y Value

When the basic assumptions of the simple linear regression model are met, a **confidence interval for $\alpha + \beta x^*$**, the average y value when x has value x^*, is

$$a + bx^* \pm (t \text{ critical value})(s_{a+bx^*})$$

where the t critical value is based on df $= n - 2$. Appendix Table 3 gives critical values corresponding to the most frequently used confidence levels.

Because s_{a+bx^*} is larger the farther x^* is from $\bar{x}$, the confidence interval widens as x^* moves away from the center of the data.

▪ Example 13.11 Shark Length and Jaw Width

Physical characteristics of sharks are of interest to surfers and scuba divers as well as to marine researchers. The following data on x = length (in feet) and y = jaw width (in inches) for 44 sharks was found in various articles appearing in the magazines *Skin Diver* and *Scuba News*:

x	y	x	y	x	y	x	y
18.7	17.5	14.6	13.9	16.7	15.2	12.2	14.8
12.3	12.3	15.8	14.7	17.8	18.2	15.2	15.9
18.6	21.8	14.9	15.1	16.2	16.7	14.7	15.3
16.4	17.2	17.6	18.5	12.6	11.6	12.4	11.9
15.7	16.2	12.1	12.0	17.8	17.4	13.2	11.6
18.3	19.9	16.4	13.8	13.8	14.2	15.8	14.3

x	y	x	y	x	y	x	y
14.3	13.3	13.6	14.2	16.2	15.7	15.7	14.3
16.6	15.8	15.3	16.9	22.8	21.2	19.7	21.3
9.4	10.2	16.1	16.0	16.8	16.3	18.7	20.8
18.2	19.0	13.5	15.9	13.6	13.0	13.2	12.2
13.2	16.8	19.1	17.9	13.2	13.3	16.8	16.9

Because it is difficult to measure jaw width in living sharks, researchers would like to determine whether it is possible to estimate jaw width from body length, which is more easily measured. A scatterplot of the data (Figure 13.19) shows a linear pattern and is consistent with use of the simple linear regression model.

Figure 13.19
Scatterplot for the data of Example 13.11.

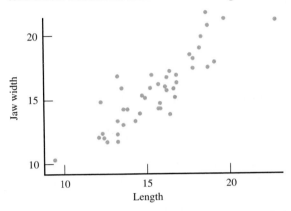

From the accompanying MINITAB output, it is easily verified that

$$a = 0.688$$
$$b = 0.96345$$
$$\text{SSResid} = 79.49$$
$$\text{SSTo} = 339.02$$
$$s_e = 1.376$$
$$r^2 = .766$$

Regression Analysis
The regression equation is
Jaw Width = 0.69 + 0.963Length

Predictor	Coef	StDev	T	P
Constant	0.688	1.299	0.53	0.599
Length	0.96345	0.08228	11.71	0.000

S = 1.376   R-Sq = 76.6%   R-Sq(adj) = 76.0%

Analysis of Variance

Source	DF	SS	MS	F	P
Regression	1	259.53	259.53	137.12	0.000
Residual Error	42	79.49	1.89		
Total	43	339.02			

From the data, we can also compute $S_{xx} = 279.8718$. Because $r^2 = .766$, the simple linear regression model explains 76.6% of the variability in jaw width. The model utility test also confirms the usefulness of this model (P-value = .000).

Let's use the data to compute a 90% confidence interval for the mean jaw width for 15-ft-long sharks. The mean jaw width when length is 15 ft is $\alpha + \beta(15)$. The point estimate is

$$a + b(15) = 0.688 + 0.96345(15) = 15.140 \text{ in.}$$

Because

$$\bar{x} = \frac{\sum x}{n} = \frac{685.80}{44} = 15.586$$

the estimated standard deviation of $a + b(15)$ is

$$s_{a+b(15)} = s_e\sqrt{\frac{1}{n} + \frac{(15 - \bar{x})^2}{S_{xx}}}$$

$$= (1.376)\sqrt{\frac{1}{44} + \frac{(15 - 15.586)^2}{279.8718}}$$

$$= 0.213$$

The t critical value for df $= 42$ is 1.68 (using the tabulated value for df $= 40$ from Appendix Table 3). We now have all the relevant quantities needed to compute a 90% confidence interval:

$$a + b(15) \pm (t \text{ critical value})(s_{a+bx^*}) = 15.140 \pm (1.68)(0.213)$$

$$= 15.140 \pm 0.358$$

$$= (14.782, 15.498)$$

Based on these sample data, we can be 90% confident that the mean jaw width for sharks whose length is 15 ft is between 14.782 and 15.498 in. As with all confidence intervals, the 90% confidence level means that we have used a *method* that has a 10% error rate to construct this interval estimate.

We have just considered estimation of the mean y value at a fixed $x = x^*$. When the basic assumptions of the simple linear regression model are met, the true value of this mean is $\alpha + \beta x^*$. The reason that our point estimate $a + bx^*$ is not exactly equal to $\alpha + \beta x^*$ is that the values of α and β are not known, so they have been estimated from sample data. As a result, the estimate $a + bx^*$ is subject to sampling variability and the extent to which the estimated line might differ from the true population line is reflected in the width of the confidence interval.

▪ Prediction Interval for a Single y

We now turn our attention to the problem of predicting a single y value at a particular $x = x^*$ (rather than estimating the mean y value when $x = x^*$). This problem is equivalent to trying to predict the y value of an individual point in a scatterplot of the population. If we use the estimated regression line to obtain a point prediction $a + bx^*$, this prediction will probably not be exactly equal to the true y value for two reasons. First, as was the case when estimating a mean y value, the estimated line is not going to be exactly equal to the true population regression line.

But, in the case of predicting a single y value, there is an additional source of error: e, the deviation from the line. Even if we knew the true population line, individual points would not fall exactly on the population line. This implies that there is more uncertainty associated with predicting a single y value at a particular x^* than with estimating the mean y value at x^*. This extra uncertainty is reflected in the width of the corresponding intervals.

An interval for a single y value, y^*, is called a **prediction interval** (to distinguish it from the confidence interval for a mean y value). The interpretation of a prediction interval is similar to the interpretation of a confidence interval. A 95% prediction interval for y^* is constructed using a method for which 95% of all possible samples would yield interval limits capturing y^*; only 5% of all samples would give an interval that did not include y^*.

Manipulation of a standardized variable similar to the one from which a confidence interval was obtained gives the following prediction interval.

▪ **Prediction Interval for a Single y Value**

When the four basic assumptions of the simple linear regression model are met, a **prediction interval for y^***, a single y observation made when $x = x^*$, has the form

$$a + bx^* \pm (t \text{ critical value}) \sqrt{s_e^2 + s_{a+bx^*}^2}$$

The prediction interval and the confidence interval are centered at exactly the same place, $a + bx^*$. The addition of s_e^2 under the square-root symbol makes the prediction interval wider — often substantially so — than the confidence interval.

▪ Example 13.12   Jaws II

In Example 13.11, we computed a 90% confidence interval for the mean jaw width of sharks whose length is 15 ft. Suppose that we are interested in predicting the jaw width of a single shark of length 15 ft. The required calculations for a 90% prediction interval for y^* are

$$a + b(15) = 0.688 + 0.96345(15) = 15.140$$
$$s_e^2 = (1.376)^2 = 1.8934$$
$$s_{a+b(15)}^2 = (0.213)^2 = 0.0454$$

The t critical value for df = 42 and a 90% prediction level is 1.68 (using the tabulated value for df = 40 from Appendix Table 3). Substitution into the prediction interval formula then gives

$$a + b(15) \pm (t \text{ critical value}) \sqrt{s_e^2 + s_{a+b(15)}^2} = 15.140 \pm (1.68)\sqrt{1.9388}$$
$$= 15.140 \pm 2.339$$
$$= (12.801, 17.479)$$

We can be 90% confident that an individual shark with length 15 ft will have a jaw width between 12.801 and 17.479 in. Notice that, as expected, this 90% prediction interval is much wider than the 90% confidence interval when $x^* = 15$ from Example 13.11.

Figure 13.20 gives MINITAB output that includes a 95% confidence interval and a 95% prediction interval when $x^* = 15$ and when $x^* = 20$. The intervals for $x^* = 20$ are wider than the corresponding intervals for $x^* = 15$ because 20 is farther from $\bar{x}$ (the center of the sample x values) than is 15. Each prediction interval is wider than the corresponding confidence interval. Figure 13.21 is a MINITAB plot that shows the estimated regression line as well as 90% confidence limits and prediction limits.

Figure 13.20 MINITAB output for the data of Example 13.12.

Regression Analysis
The regression equation is
Jaw Width = 0.69 + 0.963 Length

Predictor	Coef	StDev	T	P
Constant	0.688	1.299	0.53	0.599
Length	0.96345	0.08228	11.71	0.000

S = 1.376   R-Sq = 76.6%   R-Sq(adj) = 76.0%

Analysis of Variance

Source	DF	SS	MS	F	P
Regression	1	259.53	259.53	137.12	0.000
Residual Error	42	79.49	1.89		
Total	43	399.02			

Predicted Values

Fit	StDev Fit	95.0% CI	95.0% PI
15.140	0.213	(14.710, 15.569)	(12.330, 17.949)
19.957	0.418	(19.113, 20.801)	(17.055, 22.859)

$x^* = 15 \rightarrow$ 15.140
$x^* = 20 \rightarrow$ 19.957

Figure 13.21 MINITAB plot showing estimated regression line and 90% confidence and prediction limits for the data of Example 13.12.

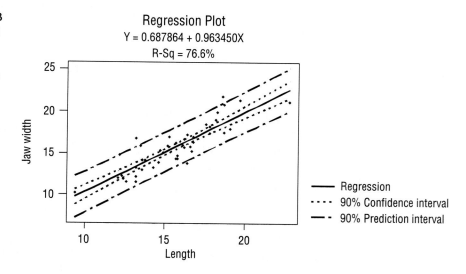

Regression Plot
Y = 0.687864 + 0.963450X
R-Sq = 76.6%

■ Exercises 13.34–13.48

13.34 Explain the difference between a confidence interval and a prediction interval. How can a prediction level of 95% be interpreted?

13.35 Suppose that a regression data set is given and that you are asked to obtain a confidence interval. How would you tell from the phrasing of the request whether the interval is for β or for $\alpha + \beta x^*$?

13.36 In Exercise 13.17, we considered a regression of y = oxygen consumption on x = time spent exercising. Summary quantities given there yield

$$n = 20 \quad \bar{x} = 2.50 \quad S_{xx} = 25$$
$$b = 97.26 \quad a = 592.10 \quad s_e = 16.486$$

a. Calculate $s_{a+b(2.0)}$, the estimated standard deviation of the statistic $a + b(2.0)$.
b. Without any further calculation, what is $s_{a+b(3.0)}$ and what reasoning did you use to obtain it?
c. Calculate the estimated standard deviation of the statistic $a + b(2.8)$.
d. For what value of x^* is the estimated standard deviation of $a + bx^*$ smallest, and why?

13.37 Example 13.3 gave data on x = ambient temperature (in °C) and y = cavity depth (in centimeters).
a. Calculate a confidence interval using a 95% confidence level for the true average change in depth associated with a 1° increase in temperature.
b. Calculate a confidence interval using a 95% confidence level for the true average depth when temperature is 1°C.

13.38 The data of Exercise 13.25, in which x = milk temperature and y = milk pH, yields

$$n = 16 \quad \bar{x} = 43.375 \quad S_{xx} = 7325.75$$
$$b = -0.00730608 \quad a = 6.843345 \quad s_e = 0.0356$$

a. Obtain a 95% confidence interval for $\alpha + \beta(40)$, the true average milk pH when the milk temperature is 40°C.
b. Calculate a 99% confidence interval for the true average milk pH when the milk temperature is 35°C.
c. Would you recommend using the data to calculate a 95% confidence interval for the true average pH when the temperature is 90°C? Why or why not?

13.39 Return to the regression of y = milk pH on x = milk temperature described in Exercise 13.38.
a. Obtain a 95% prediction interval for a single pH observation to be made when milk temperature = 40°C.
b. Calculate a 99% prediction interval for a single pH observation when milk temperature = 35°C.
c. When the milk temperature is 60°C, would a 99% prediction interval be wider than the intervals of Parts (a) and (b)? Answer without calculating the interval.

13.40 An experiment was carried out by geologists to see how the time necessary to drill a distance of 5 ft in rock (y, in minutes) depended on the depth at which the drilling began (x, in feet, between 0 and 400). We show part of the MINITAB output obtained from fitting the simple linear regression model ("Mining Information," *American Statistician* [1991]: 4–9):

The regression equation is
time = 4.79 + 0.0144depth

Predictor	Coef	Stdev	t-ratio	p
Constant	4.7896	0.6663	7.19	0.000
depth	0.014388	0.002847	5.05	0.000

s = 1.432   R-sq = 63.0%   R-sq(adj) = 60.5%

Analysis of Variance

Source	DF	SS	MS	F	p
Regression	1	52.378	52.378	25.54	0.000
Error	15	30.768	2.051		
Total	16	83.146			

a. What proportion of observed variation in time can be explained by the simple linear regression model?
b. Does the simple linear regression model appear to be useful?
c. MINITAB reported that $s_{a+b(200)} = 0.347$. Calculate a 90% confidence interval for the true average time when depth = 200 ft.
d. A single observation on time is to be made when drilling starts at a depth of 200 ft. Use a 95% prediction interval to predict the resulting value of time.
e. MINITAB gave (8.147, 10.065) as a 95% confidence interval for true average time when depth = 300. Calculate a 99% confidence interval for this average.

13.41 According to "Reproductive Biology of the Aquatic Salamander *Amphiuma tridactylum* in Louisiana" (*Journal of Herpetology* [1999]: 100–105), the size of a female salamander's snout is correlated with the number of eggs in her clutch. The following data are consistent with summary quantities reported in the article; MINITAB output is also included:

Snout-vent length	Clutch size	Snout-vent length	Clutch size
32	45	58	175
53	215	58	245
53	160	59	215
53	170	63	170
54	190	63	240
57	200	64	245
57	270	67	280

The regression equation is
y = −133 + 5.92x

Predictor	Coef	StDev	T	P
Constant	−133.02	64.30	−2.07	0.061
x	5.919	1.127	5.25	0.000

S = 33.90   R-Sq = 69.7%   R-Sq(adj) = 67.2%

Additional summary statistics are

$$n = 14 \quad \bar{x} = 56.5 \quad \bar{y} = 201.4 \quad \sum x^2 = 45{,}958$$

$$\sum y^2 = 613{,}550 \quad \sum xy = 164{,}969$$

a. What is the equation of the regression line for predicting clutch size based on snout-vent length?
b. Calculate the standard deviation of b.
c. Test the hypothesis that the slope of the population line is positive.
d. Predict the clutch size for a salamander with a snout-vent length of 65 using a 95% interval.
e. Predict the clutch size for a salamander with snout-vent length of 105.

13.42 The article first introduced in Exercise 13.28 gave data on the dimensions of 27 representative food products.
a. Use the information given in Exercise 13.28 to test the hypothesis that there is a positive linear relationship between the minimum width and the maximum width of an object.
b. Calculate and interpret s_e.
c. Calculate a 95% confidence interval for the mean maximum width of products with a minimum width of 6 cm.
d. Calculate a 95% prediction interval for the maximum width of a food package with a minimum width of 6 cm.

13.43 The shelf life of packaged food depends on many factors. Dry cereal is considered a moisture-sensitive product (no one likes soggy cereal!), with the shelf life determined primarily by moisture content. In a study of the shelf life of one particular brand of cereal, x = time on shelf (stored at 73°F and 50% relative humidity) and y = moisture content were recorded. The resulting data are from "Computer Simulation Speeds Shelf Life Assessments" (*Package Engineering* [1983]: 72–73):

x	0	3	6	8	10	13	16	20
y	2.8	3.0	3.1	3.2	3.4	3.4	3.5	3.1

x	24	27	30	34	37	41
y	3.8	4.0	4.1	4.3	4.4	4.9

a. Summary quantities are

$$\sum x = 269 \quad \sum y = 51 \quad \sum xy = 1081.5$$

$$\sum y^2 = 7745 \quad \sum x^2 = 190.78$$

Find the equation of the estimated regression line for predicting moisture content from time on the shelf.
b. Does the simple linear regression model provide useful information for predicting moisture content from knowledge of shelf time?
c. Find a 95% interval for the moisture content of an individual box of cereal that has been on the shelf for 30 days.
d. According to the article, taste tests indicated that this brand of cereal is unacceptably soggy when the moisture content exceeds 4.1. Based on your interval in Part (c), do you think that a box of cereal that has been on the shelf for 30 days will be acceptable? Explain.

13.44 For the cereal data of Exercise 13.43, the average x value is 19.21. Would a 95% confidence interval with $x^* = 20$ or $x^* = 17$ be wider? Explain. Answer the same question for a prediction interval.

13.45 High blood lead levels are associated with a number of different health problems. The article "A Study of the Relationship between Blood Lead Levels and Occupational Air Lead Levels" (*American Statistician* [1983]: 471–475) gave data on x = air lead level (in micrograms per cubic meter) and y = blood lead level (in micrograms per deciliter). Summary quantities (based on a subset of the data given in a plot in the article) are

$$n = 15 \quad \sum x = 1350 \quad \sum y = 600$$

$$\sum x^2 = 155{,}400 \quad \sum y^2 = 24{,}869.33 \quad \sum xy = 57{,}760$$

a. Find the equation of the estimated regression line.
b. Estimate the mean blood lead level for people who work where the air lead level is 100 $\mu g/m^3$ using a 90% interval.
c. Construct a 90% prediction interval for the blood lead level of a particular person who works where the air-lead level is 100 $\mu g/m^3$.
d. Explain the difference in interpretation of the intervals computed in Parts (b) and (c).

13.46 A regression of y = sunburn index for a pea plant on x = distance from an ultraviolet light source was considered in Exercise 13.22. The data and summary statistics presented there give

$$n = 15 \quad \bar{x} = 40.60 \quad \sum (x - \bar{x})^2 = 3311.60$$

$$b = -0.0565 \quad a = 4.500 \quad \text{SSResid} = 0.8430$$

a. Calculate a 95% confidence interval for the true average sunburn index when the distance from the light source is 35 cm.
b. When two 95% confidence intervals are computed, it can be shown that the simultaneous confidence level is at least $[100 - 2(5)]\% = 90\%$. That is,

if both intervals are computed for a first sample, for a second sample, yet again for a third sample, and so on, in the long run at least 90% of the samples will result in intervals both of which capture the values of the corresponding population characteristics. Calculate confidence intervals for the true mean sunburn index when the distance is 35 cm and when the distance is 45 cm in such a way that the simultaneous confidence level is at least 90%.

c. If two 99% intervals were computed, what do you think could be said about the simultaneous confidence level?

d. If a 95% confidence interval was computed for the true mean index when $x = 35$, another 95% confidence interval was computed when $x = 40$, and yet another one was computed when $x = 45$, what do you think would be the simultaneous confidence level for the three resulting intervals?

e. Return to Part (d), but assume that the three interval calculations had been done using a confidence level of 99% instead of 95%. Now what do you think the simultaneous confidence level would be for the three resulting intervals?

13.47 By analogy with the discussion in Exercise 13.46, when two different prediction intervals are computed, each at the 95% prediction level, the *simultaneous prediction level* is at least $[100 - 2(5)]\% = 90\%$.

a. Return to Exercise 13.46 and obtain prediction intervals for the sunburn index when the distance

is 35 cm and when the distance is 45 cm, so that the simultaneous prediction level is at least 90%.

b. If three different 99% prediction intervals are calculated for distances of 35, 40, and 45 cm, respectively, what can be said about the simultaneous prediction level?

13.48 The article "Performance Test Conducted for a Gas Air-Conditioning System" (*American Society of Heating, Refrigerating, and Air Conditioning Engineers* [1969]: 54) reported the following data on maximum outdoor temperature (x) and hours of chiller operation per day (y) for a 3-ton residential gas air-conditioning system:

x	72	78	80	86	88	92
y	4.8	7.2	9.5	14.5	15.7	17.9

Suppose that the system is actually a prototype model and that the manufacturer does not wish to produce this model unless the data strongly indicate that, when maximum outdoor temperature is 82°F, the true average number of hours of chiller operation is less than 12. The appropriate hypotheses are then

$$H_0: \alpha + \beta(82) = 12 \quad \text{versus} \quad H_a: \alpha + \beta(82) < 12$$

Use the statistic

$$t = \frac{a + b(82) - 12}{s_{a+b(82)}}$$

which has a t distribution based on $(n - 2)$ df when H_0 is true, to test the hypotheses at significance level .01.

▪ 13.5 Inferences About the Population Correlation Coefficient (Optional)

The sample correlation coefficient r, defined in Chapter 5, measures how strongly the x and y values in a *sample* of pairs are linearly related to one another. There is an analogous measure of how strongly x and y are related in the entire *population* of pairs from which the sample $(x_1, y_1), \ldots, (x_n, y_n)$ was obtained. It is called the **population correlation coefficient** and is denoted by ρ. As with r, ρ must be between -1 and 1, and it assesses the extent of any *linear* association in the population. To have $\rho = 1$ or -1, all (x, y) pairs in the population must lie exactly on a straight line. The value of ρ is a population characteristic and is generally unknown. The sample correlation coefficient r can be used as the basis for making inferences about ρ.

▪ Test for Independence ($\rho = 0$)

Investigators are often interested in detecting not just linear association but also association of *any* kind. When there is no association of any type between the x and y values, statisticians say that the two variables are *independent*. In general, $\rho = 0$ is not equivalent to the independence of x and y. However, there is one special — yet frequently occurring — situation in which the two conditions ($\rho = 0$ and independence) are identical. This is when the pairs in the population have what is called a **bivariate normal distribution**. The essential feature of such a distribution is that for *any* fixed x value, the distribution of associated y values is normal, *and* for any fixed y value, the distribution of x values is normal.

As an example, suppose that height x and weight y have a bivariate normal distribution in the American adult male population. (There is good empirical evidence for this.) Then, when $x = 68$ in., weight y has a normal distribution; when $x = 72$ in., weight is normally distributed; when $y = 160$ lb, height x has a normal distribution; when $y = 175$ lb, height has a normal distribution; and so on. In this example, of course, x and y are not independent, because a large height value tends to be associated with a large weight value and a small height value tends to be associated with a small weight value.

There is no easy way to check the assumption of bivariate normality, especially when the sample size n is small. A partial check can be based on the following property: If (x, y) has a bivariate normal distribution, then x alone has a normal distribution and so does y. This suggests constructing a normal probability plot of $x_1, x_2, \ldots, x_n$ and a separate normal probability plot of $y_1, y_2, \ldots, y_n$. If either plot shows a substantial departure from a straight line, then bivariate normality is a questionable assumption. If both plots are reasonably straight, then bivariate normality is plausible, although no guarantee can be given.

The test of independence (zero correlation) is a t test. The formula for the test statistic essentially involves standardizing the estimate r under the assumption that the null hypothesis $H_0: \rho = 0$ is true.

▪ **Test for Independence in a Bivariate Normal Population**

Null hypothesis: $H_0: \rho = 0$

Test statistic: $t = \dfrac{r}{\sqrt{\dfrac{1 - r^2}{n - 2}}}$

The t critical value is based on df $= n - 2$.

Alternative Hypothesis	P-Value
$H_a: \rho > 0$ (positive dependence)	Area under the appropriate t curve to the right of the computed t
$H_a: \rho < 0$ (negative dependence)	Area under the appropriate t curve to the left of the computed t
$H_a: \rho \neq 0$ (dependence)	(1) 2(area to the right of t) if t is positive, or (2) 2(area to the left of t) if t is negative

Assumption: r is the correlation coefficient for a *random sample* from a *bivariate normal population*.

▪ **Example 13.13   Age and Retirement Plan**

Is there any relationship between the age of an individual and the age at which that individual plans to retire? The authors of the article "Predictors of Planned Retirement Age" (*Psychology and Aging* [1995]: 76–83) hypothesized that younger people would plan to retire earlier than would older people. They selected a random sample of $n = 303$ workers from a wide variety of job positions and recorded x = current age and y = planned retirement age for each individual in the sample. The sample correlation coefficient was $r = .35$. Do these data support the stated hypothesis? Let's carry out a test using a significance level of .01.

1. ρ = the correlation between age and planned retirement age in the population from which the 303 workers were selected.

2. $H_0: \rho = 0$.

3. $H_a: \rho > 0$ (large values of y = planned retirement age tend to be associated with large values of x = age).

4. $\alpha = 0.01$.

5. Test statistic: $t = \dfrac{r}{\sqrt{\dfrac{1 - r^2}{n - 2}}}$

6. Assumptions: Without the actual data values, it is not possible to assess whether the assumption of bivariate normality is reasonable. For purposes of this example, we will assume that it is reasonable and proceed with the test. Had the data been available, we would have looked at normal probability plots of the x values and of the y values.

7. Calculation: $t = \dfrac{.35}{\sqrt{\dfrac{1 - (.35)^2}{301}}} = \dfrac{.35}{\sqrt{.0029153}} = 6.48$

8. *P*-value: The t curve with 301 df is essentially indistinguishable from the z curve, and Appendix Table 4 shows that the area under this curve and to the right of 6.48 is approximately 0. That is, *P*-value ≈ 0.

9. Conclusion: The *P*-value is smaller than any reasonable significance level; in particular, *P*-value $\leq .01$. We reject H_0, as did the authors of the article, and confirm their conjecture. Notice, though, that according to the guidelines described in Chapter 5, $r = .35$ suggests only a weak linear relationship. Because $r^2 = .1225$, fitting the simple linear regression model to the data would result in only about 12% of observed variation in planned retirement age being explained.

In the context of regression analysis, the hypothesis of no linear relationship ($H_0: \beta = 0$) was tested using the t ratio $t = \dfrac{b}{s_b}$. Some algebraic manipulation shows that

$$\frac{r}{\sqrt{\dfrac{1 - r^2}{n - 2}}} = \frac{b}{s_b}$$

Therefore the model utility test can also be used to test for independence in a bivariate normal population. The reason for using the formula for t that involves r is that when we are interested only in correlation, the extra effort involved in computing the regression quantities b, a, SSResid, s_e, and s_b need not be expended.

Other inferential procedures for drawing conclusions about ρ — a confidence interval or a test of hypothesis with nonzero hypothesized value — are somewhat complicated. One of the references at the end of Chapter 5 can be consulted for details.

▪ Exercises 13.49–13.57

13.49 Discuss the difference between r and ρ.

13.50 If the sample correlation coefficient is equal to 1, is it necessarily true that $\rho = 1$? If $\rho = 1$, is it necessarily true that $r = 1$?

13.51 A sample of $n = 353$ college faculty members was obtained, and the values of x = teaching evaluation index and y = annual raise were determined ("Determination of Faculty Pay: An Agency Theory Perspective," *Academy of Management Journal* [1992]: 921–955). The resulting value of r was .11. Does there appear to be a linear association between these variables in the population from which the sample was selected? Carry out a test of hypothesis using a significance level of .05. Does the conclusion surprise you? Explain.

13.52 It seems plausible that higher rent for retail space could be justified only by a higher level of sales. A random sample of $n = 53$ specialty stores in a retail chain was selected, and the values of x = annual dollar rent per square foot and y = annual dollar sales per square foot were determined, resulting in $r = .37$ ("Association of Shopping Center Anchors with Performance of a Nonanchor Specialty Chain Store," *Journal of Retailing* [1985]: 61–74). Carry out a test at significance level .05 to see whether there is in fact a positive linear association between x and y in the population of all such stores.

13.53 Television is regarded by many as a prime culprit for the difficulty many students have in performing well in school. The article "The Impact of Athletics, Part-Time Employment, and Other Activities on Academic Achievement" (*Journal of College Student Development* [1992]: 447–453) reported that for a random sample of $n = 528$ college students, the sample correlation coefficient between time spent watching television (x) and grade point average (y) was $r = -.26$.
a. Does this value of r suggest that there is a negative correlation between x and y in the population from which the 528 students were selected? Use a test with significance level .01.
b. If y were regressed on x, would the linear relationship explain a substantial percentage of the observed variation in grade point average? Explain your reasoning.

13.54 The following summary quantities for x = particulate pollution (in micrograms per cubic meter) and y = luminance (in hundredths of a candela per square meter) were calculated from a representative sample of data that appeared in the article "Luminance and Polarization of the Sky Light at Seville (Spain) Measured in White Light" (*Atmospheric Environment* [1988]: 595–599):

$$n = 15 \quad \sum x = 860 \quad \sum y = 348$$
$$\sum x^2 = 56{,}700 \quad \sum y^2 = 8954 \quad \sum xy = 22{,}265$$

a. Test whether there is a positive correlation between particulate pollution and luminance in the population from which the data were selected.
b. What proportion of observed variation in luminance can be attributed to the approximate linear relationship between luminance and particulate pollution?

13.55 In a study of bacterial concentration in surface and subsurface water ("Pb and Bacteria in a Surface Microlayer," *Journal of Marine Research* [1982]: 1200–1206), the following data on concentration (in millions of bacteria per milliliter) were obtained:

Surface	48.6	24.3	15.9	8.29	5.75
Subsurface	5.46	6.89	3.38	3.72	3.12
Surface	10.8	4.71	8.26	9.41	
Subsurface	3.39	4.17	4.06	5.16	

Summary quantities are

$$\sum x = 136.02 \quad \sum y = 39.35 \quad \sum x^2 = 3602.65$$
$$\sum y^2 = 184.27 \quad \sum xy = 673.65$$

Using a significance level of .05, determine whether the data support the hypothesis of a linear relationship between surface and subsurface concentration.

13.56 A sample of $n = 500$ (x, y) pairs was collected, and a test of $H_0: \rho = 0$ versus $H_a: \rho \neq 0$ was carried out. The resulting P-value was computed to be .00032.
a. What conclusion would be appropriate at level of significance .001?
b. Does this small P-value necessarily indicate that there is a very strong linear relationship between x

and *y* (a value of ρ that is greatly different from 0)? Explain.

13.57 A sample of *n* = 10,000 (*x*, *y*) pairs resulted in *r* = .022. Test $H_0: \rho = 0$ versus $H_a: \rho \neq 0$ at sig-

nificance level .05. Is the result statistically significant? Comment on the practical significance of your analysis.

■ 13.6 Communicating and Interpreting the Results of Statistical Analyses

Although regression analysis can be used as a tool for summarizing bivariate data, it is also widely used to enable researchers to make inferences about the way in which two variables are related.

■ What to Look For in Published Data

Here are some things to consider when you evaluate research that involves fitting a simple linear regression model:

- Which variable is the dependent variable? Is it a numerical (rather than a qualitative) variable?
- If sample data have been used to estimate the coefficients in a simple linear regression model, is it reasonable to think that the basic assumptions required for inference are met?
- Does the model appear to be useful? Are the results of a model utility test reported? What is the *P*-value associated with the test?
- Has the model been used in an appropriate way? Has the regression equation been used to predict *y* for values of the independent variable that are outside the range of the data?
- If a correlation coefficient is reported, is it accompanied by a test of significance? Are the results of the test interpreted properly?

The effects of caffeine were examined in the article "Withdrawal Syndrome After the Double-Blind Cessation of Caffeine Consumption" (*New England Journal of Medicine* [1992]: 1109–1113). The authors found that the dose of caffeine was significantly correlated with a measure of insomnia and also with latency on a test of reaction time. They reported *r* = .26 (*P*-value = .042) for insomnia and *r* = .31 (*P*-value = .014) for latency. Because both *P*-values are small, the authors concluded that the population correlation coefficients differ from 0. We should also note, however, that the reported correlation coefficients are not particularly large and do not indicate a strong linear relationship.

Regression analysis was used by the authors of the article "On the Generality of the 'Sit and Reach' Test: An Analysis of Flexibility Data for an Aging Population" (*Research Quarterly for Exercise and Sport* [1990]: 326–330). A linear regression model was used to describe the relationship between *y* = head rotation to the left (in degrees) and *x* = age (in years). The authors reported that "regression analysis showed an age effect for leftward head rotation (significant beta weight, $t_{78} = 3.10$, *P* < .003)." The term *beta weight* refers to the β term in the linear regression model. A model utility test was performed, and the value of the test statistic was 3.10. The test was based on 78 degrees of freedom (this is the meaning of the subscript in t_{78}).

The associated *P*-value leads to the rejection of $H_0: \beta = 0$ and to the authors' stated conclusion of an "age effect." The reported regression equation is

$$\hat{y} = 87.5 - 0.31x$$

The coefficient of age in this equation (-0.31) indicates that, on average, leftward head rotation decreases by about 0.31 degree with each additional year of age. (It is interesting to note that age did not turn out to be an important predictor of right-ward motion!)

The article does not provide enough information to determine whether the basic assumptions required for the validity of the test reported are reasonable, and the authors did not address this issue directly. If the raw data were available, we could check on this by looking at such things as a standardized residual plot and various normal probability plots.

▪ A Word to the Wise: Cautions and Limitations

In addition to the cautions and limitations described in Chapter 5 (which also apply here), here are a few additional things to keep in mind:

1. It doesn't make sense to use a regression line as the basis for making inferences about a population if there is no convincing evidence of a useful linear relationship between the two variables under study. It is particularly important to remember this when you are working with small samples. If the sample size is small, it is not uncommon for a weak linear pattern in the scatterplot to be due to chance rather than to a meaningful relationship in the population of interest.

2. As with all inferential procedures, it makes sense to carry out a model utility test or to construct confidence or prediction intervals only if the linear regression is based on a random sample from some larger population. It is not unusual to see a least-squares line used as a descriptive tool in situations where the data used to construct the line cannot reasonably be viewed as a sample. In this case, the inferential methods of this chapter are not appropriate.

3. Don't forget to check assumptions. If you are used to checking assumptions before doing much in the way of computation, it is sometimes easy to forget to check them in this setting because the equation of the least-squares line and then the residuals must be computed before a residual plot or a normal probability plot of the standardized residuals can be constructed. Be sure to step back and think about whether the linear regression model is appropriate and reasonable before using the model to draw inferences about a population.

▪ Activity 13.1: Are Tall Women from "Big" Families?

In this activity, you should work with a partner (or in a small group).

Consider the following data on height (in inches) and number of siblings for a random sample of 10 female students at a large university.

1. Construct a scatterplot of the given data. Does there appear to be a linear relationship between *y* = height and *x* = number of siblings?

Height (y)	Number of Siblings (x)	Height (y)	Number of Siblings (x)
64.2	2	65.5	1
65.4	0	67.2	2
64.6	2	66.4	2
66.1	6	63.3	0
65.1	3	61.7	1

2. Compute the value of the correlation coefficient. Is the value of the correlation coefficient consistent with your answer from Step 1? Explain.

3. What is the equation of the least-squares line for these data?

4. Is the slope of the least-squares regression line from Step 3 equal to 0? Does this necessarily mean that there is a meaningful relationship between height and number of siblings in the population of female students at this university? Discuss this with your partner, and then write a few sentences of explanation.

5. For the population of all female students at the university, do you think it is reasonable to assume that the distribution of heights at each particular x value is approximately normal and that the standard deviation of the height distribution at each particular x value is the same? That is, do you think it is

reasonable to assume that the distribution of heights for female students with zero siblings is approximately normal and that the distribution of heights for female students with one sibling is approximately normal with the same standard deviation as for female students with no siblings, and so on? Discuss this with your partner, and then write a few sentences of explanation.

6. Carry out the model utility test (H_0: $\beta = 0$). Explain why the conclusion from this test is consistent with your explanation in Step 4.

7. Would you recommend using the least-squares regression line as a way of predicting heights for women at this university? Explain.

8. After consulting with your partner, write a paragraph explaining why it is a good idea to include a model utility test (H_0: $\beta = 0$) as part of a regression analysis.

■ Summary of Key Concepts and Formulas

Term or Formula	Comment
Simple linear regression model, $y = \alpha + \beta x + e$	This model assumes that there is a line with slope β and y intercept α, called the population (true) regression line, such that an observation deviates from the line by a random amount e. The random deviation is assumed to have a normal distribution with mean 0 and standard deviation σ, and random deviations for different observations are assumed to be independent of one another.
Estimated regression line, $\hat{y} = a + bx$	The least-squares line introduced in Chapter 5.
$s_e = \sqrt{\dfrac{\text{SSResid}}{n-2}}$	The point estimate of the standard deviation σ, with associated degrees of freedom $n - 2$.
$s_b = \dfrac{s_e}{\sqrt{S_{xx}}}$	The estimated standard deviation of the statistic b.
$b \pm (t \text{ critical value})(s_b)$	A confidence interval for the slope β of the population regression line, where the t critical value is based on $(n - 2)$ df.
$t = \dfrac{b - \text{hypothesized value}}{s_b}$	The test statistic for testing hypotheses about β. The test is based on $(n - 2)$ df.
Model utility test, with test statistic $t = \dfrac{b}{s_b}$	A test of H_0: $\beta = 0$, which asserts that there is no useful linear relationship between x and y, versus H_a: $\beta \neq 0$, the claim that there is a useful linear relationship.
Residual analysis	Methods based on the residuals or standardized residuals for checking the assumptions of a regression model.
Standardized residual	A residual divided by its standard deviation.

Term or Formula	Comment
Standardized residual plot	A plot of the $(x$, standardized residual$)$ pairs. A pattern in this plot suggests a problem with the simple linear regression model.
$s_{a+bx^*} = s_e\sqrt{\dfrac{1}{n} + \dfrac{(x^* - \bar{x})^2}{S_{xx}}}$	The estimated standard deviation of the statistic $a + bx^*$, where x^* denotes a particular value of x.
$a + bx^* \pm (t\text{ critical value})(s_{a+bx^*})$	A confidence interval for $\alpha + \beta x^*$, the average value of y when $x = x^*$.
$a + bx^* \pm (t\text{ critical value})\sqrt{s_e^2 + s_{a+bx^*}^2}$	A prediction interval for a single y value to be observed when $x = x^*$.
Population correlation coefficient ρ	A measure of the extent to which the x and y values in an entire population are linearly related.
$t = \dfrac{r}{\sqrt{\dfrac{1 - r^2}{n - 2}}}$	The test statistic for testing H_0: $\rho = 0$, according to which (assuming a bivariate normal population distribution) x and y are independent of one another.

▪ Supplementary Exercises 13.58–13.76

13.58 The effects of grazing animals on grasslands have been the focus of numerous investigations by ecologists. One such study, reported in "The Ecology of Plants, Large Mammalian Herbivores, and Drought in Yellowstone National Park" (*Ecology* [1992]: 2043–2058), proposed using the simple linear regression model to relate y = green biomass concentration (in grams per cubic centimeter) to x = elapsed time since snowmelt (in days).
a. The estimated regression equation was given as $\hat{y} = 106.3 - 0.640x$. What is the estimate of average change in biomass concentration associated with a 1-day increase in elapsed time?
b. What value of biomass concentration would you predict when elapsed time is 40 days?
c. The sample size was $n = 58$, and the reported value of the coefficient of determination was .470. Does this suggest that there is a useful linear relationship between the two variables? Carry out an appropriate test.

13.59 A random sample of $n = 347$ students was selected, and each student was asked to complete several questionnaires, from which a Coping Humor Scale value x and a Depression Scale value y were determined ("Depression and Sense of Humor," *Psychological Reports* [1994]: 1473–1474). The resulting value of the sample correlation coefficient was $-.18$.
a. The investigators reported that P-value $< .05$. Do you agree?
b. Is the sign of r consistent with your intuition? Explain. (Higher scale values correspond to more developed sense of humor and greater extent of depression.)
c. Would the simple linear regression model give accurate predictions? Why or why not?

13.60 Data on x = depth of flooding and y = flood damage were given in Exercise 5.64. Summary quantities are

$$n = 13 \quad \sum x = 91 \quad \sum x^2 = 819$$
$$\sum y = 470 \quad \sum y^2 = 19{,}118 \quad \sum xy = 3867$$

a. Do the data suggest the existence of a positive linear relationship (one in which an increase in y tends to be associated with an increase in x)? Test using a .05 significance level.
b. Predict flood damage resulting from a claim made when depth of flooding is 3.5 ft, and do so in a way that conveys information about the precision of the prediction.

13.61 Exercise 13.8 gave data on x = treadmill run time to exhaustion and y = 20-km ski time for a sample of 11 biathletes. Use the accompanying MINITAB output to answer the questions that follow.

The regression equation is
ski = 88.8 − 2.33tread

Predictor	Coef	Stdev	t-ratio	p
Constant	88.796	5.750	15.44	0.000
tread	−2.3335	0.5911	−3.95	0.003

s = 2.188   R-sq = 63.4%   R-sq(adj) = 59.3%

Analysis of Variance

Source	DF	SS	MS	F	p
Regression	1	74.630	74.630	15.58	0.003
Error	9	43.097	4.789		
Total	10	117.727			

a. Carry out a test at significance level .01 to decide whether the simple linear regression model is useful.
b. Estimate the average change in ski time associated with a 1-min increase in treadmill time, and do so in a way that conveys information about the precision of estimation.
c. MINITAB reported that $s_{a+b(10)} = 0.689$. Predict ski time for a single biathlete whose treadmill time is 10 min, and do so in a way that conveys information about the precision of prediction.
d. MINITAB also reported that $s_{a+b(11)} = 1.029$. Why is this value larger than $s_{a+b(10)}$?

13.62 Exercise 5.41 presented data on $x =$ squaw-fish length and $y =$ maximum size of salmonid consumed, both in millimeters. Use the accompanying MINITAB output along with the values $\bar{x} = 343.27$ and $S_{xx} = 69{,}112.18$ to answer the questions that follow.

The regression equation is
size = −89.1 + 0.729length

Predictor	Coef	Stdev	t-ratio	p
Constant	−89.09	16.83	−5.29	0.000
length	0.72907	0.04778	15.26	0.000

s = 12.56   R-sq = 96.3%   R-sq(adj) = 95.9%

Analysis of Variance

Source	DF	SS	MS	F	p
Regression	1	36736	36736	232.87	0.000
Error	9	1420	158		
Total	10	38156			

a. Does there appear to be a useful linear relationship between length and size?
b. Does it appear that the average change in maximum size associated with a 1-mm increase in length is less than 0.8 mm? State and test the appropriate hypotheses.
c. Estimate average maximum size when length is 325 mm in a way that conveys information about the precision of estimation.
d. How would a similar estimate when length is 250 mm compare to the estimate of Part (c)? Answer without actually calculating the new estimate.

13.63 A sample of $n = 61$ penguin burrows was selected, and values of both $y =$ trail length (in meters) and $x =$ soil hardness (force [in kilograms] required to penetrate the substrate to a depth of 12 cm with a certain gauge) were determined for each burrow ("Effects of Substrate on the Distribution of Magellanic Penguin Burrows," *The Auk* [1991]:

923–933). The equation of the least-squares line was $\hat{y} = 11.607 − 1.4187x$, and $r^2 = .386$.
a. Does the relationship between soil hardness and trail length appear to be linear, with shorter trails associated with harder soil (as the article asserted)? Carry out an appropriate test of hypotheses.
b. Using $s_e = 2.35$, $\bar{x} = 4.5$, and $\sum(x − \bar{x})^2 = 250$, predict trail length when soil hardness is 6.0 in a way that conveys information about the reliability and precision of the prediction.
c. Would you use the simple linear regression model to predict trail length when hardness is 10.0? Explain your reasoning.

13.64 The article "Photocharge Effects in Dye Sensitized Ag[Br, I] Emulsions at Millisecond Range Exposures" (*Photographic Science and Engineering* [1981]: 138–144) gave the following data on $x =$ light absorption (percentage) and $y =$ peak photovoltage:

x	4.0	8.7	12.7	19.1	21.4	24.6
y	0.12	0.28	0.55	0.68	0.85	1.02

x	28.9	29.8	30.5
y	1.15	1.34	1.29

$$\sum x = 179.7 \quad \sum x^2 = 4334.41$$

$$\sum y = 7.28 \quad \sum y^2 = 7.4028 \quad \sum xy = 178.683$$

a. Construct a scatterplot of the data. What does it suggest?
b. Assuming that the simple linear regression model is appropriate, obtain the equation of the estimated regression line.
c. How much of the observed variation in peak photovoltage can be explained by the linear relationship?
d. Predict peak photovoltage when absorption is 19.1%, and compute the value of the corresponding residual.
e. The authors claimed that there is a useful linear relationship between the two variables. Do you agree? Carry out a formal test.
f. Give an estimate of the average change in peak photovoltage associated with a 1-percentage-point increase in light absorption. Your estimate should convey information about the precision of estimation.
g. Give an estimate of true average peak photovoltage when light absorption is 20%, and do so in a way that conveys information about precision.

13.65 Reduced visual performance with increasing age has been a much-studied phenomenon in recent years. This decline is due partly to changes in optical properties of the eye itself and partly to neural degeneration throughout the visual system. As one aspect of this problem, the article "Morphometry of Nerve Fiber Bundle Pores in the Optic Nerve Head of the Human" (*Experimental Eye Research* [1988]: 559–568) presented the following data on x = age and y = percentage of the cribriform area of the lamina scleralis occupied by pores:

x	y	x	y
22	75	48	52
25	62	50	58
27	50	57	49
39	49	58	52
42	54	63	49
43	49	63	31
44	59	74	42
46	47	74	41
46	54		

a. Suppose that the researchers had believed a priori that the average decrease in area associated with a 1-year age increase was 0.5%. Do the data contradict this prior belief? State and test the appropriate hypotheses using a .10 significance level.
b. Estimate true average area (%) covered by pores for all 50-year-olds in the population in a way that conveys information about the precision of estimation.

13.66 Occasionally an investigator may wish to compute a confidence interval for α, the y intercept of the true regression line, or test hypotheses about α. The estimated y intercept is simply the height of the estimated line when $x = 0$, because $a + b(0) = a$. This implies that s_a, the estimated standard deviation of the statistic a, results from substituting $x^* = 0$ in the formula for s_{a+bx^*}. The desired confidence interval is then

$$a \pm (t \text{ critical value})(s_a)$$

and a test statistic is

$$t = \frac{a - \text{hypothesized value}}{s_a}$$

a. The article "Comparison of Winter-Nocturnal Geostationary Satellite Infrared-Surface Temperature with Shelter-Height Temperature in Florida" (*Remote Sensing of the Environment* [1983]: 313–327) used the simple linear regression model to relate surface temperature as measured by a satellite (y) to actual air temperature (x) determined from a thermocouple placed on a traversing vehicle. Selected data are given here (they were read from a scatterplot in the article):

x	−2	−1	0	1	2	3	4	5
y	−3.9	−2.1	−2.0	−1.2	0.0	1.9	0.6	2.1

x	6	7
y	1.2	3.0

Estimate the true regression line.

b. Compute the estimated standard deviation s_a. Carry out a test at level of significance .05 to see whether the y intercept of the true regression line differs from 0.
c. Compute a 95% confidence interval for α. Does the result indicate that $\alpha = 0$ is plausible? Explain.

13.67 Example 13.3 gave data on x = ambient temperature and y = cavity depth. Use the following MINITAB output to estimate the true average depth when temperature = 0°C with a 95% confidence interval (hint: refer to Exercise 13.68):

```
The regression equation is
depth = 20.1 − 0.345temp
Predictor      Coef      Stdev
Constant     20.1251    0.9402
temp         −0.34504   0.06008
S = 2.334   R-SQ = 76.7%
```

13.68 In some studies, an investigator has n (x, y) pairs sampled from one population and m (x, y) pairs sampled from a second population. Let β and β' denote the slopes of the first and second population lines, respectively, and let b and b' denote the estimated slopes calculated from the first and second samples, respectively. The investigator may then wish to test the null hypothesis $H_0: \beta - \beta' = 0$ (i.e., $\beta = \beta'$) against an appropriate alternative hypothesis. Suppose that σ^2, the variance about the population line, is the same for both populations. Then this common variance can be estimated by

$$s^2 = \frac{\text{SSResid} + \text{SSResid}'}{(n - 2) + (m - 2)}$$

where SSResid and SSResid' are the residual sums of squares for the first and second samples, respectively. With S_{xx} and S'_{xx} denoting the quantity $\Sigma (x - \bar{x})^2$ for the first and second samples, respectively, the test statistic is

$$t = \frac{b - b'}{\sqrt{\dfrac{s^2}{S_{xx}} + \dfrac{s^2}{S'_{xx}}}}$$

When H_0 is true, this statistic has a t distribution based on $(n + m - 4)$ df.

The following data are a subset of the data in the article "Diet and Foraging Model of *Bufa marinus* and *Leptodactylus ocellatus*" (*Journal of Herpetology* [1984]: 138–146), where the independent variable x is body length (in centimeters) and the dependent variable y is mouth width (in centimeters), with $n = 9$ observations for one type of nocturnal frog and $m = 8$ observations for a second type (summary statistics are also given):

Leptodactylus ocellatus

x	3.8	4.0	4.9	7.1	8.1	8.5	8.9	9.1	9.8
y	1.0	1.2	1.7	2.0	2.7	2.5	2.4	2.9	3.2

Bufa marinus

x	3.8	4.3	6.2	6.3	7.8	8.5	9.0	10.0
y	1.6	1.7	2.3	2.5	3.2	3.0	3.5	3.8

	Leptodactylus	*Bufa*
Sample size	9	8
$\sum x$	64.2	55.9
$\sum x^2$	500.78	425.15
$\sum y$	19.6	21.6
$\sum y^2$	47.28	62.92
$\sum xy$	153.36	163.36

Use a level .05 test to determine whether the slopes of the true regression lines for the two different frog populations are equal.

13.69 Consider the following four (x, y) data sets: the first three sets have the same x values, so these values are listed only once (data from "Graphs in Statistical Analysis," *American Statistician* [1973]: 17–21):

Data Set	1–3	1	2	3	4	4
Variable	*x*	*y*	*y*	*y*	*x*	*y*
	10.0	8.04	9.14	7.46	8.0	6.58
	8.0	6.95	8.14	6.77	8.0	5.76
	13.0	7.58	8.74	12.74	8.0	7.71
	9.0	8.81	8.77	7.11	8.0	8.84
	11.0	8.33	9.26	7.81	8.0	8.47
	14.0	9.96	8.10	8.84	8.0	7.04
	6.0	7.24	6.13	6.08	8.0	5.25
	4.0	4.26	3.10	5.39	19.0	12.50
	12.0	10.84	9.13	8.15	8.0	5.56
	7.0	4.82	7.26	6.42	8.0	7.91
	5.0	5.68	4.74	5.73	8.0	6.89

For each of these data sets, the values of the summary quantities $\bar{x}$, $\bar{y}$, $\sum(x - \bar{x})^2$, and $\sum(x - \bar{x})(y - \bar{y})$ are identical, so all quantities computed from these values will be identical for the four sets: the estimated regression line, SSResid, s_e, r^2, and so on. The summary quantities provide no way of distinguishing among the four data sets.

Based on a scatterplot for each set, comment on the appropriateness or inappropriateness of fitting the simple linear regression model in each case.

13.70 The following scatterplot, based on 34 sediment samples with x = sediment depth (in centimeters) and y = oil and grease content (in milligrams per kilogram), appeared in the article "Mined Land Reclamation Using Polluted Urban Navigable Waterway Sediments" (*Journal of Environmental Quality* [1984]: 415–422):

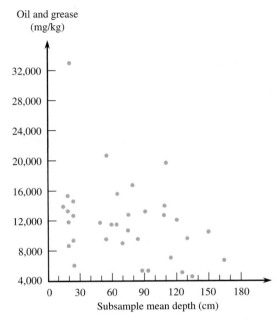

Discuss the effect that the observation (20, 33,000) will have on the estimated regression line. If this point was omitted, what would you say about the slope of the estimated regression line? What do you think will happen to the slope if this observation is included in the computations?

13.71 The article "Improving Fermentation Productivity with Reverse Osmosis" (*Food Technology* [1984]: 92–96) gave the following data (read from a scatterplot) on y = glucose concentration (in grams per liter) and x = fermentation time (in days) for a blend of malt liquor:

x	1	2	3	4	5	6	7	8
y	74	54	52	51	52	53	58	71

a. Use the data to calculate the estimated regression line.
b. Do the data indicate a linear relationship between y and x? Test using a .10 significance level.
c. Using the estimated regression line of Part (a), compute the residuals and construct a plot of the residuals versus x (i.e., of the (x, residual) pairs).
d. Based on the plot in Part (c), do you think that a linear model is appropriate for describing the relationship between y and x? Explain.

13.72 The employee relations manager of a large company was concerned that raises given to employees during a recent period might not have been based strictly on objective performance criteria. A sample of $n = 20$ employees was selected, and the values of x, a quantitative measure of productivity, and y, the percentage salary increase, were determined for each one. A computer package was used to fit the simple linear regression model, and the resulting output gave P-value $= .0076$ for the model utility test. Does the percentage raise appear to be linearly related to productivity? Explain.

13.73 The article "Statistical Comparison of Heavy Metal Concentrations in Various Louisiana Sediments" (*Environmental Monitoring and Assessment* [1984]: 163–170) gave the following data on depth (in meters), zinc concentration (in parts per million), and iron concentration (%) for 17 core samples:

Core	Depth	Zinc	Iron
1	0.2	86	3.4
2	2.0	77	2.9
3	5.8	91	3.1
4	6.5	86	3.4
5	7.6	81	3.2
6	12.2	87	2.9
7	16.4	94	3.2
8	20.8	92	3.4
9	22.5	90	3.1
10	29.0	108	4.0
11	31.7	112	3.4
12	38.0	101	3.6
13	41.5	88	3.7
14	60.0	99	3.5
15	61.5	90	3.4
16	72.0	98	3.5
17	104.0	70	4.8

a. Using a .05 significance level, test appropriate hypotheses to determine whether a correlation exists between depth and zinc concentration.
b. Using a .05 significance level, do the data strongly suggest a correlation between depth and iron concentration?
c. Calculate the slope and intercept of the estimated regression line relating $y =$ iron concentration and $x =$ depth.
d. Use the estimated regression equation to construct a 95% prediction interval for the iron concentration of a single core sample taken at a depth of 50 m.
e. Compute and interpret a 95% interval estimate for the true average iron concentration of core samples taken at 70 m.

13.74 The following figure is from the article "Root and Shoot Competition Intensity Along a Soil Depth Gradient" (*Ecology* [1995]: 673–682):

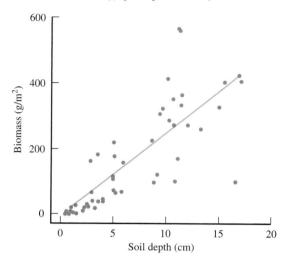

The figure shows the relationship between above-ground biomass and soil depth within experimental plots. The relationship is described by the linear equation Biomass $= -9.85 + 25.29$(Soil Depth) and $r^2 = .65; P < 0.001; n = 55$. Do you think the simple linear regression model is appropriate here? Explain. What would you expect to see in a plot of the standardized residuals versus x?

13.75 Give a brief answer, comment, or explanation for each of the following:
a. What is the difference between $e_1, e_2, \ldots, e_n$ and the n residuals?
b. The simple linear regression model states that $y = \alpha + \beta x$.
c. Does it make sense to test hypotheses about b?
d. SSResid is always positive.
e. A student reported that a data set consisting of $n = 6$ observations yielded residuals 2, 0, 5, 3, 0, and 1 from the least-squares line.
f. A research report included the following summary quantities obtained from a simple linear regression analysis: $\sum (y - \bar{y})^2 = 615$ and $\sum (y - \hat{y})^2 = 731$.

13.76 Some straightforward but slightly tedious algebra shows that

$$\text{SSResid} = (1 - r^2)\sum(y - \bar{y})^2$$

from which it follows that

$$s_e = \sqrt{\frac{n-1}{n-2}}(\sqrt{1 - r^2})s_y$$

Unless n is quite small, $(n - 1)/(n - 2) \approx 1$, so

$$s_e \approx (\sqrt{1 - r^2})s_y$$

a. For what value of r is s_e as large as s_y? What is the equation of the least-squares line in this case?

b. For what values of r will s_e be much smaller than s_y?

▪ References

See the references at the end of Chapter 5.

14 · Multiple Regression Analysis

The general objective of regression analysis is to model the relationship between a dependent variable y and one or more independent (i.e., predictor or explanatory) variables. The simple linear regression model $y = \alpha + \beta x + e$, discussed in Chapter 13, has been used successfully by many investigators in a wide variety of disciplines to relate y to a single predictor variable x. In many situations, however, the relationship between y and any single predictor variable is not strong, but knowing the values of several independent variables may considerably reduce uncertainty about the associated y value. For example, some variation in house prices in a large city can certainly be attributed to house size, but knowledge of size by itself would not usually enable a bank appraiser to accurately assess (predict) a home's value. Price is also determined to some extent by other variables, such as age, lot size, number of bedrooms and bathrooms, and distance from schools. As another example, let y denote a quantitative measure of wrinkle resistance in cotton cellulose fabric. Then the value of y may depend on such variable quantities as curing temperature, curing time, formaldehyde concentration, and concentration of the catalyst sulfur dioxide used in the production process.

In this chapter, we extend the regression methodology developed in the previous chapter to *multiple regression models*, which include at least two predictor variables. Fortunately, many of the concepts developed in the context of simple linear regression carry over to multiple regression with little or no modification. The calculations required to fit such a model and make further inferences are *much* more tedious than those for simple linear regression, so a computer is an indispensable tool for multiple regression analysis. The computer's ability to perform a huge number of computations in a short time has spurred the development of new methods for analyzing large data sets with many predictor variables. These include techniques for fitting numerous alternative models and choosing between them, tools for identifying influential observations, and both algebraic and graphical diagnostics designed to reveal potential violations of model assumptions. A single chapter can do little more than scratch the surface of this important subject area.

▪ 14.1 Multiple Regression Models

The relationship between a dependent or response variable y and two or more independent or predictor variables is deterministic if the value of y is completely determined, with no uncertainty, once values of the independent variables have been specified. Consider, for example, a school district in which teachers with no prior teaching experience and no college credits beyond a bachelor's degree start at an annual salary of $28,000. Suppose that for each year of teaching experience up to 20 years, a teacher receives an additional $800 per year and that each unit of postcollege coursework up to 75 units results in an extra $60 per year. Define three variables:

y = salary of a teacher who has at most 20 years teaching experience and at most 75 postcollege units

x_1 = number of years teaching experience

x_2 = number of postcollege units

Previously, x_1 and x_2 denoted the first two observations on the single variable x. In the usual notation for multiple regression, however, x_1 and x_2 represent two different variables.

The value of y is entirely determined by values of x_1 and x_2 through the equation

$$y = 28{,}000 + 800x_1 + 60x_2$$

Thus, if $x_1 = 10$ and $x_2 = 30$,

$$y = 28{,}000 + 800(10) + 60(30)$$
$$= 28{,}000 + 8000 + 1800$$
$$= 37{,}800$$

If two different teachers have both the same x_1 value and the same x_2 value, they will also have identical y values.

Only rarely is y deterministically related to predictors $x_1, \ldots, x_k$. A probabilistic model results from adding a random deviation e to a deterministic function of the x_i's.

▪ Definition

A **general additive multiple regression model**, which relates a dependent variable y to k predictor variables $x_1, x_2, \ldots, x_k$, is given by the model equation

$$y = \alpha + \beta_1 x_1 + \beta_2 x_2 + \cdots + \beta_k x_k + e$$

The random deviation e is assumed to be normally distributed with mean value 0 and variance σ^2 for any particular values of $x_1, \ldots, x_k$. This implies that for fixed $x_1, x_2, \ldots, x_k$ values, y has a normal distribution with variance σ^2 and

$$(\text{mean } y \text{ value for fixed } x_1, \ldots, x_k \text{ values}) = \alpha + \beta_1 x_1 + \beta_2 x_2 + \cdots + \beta_k x_k$$

The β_i's are called **population regression coefficients**; each β_i can be interpreted as the true average change in y when the predictor x_i increases by 1 unit *and* the values of all the other predictors remain fixed.

The deterministic portion $\alpha + \beta_1 x_1 + \beta_2 x_2 + \cdots + \beta_k x_k$ is called the **population regression function**.

As in simple linear regression, if σ^2 is quite close to 0, any particular observed y will tend to be quite near its mean value. When σ^2 is large, many of the y observations may deviate substantially from their mean y values.

■ **Example 14.1   Cardiorespiratory Fitness**

Cardiorespiratory fitness is widely recognized as a major component of overall physical well-being. Direct measurement of maximal oxygen uptake is the single best measure of such fitness, but direct measurement is time-consuming and expensive. It is therefore desirable to have a prediction equation for maximal oxygen uptake in terms of easily obtained quantities. Consider the variables

y = maximal oxygen uptake (in liters per minute)

x_1 = weight (in kilograms)

x_2 = age (in years)

x_3 = time necessary to walk 1 mi (in minutes)

x_4 = heart rate at the end of the walk (in beats per minute)

One possible population model for male students that is consistent with information given in the article "Validation of the Rockport Fitness Walking Test in College Males and Females" (*Research Quarterly for Exercise and Sport* [1994]: 152–158) is

$$y = 5.0 + 0.01x_1 - 0.05x_2 - 0.13x_3 - 0.01x_4 + e$$
$$\sigma = 0.4$$

The population regression function is

(mean y value for fixed $x_1, \ldots, x_4$) $= 5.0 + 0.01x_1 - 0.05x_2 - 0.13x_3 - 0.01x_4$

For individuals whose weight is 76 kg, age is 20 years, walk time is 12 min, and heart rate is 140 beats/min,

$$\begin{pmatrix} \text{mean value of} \\ \text{maximal oxygen uptake} \end{pmatrix} = 5.0 + 0.01(76) - 0.05(20) - 0.13(12) - 0.01(140)$$
$$= 1.80$$

With $2\sigma = 0.80$, it is quite likely that an actual y value will be within 0.80 of the mean value (i.e., in the interval from 1.00 to 2.60).

The true average change in maximal oxygen uptake when weight increases by 1 kg and the other predictors remain fixed is $\beta_1 = 0.01$. Similarly, because $\beta_2 = -0.05$, when age increases by 1 year and the other three predictors remain constant, we expect maximal oxygen uptake to decrease by 0.05 l/min.

■ **A Special Case: Polynomial Regression**

Consider again the case of a single independent variable x, and suppose that a scatterplot of the n sample (x, y) pairs has the appearance of Figure 14.1. The simple linear regression model is clearly not appropriate, but it does look as though a parabola (quadratic function) with equation $y = \alpha + \beta_1 x + \beta_2 x^2$ would provide a very good fit to the data for appropriately chosen values of α, β_1, and β_2. Just as the inclusion of the random deviation e in simple linear regression allowed an observation to deviate from the population regression line by a random amount, adding e to this quadratic function yields a probabilistic model in which an observation is allowed to fall above or below the parabola. The model equation is

$$y = \alpha + \beta_1 x + \beta_2 x^2 + e$$

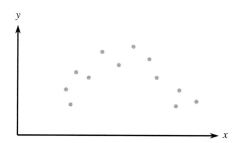

Figure 14.1 A scatter-plot that suggests the appropriateness of a quadratic probabilistic model.

Let's rewrite the model equation using x_1 to denote x and x_2 to denote x^2. The model equation then becomes

$$y = \alpha + \beta_1 x_1 + \beta_2 x_2 + e$$

This is a special case of the general multiple regression model with $k = 2$. You may wonder about the legitimacy of allowing one predictor variable to be a mathematical function of another predictor — here, $x_2 = (x_1)^2$. However, there is absolutely nothing in the general multiple regression model that prevents this. *In the model* $y = \alpha + \beta_1 x_1 + \beta_2 x_2 + \cdots + \beta_k x_k + e$, *it is permissible to have predictors that are mathematical functions of other predictors.* For example, starting with the two independent variables x_1 and x_2, we could create a model with $k = 4$ predictors in which x_1 and x_2 themselves are the first two predictor variables and $x_3 = (x_1)^2$ and $x_4 = x_1 x_2$. (We will soon discuss the consequences of using a predictor such as x_4.) In particular, the general polynomial regression model begins with a single independent variable x and creates predictors $x_1 = x, x_2 = x^2, x_3 = x^3, \ldots, x_k = x^k$ for some specified value of k.

▪ **Definition**

The **kth-degree polynomial regression model**

$$y = \alpha + \beta_1 x + \beta_2 x^2 + \cdots + \beta_k x^k + e$$

is a special case of the general multiple regression model with $x_1 = x, x_2 = x^2, x_3 = x^3, \ldots, x_k = x^k$.

The **population regression function** (mean value of y for fixed values of x) is

$$\alpha + \beta_1 x + \beta_2 x^2 + \cdots + \beta_k x^k$$

The most important special case other than simple linear regression ($k = 1$) is the **quadratic regression model**

$$y = \alpha + \beta_1 x + \beta_2 x^2 + e$$

This model replaces the line of mean values $\alpha + \beta x$ in simple linear regression with a parabolic curve of mean values $\alpha + \beta_1 x + \beta_2 x^2$. If $\beta_2 > 0$, the curve opens upward, whereas if $\beta_2 < 0$, the curve opens downward (Figure 14.2). A less frequently encountered special case is that of cubic regression, in which $k = 3$.

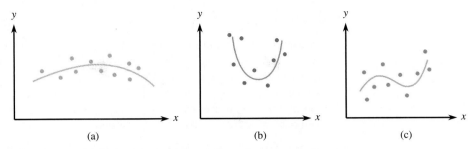

Figure 14.2 Polynomial regression: (a) quadratic regression model with $\beta_2 < 0$; (b) quadratic regression model with $\beta_2 > 0$; (c) cubic regression model with $\beta_3 > 0$.

■ **Example 14.2** Sunshine and Cloud Cover

Researchers have examined a variety of climatic variables in an attempt to gain an understanding of the mechanisms that govern rainfall runoff. The article "The Applicability of Morton's and Penman's Evapotranspiration Estimates in Rainfall Runoff Modeling" (*Water Resources Bulletin* [1991]: 611–620) reported on a study in which data on

x = cloud cover

and

y = daily sunshine (in hours)

was gathered from a number of different locations. The investigators used a cubic regression model to relate these variables (see Figure 14.2(c)). Suppose that the actual model equation for a particular location is

$$y = 11 - 0.400x - 0.250x^2 + 0.005x^3 + e$$

Then the population regression function is

$$\left(\begin{array}{c} \text{mean daily sunshine for} \\ \text{given cloud cover } x \end{array}\right) = 11 - 0.400x - 0.250x^2 + 0.005x^3$$

For example,

$$\left(\begin{array}{c} \text{mean daily sunshine} \\ \text{when cloud cover is 4} \end{array}\right) = 11 - 0.400(4) - 0.250(4)^2 + 0.005(4)^3$$

$$= 5.72$$

If $\sigma = 1$, then it is quite likely that an observation on daily sunshine made when $x = 4$ would be between 3.72 and 7.72 hr.

The interpretation of β_i previously given for the general multiple regression model cannot be applied in polynomial regression. This is because all predictors are functions of the single variable x, so $x_i = x^i$ cannot be increased by 1 unit without changing the values of all the other predictor variables as well. *In general, the interpretation of regression coefficients requires extra care when some predictor variables are mathematical functions of other variables.*

▪ Interaction Between Variables

Suppose that an industrial chemist is interested in the relationship between product yield (y) from a certain chemical reaction and two independent variables, x_1 = reaction temperature and x_2 = pressure at which the reaction is carried out. The chemist initially suggests that for temperature values between 80 and 110 in combination with pressure values ranging from 50 to 70, the relationship can be well described by the probabilistic model

$$y = 1200 + 15x_1 - 35x_2 + e$$

The regression function, which gives the mean y value for any specified values of x_1 and x_2, is then $1200 + 15x_1 - 35x_2$. Consider this mean y value for three different particular temperature values:

$x_1 = 90$:   mean y value = $1200 + 15(90) - 35x_2 = 2550 - 35x_2$

$x_1 = 95$:   mean y value = $2625 - 35x_2$

$x_1 = 100$:   mean y value = $2700 - 35x_2$

Graphs of these three mean value functions (each a function only of pressure x_2, because the temperature value has been specified) are shown in Figure 14.3(a). Each graph is a straight line, and the three lines are parallel, each one having slope -35. Because of this, the average change in yield when pressure x_2 is increased by 1 unit is -35, regardless of the fixed temperature value.

Figure 14.3 Graphs of mean y value for two different models: (a) $1200 + 15x_1 - 35x_2$; (b) $-4500 + 75x_1 + 60x_2 - x_1x_2$.

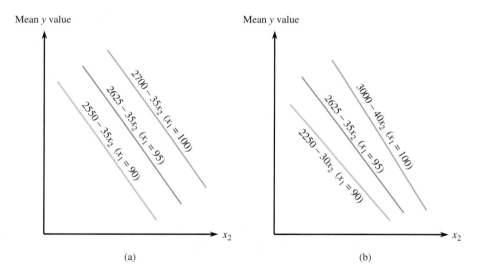

Because chemical theory suggests that the decline in average yield when pressure x_2 increases should be more rapid for a high temperature than for a low temperature, the chemist now has reason to doubt the appropriateness of the proposed model. Rather than the lines being parallel, the line for a temperature of 100 should be steeper than the line for a temperature of 95, and that line in turn should be steeper than the one for $x_1 = 90$. A model that has this property includes, in addition to predictors x_1 and x_2 separately, a third predictor variable $x_3 = x_1x_2$. One such model is

$$y = -4500 + 75x_1 + 60x_2 - x_1x_2 + e$$

which has regression function $-4500 + 75x_1 + 60x_2 - x_1x_2$. Then

$$(\text{mean } y \text{ when } x_1 = 100) = -4500 + 75(100) + 60x_2 - 100x_2$$
$$= 3000 - 40x_2$$

whereas

$$(\text{mean } y \text{ when } x_1 = 95) = 2625 - 35x_2$$
$$(\text{mean } y \text{ when } x_1 = 90) = 2250 - 30x_2$$

These functions are graphed in Figure 14.3(b), where it is clear that the three slopes are different. In fact, each different value of x_1 yields a different slope, so the average change in yield associated with a 1-unit increase in x_2 depends on the value of x_1. When this is the case, the two variables are said to *interact*.

▪ **Definition**

If the change in the mean y value associated with a 1-unit increase in one independent variable depends on the value of a second independent variable, there is **interaction** between these two variables. When the variables are denoted by x_1 and x_2, such interaction can be modeled by including x_1x_2, the product of the variables that interact, as a predictor variable.

The general equation for a multiple regression model based on two independent variables x_1 and x_2 that also includes an interaction predictor is

$$y = \alpha + \beta_1x_1 + \beta_2x_2 + \beta_3x_1x_2 + e$$

When x_1 and x_2 do interact, this model usually gives a much better fit to the resulting sample data — and thus explains more variation in y — than does the no-interaction model. Failure to consider a model with interaction often leads an investigator to conclude incorrectly that there is no strong relationship between y and a set of independent variables.

More than one interaction predictor can be included in the model when more than two independent variables are available. If, for example, there are three independent variables x_1, x_2, and x_3, one possible model is

$$y = \alpha + \beta_1x_1 + \beta_2x_2 + \beta_3x_3 + \beta_4x_4 + \beta_5x_5 + \beta_6x_6 + e$$

where

$$x_4 = x_1x_2 \qquad x_5 = x_1x_3 \qquad x_6 = x_2x_3$$

One could even include a three-way interaction predictor $x_7 = x_1x_2x_3$ (the product of all three independent variables), although in practice this is rarely done.

In applied work, quadratic terms, such as x_1^2 and x_2^2, are often included to model a curved relationship between y and several independent variables. For example, a frequently used model involving just two independent variables x_1 and x_2 but $k = 5$ predictors is the *full quadratic* or **complete second-order model**

$$y = \alpha + \beta_1x_1 + \beta_2x_2 + \beta_3x_1x_2 + \beta_4x_1^2 + \beta_5x_2^2 + e$$

This model replaces the straight lines of Figure 14.3 with parabolas (each one is the graph of the regression function for different values of x_2 when x_1 has a fixed value). With four independent variables, one could examine a model containing four quadratic predictors and six two-way interaction predictor variables. Clearly, with just a few independent variables, one could examine a great many different multiple regression models. In Section 14.5 we briefly discuss methods for selecting one model from a number of competing models.

When developing a multiple regression model, scatterplots of *y* with each potential predictor can be informative. This is illustrated in Example 14.3, which describes a model that includes a term that is a function of one of the predictor variables and also an interaction term.

▪ Example 14.3   Wind Chill Factor

The wind chill index, often included in winter weather reports, combines information on air temperature and wind speed to describe how cold it really feels. In 2001, the National Weather Service announced that it would begin using a new wind chill formula beginning in the fall of that year (*USA Today*, August 13, 2001). The following table gives the wind chill index for various combinations of air temperature and wind speed.

Wind (mph)	Temperature (°F)													
	35	30	25	20	15	10	5	0	−5	−10	−15	−20	−25	−30	−35
5	31	25	19	13	7	1	−5	−11	−16	−22	−28	−34	−40	−46	−52
10	27	21	15	9	3	−4	−10	−16	−22	−28	−35	−41	−47	−53	−59
15	25	19	13	6	0	−7	−13	−19	−26	−32	−39	−45	−51	−58	−64
20	24	17	11	4	−2	−9	−15	−22	−29	−35	−42	−48	−55	−61	−68
25	23	16	9	3	−4	−11	−17	−24	−31	−37	−44	−51	−58	−64	−71
30	22	15	8	1	−5	−12	−19	−26	−33	−39	−46	−53	−60	−67	−73
35	21	14	7	0	−7	−14	−21	−27	−34	−41	−48	−55	−62	−69	−76
40	20	13	6	−1	−8	−15	−22	−29	−36	−43	−50	−57	−64	−71	−78
45	19	12	5	−2	−9	−16	−23	−30	−37	−44	−51	−58	−65	−72	−79

Figure 14.4(a) shows a scatterplot of wind chill index versus air temperature with the different wind speeds denoted by different colors in the plot. It appears that the wind chill index increases linearly with air temperature at each of the wind speeds, but the linear patterns for the different wind speeds are not quite parallel. This suggests that to model the relationship between *y* = wind chill index and the two variables x_1 = air temperature and x_2 = wind speed, we should include both x_1 and an interaction term that involves x_2. Figure 14.4(b) shows a scatterplot of wind chill index versus wind speed with the different temperatures denoted by different colors. This plot reveals that the relationship between wind chill index and wind speed is nonlinear at each of the different temperatures, and because the pattern is more markedly curved at some temperatures than at others, an interaction is suggested.

These observations are consistent with the new model used by the National Weather Service for relating wind chill index to air temperature and wind speed. The model used is

$$y = 35.74 + 0.621x_1 - 35.75(x_2') + 0.4275x_1x_2'$$

where

$$x_2' = x_2^{0.16}$$

which incorporates a transformed x_2 (to model the nonlinear relationship between wind chill index and wind speed) and an interaction term.

Figure 14.4 Scatterplots of the wind chill index data of Example 14.3: (a) wind chill index versus air temperature; (b) wind chill index versus wind speed.

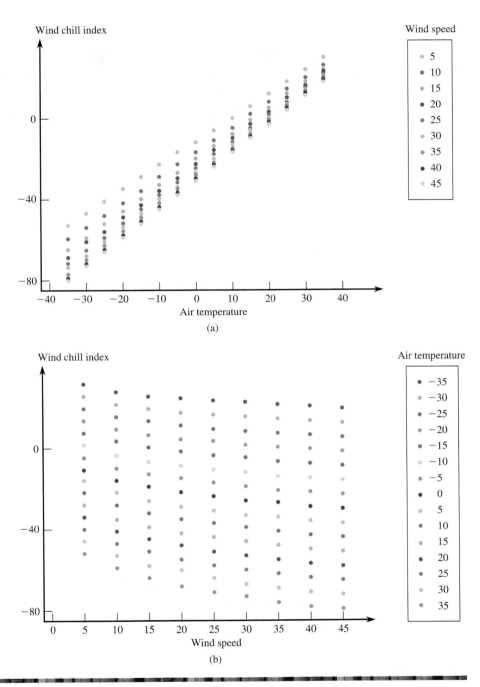

■ Qualitative Predictor Variables

Up to this point, we have explicitly considered the inclusion only of quantitative (numerical) predictor variables in a multiple regression model. Using simple numerical coding, qualitative (categorical) variables can also be incorporated into a model. Let's focus first on a dichotomous variable, one with just two possible categories: male or female, U.S. or foreign manufacture, a house with or without a view, and so on. With any such variable, we associate a numerical variable x whose possible values are 0 and 1, where 0 is identified with one category (e.g., married) and 1 is identified with the other possible category (e.g., not married). This 0–1 variable is often called a **dummy variable** or **indicator variable**.

▪ Example 14.4   Predictors of Writing Competence

The article "Grade Level and Gender Differences in Writing Self-Beliefs of Middle School Students" (*Contemporary Educational Psychology* [1999]: 390–405) considered relating writing competence score to a number of predictor variables, including perceived value of writing and gender. Both writing competence and perceived value of writing were represented by a numerically scaled variable, but gender was a qualitative predictor.

Let

$$y = \text{writing competence score}$$

$$x_1 = \begin{cases} 0 & \text{if male} \\ 1 & \text{if female} \end{cases}$$

$$x_2 = \text{perceived value of writing}$$

One possible multiple regression model is

$$y = \alpha + \beta_1 x_1 + \beta_2 x_2 + e$$

Considering the mean y value first when $x_1 = 0$ and then when $x_1 = 1$ yields

$$\text{average score} = \alpha + \beta_2 x_2 \qquad \text{when } x_1 = 0 \text{ (males)}$$
$$\text{average score} = \alpha + \beta_1 + \beta_2 x_2 \quad \text{when } x_1 = 1 \text{ (females)}$$

The coefficient β_1 is the difference in average writing competence score between males and females when perceived value of writing is held fixed.

A second possibility is a model with an interaction term:

$$y = \alpha + \beta_1 x_1 + \beta_2 x_2 + \beta_3 x_1 x_2 + e$$

The regression function for this model is $\alpha + \beta_1 x_1 + \beta_2 x_2 + \beta_3 x_3$, where $x_3 = x_1 x_2$. Now the two cases $x_1 = 0$ and $x_1 = 1$ give

$$\text{average score} = \alpha + \beta_2 x_2 \qquad\qquad \text{when } x_1 = 0 \text{ (males)}$$
$$\text{average score} = \alpha + \beta_1 + (\beta_2 + \beta_3) x_2 \quad \text{when } x_1 = 1 \text{ (females)}$$

For each model, the graph of the average writing competence score, when regarded as a function of perceived value of writing, is a line for either gender (Figure 14.5).

Figure 14.5 Regression functions for models with one qualitative variable (x_1) and one quantitative variable (x_2): (a) no interaction; (b) interaction.

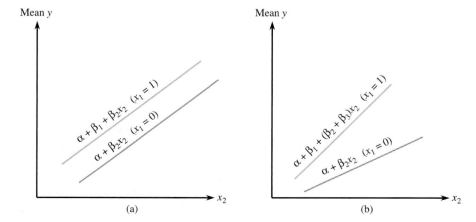

In the no-interaction model, the coefficient on x_2 is β_2 both when $x_1 = 0$ and when $x_1 = 1$, so the two lines are parallel, although their intercepts are different (unless $\beta_1 = 0$). With interaction, the lines not only have different intercepts but also have different slopes (unless $\beta_3 = 0$). For this model, the change in average writing competence score when perceived value of writing increases by 1 unit depends on gender — the two variables *perceived value* and *gender* interact.

You might think that the way to handle a three-category situation is to define a single numerical variable with coded values such as 0, 1, and 2 corresponding to the three categories. This is incorrect, because it imposes an ordering on the categories that is not necessarily implied by the problem. The correct approach to modeling a categorical variable with three categories is to define *two* different 0–1 variables, as illustrated in Example 14.5.

▪ Example 14.5   Organizational Structure

The article "The Effect of Ownership on the Organization Structure in Small Firms" (*Administrative Science Quarterly* [1984]: 232–237) considered three different management status categories — owner-successors, professional managers, and owner-founders — in relating degree of horizontal differentiation (y) to firm size. Let

$$x_1 = \begin{cases} 1 & \text{if the firm is managed by owner-successors} \\ 0 & \text{otherwise} \end{cases}$$

$$x_2 = \begin{cases} 1 & \text{if the firm is managed by professional managers} \\ 0 & \text{otherwise} \end{cases}$$

$$x_3 = \text{firm size}$$

Thus, $x_1 = 1, x_2 = 0$ indicates owner-successors; $x_1 = 0, x_2 = 1$ indicates professional managers; and $x_1 = x_2 = 0$ indicates owner-founders ($x_1 = x_2 = 1$ is not possible). The investigators suggested an interaction model of the form

$$y = \alpha + \beta_1 x_1 + \beta_2 x_2 + \beta_3 x_3 + \beta_4 x_1 x_3 + \beta_5 x_2 x_3 + e$$

This model allows the change in mean differentiation when size increases by 1 to be different for all three management status categories.

In general, incorporating a categorical variable with c possible categories into a regression model requires the use of $c - 1$ indicator variables. Thus, even one such categorical variable can add many predictors to a model.

▪ Nonlinear Multiple Regression Models

Many nonlinear relationships can be put into the form $y = \alpha + \beta_1 x_1 + \cdots + \beta_k x_k + e$ by transforming one or more of the variables. An appropriate transformation could be suggested by theory or by various plots of the data (e.g., residual plots after fitting a particular model). There are also relationships that cannot be linearized by transformations, in which case more complicated methods of analysis must be used. A general discussion of nonlinear regression is beyond the scope of this textbook; you can learn more by consulting one of the sources listed in the references to Chapter 5.

▪ Exercises 14.1–14.14

14.1 Explain the difference between a deterministic model and a probabilistic model. Give an example of a dependent variable y and two or more independent variables that might be related to y deterministically. Give an example of a dependent variable y and two or more independent variables that might be related to y in a probabilistic fashion.

14.2 A number of recent investigations have focused on the problem of assessing loads that can be manually handled in a safe manner. The article "Anthropometric, Muscle Strength, and Spinal Mobility Characteristics as Predictors in the Rating of Acceptable Loads in Parcel Sorting" (*Ergonomics* [1992]: 1033–1044) proposed using a regression model to relate the dependent variable y = individual's rating of acceptable load (in kilograms) to $k = 3$ independent (predictor) variables: x_1 = extent of left lateral bending (in centimeters), x_2 = dynamic hand grip endurance (in seconds), and x_3 = trunk extension ratio (in newtons per kilogram). Suppose that the model equation is

$$y = 30 + 0.90x_1 + 0.08x_2 - 4.50x_3 + e$$

and that $\sigma = 5$.
a. What is the population regression function?
b. What are the values of the population regression coefficients?
c. Interpret the value of β_1.
d. Interpret the value of β_3.
e. What is the mean value of rating of acceptable load when extent of left lateral bending is 25 cm, dynamic hand grip endurance is 200 sec, and trunk extension ratio is 10 N/kg?
f. If repeated observations on rating are made on different individuals, all of whom have the values of x_1, x_2, and x_3 specified as in Part (e), in the long run approximately what percentage of ratings will be between 13.5 and 33.5 kg?

14.3 The following statement appeared in the article "Dimensions of Adjustment Among College Women" (*Journal of College Student Development* [1998]: 364–372): "Regression analyses indicated that academic adjustment and race made independent contributions to academic achievement, as measured by current GPA." Suppose that y = current GPA, x_1 = academic adjustment score, and x_2 = race (with white = 0, other = 1). What multiple regression model is suggested by the statement? Did you include an interaction term in the model? Why or why not?

14.4 According to "Assessing the Validity of the Post-Materialism Index" (*American Political Science Review* [1999]: 649–664), we might be able to predict

an individual's level of support for ecology based on demographic and ideological characteristics. The multiple regression model proposed by the investigators was

$$y = 3.60 - 0.01x_1 + 0.01x_2 - 0.07x_3 + 0.12x_4 + 0.02x_5 \\ - 0.04x_6 - 0.01x_7 - 0.04x_8 - 0.02x_9 + e$$

where the variables are defined as follows: y = ecology score (higher values indicate a greater concern for ecology), x_1 = age times 10, x_2 = income (in thousands of dollars), x_3 = gender (1 = male, 0 = female), x_4 = race (1 = white, 0 = nonwhite), x_5 = education (in years), x_6 = ideology (4 = conservative, 3 = right of center, 2 = middle of the road, 1 = left of center, and 0 = liberal), x_7 = social class (4 = upper, 3 = upper middle, 2 = middle, 1 = lower middle, 0 = lower), x_8 = postmaterialist (1 if postmaterialist, 0 otherwise), and x_9 = materialist (1 if materialist, 0 otherwise).
a. Suppose that you knew a person with the following characteristics: a 25-year-old white female with a college degree (16 years of education) who has a job that pays $32,000 per year; she is from the upper middle class and considers herself left of center, but she is neither a materialist nor a postmaterialist. Predict her ecology score.
b. If the woman described in Part (a) was Hispanic rather than white, how would the prediction change?
c. Given that the other variables are the same, what is the estimated mean difference in ecology score for men and women?
d. How would you interpret the coefficient of x_2?
e. Comment on the numerical coding of the ideology and social class variables. Suggest a better way of incorporating these two variables into the model.

14.5 The article "The Influence of Temperature and Sunshine on the Alpha-Acid Contents of Hops" (*Agricultural Meteorology* [1974]: 375–382) used a multiple regression model to relate y = yield of hops to x_1 = mean temperature (°C) between date of coming into hop and date of picking and x_2 = mean percentage of sunshine during the same period. The proposed model equation is

$$y = 415.11 - 6060x_1 - 4.50x_2 + e$$

a. Suppose that this equation does indeed describe the true relationship. What mean yield corresponds to a temperature of 20 and a sunshine percentage of 40?
b. What is the mean yield when the mean temperature and percentage of sunshine are 18.9 and 43, respectively?
c. Interpret the values of the population regression coefficients.

14.6 The article "Readability of Liquid Crystal Displays: A Response Surface" (*Human Factors* [1983]: 185–190) used a multiple regression model with four independent variables, where y = error percentage for subjects reading a four-digit liquid crystal display, x_1 = level of backlight (from 0 to 122 cd/m), x_2 = character subtense (from 0.025° to 1.34°), x_3 = viewing angle (from 0° to 60°), and x_4 = level of ambient light (from 20 to 1500 lx). The model equation suggested in the article is

$$y = 1.52 + 0.02x_1 - 1.40x_2 + 0.02x_3 - 0.0006x_4 + e$$

a. Assume that this is the correct model equation. What is the mean value of y when $x_1 = 10$, $x_2 = 0.5$, $x_3 = 50$, and $x_4 = 100$?
b. What mean error percentage is associated with a backlight level of 20 cd/m, a character subtense of 0.5°, a viewing angle of 10°, and an ambient light level of 30 lx?
c. Interpret the values of β_2 and β_3.

14.7 The article "Pulp Brightness Reversion: Influence of Residual Lignin on the Brightness Reversion of Bleached Sulfite and Kraft Pulps" (*Tappi* [1964]: 653–662) proposed a quadratic regression model to describe the relationship between x = degree of delignification during the processing of wood pulp for paper and y = total chlorine content. Suppose that the actual model is

$$y = 220 + 75x - 4x^2 + e$$

a. Graph the regression function $220 + 75x - 4x^2$ over x values between 2 and 12. (Hint: Substitute $x = 2, 4, 6, 8, 10$, and 12 to find points on the graph, and connect them with a smooth curve.)
b. Would mean chlorine content be higher for a degree of delignification value of 8 or 10?
c. What is the change in mean chlorine content when the degree of delignification increases from 8 to 9? from 9 to 10?

14.8 The relationship between yield of maize, date of planting, and planting density was investigated in the article "Development of a Model for Use in Maize Replant Decisions" (*Agronomy Journal* [1980]: 459–464). Let y = maize yield (percentage), x_1 = planting date (days after April 20), and x_2 = planting density (plants per hectare). The regression model with both quadratic terms ($y = \alpha + \beta_1 x_1 + \beta_2 x_2 + \beta_3 x_3 + \beta_4 x_4 + e$, where $x_3 = x_1^2$ and $x_4 = x_2^2$) provides a good description of the relationship between y and the independent variables.
a. If $\alpha = 21.09$, $\beta_1 = 0.653$, $\beta_2 = 0.0022$, $\beta_3 = -0.0206$, and $\beta_4 = 0.00004$, what is the population regression function?
b. Use the regression function from Part (a) to determine the mean yield for a plot planted on May 6 with a density of 41,180 plants/ha.

c. Would the mean yield be higher for a planting date of May 6 or May 22 (for the same density)?
d. Is it legitimate to interpret $\beta_1 = 0.653$ as the true average change in yield when planting date increases by 1 day and the values of the other three predictors are held fixed? Why or why not?

14.9 Suppose that the variables y, x_1, and x_2 are related by the regression model

$$y = 1.8 + 0.1x_1 + 0.8x_2 + e$$

a. Construct a graph (similar to that of Figure 14.5) that shows the relationship between mean y and x_2 for fixed x_1 values 10, 20, and 30.
b. Construct a graph depicting the relationship between mean y and x_1 for fixed x_2 values 50, 55, and 60.
c. What aspect of the graphs in Parts (a) and (b) can be attributed to the lack of an interaction between x_1 and x_2?
d. Suppose that the interaction term $0.03x_3$, where $x_3 = x_1 x_2$, is added to the regression model equation. Using this new model, construct the graphs described in Parts (a) and (b). How do they differ from those obtained in Parts (a) and (b)?

14.10 A manufacturer of wood stoves collected data on y = particulate matter concentration and x_1 = flue temperature for three different air intake settings (low, medium, and high).
a. Write a model equation that includes dummy variables to incorporate intake setting, and give an interpretation of each of the β coefficients.
b. What additional predictors would be needed to incorporate interaction between temperature and intake setting?

14.11 Consider a regression analysis with three independent variables x_1, x_2, and x_3. Give the equation for the following regression models:
a. The model that includes as predictors all independent variables but no quadratic or interaction terms
b. The model that includes as predictors all independent variables and all quadratic terms
c. All models that include as predictors all independent variables, no quadratic terms, and exactly one interaction term
d. The model that includes as predictors all independent variables, all quadratic terms, and all interaction terms (the complete second-order model)

14.12 The article "The Value and the Limitations of High-Speed Turbo-Exhausters for the Removal of Tar-Fog from Carburetted Water-Gas" (*Society of Chemical Industry Journal* [1946]: 166–168) presented data on y = tar content (in grains per 100 ft^3) of a gas stream as a function of x_1 = rotor speed (in revolutions per minute) and x_2 = gas inlet temperature (in °F). A regression model using x_1, x_2, $x_3 = x_2^2$, and $x_4 = x_1 x_2$ was suggested:

mean y value $= 86.8 - 0.123x_1 + 5.09x_2$
$$- 0.0709x_3 + 0.001x_4$$

a. According to this model, what is the mean y value if $x_1 = 3200$ and $x_2 = 57$?

b. For this particular model, does it make sense to interpret the value of any individual β_i ($\beta_1, \beta_2, \beta_3,$ or β_4) in the way we have previously suggested? Explain.

14.13 Consider the dependent variable $y = $ fuel efficiency of a car (in miles per gallon).

a. Suppose that you want to incorporate size class of car, with four categories (subcompact, compact, mid-size, and large), into a regression model that also includes $x_1 = $ age of car and $x_2 = $ engine size. Define the necessary dummy variables, and write out the complete model equation.

b. Suppose that you want to incorporate interaction between age and size class. What additional predictors would be needed to accomplish this?

14.14 If we knew the width and height of cylindrical tin cans of food, could we predict the volume of these cans with precision and accuracy?

a. Give the equation that would allow us to make such predictions.

b. Is the relationship between volume and its predictors, height and width, a linear one?

c. Should we use an additive multiple regression model to predict a volume of a can from its height and width? Explain.

d. If you were to take logarithms of each side of your equation in Part (a), would the relationship be linear?

▪ 14.2 Fitting a Model and Assessing Its Utility

In Section 14.1, we introduced multiple regression models containing several different types of predictors. Let's now suppose that a particular set of k predictor variables $x_1, x_2, \ldots, x_k$ has been selected for inclusion in the model

$$y = \alpha + \beta_1 x_1 + \beta_2 x_2 + \cdots + \beta_k x_k + e$$

It is then necessary to estimate the model coefficients $\alpha, \beta_1, \ldots, \beta_k$ and the regression function $\alpha + \beta_1 x_1 + \cdots + \beta_k x_k$ (the mean y value for specified values of the predictors), assess the model's utility, and perhaps use the estimated model to make further inferences. All this, of course, requires sample data. As before, n denotes the number of observations in the sample. With just one predictor variable, the sample consisted of n (x, y) pairs. Now each observation consists of $k + 1$ numbers: a value of x_1, a value of $x_2, \ldots$, a value of x_k, and the associated value of y. The n observations are assumed to have been selected independently of one another.

▪ Example 14.6   Soil and Sediment Characteristics

Soil and sediment adsorption, the extent to which chemicals collect in a condensed form on the surface, is an important characteristic because it influences the effectiveness of pesticides and various agricultural chemicals. The article "Adsorption of Phosphates, Arsenate, Methane Arsenate, and Calcodylate by Lake and Stream Sediments: Comparisons with Soils" (*Journal of Environmental Quality* [1984]: 499–504) presented the following data consisting of $n = 13$ (x_1, x_2, y) triples and proposed the model

$$y = \alpha + \beta_1 x_1 + \beta_2 x_2 + e$$

for relating

$y = $ phosphate adsorption index

$x_1 = $ amount of extractable iron

$x_2 = $ amount of extractable aluminum

Observation	x_1	x_2	y
1	61	13	4
2	175	21	18
3	111	24	14
4	124	23	18
5	130	64	26
6	173	38	26
7	169	33	21
8	169	61	30
9	160	39	28
10	244	71	36
11	257	112	65
12	333	88	62
13	199	54	40

We will return to this example after we see how sample data are used to estimate model coefficients.

As in simple linear regression, the principle of least squares is used to estimate the coefficients $\alpha, \beta_1, \ldots, \beta_k$. For specified estimates $a, b_1, \ldots, b_k$

$$y - (a + b_1x_1 + b_2x_2 + \cdots + b_kx_k)$$

is the deviation between the observed y value for a particular observation and the predicted value using the estimated regression function $a + b_1x_1 + \cdots + b_kx_k$. The first observation in the data set of Example 14.6 is $(x_1, x_2, y) = (61, 13, 4)$. The resulting deviation between observed and predicted y values is

$$4 - [a + b_1(61) + b_2(13)]$$

Deviations corresponding to other observations are expressed in a similar manner. The principle of least squares then says to use as estimates of α, β_1, and β_2—the values of a, b_1, and b_2 that minimize the sum of these squared deviations.

▪ **Definition**

According to the principle of least squares, the fit of a particular estimated regression function $a + b_1x_1 + \cdots + b_kx_k$ to the observed data is measured by the sum of squared deviations between the observed y values and the y values predicted by the estimated function:

$$\sum[y - (a + b_1x_1 + \cdots + b_kx_k)]^2$$

The **least-squares estimates** of $\alpha, \beta_1, \ldots, \beta_k$ are those values of $a, b_1, \ldots, b_k$ that make this sum of squared deviations as small as possible.

The least-squares estimates for a given data set are obtained by solving a system of $k + 1$ equations in the $k + 1$ unknowns $a, b_1, \ldots, b_k$ (called the *normal equations*). In the case $k = 1$ (simple linear regression), there are only two equations, and we gave their general solution — the expressions for b and a — in Chapter 5. For $k \geq 2$, it is not as easy to write general expressions for the estimates without using advanced mathematical notation. Fortunately, the computer saves us! Formulas for the estimates have been programmed into all the commonly used statistical software packages.

■ Example 14.7 More on Soil and Sediment Characteristics

Figure 14.6 displays MINITAB output from a regression command requesting that the model $y = \alpha + \beta_1 x_1 + \beta_2 x_2 + e$ be fit to the phosphate adsorption data of Example 14.6. (We named the dependent variable HPO, an abbreviation for the chlorate H_2PO_4 [dihydrogen phosphate], and the two predictor variables x_1 and x_2 were named FE and AL, abbreviations derived from the chemical symbols for iron and aluminum, respectively.) Focus on the column labeled Coef (for coefficient) in the table near the top of the figure. The three numbers in this column are the estimated model coefficients:

$a = -7.351$ (the estimate of the constant term α)
$b_1 = 0.11273$ (the estimate of the coefficient β_1)
$b_2 = 0.34900$ (the estimate of the coefficient β_2)

Thus, we estimate that the average change in HPO associated with a 1-unit increase in FE while AL remains fixed is 0.11273. A similar interpretation applies to b_2. The estimated regression function is

$$\left(\begin{array}{c} \text{estimated mean value of } y \\ \text{for specified } x_1 \text{ and } x_2 \text{ values} \end{array} \right) = -7.351 - 0.11273x_1 + 0.34900x_2$$

Substituting $x_1 = 150$ and $x_2 = 60$ gives

$$-7.351 + 0.11273(150) + 0.34900(60) = 30.5$$

which can be interpreted either as a point estimate for the mean value of HPO or as a point prediction for a single HPO value when FE = 150 and AL = 60.

Figure 14.6 MINITAB output for the regression analysis of Example 14.7.

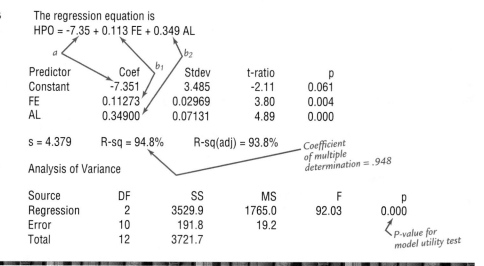

The regression equation is
HPO = -7.35 + 0.113 FE + 0.349 AL

Predictor	Coef	Stdev	t-ratio	p
Constant	-7.351	3.485	-2.11	0.061
FE	0.11273	0.02969	3.80	0.004
AL	0.34900	0.07131	4.89	0.000

s = 4.379   R-sq = 94.8%   R-sq(adj) = 93.8%

Coefficient of multiple determination = .948

Analysis of Variance

Source	DF	SS	MS	F	p
Regression	2	3529.9	1765.0	92.03	0.000
Error	10	191.8	19.2		
Total	12	3721.7			

P-value for model utility test

■ Is the Model Useful?

The utility of an estimated model can be assessed by examining the extent to which predicted y values based on the estimated regression function are close to the y values actually observed.

■ **Definition**

The first predicted value $\hat{y}_1$ is obtained by taking the values of the predictor variables $x_1, x_2, \ldots, x_k$ for the first sample observation and substituting these values into the estimated regression function. Doing this successively for the remaining observations yields the **predicted values** $\hat{y}_2, \ldots, \hat{y}_n$. The **residuals** are then the differences $y_1 - \hat{y}_1, y_2 - \hat{y}_2, \ldots, y_n - \hat{y}_n$ between the observed and predicted y values.

The predicted values and residuals are defined here exactly as they were in simple linear regression, but computation of the $\hat{y}$ values is more tedious because there is more than one predictor. Fortunately, the $\hat{y}$'s and $(y - \hat{y})$'s are automatically computed and displayed in the output of all good statistical software packages. Consider again the phosphate adsorption data discussed in Examples 14.6 and 14.7. Because the first y observation, $y_1 = 4$, was made with $x_1 = 61$ and $x_2 = 13$, the first predicted value is

$$\hat{y}_1 = -7.351 + 0.11273(61) + 0.34900(13) \approx 4.06$$

The first residual is then

$$y_1 - \hat{y}_1 = 4 - 4.06 = -0.06$$

The other predicted values and residuals are computed in a similar fashion. The sum of residuals from a least-squares fit should, except for rounding effects, be 0.

As in simple linear regression, the sum of squared residuals is the basis for several important summary quantities that are indicative of a model's utility.

■ **Definition**

The **residual (or error) sum of squares, SSResid**, and **total sum of squares, SSTo**, are given by

$$\text{SSResid} = \sum(y - \hat{y})^2 \qquad \text{SSTo} = \sum(y - \bar{y})^2$$

where $\bar{y}$ is the mean of the y observations in the sample.

The number of degrees of freedom associated with SSResid is $n - (k + 1)$, because $k + 1$ df are lost in estimating the $k + 1$ coefficients $\alpha, \beta_1, \ldots, \beta_k$.

An estimate of the random deviation variance σ^2 is given by

$$s_e^2 = \frac{\text{SSResid}}{n - (k + 1)}$$

and $s_e = \sqrt{s_e^2}$ is the estimate of σ.

The **coefficient of multiple determination**, R^2, interpreted as the proportion of variation in observed y values that is explained by the fitted model, is

$$R^2 = 1 - \frac{\text{SSResid}}{\text{SSTo}}$$

▪ **Example 14.8** **Soil Characteristics Revisited**

Looking again at Example 14.7, which contains MINITAB output for the adsorption data fit by a two-predictor model, we can find the residual sum of squares in the Error row and SS column of the table headed Analysis of Variance (see Figure 14.6): SSResid = 191.8. The associated number of degrees of freedom is $n - (k + 1) = 13 - (2 + 1) = 10$, which appears in the DF column just to the left of SSResid. The sample average y value is $\bar{y} = 29.85$, and SSTo $= \sum (y - 29.85)^2 = 3721.7$ appears in the Total row and SS column of the Analysis of Variance table just under the value of SSResid. The values of s_e, s_e^2, and R^2 are then

$$s_e^2 = \frac{\text{SSResid}}{n - (k + 1)} = \frac{191.8}{10} = 19.18 \approx 19.2$$

(in the MS column of the MINITAB output)

$$s_e = \sqrt{s_e^2} = \sqrt{19.18} = 4.379$$

(which appears just above the Analysis of Variance table)

$$R^2 = 1 - \frac{\text{SSResid}}{\text{SSTo}} = 1 - \frac{191.8}{3721.7} = 1 - .052 = .948$$

Thus, the percentage of variation explained is $100R^2 = 94.8\%$, which appears in the output as R-sq = 94.8%. The values of R^2 and s_e suggest that the chosen model has been extremely successful in modeling the relationship between y and the predictors.

In general, a desirable model is one that results in both a large R^2 value and a small s_e value. However, there is a catch. These two conditions can be achieved by fitting a model that contains a large number of predictors. Such a model might be successful in explaining y variation, but it almost always specifies a relationship that is unrealistic and difficult to interpret. What we really want is a simple model — that is, a model that has relatively few predictors whose roles are easily interpreted and that also does a good job of explaining variation in y.

All statistical software packages include R^2 and s_e in their output, and most also give SSResid. In addition, some packages compute the quantity called the **adjusted R^2**:

$$\text{adjusted } R^2 = 1 - \left[\frac{n - 1}{n - (k + 1)} \right] \left(\frac{\text{SSResid}}{\text{SSTo}} \right)$$

Because the quantity in square brackets exceeds 1, the number subtracted from 1 is larger than $\frac{\text{SSResid}}{\text{SSTo}}$, so the adjusted R^2 is smaller than R^2. The value of R^2 must be between 0 and 1, but the adjusted R^2 can be negative. If a large R^2 has been achieved by using just a few model predictors, the adjusted R^2 will differ little from R^2. However, the adjustment can be substantial when a great many predictors (relative to the number of observations) have been used or when R^2 itself is small to moderate (which could happen even when there is no relationship between y and the predictors). In Example 14.7, the adjusted $R^2 = .938$, which is not much less than R^2 because the model included only two predictor variables.

■ *F* Distributions

The model utility test in simple linear regression was based on the fact that when $H_0: \beta = 0$ is true, the test statistic $t = \dfrac{b - 0}{s_b}$ has a t distribution. The model utility test for multiple regression uses a type of probability distribution called an *F distribution*. We digress briefly to describe some general properties of *F* distributions.

An *F* distribution always arises in connection with a ratio in which the numerator involves one sum of squares and the denominator involves a second sum of squares. Each sum of squares has associated with it a specified number of degrees of freedom, so a particular *F* distribution is determined by fixing values of df_1 = numerator degrees of freedom and df_2 = denominator degrees of freedom. There is a different *F* distribution for each different df_1–df_2 combination. For example, there is an *F* distribution based on 4 numerator degrees of freedom and 12 denominator degrees of freedom, another *F* distribution based on 3 numerator degrees of freedom and 20 denominator degrees of freedom, and so on. A typical *F* curve for fixed numerator and denominator degrees of freedom appears in Figure 14.7. All *F* tests presented in this book are upper-tailed. Recall that for an upper-tailed *t* test, the *P*-value is the area under the associated *t* curve to the right of the calculated *t*. Similarly, the *P*-value for an upper-tailed *F* test is the area under the associated *F* curve to the right of the calculated *F*. Figure 14.7 illustrates this for a test based on $df_1 = 4$ and $df_2 = 6$.

Figure 14.7 A *P*-value for an upper-tailed *F* test.

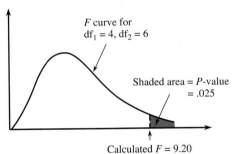

F curve for $df_1 = 4$, $df_2 = 6$

Shaded area = *P*-value = .025

Calculated *F* = 9.20

Unfortunately, tabulation of these upper-tail areas is much more cumbersome than for *t* distributions, because here 2 df are involved. For each of a number of different *F* distributions, our *F* table (Appendix Table 7) tabulates only four numbers: the values that capture tail areas .10, .05, .01, and .001. Different columns correspond to different values of df_1, and each different group of rows is for a different value of df_2. Figure 14.8 shows how this table is used to obtain *P*-value information.

Figure 14.8 Obtaining *P*-value information from the *F* table.

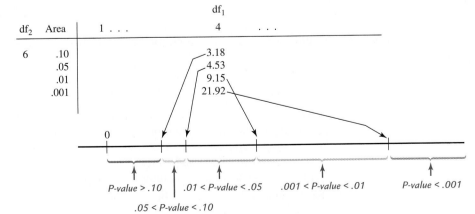

df_2	Area	1 . . .	4	. . .
6	.10		3.18	
	.05		4.53	
	.01		9.15	
	.001		21.92	

0

P-value > .10

.05 < P-value < .10

.01 < P-value < .05

.001 < P-value < .01

P-value < .001

For example, for a test with $df_1 = 4$ and $df_2 = 6$,

calculated $F = 5.70 \Rightarrow .01 < P\text{-value} < .05$

calculated $F = 2.16 \Rightarrow P\text{-value} > .10$

calculated $F = 25.03 \Rightarrow P\text{-value} < .001$

Only if calculated F equals a tabulated value do we obtain an exact P-value (e.g., if calculated $F = 4.53$, then $P\text{-value} = .05$). If $.01 < P\text{-value} < .05$, we should reject the null hypothesis at a significance level of .05 but not at a level of .01. When $P\text{-value} < .001$, H_0 would be rejected at any reasonable significance level.

■ The F Test for Model Utility

In the simple linear model with regression function $\alpha + \beta x$, if $\beta = 0$, there is no useful linear relationship between y and the single predictor variable x. Similarly, if all k coefficients $\beta_1, \beta_2, \ldots, \beta_k$ are 0 in the general k-predictor multiple regression model, there is no useful linear relationship between y and *any* of the predictor variables $x_1, x_2, \ldots, x_k$ included in the model. Before using an estimated model to make further inferences (e.g., predictions and estimates of mean values), you should confirm the model's utility through a formal test procedure.

Recall that SSTo is a measure of total variation in the observed y values and that SSResid measures the amount of total variation that has not been explained by the fitted model. The difference between total and error sums of squares is itself a sum of squares, called the **regression sum of squares**, which is denoted by SSRegr:

$$\text{SSRegr} = \text{SSTo} - \text{SSResid}$$

SSRegr is interpreted as the amount of total variation that *has* been explained by the model. Intuitively, the model should be judged useful if SSRegr is large relative to SSResid and the model uses a small number of predictors relative to the sample size. The number of degrees of freedom associated with SSRegr is k, the number of model predictors, and the number of degrees of freedom for SSResid is $n - (k + 1)$. The model utility F test is based on the following distributional result.

When all k β_i's are 0 in the model $y = \alpha + \beta_1 x_1 + \beta_2 x_2 + \cdots + \beta_k x_k + e$ and when the distribution of e is normal with mean 0 and variance σ^2 for any particular values of $x_1, x_2, \ldots, x_k$, the statistic

$$F = \frac{\text{SSRegr}/k}{\text{SSResid}/[n - (k + 1)]}$$

has an F probability distribution based on k numerator degrees of freedom and $n - (k + 1)$ denominator degrees of freedom.

The value of F tends to be larger when at least one β_i is not 0 than when all the β_i's are 0, because more variation is typically explained by the model in the former case than in the latter case. An F statistic value far out in the upper tail of the associated F distribution can be more plausibly attributed to at least one nonzero β_i than to something extremely unusual having occurred when all the β_i's are 0. This is why the F test for model utility is upper-tailed.

▪ **The F Test for Utility of the Model $y = \alpha + \beta_1 x_1 + \beta_2 x_2 + \cdots + \beta_k x_k + e$**

Null hypothesis: $H_0: \beta_1 = \beta_2 = \cdots = \beta_k = 0$
(i.e., there is no useful linear relationship between y and *any* of the predictors).

Alternative hypothesis: H_a: At least one among $\beta_1, \ldots, \beta_k$ is not 0
(i.e., there is a useful linear relationship between y and *at least one* of the predictors).

Test statistic: $F = \dfrac{\text{SSRegr}/k}{\text{SSResid}/[n - (k + 1)]}$

where SSRegr = SSTo − SSResid. An equivalent formula is

$$F = \frac{R^2/k}{(1 - R^2)/[n - (k + 1)]}$$

The test is upper-tailed, and the information in Appendix Table 7 is used to obtain a bound or bounds on the P-value using numerator degrees of freedom $df_1 = k$ and denominator degrees of freedom $df_2 = n - (k + 1)$.

Assumptions: For any particular combination of predictor variable values, the distribution of e, the random deviation, is *normal* with mean 0 and *constant variance*.

For the model utility test, the null hypothesis is the claim that the model is not useful. Unless H_0 can be rejected at a small level of significance, the model has not demonstrated its utility, in which case the investigator must search further for a model that can be judged useful. The alternative formula for F allows the test to be carried out when only R^2, k, and n are available, as is frequently the case in published articles.

▪ **Example 14.9** Soil Characteristics Revisited Again

The model fit to the phosphate adsorption data introduced in Example 14.6 involved $k = 2$ predictors. The MINITAB output in Figure 14.6 contains the relevant information for carrying out the model utility test.

1. The model is $y = \alpha + \beta_1 x_1 + \beta_2 x_2 + e$, where y = phosphate adsorption, x_1 = extractable iron, and x_2 = extractable aluminum.

2. $H_0: \beta_1 = \beta_2 = 0$.

3. H_a: At least one of the two β_i's is not 0.

4. Significance level: $\alpha = .05$.

5. Test statistic: $F = \dfrac{\text{SSRegr}/k}{\text{SSResid}/[n - (k + 1)]}$

6. Assumptions: The following table gives the residuals and standardized residuals (from MINITAB) for the model under consideration:

Obs.	FE	AL	HPO	Residual	StResid
1	61	13	4	−0.06305	−0.01730
2	175	21	18	−1.70661	−0.45867
3	111	24	14	0.46130	0.11455
4	124	23	18	3.34477	0.82601
5	130	64	26	−3.64064	−1.03827
6	173	38	26	0.58585	0.14126
7	169	33	21	−2.21821	−0.54206
8	169	61	30	−2.99022	−0.73346
9	160	39	28	3.70238	0.88505
10	244	71	36	−8.93519	−2.21611
11	257	112	65	4.29026	1.43595
12	333	88	62	1.09857	0.36646
13	199	54	40	6.07079	1.45040

A normal probability plot of the standardized residuals is shown here; the plot is quite straight, indicating that the assumption of normality of the random deviation distribution is reasonable:

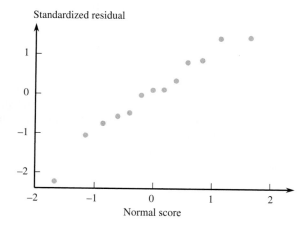

7. Reading directly from the Analysis of Variance table in Figure 14.6, we find that the SS column gives SSRegr = 3529.9 and SSResid = 191.8. Thus,

$$F = \frac{3529.9/2}{191.8/10} = \frac{1764.95}{19.18} = 92.02 \qquad (92.03 \text{ in Figure 14.6})$$

8. Appendix Table 7 shows that for a test based on $df_1 = k = 2$ and $df_2 = n - (k + 1) = 13 - (2 + 1) = 10$, the value 14.91 captures the upper-tail F curve area .001. Because calculated $F = 92.02 > 14.91$, it follows that P-value $< .001$. In fact, Figure 14.6 shows that, to three decimal places, P-value $= 0$.

9. Because P-value $< .001 \le .05 = \alpha$, H_0 should be rejected. The conclusion would be the same using $\alpha = .01$ or $\alpha = .001$. The utility of the model is resoundingly confirmed.

■ **Example 14.10** School Board Politics

A multiple regression analysis presented in the article "The Politics of Bureaucratic Discretion: Educational Access as an Urban Service" (*American Journal of*

Political Science [1991]: 155–177) considered a model in which the dependent variable was

y = percentage of school board members in a school district who are black

and the predictors were

x_1 = black-to-white income ratio in the district

x_2 = percentage of whites in the district below the poverty line

x_3 = indicator for whether district was in the South

x_4 = percentage of blacks in the district with a high school education

x_5 = black population percentage in the district

Summary quantities included $n = 140$ and $R^2 = .749$.

1. The fitted model was $y = \alpha + \beta_1 x_1 + \beta_2 x_2 + \cdots + \beta_5 x_5 + e$.

2. $H_0: \beta_1 = \beta_2 = \beta_3 = \beta_4 = \beta_5 = 0$.

3. H_a: At least one of the β_i's is not 0.

4. Significance level: $\alpha = .01$.

5. Test statistic: $F = \dfrac{R^2/k}{(1 - R^2)/[n - (k + 1)]}$

6. Assumptions: The raw data were not given in this article, so we are unable to compute standardized residuals or construct a normal probability plot. For this test to be valid, we must be willing to assume that the random deviation distribution is normal.

7. Computation: $F = \dfrac{0.749/5}{0.251/[140 - (5 + 1)]} = \dfrac{0.1498}{0.001873} = 80.0$

8. The test is based on $df_1 = k = 5$ and $df_2 = n - (k + 1) = 134$. This latter number of degrees of freedom is not included in the F table. However, the .001 cutoff value for $df_2 = 120$ is 4.42, and for $df_2 = 240$ it is 4.25; so for $df_2 = 134$, the cutoff value is roughly 4.4. Clearly, 80.0 greatly exceeds this value, implying that P-value $< .001$.

9. Because P-value $< .001 \leq .01 = \alpha$, H_0 is rejected at significance level .01. There appears to be a useful linear relationship between y and at least one of the five predictors.

In the next section, we presume that a model has been judged useful after performing an F test and then show how the estimated coefficients and regression function can be used to draw further conclusions. However, you should realize that in many applications, more than one model's utility could be confirmed by the F test. Also, just because the model utility test indicates that the multiple regression model is useful does not necessarily mean that all the predictors included in the model contribute to the usefulness of the model. This is illustrated in Example 14.11, and strategies for selecting a model are briefly considered later in Section 14.4.

▪ **Example 14.11   The Cost of Energy Bars**

What factors contribute to the price of energy bars promoted to provide endurance and promote muscle power? The article "Energy Bars, Unwrapped" (*Consumer*

Reports, June 2003, 19–21) included the following data on price, calorie content, protein content (in grams), and fat content (in grams) for a sample of 19 energy bars:

Price	Calories	Protein	Fat	Price	Calories	Protein	Fat
1.40	180	12	3.0	1.40	180	10	4.5
1.28	200	14	6.0	0.53	200	7	6.0
1.31	210	16	7.0	1.02	220	8	5.0
1.10	220	13	6.0	1.13	230	9	6.0
2.29	220	17	11.0	1.29	230	10	2.0
1.15	230	14	4.5	1.28	240	10	4.0
2.24	240	24	10.0	1.44	260	6	5.0
1.99	270	24	5.0	1.27	260	7	5.0
2.57	320	31	9.0	1.47	290	13	6.0
0.94	110	5	3.0				

Figure 14.9 displays MINITAB output from a regression for the model

$$y = \alpha + \beta_1 x_1 + \beta_2 x_2 + \beta_3 x_3 + e$$

where y = price, x_1 = calorie content, x_2 = protein content, and x_3 = fat content.

Figure 14.9 MINITAB output for the energy bar data of Example 14.11.

The regression equation is
Price = 0.251 + 0.00125 Calories + 0.0485 Protein + 0.0444 Fat

Predictor	Coef	SE Coef	T	P
Constant	0.2511	0.3524	0.71	0.487
Calories	0.001254	0.001724	0.73	0.478
Protein	0.04849	0.01353	3.58	0.003
Fat	0.04445	0.03648	1.22	0.242

S = 0.2789   R-Sq = 74.7%   R-Sq(adj) = 69.6%

Analysis of Variance

Source	DF	SS	MS	F	P
Regression	3	3.4453	1.1484	14.76	0.000
Residual Error	15	1.1670	0.0778		
Total	18	4.6122			

From the MINITAB output, $F = 14.76$, with an associated P-value of 0.000, indicating that the null hypothesis in the model utility test, H_0: $\beta_1 = \beta_2 = \beta_3 = 0$, would be rejected. We would conclude that there is a useful linear relationship between y and x_1, x_2, and x_3. However, consider the MINITAB output shown in Figure 14.10, which resulted from fitting a model that uses only x_2 = protein content as a predictor. Notice that the F test would also indicate the utility of this model and also that the R^2 and adjusted R^2 values of 71.1% and 69.4% are quite similar to those of the model that included all three predictors (74.7% and 69.6% from the MINITAB output of Figure 14.9). This suggests that protein content alone explains about the same amount of the variability in price as all three variables together, and so the simpler model with just one predictor may be preferred over the more complicated model with three predictor variables.

Figure 14.10 MINITAB output for the energy bar data of Example 14.11 when only x_2 = protein content is included as a predictor.

The regression equation is
Price = 0.607 + 0.0623 Protein

Predictor	Coef	SE Coef	T	P
Constant	0.6072	0.1419	4.28	0.001
Protein	0.062256	0.009618	6.47	0.000

S = 0.2798   R-Sq = 71.1%   R-Sq(adj) = 69.4%

Analysis of Variance

Source	DF	SS	MS	F	P
Regression	1	3.2809	3.2809	41.90	0.000
Residual Error	17	1.3313	0.0783		
Total	18	4.6122			

▪ Exercises 14.15–14.35

14.15 The article cited in Example 14.1 also recommended the following estimated regression equation for relating y = maximal oxygen uptake (in liters per minute) to the predictors x_1 = gender (female = 0, male = 1), x_2 = weight (in kilograms), x_3 = 1-mi walk time (in minutes), and x_4 = heart rate at the end of the walk (beats per minute):

$$\hat{y} = 3.5959 + 0.6566x_1 + 0.0096x_2 - 0.0996x_3 - 0.0080x_4$$

a. How would you interpret the value of $b_3 = -0.0996$?
b. How would you interpret the value of $b_1 = 0.6566$?
c. Suppose that an observation made on a male whose weight was 80 kg, walk time was 11 min, and heart rate was 140 beats/min resulted in maximal oxygen uptake $y = 3.15$. What would you have predicted for maximal oxygen uptake and what is the corresponding residual?
d. If SSResid = 30.1033 and SSTo = 102.3922, what proportion of observed variation in maximal oxygen uptake can be attributed to the model relationship?
e. The sample size was $n = 196$. Calculate an estimate of σ.

14.16 When coastal power stations take in large quantities of cooling water, it is inevitable that a number of fish are drawn in with the water. Various methods have been designed to screen out the fish. The article "Multiple Regression Analysis for Forecasting Critical Fish Influxes at Power Station Intakes" (*Journal of Applied Ecology* [1983]: 33–42) examined intake fish catch at an English power plant and several other variables thought to affect fish intake: y = fish intake (number of fish), x_1 = water temperature (in °C), x_2 = number of pumps running,

x_3 = sea state (values 0, 1, 2, or 3), and x_4 = speed (in knots). Part of the data given in the article were used to obtain the estimated regression equation

$$\hat{y} = 92 - 2.18x_1 - 19.20x_2 - 9.38x_3 + 2.32x_4$$

(based on $n = 26$). SSRegr = 1486.9 and SSResid = 2230.2 were also calculated.
a. Interpret the values of b_1 and b_4.
b. What proportion of observed variation in fish intake can be explained by the model relationship?
c. Estimate the value of σ.
d. Calculate the adjusted R^2. How does it compare to R^2 itself?

14.17 Obtain as much information as you can about the P-value for an upper-tailed F test in each of the following situations:
a. $df_1 = 3$, $df_2 = 15$, calculated $F = 4.23$
b. $df_1 = 4$, $df_2 = 18$, calculated $F = 1.95$
c. $df_1 = 5$, $df_2 = 20$, calculated $F = 4.10$
d. $df_1 = 4$, $df_2 = 35$, calculated $F = 4.58$

14.18 Obtain as much information as you can about the P-value for the F test for model utility in each of the following situations:
a. $k = 2$, $n = 21$, calculated $F = 2.47$
b. $k = 8$, $n = 25$, calculated $F = 5.98$
c. $k = 5$, $n = 26$, calculated $F = 3.00$
d. The full quadratic model based on x_1 and x_2 is fit, $n = 20$, and calculated $F = 8.25$
e. $k = 5$, $n = 100$, calculated $F = 2.33$

14.19 The ability of ecologists to identify regions of greatest species richness could have an impact on the preservation of genetic diversity, a major objective of the World Conservation Strategy. The article "Prediction of Rarities from Habitat Variables: Coastal Plain Plants on Nova Scotian Lakeshores" (*Ecology*

[1992]: 1852–1859) used a sample of $n = 37$ lakes to obtain the estimated regression equation

$$\hat{y} = 3.89 + 0.033x_1 + 0.024x_2 + 0.023x_3 - 0.008x_4 - 0.13x_5 - 0.72x_6$$

where y = species richness, x_1 = watershed area, x_2 = shore width, x_3 = drainage (percentage), x_4 = water color (total color units), x_5 = sand (percentage), and x_6 = alkalinity. The coefficient of multiple determination was reported as $R^2 = .83$. Use a test with significance level .01 to decide whether the chosen model is useful.

14.20 The article "Impacts of On-Campus and Off-Campus Work on First-Year Cognitive Outcomes" (*Journal of College Student Development* [1994]: 364–370) reported on a study in which y = spring 1992 math comprehension score was regressed against x_1 = fall 1991 test score, x_2 = fall 1991 academic motivation score, x_3 = age, x_4 = number of credit hours, x_5 = residence (1 if on campus, 0 otherwise), x_6 = hours worked on campus, and x_7 = hours worked off campus. The sample size was $n = 210$, and $R^2 = .543$. Test to see whether there is a useful linear relationship between y and at least one of the predictors.

14.21 Return to the situation described in Exercise 14.15, and carry out the model utility test to decide whether there is a useful linear relationship between y and at least one of the four predictors.

14.22 Is the model fit in Exercise 14.16 useful? Carry out a test using a significance level of .10.

14.23 The following MINITAB output results from fitting the model described in Exercise 14.12 to data:

Predictor	Coef	Stdev	t-ratio
Constant	86.85	85.39	1.02
X1	−0.12297	0.03276	−3.75
X2	5.090	1.969	2.58
X3	−0.07092	0.01799	−3.94
X4	0.0015380	0.0005560	2.77

$s = 4.784$ R-sq = 90.8%   R-sq(adj) = 89.4%

Analysis of Variance	DF	SS	MS
Regression	4	5896.6	1474.2
Error	26	595.1	22.9
Total	30	6491.7	

a. What is the estimated regression equation?
b. Using a .01 significance level, perform the model utility test.
c. Interpret the values of R^2 and s_e given in the output.

14.24 For the multiple regression model in Exercise 14.4, the value of R^2 was .06 and the adjusted R^2 was .06. The model was based on a data set with 1136 observations. Perform a model utility test for this regression.

14.25 *This exercise requires the use of a computer package.* The article "Movement and Habitat Use by Lake Whitefish During Spawning in a Boreal Lake: Integrating Acoustic Telemetry and Geographic Information Systems" (*Transactions of the American Fisheries Society* [1999]: 939–952) included the following data on 17 fish caught in 1995 and 1996:

Year	Fish Number	Weight (g)	Length (mm)	Age (years)
1995	1	776	410	9
	2	580	368	11
	3	539	357	15
	4	648	373	12
	5	538	361	9
	6	891	385	9
	7	673	380	10
	8	783	400	12
1996	9	571	407	12
	10	627	410	13
	11	727	421	12
	12	867	446	19
	13	1042	478	19
	14	804	441	18
	15	832	454	12
	16	764	440	12
	17	727	427	12

a. Fit a multiple regression model to describe the relationship between weight and the predictors *length* and *age*.
b. Carry out the model utility test to determine whether the predictors *length* and *age* together are useful for predicting weight.

14.26 *This exercise requires the use of a computer package.* The authors of the article "Absolute Versus per Unit Body Length Speed of Prey as an Estimator of Vulnerability to Predation" (*Animal Behaviour* [1999]: 347–352) found that the speed of a prey (twips per second) and the length of a prey (twips × 100) are good predictors of the time (in seconds) required to catch the prey. (A twip is a measure of distance used by programmers.) Data were collected in an experiment where subjects were asked to "catch" an animal of prey moving across his or her computer screen by clicking on it with the mouse. The investigators varied the length of the prey and the speed with which the prey moved across the screen. The following data are consistent with summary values and a graph given in the article (each value represents the average catch time over all subjects, and the order of the various speed–length combinations was randomized for each subject):

Subject	Prey Length	Prey Speed	Catch Time
1	7	20	1.10
2	6	20	1.20
3	5	20	1.23
4	4	20	1.40
5	3	20	1.50
6	3	40	1.40
7	4	40	1.36
8	6	40	1.30
9	7	40	1.28
10	7	80	1.40
11	6	60	1.38
12	5	80	1.40
13	7	100	1.43
14	6	100	1.43
15	7	120	1.70
16	5	80	1.50
17	3	80	1.40
18	6	100	1.50
19	3	120	1.90

a. Fit a multiple regression model for predicting catch time using prey length and speed as predictors.
b. Predict the catch time for an animal of prey whose length is 6 and whose speed is 50.
c. Is the multiple regression model useful for predicting catch time? Test the relevant hypotheses using $\alpha = .05$.
d. The authors of the article suggested that a simple linear regression model with the single predictor

$$x = \frac{\text{length}}{\text{speed}}$$

might be a better model for predicting catch time. Calculate the x values, and use them to fit this linear regression model.
e. Which of the two models considered (the multiple regression model from Part (a) or the simple linear regression model from Part (d)) would you recommend for predicting catch time? Justify your choice.

14.27 *This exercise requires the use of a computer package.* The article "Vital Dimensions in Volume Perception: Can the Eye Fool the Stomach?" (*Journal of Marketing Research* [1999]: 313–326) gave the following data on dimensions of 27 representative food products:

Data for Exercise 14.27

Product	Material	Height	Maximum Width	Minimum Width	Elongation	Volume
1	glass	7.7	2.50	1.80	1.50	125
2	glass	6.2	2.90	2.70	1.07	135
3	glass	8.5	2.15	2.00	1.98	175
4	glass	10.4	2.90	2.60	1.79	285
5	plastic	8.0	3.20	3.15	1.25	330
6	glass	8.7	2.00	1.80	2.17	90
7	glass	10.2	1.60	1.50	3.19	120
8	plastic	10.5	4.80	3.80	1.09	520
9	plastic	3.4	5.90	5.00	0.29	330
10	plastic	6.9	5.80	4.75	0.59	570
11	tin	10.9	2.90	2.80	1.88	340
12	plastic	9.7	2.45	2.10	1.98	175
13	glass	10.1	2.60	2.20	1.94	240
14	glass	13.0	2.60	2.60	2.50	240
15	glass	13.0	2.70	2.60	2.41	360
16	glass	11.0	3.10	2.90	1.77	310
17	cardboard	8.7	5.10	5.10	0.85	635
18	cardboard	17.1	10.20	10.20	0.84	1250
19	glass	16.5	3.50	3.50	2.36	650
20	glass	16.5	2.70	1.20	3.06	305
21	glass	9.7	3.00	1.70	1.62	315
22	glass	17.8	2.70	1.75	3.30	305
23	glass	14.0	2.50	1.70	2.80	245
24	glass	13.6	2.40	1.20	2.83	200
25	plastic	27.9	4.40	1.20	3.17	1205
26	tin	19.5	7.50	7.50	1.30	2330
27	tin	13.8	4.25	4.25	1.62	730

a. Fit a multiple regression model for predicting the volume (in milliliters) of a package based on its minimum width, maximum width, and elongation score.
b. Why should we consider adjusted R^2 instead of R^2 when attempting to determine the quality of fit of the data to our model?
c. Perform a model utility test.

14.28 The article "The Caseload Controversy and the Study of Criminal Courts" (*Journal of Criminal Law and Criminology* [1979]: 89–101) used a multiple regression analysis to help assess the impact of judicial caseload on the processing of criminal court cases. Data were collected in the Chicago criminal courts on the following variables: y = number of indictments, x_1 = number of cases on the docket, and x_2 = number of cases pending in criminal court trial system. The estimated regression equation (based on $n = 367$ observations) was

$$\hat{y} = 28 - 0.05x_1 - 0.003x_2 + 0.00002x_3$$

where $x_3 = x_1x_2$.
a. The reported value of R^2 was .16. Conduct the model utility test. Use a .05 significance level.
b. Given the results of the test in Part (a), does it surprise you that the R^2 value is so low? Can you think of a possible explanation for this?
c. How does adjusted R^2 compare to R^2?

14.29 The article "The Undrained Strength of Some Thawed Permafrost Soils" (*Canadian Geotechnical Journal* [1979]: 420–427) contained the following data on y = shear strength of sandy soil (in kilopascals), x_1 = depth (in meters), and x_2 = water content (percentage):

y	x_1	x_2	Predicted y	Residual
14.7	8.9	31.5	23.35	−8.65
48.0	36.6	27.0	46.38	1.62
25.6	36.8	25.9	27.13	−1.53
10.0	6.1	39.1	10.99	−0.99
16.0	6.9	39.2	14.10	1.90
16.8	6.9	38.3	16.54	0.26
20.7	7.3	33.9	23.34	−2.64
38.8	8.4	33.8	25.43	13.37
16.9	6.5	27.9	15.63	1.27
27.0	8.0	33.1	24.29	2.71
16.0	4.5	26.3	15.36	0.64
24.9	9.9	37.8	29.61	−4.71
7.3	2.9	34.6	15.38	−8.08
12.8	2.0	36.4	7.96	4.84

The predicted values and residuals were computed using the estimated regression equation

$$\hat{y} = -151.36 - 16.22x_1 + 13.48x_2 + 0.094x_3 - 0.253x_4 + 0.492x_5$$

where $x_3 = x_1^2$, $x_4 = x_2^2$, and $x_5 = x_1x_2$.

a. Use the given information to compute SSResid, SSTo, and SSRegr.
b. Calculate R^2 for this regression model. How would you interpret this value?
c. Use the value of R^2 from Part (b) and a .05 level of significance to conduct the appropriate model utility test.

14.30 The article "Readability of Liquid Crystal Displays: A Response Surface" (*Human Factors* [1983]: 185–190) used an estimated regression equation to describe the relationship between y = error percentage for subjects reading a four-digit liquid crystal display and the independent variables x_1 = level of backlight, x_2 = character subtense, x_3 = viewing angle, and x_4 = level of ambient light. From a table given in the article, SSRegr = 19.2, SSResid = 20.0, and $n = 30$.
a. Does the estimated regression equation specify a useful relationship between y and the independent variables? Use the model utility test with a .05 significance level.
b. Calculate R^2 and s_e for this model. Interpret these values.
c. Do you think that the estimated regression equation would provide reasonably accurate predictions of error rate? Explain.

14.31 The article "Effect of Manual Defoliation on Pole Bean Yield" (*Journal of Economic Entomology* [1984]: 1019–1023) used a quadratic regression model to describe the relationship between y = yield (in kilograms per plot) and x = defoliation level (a proportion between 0 and 1). The estimated regression equation based on $n = 24$ was

$$\hat{y} = 12.39 + 6.67x_1 - 15.25x_2$$

where $x_1 = x$ and $x_2 = x^2$. The article also reported that R^2 for this model was .902. Does the quadratic model specify a useful relationship between y and x? Carry out the appropriate test using a .01 level of significance.

14.32 Suppose that a multiple regression data set consists of $n = 15$ observations. For what values of k, the number of model predictors, would the corresponding model with $R^2 = .90$ be judged useful at significance level .05? Does such a large R^2 value necessarily imply a useful model? Explain.

14.33 *This exercise requires the use of a computer package.* Use the data given in Exercise 14.29 to verify that the true regression function

$$\text{mean } y \text{ value} = \alpha + \beta_1x_1 + \beta_2x_2 + \beta_3x_3 + \beta_4x_4 + \beta_5x_5$$

is estimated by

$$\hat{y} = -151.36 - 16.22x_1 + 13.48x_2 + 0.094x_3 - 0.253x_4 + 0.492x_5$$

14.34 *This exercise requires the use of a computer package.* The following data resulted from a study of the relationship between y = brightness of finished paper and the independent variables x_1 = hydrogen peroxide (percentage by weight), x_2 = sodium hydroxide (percentage by weight), x_3 = silicate (percentage by weight), and x_4 = process temperature (in °F) ("Advantages of CE-HDP Bleaching for High Brightness Kraft Pulp Production," *Tappi* [1964]: 107A–173A):

x_1	x_2	x_3	x_4	y
0.2	0.2	1.5	145	83.9
0.4	0.2	1.5	145	84.9
0.2	0.4	1.5	145	83.4
0.4	0.4	1.5	145	84.2
0.2	0.2	3.5	145	83.8
0.4	0.2	3.5	145	84.7
0.2	0.4	3.5	145	84.0
0.4	0.4	3.5	145	84.8
0.2	0.2	1.5	175	84.5
0.4	0.2	1.5	175	86.0
0.2	0.4	1.5	175	82.6
0.4	0.4	1.5	175	85.1
0.2	0.2	3.5	175	84.5
0.4	0.2	3.5	175	86.0
0.2	0.4	3.5	175	84.0
0.4	0.4	3.5	175	85.4
0.1	0.3	2.5	160	82.9
0.5	0.3	2.5	160	85.5
0.3	0.1	2.5	160	85.2
0.3	0.5	2.5	160	84.5
0.3	0.3	0.5	160	84.7
0.3	0.3	4.5	160	85.0
0.3	0.3	2.5	130	84.9
0.3	0.3	2.5	190	84.0
0.3	0.3	2.5	160	84.5
0.3	0.3	2.5	160	84.7
0.3	0.3	2.5	160	84.6
0.3	0.3	2.5	160	84.9
0.3	0.3	2.5	160	84.9
0.3	0.3	2.5	160	84.5
0.3	0.3	2.5	160	84.6

a. Find the estimated regression equation for the model that includes all independent variables, all quadratic terms, and all interaction terms.
b. Using a .05 significance level, perform the model utility test.
c. Interpret the values of the following quantities: SSResid, R^2, s_e.

14.35 *This exercise requires the use of a computer package.* The cotton aphid poses a threat to cotton crops in Iraq. The following data on y = infestation rate (aphids per 100 leaves), x_1 = mean temperature (in °C), and x_2 = mean relative humidity appeared in the article "Estimation of the Economic Threshold of Infestation for Cotton Aphid" (*Mesopotamia Journal of Agriculture* [1982]: 71–75):

y	x_1	x_2	y	x_1	x_2
61	21.0	57.0	25	33.5	18.5
77	24.8	48.0	67	33.0	24.5
87	28.3	41.5	40	34.5	16.0
93	26.0	56.0	6	34.3	6.0
98	27.5	58.0	21	34.3	26.0
100	27.1	31.0	18	33.0	21.0
104	26.8	36.5	23	26.5	26.0
118	29.0	41.0	42	32.0	28.0
102	28.3	40.0	56	27.3	24.5
74	34.0	25.0	60	27.8	39.0
63	30.5	34.0	59	25.8	29.0
43	28.3	13.0	82	25.0	41.0
27	30.8	37.0	89	18.5	53.5
19	31.0	19.0	77	26.0	51.0
14	33.6	20.0	102	19.0	48.0
23	31.8	17.0	108	18.0	70.0
30	31.3	21.0	97	16.3	79.5

Use the data to find the estimated regression equation and assess the utility of the multiple regression model

$$y = \alpha + \beta_1 x_1 + \beta_2 x_2 + e$$

▪ 14.3 Inferences Based on an Estimated Model

In Section 14.2, we discussed estimating the coefficients $\alpha, \beta_1, \ldots, \beta_k$ in the model $y = \alpha + \beta_1 x_1 + \beta_2 x_2 + \cdots + \beta_k x_k + e$ (using the principle of least squares) and then showed how the usefulness of the model could be confirmed by application of the F test for model utility. If H_0: $\beta_1 = \beta_2 = \cdots = \beta_k = 0$ cannot be rejected at a reasonably small level of significance, it must be concluded that the model does not specify a useful relationship between y and any of the predictor variables $x_1, x_2, \ldots, x_k$.

The investigator must then search further for a model that does describe a useful relationship, perhaps by introducing different predictors or making variable transformations. Only if H_0 can be rejected is it appropriate to proceed further with the chosen model and make inferences based on the estimated coefficients $a, b_1, \ldots, b_k$ and the estimated regression function $\hat{y} = a + b_1x_1 + b_2x_2 + \cdots + b_kx_k$. In this section, we consider two different types of inferential problems. One type involves drawing a conclusion about an individual regression coefficient β_i—either computing a confidence interval for β_i or testing a hypothesis concerning β_i. The second type of problem involves first fixing values of $x_1, x_2, \ldots, x_k$ and then computing either a point estimate or a confidence interval for the corresponding mean y value or predicting a future y value with a point prediction or a prediction interval.

▪ Inferences About a Regression Coefficient

A confidence interval for the slope coefficient β and a hypothesis test about β in a simple linear regression were based on facts about the sampling distribution of the statistic b used to obtain a point estimate of β. Similarly, in multiple regression, procedures for making inferences about β_i are derived from properties of the sampling distribution of b_i. Formulas for b_i and its standard deviation σ_{b_i} are quite complicated and cannot be stated concisely except by using some advanced mathematical notation.

▪ The Sampling Distribution of b_i

Let b_i denote the least-squares estimate of the coefficient β_i in the model $y = \alpha + \beta_1x_1 + \beta_2x_2 + \cdots + \beta_kx_k + e$. Assumptions about this model given in Section 14.1 imply the following:

1. b_i has a normal distribution.

2. $\mu_{b_i} = \beta_i$, so b_i is an unbiased statistic for estimating β_i.

3. The standard deviation of b_i, σ_{b_i}, involves σ^2 and a complicated function of all the values of $x_1, x_2, \ldots, x_k$ in the sample. The estimated standard deviation s_{b_i} results from replacing σ^2 with s_e^2 in the formula for σ_{b_i}.

A consequence of the properties of the sampling distribution of b_i is that the standardized variable

$$t = \frac{b_i - \beta_i}{s_{b_i}}$$

has a t distribution with df $= n - (k + 1)$. This leads to a confidence interval for β_i.

▪ A Confidence Interval for β_i

A confidence interval for β_i is

$$b_i \pm (t \text{ critical value})s_{b_i}$$

The t critical value is based on df $= n - (k + 1)$.

Any good statistical computer package will provide both the estimated coefficients $a, b_1, \ldots, b_k$ and their estimated standard deviations $s_a, s_{b_1}, \ldots, s_{b_k}$.

▪ Example 14.12   Soil Characteristics Continued

Reconsider the regression of phosphate adsorption (y) against extractable iron and extractable aluminum with the sample data given in Example 14.6 and the MINITAB output displayed in Figure 14.6. Suppose that the investigator had required a 95% confidence interval for β_1, the average change in phosphate adsorption when extractable iron increases by 1 unit and extractable aluminum is held fixed. The necessary quantities are

$$b_1 = 0.11273 \quad \text{and} \quad s_{b_1} = 0.02969 \quad \text{(from the Coef and Stdev columns in the output)}$$

$$\text{df} = n - (k + 1) = 13 - (2 + 1) = 10$$

$$t \text{ critical value} = 2.23$$

The resulting confidence interval is

$$\begin{aligned}
b_i \pm (t \text{ critical value})s_{b_i} &= 0.11273 \pm (2.23)(0.02969) \\
&= 0.11273 \pm 0.06621 \\
&= (0.047, 0.179)
\end{aligned}$$

The standardized variable $t = \dfrac{b_i - \beta_i}{s_{b_i}}$ is also the basis for a test statistic for testing hypotheses about β_i.

▪ Testing Hypotheses About β_i

Null hypothesis: H_0: β_i = hypothesized value

Test statistic: $t = \dfrac{b_i - \text{hypothesized value}}{s_{b_i}}$

The test is based on df $= n - (k + 1)$.

Alternative Hypothesis	P-Value
H_a: $\beta_i >$ hypothesized value	Area under the appropriate t curve to the right of the computed t
H_a: $\beta_i <$ hypothesized value	Area under the appropriate t curve to the left of the computed t
H_a: $\beta_i \neq$ hypothesized value	(1) 2(area to the right of t) if t is positive, or (2) 2(area to the left of t) if t is negative

Assumptions: For any particular combination of predictor variable values, the distribution of e, the random deviation from the model, is *normal* with mean 0 and *constant variance*.

The most frequently tested null hypothesis in this situation is H_0: $\beta_i = 0$. The interpretation of this null hypothesis is that *as long as the other predictors $x_1, \ldots, x_{i-1}, x_{i+1}, \ldots, x_k$ remain in the model*, the predictor x_i provides no useful information about y, and so it can be eliminated. The test statistic for testing H_0: $\beta_i = 0$ simplifies to the *t* **ratio**

$$t = \dfrac{b_i}{s_{b_i}}$$

▪ **Example 14.13   The Price of Fish**

Economists are always interested in studying the factors affecting the price paid for a product. The article "Testing for Imperfect Competition at the Fulton Fish Market" (*Rand Journal of Economics* [1995]: 75–92) considered a regression of $y =$ price paid for whiting on eight different predictors. (The Fulton Fish Market, located in New York City, is the largest such market in the United States; the many dealers make it highly competitive, and whiting is one of the most frequently purchased types of fish.) There were $n = 54$ observations, and the coefficient of multiple determination was $R^2 = .992$, indicating a highly useful model (model utility $F = 697.5$, P-value $= .000$). One of the predictors used, x_3, was an indicator variable for whether a purchaser was Asian. The values $b_3 = -0.0463$ and $s_{b_3} = 0.0191$ were given. Provided that all seven other predictors remain in the model, does x_3 appear to provide useful information about y?

1. β_3 is the average difference in price paid by Asian and non-Asian purchasers when all other predictors are held fixed.
2. $H_0: \beta_3 = 0$.
3. $H_a: \beta_3 \neq 0$.
4. A significance level of $\alpha = .05$ was suggested in the article.
5. Test statistic: $t = \dfrac{b_3}{s_{b_3}}$
6. Assumptions: Without raw data, we cannot assess the reasonableness of the assumptions. If we had the data, a normal probability plot or boxplot of the standardized residuals would be a good place to start. For the purposes of this example, we will assume that it is reasonable to proceed with the test.
7. The test statistic value is

$$t = \frac{b_3}{s_{b_3}} = \frac{-0.0463}{0.0191} = -2.42$$

8. The test is based on $n - (k + 1) = 54 - (8 + 1) = 45$ df. An examination of the 40 and 60 df columns of Appendix Table 4 shows that the P-value is roughly $2(.010) = .020$.
9. Because P-value $= .020 \leq .05 = \alpha$, H_0 should be rejected. The predictor that indicates whether a purchaser is Asian does appear to provide useful information about price, over and above the information contained in the other predictors. The author of the article indicated that this result is inconsistent with the model of perfect competition.

▪ **Example 14.14   Soil Characteristics Continued Again**

Our analysis of the phosphate adsorption data introduced in Example 14.6 has so far focused on the model $y = \alpha + \beta_1 x_1 + \beta_2 x_2 + e$, in which x_1 (extractable iron) and x_2 (extractable aluminum) affect the response separately. Suppose that the researcher wishes to investigate the possibility of interaction between x_1 and x_2

through fitting the model with predictors x_1, x_2, and $x_3 = x_1x_2$. We list a few of the sample values of y, x_1, and x_2 along with the corresponding values of x_3:

Observation	y	x_1	x_2	x_3
1	4	61	13	793
2	18	175	21	3675
3	14	111	24	2664
⋮	⋮	⋮	⋮	⋮
13	40	199	54	10,746

In practice, a statistical software package would automatically compute the x_3 values upon request once the x_1 and x_2 values had been entered, so hand computations would not be necessary. Figure 14.11 displays partial MINITAB output resulting from a request to fit this model. Let's use the output to see whether inclusion of the interaction predictor is justified.

Figure 14.11 MINITAB output for model with interaction fitted to the phosphate adsorption data of Example 14.14 (x_1 = FE, x_2 = AL, $x_3 = x_1x_2$ = FEAL).

The regression equation is
HPO = − 2.37 + 0.0828 FE + 0.246 AL + 0.000528 FEAL

Predictor	Coef	SE Coef	T	P
Constant	−2.368	7.179	−0.33	0.749
FE	0.08279	0.04818	1.72	0.120
AL	0.2460	0.1481	1.66	0.131
FEAL	0.0005278	0.0006610	0.80	0.445

S = 4.461   R-Sq = 95.2%   R-Sq(adj) = 93.6%

1. The model is $y = \alpha + \beta_1x_1 + \beta_2x_2 + \beta_3x_3 + e$, where $x_3 = x_1x_2$.
2. $H_0: \beta_3 = 0$.
3. $H_a: \beta_3 \neq 0$.
4. Test statistic: $t = \dfrac{b_3}{s_{b_3}}$
5. Significance level: $\alpha = .05$.
6. Assumptions: The residuals and standardized residuals (from MINITAB) for this model are as follows:

Obs.	FE	HPO	Residual	StResid
1	61	4	−2.29844	−0.94279
2	175	18	−1.22530	−0.32746
3	111	14	−0.13111	−0.03250
4	124	18	2.93937	0.71808
5	130	26	−2.52836	−0.76874
6	173	26	1.22858	0.29624
7	169	21	−1.68416	−0.40933
8	169	30	−2.06905	−0.51864
9	160	28	4.23516	1.00631
10	244	36	−8.44065	−2.07895
11	257	65	3.34931	1.19361
12	333	62	−0.31378	−0.12606
13	199	40	6.93843	1.68295

A normal probability plot of the standardized residuals is shown here:

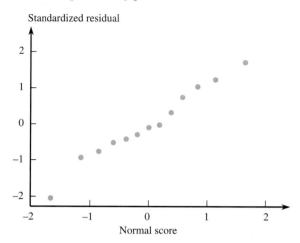

The plot is quite straight, indicating that normality of the random error distribution is plausible.

7. By reading from the t ratio column in Figure 14.11, we find that $t = 0.80$.

8. Figure 14.11 shows that the P-value is .445.

9. Because $.445 > .05$, H_0 cannot be rejected. The interaction predictor does not appear to be useful and can be removed from the model.

An interesting aspect of the computer output for the interaction model in Example 14.14 is that the t ratios for β_1, β_2, and β_3 (1.72, 1.66, and 0.80, respectively) are all relatively small and the corresponding P-values are large, yet $R^2 = .952$ is quite large. The high R^2 value suggests a useful model (this can be confirmed by the model utility test), yet the size of each t ratio and P-value might tempt us to conclude that all three β_i's are 0. This sounds like a contradiction, but it involves a misinterpretation of the t ratios. For example, the t ratio for β_3, $t = \dfrac{b_3}{s_{b_3}}$, tests the null hypothesis H_0: $\beta_3 = 0$ *when x_1 and x_2 are included in the model*. The smallness of a given t ratio suggests that the associated predictor can be dropped from the model *as long as the other predictors are retained*. The fact that all t ratios are small in this example does not, therefore, allow us to simultaneously delete all predictors. The data do suggest deleting x_3 because the model that includes x_1 and x_2 has already been found to give a very good fit to the data and is simple to interpret. In Section 14.4, we comment further on why it might happen that all t ratios are small even when the model seems quite useful.

The model utility test amounts to testing a simultaneous claim about the values of all the β_i's — that they are all 0. There is also an F test for testing a hypothesis involving a specified subset consisting of at least two β_i's. For example, we might fit the model

$$y = \alpha + \beta_1 x_1 + \beta_2 x_2 + \beta_3 x_1 x_2 + \beta_4 x_1^2 + \beta_5 x_2^2 + e$$

and then wish to test H_0: $\beta_3 = \beta_4 = \beta_5 = 0$ (which says that the second-order predictors contribute nothing to the model). Please see one of the chapter references for Chapter 5 for further details.

■ Inferences Based on the Estimated Regression Function

The estimated regression line $\hat{y} = a + bx$ in simple linear regression was used both to estimate the mean y value when x had a specified value and to predict the associated y value for a single observation made at a particular x value. The estimated regression function for the model $y = \alpha + \beta_1 x_1 + \beta_2 x_2 + \cdots + \beta_k x_k + e$ can be used in the same two ways. When fixed values of the predictor variables $x_1, x_2, \ldots, x_k$ are substituted into the estimated regression function

$$\hat{y} = a + b_1 x_1 + b_2 x_2 + \cdots + b_k x_k$$

the result can be used either as a point estimate of the corresponding mean y value or as a point prediction of the y value that will result from a single observation when the x_i's have the specified values.

■ Example 14.15   Predicting Transit Times

Precise information concerning bus transit times is important when making transportation planning decisions. The article "Factors Affecting Running Time on Transit Routes" (*Transportation Research* [1983]: 107–113) reported on an empirical study based on data gathered in Cincinnati, Ohio. The variables of interest were

y = running time per mile during the morning peak period (in seconds)
x_1 = number of passenger boardings per mile
x_2 = number of passenger alightings per mile
x_3 = number of signaled intersections per mile
x_4 = proportion of the route on which parking is allowed

The values of x_1 and x_2 were not necessarily equal, because observations were made over partial segments of routes (so not all passengers entered or exited on the segments). The estimated regression function was

$$\hat{y} = 169.50 + 5.07x_1 + 4.53x_2 + 6.61x_3 + 67.70x_4$$

Consider the predictor variable values $x_1 = 4.5$, $x_2 = 5.5$, $x_3 = 5$, and $x_4 = 0.1$. Then a point estimate for true average running time per mile when the x_k's have these values is

$$\hat{y} = 169.50 + 5.07(4.5) + 4.53(5.5) + 6.61(5) + 67.70(0.1) = 257.05$$

This value of 257.05 is also the predicted running time per mile for a single trip when x_1, x_2, x_3, and x_4 have the given values.

Remember that before the sample observations $y_1, y_2, \ldots, y_n$ are obtained, a and all the b_i's are statistics (because they all involve the y_i's). This implies that for fixed values of $x_1, x_2, \ldots, x_k$ the estimated regression function $a + b_1 x_1 + b_2 x_2 + \cdots + b_k x_k$ is a statistic (its value varies from sample to sample). To obtain a confidence interval for the mean y value for specified $x_1, x_2, \ldots, x_k$ values, we need some facts about the sampling distribution of this statistic.

▪ **Properties of the Sampling Distribution of** $\hat{y} = a + b_1x_1 + b_2x_2 + \cdots + b_kx_k$

Assumptions about the model $y = \alpha + \beta_1x_1 + \beta_2x_2 + \cdots + \beta_kx_k + e$ given in Section 14.1 imply that for fixed values of the predictors $x_1, x_2, \ldots, x_k$, the statistic $\hat{y} = a + b_1x_1 + b_2x_2 + \cdots + b_kx_k$ satisfies the following properties:

1. It has a normal distribution.

2. Its mean value is $\alpha + \beta_1x_1 + \beta_2x_2 + \cdots + \beta_kx_k$. That is, the statistic is unbiased for estimating the mean y value when $x_1, x_2, \ldots, x_k$ are fixed.

3. The standard deviation of $\hat{y}$ involves σ^2 and a complicated function of all the sample predictor variable values. The estimated standard deviation of $\hat{y}$, denoted by $s_{\hat{y}}$, results from replacing σ^2 with s_e^2 in this function.

The standardized variable

$$t = \frac{\hat{y} - (\alpha + \beta_1x_1 + \beta_2x_2 + \cdots + \beta_kx_k)}{s_{\hat{y}}}$$

then has a t distribution with $n - (k + 1)$ degrees of freedom.

The formula for $s_{\hat{y}}$ has been programmed into the most widely used statistical computer packages, and its value for specified $x_1, x_2, \ldots, x_k$ is available upon request. Manipulation of the t variable as before gives a confidence interval formula. A prediction interval formula is based on a similar standardized variable.

For fixed values of $x_1, x_2, \ldots, x_k$, a **confidence interval for the mean y value**—that is, for $\alpha + \beta_1x_1 + \beta_2x_2 + \cdots + \beta_kx_k$—is

$$\hat{y} \pm (t \text{ critical value})s_{\hat{y}}$$

A **prediction interval for a single y value** is

$$\hat{y} \pm (t \text{ critical value})\sqrt{s_e^2 + s_{\hat{y}}^2}$$

The t critical value in both intervals is based on $n - (k + 1)$ degrees of freedom.

▪ **Example 14.16** Soil Characteristics—Last Time!

Figure 14.6 shows MINITAB output from fitting the model $y = \alpha + \beta_1x_1 + \beta_2x_2 + e$ to the phosphate data introduced previously. A request for estimation and prediction information when $x_1 = 150$ and $x_2 = 40$ resulted in the following additional output:

Fit	StDev Fit	95.0% CI	95.0% PI
23.52	1.31	(20.60, 26.44)	(13.33, 33.70)

The t critical value for a 95% confidence level when df = 10 is 2.23, and the corresponding interval is

$$23.52 \pm (2.23)(1.31) = 23.52 \pm 2.92 = (20.60, 26.44)$$

Using $s_e^2 = 4.379$ along with the given values of $\hat{y}$, $s_{\hat{y}}$, and t critical value in the prediction interval formula results in (13.33, 33.70). Both intervals are centered at $\hat{y}$, but the prediction interval is much wider than the confidence interval.

The danger of extrapolation in simple linear regression is that if the x value of interest is much outside the interval of x values in the sample, the proposed model might no longer be valid, and even if it were, $\alpha + \beta x$ could still be quite poorly estimated (s_{a+bx} could be large). There is a similar danger in multiple regression, but it is not always obvious when the x_i values of interest involve an extrapolation from sample data. As an example, suppose that a single y observation is made for each of the following 13 (x_1, x_2) pairs:

Observation	1	2	3	4	5	6	7	8	9	10	11	12	13
x_1	0	0	0	0	0	−5	5	5	5	10	−5	−5	−10
x_2	10	5	0	−5	−10	−5	0	5	−5	0	0	5	0

The x_1 values range between -10 and 10, as do the x_2 values. After fitting the model $y = \alpha + \beta_1 x_1 + \beta_2 x_2 + e$, we might then want a confidence interval for the mean y value when $x_1 = 10$ and $x_2 = 10$. However, even though each of these values separately is within the range of sample x_i values, Figure 14.12 shows that the point $(x_1, x_2) = (10, 10)$ is actually far from the (x_1, x_2) pairs in the sample. Thus, drawing a conclusion about y when $x_1 = 10$ and $x_2 = 10$ involves a substantial extrapolation. In particular, the estimated standard deviation of $a + b_1(10) + b_2(10)$ would probably be quite large even if the model was valid near this point.

Figure 14.12 The danger of extrapolation: Each orange dot denotes an (x_1, x_2) pair for which the sample contains a y observation; the blue dot denotes an (x_1, x_2) pair well outside the sample region, although the individual values $x_1 = 10$ and $x_2 = 10$ are within the x_1 and x_2 sample ranges separately.

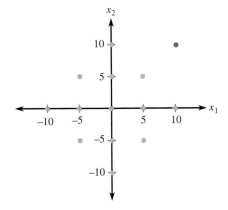

When more than two predictor variables are included in the model, we cannot tell from a plot like that of Figure 14.12 whether the $x_1, x_2, \ldots, x_k$ values of interest involve an extrapolation. It is then best to compare the $s_{\hat{y}}$'s based on $a + b_1 x_1 + b_2 x_2 + \cdots + b_k x_k$ for the values of interest with the $s_{\hat{y}}$'s corresponding to the $x_1, x_2, \ldots, x_k$ values in the sample (MINITAB gives these on request). Extrapolation is indicated by a value of the former standard deviation (at the x_i values of interest) that is much larger than the standard deviations for sampled values.

▪ Exercises 14.36–14.50

14.36 Explain why it is preferable to perform a model utility test before using an estimated regression model to make predictions or to estimate the mean y value for specified values of the independent variables.

14.37 The article "Zoning and Industrial Land Values: The Case of Philadelphia" (*AREUEA Journal* [1991]: 154–159) considered a regression model to relate the value of a vacant lot in Philadelphia to a number of different predictor variables.

a. One of the predictors was x_3 = the distance from the city's major east–west thoroughfare, for which $b_3 = -0.489$ and $s_{b_3} = 0.1044$. The model contained seven predictors, and the coefficients were estimated from a sample of 100 vacant lots. Calculate and interpret a confidence interval for β_3 using a confidence level of 95%.

b. Another predictor was x_1, an indicator variable for whether the lot was zoned for residential use, for which $b_1 = -0.183$ and $s_{b_1} = 0.3055$. Carry out a test of $H_0: \beta_1 = 0$ versus $H_a: \beta_1 \neq 0$, and interpret the conclusion in the context of the problem.

14.38 Twenty-six observations from the article "Multiple Regression Analysis for Forecasting Critical Fish Influxes at Power Station Intakes" (*Journal of Applied Ecology* [1983]: 33–42) were used to fit a multiple regression model relating y = number of fish at intake to the independent variables x_1 = water temperature (in °C), x_2 = number of pumps running, x_3 = sea state (taking values 0, 1, 2, or 3), and x_4 = speed (in knots). Partial MINITAB output follows:

The regression equation is
Y = 92.0 − 2.18X1 − 19.2X2 − 9.38X3 + 2.32X4

Predictor	Coef	Stdev	t-ratio
Constant	91.98	42.07	2.19
X1	−2.179	1.087	−2.00
X2	−19.189	9.215	−2.08
X3	−9.378	4.356	−2.15
X4	2.3205	0.7686	3.02

s = 10.53   R-sq = 39.0%   R-sq(adj) = 27.3%

a. Construct a 95% confidence interval for β_3, the coefficient of x_3 = sea state. Interpret the resulting interval.

b. Construct a 90% confidence interval for the mean change in y associated with a 1° increase in temperature when number of pumps, sea state, and speed remain fixed.

14.39 A study reported in the article "Leakage of Intracellular Substances from Alfalfa Roots at Various Subfreezing Temperatures" (*Crop Science* [1991]:

1575–1578) considered a quadratic regression model to relate y = MDH activity (a measure of the extent to which cellular membranes suffer extensive damage from freezing) to x = electrical conductivity (which describes the damage in the early stages of freezing). The estimated regression function was

$$\hat{y} = -0.1838 + 0.0272x + 0.0446x^2$$

with $R^2 = .860$.

a. Suppose that $n = 50$. Does the quadratic model appear to be useful? Test the appropriate hypotheses.

b. Using $s_{b_2} = 0.0103$, carry out a test at significance level .01 to decide whether the quadratic predictor x^2 is important.

c. If the standard deviation of the statistic $\hat{y} = a + b_1(40) + b_2(1600)$ is $s_{\hat{y}} = 0.120$, calculate a confidence interval with confidence level 90% for true average MDH activity when conductivity is 40.

14.40 The article first introduced in Exercise 14.27 gave data on the dimensions of 27 representative food products. Use the multiple regression model fit in Exercise 14.27 to answer the following questions:

a. Could any of these variables be eliminated from the regression model?

b. Predict the volume of a package with a minimum width of 2.5 cm, a maximum width of 3.0 cm, and an elongation of 1.55.

c. Calculate a 95% prediction interval for a package with a minimum width of 2.5 cm, a maximum width of 3.0 cm, and an elongation of 1.55.

14.41 Data from a random sample of 107 students taking a managerial accounting course were used to obtain the estimated regression equation $\hat{y} = 2.178 + 0.469x_1 + 3.369x_2 + 3.054x_3$, where y = student's exam score, x_1 = student's expected score on the exam, x_2 = time spent studying (hours per week), and x_3 = student's grade point average (GPA) ("Effort, Expectation, and Academic Performance in Managerial Cost Accounting," *Journal of Accounting Education* [1989]: 57–68). The value of R^2 was .686, and the estimated standard deviations of the statistics b_1, b_2, and b_3 were 0.090, 0.456, and 1.457, respectively.

a. How would you interpret the estimated coefficient $b_1 = 0.469$?

b. Does there appear to be a useful linear relationship between exam score and at least one of the three predictors?

c. Calculate a confidence interval for the mean change in exam score associated with a 1-hr increase in study time when expected score and GPA remain fixed.

d. Obtain a point prediction of the exam score for a student who expects a 75, has studied 8 hr/wk, and has a GPA of 2.8.

e. If the standard deviation of the statistic on which the prediction of Part (d) is based is 1.2 and SSTo = 10,200, calculate a 95% prediction interval for the score of the student described in Part (d).

14.42 Benevolence payments are monies collected by a church to fund activities and ministries beyond those provided by the church to its own members. The article "Optimal Church Size: The Bigger the Better" (*Journal for the Scientific Study of Religion* [1993]: 231–241) considered a regression of y = benevolence payments on x_1 = number of church members, x_2 = x_1^2, and x_3 = an indicator variable for urban versus nonurban churches.

a. The sample size was n = 300, and adjusted R^2 = .774. Does at least one of the three predictors provide useful information about y?

b. The article reported that b_3 = 101.1 and s_{b_3} = 625.8. Should the indicator variable be retained in the model? Test the relevant hypotheses.

14.43 The estimated regression equation

$$\hat{y} = 28 - 0.05x_1 - 0.003x_2 + 0.00002x_3$$

where y = number of indictments disposed of in a given month, x_1 = number of cases on judge's docket, x_2 = number of cases pending in the criminal trial court, and x_3 = x_1x_2, appeared in the article "The Caseload Controversy and the Study of Criminal Courts" (*Journal of Criminal Law and Criminology* [1979]: 89–101). This equation was based on n = 367 observations. The b_i's and their associated standard deviations are given in the following table:

i	b_i	Estimated Standard Deviation of b_i
1	−0.05	0.03
2	−0.003	0.0024
3	0.00002	0.000009

Is inclusion of the interaction predictor important? Test $H_0: \beta_3 = 0$ using a .05 level of significance.

14.44 The following data on the independent variable x = volume and the dependent variable y = amount of isobutylene converted were obtained from a study of a certain method for preparing pure alcohol from refinery streams ("Direct Hydration of Olefins," *Industrial and Engineering Chemistry* [1961]: 209–211):

x	1	1	2	4	4	4	6
y	23.0	24.5	28.0	30.9	32.0	33.6	20.0

MINITAB output — the result of fitting a quadratic regression model where x_1 = x and x_2 = x^2 — is given here:

```
The regression equation is
Y = 13.6 + 11.4X1 − 1.72X2
Predictor     Coef      Stdev     t-ratio
Constant     13.636     1.896      7.19
X1           11.406     1.356      8.41
X2           −1.7155    0.2036    −8.42
s = 1.428   R-sq = 94.7%   R-sq(adj) = 92.1%
```

Would a linear regression have sufficed? That is, is the quadratic predictor important? Use a level .05 test.

14.45 The article "Bank Full Discharge of Rivers" (*Water Resources Journal* [1978]: 1141–1154) reported data on y = discharge amount (in square meters per second), x_1 = flow area (in square meters), and x_2 = slope of the water surface (meters per meter) obtained at n = 10 floodplain stations. A multiple regression model using x_1, x_2, and x_3 = x_1x_2 was fit to these data, and partial MINITAB output appears here:

```
The regression equation is
Y = −3.14 + 1.70X1 + 96.1X2 + 8.38X3
Predictor     Coef      Stdev     t-ratio
Constant     −3.14      14.54     −0.22
X1            1.697      1.431      1.19
X2           96.1       702.7      0.14
X3            8.38      199.0       0.04
s = 17.58   R-sq = 73.2%   R-sq(adj) = 59.9%
```

a. Perform the model utility test using a significance level of .05.

b. Is the interaction term important? Test using a .05 significance level.

c. Does it bother you that the model utility test indicates a useful model but that all values in the t ratio column of the MINITAB output are small? Explain.

14.46 In the article "An Ultracentrifuge Flour Absorption Method" (*Cereal Chemistry* [1978]: 96–101), the authors discussed the relationship between water absorption for wheat flour and various characteristics of the flour. The model $y = \alpha + \beta_1x_1 + \beta_2x_2 + e$ was used to relate y = absorption (percentage) to x_1 = flour protein (percentage) and x_2 = starch damage (in farrands). MINITAB output based on n = 28 observations is given here:

```
The regression equation is
Y = 19.4 + 1.44X1 + .336X2
Predictor     Coef      Stdev     t-ratio     p
Constant     19.440     2.188      8.88     .000
X1            1.4423    0.2076     6.95     .000
X2            0.33563   0.01814   18.51     .000
s = 1.094   R-sq = 96.4%   R-sq(adj) = 96.2%
```

Analysis of Variance

SOURCE	DF	SS	MS	F	p
Regression	2	812.380	406.190	339.3	.000
Error	25	29.928	1.197		
Total	27	842.307			

Use a significance level of .05 for all tests requested.
a. Does the model appear to be useful? Test the relevant hypotheses.
b. Calculate and interpret a 95% confidence interval for β_2.
c. Conduct tests for each of the following pairs of hypotheses:
i. $H_0: \beta_1 = 0$ versus $H_a: \beta_1 \neq 0$
ii. $H_0: \beta_2 = 0$ versus $H_a: \beta_2 \neq 0$
d. Based on the results of Part (c), would you conclude that both independent variables are important? Explain.
e. An estimate of the mean water absorption for wheat with 11.7% protein and a starch damage value of 57 is desired. Compute a 90% confidence interval for $\alpha + \beta_1(11.7) + \beta_2(57)$ if the estimated standard deviation of $a + b_1(11.7) + b_2(57)$ is 0.522. Interpret the resulting interval.
f. A single shipment of wheat is received. For this particular shipment, $x_1 = 11.7$ and $x_2 = 57$. Predict the water absorption for this shipment (a single y value) using a 90% interval.

14.47 Exercise 14.25 gave data on fish weight, length, age, and year caught. A multiple regression model was fit to describe the relationship between weight and the predictors length and age.
a. Could either length or age be eliminated from the model without significantly affecting your ability to make accurate predictions? Why or why not? Give statistical evidence to support your conclusion.
b. Create a dummy (indicator) variable for year caught. Fit a multiple regression model that includes year, length, and age as predictors of weight. Is there evidence that year is a useful predictor given that length and age are included in the model? Test the relevant hypotheses using $\alpha = .05$.

14.48 The article "Predicting Marathon Time from Anaerobic Threshold Measurements" (*The Physician and Sports Medicine* [1984]: 95–98) gave data on y = maximum heart rate (beats per minute), x_1 = age, and x_2 = weight (in kilograms) for $n = 18$ marathon runners. The estimated regression equation for the model $y = \alpha + \beta_1 x_1 + \beta_2 x_2 + e$ was $\hat{y} = 179 - 0.8x_1 + 0.5x_2$, and SSRegr = 649.75 and SSResid = 538.03.
a. Is the model useful for predicting maximum heart rate? Use a significance level of .10.
b. Using $s_{b_1} = 0.280$, calculate and interpret a 95% confidence interval for β_1.

c. Predict the maximum heart rate of a particular runner who is 43 years old and weighs 65 kg, using a 99% interval. The estimated standard deviation of the statistic $a + b_1(43) + b_2(65)$ is 3.52.
d. Use a 90% interval to estimate the average maximum heart rate for all marathon runners who are 30 years old and weigh 77.2 kg. The estimated standard deviation of $a + b_1(30) + b_2(77.2)$ is 2.97.
e. Would a 90% prediction interval for a single 30-year-old runner weighing 77.2 kg be wider or narrower than the interval computed in Part (d)? Explain. (Hint: You need not compute the interval.)

14.49 The effect of manganese (Mn) on wheat growth is examined in the article "Manganese Deficiency and Toxicity Effects on Growth, Development, and Nutrient Composition in Wheat" (*Agronomy Journal* [1984]: 213–217). A quadratic regression model was used to relate y = plant height (in centimeters) to x = log(added Mn) (added Mn is measured in micromoles). The accompanying data were read from a scatterplot in the article. Also given is a MINITAB output, where $x_1 = x$ and $x_2 = x^2$.

x	-1	-0.4	0	0.2	1	2	2.8	3.2	3.4	4
y	32	37	44	45	46	42	42	40	37	30

The regression equation is
$Y = 41.7 + 6.58X1 - 2.36X2$

Predictor	Coef	Stdev	t-ratio
Constant	41.7422	0.8522	48.98
X1	6.581	1.002	6.57
X2	-2.3621	0.3073	-7.69

$s = 1.963$ R-sq = 89.8% R-sq(adj) = 86.9%

Analysis of Variance

Source	DF	SS	MS
Regression	2	237.520	118.760
Error	7	26.980	3.854
Total	9	264.500	

Use a .05 significance level for any hypothesis tests needed to answer the following questions:
a. Is the quadratic model useful for describing the relationship between y and x?
b. Are both the linear and the quadratic predictors important? Could either one be eliminated from the model? Explain.
c. Give a 95% confidence interval for the mean y value when $x = 2$. The estimated standard deviation of $a + b_1(2) + b_2(4)$ is 1.037. Interpret the resulting interval.
d. Estimate the mean height for wheat treated with 10 μmol of Mn using a 90% interval. (Note: The estimated standard deviation of $a + b_1(1) + b_2(1)$ is 1.031 and log(10) = 1.)

14.50 *This exercise requires the use of a computer package.* A study on the effect of applying fertilizer in bands is described in the article "Fertilizer Place-ment Effects on Growth, Yield, and Chemical Com-position of Burley Tobacco" (*Agronomy Journal* [1984]: 183–188). The following data were taken from a scatterplot appearing in the article, with y = plant Mn (in micrograms per gram, dry weight) and x = distance from the fertilizer band (in centimeters):

x	0	10	20	30	40
y	110	90	76	72	70

The authors suggested a quadratic regression model.

a. Use a suitable computer package to find the esti-mated quadratic regression equation.
b. Perform the model utility test.
c. Interpret the values of R^2 and s_e.
d. Are both the linear and the quadratic predictors important? Carry out the necessary hypothesis tests, and interpret the results.
e. Find a 90% confidence interval for the mean plant Mn for plants that are 30 cm from the fertilizer band.

■ 14.4 Other Issues in Multiple Regression

Primary objectives in multiple regression include estimating a mean y value, pre-dicting an individual y value, and gaining insight into how changes in predictor vari-able values affect y. Often an investigator has data on a number of predictor vari-ables that might be incorporated into a model to be used for such purposes. Some of these predictor variables may actually be unrelated or only weakly related to y, or they may contain information that duplicates information provided by other pre-dictors. If all these predictors are included in the model, many model coefficients will have to be estimated. This reduces the number of degrees of freedom associ-ated with SSResid, leading to a deterioration in the degree of precision associated with other inferences (e.g., wide confidence and prediction intervals). A model with many predictors can also be cumbersome to use and difficult to interpret.

In this section, we first introduce some guidelines and procedures for selecting a set of useful predictors. In choosing a model, the analyst should examine the data carefully for evidence of unusual observations or potentially troublesome patterns. It is important to identify unusually deviant or influential observations and to look for possible inconsistencies with model assumptions. Our discussion of multiple re-gression closes with a brief mention of some diagnostic methods designed for these purposes.

■ Variable Selection

Suppose that an investigator has data on p predictors $x_1, x_2, \ldots, x_p$, which are can-didates for use in building a model. Some of these predictors might be specified functions of others, for example, $x_3 = x_1 x_2$, $x_4 = x_1^2$, and so on. The objective is then to select a set of these predictors that in some sense specifies a best model (of the general additive form considered in Sections 14.2 and 14.3). Fitting a model that uses a specified k predictor requires that $k + 1$ model coefficients (α and the k cor-responding β's) be estimated. In general, the number of observations, n, should be at least twice the number of predictors in the largest model under consideration to ensure reasonably accurate coefficient estimates and a sufficient number of degrees of freedom associated with SSResid.

If p is not too large, a good statistical computer package can quickly fit a model based on each different subset of the p predictors. Consider the case $p = 4$. There are two possibilities for each predictor — it could be included or not included in a model — so the number of possible models in this case is $2(2)(2)(2) = 2^4 = 16$ (including the model with all four predictors and the model with only the constant term and none of the four predictors). These 16 possibilities are displayed in the following table:

Predictors Included	Number of Predictors in Model	Predictors Included	Number of Predictors in Model
None	0	x_2, x_3	2
x_1	1	x_2, x_4	2
x_2	1	x_3, x_4	2
x_3	1	x_1, x_2, x_3	3
x_4	1	x_1, x_2, x_4	3
x_1, x_2	2	x_1, x_3, x_4	3
x_1, x_3	2	x_2, x_3, x_4	3
x_1, x_4	2	x_1, x_2, x_3, x_4	4

More generally, when there are p candidate predictors, the number of possible models is 2^p. The number of possible models is therefore substantial if p is even moderately large — for example, 1024 possible models when $p = 10$ and 32,768 possibilities when $p = 15$.

Model selection methods can be divided into two types. First, there are methods based on fitting every possible model, computing one or more summary quantities from each fit, and comparing these quantities to identify the most satisfactory models. Several of the most powerful statistical computer packages have an **all-subsets** option, which gives limited output from fitting each possible model. Methods of the second type are appropriate when p is so large that it is not feasible to examine all subsets. These methods are often referred to as **automatic selection** or **stepwise procedures**. The general idea is either to begin with the p predictor model and delete predictors one by one until all remaining predictors are judged important or to begin with no predictors and add predictors until no predictor not in the model seems important. With present-day computing power, the value of p for which examination of all subsets is feasible is surprisingly large, so automatic selection procedures are not as important as they once were.

Suppose, then, that p is small enough that all subsets can be fit. What characteristic(s) of the estimated models should be examined in the search for a best model? An obvious and appealing candidate is the coefficient of multiple determination, R^2, which measures the proportion of observed y variation explained by the model. Certainly a model with a large R^2 value is preferable to another model that contains the same number of predictors but has a much smaller R^2 value. Thus, if the model with predictors x_1 and x_2 has $R^2 = .765$ and the model with predictors x_1 and x_3 has $R^2 = .626$, the second model would almost surely be eliminated from further consideration.

However, using R^2 to choose between models containing different numbers of predictors is not so straightforward, because adding a predictor to a model can never decrease the value of R^2. Let

$R^2_{(1)}$ = largest R^2 for any one-predictor model
$R^2_{(2)}$ = largest R^2 for any two-predictor model

and so on. Then $R^2_{(1)} \leq R^2_{(2)} \leq \cdots \leq R^2_{(p-1)} \leq R^2_{(p)}$. When statisticians base model selection on R^2, the objective is not simply to find the model with the largest R^2 value; the model with p predictors does that. Instead, we should look for a model that contains relatively few predictors but has a large R^2 value and is such that no other model containing more predictors gives much of an improvement in R^2. Suppose, for example, that $p = 5$ and that

$$R^2_{(1)} = .427 \qquad R^2_{(2)} = .733 \qquad R^2_{(3)} = .885 \qquad R^2_{(4)} = .898 \qquad R^2_{(5)} = .901$$

Then the best three-predictor model appears to be a good choice, because it substantially improves on the best one- and two-predictor models, whereas little is gained by using the best four-predictor model or all five predictors.

A small increase in R^2 resulting from the addition of a predictor to a model can be offset by the increased complexity of the new model and the reduction in the number of degrees of freedom associated with SSResid. This has led statisticians to consider an adjusted R^2, which can either decrease or increase when a predictor is added to a model. It follows that the adjusted R^2 for the best k-predictor model (i.e., the model with coefficient of multiple determination $R^2_{(k)}$) may be larger than the adjusted R^2 for the best model based on $k + 1$ predictors. The adjusted R^2 formalizes the notion of diminishing returns as more predictors are added: Small increases in R^2 are outweighed by corresponding decreases in the number of degrees of freedom associated with SSResid. A reasonable strategy in model selection is to identify the model with the largest value of adjusted R^2 (the corresponding number of predictors k is often much smaller than p) and then consider only that model and any others whose adjusted R^2 values are nearly as large.

■ Example 14.17   Modeling the Price of Industrial Properties

The paper "Using Multiple Regression Analysis in Real Estate Appraisal" (*The Appraisal Journal* [2001]: 424–430) reported on a study that aimed to relate the price of a property to various other characteristics of the property. The variables were

y = price per square foot

x_1 = size of building (in square feet)

x_2 = age of building (in years)

x_3 = quality of location (measured on a scale of 1 [very poor location] to 4 [very good location])

x_4 = land to building ratio

The study reported the following data for a random sample of nine large industrial properties:

y	x_1	x_2	x_3	x_4	y	x_1	x_2	x_3	x_4
4.89	2,166,600	30	4.0	2.01	2.79	2,866,526	35	4.0	2.27
3.49	751,658	30	2.0	3.54	5.89	1,698,161	28	3.0	3.12
4.33	2,422,650	28	3.0	3.63	6.38	1,046,260	33	4.0	4.77
8.24	224,573	25	1.5	4.65	5.25	1,108,828	28	4.0	7.56
5.10	3,917,800	26	4.0	1.71					

Consider x_1, x_2, x_3, and x_4 as the set of potential predictors ($p = 4$, with no derived predictors, such as squares or interaction terms, as candidates for inclusion). Then there are $2^4 = 16$ possible models, among which 4 consist of a single predictor, 6 involve two predictors, 4 others use three predictors, and 1 includes all four predictor variables. We used a statistical computer package to fit each possible model and extracted both R^2 and adjusted R^2 from the output. These values are as follows:

Models with One Predictor

Predictor	x_2	x_1	x_4	x_3
R^2	32.4	24.7	13.2	11.9
Adjusted R^2	22.8	14.0	0.8	0.0

Models with Two Predictors

Predictors	x_1, x_2	x_2, x_4	x_2, x_3	x_1, x_3	x_1, x_4	x_3, x_4
R^2	52.3	40.2	33.3	25.0	24.9	22.8
Adjusted R^2	36.3	20.2	11.1	0.0	0.0	0.0

Models with Three Predictors

Predictors	x_1, x_2, x_3	x_1, x_2, x_4	x_2, x_3, x_4	x_1, x_3, x_4
R^2	58.6	52.3	40.9	25.7
Adjusted R^2	33.7	23.7	5.4	0.0

Models with All Four Predictors

Predictors	x_1, x_2, x_3, x_4
R^2	65.3
Adjusted R^2	30.6

It is clear that the best two-predictor model offers considerable improvement with respect to both R^2 (52.3%) and adjusted R^2 (36.3%) over any model with just a single predictor. The best two-predictor model also has the largest value of adjusted R^2 — even larger than adjusted R^2 for the three-predictor models and the model that uses all four predictors.

This suggests selecting the model that uses x_1 and x_2 (building size and age) as predictors of price. The MINITAB output resulting from fitting this model is given in Figure 14.13. Note that, although this model may be the best choice among those considered here, the R^2 value is not particularly large and the value of $s_e = 1.283$ (in dollars per square foot) is quite large given the range of y values in the data set.

Figure 14.13 MINITAB output for the data of Example 14.17.

Regression Analysis: Price versus Size, Age
The regression equation is
Price = 14.0 − 0.000001 Size − 0.265 Age

Predictor	Coef	SE Coef	T	P
Constant	14.022	4.185	3.35	0.015
Size	−0.00000062	0.00000039	−1.58	0.166
Age	−0.2654	0.1427	−1.86	0.112

S = 1.283   R-Sq = 52.3%   R-Sq(adj) = 36.3%

Analysis of Variance

Source	DF	SS	MS	F	P
Regression	2	10.810	5.405	3.28	0.109
Residual Error	6	9.876	1.646		
Total	8	20.686			

Various other criteria have been proposed and used for model selection after fitting all subsets. The references at the end of this chapter can be consulted for more details.

When using particular criteria as a basis for model selection, many of the 2^p possible subset models are not serious candidates because of poor criteria values. For example, if $p = 15$, there are 5005 different models consisting of 6 predictor variables, many of which typically have small R^2 and adjusted R^2 values. An investigator usually wishes to consider only a few of the best models of each different size (a model whose criteria value is close to the best one may be easier to interpret than the best model or may include a predictor that the investigator thinks should be in the selected model). In recent years, statisticians have developed computer programs to select the best models of a given size without actually fitting all possible models. One version of such a program has been implemented in MINITAB and can be used as long as $p \leq 20$. (There are roughly 1 million possible models when $p = 20$, so fitting them all would be out of the question.) The user specifies a number between 1 and 10 as the number of models of each given size for which the output will be provided. In addition to R^2 and adjusted R^2, values of another criterion, called *Mallow's* C_p, are included. A good model according to this criterion is one that has small C_p (for accurate predictions) and for which $C_p \approx k + 1$ (for unbiasedness in estimating model coefficients). After the choice of models is narrowed, the analyst can request more detail on each finalist.

▪ Example 14.18   Durable Press Rating of Cotton Fabric

In the article "Applying Stepwise Multiple Regression Analysis to the Reaction of Formaldehyde with Cotton Cellulose" (*Textile Research Journal* [1984]: 157–165), the investigators looked at the dependent variable

y = durable press rating

which is a quantitative measure of wrinkle resistance. The four independent variables used in the model building process were

x_1 = formaldehyde concentration

x_2 = catalyst ratio

x_3 = curing temperature

x_4 = curing time

In addition to these variables, the investigators considered as potential predictors $x_1^2, x_2^2, x_3^2, x_4^2$ and all six interactions $x_1 x_2, \ldots, x_3 x_4$, a total of $p = 14$ candidates.

The following data were taken from the article:

Observation	y	x_1	x_2	x_3	x_4
1	1.4	8	4	100	1
2	2.2	2	4	180	7
3	4.6	7	4	180	1
4	4.9	10	7	120	5
5	4.6	7	4	180	5
6	4.7	7	7	180	1

(*continued*)

(*continued*)

Observation	y	x_1	x_2	x_3	x_4
7	4.6	7	13	140	1
8	4.5	5	4	160	7
9	4.8	4	7	140	3
10	1.4	5	1	100	7
11	4.7	8	10	140	3
12	1.6	2	4	100	3
13	4.5	4	10	180	3
14	4.7	6	7	120	7
15	4.8	10	13	180	3
16	4.6	4	10	160	5
17	4.3	4	13	100	7
18	4.9	10	10	120	7
19	1.7	5	4	100	1
20	4.6	8	13	140	1
21	2.6	10	1	180	1
22	3.1	2	13	140	1
23	4.7	6	13	180	7
24	2.5	7	1	120	7
25	4.5	5	13	140	1
26	2.1	8	1	160	7
27	1.8	4	1	180	7
28	1.5	6	1	160	1
29	1.3	4	1	100	1
30	4.6	7	10	100	7

Figure 14.14 displays output for the best three subset models of each size from $k = 1$ to $k = 9$ predictor variables and also for the model with all 14 predictors.

The choice of a best model here, as often happens, is not clear-cut. We certainly don't see the benefit of including more than $k = 8$ predictor variables (after that, the adjusted R^2 begins to decrease), nor would we suggest a model with fewer than five predictors (the adjusted R^2 is still increasing and C_p is large). Based only on this output, the best six-predictor model is a reasonable choice. The corresponding estimated regression function is

$$y = -1.218 + 0.9599x_2 - 0.0373x_1^2 - 0.0389x_2^2 + 0.0037x_1x_3 + 0.019x_1x_4$$
$$- 0.0013x_2x_3$$

Another good candidate is the best seven-predictor model. Although this model includes one more predictor than the six-predictor model just suggested, only one of the seven predictors is an interaction term (x_1x_3), so model interpretation is somewhat easier. (Notice, though, that none of the best three models with seven predictors results simply from adding a single predictor to the best six-predictor model.)

FIGURE 14.14
MINITAB output for the data of Example 14.18.

```
                                     c c t                     c c c c t
                                     o a e                   c o o a a e
                                     n t m         t t o n n t t m
                                     c a p     c c e l n * * * * p
                                     e l e t o a m m * t t t t *
                                     n y r l n t p e c e l e l t
              R-Sq                   t s a m s s s s a m m m m i
Vars  R-Sq  (adj)  C-p      S        r t t e q q q q t p e p e m
 1    52.5  50.8  64.5   0.98251                                       X
 1    52.1  50.4  65.2   0.98614                       X
 1    51.6  49.9  66.2   0.99168     X
 2    64.6  62.0  43.4   0.86298     X               X
 2    63.3  60.6  45.9   0.87898     X                           X
 2    61.9  59.1  48.5   0.89548                                 X X
 3    78.7  76.2  18.6   0.68304     X           X               X
 3    72.8  69.7  29.8   0.77112     X X         X
 3    72.8  69.6  29.9   0.77202   X X           X
 4    82.2  79.4  13.9   0.63607     X     X     X           X
 4    82.1  79.2  14.1   0.63843     X           X     X     X
 4    81.7  78.8  14.8   0.64455     X           X           X               X
 5    86.7  83.9   7.4   0.56196     X        X X           X X
 5    85.5  82.5   9.6   0.58571     X      X X X           X
 5    85.5  82.5   9.6   0.58614     X        X X     X     X
 6    88.5  85.5   5.9   0.53337     X        X X           X X X
 6    88.2  85.1   6.5   0.54094     X      X X X           X               X
 6    88.0  84.9   6.8   0.54386     X        X X           X X             X
 7    89.9  86.7   5.2   0.51045     X X X X X X           X
 7    89.9  86.7   5.3   0.51140     X X      X X X X       X
 7    89.7  86.4   5.7   0.51683     X X      X X X         X X
 8    91.5  88.3   4.2   0.47933     X X X X X X           X               X
 8    90.6  87.0   6.0   0.50535     X X      X X X X       X               X
 8    90.5  86.8   6.1   0.50793     X X      X X X         X X X
 9    91.7  87.9   5.9   0.48650   X X X X X X X           X               X
 9    91.6  87.8   6.0   0.48853   X X X X X X             X     X         X
 9    91.6  87.8   6.0   0.48860   X X X X X X             X X             X
14    92.1  84.8  15.0   0.54639   X X X X X X X X X X X X X X
```

Because every good model includes x_1, x_2, x_3, and x_4 in some predictor, it appears that formaldehyde concentration, catalyst ratio, curing time, and curing temperature are all important determinants of durable press rating.

The most easily understood and implemented automatic selection procedure is referred to as **backward elimination**. It involves starting with the model that contains all p potential predictors and then deleting them one by one until all remaining predictors seem important. The first step is to specify the value of a positive constant t_{out}, which is used to decide whether deletion should be continued. After fitting the p predictor model, the t ratios $\frac{b_1}{s_{b_1}}, \frac{b_2}{s_{b_2}}, \ldots, \frac{b_p}{s_{b_p}}$ are examined. The predictor variable whose t ratio is closest to 0, whether positive or negative, is the obvious candidate for deletion. The corresponding predictor is eliminated from the model if this t ratio satisfies the inequalities $-t_{out} \le t$ ratio $\le t_{out}$. Suppose that this is the case. The model with the remaining $p - 1$ predictors is then fit, and again the predictor with the t ratio closest to 0 is eliminated, provided that it satisfies $-t_{out} \leq t$ ratio $\leq t_{out}$. The procedure continues until, at some stage, no t ratio satisfies $-t_{out} \leq t$ ratio $\leq t_{out}$ (all are either greater than t_{out} or less than $-t_{out}$). The chosen model is then the last one fit (although some analysts recommend examining other models of the same size). It is customary to use $t_{out} = 2$, because for many different numbers of degrees of freedom, $t_{out} = 2$ corresponds to a two-tailed test with approximate significance level .05.*

▪ Example 14.19 More on Price of Industrial Properties

Figure 14.15 shows a MINITAB output resulting from the application of the backward elimination procedure with $t_{out} = 2$ to the industrial property price data of Example 14.17. The t ratio closest to 0 when the model with all four predictors was fit was $-.88$, so the corresponding predictor, x_4, was deleted from the model. When the model with the three remaining predictors was fit, the t ratio closest to 0 was 0.87, which satisfies $-2 \leq 0.87 \leq 2$. As a consequence, the corresponding predictor, x_3, was eliminated, leaving x_1 and x_2. The next predictor to be dropped was x_1, because its t ratio was -1.58, and $-2 \leq -1.58 \leq 2$. When the model with just $x_2 =$ age was fit, it was also eliminated! This illustrates that automatic selection procedures don't always work well.

Figure 14.15
MINITAB output for Example 14.19.

Step	1	2	3	4	5
Constant	18.428	14.757	14.022	13.533	5.151
Size	−0.00000	−0.00000	−0.00000		
T-Value	−1.68	−1.75	−1.58		
P-Value	0.169	0.141	0.166		
Age	−0.43	−0.34	−0.27	−0.29	
T-Value	−2.14	−2.01	−1.86	−1.83	
P-Value	0.100	0.100	0.112	0.109	
Location	1.19	0.61			
T-Value	1.22	0.87			
P-Value	0.288	0.422			
Land/Bui	−0.42				
T-Value	−0.88				
P-Value	0.429				
S	1.34	1.31	1.28	1.41	1.61
R-Sq	65.29	58.58	52.26	32.43	−0.00
R-Sq(adj)	30.58	33.72	36.34	22.78	0.00

Unfortunately, the backward elimination method does not always terminate with a model that is the best of its size, and this is also true of other automatic selection procedures. For example, the authors of the article mentioned in Example 14.18 used an automatic procedure to obtain a six-predictor model with

---

*Some computer packages base the procedure on the squares of the t ratios, which are F ratios, and continue to delete as long as F ratio $\leq F_{out}$ for at least one predictor. The predictor with the smallest F ratio is eliminated. $F_{out} = 4$ corresponds to $t_{out} = 2$.

$R^2 = .77$, whereas all of the 10 best six-predictor models have R^2 values of at least .87. Because of this, we recommend using a statistical software package that will identify best subsets of different sizes whenever possible.

■ Checks on Model Adequacy

In Chapter 13 we discussed some informal techniques for checking the adequacy of the simple linear regression model. Most of these were based on plots involving the standardized residuals. The formula for standardizing residuals in multiple regression is quite complicated, but it has been programmed into many statistical computer packages. Once the standardized residuals resulting from the fit of a particular model have been computed, plots similar to those discussed previously are useful in diagnosing model defects. A normal probability plot of the standardized residuals that departs too much from a straight line casts doubt on the assumption that the random deviation e has a normal distribution. Plots of the standardized residuals against each predictor variable in the model — that is, a plot of $(x_1,$ standardized residual) pairs, another of $(x_2,$ standardized residual) pairs, and so on — are analogous to the standardized residual versus x plot discussed and illustrated in Chapter 13. The appearance of any discernible pattern in these plots (e.g., curvature or increasing spread from left to right) points to the need for model modification. If observations have been made over time, a periodic pattern when the standardized residuals are plotted in time order suggests that successive observations were not independent. Models that incorporate dependence of successive observations are substantially more complicated than those we have presented here. They are especially important in econometrics, which involves using statistical methods to model economic data. Please consult one of the references in Chapter 5 for more information.

One other aspect of model adequacy that has received much attention from statisticians in recent years is the identification of any observations in the data set that may have been highly influential in estimating model coefficients. Recall that in simple linear regression, an observation with potentially high influence is one whose x value places it far to the right or left of the other points in the scatterplot or standardized residual plot. If a multiple regression model involves only two predictor variables x_1 and x_2, an observation with potentially large influence can be revealed by examining a plot of (x_1, x_2) pairs. Any point in this plot that is far away from the others corresponds to an observation that, if deleted from the sample, may cause coefficient estimates and other quantities to change considerably. Detecting influential observations when the model contains three predictors or more is more difficult. Recent research has yielded several helpful diagnostic quantities. One of these has been implemented in MINITAB, and a large value of this quantity automatically results in the corresponding observation being identified as one that might have great influence. Deleting the observation and refitting the model reveals the extent of actual influence (which depends on how consistent the corresponding y observation is with the rest of the data).

■ Multicollinearity

When the simple linear regression model is fit using sample data in which the x values are all close to one another, small changes in observed y values can cause the values of the estimated coefficients a and b to change considerably. This may well result in standard deviations σ_b and σ_a that are quite large, so that estimates of β and α from any given sample are likely to differ greatly from the true values. There is an analogous condition that leads to this same type of behavior in multiple regression: a configuration of predictor variable values that is likely to result in poorly estimated model coefficients.

When the model to be fit includes k predictors $x_1, x_2, \ldots, x_k$, there is said to be **multicollinearity** if there is a strong linear relationship between values of the predictors. Severe multicollinearity leads to instability of estimated coefficients and to various other problems. Such a relationship is difficult to visualize when $k > 2$, so statisticians have developed various quantitative indicators to measure the extent of multicollinearity in a data set. The most straightforward approach involves computing R^2 values for regressions in which the dependent variable is taken to be one of the k x's and the predictors are the remaining $(k - 1)$ x's. For example, when $k = 3$, there are three relevant regressions:

1. Dependent variable = x_1, predictor variables = x_2 and x_3
2. Dependent variable = x_2, predictor variables = x_1 and x_3
3. Dependent variable = x_3, predictor variables = x_1 and x_2

Each regression yields an R^2 value. In general, there are k such regressions and therefore k resulting R^2 values. If one or more of these R^2 values is large (close to 1), multicollinearity is present. MINITAB prints a message saying that the predictors are highly correlated when at least one of these R^2 values exceeds .99 and refuses to include a predictor in the model if the corresponding R^2 value is larger than .9999. Other analysts are more conservative and would judge multicollinearity to be a potential problem if any of these R^2 values exceeded .9.

When the values of predictor variables are under the control of the investigator, which often happens in scientific experimentation, a careful choice of values precludes multicollinearity. Multicollinearity does frequently occur in social science and business applications of regression analysis, where data result simply from observation rather than from intervention by an experimenter. Statisticians have proposed various remedies for the problems associated with multicollinearity in such situations, but a discussion would take us beyond the scope of this book. (After all, we want to leave something for your next statistics course!)

■ Exercises 14.51–14.62

14.51 The article "The Caseload Controversy and the Study of Criminal Courts" (*Journal of Criminal Law and Criminology* [1979]: 89–101) used multiple regression to analyze a data set consisting of observations on the variables y = length of sentence in trial case (in months), x_1 = seriousness of first offense, x_2 = dummy variable indicating type of trial (bench or jury), x_3 = number of legal motions, x_4 = measure of delay for confined defendants (0 for those not confined), and x_5 = measure of judge's caseload. The estimated regression equation proposed by the authors is

$$\hat{y} = 12.6 + 0.59x_1 - 70.8x_2 - 33.6x_3 - 15.5x_4 + 0.0007x_6 + 3x_7 - 41.5x_8$$

where $x_6 = x_4^2$, $x_7 = x_1 x_2$, and $x_8 = x_3 x_5$. How do you think the authors might have arrived at this particular model?

14.52 The article "Histologic Estimation of Age at Death Using the Anterior Cortex of the Femur" (*American Journal of Physical Anthropology* [1991]: 171–179) developed multiple regression models to relate y = age at death to a number of different predictor variables. Stepwise regression (an automatic selection procedure favored over the backward elimination method by many statisticians) was used to identify models, shown in the following table, for females:

Predictors	R^2	Adjusted R^2
x_6	.60	.60
x_3, x_6	.66	.66
x_3, x_5	.68	.68
x_2, x_3, x_5	.69	.68
x_2, x_3, x_5, x_6	.70	.69
x_2, x_3, x_4, x_5, x_6	.71	.70
$x_1, x_2, x_3, x_4, x_5, x_6$	.71	.70
$x_1, x_2, x_3, x_4, x_5, x_6, x_7$	.71	.70
$x_1, x_2, x_3, x_5, x_6, x_7, x_8$	.72	.71
$x_1, x_2, x_3, x_4, x_5, x_6, x_7, x_8$	.72	.70

Based on the given information, which of these models would you recommend using, and why?

14.53 The article "The Analysis and Selection of Variables in Linear Regression" (*Biometrics* [1976]: 1–49) reported on an analysis of data taken from issues of *Motor Trend* magazine. The dependent variable y was gas mileage, there were $n = 32$ observations, and the independent variables were x_1 = engine type (1 = straight, 0 = V), x_2 = number of cylinders, x_3 = transmission type (1 = manual, 0 = automatic), x_4 = number of transmission speeds, x_5 = engine size, x_6 = horsepower, x_7 = number of carburetor barrels, x_8 = final drive ratio, x_9 = weight, and x_{10} = quarter-mile time. The R^2 and adjusted R^2 values given in the following table are for the best model using k predictors for $k = 1, \ldots, 10$:

k	Variables Included	R^2	Adjusted R^2
1	x_9	.756	.748
2	x_2, x_9	.833	.821
3	x_3, x_9, x_{10}	.852	.836
4	x_3, x_6, x_9, x_{10}	.860	.839
5	$x_3, x_5, x_6, x_9, x_{10}$	.866	.840
6	$x_3, x_5, x_6, x_8, x_9, x_{10}$	.869	.837
7	$x_3, x_4, x_5, x_6, x_8, x_9, x_{10}$	.870	.832
8	$x_3, x_4, x_5, x_6, x_7, x_8, x_9, x_{10}$	.871	.826
9	$x_1, x_3, x_4, x_5, x_6, x_7, x_8, x_9, x_{10}$	.871	.818
10	All independent variables	.871	.809

Which model would you select? Explain your choice and the criteria used to reach your decision.

14.54 The article "Estimation of the Economic Threshold of Infestation for Cotton Aphid" (*Mesopotamia Journal of Agriculture* [1982]: 71–75) gave $n = 34$ observations on y = infestation rate (number of aphids per 100 leaves), x_1 = mean temperature (in °C), and x_2 = mean relative humidity. Partial SAS computer output resulting from fitting the model $y = \alpha + \beta_1 x_1 + \beta_2 x_2 + e$ to the data given in the article is shown:

		R-SQUARE	0.5008
		ADJ R-SQ	0.4533

VARIABLE	DF	PARAMETER ESTIMATE	STANDARD ERROR	T FOR H0: PARAMETER = 0
INTERCEP	1	15.667014	84.157420	0.186
TEMP	1	−0.360928	2.330835	−0.155
RH	1	1.959715	0.576946	2.877

If the method of backward elimination is to be used, which variable would be the first candidate for elimination? By using the criterion of eliminating a variable if its t ratio satisfies $-2 \le t$ ratio ≤ 2, can this variable be eliminated from the model?

14.55 The following statement is from the article "Blood Cadmium Levels in Nonexposed Male Subjects Living in the Rome Area: Relationship to Selected Cardiovascular Risk Factors" (*Microchemical Journal* [1998]: 173–179): "A multiple regression analyzed ln(blood cadmium level) as a dependent variable, with a model including age, alcohol consumption, daily cigarette consumption, driving habits, body mass index, skinfold thickness, high density lipoprotein cholesterol, non–high density lipoprotein cholesterol, and ln(triglyceride level) as possible predictors. This model explained 30.92% of the total variance in ln(blood cadmium level)."
a. A total of $n = 1856$ men participated in this study. Perform a model utility test to determine whether the predictor variables, taken as a group, are useful in predicting ln(blood cadmium level).
b. The article went on to state: "Smoking habits, the most important predictor of ln(blood cadmium level), explained 30.26% of the total variation, and alcohol consumption was the only other variable influencing ln(blood cadmium level)." These statements were based on the result of a forward stepwise regression. In forward stepwise regression, variables are added to the model one at a time. At each step, the variable that results in the largest improvement is added to the model. This continues until there is no variable whose addition will result in a significant improvement. Describe what you think happened at each step of the stepwise procedure.

14.56 For the multiple regression model in Exercise 14.4, the estimated standard deviations for each of the coefficients (excluding the constant) are (in order) 0.01, 0.01, 0.03, 0.06, 0.01, 0.01, 0.03, 0.04, and 0.05.
a. If a backward elimination variable selection process was to be used, which of the nine predictors would be the first candidate for elimination? Would it be eliminated?
b. Based on the given information, can you tell what the second candidate for elimination would be? Explain.

14.57 The accompanying table showing the results of two multiple regressions appeared in "Does the Composition of the Compensation Committee Influence CEO Compensation Practices?" (*Financial Management* [1999]: 41–53). The dependent variable is log(compensation).

Dependent Variable	Independent Variable	b	t
log(1992 Compensation)	Intercept	−2.508	−4.80
	Stock ownership	−0.002	−0.58
	Log(Sales)	0.372	6.56
	CEO tenure	0.005	0.56
	Stock returns	0.003	2.12
	Return on equity	0.005	2.79
	Insider	0.135	1.35
log(1991 Compensation)	Intercept	−2.098	−3.92
	Stock ownership	−0.005	−1.73
	Log(Sales)	0.286	4.92
	CEO tenure	0.020	2.16
	Stock returns	0.006	3.88
	Return on equity	0.001	0.33
	Insider	0.061	0.57

a. Do both regressions show a similar pattern?
b. Calculate the standard error for the coefficient corresponding to log(Sales) for the 1992 regression.
c. For the 1991 regression, which variable would be considered first in a backward elimination variable selection procedure? Would it be eliminated?
d. Would the same variable identified in Part (c) be the first to be considered in a backward elimination process for the 1992 regression?
e. Assuming that the number of observations is 160, determine the P-value for a test of the hypothesis that β_1 in the 1991 model is negative.

14.58 Suppose that you were considering a multiple regression analysis with y = house price, x_1 = number of bedrooms, x_2 = number of bathrooms, and x_3 = total number of rooms. Do you think that multicollinearity might be a problem? Explain.

14.59 *This exercise requires use of a computer package.* Compare the best one-, two-, three-, and four-predictor models for the data given in Exercise 14.60. Does this variable selection procedure lead you to the same model as the backward elimination used in Exercise 14.60?

14.60 In the article "Breeding Success of the Common Puffin on Different Habitats at Great Island, Newfoundland" (*Ecology Monographs* [1972]: 246–252), the variables considered were y = nesting frequency (burrows per 9 m^2), x_1 = grass cover (percentage), x_2 = mean soil depth (in centimeters), x_3 = angle of slope (in degrees), and x_4 = distance from cliff edge (in meters). The following data are from the article:

y	x_1	x_2	x_3	x_4
16	45	39.2	38	3
15	65	47.0	36	12
10	40	24.3	14	18
7	20	30.0	16	21
11	40	47.6	6	27
7	80	47.6	9	36
4	80	45.6	7	39
0	15	27.8	8	45
0	0	41.9	8	54
0	20	36.8	5	60
15	40	34.9	31	3
21	60	45.2	37	12
12	95	32.9	24	18
8	50	26.6	11	24
9	80	32.7	10	30
6	80	38.1	5	36
16	60	37.1	35	6
25	60	47.1	35	12
13	85	34.0	23	18
13	90	43.6	12	21
11	20	30.8	9	27
3	85	34.6	6	33
0	30	37.7	8	42
0	75	45.5	5	48
0	15	51.4	8	54
18	40	32.1	36	6
19	40	35.4	37	9
8	90	30.2	11	18
12	80	33.9	9	24
10	80	40.2	11	30
3	75	33.5	7	36
0	65	40.3	10	42
0	60	31.4	5	39
0	70	32.7	2	48
0	35	38.1	8	51
0	80	40.3	12	45
0	50	43.1	13	51
0	50	42.0	3	57

MINITAB output resulting from application of the backward elimination procedure is also given in what follows. Explain what action was taken at each step and why.

Step	1	2	3
Constant	12.29	11.45	13.96
X1	−0.019		
T-Ratio	−0.70		
X2	−0.043	−0.043	
T-Ratio	−1.42	−1.42	

X3	0.224	0.225	0.176
T-Ratio	2.91	2.94	2.54
X4	−0.182	−0.203	−0.293
T-Ratio	−2.12	−2.55	−6.01
S	2.83	2.81	2.85
R-SQ	86.20	85.99	85.16

14.61 *This exercise requires use of a computer package.* The accompanying $n = 25$ observations on $y =$ catch at intake (number of fish), $x_1 =$ water temperature (in °C), $x_2 =$ minimum tide height (in meters), $x_3 =$ number of pumps running, $x_4 =$ speed (in knots), and $x_5 =$ wind-range of direction (in degrees) constitute a subset of the data that appeared in the article "Multiple Regression Analysis for Forecasting Critical Fish Influxes at Power Station Intakes" (*Journal of Applied Ecology* [1983]: 33–42):

y	x_1	x_2	x_3	x_4	x_5
17	6.7	0.5	4	10	50
42	7.8	1.0	4	24	30
1	9.9	1.2	4	17	120
11	10.1	0.5	4	23	30
8	10.0	0.9	4	18	20
30	8.7	0.8	4	9	160
2	10.3	1.5	4	13	40
6	10.5	0.3	4	10	150
11	11.0	1.2	3	9	50

y	x_1	x_2	x_3	x_4	x_5
14	11.2	0.6	3	7	100
53	12.9	1.8	3	10	90
9	13.2	0.2	3	12	50
4	16.2	0.7	3	6	80
3	15.8	1.6	3	7	120
7	16.2	0.4	3	10	50
9	15.8	1.2	3	9	60
10	16.0	0.8	3	12	90
7	16.2	1.2	3	5	160
12	17.1	0.7	3	10	90
12	17.5	0.8	3	12	110
26	17.5	1.2	3	18	130
14	17.4	0.8	3	9	60
18	17.4	1.1	3	13	30
14	17.8	0.5	3	8	160
5	18.0	1.6	3	10	40

Use the variable selection procedures discussed in this section to formulate a model.

14.62 Suppose that $R^2 = .723$ for a model containing predictors $x_1, x_4, x_5,$ and x_8 and $R^2 = .689$ for a model with predictors x_1, x_3, x_5 and x_6.
a. What can you say about R^2 for the model containing predictors $x_1, x_3, x_4, x_5, x_6,$ and x_8? Explain.
b. What can you say about R^2 for the model containing predictors x_1 and x_4? Explain.

▪ 14.5 Communicating and Interpreting the Results of Statistical Analyses

Multiple regression is a powerful tool for analyzing multivariate data sets. Unfortunately, the results of such an analysis are often reported in a compact way, and it is often difficult to extract the relevant information from the summary.

▪ What to Look For in Published Data

Here are some things to consider when evaluating a study that has used data to fit a multiple regression model:

- What variables are being used as predictors? Are any of them categorical? If so, have they been incorporated in an appropriate way (using indicator variables)?

- Has a model utility test been performed? Does the conclusion of this test indicate that the reported model is useful?

- Are the values of R^2 and adjusted R^2 reported? Is there a large difference between these two values? If so, this may indicate that the number of predictors used in the model is excessive, given the number of observations in the data set.

▪ Has a variable selection procedure been used? If so, what procedure was used? Which variables were selected for inclusion in the final model?

The following example illustrates how the results of a multiple regression analysis might be described. Factors influencing teacher referrals to special education programs were examined in the paper "Stress, Biases, or Professionalism: What Drives Teachers' Referral Judgments of Students with Challenging Behaviors" (*Journal of Emotional and Behavioral Disorders* [2002]: 204–212). Teachers were asked to review a student record and to rate on a scale of 1 to 10 the likelihood that they would refer the student for a psycho-educational assessment (1 = very likely; 10 = very unlikely). The following variables were considered in a multiple regression analysis relating teacher likelihood of referral to various student characteristics:

y = teacher likelihood of referral

x_1 = a measure of off-task behavior

x_2 = a measure of problem behavior

x_3 = a measure of academic competence

The paper reported that "the final model included three predictor variables (i.e., off-task behavior, problem behavior, and academic competence) that jointly accounted for 51% of the variance in likelihood of student referral." Other variables, including demographic characteristics (such as gender, race, and age) of both students and teachers, were also considered as potential predictors of y = likelihood of referral, but they were eliminated from consideration using a variable selection procedure described in the paper.

▪ Activity 14.1: Exploring the Relationship Between Number of Predictors and Sample Size

This activity requires the use of a statistical computer package capable of fitting multiple regression models.

Background: The given data on y, x_1, x_2, x_3, and x_4 were generated using a computer package capable of producing random observations from any specified normal distribution. Because the data were generated at random, there is no reason to believe that y is related to any of the proposed predictor variables x_1, x_2, x_3, and x_4.

y	x_1	x_2	x_3	x_4
20.5	18.6	22.0	17.1	18.5
20.1	23.9	19.1	21.1	21.3
20.0	20.9	20.7	19.4	20.6
21.7	18.7	18.1	20.9	18.1
20.7	21.1	21.7	23.7	17.0

1. Construct four scatterplots — one of y versus each of x_1, x_2, x_3, and x_4. Do the scatterplots look the way you expected based on the way the data were generated? Explain.

2. Fit each of the following regression models:
i. y with x_1
ii. y with x_1 and x_2
iii. y with x_1 and x_2 and x_3
iv. y with x_1 and x_2 and x_3 and x_4

3. Make a table that gives the R^2, the adjusted R^2, and s_e values for each of the models fit in Step 2. Write a few sentences describing what happens to each of these three quantities as additional variables are added to the multiple regression model.

4. Given the manner in which these data were generated, what is the implication of what you observed in Step 3? What does this suggest about the relationship between number of predictors and sample size?

■ Summary of Key Concepts and Formulas

Term or Formula	Comment
Additive multiple regression model, $y = \alpha + \beta_1 x_1 + \beta_2 x_2 + \cdots + \beta_k x_k + e$	This equation specifies a general probabilistic relationship between y and k predictor variables $x_1, x_2, \ldots, x_k$, where $\beta_1, \ldots, \beta_k$ are population regression coefficients and $\alpha + \beta_1 x_1 + \beta_2 x_2 + \cdots + \beta_k x_k$ is the population regression function (the mean value of y for fixed values of $x_1, x_2, \ldots, x_k$).
Estimated regression function, $\hat{y} = a + b_1 x_1 + b_2 x_2 + \cdots + b_k x_k$	The estimates $a, b_1, \ldots, b_k$ of $\alpha, \beta_1, \ldots, \beta_k$ result from applying the principle of least squares.
Coefficient of multiple determination, $R^2 = 1 - \dfrac{\text{SSResid}}{\text{SSTo}}$	The proportion of observed y variation that can be explained by the model relationship, where SSResid is defined as it is in simple linear regression but is now based on $n - (k + 1)$ degrees of freedom.
Adjusted R^2	A downward adjustment of R^2 that depends on the number of predictors k relative to the sample size n.
F distribution	A type of probability distribution used in many different inferential procedures. A particular F distribution results from specifying both numerator degrees of freedom and denominator degrees of freedom.
$F = \dfrac{R^2/k}{(1 - R^2)/[n - (k + 1)]}$ or $F = \dfrac{\text{SSRegr}/k}{\text{SSResid}/[n - (k + 1)]}$	The test statistic for testing $H_0: \beta_1 = \beta_2 = \cdots = \beta_k = 0$, which asserts that there is no useful linear relationship between y and any of the model predictors. The F test is upper-tailed and is based on numerator degrees of freedom $df_1 = k$ and denominator degrees of freedom $df_2 = n - (k + 1)$.
$b_i \pm (t \text{ critical value}) s_{b_i}$	A confidence interval for the regression coefficient β_i (based on $n - (k + 1)$ df).
$t = \dfrac{b_i - \text{hypothesized value}}{s_{b_i}}$	The test statistic for testing hypotheses about β_i (based on $n - (k + 1)$ df). The most important case is $H_0: \beta_i = 0$, according to which the predictor is not useful as long as all other predictors remain in the model.
$\hat{y} \pm (t \text{ critical value}) s_{\hat{y}}$ $\hat{y} \pm (t \text{ critical value}) \sqrt{s_e^2 + s_{\hat{y}}^2}$	A confidence interval for a mean y value and a prediction interval for a single y value when $x_1, x_2, \ldots, x_k$ have specified values, where $\hat{y} = a + b_1 x_1 + b_2 x_2 + \cdots + b_k x_k$ and $s_{\hat{y}}$ denotes the estimated standard deviation of $\hat{y}$.
Model selection	An investigator may have data on many predictors that could be included in a model. There are two different approaches to choosing a model: (1) Use an *all-subsets procedure* to identify the best models of each different size (one-predictor models, two-predictor models, etc.) and then compare these according to criteria such as R^2 and adjusted R^2; and (2) use an *automatic selection procedure*, which either successively eliminates predictors until a stopping point is reached (backward elimination) or starts with no predictors and adds them successively until the inclusion of additional predictors cannot be justified.

▪ Supplementary Exercises 14.63–14.73

14.63 What factors affect the quality of education in our schools? The article "How Much Is Enough? Applying Regression to a School Finance Case" (in *Statistics and the Law* [New York: Wiley, 1986]: 257–287) used data from $n = 88$ school districts to carry out a regression analysis based on the following variables: y = average language score for fourth-grade students; the background variables x_1 = occupational index (percentages of managerial and professional workers in the community) and x_2 = median income; the peer group variables x_3 = Title I enrollment (percentage) and x_4 = Hispanic enrollment ($= 1$ if $>5\%$ and 0 otherwise), x_5 = logarithm of fourth-grade enrollment, x_6 = prior test score (from 4 years previous to year under study); and the school variables x_7 = administrator–teacher ratio, x_8 = pupil–teacher ratio, x_9 = certified staff–pupil ratio, and x_{10}, x_{11} = indicator variables for average teaching experience ($x_{10} = 1$ if less than 3 years and 0 otherwise; $x_{11} = 1$ if more than 6 years and 0 otherwise). The following table gives estimated coefficients, values of t ratios $\left(\dfrac{b_1}{s_{b_1}}, \dfrac{b_2}{s_{b_2}}, \text{and so on}\right)$, and other relevant information:

Variable	Estimated Coefficient	t Ratio
Constant	70.67	6.08
x_1	0.458	3.08
x_2	-0.000141	-0.59
x_3	-58.47	-4.02
x_4	-3.668	-1.73
x_5	-2.942	-2.34
x_6	0.200	3.60
x_7	17.93	0.42
x_8	0.689	0.42
x_9	-0.403	-0.27
x_{10}	-0.614	-1.30
x_{11}	-0.382	-0.58
$s_e = 5.57$	$R^2 = .64$	

a. Is there a useful linear relationship between y and at least one of the predictors? Test the relevant hypotheses using a .01 significance level.
b. What is the value of the adjusted R^2?
c. Calculate and interpret a 95% confidence interval for β_1.
d. Does it appear that one or more predictors could be eliminated from the model without discarding useful information? If so, which one would you eliminate first? Explain your reasoning.
e. Suppose that you wanted to decide whether any of the peer group variables provided useful information about y (provided that all other predictors remained in the model). What hypotheses would you test (a single H_0 and H_a)? Could one of the test procedures presented in this chapter be used? Explain.

14.64 The article "Innovation, R&D Productivity, and Global Market Share in the Pharmaceutical Industry" (*Review of Industrial Organization* [1995]: 197–207) presented the results of a multiple regression analysis in which the dependent variable was y = the number of patented R&D compounds per R&D employee. The number of observations was $n = 78$, the model contained six predictors, and SSResid = 1.02 and $R^2 = .50$.
a. Is the chosen model useful? Test the relevant hypotheses.
b. What is the value of the adjusted R^2?
c. The predictor x_6 was an indicator variable with a value of 1 if a firm was headquartered outside the United States. The article reported that $b_6 = 0.001$ and that the corresponding t ratio was $-.92$. Provided that the other five predictors remain in the model, is x_6 useful? Test the relevant hypotheses.

14.65 The following data on y = glucose concentration (in grams per liter) and x = fermentation time (in days) for a particular blend of malt liquor was read from a scatterplot in the article "Improving Fermentation Productivity with Reverse Osmosis" (*Food Technology* [1984]: 92–96):

x	1	2	3	4	5	6	7	8
y	74	54	52	51	52	53	58	71

a. Construct a scatterplot for these data. Based on the scatterplot, what type of model would you suggest?
b. MINITAB output resulting from fitting a multiple regression model with $x_1 = x$ and $x_2 = x^2$ is shown here. Does this quadratic model specify a useful relationship between y and x?

```
The regression equation is
Y = 84.5 − 15.9X1 + 1.77X2
Predictor    Coef      Stdev    t-ratio
Constant     84.482    4.904    17.23
X1          −15.875    2.500    −6.35
X2            1.7679   0.2712    6.52
s = 3.515   R-sq = 89.5%   R-sq(adj) = 85.3%
```

c. Could the quadratic term have been eliminated? That is, would a simple linear model have sufficed? Test using a .05 significance level.

14.66 Much interest in management circles has recently focused on how employee compensation is related to various company characteristics. The article "Determinants of R and D Compensation Strategies" (*Personnel Psychology* [1984]: 635–650) proposed a quantitative scale for y = base salary for employees of high-tech companies. The following estimated multiple regression equation was then presented:

$$\hat{y} = 2.60 + 0.125x_1 + 0.893x_2 + 0.057x_3 - 0.014x_4$$

where x_1 = sales volume (in millions of dollars), x_2 = stage in product life cycles (1 = growth, 0 = mature), x_3 = profitability (percentage), and x_4 = attrition rate (percentage).

a. There were $n = 33$ firms in the sample and $R^2 = .69$. Is the fitted model useful?

b. Predict base compensation for a growth-stage firm with sales volume $50 million, profitability 8%, and attrition rate 12%.

c. The estimated standard deviations for the coefficient estimates were 0.064, 0.141, 0.014, and 0.005 for b_1, b_2, b_3, and b_4, respectively. Should any of the predictors be deleted from the model? Explain.

d. β_2 is the difference between average base compensation for growth-stage and mature-stage firms when all other predictors are held fixed. Use the information in Part (c) to calculate a 95% confidence interval for β_2.

14.67 If $n = 21$ and $k = 10$, for what values of R^2 would the adjusted R^2 be negative?

14.68 A study of total body electrical conductivity was described in the article "Measurement of Total Body Electrical Conductivity: A New Method for Estimation of Body Composition" (*American Journal of Clinical Nutrition* [1983]: 735–739). Nineteen observations were given for the variables y = total body electrical conductivity, x_1 = age (in years), x_2 = sex (0 = male, 1 = female), x_3 = body mass (in kilograms per square meter), x_4 = body fat (in kilograms), and x_5 = lean body mass (in kilograms).

a. The backward elimination method of variable selection was used, and the resulting MINITAB output is given here. Explain what occurred at each step in the process.

Stepwise regression of Y on 5
predictors, with N = 19

Step	1	2	3
Constant	−6.193	−13.285	−15.175
X1	0.31	0.36	0.38
T-Ratio	1.90	2.36	2.96
X2	−7.9	−7.5	−7.0
T-Ratio	−1.93	−1.89	−2.04
X3	−0.43		
T-Ratio	−0.72		
X4	0.22	0.03	
T-Ratio	0.78	0.29	
X5	0.365	0.339	0.378
T-Ratio	2.16	2.09	4.18
S	5.16	5.07	4.91
R-SQ	79.53	78.70	78.57

b. MINITAB output for the multiple regression model relating y to x_1, x_2, and x_5 is shown here. Interpret the values of R^2 and s_e.

The regression equation is
Y = −15.2 + 0.377X1 − 6.99X2 + 0.378X5

Predictor	Coef	Stdev	t-ratio
Constant	−15.175	9.620	−1.58
X1	0.3771	0.1273	2.96
X2	−6.988	3.425	−2.04
X5	0.37779	0.09047	4.18

s = 4.914 R-sq = 78.6% R-sq(adj) = 74.3%

c. Interpret the value of b_2 in the estimated regression equation.

d. If the estimated standard deviation of the statistic $a + b_1(31) + b_2(1) + b_5(52.7)$ is 1.42, give a 95% confidence interval for the true mean total body electrical conductivity of all 31-year-old females whose lean body mass is 52.7 kg.

14.69 The article "Creep and Fatigue Characteristics of Ferrocement Slabs" (*Journal of Ferrocement* [1984]: 309–322) reported data on y = tensile strength (in megapascals), x_1 = slab thickness (in centimeters), x_2 = load (in kilograms), x_3 = age at loading (in days), and x_4 = time under test (in days) resulting from stress tests of $n = 9$ reinforced concrete slabs. The backward elimination method of variable selection was applied. Partial MINITAB output follows.

Step	1	2	3
Constant	8.496	12.670	12.989
X1	−0.29	−0.42	−0.49
T-Ratio	−1.33	−2.89	−3.14
X2	0.0104	0.0110	0.0116
T-Ratio	6.30	7.40	7.33
X3	0.0059		
T-Ratio	0.83		
X4	−0.023	−0.023	
T-Ratio	−1.48	−1.53	
S	0.533	0.516	0.570
R-SQ	95.81	95.10	92.82

The regression equation is
Y = 13.0 − 0.487X1 + 0.116X2

Predictor	Coef	Stdev	T-ratio
Constant	12.989	3.640	3.57
X1	−0.4867	0.1549	−3.14
X2	0.011569	0.001579	7.33

s = 0.5698 R-sq = 92.8% R-sq(adj) = 90.4%

Explain what action was taken at each step in the process. Use the estimated regression equation for the selected model to predict tensile strength for a slab that is 25 cm thick, 150 days old, and subjected to a load of 200 kg for 50 days.

14.70 A study of pregnant grey seals involved $n = 25$ observations on the variables y = fetus progesterone level (in milligrams), x_1 = fetus sex (0 = male, 1 = female), x_2 = fetus length (in centimeters), and x_3 = fetus weight (in grams). MINITAB output for the model using all three independent variables is given ("Gonadotrophin and Progesterone Concentration in Placenta of Grey Seals," *Journal of Reproduction and Fertility* [1984]: 521–528):

```
The regression equation is
Y = -1.98 - 1.87X1 + .234X2 + .0001X3
Predictor    Coef       Stdev     t-ratio
Constant    -1.982      4.290     -0.46
X1          -1.871      1.709     -1.09
X2           0.2340     0.1906     1.23
X3           0.000060   0.002020   0.03
s = 4.189   R-sq = 55.2%   R-sq(adj) = 48.8%
Analysis of Variance
Source       DF    SS       MS      F      p
Regression    3   454.63   151.54  8.63   .001
Error        21   368.51    17.55
Total        24   823.15
```

a. Use information from the MINITAB output to test the hypothesis H_0: $\beta_1 = \beta_2 = \beta_3 = 0$.
b. Using an elimination criterion of $-2 \le t$ ratio ≤ 2, should any variable be eliminated? If so, which one?
c. MINITAB output for the regression using only x_1 = sex and x_2 = length is given here:

```
The regression equation is
Y = -2.09 - 1.87X1 + 0.240X2
Predictor    Coef      Stdev     t-ratio
Constant    -2.090     2.212     -0.94
X1          -1.865     1.661     -1.12
X2           0.23952   0.04604    5.20
s = 4.093   R-sq = 55.2%   R-sq(adj) = 51.2%
```

Would you recommend keeping both x_1 and x_2 in the model? Explain.
d. After elimination of both x_3 and x_1, the estimated regression equation is $\hat{y} = -2.61 + 0.231x_2$. The corresponding values of R^2 and s_e are .527 and 4.116, respectively. Interpret these values.
e. Referring to Part (d), how would you interpret the value of $b_2 = 0.231$? Does it make sense to interpret the value of a as the estimate of average progesterone level when length is 0? Explain.

14.71 The authors of the article "Influence of Temperature and Salinity on Development of White Perch Eggs" (*Transactions of the American Fisheries Society* [1982]: 396–398) used a quadratic regression model to describe the relationship between y = percentage hatch of white perch eggs and x = water temperature (in °C). The estimated regression equation was given as $\hat{y} = -41.0 + 14.6x - 0.55x^2$.

a. Graph the curve corresponding to this equation by plotting the (x, y) pairs using x values of 10, 12, 14, 16, 18, 20, 22, and 24 (these are the temperatures used in the article). Draw a smooth curve through the points in your plot.
b. The authors make the claim that the optimal temperature for hatching is 14°C. Based on your graph in Part (a), does this statement seem reasonable? Explain.

14.72 A sample of $n = 20$ companies was selected, and the values of y = stock price and $k = 15$ predictor variables (such as quarterly dividend, previous year's earnings, and debt ratio) were determined. When the multiple regression model with these 15 predictors was fit to the data, $R^2 = .90$ resulted.
a. Does the model appear to specify a useful relationship between y and the predictor variables? Carry out a test at a significance level of .05. (The value 5.86 captures upper-tail area .05 under the F curve with 15 numerator degrees of freedom and 4 denominator degrees of freedom.)
b. Based on the result of Part (a), does a high R^2 value by itself imply that a model is useful? Under what circumstances might you be suspicious of a model with a high R^2 value?
c. With n and k as given here, how large would R^2 have to be for the model to be judged useful at the .05 level of significance?

14.73 *This exercise requires the use of a computer package.* The article "Entry and Profitability in a Rate-Free Savings and Loan Market" (*Quarterly Review of Economics and Business* [1978]: 87–95) gave the following data on y = profit margin of savings and loan companies in a given year, x_1 = net revenues in that year, and x_2 = number of savings and loan branch offices:

x_1	x_2	y	x_1	x_2	y
3.92	7298	.75	3.78	6672	.84
3.61	6855	.71	3.82	6890	.79
3.32	6636	.66	3.97	7115	.70
3.07	6506	.61	4.07	7327	.68
3.06	6450	.70	4.25	7546	.72
3.11	6402	.72	4.41	7931	.55
3.21	6368	.77	4.49	8097	.63
3.26	6340	.74	4.70	8468	.56
3.42	6349	.90	4.58	8717	.41
3.42	6352	.82	4.69	8991	.51
3.45	6361	.75	4.71	9179	.47
3.58	6369	.77	4.78	9318	.32
3.66	6546	.78			

a. Fit a multiple regression model using both independent variables.

b. Use the *F* test to determine whether the model provides useful information for predicting profit margin.

c. Interpret the values of R^2 and s_e.

d. Would a regression model using a single independent variable (x_1 alone or x_2 alone) have sufficed? Explain.

e. Plot the (x_1, x_2) pairs. Does the plot indicate any sample observation that may have been highly influential in estimating the model coefficients? Explain. Do you see any evidence of multicollinearity? Explain.

■ References

Additional references are listed at the end of Chapter 5.

Kerlinger, Fred N., and Elazar J. Pedhazar, *Multiple Regression in Behavioral Research*. Austin, Texas: Holt, Rinehart & Winston, 1973. (A readable introduction to multiple regression.)

15 · Analysis of Variance

In Chapter 11 we discussed methods for testing $H_0: \mu_1 - \mu_2 = 0$ (i.e., $\mu_1 = \mu_2$), where μ_1 and μ_2 are the means of two different populations or the true mean responses when two different treatments are applied. Many investigations involve a comparison of more than two population or treatment means. For example, an investigation carried out to study purchasers of luxury automobiles reported data on a number of different attributes that might affect purchase decisions, including comfort, safety, styling, durability, and reliability ("Measuring Values Can Sharpen Segmentation in the Luxury Car Market," *Journal of Advertising Research* [1995]: 9–22). Here is summary information on the level of importance of speed, rated on a seven-point scale:

Type of Car	American	German	Japanese
Sample size	58	38	59
Sample mean rating	3.05	2.87	2.67

Let μ_1, μ_2, and μ_3 denote the true average (i.e., population mean) rating on this attribute for owners of American, German, and Japanese luxury cars, respectively. Do the data support the claim that $\mu_1 = \mu_2 = \mu_3$, or does it appear that at least two of the μ's are different from one another? This is an example of a **single-factor analysis of variance (ANOVA)** problem, in which the objective is to decide whether the means for more than two populations or treatments are identical. The first two sections of this chapter discuss various aspects of single-factor ANOVA. In Sections 15.3 and 15.4, we consider more complex ANOVA situations and methods.

■ 15.1 Single-Factor ANOVA and the *F* Test

When two or more populations or treatments are being compared, the characteristic that distinguishes the populations or treatments from one another is called the **factor** under investigation. For example, an experiment might be carried out to compare three different methods for teaching reading (three different treatments), in which case the factor of interest would be *teaching method*, a qualitative factor. If the growth of fish raised in waters having different salinity levels — 0%, 10%, 20%, and 30% — is of interest, the factor *salinity level* is quantitative.

A **single-factor analysis of variance (ANOVA)** problem involves a comparison of k population or treatment means $\mu_1, \mu_2, \ldots, \mu_k$. The objective is to test

$$H_0: \quad \mu_1 = \mu_2 = \cdots = \mu_k$$

against

$$H_a: \quad \text{At least two of the } \mu\text{'s are different.}$$

The analysis is based on k independently selected samples, one from each population or for each treatment. For populations a random sample from each population is selected independently of the random sample from any other population. When comparing treatments, the experimental units (subjects or objects) that receive any particular treatment are chosen at random from all the units available for the experiment. A comparison of treatments based on independently selected experimental units is often referred to as a **completely randomized design**.

Whether the null hypothesis of a single-factor ANOVA should be rejected depends on how substantially the samples from the different populations or treatments differ from one another. Figure 15.1 displays two possible data sets that might result when observations are selected from each of three populations under study. Each data set consists of five observations from the first population, four observations from the second population, and six observations from the third population. For both data sets, the three sample means are located by arrows. The means of the two samples from Population 1 are identical, and a similar statement holds for the two samples from Population 2 and those from Population 3.

Figure 15.1 Two possible ANOVA data sets when three populations are under investigation: green circle = observation from Population 1; orange circle = observation from Population 2; blue circle = observation from Population 3.

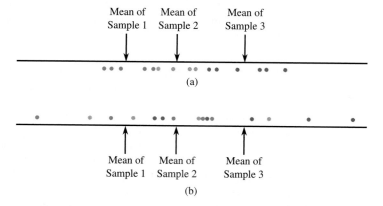

After looking at the data set in Figure 15.1(a), almost anyone would readily agree that the claim $\mu_1 = \mu_2 = \mu_3$ appears to be false. Not only are the three sample means different, but also the three samples are clearly separated. In other words, differences between the three sample means are quite large relative to the variability within each sample. (If all data sets gave such obvious messages, statisticians would not be in such great demand!)

The situation pictured in Figure 15.1(b) is much less clear-cut. The sample means are as different as they were in the first data set, but now there is considerable overlap among the three samples. The separation between sample means here can plausibly be attributed to substantial variability in the populations (and therefore the samples) rather than to differences between μ_1, μ_2, and μ_3. The phrase *analysis of variance* comes from the idea of analyzing variability in the data to see how much can be attributed to differences in the μ's and how much is due to variability in the individual populations. In Figure 15.1(a), there is little within-sample variability relative to the amount of between-sample variability, whereas in Figure 15.1(b), a great deal more of the total variability is due to variation within each sample. If differences between the sample means can be explained by within-sample variability, there is no compelling reason to reject H_0.

▪ Notation and Assumptions

Notation in single-factor ANOVA is a natural extension of the notation used in Chapter 11 for comparing two population or treatment means.

▪ ANOVA Notation

k = number of populations or treatments being compared.

Population or treatment	1	2	...	k
Population or treatment mean	μ_1	μ_2	...	μ_k
Population or treatment variance	σ_1^2	σ_2^2	...	σ_k^2
Sample size	n_1	n_2	...	n_k
Sample mean	$\bar{x}_1$	$\bar{x}_2$	...	$\bar{x}_k$
Sample variance	s_1^2	s_2^2	...	s_k^2

The total number of observations in the data set is $N = n_1 + n_2 + \cdots + n_k$.

The sum of all N observations — the grand total — is denoted by $T = n_1\bar{x}_1 + n_2\bar{x}_2 + \cdots + n_k\bar{x}_k$.

The grand mean is $\bar{\bar{x}} = \dfrac{T}{N}$.

A decision between H_0 and H_a is based on examining the $\bar{x}$ values to see whether observed discrepancies are small enough to be attributable simply to sampling variability or whether an alternative explanation for the differences is necessary.

▪ Example 15.1 Strength of Cardboard

The article "Compression of Single-Wall Corrugated Shipping Containers Using Fixed and Floating Text Platens" (*Journal of Testing and Evaluation* [1992]: 318–320) described an experiment in which several different types of boxes were compared with respect to compression strength (in pounds). Table 15.1 presents the results of a single-factor experiment involving $k = 4$ types of boxes (the sample means and standard deviations are in close agreement with values given in the paper).

With μ_i denoting the true average compression strength for boxes of type i ($i = 1, 2, 3, 4$), the null hypothesis is $H_0: \mu_1 = \mu_2 = \mu_3 = \mu_4$. Figure 15.2(a) shows a comparative boxplot for the four samples. The boxplots for the first three types of boxes show a substantial amount of overlap, but compression strengths for the fourth type appear considerably smaller than those for the other types. This suggests that H_0 is not true. If the comparative boxplot looked instead like the plot in Figure 15.2(b), it would not be so obvious whether H_0 was true or false. In situations such as this, we need a formal test procedure.

Table 15.1 ▪ Data and Summary Quantities for Example 15.1

Type of Box	Compression Strength (lb)						Sample Mean	Sample SD
1	655.5	788.3	734.3	721.4	679.1	699.4	713.00	46.55
2	789.2	772.5	786.9	686.1	732.1	774.8	756.93	40.34
3	737.1	639.0	696.3	671.7	717.2	727.1	698.07	37.20
4	535.1	628.7	542.4	559.0	586.9	520.0	562.02	39.87

$$\overline{\overline{x}} = 682.50$$

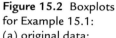
Figure 15.2 Boxplots for Example 15.1: (a) original data; (b) altered data.

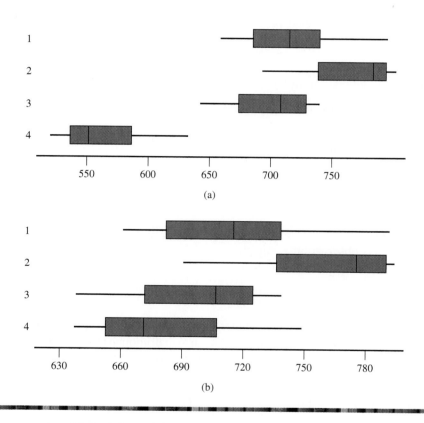

As with the inferential methods of Chapters 10–12, the validity of the ANOVA test for $H_0: \mu_1 = \mu_2 = \cdots = \mu_k$ requires some assumptions.

▪ **Assumptions for ANOVA**

1. Each of the k population or treatment response distributions is normal.

2. $\sigma_1 = \sigma_2 = \cdots = \sigma_k$ (i.e., the k normal distributions have identical standard deviations).

3. The observations in the sample from any particular one of the k populations or treatments are independent of one another.

4. When comparing population means, k random samples are selected independently of one another. When comparing treatment means, treatments are assigned at random to subjects or objects (or subjects are assigned at random to treatments).

In practice, the test based on these assumptions works well as long as the assumptions are not too badly violated. If the sample sizes are reasonably large, normal probability plots or boxplots of the data in each sample are helpful in checking the assumption of normality. Often, however, sample sizes are so small that a separate normal probability plot for each sample is of little value in checking normality. A single combined plot results from first subtracting $\bar{x}_1$ from each observation in the first sample, $\bar{x}_2$ from each value in the second sample, and so on and then constructing a normal probability plot of all N deviations from their respective means. The plot should be reasonably straight. Figure 15.3 shows such a plot for the data of Example 15.1.

Figure 15.3 A normal probability plot based on the data of Example 15.1.

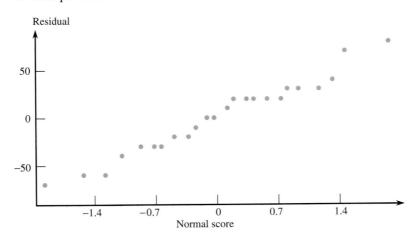

There is a formal procedure for testing the equality of population standard deviations. Unfortunately, it is quite sensitive to even a small departure from the normality assumption, so we do not recommend its use. Instead, we suggest that the ANOVA *F* test (to be described subsequently) can safely be used if the largest of the sample standard deviations is at most twice the smallest one. The largest standard deviation in Example 15.1 is $s_1 = 46.55$, which is only about 1.25 times the smallest standard deviation ($s_3 = 37.20$). The book *Beyond ANOVA: The Basics of Applied Statistics* (see the references at the end of the chapter) is a good source for alternative methods of analysis if there appears to be a violation of assumptions. The test procedure is based on the following measures of variation in the data.

■ **Definition**

A measure of disparity among the sample means is the **treatment sum of squares**, denoted by **SSTr** and given by

$$\text{SSTr} = n_1(\bar{x}_1 - \bar{\bar{x}})^2 + n_2(\bar{x}_2 - \bar{\bar{x}})^2 + \cdots + n_k(\bar{x}_k - \bar{\bar{x}})^2$$

A measure of variation within the k samples, called the **error sum of squares** and denoted by **SSE**, is

$$\text{SSE} = (n_1 - 1)s_1^2 + (n_2 - 1)s_2^2 + \cdots + (n_k - 1)s_k^2$$

Each sum of squares has an associated number of degrees of freedom:

   treatment df $= k - 1$

   error df $= N - k$

(continued)

A **mean square** is a sum of squares divided by its number of degrees of freedom. In particular,

$$\text{mean square for treatments} = \text{MSTr} = \frac{\text{SSTr}}{k-1}$$

$$\text{mean square for error} = \text{MSE} = \frac{\text{SSE}}{N-k}$$

The number of error degrees of freedom comes from adding the number of degrees of freedom associated with each of the sample variances:

$$(n_1 - 1) + (n_2 - 1) + \cdots + (n_k - 1) = n_1 + n_2 + \cdots + n_k - 1 - 1 - \cdots - 1 = N - k$$

Example 15.2   Cardboard Calculations

Let's return to the compression strength data of Example 15.1. With $\bar{x}_1 = 713.00$, $\bar{x}_2 = 756.93$, $\bar{x}_3 = 698.07$, $\bar{x}_4 = 562.02$, $\bar{\bar{x}} = 682.50$, and $n_1 = n_2 = n_3 = n_4 = 6$, we can calculate

$$\begin{aligned} \text{SSTr} &= n_1(\bar{x}_1 - \bar{\bar{x}})^2 + n_2(\bar{x}_2 - \bar{\bar{x}})^2 + \cdots + n_k(\bar{x}_k - \bar{\bar{x}})^2 \\ &= 6(713.00 - 682.50)^2 + 6(756.93 - 682.50)^2 + 6(698.07 - 682.50)^2 \\ &\quad + 6(562.02 - 682.50)^2 \\ &= 5581.50 + 33{,}238.95 + 1454.55 + 87{,}092.58 \\ &= 127{,}367.58 \end{aligned}$$

Because $s_1 = 46.55$, $s_2 = 40.34$, $s_3 = 37.20$, and $s_4 = 39.87$, we can compute

$$\begin{aligned} \text{SSE} &= (n_1 - 1)s_1^2 + (n_2 - 1)s_2^2 + \cdots + (n_k - 1)s_k^2 \\ &= (6 - 1)(46.55)^2 + (6 - 1)(40.34)^2 + (6 - 1)(37.20)^2 + (6 - 1)(39.87)^2 \\ &= 33{,}838.375 \end{aligned}$$

The number of degrees of freedom are

treatment df $= k - 1 = 3$

error df $= N - k = 6 + 6 + 6 + 6 - 4 = 20$

from which

$$\text{MSTr} = \frac{\text{SSTr}}{k-1} = \frac{127{,}367.58}{3} = 42{,}455.86$$

$$\text{MSE} = \frac{\text{SSE}}{N-k} = \frac{33{,}838.375}{20} = 1691.92$$

Both MSTr and MSE are quantities whose values can be calculated once sample data are available; that is, they are statistics. Each of these statistics varies in value from data set to data set. For example, repeating the experiment of Example 15.1 with the same sample sizes might yield

$\bar{x}_1 = 703.24$	$\bar{x}_2 = 769.15$	$\bar{x}_3 = 678.83$	$\bar{x}_4 = 575.50$
$s_1 = 42.69$	$s_2 = 41.35$	$s_3 = 35.02$	$s_4 = 41.20$

from which

$$\text{MSTr} = 38{,}796.30 \quad \text{(previously } 42{,}455.86\text{)}$$
$$\text{MSE} = 1614.02 \quad \text{(previously } 1691.92\text{)}$$

Both statistics MSTr and MSE have sampling distributions, and these sampling distributions have mean values. The following box describes the key relationship between MSTr and MSE and the mean values of these two statistics.

When H_0 is true ($\mu_1 = \mu_2 = \cdots = \mu_k$),

$$\mu_{\text{MSTr}} = \mu_{\text{MSE}}$$

However, when H_0 is false,

$$\mu_{\text{MSTr}} > \mu_{\text{MSE}}$$

and the greater the differences among the μ's, the larger μ_{MSTr} will be relative to μ_{MSE}.

According to this result, when H_0 is true, we expect the two mean squares to be close to one another, whereas we expect MSTr to substantially exceed MSE when some μ's differ greatly from others. Thus, a calculated MSTr that is much larger than MSE casts doubt on H_0. In Example 15.2, MSTr = 42,455.86 and MSE = 1691.92, so MSTr is about 25 times as large as MSE. Can this difference be attributed solely to sampling variability, or is the ratio $\dfrac{\text{MSTr}}{\text{MSE}}$ of sufficient magnitude to suggest that H_0 is false? Before we can describe a formal test procedure, it is necessary to revisit *F* distributions, first introduced in multiple regression analysis (Chapter 14).

Many ANOVA test procedures are based on probability distributions called *F* distributions. As explained in Chapter 14, an *F* distribution always arises in connection with a ratio. A particular *F* distribution is obtained by specifying both numerator degrees of freedom (df_1) and denominator degrees of freedom (df_2). Figure 15.4 shows an *F* curve for a particular choice of df_1 and df_2. All *F* tests in this book are upper-tailed, so the *P*-value is the area under the *F* curve to the right of the calculated *F*.

Tabulation of these upper-tail areas is cumbersome, because there are two degrees of freedom rather than just one (as in the case of *t* distributions). For selected (df_1, df_2) pairs, our *F* table (Appendix Table 7) gives only the four numbers that capture tail areas .10, .05, .01, and .001, respectively. Here are the four numbers for $\text{df}_1 = 4$, $\text{df}_2 = 10$ along with the statements that can be made about the *P*-value:

Figure 15.4 An *F* curve and *P*-value for an upper-tailed test.

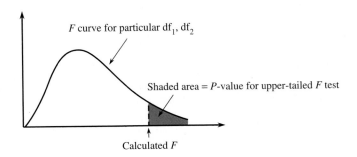

F curve for particular df_1, df_2

Shaded area = *P*-value for upper-tailed *F* test

Calculated *F*

Tail area   .10   .05   .01   .001
Value   2.61   3.48   5.99   11.28
              ↑   ↑   ↑   ↑   ↑
              a   b   c   d   e

a. $F < 2.61 \Rightarrow P\text{-value} > .10$.

b. $2.61 < F < 3.48 \Rightarrow .05 < P\text{-value} < .10$.

c. $3.48 < F < 5.99 \Rightarrow .01 < P\text{-value} < .05$.

d. $5.99 < F < 11.28 \Rightarrow .001 < P\text{-value} < .01$.

e. $F > 11.28 \Rightarrow P\text{-value} < .001$.

Thus, if $F = 7.12$, then $.001 < P\text{-value} < .01$. If a test with $\alpha = .05$ is used, H_0 should be rejected, because $P\text{-value} \leq \alpha$. The most frequently used statistical computer packages can provide exact P-values for F tests.

■ The Single-Factor ANOVA F Test

Null hypothesis: $H_0: \mu_1 = \mu_2 = \cdots = \mu_k$

Test statistic: $F = \dfrac{\text{MSTr}}{\text{MSE}}$

When H_0 is true and the ANOVA assumptions are reasonable, F has an F distribution with $df_1 = k - 1$ and $df_2 = N - k$.

Values of F more contradictory to H_0 than what was calculated are values even farther out in the upper tail, so the P-value is the area captured in the upper tail of the corresponding F curve.

■ Example 15.3   Cardboard Strength Revisited

The two mean squares for the compression strength data given in Example 15.1 were calculated in Example 15.2 as

$MSTr = 42{,}455.86$

$MSE = 1691.92$

The value of the F statistic is then

$$F = \frac{\text{MSTr}}{\text{MSE}} = \frac{42{,}455.86}{1691.92} = 25.09$$

With $df_1 = k - 1 = 3$ and $df_2 = N - k = 24 - 4 = 20$, Appendix Table 7 shows that 8.10 captures tail area .001. Because $25.09 > 8.10$, it follows that $P\text{-value} = \text{captured tail area} < .001$. The P-value is smaller than any reasonable α, so there is compelling evidence for rejecting $H_0: \mu_1 = \mu_2 = \cdots = \mu_k$. True average compression strength appears to depend in some way on the type of box.

■ Example 15.4   Musical Preferences and Reckless Behavior

The article "The Soundtrack of Recklessness: Musical Preferences and Reckless Behavior Among Adolescents" (*Journal of Adolescent Research* [1992]: 313–331) described a study whose purpose was to determine whether adolescents who preferred certain types of music reported higher rates of reckless behaviors, such as speeding, drug use, shoplifting, and unprotected sex. Independently chosen random samples were selected from each of four groups of students with different musical preferences at a large high school: (1) acoustic/pop, (2) mainstream rock, (3) hard rock, and (4) heavy metal. Each student in these samples was asked how many times he or she had engaged in various reckless activities during the last year. The following table lists data and summary quantities on driving over 80 mph that is consistent with summary quantities given in the article (the sample sizes in the article were much larger, but for the purposes of this example, we use $n_1 = n_2 = n_3 = n_4 = 20$):

Number of Times Driving over 80 mph

	Acoustic/ Pop	Mainstream Rock	Hard Rock	Heavy Metal
		Musical Preference		
	2	3	3	4
	3	2	4	3
	4	1	3	4
	1	2	1	3
	3	3	2	3
	3	4	1	3
	3	3	4	3
	3	2	2	3
	2	4	2	2
	2	4	2	4
	1	4	3	4
	3	4	3	5
	2	2	4	4
	2	3	3	5
	2	2	3	3
	3	2	2	4
	2	2	3	5
	2	3	4	4
	3	1	2	2
	4	3	4	3
n	20	20	20	20
$\bar{x}$	2.50	2.70	2.75	3.55
s	0.827	0.979	0.967	0.887
s^2	0.6839	0.9584	0.9351	0.7868

Also, $N = 80$, $T = $ grand total $= 230.0$, and $\bar{\bar{x}} = \dfrac{230.0}{80} = 2.875$.

Let's carry out an F test to see whether true mean number of times driving over 80 mph varies with musical preference.

1. Let μ_1, μ_2, μ_3, and μ_4 denote the true mean number of times driving over 80 mph for individuals preferring acoustic/pop, mainstream rock, hard rock, and heavy metal, respectively.

2. $H_0: \mu_1 = \mu_2 = \mu_3 = \mu_4$.

3. H_a: At least two among μ_1, μ_2, μ_3, and μ_4 are different.

4. Significance level: $\alpha = .01$.

5. Test statistic: $F = \dfrac{\text{MSTr}}{\text{MSE}}$

6. Assumptions: Figure 15.5 shows boxplots for the four samples. The boxplots are roughly symmetric, and there are no outliers. The largest standard deviation ($s_2 = 0.979$) is not more than twice as big as the smallest standard deviation ($s_1 = 0.827$). The data consist of independently chosen random samples. The assumptions of ANOVA are reasonable.

Figure 15.5 Boxplots for the data of Example 15.4.

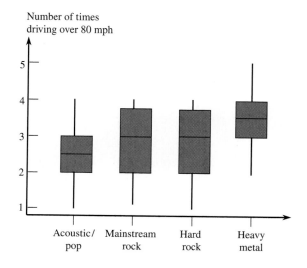

7. Computation:

$$\text{SSTr} = n_1(\bar{x}_1 - \bar{\bar{x}})^2 + n_2(\bar{x}_2 - \bar{\bar{x}})^2 + \cdots + n_k(\bar{x}_k - \bar{\bar{x}})^2$$
$$= 20(2.50 - 2.875)^2 + 20(2.70 - 2.875)^2 + 20(2.75 - 2.875)^2$$
$$+ 20(3.55 - 2.875)^2$$
$$= 12.850$$

treatment df $= k - 1 = 3$

$$\text{SSE} = (n_1 - 1)s_1^2 + (n_2 - 1)s_2^2 + \cdots + (n_k - 1)s_k^2$$
$$= 19(0.6839) + 19(0.9584) + 19(0.9351) + 19(0.7868)$$
$$= 63.9020$$

error df $= N - k = 80 - 4 = 76$

Thus,

$$F = \frac{\text{MSTr}}{\text{MSE}} = \frac{\text{SSTr/df}}{\text{SSE/df}} = \frac{12.850/3}{63.9020/76} = \frac{4.283}{0.841} = 5.09$$

8. *P*-value: Appendix Table 7 shows that for $df_1 = 3$ and $df_2 = 90$ (the closest tabulated degrees of freedom to df $= 76$), the value 4.01 captures upper-tail area .01 and 5.91 captures upper-tail area .001. Because $4.01 < F = 5.09 < 5.91$, it follows that $.001 < P\text{-value} < .01$.

9. Conclusion: Because *P*-value $\leq \alpha$, we reject H_0. The mean number of times driving over 80 mph does not appear to be the same for all four musical preference groups.

■ Summarizing an ANOVA

ANOVA calculations are often summarized in a tabular format called an ANOVA table. To understand such a table, we must define one more sum of squares.

Total sum of squares, denoted by **SSTo**, is given by

$$SSTo = \sum_{\text{all } N \text{ obs.}} (x - \bar{\bar{x}})^2$$

with associated df = $N - 1$.

The relationship between the three sums of squares is

$$SSTo = SSTr + SSE$$

which is often called the *fundamental identity for single-factor ANOVA*.

The quantity SSTo, the sum of squared deviations about the grand mean, is a measure of total variability in the data set consisting of all *k* samples. The quantity SSE results from measuring variability separately within each sample and then combining as indicated in the formula for SSE. Such within-sample variability is present regardless of whether or not H_0 is true. The magnitude of SSTr, on the other hand, has much to do with the status of H_0 (whether it is true or false). The more the μ's differ from one another, the larger SSTr will tend to be. Thus, SSTr represents variation that can (at least to some extent) be explained by any differences between means. An informal paraphrase of the fundamental identity is

total variation = explained variation + unexplained variation

Once any two of the sums of squares have been calculated, the remaining one is easily obtained from the fundamental identity. Often SSTo and SSTr are calculated first (using computational formulas given in the appendix to this chapter), and then SSE is obtained by subtraction: SSE = SSTo − SSTr. All the degrees of freedom, sums of squares, and mean squares are entered in an ANOVA table, as displayed in Table 15.2. The *P*-value usually appears to the right of *F* when the analysis is done by a statistical software package.

Table 15.2 ■ General Format for a Single-Factor ANOVA Table

Source of Variation	df	Sum of Squares	Mean Square	F
Treatments	$k - 1$	SSTr	$MSTr = \dfrac{SSTr}{k-1}$	$F = \dfrac{MSTr}{MSE}$
Error	$N - k$	SSE	$MSE = \dfrac{SSE}{N-k}$	
Total	$N - 1$	SSTo		

An ANOVA table from MINITAB for the reckless driving data of Example 15.4 is shown in Table 15.3. The P-value is .003, consistent with our previous conclusion that $.001 < P\text{-value} < .01$.

Table 15.3 ▪ **An ANOVA Table from MINITAB for the Data of Example 15.4**

One-way Analysis of Variance
Analysis of Variance

Source	DF	SS	MS	F	P
Factor	3	12.850	4.283	5.09	0.003
Error	76	63.900	0.841		
Total	79	76.750			

▪ Exercises 15.1–15.18

15.1 Give as much information as you can about the P-value for an upper-tailed F test in each of the following situations:
a. $df_1 = 4, df_2 = 15, F = 5.37$
b. $df_1 = 4, df_2 = 15, F = 1.90$
c. $df_1 = 4, df_2 = 15, F = 4.89$
d. $df_1 = 3, df_2 = 20, F = 14.48$
e. $df_1 = 3, df_2 = 20, F = 2.69$
f. $df_1 = 4, df_2 = 50, F = 3.24$

15.2 Give as much information as you can about the P-value of the single-factor ANOVA F test in each of the following situations:
a. $k = 5, n_1 = n_2 = n_3 = n_4 = n_5 = 4, F = 5.37$
b. $k = 5, n_1 = n_2 = n_3 = 5, n_4 = n_5 = 4, F = 2.83$
c. $k = 3, n_1 = 4, n_2 = 5, n_3 = 6, F = 5.02$
d. $k = 3, n_1 = n_2 = 4, n_3 = 6, F = 15.90$
e. $k = 4, n_1 = n_2 = 15, n_3 = 12, n_4 = 10, F = 1.75$

15.3 Employees of a certain state university system can choose from among four different health plans. Each plan differs somewhat from the others in terms of hospitalization coverage. Four samples of recently hospitalized individuals were selected, with each sample consisting of people covered by a different health plan. The length of the hospital stay (number of days) was determined for each individual selected.
a. What hypotheses would you test to decide whether average length of stay was related to health plan? (Note: Carefully define the population characteristics of interest.)
b. If each sample consisted of eight individuals and the value of the ANOVA F statistic was $F = 4.37$, what conclusion would be appropriate for a test with $\alpha = .01$?
c. Answer the question posed in Part (b) if the F value given there resulted from sample sizes $n_1 = 9$, $n_2 = 8, n_3 = 7$, and $n_4 = 8$.

15.4 The article "Heavy Drinking and Problems Among Wine Drinkers" (*Journal of Studies on Alcohol* [1999]: 467–471) analyzed drinking problems among Canadians. For each of several different groups of drinkers, the mean and standard deviation of the "highest number of drinks consumed" were calculated:

	$\bar{x}$	s	n
Beer only	7.52	6.41	1256
Wine only	2.69	2.66	1107
Spirits only	5.51	6.44	759
Beer and wine	5.39	4.07	1334
Beer and spirits	9.16	7.38	1039
Wine and spirits	4.03	3.03	1057
Beer, wine, and spirits	6.75	5.49	2151

Assume that each of the seven samples can be viewed as a random sample from the respective group. Is there sufficient evidence to conclude that the mean value of highest number of drinks consumed is not the same for all seven groups?

15.5 Suppose that a random sample of size $n = 5$ was selected from the vineyard properties for sale in Sonoma County, California, in each of three years. The following data are consistent with summary information on price per acre (in dollars, rounded to the nearest thousand) for disease-resistant grape vineyards in Sonoma County (*Wines and Vines*, November 1999):

1996	30,000	34,000	36,000	38,000	40,000
1997	30,000	35,000	37,000	38,000	40,000
1998	40,000	41,000	43,000	44,000	50,000

a. Construct boxplots for each of the three years on a common axis, and label each by year. Comment on the similarities and differences.

b. Carry out an ANOVA to determine whether there is evidence to support the claim that the mean price per acre for vineyard land in Sonoma County was not the same for the three years considered. Use a significance level of .05 for your test.

15.6 The article "An Analysis of Job Sharing, Full-Time, and Part-Time Arrangements" (*American Business Review* [1989]: 34–40) reported on a study of hospital employees from three different groups. Each employee reported a level of satisfaction with his or her work schedule (1 = very dissatisfied to 7 = very satisfied), as summarized in the following table:

Group	Sample Size	Sample Mean
1. Job sharers	24	6.60
2. Full-time employees	24	5.37
3. Part-time employees	20	5.20

a. What are numerator and denominator degrees of freedom for the *F* test?
b. The test statistic value is $F = 6.62$. Use a test with significance level .05 to decide whether it is plausible that true average satisfaction levels are identical for the three groups.
c. What is the value of MSE?

15.7 High productivity and carbohydrate storage ability of the Jerusalem artichoke make it a promising agricultural crop. The article "Leaf Gas Exchange and Tuber Yield in Jerusalem Artichoke Cultivars" (*Field Crops Research* [1991]: 241–252) reported on various plant characteristics. Consider the following data on chlorophyll concentration (in grams per square meter) for four varieties of Jerusalem artichoke:

Variety	1	2	3	4
Sample mean	0.30	0.24	0.41	0.33

Suppose that the sample sizes were 5, 5, 4, and 6, respectively, and also that $MSE = 0.0130$. Do the data suggest that true average chlorophyll concentration depends on the variety? State and test the appropriate hypotheses using a significance level of .05.

15.8 It has been reported that varying work schedules can lead to a variety of health problems for workers. The article "Nutrient Intake in Day Workers and Shift Workers" (*Work and Stress* [1994]: 332–342) reported on blood glucose levels (in millimoles per liter) for day-shift workers and for workers on two different types of rotating shifts. The sample sizes were $n_1 = 37$ for the day shift, $n_2 = 34$ for the second shift, and $n_3 = 25$ for the third shift. A single-factor ANOVA resulted in $F = 3.834$. At a significance level of .05, does true average blood glucose level appear to depend on the type of shift?

15.9 In the introduction to this chapter we discussed a study in which luxury car owners were asked to rate the importance of various attributes on a quantitative scale. Summary information for the importance of speed is as follows:

Type of Car	American	German	Japanese
Sample size	58	38	59
Sample mean	3.05	2.87	2.67

In addition, $SSE = 459.04$. Carry out a test of hypothesis to decide whether true average importance rating of speed is the same for owners of the three types of cars.

15.10 The article "Utilizing Feedback and Goal Setting to Increase Performance Skills of Managers" (*Academy of Management Journal* [1979]: 516–526) reported the results of an experiment to compare three different interviewing techniques for employee evaluations. One method allowed the employee being evaluated to discuss previous evaluations, the second method involved setting goals for the employee, and the third method did not allow either feedback or goal setting. After the interviews were concluded, the evaluated employee was asked to indicate how satisfied he or she was with the interview. (A numerical scale was used to quantify level of satisfaction.) The investigators used an ANOVA to compare the three interview techniques. An *F* statistic value of 4.12 was reported.
a. Suppose that 33 subjects had been used, with each technique applied to 11 of them. Use this information to conduct a level .05 test of the null hypothesis of no difference in mean satisfaction level for the three interview techniques.
b. The actual number of subjects on which each technique was used was 45. After studying the *F* table, explain why the conclusion in Part (a) still holds.

15.11 An investigation carried out to study the toxic effects of mercury was described in the article "Comparative Responses of the Action of Different Mercury Compounds on Barley" (*International Journal of Environmental Studies* [1983]: 323–327). Ten different concentrations of mercury (0, 1, 5, 10, 50, 100, 200, 300, 400, and 500 mg/l) were compared with respect to their effects on average dry weight (per 100 seven-day-old seedlings). The basic experiment was replicated four times for a total of 40 dry-weight observations (four for each treatment level). The article reported an ANOVA *F* statistic value of 1.895. Using a significance level of .05, test the null hypothesis that the true mean dry weight is the same for all 10 concentration levels.

15.12 The following data on calcium content of wheat are consistent with summary quantities that appeared in the article "Mineral Contents of Cereal Grains as Affected by Storage and Insect Infestation" (*Journal of Stored Products Research* [1992]: 147–151); four different storage times were considered:

Storage Period	Observations					
0 months	58.75	57.94	58.91	56.85	55.21	57.30
1 month	58.87	56.43	56.51	57.67	59.75	58.48
2 months	59.13	60.38	58.01	59.95	59.51	60.34
4 months	62.32	58.76	60.03	59.36	59.61	61.95

Partial output from the SAS computer package is also shown:

Dependent Variable: CALCIUM

Source	DF	Sum of Squares	Mean Square	F Value	Pr > F
Model	3	32.13815000	10.71271667	6.51	0.0030
Error	20	32.90103333	1.64505167		
Corrected Total	23	65.03918333			

R-Square	C.V.	Root MSE	CALCIUMMean
0.494135	2.180018	1.282596	58.8341667

a. Verify that the sums of squares and the numbers of degrees of freedom are as given in the ANOVA table.

b. Is there sufficient evidence to conclude that the mean calcium content is not the same for the four different storage times? Use the value of F from the ANOVA table to test the appropriate hypotheses at significance level .05.

15.13 In an experiment to investigate the performance of four different brands of spark plugs intended for use on a 125-cc motorcycle, five plugs of each brand were tested, and the number of miles (at a constant speed) until failure was observed. A partially completed ANOVA table is given. Fill in the missing entries, and test the relevant hypotheses using a .05 level of significance.

Source of Variation	df	Sum of Squares	Mean Square	F
Treatments				
Error		235,419.04		
Total		310,500.76		

15.14 The partially completed ANOVA table given in this problem is taken from the article "Perception of Spatial Incongruity" (*Journal of Nervous and Mental Disease* [1961]: 222), in which the abilities of three

different groups to identify a perceptual incongruity were assessed and compared. All individuals in the experiment had been hospitalized to undergo psychiatric treatment. There were 21 individuals in the depressive group, 32 individuals in the functional "other" group, and 21 individuals in the brain-damaged group. Complete the ANOVA table, and then carry out the appropriate test of hypothesis (use $\alpha = .01$) and interpret your results.

Source of Variation	df	Sum of Squares	Mean Square	F
Treatments			76.09	
Error				
Total		1123.14		

15.15 Research carried out to investigate the relationship between smoking status of workers and short-term absenteeism rate (in hours per month) yielded the following summary information ("Work-Related Consequences of Smoking Cessation," *Academy of Management Journal* [1989]: 606–621):

Status	Sample Size	Sample Mean
Continuous smoker	96	2.15
Recent ex-smoker	34	2.21
Long-term ex-smoker	86	1.47
Never smoked	206	1.69

In addition, $F = 2.56$. Construct an ANOVA table, and then state and test the appropriate hypotheses using a .01 significance level.

15.16 Parents are frequently concerned when their child seems slow to begin walking (although when the child finally walks, the resulting havoc sometimes has the parents wishing they could turn back the clock!). The article "Walking in the Newborn" (*Science* [1972]: 314–315) reported on an experiment in which the effects of several different treatments on the age at which a child first walks were compared. Children in the first group were given special walking exercises for 12 min/day beginning at age 1 week and lasting 7 weeks. The second group of children received daily exercises but not the walking exercises administered to the first group. The third and fourth groups were control groups: They received no special treatment and differed only in that the third group's progress was checked weekly, whereas the fourth group's progress was checked just once at the end of the study. Observations on age (in months) when the children first walked are shown in the accompanying table. Also given is the ANOVA table, obtained from the SPSS computer package.

	Age (months)			n	Total
Treatment 1	9.00	9.50	9.75	6	**60.75**
	10.00	13.00	9.50		
Treatment 2	11.00	10.00	10.00	6	**68.25**
	11.75	10.50	15.00		
Treatment 3	11.50	12.00	9.00	6	**70.25**
	11.50	13.25	13.00		
Treatment 4	13.25	11.50	12.00	5	**61.75**
	13.50	11.50			

Analysis of Variance

Source	df	Sum of sq.	Mean Sq.	F Ratio	F Prob
Between Groups	3	14.778	4.926	2.142	.129
Within Groups	19	43.690	2.299		
Total	**22**	**58.467**			

a. Verify the entries in the ANOVA table.
b. State and test the relevant hypotheses using a significance level of .05.

15.17 The Fog Index is a measure of reading difficulty based on the average number of words per sentence and the percentage of words with three or more syllables. High values of the Fog Index are associated with difficult reading levels. Independent random samples of six advertisements were taken from three different magazines, and Fog Indexes were computed to obtain the data given in the following table ("Readability Levels of Magazine Advertisements," *Journal of Advertising Research* [1981]: 45–50):

Scientific						
American	15.75	11.55	11.16	9.92	9.23	8.20
Fortune	12.63	11.46	10.77	9.93	9.87	9.42
New Yorker	9.27	8.28	8.15	6.37	6.37	5.66

Construct an ANOVA table, and then use a significance level of .01 to test the null hypothesis of no difference between the mean Fog Index levels for advertisements appearing in the three magazines.

15.18 Some investigators think that the concentration (in milligrams per milliliter) of a particular antigen in supernatant fluids could be related to the onset of meningitis in infants. The following data are typical of the data given in plots in the article "Type-Specific Capsular Antigen Is Associated with Virulence in Late-Onset Group B Streptococcal Type III Disease" (*Infection and Immunity* [1984]: 124–129):

Asymptomatic infants	1.56	1.06	0.87	1.39	0.71	0.87		
Infants with late-onset sepsis	1.51	1.78	1.45	1.13	1.87	1.89	1.07	1.72
Infants with late-onset meningitis	1.21	1.34	1.95	2.27	0.88	1.67	2.57	

Construct an ANOVA table, and use it to test the null hypothesis of no difference in mean antigen concentrations for the three groups.

■ 15.2 Multiple Comparisons

When $H_0: \mu_1 = \mu_2 = \cdots = \mu_k$ is rejected by the F test, we believe that there are differences among the k population or treatment means. A natural question to ask at this point is, Which means differ? For example, with $k = 4$, it might be the case that $\mu_1 = \mu_2 = \mu_4$, with μ_3 different from the other three means. Another possibility is that $\mu_1 = \mu_4$ and $\mu_2 = \mu_3$. Still another possibility is that all four means are different from one another. A **multiple comparisons procedure** is a method for identifying differences among the μ's once the hypothesis of overall equality has been rejected. We present one such method, the **Tukey–Kramer** (T–K) multiple comparisons procedure.

The T–K procedure is based on computing confidence intervals for the difference between each possible pair of μ's. For $k = 3$, there are three differences to consider:

$$\mu_1 - \mu_2 \qquad \mu_1 - \mu_3 \qquad \mu_2 - \mu_3$$

(The difference $\mu_2 - \mu_1$ is not considered, because the interval for $\mu_1 - \mu_2$ provides the same information. Similarly, intervals for $\mu_3 - \mu_1$ and $\mu_3 - \mu_2$ are not necessary.) Once all confidence intervals have been computed, each is examined to determine whether the interval includes 0. If a particular interval does not include 0, the two means are declared "significantly different" from one another. An interval that does include 0 supports the conclusion that there is no significant difference between the means involved.

Suppose, for example, that $k = 3$ and that the three confidence intervals are

Difference	T–K Interval
$\mu_1 - \mu_2$	$(-0.9, 3.5)$
$\mu_1 - \mu_3$	$(2.6, 7.0)$
$\mu_2 - \mu_3$	$(1.2, 5.7)$

Because the interval for $\mu_1 - \mu_2$ includes 0, we judge that μ_1 and μ_2 do not differ significantly. The other two intervals do not include 0, so we conclude that $\mu_1 \neq \mu_3$ and $\mu_2 \neq \mu_3$.

The T–K intervals are based on critical values for a probability distribution called the *Studentized range distribution*. These critical values appear in Appendix Table 8. To find a critical value, enter the table at the column corresponding to the number of populations or treatments being compared, move down to the rows corresponding to the number of error degrees of freedom, and select either the value for a 95% confidence level or the one for a 99% level.

■ **Tukey–Kramer Multiple Comparisons Procedure**

When there are k populations or treatments being compared, $\dfrac{k(k-1)}{2}$ confidence intervals must be computed. If we denote the relevant Studentized range critical value by q, the intervals are as follows:

For $\mu_i - \mu_j$: $(\bar{x}_i - \bar{x}_j) \pm q\sqrt{\dfrac{MSE}{2}\left(\dfrac{1}{n_i} + \dfrac{1}{n_j}\right)}$

Two means are judged to differ significantly if the corresponding interval does not include 0.

If the sample sizes are all the same, let n denote the common value of $n_1, \ldots, n_k$. In this case, the $\pm$ factor for each interval is the same quantity

$$q\sqrt{\dfrac{MSE}{n}}$$

■ **Example 15.5 Imagining Golf**

The degree of success at mastering a skill often depends on the method used to learn the skill. The article "Effects of Occluded Vision and Imagery on Putting Golf Balls" (*Perceptual and Motor Skills* [1995]: 179–186) reported on a study involving the following four learning methods: (1) visual contact and imagery, (2) nonvisual contact and imagery, (3) visual contact, and (4) control. There were 20 subjects randomly assigned to each method. The following summary data on putting performance were reported:

Method	1	2	3	4
$\bar{x}$	16.30	15.25	12.05	9.30
s	2.03	3.23	2.91	2.85

It is easily verified that MSTr = 202.283, MSE = 7.7861, and F = 25.98. With $df_1 = 3$ and $df_2 = 76$, Appendix Table 7 shows that P-value < .001. The investigators concluded that the mean performance is not the same for the four methods.

Appendix Table 8 gives the 95% Studentized range critical value $q = 3.74$ (using error df = 60, the closest tabulated value to df = 76). The first two T–K intervals are

$$\mu_1 - \mu_2: \quad (16.30 - 15.25) \pm 3.74\sqrt{\left(\frac{7.7861}{2}\right)\left(\frac{1}{20} + \frac{1}{20}\right)}$$

$$= 1.05 \pm 2.334$$

$$= (-1.284, 3.384) \leftarrow \textit{Includes 0}$$

$$\mu_1 - \mu_3: \quad (16.30 - 12.05) \pm 3.74\sqrt{\left(\frac{7.7861}{2}\right)\left(\frac{1}{20} + \frac{1}{20}\right)}$$

$$= 4.25 \pm 2.334$$

$$= (1.916, 6.584) \leftarrow \textit{Does not include 0}$$

The remaining intervals are

$\mu_1 - \mu_4:$ $7.00 \pm 2.334 = (4.666, 9.334)$ ← *Does not include 0*
$\mu_2 - \mu_3:$ $3.20 \pm 2.334 = (0.866, 5.534)$ ← *Does not include 0*
$\mu_2 - \mu_4:$ $5.95 \pm 2.334 = (3.616, 8.284)$ ← *Does not include 0*
$\mu_3 - \mu_4:$ $2.75 \pm 2.334 = (0.416, 5.084)$ ← *Does not include 0*

We would conclude that μ_1 is not significantly different from μ_2 but that all other means differ from one another. That is, there is no significant difference between the visual contact and imagery treatment (Treatment 1) and the nonvisual contact and imagery treatment (Treatment 2). However, the visual contact treatment and the control treatment (Treatments 3 and 4) differ from each other and from Treatments 1 and 2.

MINITAB can be used to construct T–K intervals if raw data are available. A typical output (based on Example 15.5) is shown in Figure 15.6. From the output, we see that the confidence interval for $\mu_1 - \mu_2$ is $(-1.284, 3.384)$, that for $\mu_2 - \mu_4$ is $(3.616, 8.284)$, and so on.

Figure 15.6 The Tukey–Kramer intervals for Example 15.5 (from MINITAB).

Tukey's pairwise comparisons
    Family error rate = 0.0500
    Individual error rate = 0.0111
    Critical value = 3.74
    Intervals for (column level mean) − (row level mean)

	1	2	3
2	−1.284		
	3.384		
3	1.916	.866	
	6.584	5.534	
4	4.666	3.616	.416
	9.334	8.284	5.084

Why calculate the T–K intervals rather than use the t confidence interval for a difference between μ's (from Chapter 11)? The answer is that the T–K intervals control the **simultaneous confidence level** at approximately 95% (or 99%). That is, if the procedure is used repeatedly on many different data sets, in the long run only about 5% (or 1%) of the time would at least one of the intervals not include the value of what the interval is estimating. Consider using separate 95% t intervals, each one having a 5% error rate. Then the chance that at least one interval would make an incorrect statement about a difference in μ's increases dramatically with the number of intervals calculated. The MINITAB output in Figure 15.6 shows that to achieve a simultaneous confidence level of about 95% (experimentwise or "family" error rate of 5%) when $k = 4$ and error df = 76, the individual confidence level must be 98.89% (individual error rate 1.11%).

An effective display for summarizing the results of any multiple comparisons procedure involves listing the $\bar{x}$'s and underscoring pairs judged to be not significantly different. The process for constructing such a display is described in the accompanying box.

■ **Summarizing the Results of the Tukey–Kramer Procedure**

1. List the sample means in increasing order, identifying the corresponding population just above the value of each $\bar{x}$.

2. Use the T–K intervals to determine the group of means that do not differ significantly from the first (smallest) sample mean in the list. Draw a horizontal line extending from the smallest mean to the last mean in the group identified. For example, if there are five means arranged in the following order:

Population	3	2	1	4	5
Sample mean	$\bar{x}_3$	$\bar{x}_2$	$\bar{x}_1$	$\bar{x}_4$	$\bar{x}_5$

and if μ_3 is judged to be not significantly different from μ_2 or μ_1 but is judged to be significantly different from μ_4 and μ_5, then draw the following line:

Population	3	2	1	4	5
Sample mean	$\bar{x}_3$	$\bar{x}_2$	$\bar{x}_1$	$\bar{x}_4$	$\bar{x}_5$

3. Use the T–K intervals to determine the group of means that are not significantly different from the second smallest mean. (You need to consider only the means that appear to the right of the mean under consideration.) If there is already a line connecting the second smallest mean with all means in the new group identified, then no new line need be drawn. If this entire group of means is not underscored with a single line, draw a line extending from the second smallest mean to the last mean in the new group. Continuing with our example, if μ_2 is not significantly different from μ_1 but is significantly different from μ_4 and μ_5, no new line need be drawn. However, if μ_2 is not significantly different from either μ_1 or μ_4 but is judged to be different from μ_5, a second line is drawn, as shown:

Population	3	2	1	4	5
Sample mean	$\bar{x}_3$	$\bar{x}_2$	$\bar{x}_1$	$\bar{x}_4$	$\bar{x}_5$

4. Continue considering the means in the order listed, adding new lines as needed.

To illustrate this summary procedure, suppose that four samples with $\bar{x}_1 = 19$, $\bar{x}_2 = 27$, $\bar{x}_3 = 24$, and $\bar{x}_4 = 10$ are used to test $H_0: \mu_1 = \mu_2 = \mu_3 = \mu_4$ and that this hypothesis is rejected. Suppose that the T–K confidence intervals indicate that μ_2 is significantly different from both μ_1 and μ_4 and that there are no other significant differences. The resulting summary display would then be

Population	4	1	3	2
Sample mean	10	19	24	27

Example 15.6 Sleep Time

A biologist wished to study the effects of ethanol on sleep time. A sample of 20 rats, matched for age and other characteristics, was selected, and each rat was given an oral injection having a particular concentration of ethanol per body weight. The rapid eye movement (REM) sleep time for each rat was then recorded for a 24-hr period, with the results shown in the following table:

Treatment	Observations					$\bar{x}$
1. 0 (control)	88.6	73.2	91.4	68.0	75.2	79.28
2. 1 g/kg	63.0	53.9	69.2	50.1	71.5	61.54
3. 2 g/kg	44.9	59.5	40.2	56.3	38.7	47.92
4. 4 g/kg	31.0	39.6	45.3	25.2	22.7	32.76

Table 15.4 (an ANOVA table from SAS) leads to the conclusion that true average REM sleep time depends on the treatment used; the P-value for the F test is .0001.

Table 15.4 ▪ SAS ANOVA Table for Example 15.6

Analysis of Variance Procedure
Dependent Variable: TIME

Source	DF	Sum of Squares	Mean Square	F Value	Pr > F
Model	3	5882.35750	1960.78583	21.09	0.0001
Error	16	1487.40000	92.96250		
Total	19	7369.75750			

The T–K intervals are

Difference	Interval	Includes 0?
$\mu_1 - \mu_2$	17.74 ± 17.446	no
$\mu_1 - \mu_3$	31.36 ± 17.446	no
$\mu_1 - \mu_4$	46.24 ± 17.446	no
$\mu_2 - \mu_3$	13.08 ± 17.446	yes
$\mu_2 - \mu_4$	28.78 ± 17.446	no
$\mu_3 - \mu_4$	15.16 ± 17.446	yes

The only T–K intervals that include 0 are those for $\mu_2 - \mu_3$ and $\mu_3 - \mu_4$. The corresponding underscoring pattern is

$\overline{x}_4$	$\overline{x}_3$	$\overline{x}_2$	$\overline{x}_1$
32.76	47.92	61.54	79.28

Figure 15.7 displays the SAS output that agrees with our underscoring; letters are used to indicate groupings in place of the underscoring.

Figure 15.7 SAS output for Example 15.6.

Alpha = 0.05 df = 16 MSE = 92.9625
Critical Value of Studentized Range = 4.046
Minimum Signfi cant Difference = 17.446
Means with the same letter are not signfi cantly different.

Tukey Grouping		Mean	N	Treatment
	A	79.280	5	0 (control)
	B	61.540	5	1 g/kg
C	B	47.920	5	2 g/kg
C		32.760	5	4 g/kg

■ Example 15.7   Roommate Satisfaction

How satisfied are college students with dormitory roommates? The article "Roommate Satisfaction and Ethnic Identity in Mixed-Race and White University Roommate Dyads" (*Journal of College Student Development* [1998]: 194–199) investigated differences among randomly assigned African American/white, Asian/white, Hispanic/white, and white/white roommate pairs. The researchers used a one-way ANOVA to analyze scores on the Roommate Relationship Inventory to see whether a difference in mean score existed for the four types of roommate pairs. They reported "significant differences among the means ($P < .01$). Follow-up Tukey [intervals] . . . indicated differences between White dyads ($M = 77.49$) and African American/White dyads ($M = 71.27$). . . . No other significant differences were found."

Although the mean satisfaction score for the Asian/white and Hispanic/white groups were not given, they must have been between 77.49 (the mean for the white/white pairs) and 71.27 (the mean for the African American/white pairs). (If they had been larger than 77.49, they would have been significantly different from the African American/white pairs mean, and if they had been smaller than 71.27, they would have been significantly different from the white/white pairs mean.) An underscoring consistent with the reported information is

White/White	Hispanic/White and Asian/White	African American/ White

■ Exercises 15.19–15.29

15.19 Leaf surface area is an important variable in plant gas-exchange rates. The article "Fluidized Bed Coating of Conifer Needles with Glass Beads for Determination of Leaf Surface Area" (*Forest Science* [1980]: 29–32) included an analysis of dry matter per unit surface area (in milligrams per cubic centimeter) for trees raised under three different growing conditions. Let μ_1, μ_2, and μ_3 represent the true mean dry matter per unit surface area for the growing conditions 1, 2, and 3, respectively. The following 95%

simultaneous confidence intervals are based on summary quantities that appeared in the article:

Difference	$\mu_1 - \mu_2$	$\mu_1 - \mu_3$	$\mu_2 - \mu_3$
Interval	$(-3.11, -1.11)$	$(-4.06, -2.06)$	$(-1.95, 0.05)$

Which of the following four statements do you think describes the relationship between μ_1, μ_2, and μ_3. Explain your choice.

a. $\mu_1 = \mu_2$, and μ_3 differs from μ_1 and μ_2.
b. $\mu_1 = \mu_3$, and μ_2 differs from μ_1 and μ_3.
c. $\mu_2 = \mu_3$, and μ_1 differs from μ_2 and μ_3.
d. All three μ's are different from one another.

15.20 The article referenced in Example 15.5 reported on a study involving the following four learning methods for golf: (1) visual contact and imagery, (2) nonvisual contact and imagery, (3) visual contact, and (4) control. There were 20 subjects assigned to each method. The following summary data on putting performance were reported:

Method	1	2	3	4
$\bar{x}$	16.30	15.25	12.05	9.30
s	2.03	3.23	2.91	2.85

These summary statistics, in turn, give MSE = 7.7861 and F = 25.98, from which P-value < .001, so H_0 can be rejected at significance level .01. Calculate the 99% T–K intervals, indicate which methods differ significantly from one another, and summarize the results by underscoring. How do the 99% T–K intervals compare to the 95% T–K intervals of Example 15.5?

15.21 The following data resulted from a flammability study in which specimens of five different fabrics were tested to determine burn times:

Fabric 1	17.8	16.2	15.9	15.5	
Fabric 2	13.2	10.4	11.3		
Fabric 3	11.8	11.0	9.2	10.0	
Fabric 4	16.5	15.3	14.1	15.0	13.9
Fabric 5	13.9	10.8	12.8	11.7	

The study also gave MSTr = 23.67, MSE = 1.39, F = 17.08, and P-value = .000. The following output gives the T–K intervals as calculated by MINITAB:

```
Individual error rate = 0.00750
Critical value = 4.37
Intervals for (column level mean) − (row level mean)
         1        2        3        4
      1.938
2     7.495
      3.278   −1.645
3     8.422    3.912
     −1.050   −5.983   −6.900
4     3.830   −0.670   −2.020
      1.478   −3.445   −4.372   0.220
5     6.622    2.112    0.772   5.100
```

Identify significant differences, and give the underscoring pattern.

15.22 Exercise 15.6 presented the following summary information on satisfaction levels for employees on three different work schedules:

$n_1 = 24$	$n_2 = 24$	$n_3 = 20$
$\bar{x}_1 = 6.60$	$\bar{x}_2 = 5.37$	$\bar{x}_3 = 5.20$

MSE was computed to be 2.028. Calculate the 95% T–K intervals, and identify any significant differences.

15.23 Sample mean chlorophyll concentrations for the four Jerusalem artichoke varieties introduced in Exercise 15.7 were 0.30, 0.24, 0.41, and 0.33, with corresponding sample sizes of 5, 5, 4, and 6, respectively. In addition, MSE = 0.0130. Calculate the 95% T–K intervals, and then use the underscoring procedure described in this section to identify significant differences among the varieties.

15.24 Do lizards play a role in spreading plant seeds? Research carried out in South Africa suggests that they do ("Dispersal of Namaqua Fig [*Ficus cordata cordata*] Seeds by the Augrabies Flat Lizard [*Platysaurus broadleyi*]," *Journal of Herpetology* [1999]: 328–330). The researchers collected 400 seeds of the Namaqua fig, 100 of which were from each of the following treatments: lizard dung, bird dung, rock hyrax dung, and uneaten figs. They planted these seeds in batches of 5, and for each group of 5 they recorded how many of the seeds germinated. This resulted in 20 observations for each treatment. The treatment means and standard deviations are given in the following table:

Treatment	n	$\bar{x}$	s
Uneaten figs	20	2.40	0.30
Lizard dung	20	2.35	0.33
Bird dung	20	1.70	0.34
Hyrax dung	20	1.45	0.28

a. Construct the appropriate ANOVA table, and test the hypothesis that there is no difference between mean number of seeds germinating for the four treatments.
b. Is there evidence that seeds eaten and then excreted by lizards germinate at a higher rate than those eaten and then excreted by birds? Give statistical evidence to support your answer.

15.25 The article "Growth Response in Radish to Sequential and Simultaneous Exposures of NO_2 and SO_2" (*Environmental Pollution* [1984]: 303–325) compared a control group (no exposure), a sequential-exposure group (plants exposed to one pollutant followed by exposure to a second pollutant 4 weeks later), and a simultaneous-exposure group (plants exposed to both pollutants at the same time). The article stated that "sequential exposure to the two pollutants had no effect on growth compared to the control. Simultaneous exposure to the gases significantly reduced plant growth." Let $\bar{x}_1$, $\bar{x}_2$, and $\bar{x}_3$ represent the sample means for the control, sequential-exposure, and simultaneous-exposure groups, respectively. Suppose that $\bar{x}_1 > \bar{x}_2 > \bar{x}_3$. Use the given information to construct a table in which the sample means are listed in increasing order, with those that are judged to be not significantly different underscored.

15.26 The nutritional quality of shrubs commonly used for feed by rabbits was the focus of a study summarized in the article "Estimation of Browse by Size Classes for Snowshoe Hare" (*Journal of Wildlife Management* [1980]: 34–40). The energy content (in calories per gram) of three sizes (4 mm or less, 5–7 mm, and 8–10 mm) of serviceberries was studied. Let μ_1, μ_2, and μ_3 denote the true energy content for the three size classes. Suppose that 95% simultaneous confidence intervals for $\mu_1 - \mu_2$, $\mu_1 - \mu_3$, and $\mu_2 - \mu_3$ are (−10, 290), (150, 450), and (10, 310), respectively. How would you interpret these intervals?

15.27 The following underscoring pattern appeared in the article "Effect of SO_2 on Transpiration, Chlorophyll Content, Growth, and Injury in Young Seedlings of Woody Angiosperms" (*Canadian Journal of Forest Research* [1980]: 78–81), where water loss of plants (*Acer saccharinum*) exposed to 0, 2, 4, 8, and 16 hr of fumigation was recorded:

Duration of fumigation	16	0	8	2	4
Sample mean water loss	27.57	28.23	30.21	31.16	36.21

A multiple comparisons procedure was used to detect differences among the mean water losses for the different fumigation durations. How would you interpret this underscoring pattern?

15.28 Samples of six different brands of diet or imitation margarine were analyzed to determine the level of physiologically active polyunsaturated fatty acids (PAPUFA, in percent), resulting in the data shown in the following table (the data are fictitious, but the sample means agree with data reported in *Consumer Reports*):

Imperial	14.1	13.6	14.4	14.3	
Parkay	12.8	12.5	13.4	13.0	12.3
Blue Bonnet	13.5	13.4	14.1	14.3	
Chiffon	13.2	12.7	12.6	13.9	
Mazola	16.8	17.2	16.4	17.3	18.0
Fleischmann's	18.1	17.2	18.7	18.4	

a. Test for differences among the true mean PAPUFA percentages for the different brands. Use $\alpha = .05$.
b. Use the T–K procedure to compute 95% simultaneous confidence intervals for all differences between means and interpret the resulting intervals.

15.29 Consider the following data on plant growth after the application of different types of growth hormone:

Hormone 1	13	17	7	14
Hormone 2	21	13	20	17
Hormone 3	18	14	17	21
Hormone 4	7	11	18	10
Hormone 5	6	11	15	8

a. Carry out the *F* test at level $\alpha = .05$.
b. What happens when the T–K procedure is applied? (Note: This "contradiction" can occur when H_0 is "barely" rejected. It happens because the test and the multiple comparisons method are based on different distributions. Consult your friendly neighborhood statistician for more information.)

▪ 15.3 The *F* Test for a Randomized Block Experiment

We saw in Chapter 11 that when two treatments are to be compared, a paired experiment is often more effective than one involving two independent samples. This is because pairing can considerably reduce the extraneous variation in subjects or experimental units. A similar result can be achieved when more than two treatments are to be compared. Suppose that four different pesticides (the treatments) are being considered for application to a particular crop. There are 20 plots of land available for planting. If 5 of these plots are randomly selected to receive Pesticide 1, 5 of the remaining 15 randomly selected for Pesticide 2, and so on, the result is a *completely randomized experiment*, and the data should be analyzed using a single-factor ANOVA. The disadvantage of this experiment is that if there are any substantial differences in characteristics of the plots that could affect yield, a separate assessment of any differences between treatments will not be possible (plot effects are confounded with treatment effects).

Here is an alternative experiment. Consider separating the 20 plots into 5 groups, each consisting of 4 plots. Within each group, the plots are as much alike as possible with respect to characteristics affecting yield. Then, within each group one plot is randomly selected for Pesticide 1, a second plot is randomly chosen to receive Pesticide 2, and so on. The homogeneous groups are called *blocks*, and the random allocation of treatments within each block as described results in a *randomized block experiment*.

■ Definition

Suppose that experimental units (individuals or objects to which the treatments are applied) are first separated into groups consisting of *k* units in such a way that the units within each group are as similar as possible. Within any particular group, the treatments are then randomly allocated so that each unit in a group receives a different treatment. The groups are called **blocks**, and the experimental design is referred to as a **randomized block design**.

■ Example 15.8   Cost of Air-Conditioning: Type of System Blocked by House Type

High energy costs have made consumers and home builders increasingly aware of whether household appliances are energy efficient. A large developer carried out a study to compare electricity usage for four different residential air-conditioning systems being considered for tract homes. Each system was installed in five homes, and the resulting electricity usage (in kilowatt-hours) was monitored for a 1-month period. Because the developer realized that many characteristics of a home could affect usage (e.g., floor space, type of insulation, directional orientation, and type of roof and exterior), care was taken to ensure that extraneous variation in such characteristics did not influence the conclusions. Homes selected for the experiment were grouped into five blocks consisting of four homes each in such a way that the four homes within any given block were as similar as possible. The resulting data are displayed in Table 15.5, in which rows correspond to the different treatments (air-conditioning systems) and columns correspond to the different blocks. Later in

Table 15.5 ■ Data from the Randomized Block Experiment of Example 15.8

Treatment	Block 1	Block 2	Block 3	Block 4	Block 5	Treatment Average
1	116	118	97	101	115	109.40
2	171	131	105	107	129	128.60
3	138	131	115	93	110	117.40
4	144	141	115	93	99	118.40
Block average	142.25	130.25	108.00	98.50	113.25	**Grand mean** 118.45

this section we will analyze these data to see whether electricity usage depends on which system is used.

The hypotheses of interest and assumptions underlying the analysis of a randomized block design are similar to those for a completely randomized design.

▪ Assumptions and Hypotheses

The single observation made on any particular treatment in a given block is assumed to be selected from a normal distribution. The variance of this distribution, σ^2, is the same for each block–treatment combination. However, the mean value may depend separately on both the treatment applied and the block.

The hypotheses of interest are as follows:

H_0: The mean value does not depend on which treatment is applied.

H_a: The mean value does depend on which treatment is applied.

The μ notation used previously is no longer adequate for stating hypotheses, because an observation's mean value may depend on both the treatment applied and the block used.

▪ The F Test

The key to analyzing data from a randomized block experiment is to represent SSTo, which measures total variation, as a sum of three pieces: SSTr, SSE, and a block sum of squares SSBl. SSBl incorporates any variation resulting from differences between the blocks; these differences can be substantial if, before creating the blocks, there was great heterogeneity in experimental units (e.g., plots). Once the four sums of squares have been computed, the test statistic is an F ratio, $\dfrac{\text{MSTr}}{\text{MSE}}$, but the number of error degrees of freedom is no longer $N - k$ as it was in single-factor ANOVA.

▪ Summary of the Randomized Block F Test

Notation: Let

   k = number of treatments

   l = number of blocks

   $\overline{x}_i$ = average of all observations for treatment i

   $\overline{b}_j$ = average of all observations in block j

   $\overline{\overline{x}}$ = average of all kl observations in the experiment (the grand mean)

Sums of squares and associated degrees of freedom are as follows:

Sum of Squares	Symbol	df	Formula
Treatments	SSTr	$k - 1$	$l[(\bar{x}_1 - \bar{\bar{x}})^2 + (\bar{x}_2 - \bar{\bar{x}})^2 + \cdots + (\bar{x}_k - \bar{\bar{x}})^2]$
Blocks	SSBl	$l - 1$	$k[(\bar{b}_1 - \bar{\bar{x}})^2 + (\bar{b}_2 - \bar{\bar{x}})^2 + \cdots + (\bar{b}_l - \bar{\bar{x}})^2]$
Error	SSE	$(k - 1)(l - 1)$	by subtraction
Total	**SSTo**	$kl - 1$	$\sum_{\text{all } kl \text{ obs.}} (x - \bar{\bar{x}})^2$

SSE is obtained by subtraction through the use of the fundamental identity

$$\text{SSTo} = \text{SSTr} + \text{SSBl} + \text{SSE}$$

Test statistic: $F = \dfrac{\text{MSTr}}{\text{MSE}}$

where

$$\text{MSTr} = \frac{\text{SSTr}}{k - 1}$$

$$\text{MSE} = \frac{\text{SSE}}{(k - 1)(l - 1)}$$

The test is based on $\text{df}_1 = k - 1$ and $\text{df}_2 = (k - 1)(l - 1)$. *P*-value information comes from Appendix Table 7.

Alternative formulas for the sums of squares appropriate for efficient hand computation appear in the appendix to this chapter.

Calculations for this *F* test are usually summarized in an ANOVA table. The table is similar to the one for a single-factor ANOVA except that blocks are an extra source of variation, so four rows are included rather than just three.

Table 15.6 shows a mean square for blocks and the mean squares for treatments and error. Sometimes the *F* ratio $\dfrac{\text{MSBl}}{\text{MSE}}$ is also computed. A large value of this ratio suggests that blocking was effective in filtering out extraneous variation.

Table 15.6 ▪ ANOVA Table for a Randomized Block Experiment

Source of Variation	df	Sum of Squares	Mean Square	F
Treatments	$k - 1$	SSTr	$\text{MSTr} = \dfrac{\text{SSTr}}{k - 1}$	$F = \dfrac{\text{MSTr}}{\text{MSE}}$
Blocks	$l - 1$	SSBl	$\text{MSBl} = \dfrac{\text{SSBl}}{l - 1}$	
Error	$(k - 1)(l - 1)$	SSE	$\text{MSE} = \dfrac{\text{SSE}}{(k - 1)(l - 1)}$	
Total	$kl - 1$	**SSTo**		

▪ **Example 15.9** **Electricity Cost of Air-Conditioning Revisited**

Reconsider the electricity usage data given in Example 15.8.

H_0: Mean electricity usage does not depend on which air-conditioning system is used.

H_a: Mean electricity usage does depend on which system is used.

Test statistic: $F = \dfrac{\text{MSTr}}{\text{MSE}}$

From Example 15.8,

$$\bar{\bar{x}} = 118.45$$

$$\bar{x}_1 = 109.40 \qquad \bar{x}_2 = 128.60 \qquad \bar{x}_3 = 117.40 \qquad \bar{x}_4 = 118.40$$

(these are the four row averages), and

$$\bar{b}_1 = 142.25 \qquad \bar{b}_2 = 130.25 \qquad \bar{b}_3 = 108.00 \qquad \bar{b}_4 = 98.50 \qquad \bar{b}_5 = 113.25$$

(these are the five column averages). Using the individual observations given previously, we calculate

$$\text{SSTo} = \sum_{\text{all } kl \text{ obs.}} (x - \bar{\bar{x}})^2$$
$$= (116 - 118.45)^2 + (118 - 118.45)^2 + \cdots + (99 - 118.45)^2$$
$$= 7594.95$$

The other sums of squares are

$$\text{SSTr} = l[(\bar{x}_1 - \bar{\bar{x}})^2 + (\bar{x}_2 - \bar{\bar{x}})^2 + \cdots + (\bar{x}_k - \bar{\bar{x}})^2]$$
$$= 5[(109.4 - 118.45)^2 + (128.6 - 118.45)^2$$
$$+ (117.4 - 118.45)^2 + (118.4 - 118.45)^2]$$
$$= 930.15$$

$$\text{SSBl} = k[(\bar{b}_1 - \bar{\bar{x}})^2 + (\bar{b}_2 - \bar{\bar{x}})^2 + \cdots + (\bar{b}_l - \bar{\bar{x}})^2]$$
$$= 4[(142.25 - 118.45)^2 + (130.25 - 118.45)^2 + \cdots + (113.25 - 118.45)^2]$$
$$= 4959.70$$

$$\text{SSE} = \text{SSTo} - \text{SSTr} - \text{SSBl}$$
$$= 7594.95 - 930.15 - 4959.70$$
$$= 1705.10$$

The remaining calculations are displayed in the following ANOVA table:

Source of Variation	df	Sum of Squares	Mean Square	F
Treatments	3	930.15	310.05	$\dfrac{310.05}{142.09} = 2.18$
Blocks	4	4959.70	1239.93	
Error	12	1705.10	142.09	
Total	19	7594.95		

In Appendix Table 7 with $df_1 = 3$ and $df_2 = 12$, the value 2.61 corresponds to a *P*-value of .10. Because $2.18 < 2.61$, *P*-value $> .10$ and H_0 cannot be rejected. Mean electricity usage does not seem to depend on which of the four air-conditioning systems is used.

In many studies, all k treatments can be applied to the same experimental unit, so there is no need to group different experimental units to form blocks. For example, an experiment to compare the effects of four different gasoline additives on automobile engine efficiency could be carried out by selecting just 5 engines and using all four treatments on each one rather than using 20 engines and blocking them. Each engine by itself then constitutes a block. As another example, a manufacturing company might wish to compare outputs for three different packaging machines. Because output could be affected by which operator is using the machine, a design that controls for the effects of operator variation is desirable. One possibility is to use 15 operators grouped into homogeneous blocks of 5 operators each, but such homogeneity within each block may be difficult to achieve. An alternative approach is to use only five operators and to have each one operate all three machines. There are then three observations in each block, all three with the same operator.

■ Example 15.10   Comparing Four Stool Designs

In the article "The Effects of a Pneumatic Stool and a One-Legged Stool on Lower Limb Joint Load and Muscular Activity During Sitting and Rising" (*Ergonomics* [1993]: 519–535), the accompanying data were given on the effort (measured on the Borg Scale) required by a subject to rise from a sitting position for each of four different stools. Because it was suspected that different people could exhibit large differences in effort, even for the same type of stool, a sample of nine people was selected and each person was tested on all four stools, with the following results:

	1	2	3	4	5	6	7	8	9
					Subject				
Stool A	12	10	7	7	8	9	8	7	9
Stool B	15	14	14	11	11	11	12	11	13
Stool C	12	13	13	10	8	11	12	8	10
Stool D	10	12	9	9	7	10	11	7	8

For each person, the order in which the stools were tested was randomized. This is a randomized block experiment, with subjects playing the role of blocks. The test consists of these hypotheses:

H_0: Mean effort does not depend on type of stool.

H_a: Mean effort does depend on type of stool.

Test statistic: $F = \dfrac{\text{MSTr}}{\text{MSE}}$

Computations are summarized in Table 15.7, an ANOVA table from MINITAB.

Table 15.7 ■ ANOVA Table for Example 15.10

Two-way Analysis of Variance
Analysis of Variance for Effort

Source	DF	SS	MS	F	P
Stool	3	81.19	27.06	22.36	0.000
Block	8	66.50	8.31	6.87	0.000
Error	24	29.06	1.21		
Total	**35**	**176.75**			

The test statistic value is 22.36, with P-value = .000. Because P-value $< \alpha$, we reject H_0. There is sufficient evidence to conclude that the mean effort required is not the same for all four stool types.

Experiments such as the one described in Example 15.10, in which repeated observations are made on the same experimental unit, are sometimes called *repeated-measures designs*. Such designs should not be used when application of the first several treatments somehow affects responses to later treatments. This would be the case if treatments were different methods for learning the same skill, so that if all treatments were given to the same subject, the response to the treatment given last would presumably be much better than the response to the treatment initially applied.

■ Multiple Comparisons

As in single-factor ANOVA, once H_0 has been rejected, further analysis of the data is appropriate to identify significant differences among the treatments. The Tukey–Kramer method is easily adapted for this purpose.

Declare that treatments i and j differ significantly if the interval

$$(\bar{x}_i - \bar{x}_j) \pm q\sqrt{\frac{\text{MSE}}{l}}$$

does not include 0, where q is based on a comparison of k treatments and error df $= (k-1)(l-1)$.

■ Example 15.11   Multiple Comparisons for Stool Designs

In Example 15.10, we had $k = 4$ and error df $= 24$, from which $q = 4.91$ for a 99% simultaneous confidence level. The $\pm$ factor for each interval is

$$q\sqrt{\frac{\text{MSE}}{l}} = 4.91\sqrt{\frac{1.21}{9}} = 1.80$$

The four treatment means arranged in order are

Treatment	A	D	C	B
Sample mean	8.556	9.222	10.778	12.444

It is easily verified that the corresponding underscoring pattern is as follows:

Treatment	A	D	C	B
Sample mean	8.556	9.222	10.778	12.444

■ Exercises 15.30–15.38

15.30 A particular county employs three assessors, who are responsible for determining the value of residential property in the county. To see whether these assessors differ systematically in their appraisals, five houses are selected, and each assessor is asked to determine the market value of each house. Explain why a randomized block experiment (with blocks corresponding to the 5 houses) was used rather than a completely randomized experiment involving a total of 15 houses with each assessor asked to appraise 5 different houses (a different group of 5 for each assessor).

15.31 The following display is a partially completed ANOVA table for the experiment described in Exercise 15.30 (with houses representing blocks and assessors representing treatments):

Source of Variation	df	Sum of Squares	Mean Square	F
Treatments		11.7		
Blocks		113.5		
Error				
Total		250.8		

a. Fill in the missing entries in the ANOVA table.
b. Use the ANOVA *F* statistic and a .05 level of significance to test the null hypothesis of no difference between assessors.

15.32 Land-treatment wastewater-processing systems work by removing nutrients and thereby discharging water of better quality. The land used is often planted with a crop such as corn, because plant uptake removes nitrogen from the water and sale of the crop helps reduce the costs of wastewater treatment. The concentration of nitrogen in treated water was observed from 1975 to 1979 under wastewater application rates of 0, 0.05, and 0.1 m/wk. A randomized block ANOVA was performed, with the 5 years serving as blocks. The following display is a partially completed ANOVA table from the article "Quality of Percolate Water After Treatment of a Municipal Wastewater Effluent by a Crop Irrigation System" (*Journal of Environmental Quality* [1984]: 256–264):

Source of Variation	df	Sum of Squares	Mean Square	F
Treatments		1835.2		
Blocks				
Error		206.1		
Total	14	2134.1		

a. Complete the ANOVA table.
b. Is there sufficient evidence to reject the null hypothesis of no difference between the true mean nitrogen concentrations for the three application rates? Use $\alpha = .05$.

15.33 In a comparison of the energy efficiency of three types of ovens (conventional, biradiant, and convection), the energy used in cooking was measured for eight different foods (one-layer cake, two-layer cake, biscuits, bread, frozen pie, baked potatoes, lasagna, and meat loaf). Because a comparison between the three types of ovens is desired, a randomized block ANOVA (with the eight foods serving as blocks) will be used. Suppose that calculations result in the quantities SSTo = 4.57, SSTr = 3.97, and SSBl = 0.2503. (A similar study is described in the article "Optimizing Oven Radiant Energy Use," *Home Economics Research Journal* [1980]: 242–251.) Construct an ANOVA table and test the null hypothesis of no difference in mean energy use for the three types of ovens. Use a .01 significance level.

15.34 The article "Rate of Stuttering Adaptation Under Two Electro-Shock Conditions" (*Behavior Research Therapy* [1967]: 49–54) gave adaptation scores for three different treatments: no shock (Treatment 1), shock following each stuttered word (Treatment 2), and shock during each moment of stuttering (Treatment 3). These treatments were used on each of 18 stutterers. The 18 subjects were viewed as blocks, and the data were analyzed using a randomized block ANOVA. Summary quantities are SSTr = 28.78, SSBl = 2977.67, and SSE = 469.55. Construct the ANOVA table, and test at significance level .05 to see whether true average adaptation score depends on the treatment given.

15.35 The following table shows average height of cotton plants during 1978–1980 under three different effluent application rates (350, 440, and 515 mm) ("Drip Irrigation of Cotton with Treated Municipal Effluents: Yield Response," *Journal of Environmental Quality* [1984]: 231–238):

Year	Application Rate		
	350	440	515
1978	166	176	177
1979	109	126	136
1980	140	155	156

a. These data were analyzed using a randomized block ANOVA, with years serving as blocks. Explain why this analysis would be better for comparing treatments than a completely randomized ANOVA.
b. With Treatments 1, 2, and 3 denoting the application rates 350, 440, and 515 mm, respectively, the summary quantities are

$$\sum(x - \bar{\bar{x}})^2 = 4266.0 \qquad \bar{\bar{x}} = 149.00$$

$$\bar{x}_1 = 138.33 \qquad \bar{x}_2 = 152.33 \qquad \bar{x}_3 = 156.33$$
$$\bar{b}_1 = 173.00 \qquad \bar{b}_2 = 123.67 \qquad \bar{b}_3 = 150.33$$

Construct an ANOVA table. Using $\alpha = .05$, determine whether the true mean height differs for the three effluent rates.
c. Identify significant differences among the rates. (Hint: $q = 5.04$.)

15.36 The article "Responsiveness of Food Sales to Shelf Space Changes in Supermarkets" (*Journal of Marketing Research* [1964]: 63–67) described an experiment to assess the effect of allotted shelf space on product sales. Two of the products studied were baking soda (a staple product) and Tang (considered an impulse product). Six stores (blocks) were used in the experiment, and six different shelf-space allotments were tried for 1 week each. Space allotments of 2, 4, 6, 8, 10, and 12 ft were used for baking soda, and space allotments of 6, 9, 12, 15, 18, and 21 ft were used for Tang. The investigator speculated that sales of staple goods would not be sensitive to changes in shelf space, whereas sales of impulse products would be affected by changes in shelf space. Data on number of boxes of baking soda and on number of containers of Tang sold during a 1-week period are given in the following tables:

Store	Baking Soda Shelf Space					
	2	4	6	8	10	12
1	36	42	36	40	30	22
2	74	61	65	67	83	84
3	40	58	42	73	69	63
4	43	65	65	41	43	47
5	27	33	35	17	40	26
6	23	31	36	38	42	37

Store	Tang Shelf Space					
	6	9	12	15	18	21
1	30	35	25	25	38	31
2	47	59	43	62	65	48
3	47	55	48	54	36	54
4	29	19	41	27	33	39
5	17	11	25	23	24	26
6	22	9	19	18	25	22

Construct an ANOVA table for each product, and test the appropriate hypotheses. Use a significance level of .05. Was the investigator correct in his speculation that sales of the staple product would not be affected by shelf space allocation, whereas sales of the impulse product would be affected? Explain.

15.37 The article "Measuring Treatment Effects Through Comparisons Along Plot Boundaries" (*Forest Science* [1980]: 704–709) reported the results of a randomized block experiment. Five different sources of pine seed were used in each of four blocks. The following table gives data on plant height (in meters):

Source	Block			
	I	II	III	IV
1	7.1	5.8	7.2	6.9
2	6.2	5.3	7.7	4.7
3	7.9	5.4	8.6	6.2
4	9.0	5.9	5.7	7.3
5	7.0	6.3	4.4	6.1

Do the data provide sufficient evidence to conclude that the true mean height is not the same for all five seed sources? Use a .05 level of significance.

15.38 The article cited in Exercise 15.37 also gave the following data on survival rate for five different seed sources:

Source	Block			
	I	II	III	IV
1	62.5	87.5	50.0	70.3
2	50.0	50.0	54.7	59.4
3	93.8	92.2	87.5	87.5
4	96.9	76.6	70.3	65.6
5	56.3	50.0	45.3	56.3

a. Construct an ANOVA table, and test the null hypothesis of no difference in true mean survival rates for the five seed sources. Use $\alpha = .05$.

b. Use multiple comparisons to identify significant differences among the seed sources. (Hint: $q = 5.04$.)

▪ 15.4 Two-Factor ANOVA

An investigator is often interested in assessing the effects of two different factors on a response variable. Consider the following examples.

1. A physical education researcher wishes to know how body density of football players varies with position played (a categorical factor with categories defensive back, offensive back, defensive lineman, and offensive lineman) and level of play (a second categorical factor with categories professional, college Division I, college Division II, and college Division III).

2. An agricultural scientist is interested in seeing how yield of tomatoes is affected by choice of variety planted (a categorical factor, with each category corresponding to a different variety) and planting density (a quantitative factor, with a level corresponding to each planting density being considered).

3. An applied chemist might wish to investigate how shear strength of a particular adhesive varies with application temperature (a quantitative factor with levels 250°F, 260°F, and 270°F) and application pressure (a quantitative factor with levels 110, 120, 130, and 140 lb/in.^{2}).

Let's label the two factors under study Factor A and Factor B. Even when a factor is categorical, it simplifies terminology to refer to the categories as **levels**. Thus, in the first example, the categorical factor *position played* has four levels. The number of levels of Factor A is denoted by k, and l denotes the number of levels of Factor B, as shown in Table 15.8. This rectangular table contains a row corresponding to each level of Factor A and a column corresponding to each level of Factor B. Each cell in the table corresponds to a particular level of Factor A in combination with a particular level of Factor B. Because there are l cells in each row and k rows, there are kl cells in the table. The kl different combinations of Factor A and Factor B levels are often referred to as **treatments**. For example, if there are three

Table 15.8 ▪ A Table of Factor Combinations for a Two-Way ANOVA Experiment

tomato varieties and four different planting densities under consideration, the number of treatments is 12.

Suppose that an experiment is carried out, resulting in a data set that contains some number of observations for each of the kl treatments. In general, there could be more observations for some treatments than for others, and there may even be a few treatments for which no observations are available. An experimenter may set out to make the same number of observations on each treatment, but, occasionally, forces beyond the experimenter's control — the death of an experimental subject, malfunctioning equipment, and so on — result in different sample sizes for some treatments. Such imbalances in sample sizes makes analysis of the data rather difficult. We will restrict our discussion to consideration of data sets containing the same number of observations for each treatment, and we will let m denote this number.

▪ **Notation**

k = number of levels of Factor A

l = number of levels of Factor B

kl = number of treatments (each one a combination of a Factor A level and a Factor B level)

m = number of observations on each treatment

▪ Example 15.12   Tomato Yield and Planting Density

An experiment was carried out to assess the effects of tomato variety (Factor A, with $k = 3$ levels) and planting density (Factor B, with $l = 4$ levels of 10,000, 20,000, 30,000, and 40,000 plants per hectare) on yield. Each of the $kl = 12$ treatments was used on $m = 3$ plots, resulting in the data set consisting of $klm = 36$ observations shown in Table 15.9 (adapted from "Effects of Plant Density on Tomato Yields in Western Nigeria," *Experimental Agriculture* [1976]: 43–47).

Table 15.9 ▪ Data from the Two-Factor Experiment of Example 15.12

Variety (Factor A)	Density (Factor B)			
	1	2	3	4
1	7.9, 9.2, 10.5	11.2, 12.8, 13.3	12.1, 12.6, 14.0	9.1, 10.8, 12.5
2	8.1, 8.6, 10.1	11.5, 12.7, 13.7	13.7, 14.4, 15.4	11.3, 12.5, 14.5
3	15.3, 16.1, 17.5	16.6, 18.5, 19.2	18.0, 20.8, 21.0	17.2, 18.4, 18.9

Sample average yields for each treatment, each level of Factor A, and each level of Factor B are important summary quantities. These can be displayed in a rectangular table (see Table 15.10). A plot of these sample averages is also quite informative. First, construct horizontal and vertical axes, and scale the vertical axis in units of the response variable (yield). Then mark a point on the horizontal axis for each level of one of the factors (either Factor A or Factor B can be chosen). Now above each such mark, plot a point for the sample average response for each level of the other factor. Finally, connect all points corresponding to the same level of the other factor using straight line segments.

Table 15.10 ▪ **Sample Means for the 12 Treatments of Example 15.12**

Factor A (Variety)	Factor B (Planting Density)				Sample Average Yield for Each Level of Factor A
	1	2	3	4	
1	9.20	12.43	12.90	10.80	11.33
2	8.93	12.63	14.50	12.77	12.21
3	16.30	18.10	19.93	18.17	18.13
Sample Average Yield for Each Level of Factor B	11.48	14.39	15.78	13.91	**Grand mean** = 13.89

Figure 15.8 displays two plots: one in which Factor A levels mark the horizontal axis and one in which Factor B levels mark the horizontal axis; usually only one of the two plots is constructed.

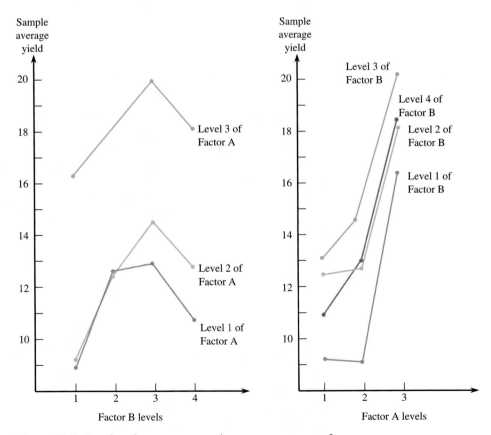

Figure 15.8 Graphs of treatment sample average responses for the data of Example 15.12.

▪ Interaction

An important aspect of two-factor studies involves assessing how simultaneous changes in the levels of both factors affect the response. As a simple example, suppose that an automobile manufacturer is studying engine efficiency (measured in miles per gallon) for two different engine sizes (Factor A, with $k = 2$ levels) in combination with two different carburetor designs (Factor B, with $l = 2$ levels). Consider the two possible sets of true average responses displayed in Figure 15.9. In Figure 15.9(a), when Factor A changes from Level 1 to Level 2 and Factor B remains at Level 1 (the change within the first column), the true average response increases by 2. Similarly, when Factor B changes from Level 1 to Level 2 and Factor A is fixed at Level 1 (the change within the first row), the true average response increases by 3. And when the levels of both factors are changed from 1 to 2, the true average response increases by 5, which is the sum of the two "one-at-a-time" increases. This is because the change in true average response when the level of either factor changes from 1 to 2 is the same for each level of the other factor: The change within either row is 3, and the change within either column is 2. In this case, changes in the levels of the two factors affect the true average response separately or in an additive manner.

Figure 15.9 Two possible sets of true average responses when $k = 2$ and $l = 2$.

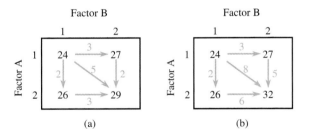

(a)   (b)

The changes in true average responses in the first row and in the first column of Figure 15.9(b) are 3 and 2, respectively, exactly as in Figure 15.9(a). However, the change in true average response when the levels of both factors change simultaneously from 1 to 2 is 8, which is much larger than the separate changes suggest. In this case, there is interaction between the two factors, so that the effect of simultaneous changes cannot be determined from the individual effects of separate changes. This is because in Figure 15.9(b), the change in going from the first to the second column is different for the two rows, and the change in going from the first to the second row is different for the two columns. That is, *the change in true average response when the level of one factor changes depends on the level of the other factor*. This is not true in Figure 15.9(a).

When there are more than two levels of either factor, a graph of true average responses, similar to that for sample average responses in Figure 15.8, provides insight into how changes in the level of one factor depend on the level of the other factor. Figure 15.10 shows several possible such graphs when $k = 4$ and $l = 3$. The most general situation is pictured in Figure 15.10(a). There, the change in true average response when the level of Factor B is changed (a vertical distance) depends on the level of Factor A. An analogous property would hold if the picture were redrawn so that levels of Factor B were marked on the horizontal axis. This is a prototypical picture suggesting **interaction** between the factors — the change in true average response when the level of one factor changes depends on the level of the other factor.

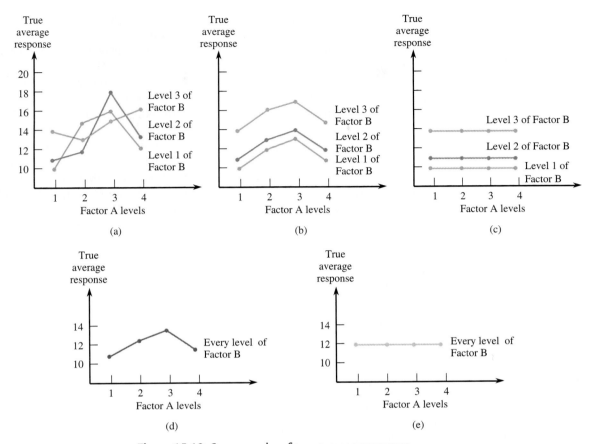

Figure 15.10 Some graphs of true average responses.

There is no interaction between the factors when the connected line segments are parallel, as in Figure 15.10(b). Then the change in true average response when the level of one factor changes is the same for each level of the other factor (the vertical distances are the same for each level of Factor A). Figure 15.10(c) illustrates an even more restrictive situation — there is no interaction between factors; in addition, the true average response does not depend on the level of Factor A. Only when the graph looks like this can it be said that Factor A has no effect on the responses. Similarly, the graph in Figure 15.10(d) indicates no interaction and no dependence on the level of Factor B. A final case, illustrated in Figure 15.10(e), shows a single set of four points connected by horizontal line segments, which indicates that the true average response is identical for every level of both factors.

If the graphs of true average responses are connected line segments that are parallel, there is no interaction between the factors. In this case, the change in true average response when the level of one factor is changed is the same for each level of the other factor. Special cases of no interaction are as follows:

1. The true average response is the same for each level of Factor A (no Factor A main effects).

2. The true average response is the same for each level of Factor B (no factor B main effects).

The graphs in Figure 15.10 depict true average responses — that is, quantities whose values are fixed but unknown to an investigator. Figure 15.8 contains graphs of the sample average responses based on data resulting from an experiment. These sample averages are, of course, subject to variability because there is sampling variation in the individual observations. If the experiment discussed in Example 15.12 was repeated, the resulting graphs of sample averages would probably look somewhat different from the graph in Figure 15.8 — perhaps a great deal different if there was substantial underlying variability in responses. Even when there is no interaction among factors, the connected line segments in the sample mean picture will not typically be exactly parallel, and they may deviate quite a bit from parallelism in the presence of substantial underlying variability. Similarly, there might actually be no Factor A effects (Figure 15.10(c)), yet the sample graphs would not usually be exactly horizontal. The sample graphs give us insight, but formal inferential procedures are necessary to draw sound conclusions about the nature of the true average responses for different factor levels.

▪ Hypotheses and *F* Tests

ANOVA procedures can used to test hypotheses about the effects of two different factors on a response.

▪ Basic Assumptions for Two-Factor ANOVA

The observations on any particular treatment are independently selected from a normal distribution with variance σ^2 (the same variance for each treatment), and samples from different treatments are independent of one another.

Because of the normality assumption, tests based on F statistics and distributions are appropriate. The necessary sums of squares result from breaking up $\text{SSTo} = \sum (x - \bar{\bar{x}})^2$ into four parts, which reflect random variation and variation attributable to various factor effects:

$$\text{SSTo} = \text{SSA} + \text{SSB} + \text{SSAB} + \text{SSE}$$

where

1. SSTo is total sum of squares, with associated df $= klm - 1$.
2. SSA is the Factor A main effect sum of squares, with associated df $= k - 1$.
3. SSB is the Factor B main effect sum of squares, with associated df $= l - 1$.
4. SSAB is the interaction sum of squares, with associated df $= (k-1)(l-1)$.
5. SSE is error sum of squares, with associated df $= kl(m-1)$.

The formulas for these sums of squares are similar to those given in previous sections, so we will not give them here. The standard statistical computer packages can calculate all sums of squares and other necessary quantities. The magnitude of SSE is related entirely to the amount of underlying variability (as specified by σ^2)

in the distributions being sampled. It has nothing to do with values of the various true average responses. SSAB reflects in part underlying variability, but its value is also affected by whether there is interaction between the factors. In general, the more extensive the amount of interaction (i.e., the further the graphs of true average responses are from being parallel), the larger the value of SSAB tends to be. The test statistic for testing the null hypothesis that there is no interaction between factors is the ratio $\dfrac{\text{MSAB}}{\text{MSE}}$. A large value of this statistic suggests that interaction effects are present.

Both the absence of Factor A effects and the absence of Factor B effects are special cases of no-interaction situations. If the data suggest that interaction is present, it does not make sense to investigate effects of one factor without reference to the other factor. Our recommendation is that hypotheses concerning the presence or absence of separate factor effects be tested only if the hypothesis of no interaction is not rejected. Then, the Factor A main effect sum of squares, SSA, will reflect random variation as well as any differences between true average responses for different levels of Factor A. The same applies to SSB.

■ **Two-Factor ANOVA Hypotheses and Tests**

1. H_0: There is no interaction between factors.

   H_a: There is interaction between factors.

   Test statistic: $F = \dfrac{\text{MSAB}}{\text{MSE}}$

   based on $\text{df}_1 = (k - 1)(l - 1)$ and $\text{df}_2 = kl(m - 1)$

The following two hypotheses should be tested only if the hypothesis of no interaction is not rejected:

2. H_0: There are no Factor A main effects (true average response is the same for each level of Factor A).

   H_a: H_0 is not true.

   Test statistic: $F = \dfrac{\text{MSA}}{\text{MSE}}$

   based on $\text{df}_1 = k - 1$ and $\text{df}_2 = kl(m - 1)$

3. H_0: There are no Factor B main effects.

   H_a: H_0 is not true.

   Test statistic: $F = \dfrac{\text{MSB}}{\text{MSE}}$

   based on $\text{df}_1 = l - 1$ and $\text{df}_2 = kl(m - 1)$

Computations are typically summarized in an ANOVA table, as shown in Table 15.11.

Table 15.11   ▪   Analysis of Variance Table

Source of Variation	df	Sum of Squares	Mean Square	F
Factor A main effects	$k - 1$	SSA	$MSA = \dfrac{SSA}{k - 1}$	$F = \dfrac{MSA}{MSE}$
Factor B main effects	$l - 1$	SSB	$MSB = \dfrac{SSB}{l - 1}$	$F = \dfrac{MSB}{MSE}$
AB Interaction	$(k - 1)(l - 1)$	SSAB	$MSAB = \dfrac{SSAB}{(k - 1)(l - 1)}$	$F = \dfrac{MSAB}{MSE}$
Error	$kl(m - 1)$	SSE	$MSE = \dfrac{SSE}{kl(m - 1)}$	
Total	$klm - 1$	SSTo		

▪ **Example 15.13   More on Tomato Yield**

An ANOVA table for the tomato yield data of Example 15.12 is given in Table 15.12.

Table 15.12   ▪   Analysis of Variance on Tomato Yield for Example 15.13

Source of Variation	df	Sum of Squares	Mean Square	F
Variety	2	327.60	163.80	103.70
Density	3	86.69	28.90	18.30
Interaction	6	8.03	1.34	0.85
Error	24	38.04	1.58	
Total	35	460.36		

1. Test for interaction:

   H_0:   no interaction between variety and density

   Calculated $F_{AB} = 0.85$, based on $df_1 = 6$ and $df_2 = 24$

   From Appendix Table 7, the smallest value for these degrees of freedom is 2.04, so P-value $> .10$. There is no evidence of interaction, so it is appropriate to carry out further tests concerning the presence of main effects.

2. Test for Factor A effects:

   H_0:   Factor A (variety) main effects are absent

   Calculated $F_A = 103.7$, based on $df_1 = 2$ and $df_2 = 24$

   Appendix Table 7 shows that P-value $< .001$. We therefore reject H_0 and conclude that true average yield does depend on variety.

3. Test for Factor B effects:

   H_0:   Factor B (density) main effects are absent

   Calculated $F_B = 18.3$, based on $df_1 = 3$ and $df_2 = 24$

Again, P-value $< .001$, so we reject H_0 and conclude that true average yield does depend on planting density.

After the null hypothesis of no Factor A main effects has been rejected, significant differences in Factor A levels can be identified by using a multiple comparisons procedure. In particular, the Tukey–Kramer method described previously can be applied. The quantities $\bar{x}_1, \bar{x}_2, \ldots, \bar{x}_k$ are now the sample average responses for levels $1, \ldots, k$ of Factor A, and the number of error degrees of freedom is $kl(m - 1)$. A similar comment applies to Factor B main effects and significant differences in Factor B levels.

▪ The Case $m = 1$

There is a problem with the foregoing analysis when $m = 1$ (one observation on each treatment). Although we did not give the formula, MSE is an estimate of σ^2 obtained by computing a separate sample variance s^2 for the m observations on each treatment and then averaging these kl sample variances. With only one observation on each treatment, there is no way to estimate σ^2 separately from each of the treatments.

One way out of this dilemma is to assume a priori that there is no interaction between factors. This should, of course, be done only when the investigator has sound reasons, based on a thorough understanding of the problem, for believing that the factors contribute separately to the response. Having made this assumption, the investigator can then use what would otherwise be an interaction sum of squares for SSE. The fundamental identity becomes

SSTo = SSA + SSB + SSE

with the four associated degrees of freedom $kl - 1, k - 1, l - 1$, and $(k - 1)(l - 1)$.

Table 15.13 gives the corresponding ANOVA table. F_A is the test statistic for testing the null hypothesis that true average responses are identical for all Factor A levels. F_B plays a similar role for Factor B main effects. The analysis of data from a randomized block experiment in fact assumed no interaction between treatments and blocks. If SSTr is relabeled SSA and if SSBl is relabeled SSB, the formulas for all sums of squares given in Section 15.3 are valid here (and are now the sample average responses for Factor A levels and Factor B levels, respectively).

Table 15.13 ▪ ANOVA Table for Two-Factor Experiment with $m = 1$

Source of Variation	df	Sum of Squares	Mean Square	F
Factor A	$k - 1$	SSA	$MSA = \dfrac{SSA}{k - 1}$	$F = \dfrac{MSA}{MSE}$
Factor B	$l - 1$	SSB	$MSB = \dfrac{SSB}{l - 1}$	$F = \dfrac{MSB}{MSE}$
Error	$(k - 1)(l - 1)$	SSE	$MSE = \dfrac{SSE}{(k - 1)(m - 1)}$	
Total	$kl - 1$	SSTo		

▪ Example 15.14   Effect of Soil Type and Pipe Coating on Corrosion

When metal pipe is buried in soil, it is desirable to apply a coating to retard corrosion. Four different coatings are under consideration for use with pipe that will ultimately be buried in three types of soil. An experiment to investigate the effects of these coatings and soils was carried out by first selecting 12 pipe segments and applying each coating to 3 segments. The segments were then buried in soil for a specified period in such a way that each soil type received one piece with each coating. The resulting data (depth of corrosion) and ANOVA table are given in Table 15.14. Assuming that there is no interaction between coating type and soil type, let's test at level .05 for the presence of separate Factor A (coating) and Factor B (soil) effects.

Table 15.14 ▪ Data and ANOVA Table for Example 15.14

Factor A (Coating)	Factor B (Soil) 1	2	3	Sample Average
1	64	49	50	54.33
2	53	51	48	50.67
3	47	45	50	47.33
4	51	43	52	48.67
Sample average	53.75	47.00	50.00	$\bar{\bar{x}} = 50.25$

Source of Variation	df	Sum of Squares	Mean Square	F
Factor A	3	83.5	27.8	$F_A = \dfrac{27.8}{20.6} = 1.3$
Factor B	2	91.5	45.8	$F_B = \dfrac{45.8}{20.6} = 2.2$
Error	6	123.3	20.6	
Total	11	298.3		

Appendix Table 7 shows that P-value $> .10$ for both tests. It appears that the true average response (amount of corrosion) depends on neither the coating used nor the type of soil in which the pipe is buried.

▪ Exercises 15.39–15.51

15.39 The behavior of undergraduate students when they were exposed to various odors was examined by the authors of the article "Effects of Environmental Odor and Coping Style on Negative Affect, Anger, Arousal, and Escape" (*Journal of Applied Social Psychology* [1999]: 245–260). The following table was constructed using data on reported discomfort level (measured on a scale from 1 to 5); there were 24 students in each odor–gender combination:

Type of Odor	Male $\bar{x}$	Female $\bar{x}$
No odor	1.36	1.68
Rotten egg	1.83	2.33
Skunk	2.42	2.69
Cigarette ash	2.74	3.16

a. Construct a graph (similar to those in Figure 15.8) that shows the mean discomfort level on the vertical axis. Mark the four odor categories on the horizontal axis. Then plot the four means for the males and connect them with line segments. Plot the four means for females and connect them with line segments.

b. Interpret the interaction plot. Do you think that there is an interaction between gender and type of odor?

15.40 Many students report that test anxiety affects their performance on exams. A study of the effect of anxiety and instructional mode (lecture versus independent study) on test performance was described in the article "Interactive Effects of Achievement Anxiety, Academic Achievement, and Instructional Mode on Performance and Course Attitudes" (*Home Economics Research Journal* [1980]: 216–227). Students classified as belonging to either high- or low-achievement anxiety groups (Factor A) were assigned to one of the two instructional modes (Factor B). Mean test scores for the four treatments (factor–level combinations) are given in the following table:

Anxiety Group	Instructional Mode	
	Lecture	Independent Group
High	145.8	144.3
Low	142.9	144.8

Use these means to construct a graph (similar to those in Figure 15.8) of the treatment sample averages. Does the picture suggest the existence of an interaction between factors? Explain.

15.41 The following plot, in which the response variable was tension (a measure of job strain) and the two factors of interest were peer group interaction (with two levels—high and low) and self-esteem (also with two levels—high and low) appeared in the article "Group Process–Work Outcome Relationships: A Note on the Moderating Impact of Self-Esteem" (*Academy of Management Journal* [1982]: 575–585):

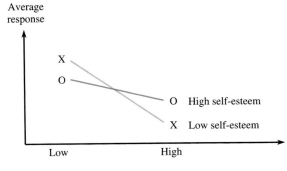

a. Does this plot suggest an interaction between peer group interaction and self-esteem? Explain.

b. The authors of the article stated that "peer group interaction had a stronger effect on individuals with lower self-esteem than on those with higher self-esteem." Do you agree with this statement? Explain.

15.42 Explain why the individual effects of Factor A or Factor B cannot be interpreted when an interaction between Factors A and B is present.

15.43 The article "Experimental Analysis of Prey Selection by Largemouth Bass" (*Transactions of the American Fisheries Society* [1991]: 500–508) gave an ANOVA summary in which the response variable was a certain preference index, there were three sizes of bass, and there were two different species of prey. Three observations were made for each size–species combination. Sums of squares for size, species, and interaction were reported as 0.088, 0.048, and 0.048, respectively, and SSTo = 0.316. Test the relevant hypotheses using a significance level of .01.

15.44 The study described in the article "From Here to Equity: The Influence of Status on Student Access to and Understanding of Science" (*Science Education* [1999]: 577–602) attempted to quantify the effect of socioeconomic status on learning science. The response variable was "rate of talk" (the number of on-task talk speech acts per minute) during group work. Data were also collected on the variable *socioeconomic status* (low, middle, high) and the variable *gender* (female, male). The investigator allowed for an interaction between gender and status. Assume that there were 12 students in each status–gender combination, for a total of 72 subjects. The following partially completed ANOVA table approximately matches summary statistics given in the article:

Source of Variation	df	Sum of Squares	Mean Square	F
Status	2			14.49
Gender	1			0.15
Interaction	2			0.95
Error	66		0.0120	
Total	**71**			

a. Fill in the missing numbers in the ANOVA table.
b. Is there a significant interaction between socioeconomic status and gender?
c. Is there a difference between the mean rate of talk scores for males and females?
d. Is there a difference between the mean rate of talk" scores across the three different status groups?

15.45 Is self-esteem related to year in college or to fraternity membership? The authors of the article

"Self-Esteem Among College Men as a Function of Greek Affiliation and Year in College" (*Journal of College Student Development* [1998]: 611–613) interviewed 75 college men who were members of fraternities and 57 who were not. In addition to Greek status (in a fraternity or not in a fraternity), year in college (freshman, sophomore, junior, senior) was also noted. Subjects completed a 10-item self-esteem inventory, which was used to compute a self-esteem score. The article reported the following F statistic values from a two-way ANOVA:

Source	F
Greek status	20.53
Year	2.59
Interaction	0.70

a. Is there evidence of a significant interaction between year and Greek status?
b. Is the main effect for Greek status significant?
c. Is the main effect for year significant?

15.46 Three ultrasonic devices (Factor A, with levels 20, 30, and 40 kHz) were tested for effectiveness under two test conditions (Factor B, with levels plentiful food supply and restricted food supply). Daily food consumption was recorded for three rats under each factor–level combination for a total of 18 observations. Data compatible with summary values given in the article "Variables Affecting Ultrasound Repellency in Philippine Rats" (*Journal of Wildlife Management* [1982]: 148–155) were used to obtain the sums of squares given in the accompanying ANOVA table. Complete the table, and use it to test the relevant hypotheses.

Source of Variation	df	Sum of Squares	Mean Square	F
Factor A main effects		4206		
Factor B main effects		1782		
Interaction between Factors A and B				
Error		2911		
Total		**10,846**		

15.47 The article "Learning, Opportunity to Cheat, and Amount of Reward" (*Journal of Experimental Education* [1977]: 30–40) described a study to determine the effects of expectations concerning cheating during a test and perceived payoff on test performance. Subjects, students at UCLA, were randomly assigned to a particular factor–level combination. Factor A was expectation of opportunity to cheat, with levels high, medium, and low. Those in the high-expectation group were asked to study and then re-call a list of words. For the first four lists, the students were left alone in a room with the door closed, so they could look at the original list of words if they wanted to. The medium-expectation group was asked to study and recall the list while left alone but with the door open. For the low-expectation group, the experimenter remained in the room. For study and recall of a fifth list, the experimenter stayed in the room for all three groups, thus precluding any cheating on the fifth list. Score on the fifth test was the response variable. The second factor (Factor B) under study was the perceived payoff, with a high and a low level. The high-payoff group was told that if they scored above average on the test, they would receive 2 hr of credit rather than just 1 hr (subjects were fulfilling a course requirement by participating in experiments for a certain number of hours during the course). The low-payoff group was not given any extra incentive for scoring above the average. The article gave the following statistics: $F_A = 4.99$, $F_B = 4.81$, $F_{AB} < 1$, error df ≈ 120. Test the null hypothesis of no interaction between the factors. If appropriate, test the null hypotheses of no Factor A and no Factor B effects. Use $\alpha = .05$.

15.48 The following (slightly modified) partial ANOVA table, for which the response variable was territory size, appeared in the article "An Experimental Test of Mate Defense in an Iguanid Lizard" (*Ecology* [1991]: 1218–1224):

Source of Variation	df	Sum of Squares
Age	1	0.614
Sex	1	1.754
Interaction	1	0.146
Error	80	5.624

a. How many age classes were there?
b. How many observations were made for each age–sex combination?
c. What conclusions can be drawn about how the factors affect the response variable?

15.49 Identification of sex in human skeletons is an important part of many anthropological studies. An experiment conducted to determine whether measurements of the sacrum could be used to determine sex was described in the article "Univariate and Multivariate Methods for Sexing the Sacrum" (*American Journal of Physical Anthropology* [1978]: 103–110). Sacra from skeletons of individuals of known ethnic group (Factor A, with two levels — Caucasian and black) and sex (Factor B, with two levels — male and female) were measured and the lengths recorded. Data compatible with summary quantities given in the article were used to compute the following: SSA = 857, SSB = 291, SSAB = 32, SSE = 5541, and error df = 36.

a. Use a significance level of .01 to test the null hypothesis of no interaction between ethnic group and sex.

b. Using a .01 significance level, test to determine whether the true average length differs for the two ethnic groups.

c. Using a .01 significance level, test to determine whether the true average length differs for males and females.

15.50 The article "Food Consumption and Energy Requirements of Captive Bald Eagles" (*Journal of Wildlife Management* [1982]: 646–654) investigated mean gross daily energy intake (the response variable) for different diet types (Factor A, with three levels—fish, rabbit, and mallard) and temperatures (Factor B, with three levels). Summary quantities given in the article were used to generate data, resulting in SSA = 18,138, SSB = 5182, SSAB = 1737, SSE = 11,291, and error df = 36. Construct an ANOVA table, and test the relevant hypotheses.

15.51 The effect of three different soil types and three phosphate application rates on total phosphorus uptake (in milligrams) of white clover was examined in the article "A Glasshouse Comparison of Six Phosphate Fertilisers" (*New Zealand Journal of Experimental Agriculture* [1984]: 131–140). Only one observation was obtained for each factor–level combination. Assuming that there is no interaction between soil type and application rate, use the accompanying data to construct an ANOVA table and to test the null hypotheses of no soil type main effects and of no application rate main effects. Also identify significant differences among soil types.

Phosphate Application Rate (kg/ha)	Soil Type		
	Ramiha	Konini	Wainui
0	1.29	10.42	17.10
75	11.73	21.08	23.69
150	17.63	31.37	32.88

▪ 15.5 Communicating and Interpreting the Results of Statistical Analyses

The ANOVA procedures introduced in this chapter are used to compare more than two population or treatment means. When a single-factor ANOVA has been used to test the null hypothesis of equal population or treatment means, the value of the *F* statistic and the associated *P*-value usually are reported. It is also fairly common to see the supporting calculations summarized in an ANOVA table, although this is not always the case.

▪ What to Look For in Published Data

Here are some questions to ask when you read an article that includes a description of a single-factor ANOVA:

- Are the assumptions required for the validity of the ANOVA procedure reasonable? Specifically, are the samples independently chosen, or is there random assignment of treatments? Is it reasonable to think that the population or treatment response distributions are normal in shape? Are the reported sample standard deviations consistent with the assumption of equal population or treatment variances?

- What is the *P*-value associated with the test? Does the *P*-value lead to rejection of the null hypothesis?

- If the ANOVA F test led to rejection of the null hypothesis, was a multiple comparisons procedure used to identify differences in the means? Are the results of the multiple comparisons procedure interpreted properly?
- Are the conclusions drawn consistent with the results of the hypothesis test and the multiple comparisons procedure? If H_0 was rejected, does this indicate practical significance or only statistical significance?

As an example, consider an ANOVA performed by the authors of "Perceived Age and Attractiveness of Models in Cigarette Advertisements" (*Journal of Marketing* [1992]: 22–37). The investigators hypothesized that certain cigarette brands aimed their advertising at the young and that the average age of readers of magazines in which the cigarette ads appeared was not the same for the 12 brands of cigarettes studied. This hypothesis was tested using a single-factor ANOVA. The following ANOVA table is from the article:

Source of Variation	df	Sum of Squares	Mean Square	F
Treatments	11	592.66	53.88	4.13
Error	175	2282.98	13.05	
Total	**186**	**2875.64**		

It was reported that the hypothesis of equal mean ages for the 12 brands considered in the study was rejected ($F = 4.13$, P-value $< .001$). This was interpreted as support for the researchers' hypothesis that some brands targeted the young. (Does this conclusion necessarily follow from the result of the ANOVA F test?)

To clarify these results, a multiple comparisons procedure could be used to identify significant differences among brands, although this was not done in the reported research. The article did include sample sizes and sample means for the 12 brands, and this information is reproduced in the following table:

Brand	Sample Size	Average Audience Age for Magazine in Which Ad Appears
Lucky Strike Lights	3	28.5
Newport Lights	5	31.1
Camel Lights	21	31.3
Kool Milds	24	31.3
Newport	12	31.3
Winston Lights	15	31.3
Winston	8	31.6
Virginia Slims Lights	32	33.9
Marlboro	26	34.3
More	17	34.8
Benson Hedges Lights	23	35.1
Carlton	1	41.0

Notice that the sample sizes vary greatly. We might also worry about the validity of the assumptions required for the ANOVA F test. It would be of interest to examine the sample standard deviations and, if the original data were available, to look at various plots of the data.

■ A Word to the Wise: Cautions and Limitations

When using analysis of variance methods to test hypotheses about the differences between population or treatment means, keep the following in mind:

1. In single-factor analysis of variance, the alternative hypothesis is that not all population or treatment means are the same. When we reject the null hypothesis, it does not mean that we have evidence that *all* the population means are different. Remember, the alternative hypothesis is not $\mu_1 \neq \mu_2 \neq \cdots \neq \mu_k$. A multiple comparisons procedure, such as the Tukey–Kramer method presented in Section 15.2, can be used to identify which means differ.

2. As was the case for the two-sample t test of Chapter 11, when the sample sizes are small, we must be willing to assume that the population distributions are at least approximately normal in order for analysis of variance to be an appropriate method of analysis. However, there is an additional assumption that is necessary for ANOVA — that the population or treatment variances are equal. When this assumption is not reasonable, it is sometimes possible to re-express the data (by using a transformation such as a logarithm or the square root) to obtain data for which the ANOVA assumptions are more plausible. This is why it is not uncommon to see an analysis of variance performed using transformed data.

■ Activity 15.1: Exploring Single-Factor ANOVA ──────────

Working with a partner, consider the following:

1. Each of the four accompanying graphs shows a dotplot of data from three separate random samples. For each of the four dotplots, indicate whether you think that the basic assumptions for single-factor ANOVA are plausible. Write a sentence or two justifying your answer.

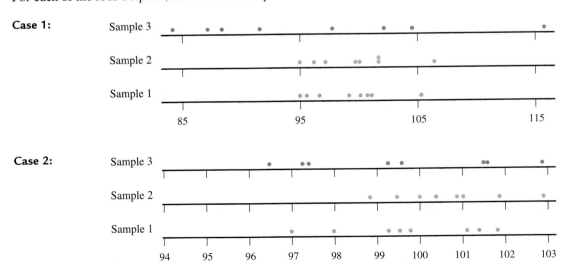

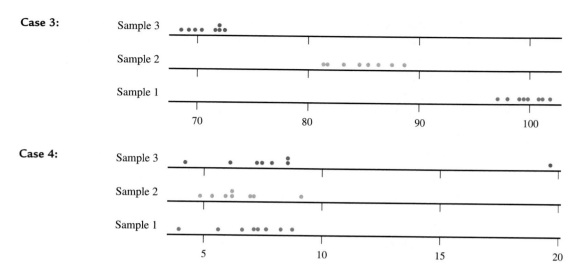

2. Each of the three accompanying graphs shows a dotplot of data from three separate random samples. For each of the three dotplots, indicate whether you think that the three population means are probably not all the same, you think that the three population means might be the same, or you are unsure whether the population means could be the same. Write a sentence or two explaining your reasoning.

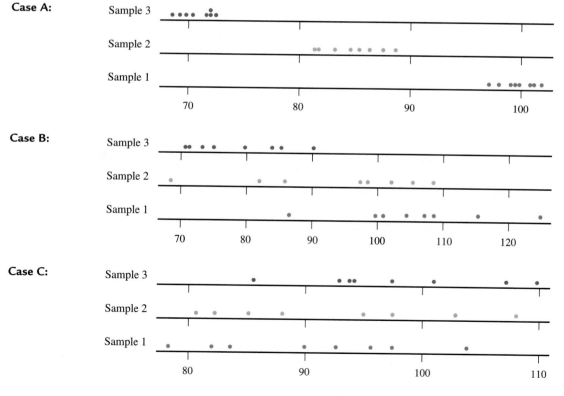

3. Sample data for each of the three cases in Step 2 are shown in the table at the top of the next page. For each of the three cases, carry out a single-factor ANOVA. Is the result of the *F* test in each case consistent with your answers in Step 2? Explain.

Table for Activity Exercise 3

	Case A			Case B			Case C	
Sample 1	Sample 2	Sample 3	Sample 1	Sample 2	Sample 3	Sample 1	Sample 2	Sample 3
99.7	91.3	69.3	104.2	81.9	71.7	82.3	82.4	94.2
98.0	82.0	72.1	107.0	105.4	79.7	97.4	87.5	109.8
101.4	83.6	71.7	88.6	98.4	70.9	83.7	97.3	94.9
99.2	84.8	69.9	99.6	108.4	76.6	103.6	102.6	85.8
101.0	86.5	68.8	124.3	102.1	85.3	78.6	94.8	97.4
101.8	91.5	70.7	100.7	68.9	90.3	90.1	81.3	101.0
99.5	81.8	72.7	108.3	85.8	84.2	92.8	85.2	93.0
97.0	85.5	72.2	116.5	97.5	74.6	95.5	107.8	107.1

■ Summary of Key Concepts and Formulas

Term or Formula	Comment
Single-factor analysis of variance (ANOVA)	A test procedure for determining whether there are significant differences among k population or treatment means. The hypotheses tested are $H_0: \mu_1 = \mu_2 = \cdots = \mu_k$ versus H_a: at least two μ's differ.
Treatment sum of squares: $\text{SSTr} = n_1(\bar{x}_1 - \bar{\bar{x}})^2 + \cdots + n_k(\bar{x}_k - \bar{\bar{x}})^2$	A measure of how different the k sample means $\bar{x}_1, \bar{x}_2, \ldots, \bar{x}_k$ are from one another; associated df $= k - 1$.
Error sum of squares: $\text{SSE} = (n_1 - 1)s_1^2 + \cdots + (n_k - 1)s_k^2$	A measure of the amount of variability within the individual samples; associated df $= N - k$, where $N = n_1 + \cdots + n_k$.
Mean square	A sum of squares divided by its number of degrees of freedom. For single-factor ANOVA, $\text{MSTr} = \dfrac{\text{SSTR}}{(k-1)}$ and $\text{MSE} = \dfrac{\text{SSE}}{(N-k)}$.
$F = \dfrac{\text{MSTr}}{\text{MSE}}$	The test statistic for testing $H_0: \mu_1 = \mu_2 = \cdots = \mu_k$ in a single-factor ANOVA. When H_0 is true, F has an F distribution with numerator df $= k - 1$ and denominator df $= N - k$.
$\text{SSTo} = \text{SSTr} + \text{SSE}$	The fundamental identity in single-factor ANOVA, where $\text{SSTo} = $ total sum of squares $= \Sigma(x - \bar{\bar{x}})^2$.
Tukey–Kramer multiple comparisons procedure	A procedure for identifying significant differences among the μ's once the hypothesis $H_0: \mu_1 = \mu_2 = \cdots = \mu_k$ has been rejected by the ANOVA F test.
Randomized block design	An experimental design that controls for extraneous variation when comparing treatments. The experimental units are grouped into homogeneous *blocks* so that within each block, the units are as similar as possible. Then each treatment is used on exactly one experimental unit in every block (each treatment appears once in every block).

Term or Formula	Comment
Randomized block F test	The four sums of squares for a randomized block design — SSTo, SSTr, SSBl, and SSE (with df $= kl - 1, k - 1, l - 1$, and $(k - 1)(l - 1)$, respectively) — are related by SSTo = SSTr + SSBl + SSE. Usually SSE is obtained by subtraction once the other three sums of squares have been calculated. The null hypothesis is that the true average response does not depend on which treatment is applied. With mean squares $MSTr = \dfrac{SSTr}{(k-1)}$ and $MSE = \dfrac{SSE}{(k-1)(l-1)}$, the test statistic is $F = \dfrac{MSTr}{MSE}$, based on $df_1 = k - 1$ and $df_2 = (k-1)(l-1)$.
Interaction between factors	Two factors are said to interact if the average change in response associated with changing the level of one factor depends on the level of the other factor.
Two-factor ANOVA	When there are k levels of Factor A and l levels of Factor B and m (>1) observations made for each combination of Factor A–B levels, the total sum of squares SSTo can be decomposed into SSA (sum of squares for Factor A main effects), SSB, SSAB (interaction sum of squares), and SSE. The associated numbers of degrees of freedom are $klm - 1, k - 1$, $l - 1, (k-1)(l-1)$, and $kl(m-1)$, respectively. The null hypothesis of no interaction between the two factors is tested using $F_{AB} = \dfrac{MSAB}{MSE}$, where $MSAB = \dfrac{SSAB}{(k-1)(l-1)}$ and $MSE = \dfrac{SSE}{kl(m-1)}$. If this null hypothesis cannot be rejected, tests for Factor A and B main effects are based on $F_A = \dfrac{MSA}{MSE}$ and $F_B = \dfrac{MSB}{MSE}$, respectively.

▪ APPENDIX: ANOVA Computations

▪ Single-Factor ANOVA

Let T_1 denote the sum of the observations in the sample from the first population or treatment, and let $T_2, \ldots, T_k$ denote the other sample totals. Also, let T represent the sum of all N observations — the **grand total** — and

$$CF = \text{correction factor} = \frac{T^2}{N}$$

Then

$$SSTo = \sum_{\text{all } N \text{ obs.}} x^2 - CF$$

$$SSTr = \frac{T_1^2}{n_1} + \frac{T_2^2}{n_2} + \cdots + \frac{T_k^2}{n_k} - CF$$

$$SSE = SSTo - SSTr$$

▪ **Example** 15.A1

Treatment 1	4.2	3.7	5.0	4.8	$T_1 = 17.7$	$n_1 = 4$
Treatment 2	5.7	6.2	6.4		$T_2 = 18.3$	$n_2 = 3$
Treatment 3	4.6	3.2	3.5	3.9	$T_3 = 15.2$	$n_3 = 4$
					$T = 51.2$	$N = 11$

$$\text{CF} = \text{correction factor} = \frac{T^2}{N} = \frac{(51.2)^2}{11} = 238.31$$

$$\text{SSTr} = \frac{T_1^2}{n_1} + \frac{T_2^2}{n_2} + \cdots + \frac{T_k^2}{n_k} - \text{CF}$$

$$= \frac{(17.7)^2}{4} + \frac{(18.3)^2}{3} + \frac{(15.2)^2}{4} - 238.31 = 9.40$$

$$\text{SSTo} = \sum_{\text{all } N \text{ obs.}} x^2 - \text{CF} = (4.2)^2 + (3.7)^2 + \cdots + (3.9)^2 - 238.31 = 11.81$$

$$\text{SSE} = \text{SSTo} - \text{SSTr} = 11.81 - 9.40 = 2.41$$

▪ Randomized Block Experiment

Let $T_1, T_2, \ldots, T_k$ denote the treatment totals and $B_1, B_2, \ldots, B_l$ represent the block totals. Also, let T be the grand total of all kl observations and

$$\text{CF} = \text{correction factor} = \frac{T^2}{kl}$$

Then

$$\text{SSTo} = \sum_{\text{all } kl \text{ obs.}} x^2 - \text{CF}$$

$$\text{SSTr} = \frac{1}{l}(T_1^2 + T_2^2 + \cdots + T_k^2) - \text{CF}$$

$$\text{SSBl} = \frac{1}{k}(B_1^2 + B_2^2 + \cdots + B_l^2) - \text{CF}$$

$$\text{SSE} = \text{SSTo} - \text{SSTr} - \text{SSBl}$$

▪ **Example** 15.A2

Treatment	Block 1	2	3	4	
1	4.2	3.7	5.0	4.8	$T_1 = 17.7$
2	5.2	4.5	6.7	5.4	$T_2 = 21.8$
3	3.4	3.2	5.1	3.9	$T_3 = 15.6$
	$B_1 = 12.8$	$B_2 = 11.4$	$B_3 = 16.8$	$B_4 = 14.1$	$T = 55.1$

$$CF = \text{correction factor} = \frac{T^2}{kl} = \frac{(55.1)^2}{(3)(4)} = \frac{3036.01}{12} = 253.00$$

$$SSTr = \frac{1}{l}(T_1^2 + T_2^2 + \cdots + T_k^2) - CF$$

$$= \frac{1}{4}[(17.7)^2 + (21.8)^2 + (15.6)^2] - 253.00 = 4.97$$

$$SSBl = \frac{1}{k}(B_1^2 + B_2^2 + \cdots + B_l^2) - CF$$

$$= \frac{1}{3}[(12.8)^2 + (11.4)^2 + (16.8)^2 + (14.1)^2] - 253.00 = 5.28$$

$$SSTo = \sum_{\text{all } kl \text{ obs.}} x^2 - CF = (4.2)^2 + \cdots + (3.9)^2 - 253.00 = 10.73$$

$$SSE = SSTo - SSTr - SSBl = 10.73 - 4.97 - 5.28 = 0.48$$

▪ Supplementary Exercises 15.52–15.63

15.52 A study was carried out to compare aptitudes and achievements of three different groups of college students ("A Comparison of Three Groups of Academically At-Risk College Students," *Journal of College Student Development* [1995]: 270–279): (1) students diagnosed as learning disabled, (2) students who identified themselves as learning disabled, and (3) students who were low achievers. The Scholastic Abilities Test for Adults was given to each student. Consider the following summary data on writing composition score:

$n_1 = 30$	$n_2 = 30$	$n_3 = 30$
$\bar{x}_1 = 9.40$	$\bar{x}_2 = 11.63$	$\bar{x}_3 = 11.00$
SSE = 749.85		

Does it appear that the population mean score is not the same for the three types of students? Carry out a test of hypothesis. Does your conclusion depend on whether a significance level of .05 or .01 is used?

15.53 Most large companies have established grievance procedures for their employees. One question of interest to employers is why certain groups within a company have higher grievance rates than other groups. The study described in the article "Grievance Rates and Technology" (*Academy of Management Journal* [1979]: 810–815) distinguished four types of work groups. These types were labeled apathetic, erratic, strategic, and conservative. Suppose that 52 work groups were selected (13 of each type) and that a measure of grievance rate was determined for each group. Here are the resulting $\bar{x}$ values:

Group type	Apathetic	Erratic	Strategic	Conservative
Sample mean	2.96	5.05	8.74	4.91

In addition, SSTo = 682.10 and SSE = 506.19. Test at significance level .05 to see whether true average grievance rate depends on group type. If appropriate, use the T–K method to identify significant differences among group types.

15.54 The results of a study on the effectiveness of line drying on the smoothness and stiffness of fabric were summarized in the article "Line-Dried vs. Machine-Dried Fabrics: Comparison of Appearance, Hand, and Consumer Acceptance" (*Home Economics Research Journal* [1984]: 27–35). Smoothness scores were given for nine different types of fabric and five different drying methods: (1) machine dry, (2) line dry, (3) line dry followed by 15-min tumble, (4) line dry with softener, and (5) line dry with air movement. The scores are given in the following table:

Fabric	Drying Method				
	1	2	3	4	5
Crepe	3.3	2.5	2.8	2.5	1.9
Doubleknit	3.6	2.0	3.6	2.4	2.3
Twill	4.2	3.4	3.8	3.1	3.1
Twill mix	3.4	2.4	2.9	1.6	1.7
Terry	3.8	1.3	2.8	2.0	1.6
Broadcloth	2.2	1.5	2.7	1.5	1.9
Sheeting	3.5	2.1	2.8	2.1	2.2
Corduroy	3.6	1.3	2.8	1.7	1.8
Denim	2.6	1.4	2.4	1.3	1.6

Using a .05 significance level, test to see whether there is a difference in the true mean smoothness scores for the drying methods. (Note: There may be "extraneous" variation because of the different fabrics.)

15.55 In many countries, grains and cereals are the primary food source. The authors of the article "Mineral Contents of Cereal Grains as Affected by Storage and Insect Infestation" (*Journal of Stored Products Research* [1992]: 147–151) investigated the effects of storage period on the mineral content of maize. Four storage periods were considered: 0 months (no storage), 1 month, 2 months, and 4 months. Twenty-four containers of maize were randomly divided into four groups of six each. The iron content (milligrams per 100 g dry weight) of the first group of six was measured immediately (0 months of storage), the second six containers were measured after 1 month in storage, and so on. The following summary quantities are consistent with information in the article:

Storage Period	$\bar{x}$	s^2
0 months	4.923	0.000107
1 month	4.923	0.000067
2 months	4.917	0.000147
4 months	4.902	0.000057

a. Use a test with $\alpha = .05$ to decide whether true average iron content is the same for all four storage periods.
b. If appropriate, carry out a multiple comparisons analysis.

15.56 Controlling a filling operation with multiple fillers requires adjustment of the individual units. Data resulting from a sample of size 5 from each pocket of a 12-pocket filler were given in the article "Evaluating Variability of Filling Operations" (*Food Technology* [1984]: 51–55). Data for the first five pockets are given in the following table:

Pocket	Fill (oz)				
1	10.2	10.0	9.8	10.4	10.0
2	9.9	10.0	9.9	10.1	10.0
3	10.1	9.9	9.8	9.9	9.7
4	10.0	9.7	9.9	9.7	9.6
5	10.2	9.8	9.9	9.7	9.8

Use the ANOVA F test to determine whether the null hypothesis of no difference in the mean fill weight of the five pockets can be rejected. If so, use an appropriate technique to determine where the differences lie.

15.57 *This problem requires the use of a computer package.* The effect of oxygen concentration on fer-

mentation end products was examined in the article "Effects of Oxygen on Pyruvate Formate Lyase in Situ and Sugar Metabolism of *Streptococcus mutans* and *Streptococcus samguis*" (*Infection and Immunity* [1985]: 129–134). Four oxygen concentrations (0, 46, 92, and 138 μmol) and two types of sugar (galactose and glucose) were used. The following table gives two observations on amount of ethanol (in millimoles per milligram) for each sugar–oxygen concentration combination:

Oxygen Concentration	Galactose	Glucose
0	0.59, 0.30	0.25, 0.03
46	0.44, 0.18	0.13, 0.02
92	0.22, 0.23	0.07, 0.00
138	0.12, 0.13	0.00, 0.01

Construct an ANOVA table, and test the relevant hypotheses.

15.58 A water specimen was taken every month for a year at each of 15 designated locations on the Lago di Piediluco in Italy. The ammonia-nitrogen concentration was determined for each specimen, and the resulting data were analyzed using a two-way ANOVA. The researchers were willing to assume that there was no interaction between the two factors *location* and *month*. The following ANOVA table is from the article "Bacteriological and Chemical Variations and Their Inter-Relationships in a Slightly Polluted Water-Body" (*International Journal of Environmental Studies* [1984]: 121–129):

Source of Variation	df	Sum of Squares	Mean Square	F
Location		0.6		
Month	11	2.3		
Error				
Total	179	6.4		

Complete the ANOVA table, and use it to perform the tests required to determine whether the true mean concentration differs by location or by month of year. Use a .05 significance level for each test.

15.59 The water absorption of two types of mortar used to repair damaged cement was discussed in the article "Polymer Mortar Composite Matrices for Maintenance-Free, Highly Durable Ferrocement" (*Journal of Ferrocement* [1984]: 337–345). Specimens of ordinary cement mortar and polymer cement mortar were submerged for varying lengths of time (5, 9, 24, or 48 hr), and water absorption (percent, by weight) was recorded. With mortar type as Factor A (with two levels) and submersion period as Factor B (with four levels), three observations were made for each factor–level combination. Data included in the article were used to compute the following sums of squares: SSA = 322.667, SSB = 35.623, SSAB = 8.557, SSTo = 372.113. Use this information to construct an ANOVA table, and then use a .05 significance level to test the appropriate hypotheses.

15.60 Suppose that each observation in a single-factor ANOVA data set is multiplied by a constant c (a change in units; e.g., $c = 2.54$ changes observations from inches to centimeters). How does this affect MSTr, MSE, and the test statistic F? Is this reasonable? Explain.

15.61 Three different brands of automobile batteries, each one having a 42-month warranty, were included in a study of battery lifetime. A random sample of batteries of each brand was selected, and lifetime (in months) was determined, resulting in the following data:

Brand 1	45	38	52	47	45	42	43
Brand 2	39	44	50	54	48	46	40
Brand 3	50	46	43	48	57	44	48

State and test the appropriate hypotheses using a significance level of .05. Be sure to summarize your calculations in an ANOVA table.

15.62 Let $c_1, c_2, \ldots, c_k$ denote k specified numbers, and consider the quantity θ defined by

$$\theta = c_1\mu_1 + c_2\mu_2 + \cdots + c_k\mu_k$$

A confidence interval for θ is then

$$c_1\bar{x}_1 + \cdots + c_k\bar{x}_k$$
$$\pm \ (t \text{ critical value})\sqrt{\text{MSE}\left(\frac{c_1^2}{n_1} + \cdots + \frac{c_k^2}{n_k}\right)}$$

where the t critical value is based on the error df = $N - k$. For example, in a study carried out to compare pain relievers with respect to true average time to relief, suppose that Brands 1, 2, and 3 are nationally available, whereas Brands 4 and 5 are sold only

by two large chains of drug stores. An investigator might then wish to consider

$$\theta = \frac{1}{3}\mu_1 + \frac{1}{3}\mu_2 + \frac{1}{3}\mu_3 - \frac{1}{2}\mu_4 - \frac{1}{2}\mu_5$$

which, in essence, compares the average time to relief of the national brands to the average for the house brands. Refer to Exercise 15.61, and suppose that Brand 1 is a store brand battery and that Brands 2 and 3 are national brands. Obtain a 95% confidence interval for $\theta = \mu_1 - \frac{1}{2}\mu_2 - \frac{1}{2}\mu_3$.

15.63 One of the assumptions that underlies the validity of the ANOVA F test is that the population or treatment response variances $\sigma_1^2, \sigma_2^2, \ldots, \sigma_k^2$ should be identical regardless of whether or not H_0 is true; this is the assumption of constant variance across populations or treatments. In some situations the x values themselves may not satisfy this assumption, yet a transformation using some specified mathematical function (e.g., taking the logarithm or the square root) will give observations that have (approximately) constant variance. The ANOVA F test can then be applied to the transformed data. When observations are made on a counting variable (x = number of something), statisticians have found that taking the square root frequently "stabilizes the variance." In an experiment to compare the quality of four different brands of videotape, cassettes of a specified length were selected, and the number of flaws in each was determined:

Brand 1	10	14	5	12	8
Brand 2	17	14	8	9	12
Brand 3	13	18	15	18	10
Brand 4	14	22	12	16	17

Make a square-root transformation, and analyze the resulting data by using the ANOVA F test at significance level .01.

▪ References

Miller, Rupert. *Beyond ANOVA: The Basics of Applied Statistics.* New York: Wiley, 1986. (This book contains a wealth of information concerning violations of basic assumptions and alternative methods of analysis.)

Appendices

■ Appendix 1: The Binomial Distribution

Suppose that we decide to record the gender of each of the next 25 newborn children at a particular hospital. What is the chance that at least 15 are female? What is the chance that between 10 and 15 are female? How many among the 25 can we expect to be female? These and other similar questions can be answered by studying the *binomial probability distribution*. This distribution arises when the experiment of interest is a *binomial experiment*—that is, an experiment having the following characteristics.

■ Properties of a Binomial Experiment

1. It consists of a fixed number of observations, called trials.
2. Each trial can result in one of only two mutually exclusive outcomes labeled success (S) and failure (F).
3. Outcomes of different trials are independent.
4. The probability that a trial results in a success is the same for each trial.

The **binomial random variable** x is defined as

x = number of successes observed when experiment is performed

The probability distribution of x is called the *binomial probability distribution*.

The term *success* here does not necessarily have any of its usual connotations. Which of the two possible outcomes is labeled "success" is determined by the random variable of interest. For example, if the variable counts the number of female births among the next 25 births at a particular hospital, a female birth would be labeled a success (because this is what the variable counts). This labeling is arbitrary: If male births had been counted instead, a male birth would have been labeled a success and a female birth a failure.

For example, suppose that each of five randomly selected customers purchasing a hot tub at a certain store chooses either an electric model or a gas model. Assume that these customers make their choices independently of one another and that 40% of all customers select an electric model. Let's define the variable

x = number among the five customers who selected an electric hot tub

This experiment is a binomial experiment with

number of trials = 4 $P(S) = P(E) = .4$

where success (S) is defined as a customer who purchased an electric model.

The binomial distribution tells us how much probability is associated with each of the possible x values 0, 1, 2, 3, 4, and 5. There are 32 possible outcomes, and 5 of them yield $x = 1$:

SFFFF   FSFFF   FFSFF   FFFSF   FFFFS

By independence, the first of these possible outcomes has probability

$$P(\text{SFFFF}) = P(S)P(F)P(F)P(F)P(F)$$
$$= (.4)(.6)(.6)(.6)(.6)$$
$$= (.4)(.6)^4$$
$$= .05184$$

The probability calculation is the same for any outcome with only one success ($x = 1$). It does not matter where in the sequence the single success occurs. Thus

$$p(1) = P(x = 1)$$
$$= P(\text{SFFFF } or \text{ FSFFF } or \text{ FFSFF } or \text{ FFFSF } or \text{ FFFFS})$$
$$= .05184 + .05184 + .05184 + .05184 + .05184$$
$$= (5)(.05184)$$
$$= .25920$$

Similarly, there are 10 outcomes for which $x = 2$, because there are 10 ways to select 2 outcomes from among the 5 trials to be the successes: SSFFF, SFSFF, ..., and FFFSS. The probability of each results from multiplying together (.4) two times and (.6) three times. For example,

$$P(\text{SSFFF}) = (.4)(.4)(.6)(.6)(.6)$$
$$= (.4)^2(.6)^3$$
$$= .03456$$

and so

$$p(2) = P(x = 2)$$
$$= P(\text{SSFFF}) + \cdots + P(\text{FFFSS})$$
$$= (10)(.4)^2(.6)^3$$
$$= .34560$$

The general form of the distribution here is

$$p(x) = P(x \text{ successes among the 5 trials})$$
$$= \left(\begin{array}{c}\text{number of outcomes}\\\text{with } x \text{ successes}\end{array}\right)\left(\begin{array}{c}\text{probability of any particular}\\\text{outcome with } x \text{ successes}\end{array}\right)$$
$$= (\text{number of outcomes with } x \text{ successes})(.4)^x(.6)^{5-x}$$

This form was seen previously, where $p(2) = 10(.4)^2(.6)^3$.

Let n denote the number of trials in the experiment. Then the number of outcomes with x successes is the number of ways of selecting x from among the n trials to be the success trials. A simple expression for this quantity is

$$\text{number of outcomes with } x \text{ successes} = \frac{n!}{x!(n-x)!}$$

where, for any positive whole number m, the symbol $m!$ (read "m factorial") is defined by

$$m! = m(m-1)(m-2)\ldots(2)(1)$$

and $0! = 1$.

▪ The Binomial Distribution

Let

n = number of independent trials in a binomial experiment

π = constant probability that any particular trial results in a success *

Then

$$p(x) = P(x \text{ successes among } n \text{ trials})$$

$$= \frac{n!}{x!(n-x)!}\,\pi^x(1-\pi)^{n-x} \qquad x = 0, 1, 2, \ldots, n$$

The expressions $\binom{n}{x}$ or $_nC_x$ are sometimes used in place of $\frac{n!}{x!(n-x)!}$. Both are read as "n choose x," and they represent the number of ways of choosing x items from a set of n. The binomial probability function can then be written as

$$p(x) = \binom{n}{x}\pi^x(1-\pi)^{n-x} \qquad x = 0, 1, 2, \ldots, n$$

or

$$p(x) = {_nC_x}\pi^x(1-\pi)^{n-x} \qquad x = 0, 1, 2, \ldots, n$$

*Some sources use p to represent the probability of success rather than π. We prefer the use of Greek letters for characteristics of a population or probability distribution, thus our use of π here.

Notice that the probability distribution is being specified by a formula that allows calculation of the various probabilities rather than by giving a table or a probability histogram.

▪ Example A.1 Computer Monitors

Sixty percent of all computer monitors sold by a large computer retailer have a flat panel display and 40% have a CRT display. The type of monitor purchased by each of the next 12 customers will be noted. Define a random variable x by

x = number of monitors among these 12 that have a flat panel display

Because x counts the number of flat panel displays, we use S to denote the sale of a flat panel monitor. Then x is a binomial random variable with $n = 12$ and $\pi = P(\text{S}) = .60$. The probability distribution of x is given by

$$p(x) = \frac{12!}{x!(12-x)!}(.6)^x(.4)^{n-x} \qquad x = 0, 1, 2, \ldots, 12$$

The probability that exactly four monitors are digital is

$$p(4) = P(x = 4)$$
$$= \frac{12!}{4!8!}(.6)^4(.4)^8$$
$$= (495)(.6)^4(.4)^8$$
$$= .042$$

If group after group of 12 purchases is examined, the long-run percentage of those with exactly 4 flat panel monitors will be 4.2%. According to this calculation, 495 of the possible outcomes (there are $2^{12} = 4096$ possible outcomes) have $x = 4$.

The probability that between four and seven (inclusive) monitors are flat panel is

$$P(4 \leq x \leq 7) = P(x = 4 \; or \; x = 5 \; or \; x = 6 \; or \; x = 7)$$

Because these outcomes are disjoint, this is equal to

$$
\begin{aligned}
P(4 \leq x \leq 7) &= p(4) + p(5) + p(6) + p(7) \\
&= \frac{12!}{4!8!}(.6)^4(.4)^8 + \cdots + \frac{12!}{7!5!}(.6)^7(.4)^5 \\
&= .042 + .101 + .177 + .227 \\
&= .547
\end{aligned}
$$

Notice that

$$
\begin{aligned}
P(4 < x < 7) &= P(x = 5 \; or \; x = 6) \\
&= p(5) + p(6) \\
&= .278
\end{aligned}
$$

so the probability depends on whether $<$ or $\leq$ appears. (This is typical of *discrete* random variables.)

The binomial distribution formula can be tedious to use unless n is very small. Appendix Table 10 gives binomial probabilities for selected n in combination with various values of π. This should help you practice using the binomial distribution without getting bogged down in arithmetic.

▪ **Using Appendix Table 10**

To find $p(x)$ for any particular value of x:

1. Locate the part of the table corresponding to your value of n (5, 10, 15, 20, or 25).
2. Move down to the row labeled with your value of x.
3. Go across to the column headed by the specified value of π.

The desired probability is at the intersection of the designated x row and π column. For example, when $n = 20$ and $\pi = .8$,

$p(15) = P(x = 15) =$ (entry at intersection of $x = 15$ row and $\pi = .8$ column) $= .175$

Although $p(x)$ is positive for every possible x value, many probabilities are 0 to three decimal places, so they appear as .000 in the table. There are much more extensive binomial tables available. Alternatively, most statistics computer packages and graphing calculators are programmed to calculate these probabilities.

■ Notation and Assumptions

Suppose that a population consists of N individuals or objects, each one classified as a success or a failure. Usually, sampling is carried out without replacement; that is, once an element has been selected into the sample, it is not a candidate for future selection. If the sampling was accomplished by selecting an element from the population, observing whether it is a success or a failure, and then returning it to the population before the next selection is made, the variable x = number of successes observed in the sample would fit all the requirements of a binomial random variable. When sampling is done without replacement, the trials (individual selections) are not independent. In this case, the number of successes observed in the sample does not have a binomial distribution but rather a different type of distribution called a *hypergeometric distribution*. Not only does the name of this distribution sound forbidding, but also probability calculations for this distribution are even more tedious than for the binomial distribution. Fortunately, when the sample size n is much smaller than N, the population size, probabilities calculated using the binomial distribution and the hypergeometric distribution are very close in value. They are so close, in fact, that statisticians often ignore the difference and use the binomial probabilities in place of the hypergeometric probabilities. Most statisticians recommend the following guideline for determining whether the binomial probability distribution is appropriate when sampling without replacement.

Let x denote the number of successes in a sample of size n selected without replacement from a population consisting of N individuals or objects. If $\frac{n}{N} \leq 0.05$ (i.e., if at most 5% of the population is sampled), then the binomial distribution gives a good approximation to the probability distribution of x.

■ Example A.2 Security Systems

In recent years, homeowners have become increasingly security conscious. A *Los Angeles Times* poll (November 10, 1991) reported that almost 20% of Southern California homeowners questioned had installed a home security system. Suppose that exactly 20% of all such homeowners have a system. Consider a random sample of $n = 20$ homeowners (much less than 5% of the population). Then x, the number of homeowners in the sample who have a security system, has (approximately) a binomial distribution with $n = 20$ and $\pi = .20$. The probability that five of those sampled have a system is

$$p(5) = P(x = 5)$$
$$= \text{(entry in } x = 5 \text{ row and } \pi = .20 \text{ column in Appendix Table 10 } (n = 20))$$
$$= .175$$

The probability that at least 40% of those in the sample — that is, 8 or more — have a system is

$$P(x \geq 8) = P(x = 8, 9, 10, \ldots, 19, \text{ or } 20)$$
$$= p(8) + p(9) + \cdots + p(20)$$
$$= .022 + .007 + .002 + .000 + \cdots + .000$$
$$= .031$$

Figure A.1 The binomial probability histogram when $n = 20$ and $\pi = .20$.

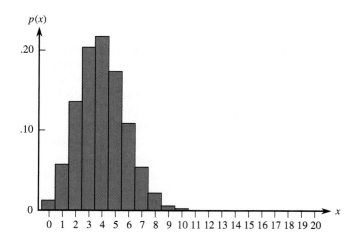

If, in fact, $\pi = .20$, only about 3% of all samples of size 20 would result in at least 8 homeowners having a security system. Because $P(x \geq 8)$ is so small when $\pi = .20$, if $x \geq 8$ were actually observed, we would have to wonder whether the reported value of $\pi = .20$ was correct. Although it is possible that we would observe $x \geq 8$ when $\pi = .20$ (this would happen about 3% of the time in the long run), it might also be the case that π is actually greater than .20. In Chapter 10, we showed how hypothesis-testing methods could be used to decide which of two contradictory claims about a population (e.g., $\pi = .20$ or $\pi > .20$) is more plausible.

The binomial formula or tables can be used to compute each of the 21 probabilities $p(0), p(1), \ldots, p(20)$. Figure A.1 shows the probability histogram for the binomial distribution with $n = 20$ and $\pi = .20$. Notice that the distribution is skewed to the right. (The binomial distribution is symmetric only when $\pi = .5$.)

▪ Mean and Standard Deviation of a Binomial Random Variable

A binomial random variable x based on n trials has possible values $0, 1, 2, \ldots, n$, so the mean value is

$$\mu_x = \sum (x)p(x)$$
$$= (0)p(0) + (1)p(1) + \cdots + (n)p(n)$$

and the variance of x is

$$\sigma_x^2 = \sum (x - \mu_x)^2 p(x)$$
$$= (0 - \mu_x)^2 p(0) + (1 - \mu_x)^2 p(1) + \cdots + (n - \mu_x)^2 p(n)$$

These expressions appear to be tedious to evaluate for any particular values of n and π. Fortunately, algebraic manipulation results in considerable simplification, making summation unnecessary.

The mean value and the standard deviation of a binomial random variable are

$$\mu_x = n\pi$$

and

$$\sigma_x = \sqrt{n\pi(1 - \pi)}$$

respectively.

▪ Example A.3 Credit Cards Paid in Full

It has been reported (*Newsweek*, December 2, 1991) that one-third of all credit card users pay their bills in full each month. This figure is, of course, an average across different cards and issuers. Suppose that 30% of all individuals holding Visa cards issued by a certain bank pay in full each month. A random sample of $n = 25$ cardholders is to be selected. The bank is interested in the variable x = number in the sample who pay in full each month. Even though sampling is done without replacement, the sample size $n = 25$ is most likely very small compared to the total number of credit card holders, so we can approximate the probability distribution of x by using a binomial distribution with $n = 25$ and $\pi = .3$. We have defined "paid in full" as a success because this is the outcome counted by the random variable x. The mean value of x is then

$$\mu_x = n\pi = 25(.30) = 7.5$$

and the standard deviation is

$$\sigma_x = \sqrt{n\pi(1 - \pi)}$$
$$= \sqrt{25(.30)(.70)}$$
$$= \sqrt{5.25}$$
$$= 2.29$$

The probability that x is farther than 1 standard deviation from its mean value is

$$P(x < \mu_x - \sigma_x \text{ or } x > \mu_x + \sigma_x) = P(x < 5.21 \text{ or } x > 9.79)$$
$$= P(x \le 5) + P(x \ge 10)$$
$$= p(0) + \cdots + p(5) + p(10) + \cdots + p(25)$$
$$= .382 \quad \text{(using Appendix Table 10)}$$

The value of σ_x is 0 when $\pi = 0$ or $\pi = 1$. In these two cases, there is no uncertainty in x: We are sure to observe $x = 0$ when $\pi = 0$ and $x = n$ when $\pi = 1$. It is also easily verified that $\pi(1 - \pi)$ is largest when $\pi = .5$. Thus the binomial distribution spreads out the most when sampling from a 50–50 population. The farther π is from .5, the less spread out and the more skewed the distribution.

▪ Exercises A.1–A.16

A.1 Consider the following two binomial experiments.
a. In a binomial experiment consisting of six trials, how many outcomes have exactly one success, and what are these outcomes?
b. In a binomial experiment consisting of 20 trials, how many outcomes have exactly 10 successes? exactly 15 successes? exactly 5 successes?

A.2 Suppose that in a certain metropolitan area, 9 out of 10 households have a VCR. Let x denote the number among four randomly selected households that have a VCR, so x is a binomial random variable with $n = 4$ and $\pi = .9$.
a. Calculate $p(2) = P(x = 2)$, and interpret this probability.
b. Calculate $p(4)$, the probability that all four selected households have a VCR.
c. Determine $P(x \le 3)$.

A.3 The *Los Angeles Times* (December 13, 1992) reported that what airline passengers like to do most on long flights is rest or sleep; in a survey of 3697 passengers, almost 80% rested or slept. Suppose that for a particular route, the actual percentage is exactly 80%, and consider randomly selecting six passengers. Then x, the number among the selected six who rested or slept, is a binomial random variable with $n = 6$ and $\pi = .8$.
a. Calculate $p(4)$, and interpret this probability.
b. Calculate $p(6)$, the probability that all six selected passengers rested or slept.
c. Determine $P(x \ge 4)$.

A.4 Refer to Exercise A.3, and suppose that 10 rather than 6 passengers are selected ($n = 10$, $\pi = .8$), so that Appendix Table 10 can be used.
a. What is $p(8)$?
b. Calculate $P(x \le 7)$.
c. Calculate the probability that more than half of the selected passengers rested or slept.

A.5 Twenty-five percent of the customers entering a grocery store between 5 P.M. and 7 P.M. use an express checkout. Consider five randomly selected customers, and let x denote the number among the five who use the express checkout.
a. What is $p(2)$, that is, $P(x = 2)$?
b. What is $P(x \le 1)$?
c. What is $P(2 \le x)$? (Hint: Make use of your computation in Part (b).)
d. What is $P(x \ne 2)$?

A.6 A breeder of show dogs is interested in the number of female puppies in a litter. If a birth is equally likely to result in a male or a female puppy,

give the probability distribution of the variable x = number of female puppies in a litter of size 5.

A.7 The article "FBI Says Fewer Than 25 Failed Polygraph Test" (*San Luis Obispo Tribune*, July 29, 2001) described the impact of a new program that requires top FBI officials to pass a polygraph test. The article states that false positives (i.e., tests in which an individual fails even though he or she is telling the truth) are relatively common and occur about 15% of the time. Suppose that such a test is given to 10 trustworthy individuals.
a. What is the probability that all 10 pass?
b. What is the probability that more than 2 fail, even though all are trustworthy?
c. The article indicated that 500 FBI agents were tested. Consider the random variable x = number of the 500 tested who fail. If all 500 agents tested are trustworthy, what are the mean and the standard deviation of x?
d. The headline indicates that fewer than 25 of the 500 agents tested failed the test. Is this a surprising result if all 500 are trustworthy? Answer based on the values of the mean and standard deviation from Part (c).

A.8 Industrial quality control programs often include inspection of incoming materials from suppliers. If parts are purchased in large lots, a typical plan might be to select 20 parts at random from a lot and inspect them. A lot might be judged acceptable if one or fewer defective parts are found among those inspected. Otherwise, the lot is rejected and returned to the supplier. Use Appendix Table 10 to find the probability of accepting lots that have each of the following (Hint: Identify success with a defective part):
a. 5% defective parts
b. 10% defective parts
c. 20% defective parts

A.9 An experiment was conducted to investigate whether a graphologist (handwriting analyst) could distinguish a normal person's handwriting from the handwriting of a psychotic. A well-known expert was given 10 files, each containing handwriting samples from a normal person and from a person diagnosed as psychotic. The graphologist was then asked to identify the psychotic's handwriting. The graphologist made correct identifications in 6 of the 10 trials (data taken from *Statistics in the Real World*, by R. J. Larsen and D. F. Stroup [New York: Macmillan, 1976]). Does this evidence indicate that the graphologist has an ability to distinguish the handwriting of psychotics? (Hint: What is the probability of correctly guessing

6 or more times out of 10? Your answer should depend on whether this probability is relatively small or relatively large.)

A.10 If the temperature in Florida falls below 32°F during certain periods of the year, there is a chance that the citrus crop will be damaged. Suppose that the probability is .1 that any given tree will show measurable damage when the temperature falls to 30°F. If the temperature does drop to 30°F, what is the expected number of trees showing damage in orchards of 2000 trees? What is the standard deviation of the number of trees that show damage?

A.11 Thirty percent of all automobiles undergoing an emissions inspection at a certain inspection station fail the inspection.
a. Among 15 randomly selected cars, what is the probability that at most 5 fail the inspection?
b. Among 15 randomly selected cars, what is the probability that between 5 and 10 (inclusive) fail to pass inspection?
c. Among 25 randomly selected cars, what is the mean value of the number that pass inspection, and what is the standard deviation of the number that pass inspection?
d. What is the probability that among 25 randomly selected cars, the number that pass is within 1 standard deviation of the mean value?

A.12 You are to take a multiple-choice exam consisting of 100 questions with 5 possible responses to each question. Suppose that you have not studied and so must guess (select 1 of the 5 answers in a completely random fashion) on each question. Let x represent the number of correct responses on the test.
a. What kind of probability distribution does x have?
b. What is your expected score on the exam? (Hint: Your expected score is the mean value of the x distribution.)
c. Compute the variance and standard deviation of x.
d. Based on your answers to Parts (b) and (c), is it likely that you would score over 50 on this exam? Explain the reasoning behind your answer.

A.13 Suppose that 20% of the 10,000 signatures on a certain recall petition are invalid. Would the number of invalid signatures in a sample of size 1000 have (approximately) a binomial distribution? Explain.

A.14 A coin is to be spun 25 times. Let x = the number of spins that result in heads (H). Consider the following rule for deciding whether or not the coin is fair:

Judge the coin to be fair if $8 \leq x \leq 17$
Judge the coin to be biased if either $x \leq 7$ or $x \geq 18$

a. What is the probability of judging the coin to be biased when it is actually fair?

b. What is the probability of judging the coin to be fair when $P(H) = .9$, so that there is a substantial bias? Repeat for $P(H) = .1$.
c. What is the probability of judging the coin to be fair when $P(H) = .6$? when $P(H) = .4$? Why are the probabilities so large compared to the probabilities in Part (b)?
d. What happens to the "error probabilities" of Parts (a) and (b) if the decision rule is changed so that the coin is judged fair if $7 \leq x \leq 18$ and judged unfair otherwise? Is this a better rule than the one first proposed?

A.15 A city ordinance requires that a smoke detector be installed in all residential housing. There is concern that too many residences are still without detectors, so a costly inspection program is being contemplated. Let π = the proportion of all residences that have a detector. A random sample of 25 residences will be selected. If the sample strongly suggests that $\pi < .80$ (fewer than 80% have detectors), as opposed to $\pi \geq .80$, the program will be implemented. Let x = the number of residences among the 25 that have a detector, and consider the following decision rule:

Reject the claim that $\pi = .8$ and implement the program if $x \leq 15$

a. What is the probability that the program is implemented when $\pi = .80$?
b. What is the probability that the program is not implemented if $\pi = .70$? if $\pi = .60$?
c. How do the "error probabilities" of Parts (a) and (b) change if the value 15 in the decision rule is changed to 14?

A.16 Exit polling has been a controversial practice in recent elections, because early release of the resulting information appears to affect whether or not those who have not yet voted will do so. Suppose that 90% of all registered California voters favor banning the release of information from exit polls in presidential elections until after the polls in California close. A random sample of 25 California voters will be selected.
a. What is the probability that more than 20 will favor the ban?
b. What is the probability that at least 20 will favor the ban?
c. What are the mean value and standard deviation of the number who favor the ban?
d. If fewer than 20 in the sample favor the ban, is this at odds with the assertion that (at least) 90% of the populace favors the ban? (Hint: Consider $P(x < 20)$ when $\pi = .9$.)

Appendix 2: Statistical Tables

Table 1 ▪ Random Numbers

Row																				
1	4	5	1	8	5	0	3	3	7	1	2	8	4	5	1	1	0	9	5	7
2	4	2	5	5	8	0	4	5	7	0	7	0	3	6	6	1	3	1	3	1
3	8	9	9	3	4	3	5	0	6	3	9	1	1	8	2	6	9	2	0	9
4	8	9	0	7	2	9	9	0	4	7	6	7	4	7	1	3	4	3	5	3
5	5	7	3	1	0	3	7	4	7	8	5	2	0	1	3	7	7	6	3	6
6	0	9	3	8	7	6	7	9	9	5	6	2	5	6	5	8	4	2	6	4
7	4	1	0	1	0	2	2	0	4	7	5	1	1	9	4	7	9	7	5	1
8	6	4	7	3	6	3	4	5	1	2	3	1	1	8	0	0	4	8	2	0
9	8	0	2	8	7	9	3	8	4	0	4	2	0	8	9	1	2	3	3	2
10	9	4	6	0	6	9	7	8	8	2	5	2	9	6	0	1	4	6	0	5
11	6	6	9	5	7	4	4	6	3	2	0	6	0	8	9	1	3	6	1	8
12	0	7	1	7	7	7	2	9	7	8	7	5	8	8	6	9	8	4	1	0
13	6	1	3	0	9	7	3	3	6	6	0	4	1	8	3	2	6	7	6	8
14	2	2	3	6	2	1	3	0	2	2	6	6	9	7	0	2	1	2	5	8
15	0	7	1	7	4	2	0	0	0	1	3	1	2	0	4	7	8	4	1	0
16	6	6	5	1	6	1	8	1	5	5	2	6	2	0	1	1	5	2	3	6
17	9	9	6	2	5	3	5	9	8	3	7	5	0	1	3	9	3	8	0	8
18	9	9	9	6	1	2	9	3	4	6	5	6	4	6	5	8	2	7	4	0
19	2	5	6	3	1	9	8	1	1	0	3	5	6	7	9	1	4	5	2	0
20	5	1	1	9	8	1	2	1	1	6	9	8	1	8	1	9	9	1	2	0
21	1	9	8	0	7	4	6	8	4	0	3	0	8	1	1	0	6	2	3	2
22	9	7	0	9	6	3	8	9	9	7	0	6	5	4	3	6	5	0	3	2
23	1	7	6	4	8	2	0	3	9	6	3	6	2	1	0	7	7	3	1	7
24	6	2	5	8	2	0	7	8	6	4	6	6	8	9	2	0	6	9	0	4
25	1	5	7	1	1	1	9	5	1	4	5	2	8	3	4	3	0	7	3	5
26	1	4	6	6	5	6	0	1	9	4	0	5	2	7	6	4	3	6	8	8
27	1	8	5	0	2	1	6	8	0	7	7	2	6	2	6	7	5	4	8	7
28	7	8	7	4	6	5	4	3	7	9	3	9	2	7	9	5	4	2	3	1
29	1	6	3	2	8	3	7	3	0	7	2	4	8	0	9	9	9	4	7	0
30	2	8	9	0	8	1	6	8	1	7	3	1	3	0	9	7	2	5	7	9
31	0	7	8	8	6	5	7	5	5	4	0	0	3	4	1	2	7	3	7	9
32	8	4	0	1	4	5	1	9	1	1	2	1	5	3	2	8	5	5	7	5
33	7	3	5	9	7	0	4	9	1	2	1	3	2	5	1	9	3	3	8	3
34	4	7	2	6	7	6	9	9	2	7	8	7	5	5	5	2	4	4	3	4
35	9	3	3	7	0	7	0	5	7	5	6	9	5	4	3	1	4	6	6	8
36	0	2	4	9	7	8	1	6	3	8	7	8	0	5	6	7	2	7	5	0
37	7	1	0	1	8	4	7	1	2	9	3	8	0	0	8	7	9	2	8	6
38	9	7	9	4	4	5	3	1	9	3	4	5	0	6	3	5	9	6	9	8
39	0	4	2	5	0	0	9	9	6	4	0	6	9	0	3	8	3	5	7	2
40	0	7	1	2	3	6	1	7	9	3	9	5	4	6	8	4	8	8	0	6
41	3	5	6	6	2	4	4	5	6	3	7	8	7	6	5	2	0	4	3	2
42	6	6	8	5	5	2	9	7	9	3	3	1	6	9	5	9	7	1	1	2
43	9	5	0	4	3	1	1	7	3	9	2	7	7	4	7	0	3	1	2	8
44	5	1	7	8	9	4	7	2	9	2	8	9	9	8	0	6	3	7	2	1
45	1	6	3	9	4	1	3	2	1	1	8	5	6	3	4	1	9	3	1	7
46	4	4	8	6	4	0	3	8	3	8	3	5	9	5	9	4	8	3	9	4
47	7	7	6	6	4	5	4	4	8	4	4	0	3	9	8	5	2	0	2	3
48	2	5	6	6	3	7	0	6	5	6	9	0	1	9	5	2	6	9	1	2
49	9	4	0	4	7	5	3	2	8	7	2	7	4	9	3	9	6	5	5	6
50	7	3	1	5	6	6	5	0	3	5	3	7	2	8	6	2	4	1	8	7

Table 1 ▪ Random Numbers (*continued*)

Row																				
51	7	5	8	2	8	8	8	7	6	4	1	1	0	2	3	1	9	3	6	0
52	3	3	6	0	9	1	1	0	3	2	7	8	2	0	5	3	4	8	9	8
53	0	2	9	6	9	8	9	3	8	1	5	3	9	9	7	0	7	7	1	6
54	8	5	9	6	2	9	6	8	2	1	2	4	7	0	6	8	3	4	6	1
55	5	4	7	6	1	0	0	1	0	4	6	1	4	1	5	0	9	6	5	5
56	5	0	3	6	4	1	9	8	4	4	1	2	0	2	5	1	8	1	2	1
57	0	2	6	3	7	5	1	1	6	6	0	5	8	1	2	3	3	6	1	3
58	3	8	1	6	3	8	1	4	5	2	9	4	2	5	7	3	2	3	1	8
59	9	1	5	6	0	6	5	6	6	3	6	2	3	0	0	0	1	8	5	9
60	5	3	5	6	3	9	5	4	7	3	6	6	7	5	0	1	5	6	7	3
61	9	6	6	4	5	7	7	6	1	5	4	4	8	0	6	5	7	6	3	0
62	6	3	0	6	7	9	5	5	4	6	2	2	8	4	4	0	0	9	9	8
63	8	5	8	3	5	2	0	6	6	0	0	6	0	6	3	0	1	7	0	5
64	3	8	2	4	9	0	9	2	6	2	9	5	1	9	1	9	0	8	3	3
65	1	4	4	1	1	7	4	6	3	6	5	6	5	5	7	7	0	3	5	8
66	5	9	9	5	3	7	2	5	1	7	1	1	0	7	1	0	9	2	8	8
67	8	7	1	7	5	2	5	6	8	7	9	9	1	3	9	6	4	9	3	0
68	6	7	2	3	1	4	9	2	1	7	0	8	6	7	8	9	9	4	7	4
69	2	3	2	8	7	0	9	7	1	1	1	2	8	2	9	1	0	6	7	7
70	2	9	5	7	8	4	7	9	0	3	6	9	2	0	6	0	6	2	6	8
71	4	8	9	8	3	2	7	6	9	1	9	8	6	9	5	2	4	9	9	9
72	1	5	6	5	7	7	5	4	3	4	3	8	1	8	9	9	4	4	1	1
73	1	8	1	1	7	2	8	5	5	8	9	9	9	6	2	0	1	6	6	7
74	5	7	7	0	9	5	5	6	8	6	8	2	2	6	0	5	5	1	8	7
75	1	8	6	0	5	4	8	3	4	5	3	5	8	7	7	7	8	5	7	0
76	2	6	6	7	9	4	2	2	8	7	4	3	4	9	6	1	9	4	3	9
77	3	6	6	4	5	7	8	3	0	2	8	4	6	7	2	1	4	5	2	3
78	0	7	8	0	1	2	1	1	3	4	2	1	6	9	3	3	5	4	0	4
79	8	3	6	0	5	7	7	9	1	5	8	8	4	9	5	7	2	2	7	6
80	5	3	6	9	0	6	3	8	7	5	9	5	9	7	4	2	5	6	2	9
81	0	9	3	7	7	2	8	6	4	3	2	9	4	8	2	9	9	6	9	9
82	9	4	7	4	0	0	0	3	5	4	6	6	2	6	2	3	6	1	1	4
83	5	5	4	1	7	8	6	4	2	3	2	9	8	4	6	3	8	3	0	5
84	5	3	0	0	5	4	8	0	7	4	7	6	2	1	1	2	1	2	6	9
85	3	3	0	9	3	2	9	4	0	5	5	4	8	7	5	7	5	3	8	8
86	3	0	5	7	1	9	5	8	0	0	4	5	3	0	3	0	2	7	6	7
87	5	0	8	6	0	8	1	6	2	0	8	6	5	4	0	7	2	9	1	0
88	3	6	4	7	8	2	3	5	7	9	8	5	2	7	6	9	0	2	4	9
89	9	0	4	4	9	1	6	8	5	2	8	9	0	7	5	7	2	5	1	8
90	9	5	2	6	9	3	9	6	5	1	8	8	7	8	2	0	4	4	7	9
91	9	4	5	7	0	3	4	6	4	2	5	4	8	6	1	1	9	1	8	8
92	8	1	1	8	0	5	4	2	8	5	3	3	3	0	1	1	4	4	8	3
93	6	9	4	7	8	3	3	9	1	2	5	0	1	2	3	0	1	1	2	5
94	0	0	6	8	8	7	2	4	4	7	6	6	0	3	4	7	5	6	8	2
95	5	3	3	9	3	8	4	9	1	9	1	7	8	4	5	2	2	5	4	4
96	2	5	6	2	7	6	0	3	8	1	4	4	2	6	8	3	6	3	2	8
97	7	4	3	7	9	6	8	6	2	8	3	8	4	2	2	0	7	0	5	3
98	1	9	0	8	8	0	1	2	2	2	7	5	6	5	5	7	8	7	2	6
99	2	4	8	0	2	5	2	7	0	5	9	6	6	1	5	8	7	9	7	5
100	4	1	7	8	6	7	1	1	5	8	9	4	8	9	8	3	0	9	0	7

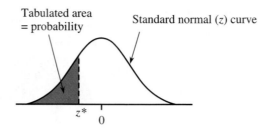

Tabulated area = probability

Standard normal (z) curve

z^* 0

Table 2 ▪ Standard Normal Probabilities (Cumulative z Curve Areas)

z^*	.00	.01	.02	.03	.04	.05	.06	.07	.08	.09
−3.8	.0001	.0001	.0001	.0001	.0001	.0001	.0001	.0001	.0001	.0000
−3.7	.0001	.0001	.0001	.0001	.0001	.0001	.0001	.0001	.0001	.0001
−3.6	.0002	.0002	.0001	.0001	.0001	.0001	.0001	.0001	.0001	.0001
−3.5	.0002	.0002	.0002	.0002	.0002	.0002	.0002	.0002	.0002	.0002
−3.4	.0003	.0003	.0003	.0003	.0003	.0003	.0003	.0003	.0003	.0002
−3.3	.0005	.0005	.0005	.0004	.0004	.0004	.0004	.0004	.0004	.0003
−3.2	.0007	.0007	.0006	.0006	.0006	.0006	.0006	.0005	.0005	.0005
−3.1	.0010	.0009	.0009	.0009	.0008	.0008	.0008	.0008	.0007	.0007
−3.0	.0013	.0013	.0013	.0012	.0012	.0011	.0011	.0011	.0010	.0010
−2.9	.0019	.0018	.0018	.0017	.0016	.0016	.0015	.0015	.0014	.0014
−2.8	.0026	.0025	.0024	.0023	.0023	.0022	.0021	.0021	.0020	.0019
−2.7	.0035	.0034	.0033	.0032	.0031	.0030	.0029	.0028	.0027	.0026
−2.6	.0047	.0045	.0044	.0043	.0041	.0040	.0039	.0038	.0037	.0036
−2.5	.0062	.0060	.0059	.0057	.0055	.0054	.0052	.0051	.0049	.0048
−2.4	.0082	.0080	.0078	.0075	.0073	.0071	.0069	.0068	.0066	.0064
−2.3	.0107	.0104	.0102	.0099	.0096	.0094	.0091	.0089	.0087	.0084
−2.2	.0139	.0136	.0132	.0129	.0125	.0122	.0119	.0116	.0113	.0110
−2.1	.0179	.0174	.0170	.0166	.0162	.0158	.0154	.0150	.0146	.0143
−2.0	.0228	.0222	.0217	.0212	.0207	.0202	.0197	.0192	.0188	.0183
−1.9	.0287	.0281	.0274	.0268	.0262	.0256	.0250	.0244	.0239	.0233
−1.8	.0359	.0351	.0344	.0336	.0329	.0322	.0314	.0307	.0301	.0294
−1.7	.0446	.0436	.0427	.0418	.0409	.0401	.0392	.0384	.0375	.0367
−1.6	.0548	.0537	.0526	.0516	.0505	.0495	.0485	.0475	.0465	.0455
−1.5	.0668	.0655	.0643	.0630	.0618	.0606	.0594	.0582	.0571	.0559
−1.4	.0808	.0793	.0778	.0764	.0749	.0735	.0721	.0708	.0694	.0681
−1.3	.0968	.0951	.0934	.0918	.0901	.0885	.0869	.0853	.0838	.0823
−1.2	.1151	.1131	.1112	.1093	.1075	.1056	.1038	.1020	.1003	.0985
−1.1	.1357	.1335	.1314	.1292	.1271	.1251	.1230	.1210	.1190	.1170
−1.0	.1587	.1562	.1539	.1515	.1492	.1469	.1446	.1423	.1401	.1379
−0.9	.1841	.1814	.1788	.1762	.1736	.1711	.1685	.1660	.1635	.1611
−0.8	.2119	.2090	.2061	.2033	.2005	.1977	.1949	.1922	.1894	.1867
−0.7	.2420	.2389	.2358	.2327	.2296	.2266	.2236	.2206	.2177	.2148
−0.6	.2743	.2709	.2676	.2643	.2611	.2578	.2546	.2514	.2483	.2451
−0.5	.3085	.3050	.3015	.2981	.2946	.2912	.2877	.2843	.2810	.2776
−0.4	.3446	.3409	.3372	.3336	.3300	.3264	.3228	.3192	.3156	.3121
−0.3	.3821	.3783	.3745	.3707	.3669	.3632	.3594	.3557	.3520	.3483
−0.2	.4207	.4168	.4129	.4090	.4052	.4013	.3974	.3936	.3897	.3859
−0.1	.4602	.4562	.4522	.4483	.4443	.4404	.4364	.4325	.4286	.4247
−0.0	.5000	.4960	.4920	.4880	.4840	.4801	.4761	.4721	.4681	.4641

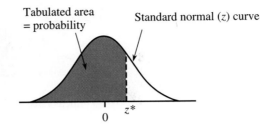

Tabulated area = probability Standard normal (z) curve

0   z*

Table 2 ▪ Standard Normal Probabilities (*continued*)

z*	.00	.01	.02	.03	.04	.05	.06	.07	.08	.09
0.0	.5000	.5040	.5080	.5120	.5160	.5199	.5239	.5279	.5319	.5359
0.1	.5398	.5438	.5478	.5517	.5557	.5596	.5636	.5675	.5714	.5753
0.2	.5793	.5832	.5871	.5910	.5948	.5987	.6026	.6064	.6103	.6141
0.3	.6179	.6217	.6255	.6293	.6331	.6368	.6406	.6443	.6480	.6517
0.4	.6554	.6591	.6628	.6664	.6700	.6736	.6772	.6808	.6844	.6879
0.5	.6915	.6950	.6985	.7019	.7054	.7088	.7123	.7157	.7190	.7224
0.6	.7257	.7291	.7324	.7357	.7389	.7422	.7454	.7486	.7517	.7549
0.7	.7580	.7611	.7642	.7673	.7704	.7734	.7764	.7794	.7823	.7852
0.8	.7881	.7910	.7939	.7967	.7995	.8023	.8051	.8078	.8106	.8133
0.9	.8159	.8186	.8212	.8238	.8264	.8289	.8315	.8340	.8365	.8389
1.0	.8413	.8438	.8461	.8485	.8508	.8531	.8554	.8577	.8599	.8621
1.1	.8643	.8665	.8686	.8708	.8729	.8749	.8770	.8790	.8810	.8830
1.2	.8849	.8869	.8888	.8907	.8925	.8944	.8962	.8980	.8997	.9015
1.3	.9032	.9049	.9066	.9082	.9099	.9115	.9131	.9147	.9162	.9177
1.4	.9192	.9207	.9222	.9236	.9251	.9265	.9279	.9292	.9306	.9319
1.5	.9332	.9345	.9357	.9370	.9382	.9394	.9406	.9418	.9429	.9441
1.6	.9452	.9463	.9474	.9484	.9495	.9505	.9515	.9525	.9535	.9545
1.7	.9554	.9564	.9573	.9582	.9591	.9599	.9608	.9616	.9625	.9633
1.8	.9641	.9649	.9656	.9664	.9671	.9678	.9686	.9693	.9699	.9706
1.9	.9713	.9719	.9726	.9732	.9738	.9744	.9750	.9756	.9761	.9767
2.0	.9772	.9778	.9783	.9788	.9793	.9798	.9803	.9808	.9812	.9817
2.1	.9821	.9826	.9830	.9834	.9838	.9842	.9846	.9850	.9854	.9857
2.2	.9861	.9864	.9868	.9871	.9875	.9878	.9881	.9884	.9887	.9890
2.3	.9893	.9896	.9898	.9901	.9904	.9906	.9909	.9911	.9913	.9916
2.4	.9918	.9920	.9922	.9925	.9927	.9929	.9931	.9932	.9934	.9936
2.5	.9938	.9940	.9941	.9943	.9945	.9946	.9948	.9949	.9951	.9952
2.6	.9953	.9955	.9956	.9957	.9959	.9960	.9961	.9962	.9963	.9964
2.7	.9965	.9966	.9967	.9968	.9969	.9970	.9971	.9972	.9973	.9974
2.8	.9974	.9975	.9976	.9977	.9977	.9978	.9979	.9979	.9980	.9981
2.9	.9981	.9982	.9982	.9983	.9984	.9984	.9985	.9985	.9986	.9986
3.0	.9987	.9987	.9987	.9988	.9988	.9989	.9989	.9989	.9990	.9990
3.1	.9990	.9991	.9991	.9991	.9992	.9992	.9992	.9992	.9993	.9993
3.2	.9993	.9993	.9994	.9994	.9994	.9994	.9994	.9995	.9995	.9995
3.3	.9995	.9995	.9995	.9996	.9996	.9996	.9996	.9996	.9996	.9997
3.4	.9997	.9997	.9997	.9997	.9997	.9997	.9997	.9997	.9997	.9998
3.5	.9998	.9998	.9998	.9998	.9998	.9998	.9998	.9998	.9998	.9998
3.6	.9998	.9998	.9999	.9999	.9999	.9999	.9999	.9999	.9999	.9999
3.7	.9999	.9999	.9999	.9999	.9999	.9999	.9999	.9999	.9999	.9999
3.8	.9999	.9999	.9999	.9999	.9999	.9999	.9999	.9999	.9999	1.0000

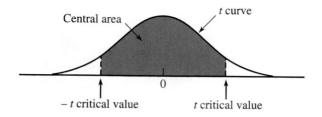

Table 3 ▪ *t* Critical Values

Central area captured:		.80	.90	.95	.98	.99	.998	.999
Confidence level:		80%	90%	95%	98%	99%	99.8%	99.9%
	1	3.08	6.31	12.71	31.82	63.66	318.31	636.62
	2	1.89	2.92	4.30	6.97	9.93	23.33	31.60
	3	1.64	2.35	3.18	4.54	5.84	10.21	12.92
	4	1.53	2.13	2.78	3.75	4.60	7.17	8.61
	5	1.48	2.02	2.57	3.37	4.03	5.89	6.86
	6	1.44	1.94	2.45	3.14	3.71	5.21	5.96
	7	1.42	1.90	2.37	3.00	3.50	4.79	5.41
	8	1.40	1.86	2.31	2.90	3.36	4.50	5.04
	9	1.38	1.83	2.26	2.82	3.25	4.30	4.78
	10	1.37	1.81	2.23	2.76	3.17	4.14	4.59
	11	1.36	1.80	2.20	2.72	3.11	4.03	4.44
	12	1.36	1.78	2.18	2.68	3.06	3.93	4.32
	13	1.35	1.77	2.16	2.65	3.01	3.85	4.22
	14	1.35	1.76	2.15	2.62	2.98	3.79	4.14
	15	1.34	1.75	2.13	2.60	2.95	3.73	4.07
	16	1.34	1.75	2.12	2.58	2.92	3.69	4.02
Degrees of	17	1.33	1.74	2.11	2.57	2.90	3.65	3.97
freedom	18	1.33	1.73	2.10	2.55	2.88	3.61	3.92
	19	1.33	1.73	2.09	2.54	2.86	3.58	3.88
	20	1.33	1.73	2.09	2.53	2.85	3.55	3.85
	21	1.32	1.72	2.08	2.52	2.83	3.53	3.82
	22	1.32	1.72	2.07	2.51	2.82	3.51	3.79
	23	1.32	1.71	2.07	2.50	2.81	3.49	3.77
	24	1.32	1.71	2.06	2.49	2.80	3.47	3.75
	25	1.32	1.71	2.06	2.49	2.79	3.45	3.73
	26	1.32	1.71	2.06	2.48	2.78	3.44	3.71
	27	1.31	1.70	2.05	2.47	2.77	3.42	3.69
	28	1.31	1.70	2.05	2.47	2.76	3.41	3.67
	29	1.31	1.70	2.05	2.46	2.76	3.40	3.66
	30	1.31	1.70	2.04	2.46	2.75	3.39	3.65
	40	1.30	1.68	2.02	2.42	2.70	3.31	3.55
	60	1.30	1.67	2.00	2.39	2.66	3.23	3.46
	120	1.29	1.66	1.98	2.36	2.62	3.16	3.37
z critical values	∞	1.28	1.645	1.96	2.33	2.58	3.09	3.29

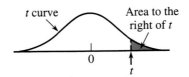

Table 4 ■ Tail Areas for *t* Curves

df / t	1	2	3	4	5	6	7	8	9	10	11	12
0.0	.500	.500	.500	.500	.500	.500	.500	.500	.500	.500	.500	.500
0.1	.468	.465	.463	.463	.462	.462	.462	.461	.461	.461	.461	.461
0.2	.437	.430	.427	.426	.425	.424	.424	.423	.423	.423	.423	.422
0.3	.407	.396	.392	.390	.388	.387	.386	.386	.386	.385	.385	.385
0.4	.379	.364	.358	.355	.353	.352	.351	.350	.349	.349	.348	.348
0.5	.352	.333	.326	.322	.319	.317	.316	.315	.315	.314	.313	.313
0.6	.328	.305	.295	.290	.287	.285	.284	.283	.282	.281	.280	.280
0.7	.306	.278	.267	.261	.258	.255	.253	.252	.251	.250	.249	.249
0.8	.285	.254	.241	.234	.230	.227	.225	.223	.222	.221	.220	.220
0.9	.267	.232	.217	.210	.205	.201	.199	.197	.196	.195	.194	.193
1.0	.250	.211	.196	.187	.182	.178	.175	.173	.172	.170	.169	.169
1.1	.235	.193	.176	.167	.162	.157	.154	.152	.150	.149	.147	.146
1.2	.221	.177	.158	.148	.142	.138	.135	.132	.130	.129	.128	.127
1.3	.209	.162	.142	.132	.125	.121	.117	.115	.113	.111	.110	.109
1.4	.197	.148	.128	.117	.110	.106	.102	.100	.098	.096	.095	.093
1.5	.187	.136	.115	.104	.097	.092	.089	.086	.084	.082	.081	.080
1.6	.178	.125	.104	.092	.085	.080	.077	.074	.072	.070	.069	.068
1.7	.169	.116	.094	.082	.075	.070	.066	.064	.062	.060	.059	.057
1.8	.161	.107	.085	.073	.066	.061	.057	.055	.053	.051	.050	.049
1.9	.154	.099	.077	.065	.058	.053	.050	.047	.045	.043	.042	.041
2.0	.148	.092	.070	.058	.051	.046	.043	.040	.038	.037	.035	.034
2.1	.141	.085	.063	.052	.045	.040	.037	.034	.033	.031	.030	.029
2.2	.136	.079	.058	.046	.040	.035	.032	.029	.028	.026	.025	.024
2.3	.131	.074	.052	.041	.035	.031	.027	.025	.023	.022	.021	.020
2.4	.126	.069	.048	.037	.031	.027	.024	.022	.020	.019	.018	.017
2.5	.121	.065	.044	.033	.027	.023	.020	.018	.017	.016	.015	.014
2.6	.117	.061	.040	.030	.024	.020	.018	.016	.014	.013	.012	.012
2.7	.113	.057	.037	.027	.021	.018	.015	.014	.012	.011	.010	.010
2.8	.109	.054	.034	.024	.019	.016	.013	.012	.010	.009	.009	.008
2.9	.106	.051	.031	.022	.017	.014	.011	.010	.009	.008	.007	.007
3.0	.102	.048	.029	.020	.015	.012	.010	.009	.007	.007	.006	.006
3.1	.099	.045	.027	.018	.013	.011	.009	.007	.006	.006	.005	.005
3.2	.096	.043	.025	.016	.012	.009	.008	.006	.005	.005	.004	.004
3.3	.094	.040	.023	.015	.011	.008	.007	.005	.005	.004	.004	.003
3.4	.091	.038	.021	.014	.010	.007	.006	.005	.004	.003	.003	.003
3.5	.089	.036	.020	.012	.009	.006	.005	.004	.003	.003	.002	.002
3.6	.086	.035	.018	.011	.008	.006	.004	.004	.003	.002	.002	.002
3.7	.084	.033	.017	.010	.007	.005	.004	.003	.002	.002	.002	.002
3.8	.082	.031	.016	.010	.006	.004	.003	.003	.002	.002	.001	.001
3.9	.080	.030	.015	.009	.006	.004	.003	.002	.002	.001	.001	.001
4.0	.078	.029	.014	.008	.005	.004	.003	.002	.002	.001	.001	.001

(*continued*)

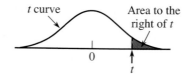

Table 4 ▪ Tail Areas for *t* Curves (*continued*)

t \ df	13	14	15	16	17	18	19	20	21	22	23	24
0.0	.500	.500	.500	.500	.500	.500	.500	.500	.500	.500	.500	.500
0.1	.461	.461	.461	.461	.461	.461	.461	.461	.461	.461	.461	.461
0.2	.422	.422	.422	.422	.422	.422	.422	.422	.422	.422	.422	.422
0.3	.384	.384	.384	.384	.384	.384	.384	.384	.384	.383	.383	.383
0.4	.348	.347	.347	.347	.347	.347	.347	.347	.347	.347	.346	.346
0.5	.313	.312	.312	.312	.312	.312	.311	.311	.311	.311	.311	.311
0.6	.279	.279	.279	.278	.278	.278	.278	.278	.278	.277	.277	.277
0.7	.248	.247	.247	.247	.247	.246	.246	.246	.246	.246	.245	.245
0.8	.219	.218	.218	.218	.217	.217	.217	.217	.216	.216	.216	.216
0.9	.192	.191	.191	.191	.190	.190	.190	.189	.189	.189	.189	.189
1.0	.168	.167	.167	.166	.166	.165	.165	.165	.164	.164	.164	.164
1.1	.146	.144	.144	.144	.143	.143	.143	.142	.142	.142	.141	.141
1.2	.126	.124	.124	.124	.123	.123	.122	.122	.122	.121	.121	.121
1.3	.108	.107	.107	.106	.105	.105	.105	.104	.104	.104	.103	.103
1.4	.092	.091	.091	.090	.090	.089	.089	.089	.088	.088	.087	.087
1.5	.079	.077	.077	.077	.076	.075	.075	.075	.074	.074	.074	.073
1.6	.067	.065	.065	.065	.064	.064	.063	.063	.062	.062	.062	.061
1.7	.056	.055	.055	.054	.054	.053	.053	.052	.052	.052	.051	.051
1.8	.048	.046	.046	.045	.045	.044	.044	.043	.043	.043	.042	.042
1.9	.040	.038	.038	.038	.037	.037	.036	.036	.036	.035	.035	.035
2.0	.033	.032	.032	.031	.031	.030	.030	.030	.029	.029	.029	.028
2.1	.028	.027	.027	.026	.025	.025	.025	.024	.024	.024	.023	.023
2.2	.023	.022	.022	.021	.021	.021	.020	.020	.020	.019	.019	.019
2.3	.019	.018	.018	.018	.017	.017	.016	.016	.016	.016	.015	.015
2.4	.016	.015	.015	.014	.014	.014	.013	.013	.013	.013	.012	.012
2.5	.013	.012	.012	.012	.011	.011	.011	.011	.010	.010	.010	.010
2.6	.011	.010	.010	.010	.009	.009	.009	.009	.008	.008	.008	.008
2.7	.009	.008	.008	.008	.008	.007	.007	.007	.007	.007	.006	.006
2.8	.008	.007	.007	.006	.006	.006	.006	.006	.005	.005	.005	.005
2.9	.006	.005	.005	.005	.005	.005	.005	.004	.004	.004	.004	.004
3.0	.005	.004	.004	.004	.004	.004	.004	.004	.003	.003	.003	.003
3.1	.004	.004	.004	.003	.003	.003	.003	.003	.003	.003	.003	.002
3.2	.003	.003	.003	.003	.003	.002	.002	.002	.002	.002	.002	.002
3.3	.003	.002	.002	.002	.002	.002	.002	.002	.002	.002	.002	.001
3.4	.002	.002	.002	.002	.002	.002	.002	.001	.001	.001	.001	.001
3.5	.002	.002	.002	.001	.001	.001	.001	.001	.001	.001	.001	.001
3.6	.002	.001	.001	.001	.001	.001	.001	.001	.001	.001	.001	.001
3.7	.001	.001	.001	.001	.001	.001	.001	.001	.001	.001	.001	.001
3.8	.001	.001	.001	.001	.001	.001	.001	.001	.001	.000	.000	.000
3.9	.001	.001	.001	.001	.001	.001	.000	.000	.000	.000	.000	.000
4.0	.001	.001	.001	.001	.000	.000	.000	.000	.000	.000	.000	.000

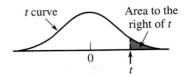

Table 4 ▪ Tail Areas for t Curves (*continued*)

t \ df	25	26	27	28	29	30	35	40	60	120	∞(= z)
0.0	.500	.500	.500	.500	.500	.500	.500	.500	.500	.500	.500
0.1	.461	.461	.461	.461	.461	.461	.460	.460	.460	.460	.460
0.2	.422	.422	.421	.421	.421	.421	.421	.421	.421	.421	.421
0.3	.383	.383	.383	.383	.383	.383	.383	.383	.383	.382	.382
0.4	.346	.346	.346	.346	.346	.346	.346	.346	.345	.345	.345
0.5	.311	.311	.311	.310	.310	.310	.310	.310	.309	.309	.309
0.6	.277	.277	.277	.277	.277	.277	.276	.276	.275	.275	.274
0.7	.245	.245	.245	.245	.245	.245	.244	.244	.243	.243	.242
0.8	.216	.215	.215	.215	.215	.215	.215	.214	.213	.213	.212
0.9	.188	.188	.188	.188	.188	.188	.187	.187	.186	.185	.184
1.0	.163	.163	.163	.163	.163	.163	.162	.162	.161	.160	.159
1.1	.141	.141	.141	.140	.140	.140	.139	.139	.138	.137	.136
1.2	.121	.120	.120	.120	.120	.120	.119	.119	.117	.116	.115
1.3	.103	.103	.102	.102	.102	.102	.101	.101	.099	.098	.097
1.4	.087	.087	.086	.086	.086	.086	.085	.085	.083	.082	.081
1.5	.073	.073	.073	.072	.072	.072	.071	.071	.069	.068	.067
1.6	.061	.061	.061	.060	.060	.060	.059	.059	.057	.056	.055
1.7	.051	.051	.050	.050	.050	.050	.049	.048	.047	.046	.045
1.8	.042	.042	.042	.041	.041	.041	.040	.040	.038	.037	.036
1.9	.035	.034	.034	.034	.034	.034	.033	.032	.031	.030	.029
2.0	.028	.028	.028	.028	.027	.027	.027	.026	.025	.024	.023
2.1	.023	.023	.023	.022	.022	.022	.022	.021	.020	.019	.018
2.2	.019	.018	.018	.018	.018	.018	.017	.017	.016	.015	.014
2.3	.015	.015	.015	.015	.014	.014	.014	.013	.012	.012	.011
2.4	.012	.012	.012	.012	.012	.011	.011	.011	.010	.009	.008
2.5	.010	.010	.009	.009	.009	.009	.009	.008	.008	.007	.006
2.6	.008	.008	.007	.007	.007	.007	.007	.007	.006	.005	.005
2.7	.006	.006	.006	.006	.006	.006	.005	.005	.004	.004	.003
2.8	.005	.005	.005	.005	.005	.004	.004	.004	.003	.003	.003
2.9	.004	.004	.004	.004	.004	.003	.003	.003	.003	.002	.002
3.0	.003	.003	.003	.003	.003	.003	.002	.002	.002	.002	.001
3.1	.002	.002	.002	.002	.002	.002	.002	.002	.001	.001	.001
3.2	.002	.002	.002	.002	.002	.002	.001	.001	.001	.001	.001
3.3	.001	.001	.001	.001	.001	.001	.001	.001	.001	.001	.000
3.4	.001	.001	.001	.001	.001	.001	.001	.001	.001	.000	.000
3.5	.001	.001	.001	.001	.001	.001	.001	.001	.000	.000	.000
3.6	.001	.001	.001	.001	.001	.001	.000	.000	.000	.000	.000
3.7	.001	.001	.000	.000	.000	.000	.000	.000	.000	.000	.000
3.8	.000	.000	.000	.000	.000	.000	.000	.000	.000	.000	.000
3.9	.000	.000	.000	.000	.000	.000	.000	.000	.000	.000	.000
4.0	.000	.000	.000	.000	.000	.000	.000	.000	.000	.000	.000

Table 5 ▪ Curves of $\beta = P$(Type II Error) for t Tests

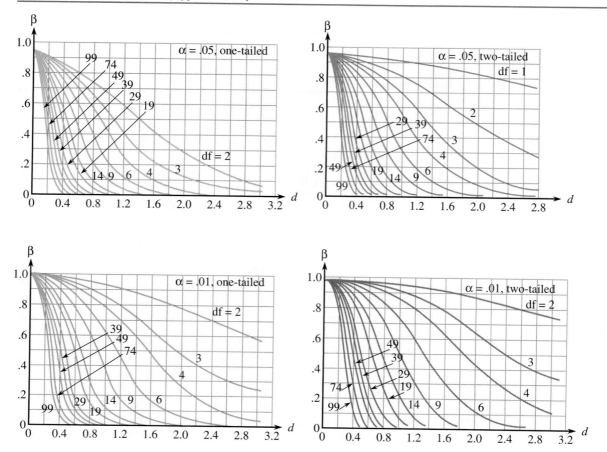

Table 6 ▪ *P*-Value Information for the Rank-Sum Test

n_1	n_2	Upper-tailed test		Lower-tailed test		Two-tailed test	
		P-value < .05 if rank sum is greater than or equal to	*P*-value < .01 if rank sum is greater than or equal to	*P*-value < .05 if rank sum is less than or equal to	*P*-value < .01 if rank sum is less than or equal to	*P*-value < .05 if rank sum is not between*	*P*-value < .01 if rank sum is not between
3	3	15	–	6	–	–	–
3	4	17	–	7	–	18,6	–
3	5	20	21	7	6	21,6	–
3	6	22	24	8	6	23,7	–
3	7	24	27	9	6	26,7	27,6
3	8	27	29	9	7	28,8	30,6
4	3	21	–	11	–	–	–
4	4	24	26	12	10	25,11	–
4	5	27	30	13	10	29,11	30,10
4	6	30	33	14	11	32,12	34,10
4	7	33	36	15	12	35,13	37,11
4	8	36	40	16	12	38,14	41,11
5	3	29	30	16	15	30,15	–
5	4	32	35	18	15	34,16	35,15
5	5	36	39	19	16	37,18	39,16
5	6	40	43	20	17	41,19	44,16
5	7	43	47	22	18	45,20	48,17
5	8	47	51	23	19	49,21	52,18
6	3	37	39	23	21	38,22	–
6	4	41	44	25	22	43,23	45,21
6	5	46	49	26	23	47,25	50,22
6	6	50	54	28	24	52,26	55,23
6	7	54	58	30	26	56,28	60,24
6	8	58	63	32	27	61,29	65,25
7	3	46	49	31	28	48,29	49,28
7	4	51	54	33	30	53,31	55,29
7	5	56	60	35	31	58,33	61,30
7	6	61	65	37	33	63,35	67,31
7	7	66	71	39	34	68,37	72,33
7	8	71	76	41	36	73,39	78,34
8	3	57	59	39	37	58,38	60,36
8	4	62	66	42	38	64,40	67,37
8	5	68	72	44	40	70,42	73,39
8	6	73	78	47	42	76,44	80,40
8	7	79	84	49	44	81,47	86,42
8	8	84	90	52	46	87,49	92,44

*Including endpoints. For example, when $n_1 = 3$ and $n_2 = 4$, *P*-value ≥ .05 if 6 ≤ rank sum ≤ 18.

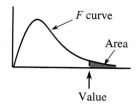

Table 7 ▪ Values That Capture Specified Upper-Tail *F* Curve Areas

df₂	Area					df₁					
		1	2	3	4	5	6	7	8	9	10
1.	1039.86	49.50	53.59	55.83	57.24	58.20	58.91	59.44	59.86	60.19	
	.05	161.40	199.50	215.70	224.60	230.20	234.00	236.80	238.90	240.50	241.90
	.01	4052.00	5000.00	5403.00	5625.00	5764.00	5859.00	5928.00	5981.00	6022.00	6056.00
2	.10	8.53	9.00	9.16	9.24	9.29	9.33	9.35	9.37	9.38	9.39
	.05	18.51	19.00	19.16	19.25	19.30	19.33	19.35	19.37	19.38	19.40
	.01	98.50	99.00	99.17	99.25	99.30	99.33	99.36	99.37	99.39	99.40
	.001	998.50	999.00	999.20	999.20	999.30	999.30	999.40	999.40	999.40	999.40
3	.10	5.54	5.46	5.39	5.34	5.31	5.28	5.27	5.25	5.24	5.23
	.05	10.13	9.55	9.28	9.12	9.01	8.94	8.89	8.85	8.81	8.79
	.01	34.12	30.82	29.46	28.71	28.24	27.91	27.67	27.49	27.35	27.23
	.001	167.00	148.50	141.10	137.10	134.60	132.80	131.60	130.60	129.90	129.20
4	.10	4.54	4.32	4.19	4.11	4.05	4.01	3.98	3.95	3.94	3.92
	.05	7.71	6.94	6.59	6.39	6.26	6.16	6.09	6.04	6.00	5.96
	.01	21.20	18.00	16.69	15.98	15.52	15.21	14.98	14.80	14.66	14.55
	.001	74.14	61.25	56.18	53.44	51.71	50.53	49.66	49.00	48.47	48.05
5	.10	4.06	3.78	3.62	3.52	3.45	3.40	3.37	3.34	3.32	3.30
	.05	6.61	5.79	5.41	5.19	5.05	4.95	4.88	4.82	4.77	4.74
	.01	16.26	13.27	12.06	11.39	10.97	10.67	10.46	10.29	10.16	10.05
	.001	47.18	37.12	33.20	31.09	29.75	28.83	28.16	27.65	27.24	26.92
6	.10	3.78	3.46	3.29	3.18	3.11	3.05	3.01	2.98	2.96	2.94
	.05	5.99	5.14	4.76	4.53	4.39	4.28	4.21	4.15	4.10	4.06
	.01	13.75	10.92	9.78	9.15	8.75	8.47	8.26	8.10	7.98	7.87
	.001	35.51	27.00	23.70	21.92	20.80	20.03	19.46	19.03	18.69	18.41
7	.10	3.59	3.26	3.07	2.96	2.88	2.83	2.78	2.75	2.72	2.70
	.05	5.59	4.74	4.35	4.12	3.97	3.87	3.79	3.73	3.68	3.64
	.01	12.25	9.55	8.45	7.85	7.46	7.19	6.99	6.84	6.72	6.62
	.001	29.25	21.69	18.77	17.20	16.21	15.52	15.02	14.63	14.33	14.08
8	.10	3.46	3.11	2.92	2.81	2.73	2.67	2.62	2.59	2.56	2.54
	.05	5.32	4.46	4.07	3.84	3.69	3.58	3.50	3.44	3.39	3.35
	.01	11.26	8.65	7.59	7.01	6.63	6.37	6.18	6.03	5.91	5.81
	.001	25.41	18.49	15.83	14.39	13.48	12.86	12.40	12.05	11.77	11.54
9	.10	3.36	3.01	2.81	2.69	2.61	2.55	2.51	2.47	2.44	2.42
	.05	5.12	4.26	3.86	3.63	3.48	3.37	3.29	3.23	3.18	3.14
	.01	10.56	8.02	6.99	6.42	6.06	5.80	5.61	5.47	5.35	5.26
	.001	22.86	16.39	13.90	12.56	11.71	11.13	10.70	10.37	10.11	9.89

Table 7 ■ Values That Capture Specified Upper-Tail *F* Curve Areas (*continued*)

df₂	Area	1	2	3	4	5	6	7	8	9	10
							df_1				
10	.10	3.29	2.92	2.73	2.61	2.52	2.46	2.41	2.38	2.35	2.32
	.05	4.96	4.10	3.71	3.48	3.33	3.22	3.14	3.07	3.02	2.98
	.01	10.04	7.56	6.55	5.99	5.64	5.39	5.20	5.06	4.94	4.85
	.001	21.04	14.91	12.55	11.28	10.48	9.93	9.52	9.20	8.96	8.75
11	.10	3.23	2.86	2.66	2.54	2.45	2.39	2.34	2.30	2.27	2.25
	.05	4.84	3.98	3.59	3.36	3.20	3.09	3.01	2.95	2.90	2.85
	.01	9.65	7.21	6.22	5.67	5.32	5.07	4.89	4.74	4.63	4.54
	.001	19.69	13.81	11.56	10.35	9.58	9.05	8.66	8.35	8.12	7.92
12	.10	3.18	2.81	2.61	2.48	2.39	2.33	2.28	2.24	2.21	2.19
	.05	4.75	3.89	3.49	3.26	3.11	3.00	2.91	2.85	2.80	2.75
	.01	9.33	6.93	5.95	5.41	5.06	4.82	4.64	4.50	4.39	4.30
	.001	18.64	12.97	10.80	9.63	8.89	8.38	8.00	7.71	7.48	7.29
13	.10	3.14	2.76	2.56	2.43	2.35	2.28	2.23	2.20	2.16	2.14
	.05	4.67	3.81	3.41	3.18	3.03	2.92	2.83	2.77	2.71	2.67
	.01	9.07	6.70	5.74	5.21	4.86	4.62	4.44	4.30	4.19	4.10
	.001	17.82	12.31	10.21	9.07	8.35	7.86	7.49	7.21	6.98	6.80
14	.10	3.10	2.73	2.52	2.39	2.31	2.24	2.19	2.15	2.12	2.10
	.05	4.60	3.74	3.34	3.11	2.96	2.85	2.76	2.70	2.65	2.60
	.01	8.86	6.51	5.56	5.04	4.69	4.46	4.28	4.14	4.03	3.94
	.001	17.14	11.78	9.73	8.62	7.92	7.44	7.08	6.80	6.58	6.40
15	.10	3.07	2.70	2.49	2.36	2.27	2.21	2.16	2.12	2.09	2.06
	.05	4.54	3.68	3.29	3.06	2.90	2.79	2.71	2.64	2.59	2.54
	.01	8.68	6.36	5.42	4.89	4.56	4.32	4.14	4.00	3.89	3.80
	.001	16.59	11.34	9.34	8.25	7.57	7.09	6.74	6.47	6.26	6.08
16	.10	3.05	2.67	2.46	2.33	2.24	2.18	2.13	2.09	2.06	2.03
	.05	4.49	3.63	3.24	3.01	2.85	2.74	2.66	2.59	2.54	2.49
	.01	8.53	6.23	5.29	4.77	4.44	4.20	4.03	3.89	3.78	3.69
	.001	16.12	10.97	9.01	7.94	7.27	6.80	6.46	6.19	5.98	5.81
17	.10	3.03	2.64	2.44	2.31	2.22	2.15	2.10	2.06	2.03	2.00
	.05	4.45	3.59	3.20	2.96	2.81	2.70	2.61	2.55	2.49	2.45
	.01	8.40	6.11	5.18	4.67	4.34	4.10	3.93	3.79	3.68	3.59
	.001	15.72	10.66	8.73	7.68	7.02	6.56	6.22	5.96	5.75	5.58
18	.10	3.01	2.62	2.42	2.29	2.20	2.13	2.08	2.04	2.00	1.98
	.05	4.41	3.55	3.16	2.93	2.77	2.66	2.58	2.51	2.46	2.41
	.01	8.29	6.01	5.09	4.58	4.25	4.01	3.84	3.71	3.60	3.51
	.001	15.38	10.39	8.49	7.46	6.81	6.35	6.02	5.76	5.56	5.39
19	.10	2.99	2.61	2.40	2.27	2.18	2.11	2.06	2.02	1.98	1.96
	.05	4.38	3.52	3.13	2.90	2.74	2.63	2.54	2.48	2.42	2.38
	.01	8.18	5.93	5.01	4.50	4.17	3.94	3.77	3.63	3.52	3.43
	.001	15.08	10.16	8.28	7.27	6.62	6.18	5.85	5.59	5.39	5.22

(*continued*)

Table 7 ▪ Values That Capture Specified Upper-Tail F Curve Areas (*continued*)

df_2	Area	df_1 1	2	3	4	5	6	7	8	9	10
20	.10	2.97	2.59	2.38	2.25	2.16	2.09	2.04	2.00	1.96	1.94
	.05	4.35	3.49	3.10	2.87	2.71	2.60	2.51	2.45	2.39	2.35
	.01	8.10	5.85	4.94	4.43	4.10	3.87	3.70	3.56	3.46	3.37
	.001	14.82	9.95	8.10	7.10	6.46	6.02	5.69	5.44	5.24	5.08
21	.10	2.96	2.57	2.36	2.23	2.14	2.08	2.02	1.98	1.95	1.92
	.05	4.32	3.47	3.07	2.84	2.68	2.57	2.49	2.42	2.37	2.32
	.01	8.02	5.78	4.87	4.37	4.04	3.81	3.64	3.51	3.40	3.31
	.001	14.59	9.77	7.94	6.95	6.32	5.88	5.56	5.31	5.11	4.95
22	.10	2.95	2.56	2.35	2.22	2.13	2.06	2.01	1.97	1.93	1.90
	.05	4.30	3.44	3.05	2.82	2.66	2.55	2.46	2.40	2.34	2.30
	.01	7.95	5.72	4.82	4.31	3.99	3.76	3.59	3.45	3.35	3.26
	.001	14.38	9.61	7.80	6.81	6.19	5.76	5.44	5.19	4.99	4.83
23	.10	2.94	2.55	2.34	2.21	2.11	2.05	1.99	1.95	1.92	1.89
	.05	4.28	3.42	3.03	2.80	2.64	2.53	2.44	2.37	2.32	2.27
	.01	7.88	5.66	4.76	4.26	3.94	3.71	3.54	3.41	3.30	3.21
	.001	14.20	9.47	7.67	6.70	6.08	5.65	5.33	5.09	4.89	4.73
24	.10	2.93	2.54	2.33	2.19	2.10	2.04	1.98	1.94	1.91	1.88
	.05	4.26	3.40	3.01	2.78	2.62	2.51	2.42	2.36	2.30	2.25
	.01	7.82	5.61	4.72	4.22	3.90	3.67	3.50	3.36	3.26	3.17
	.001	14.03	9.34	7.55	6.59	5.98	5.55	5.23	4.99	4.80	4.64
25	.10	2.92	2.53	2.32	2.18	2.09	2.02	1.97	1.93	1.89	1.87
	.05	4.24	3.39	2.99	2.76	2.60	2.49	2.40	2.34	2.28	2.24
	.01	7.77	5.57	4.68	4.18	3.85	3.63	3.46	3.32	3.22	3.13
	.001	13.88	9.22	7.45	6.49	5.89	5.46	5.15	4.91	4.71	4.56
26	.10	2.91	2.52	2.31	2.17	2.08	2.01	1.96	1.92	1.88	1.86
	.05	4.23	3.37	2.98	2.74	2.59	2.47	2.39	2.32	2.27	2.22
	.01	7.72	5.53	4.64	4.14	3.82	3.59	3.42	3.29	3.18	3.09
	.001	13.74	9.12	7.36	6.41	5.80	5.38	5.07	4.83	4.64	4.48
27	.10	2.90	2.51	2.30	2.17	2.07	2.00	1.95	1.91	1.87	1.85
	.05	4.21	3.35	2.96	2.73	2.57	2.46	2.37	2.31	2.25	2.20
	.01	7.68	5.49	4.60	4.11	3.78	3.56	3.39	3.26	3.15	3.06
	.001	13.61	9.02	7.27	6.33	5.73	5.31	5.00	4.76	4.57	4.41
28	.10	2.89	2.50	2.29	2.16	2.06	2.00	1.94	1.90	1.87	1.84
	.05	4.20	3.34	2.95	2.71	2.56	2.45	2.36	2.29	2.24	2.19
	.01	7.64	5.45	4.57	4.07	3.75	3.53	3.36	3.23	3.12	3.03
	.001	13.50	8.93	7.19	6.25	5.66	5.24	4.93	4.69	4.50	4.35
29	.10	2.89	2.50	2.28	2.15	2.06	1.99	1.93	1.89	1.86	1.83
	.05	4.18	3.33	2.93	2.70	2.55	2.43	2.35	2.28	2.22	2.18
	.01	7.60	5.42	4.54	4.04	3.73	3.50	3.33	3.20	3.09	3.00
	.001	13.39	8.85	7.12	6.19	5.59	5.18	4.87	4.64	4.45	4.29

Table 7 ▪ Values That Capture Specified Upper-Tail *F* Curve Areas (*continued*)

df₂	Area	df₁ 1	2	3	4	5	6	7	8	9	10
30	.10	2.88	2.49	2.28	2.14	2.05	1.98	1.93	1.88	1.85	1.82
	.05	4.17	3.32	2.92	2.69	2.53	2.42	2.33	2.27	2.21	2.16
	.01	7.56	5.39	4.51	4.02	3.70	3.47	3.30	3.17	3.07	2.98
	.001	13.29	8.77	7.05	6.12	5.53	5.12	4.82	4.58	4.39	4.24
40	.10	2.84	2.44	2.23	2.09	2.00	1.93	1.87	1.83	1.79	1.76
	.05	4.08	3.23	2.84	2.61	2.45	2.34	2.25	2.18	2.12	2.08
	.01	7.31	5.18	4.31	3.83	3.51	3.29	3.12	2.99	2.89	2.80
	.001	12.61	8.25	6.59	5.70	5.13	4.73	4.44	4.21	4.02	3.87
60	.10	2.79	2.39	2.18	2.04	1.95	1.87	1.82	1.77	1.74	1.71
	.05	4.00	3.15	2.76	2.53	2.37	2.25	2.17	2.10	2.04	1.99
	.01	7.08	4.98	4.13	3.65	3.34	3.12	2.95	2.82	2.72	2.63
	.001	11.97	7.77	6.17	5.31	4.76	4.37	4.09	3.86	3.69	3.54
90	.10	2.76	2.36	2.15	2.01	1.91	1.84	1.78	1.74	1.70	1.67
	.05	3.95	3.10	2.71	2.47	2.32	2.20	2.11	2.04	1.99	1.94
	.01	6.93	4.85	4.01	3.53	3.23	3.01	2.84	2.72	2.61	2.52
	.001	11.57	7.47	5.91	5.06	4.53	4.15	3.87	3.65	3.48	3.34
120	.10	2.75	2.35	2.13	1.99	1.90	1.82	1.77	1.72	1.68	1.65
	.05	3.92	3.07	2.68	2.45	2.29	2.18	2.09	2.02	1.96	1.91
	.01	6.85	4.79	3.95	3.48	3.17	2.96	2.79	2.66	2.56	2.47
	.001	11.38	7.32	5.78	4.95	4.42	4.04	3.77	3.55	3.38	3.24
240	.10	2.73	2.32	2.10	1.97	1.87	1.80	1.74	1.70	1.65	1.63
	.05	3.88	3.03	2.64	2.41	2.25	2.14	2.04	1.98	1.92	1.87
	.01	6.74	4.69	3.86	3.40	3.09	2.88	2.71	2.59	2.48	2.40
	.001	11.10	7.11	5.60	4.78	4.25	3.89	3.62	3.41	3.24	3.09
∞	.10	2.71	2.30	2.08	1.94	1.85	1.77	1.72	1.67	1.63	1.60
	.05	3.84	3.00	2.60	2.37	2.21	2.10	2.01	1.94	1.88	1.83
	.01	6.63	4.61	3.78	3.32	3.02	2.80	2.64	2.51	2.41	2.32
	.001	10.83	6.91	5.42	4.62	4.10	3.74	3.47	3.27	3.10	2.96

Table 8 ▪ Critical Values of q for the Studentized Range Distribution

Error df	Confidence level	Number of populations, treatments, or levels being compared							
		3	4	5	6	7	8	9	10
5	95%	4.60	5.22	5.67	6.03	6.33	6.58	6.80	6.99
	99%	6.98	7.80	8.42	8.91	9.32	9.67	9.97	10.24
6	95%	4.34	4.90	5.30	5.63	5.90	6.12	6.32	6.49
	99%	6.33	7.03	7.56	7.97	8.32	8.61	8.87	9.10
7	95%	4.16	4.68	5.06	5.36	5.61	5.82	6.00	6.16
	99%	5.92	6.54	7.01	7.37	7.68	7.94	8.17	8.37
8	95%	4.04	4.53	4.89	5.17	5.40	5.60	5.77	5.92
	99%	5.64	6.20	6.62	6.96	7.24	7.47	7.68	7.86
9	95%	3.95	4.41	4.76	5.02	5.24	5.43	5.59	5.74
	99%	5.43	5.96	6.35	6.66	6.91	7.13	7.33	7.49
10	95%	3.88	4.33	4.65	4.91	5.12	5.30	5.46	5.60
	99%	5.27	5.77	6.14	6.43	6.67	6.87	7.05	7.21
11	95%	3.82	4.26	4.57	4.82	5.03	5.20	5.35	5.49
	99%	5.15	5.62	5.97	6.25	6.48	6.67	6.84	6.99
12	95%	3.77	4.20	4.51	4.75	4.95	5.12	5.27	5.39
	99%	5.05	5.50	5.84	6.10	6.32	6.51	6.67	6.81
13	95%	3.73	4.15	4.45	4.69	4.88	5.05	5.19	5.32
	99%	4.96	5.40	5.73	5.98	6.19	6.37	6.53	6.67
14	95%	3.70	4.11	4.41	4.64	4.83	4.99	5.13	5.25
	99%	4.89	5.32	5.63	5.88	6.08	6.26	6.41	6.54
15	95%	3.67	4.08	4.37	4.59	4.78	4.94	5.08	5.20
	99%	4.84	5.25	5.56	5.80	5.99	6.16	6.31	6.44
16	95%	3.65	4.05	4.33	4.56	4.74	4.90	5.03	5.15
	99%	4.79	5.19	5.49	5.72	5.92	6.08	6.22	6.35
17	95%	3.63	4.02	4.30	4.52	4.70	4.86	4.99	5.11
	99%	4.74	5.14	5.43	5.66	5.85	6.01	6.15	6.27
18	95%	3.61	4.00	4.28	4.49	4.67	4.82	4.96	5.07
	99%	4.70	5.09	5.38	5.60	5.79	5.94	6.08	6.20
19	95%	3.59	3.98	4.25	4.47	4.65	4.79	4.92	5.04
	99%	4.67	5.05	5.33	5.55	5.73	5.89	6.02	6.14
20	95%	3.58	3.96	4.23	4.45	4.62	4.77	4.90	5.01
	99%	4.64	5.02	5.29	5.51	5.69	5.84	5.97	6.09
24	95%	3.53	3.90	4.17	4.37	4.54	4.68	4.81	4.92
	99%	4.55	4.91	5.17	5.37	5.54	5.69	5.81	5.92
30	95%	3.49	3.85	4.10	4.30	4.46	4.60	4.72	4.82
	99%	4.45	4.80	5.05	5.24	5.40	5.54	5.65	5.76
40	95%	3.44	3.79	4.04	4.23	4.39	4.52	4.63	4.73
	99%	4.37	4.70	4.93	5.11	5.26	5.39	5.50	5.60
60	95%	3.40	3.74	3.98	4.16	4.31	4.44	4.55	4.65
	99%	4.28	4.59	4.82	4.99	5.13	5.25	5.36	5.45
120	95%	3.36	3.68	3.92	4.10	4.24	4.36	4.47	4.56
	99%	4.20	4.50	4.71	4.87	5.01	5.12	5.21	5.30
∞	95%	3.31	3.63	3.86	4.03	4.17	4.29	4.39	4.47
	99%	4.12	4.40	4.60	4.76	4.88	4.99	5.08	5.16

Table 9 ▪ Upper-Tail Areas for Chi-Square Distributions

Right-tail area	df = 1	df = 2	df = 3	df = 4	df = 5
>0.100	< 2.70	< 4.60	< 6.25	< 7.77	< 9.23
0.100	2.70	4.60	6.25	7.77	9.23
0.095	2.78	4.70	6.36	7.90	9.37
0.090	2.87	4.81	6.49	8.04	9.52
0.085	2.96	4.93	6.62	8.18	9.67
0.080	3.06	5.05	6.75	8.33	9.83
0.075	3.17	5.18	6.90	8.49	10.00
0.070	3.28	5.31	7.06	8.66	10.19
0.065	3.40	5.46	7.22	8.84	10.38
0.060	3.53	5.62	7.40	9.04	10.59
0.055	3.68	5.80	7.60	9.25	10.82
0.050	3.84	5.99	7.81	9.48	11.07
0.045	4.01	6.20	8.04	9.74	11.34
0.040	4.21	6.43	8.31	10.02	11.64
0.035	4.44	6.70	8.60	10.34	11.98
0.030	4.70	7.01	8.94	10.71	12.37
0.025	5.02	7.37	9.34	11.14	12.83
0.020	5.41	7.82	9.83	11.66	13.38
0.015	5.91	8.39	10.46	12.33	14.09
0.010	6.63	9.21	11.34	13.27	15.08
0.005	7.87	10.59	12.83	14.86	16.74
0.001	10.82	13.81	16.26	18.46	20.51
<0.001	>10.82	>13.81	>16.26	>18.46	>20.51

Right-tail area	df = 6	df = 7	df = 8	df = 9	df = 10
>0.100	<10.64	<12.01	<13.36	<14.68	<15.98
0.100	10.64	12.01	13.36	14.68	15.98
0.095	10.79	12.17	13.52	14.85	16.16
0.090	10.94	12.33	13.69	15.03	16.35
0.085	11.11	12.50	13.87	15.22	16.54
0.080	11.28	12.69	14.06	15.42	16.75
0.075	11.46	12.88	14.26	15.63	16.97
0.070	11.65	13.08	14.48	15.85	17.20
0.065	11.86	13.30	14.71	16.09	17.44
0.060	12.08	13.53	14.95	16.34	17.71
0.055	12.33	13.79	15.22	16.62	17.99
0.050	12.59	14.06	15.50	16.91	18.30
0.045	12.87	14.36	15.82	17.24	18.64
0.040	13.19	14.70	16.17	17.60	19.02
0.035	13.55	15.07	16.56	18.01	19.44
0.030	13.96	15.50	17.01	18.47	19.92
0.025	14.44	16.01	17.53	19.02	20.48
0.020	15.03	16.62	18.16	19.67	21.16
0.015	15.77	17.39	18.97	20.51	22.02
0.010	16.81	18.47	20.09	21.66	23.20
0.005	18.54	20.27	21.95	23.58	25.18
0.001	22.45	24.32	26.12	27.87	29.58
<0.001	>22.45	>24.32	>26.12	>27.87	>29.58

(*continued*)

Table 9 ▪ Upper-Tail Areas for Chi-Square Distributions (*continued*)

Right-tail area	df = 11	df = 12	df = 13	df = 14	df = 15
>0.100	<17.27	<18.54	<19.81	<21.06	<22.30
0.100	17.27	18.54	19.81	21.06	22.30
0.095	17.45	18.74	20.00	21.26	22.51
0.090	17.65	18.93	20.21	21.47	22.73
0.085	17.85	19.14	20.42	21.69	22.95
0.080	18.06	19.36	20.65	21.93	23.19
0.075	18.29	19.60	20.89	22.17	23.45
0.070	18.53	19.84	21.15	22.44	23.72
0.065	18.78	20.11	21.42	22.71	24.00
0.060	19.06	20.39	21.71	23.01	24.31
0.055	19.35	20.69	22.02	23.33	24.63
0.050	19.67	21.02	22.36	23.68	24.99
0.045	20.02	21.38	22.73	24.06	25.38
0.040	20.41	21.78	23.14	24.48	25.81
0.035	20.84	22.23	23.60	24.95	26.29
0.030	21.34	22.74	24.12	25.49	26.84
0.025	21.92	23.33	24.73	26.11	27.48
0.020	22.61	24.05	25.47	26.87	28.25
0.015	23.50	24.96	26.40	27.82	29.23
0.010	24.72	26.21	27.68	29.14	30.57
0.005	26.75	28.29	29.81	31.31	32.80
0.001	31.26	32.90	34.52	36.12	37.69
<0.001	>31.26	>32.90	>34.52	>36.12	>37.69

Right-tail area	df = 16	df = 17	df = 18	df = 19	df = 20
>0.100	<23.54	<24.77	<25.98	<27.20	<28.41
0.100	23.54	24.76	25.98	27.20	28.41
0.095	23.75	24.98	26.21	27.43	28.64
0.090	23.97	25.21	26.44	27.66	28.88
0.085	24.21	25.45	26.68	27.91	29.14
0.080	24.45	25.70	26.94	28.18	29.40
0.075	24.71	25.97	27.21	28.45	29.69
0.070	24.99	26.25	27.50	28.75	29.99
0.065	25.28	26.55	27.81	29.06	30.30
0.060	25.59	26.87	28.13	29.39	30.64
0.055	25.93	27.21	28.48	29.75	31.01
0.050	26.29	27.58	28.86	30.14	31.41
0.045	26.69	27.99	29.28	30.56	31.84
0.040	27.13	28.44	29.74	31.03	32.32
0.035	27.62	28.94	30.25	31.56	32.85
0.030	28.19	29.52	30.84	32.15	33.46
0.025	28.84	30.19	31.52	32.85	34.16
0.020	29.63	30.99	32.34	33.68	35.01
0.015	30.62	32.01	33.38	34.74	36.09
0.010	32.00	33.40	34.80	36.19	37.56
0.005	34.26	35.71	37.15	38.58	39.99
0.001	39.25	40.78	42.31	43.81	45.31
<0.001	>39.25	>40.78	>42.31	>43.81	>45.31

Table 10 ▪ Binomial Probabilities

n = 5

							π						
x	0.05	0.1	0.2	0.25	0.3	0.4	0.5	0.6	0.7	0.75	0.8	0.9	0.95
0	.774	.590	.328	.237	.168	.078	.031	.010	.002	.001	.000	.000	.000
1	.203	.329	.409	.396	.360	.259	.157	.077	.029	.015	.007	.000	.000
2	.022	.072	.205	.263	.309	.346	.312	.230	.132	.088	.051	.009	.001
3	.001	.009	.051	.088	.132	.230	.312	.346	.309	.263	.205	.072	.022
4	.000	.000	.007	.015	.029	.077	.157	.259	.360	.396	.409	.329	.203
5	.000	.000	.000	.001	.002	.010	.031	.078	.168	.237	.328	.590	.774

n = 10

							π						
x	0.05	0.1	0.2	0.25	0.3	0.4	0.5	0.6	0.7	0.75	0.8	0.9	0.95
0	.599	.349	.107	.056	.028	.006	.001	.000	.000	.000	.000	.000	.000
1	.315	.387	.268	.188	.121	.040	.010	.002	.000	.000	.000	.000	.000
2	.075	.194	.302	.282	.233	.121	.044	.011	.001	.000	.000	.000	.000
3	.010	.057	.201	.250	.267	.215	.117	.042	.009	.003	.001	.000	.000
4	.001	.011	.088	.146	.200	.251	.205	.111	.037	.016	.006	.000	.000
5	.000	.001	.026	.058	.103	.201	.246	.201	.103	.058	.026	.001	.000
6	.000	.000	.006	.016	.037	.111	.205	.251	.200	.146	.088	.011	.001
7	.000	.000	.001	.003	.009	.042	.117	.215	.267	.250	.201	.057	.010
8	.000	.000	.000	.000	.001	.011	.044	.121	.233	.282	.302	.194	.075
9	.000	.000	.000	.000	.000	.002	.010	.040	.121	.188	.268	.387	.315
10	.000	.000	.000	.000	.000	.000	.001	.006	.028	.056	.107	.349	.599

(continued)

Table 10 ▪ Binomial Probabilities (*continued*)

$n = 15$

								π						
x	0.05	0.1	0.2	0.25	0.3	0.4	0.5	0.6	0.7	0.75	0.8	0.9	0.95
0	.463	.206	.035	.013	.005	.000	.000	.000	.000	.000	.000	.000	.000
1	.366	.343	.132	.067	.030	.005	.000	.000	.000	.000	.000	.000	.000
2	.135	.267	.231	.156	.092	.022	.004	.000	.000	.000	.000	.000	.000
3	.031	.128	.250	.225	.170	.064	.014	.002	.000	.000	.000	.000	.000
4	.004	.043	.188	.225	.218	.126	.041	.007	.001	.000	.000	.000	.000
5	.001	.011	.103	.166	.207	.196	.092	.025	.003	.001	.000	.000	.000
6	.000	.002	.043	.091	.147	.207	.153	.061	.011	.003	.001	.000	.000
7	.000	.000	.014	.040	.081	.177	.196	.118	.035	.013	.003	.000	.000
8	.000	.000	.003	.013	.035	.118	.196	.177	.081	.040	.014	.000	.000
9	.000	.000	.001	.003	.011	.061	.153	.207	.147	.091	.043	.002	.000
10	.000	.000	.000	.001	.003	.025	.092	.196	.207	.166	.103	.011	.001
11	.000	.000	.000	.000	.001	.007	.041	.126	.218	.225	.188	.043	.004
12	.000	.000	.000	.000	.000	.002	.014	.064	.170	.225	.250	.128	.031
13	.000	.000	.000	.000	.000	.000	.004	.022	.092	.156	.231	.267	.135
14	.000	.000	.000	.000	.000	.000	.000	.005	.030	.067	.132	.343	.366
15	.000	.000	.000	.000	.000	.000	.000	.000	.005	.013	.035	.206	.463

Table 10 ■ Binomial Probabilities (*continued*)

n = 20

x	0.05	0.1	0.2	0.25	0.3	0.4	0.5	0.6	0.7	0.75	0.8	0.9	0.95
0	.358	.122	.012	.003	.001	.000	.000	.000	.000	.000	.000	.000	.000
1	.377	.270	.058	.021	.007	.000	.000	.000	.000	.000	.000	.000	.000
2	.189	.285	.137	.067	.028	.003	.000	.000	.000	.000	.000	.000	.000
3	.060	.190	.205	.134	.072	.012	.001	.000	.000	.000	.000	.000	.000
4	.013	.090	.218	.190	.130	.035	.005	.000	.000	.000	.000	.000	.000
5	.002	.032	.175	.202	.179	.075	.015	.001	.000	.000	.000	.000	.000
6	.000	.009	.109	.169	.192	.124	.037	.005	.000	.000	.000	.000	.000
7	.000	.002	.055	.112	.164	.166	.074	.015	.001	.000	.000	.000	.000
8	.000	.000	.022	.061	.114	.180	.120	.035	.004	.001	.000	.000	.000
9	.000	.000	.007	.027	.065	.160	.160	.071	.012	.003	.000	.000	.000
10	.000	.000	.002	.010	.031	.117	.176	.117	.031	.010	.002	.000	.000
11	.000	.000	.000	.003	.012	.071	.160	.160	.065	.027	.007	.000	.000
12	.000	.000	.000	.001	.004	.035	.120	.180	.114	.061	.022	.000	.000
13	.000	.000	.000	.000	.001	.015	.074	.166	.164	.112	.055	.002	.000
14	.000	.000	.000	.000	.000	.005	.037	.124	.192	.169	.109	.009	.000
15	.000	.000	.000	.000	.000	.001	.015	.075	.179	.202	.175	.032	.002
16	.000	.000	.000	.000	.000	.000	.005	.035	.130	.190	.218	.090	.013
17	.000	.000	.000	.000	.000	.000	.001	.012	.072	.134	.205	.190	.060
18	.000	.000	.000	.000	.000	.000	.000	.003	.028	.067	.137	.285	.189
19	.000	.000	.000	.000	.000	.000	.000	.000	.007	.021	.058	.270	.377
20	.000	.000	.000	.000	.000	.000	.000	.000	.001	.003	.012	.122	.358

(*continued*)

Table 10 ▪ Binomial Probabilities (*continued*)

n = 25

x	0.05	0.1	0.2	0.25	0.3	0.4	0.5	0.6	0.7	0.75	0.8	0.9	0.95
0	.277	.072	.004	.001	.000	.000	.000	.000	.000	.000	.000	.000	.000
1	.365	.199	.023	.006	.002	.000	.000	.000	.000	.000	.000	.000	.000
2	.231	.266	.071	.025	.007	.000	.000	.000	.000	.000	.000	.000	.000
3	.093	.227	.136	.064	.024	.002	.000	.000	.000	.000	.000	.000	.000
4	.027	.138	.187	.118	.057	.007	.000	.000	.000	.000	.000	.000	.000
5	.006	.065	.196	.164	.103	.020	.002	.000	.000	.000	.000	.000	.000
6	.001	.024	.163	.183	.148	.045	.005	.000	.000	.000	.000	.000	.000
7	.000	.007	.111	.166	.171	.080	.015	.001	.000	.000	.000	.000	.000
8	.000	.002	.062	.124	.165	.120	.032	.003	.000	.000	.000	.000	.000
9	.000	.000	.030	.078	.134	.151	.061	.009	.000	.000	.000	.000	.000
10	.000	.000	.011	.042	.091	.161	.097	.021	.002	.000	.000	.000	.000
11	.000	.000	.004	.019	.054	.146	.133	.044	.004	.001	.000	.000	.000
12	.000	.000	.002	.007	.027	.114	.155	.076	.011	.002	.000	.000	.000
13	.000	.000	.000	.002	.011	.076	.155	.114	.027	.007	.002	.000	.000
14	.000	.000	.000	.001	.004	.044	.133	.146	.054	.019	.004	.000	.000
15	.000	.000	.000	.000	.002	.021	.097	.161	.091	.042	.011	.000	.000
16	.000	.000	.000	.000	.000	.009	.061	.151	.134	.078	.030	.000	.000
17	.000	.000	.000	.000	.000	.003	.032	.120	.165	.124	.062	.002	.000
18	.000	.000	.000	.000	.000	.001	.015	.080	.171	.166	.111	.007	.000
19	.000	.000	.000	.000	.000	.000	.005	.045	.148	.183	.163	.024	.001
20	.000	.000	.000	.000	.000	.000	.002	.020	.103	.164	.196	.065	.006
21	.000	.000	.000	.000	.000	.000	.000	.007	.057	.118	.187	.138	.027
22	.000	.000	.000	.000	.000	.000	.000	.002	.024	.064	.136	.227	.093
23	.000	.000	.000	.000	.000	.000	.000	.000	.007	.025	.071	.266	.231
24	.000	.000	.000	.000	.000	.000	.000	.000	.002	.006	.023	.199	.365
25	.000	.000	.000	.000	.000	.000	.000	.000	.000	.001	.004	.072	.277

Answers to Selected Odd-Numbered Exercises

CHAPTER 1

1.1 Descriptive statistics is made up of those methods whose purpose is to organize and summarize the values in a data set. Inferential statistics refers to those procedures whose purpose is to allow conclusions to be drawn about a population based on information from a sample. **1.3** The population is the entire student body — 15,000 students. The sample consists of the 200 students interviewed. **1.5** The population consists of all single-family homes in Northridge. The sample consists of the 100 homes selected for inspection. **1.7** The population consists of all 5000 bricks in the lot. The sample consists of the 100 bricks selected for inspection. **1.9 a.** Categorical. **b.** Categorical. **c.** Numerical (discrete). **d.** Numerical (continuous). **e.** Categorical. **f.** Numerical (continuous). **1.11 a.** For example, General Motors, Toyota, Aston Martin, Ford, Jaguar. **b.** For example, 3.23, 2.92, 4.0, 2.8 **c.** For example, 2, 0, 1, 4, 3 **d.** For example, 49.2, 48.84, 50.3, 50.23 **e.** For example, 10, 15.5, 17, 3, 3, 6.5 **1.13 a.** Two-thirds of the beaches are rated B or above. **b.** No. **1.15** MCL tear and meniscus tear are more than twice as common as ACL tear and patella dislocation in this sample of women collegiate rugby players. **1.19** Most business schools have acceptance rates between 15% and 40%. There are three schools that have acceptance rates greater than 40%. **1.23** The 20 observations are quite spread out. There is a gap that separates the three smallest and the three largest values from the rest of the data.

CHAPTER 2

2.1 Two *examples* of many possible answers: (1) Write the graduates' names on slips of paper (one name per slip). Place the slips of paper in a container and thoroughly mix. Then, select 20 slips of paper without replacement. The graduates whose names are on the slips constitute the random sample. (2) Associate each graduate with an individual number between 001 and 140. Then, use a random number generator on a computer to select 20 distinct random numbers between 1 and 140. Graduates whose numbers are selected are included in the sample. **2.3** The names appear on the petition in a list ordered from first to last. Using a computer random number generator, select 30 random numbers between 1 and 500. Signatures corresponding to the 30 selected numbers constitute the random sample. **2.5** Researcher B's plan would result in a simple random sample of

trees and would avoid possible confounding factors related to location in the field. **2.7** Because the students are already in a numbered list, we can assign to each of them their number on the list. As the largest number used has 4 digits, we will express all the numbers with 4 digits. For example, Student 5 on the list will be assigned the number 0005. Student 342 on the list will be assigned the number 0342. By using a random number table, we can go along a row, taking each 4-digit number and discarding any repeats and any numbers greater than 4500, until a sample of 20 numbers has been obtained. The 20 students corresponding to these numbers on the list will become the sample. Individual results will vary. **2.9** Using a convenience sample may introduce selection bias. Suppose that I wanted to know the average height of students in all statistics classes and that I used the students in my 10 o'clock class as my subjects. It may be that all the basketball players (who are generally tall) are in my 10 o'clock class because they have a 7 o'clock practice. **2.11 a.** Using a random number generator, create three-digit numbers. The three digits represent the page number, and the number of correctly defined words can be counted. Suppose that we want to choose 10 pages in the book. Repeat this process until 10 page numbers are selected, and count the number of "words" on each page. **b.** Because the number of words on a page may be influenced by which topic is being discussed, it makes sense to stratify by chapter. To sample three pages in each chapter, generate a two-digit number, select a page within the chapter, and repeat until three pages are chosen. Count the number of defined words on those pages, and repeat for each chapter. **c.** Choose the 10th page and then every 15th page after that. **d.** Randomly choose one chapter, and count the words on each page in that chapter. **2.13** Randomization is used in selecting the households and available subjects in each family. However, this method will eliminate households without a phone. It may be that the person who has recently had a birthday may be busy or too grumpy to talk to someone and that someone else (who may not be of voting age) responds to the questions. Someone may lie and give responses that are not true. **2.15** The best method to use would be a simple random sample where every different sample has an equal chance of being selected — for example, numbering every student in your school, randomly

selecting a sample of them, finding them, and asking them their opinion. **2.17** Any survey conducted by mail is subject to selection bias (eliminating anyone without an address) and nonresponse bias (only a few people reply). **2.19** Selection bias. **2.21 a.** The population consists of all 16–24-year-olds who live in New York state but not in New York City. **2.23** The sample consisted of volunteers, included only women, included only students from a private university, and included only students attending school in Texas. **2.25** Selection bias. **2.27** Observational study. **2.29** No cause-and-effect conclusion can be made on the basis of an observational study. **2.31** If we are considering all people 12 years of age or older, there are probably more single people who are young and more widowed people who are old. It tends to be the young who are at higher risk of being victims of violent crimes (e.g., because they stay out late at night). Hence, age could be a potential confounding variable. **2.33 a.** If the definition of affluent Americans was having a household income of over $75,000 and if the sample was a simple random sample. **b.** No. This sample is not representative of all Americans. **2.35** Cause-and-effect conclusions cannot be drawn from this study. **2.37** No, it is an observational study. Cause-and-effect conclusions cannot be made on the basis of an observational study. **2.39** The response variable is the amount of time it takes a car to exit the parking space. Extraneous factors that may affect the response variable could include location of the parking spaces, time of day, and normal driving behaviors of individuals, such as whether they have children in the car. **2.41** So many other factors could have contributed to the difference in the pregnancy rate. The only way that the difference between the two groups could have been attributed to the program was if the two groups had been originally formed by dividing all the students completely randomly. This would have minimized the effects of any other factors, leaving the program as the only *big* difference between the two groups. **2.43** An experiment that fairly compares the word processing programs while simultaneously accounting for possible differences in the user's computer proficiency would have the same user testing each one of the four new word processing programs. The design would randomly select the order in which each user tests the software. This would minimize the effects that any extraneous factors (such as possible tiredness or boredom or frustration with the software) might have on speed with which the tasks are completed. **2.45** Gender is a useful blocking variable if men and women of the same status tend to differ with respect to rate of talk. Blocking on gender would make it easier to assess the effect of status on rate of talk. **2.47** It is unlikely that a person who is described as a moderate drinker would be unable to tell whether he or she was drinking a vodka tonic or a virgin tonic. **2.49** Using a control group and a placebo treatment enables researchers to find out whether the active ingredient is having an effect or whether just the "power of suggestion" is having the effect. **2.51** Because some people often have their own personal beliefs about the effectiveness of various treatments and their response can be influenced by their personal perception. **2.53 a.** Age, weight, lean body mass, and capacity to lift weights were all dealt with by direct control. **b.** Yes. Knowing whether they were receiving creatine might have influenced the effort a subject put into his workout. **c.** Yes, especially if the trainer is the one taking the measurements. If the trainer knows which men are receiving creatine, it might influence the way in which the trainer works with the subjects. **2.55 a.** No, the judges wanted to show that one of Pismo's restaurants made the best chowder. **b.** So that the judging is fair! **2.57 a.** Randomly divide the volunteers into 2 groups of 50; one group gets PH80, and the other group gets a placebo nasal spray. The volunteers are assessed before and after the treatment, and any improvement in PMS symptoms is measured. **b.** A placebo treatment is needed to see whether any improvement is due to the PH80 or to any other influence that occurs during the time of testing. **c.** Single-blinding is strongly recommended. **2.65 a.** Alcohol consumption may differ by province. **b.** Examples include occupation and age. **2.67** Stratifying by room type would probably be a good strategy. **2.69** Some confounding variables could include exposure to sun or shade, proximity to water, slope of land, or density of insects in the soil. This study is an experiment.

CHAPTER 3

3.3 d. The majority of both men and women think that men earn more money for the same work. Very few men or women think that a woman earns more. More men than women

think that they earn the same. **3.5 c.** It is easier to see the difference in the proportions with a pie chart; however, it is easier to estimate the percentage of each response with a segmented bar chart. **3.7 a.** Given the choice, most people don't mind whether they work for a man or a woman. Of those who do have an opinion, most would prefer to have a male boss. **3.9** The pie charts show that a very large proportion of college students with medical career aspirations have at least one parent who is a physician. The charts also show that the female college students in this study have more mothers who are physicians than the male college students in the study have fathers who are physicians. **3.11 b.** In 1987, 70% of the top 10 movies that made the most money were rated R. The remaining 30% were rated G. By 1992, 40% were rated R, 50% were rated PG or PG-13, and 10% were rated G. **3.13** The bar graph shows a steady increase in the percentage of U.S. gross domestic product (GDP) spent on health care over the 1960–1980 time period. This was followed by a sharp drop of about 8.5% in 1985 and then by an increase over the next 10 years. In 1995 the percentage of GDP spent on health care was still less than that spent in 1960. **3.15** A bar chart would be better. There are too many different categories to use a pie chart. **3.17 a.** 2.2 **b.** In the row with the stem of 8. The leaf of 9 would be placed to the right of the other leaves. **c.** A large number of flow rates are between 6.0 and 8.0; a typical flow rate would be 6.9 or 7.0. **d.** There appears to be a lot of variability in the flow rates. **e.** The distribution is not symmetric. **f.** 18.9 appears to be an outlier.

3.19

Calorie Content
(cal/100 ml)
of 26 Brands of Light Beers

```
1L |
1H | 9                    stem: tens
2L | 1 2 3                 leaf: ones
2H | 7 8 8 9 9
3L | 0 0 1 1 1 2 2 3 3 4 5
3H | 9
4L | 0 1 2 3
4H |
```

3.21 a.

Percentage of Fully Credentialed Teachers in California Counties

```
7  | 5
8  | 0 3 3 4 5 5 5 5 5 6 7 8 8 8 9 9          stem: tens
9  | 0 0 1 1 1 1 2 2 3 4 4 4 4 5 5 5 5 5 5 5 5 5 6 6 7 7 7 7 7 7 7 7 7 8 8 8 8 8 8 9 9   leaf: ones
10 | 0
```

Most counties in California have more than 90% of their teachers fully credentialed. Los Angeles County is the lowest, with only 75%, and Alpine is the only county where 100% of the teachers are fully credentialed.
b. Percentage of Fully Credentialed Teachers in California Counties

```
7L  |
7H  | 5                          stem: tens
8L  | 0 3 3 4                    leaf: ones
8H  | 5 5 5 5 5 6 7 8 8 8 8 9 9
9L  | 0 0 1 1 1 1 2 2 3 4 4 4 4
9H  | 5 5 5 5 5 5 5 6 6 7 7 7 7 7 7 7 7 7 8 8 8 8 8 9 9
10L | 0
```

We can now see that there are only three counties with less than 85% of their teachers fully credentialed.
3.23 a. Percentage Increase in Population from 1990 to 2000

```
0 | 0 1 3 4 4 4 5 5 5 5 6 7 8 8 8 8 9 9 9 9 9
1 | 0 0 0 0 0 1 1 2 3 4 4 4 4 4 5 7 8
2 | 0 0 1 1 3 3 6 8
3 | 0 1
4 | 0
5 |                    stem: tens
6 | 6                  leaf: ones
```

b. Forty-eight of the states have an increase in population of 31% or less, and most of these are under 12%. There are two states that have a much larger percentage increase: Nevada (66%) and Arizona (40%).
c. Percentage Increase in Population from 1990 to 2000

```
    West   |    | East
9 9 8 8 8 0 | 0  | 1 3 4 4 4 5 5 5 5 6 7 8 9 9 9
    4 4 3 0 | 1  | 0 0 0 0 0 1 1 2 4 4 5 7 8
  8 3 1 0 0 | 2  | 1 3 6
      1 0 0 | 3  |
          0 | 4  |        stem: tens
            | 5  |        leaf: ones
          6 | 6  |
```

The states that show a large percentage increase in population are in the West. There are 5 states in the West (out of 19) that have a percentage increase greater than the maximum increase in the East. **3.25 a.** Relative frequencies: .4167, .1792, .1500, .0708, .1000, .0375, .0458 **b.** .7459 **c.** .2541 **d.** .1833 **e.** The frequencies (and relative frequencies) tend to decrease as the number of impairments increases. **3.27 a.** Most men average between 0 and 200 min/month. Far fewer average between 400 and 800 min/month. **b.** The distribution for men and women is similar in that most women average between 0 and 200 min, as do men. Fewer women than men average 600 to 800 min. **c.** .74 **d.** .79 **e.** .1775 **3.29 a.** If a commute is longer than 45 min, then the time is often rounded to the nearest 15 or 30 min. **3.33 a.** Negatively skewed. **b.** Positively skewed. **c.** Bimodal. **3.35 d. i.** .8651 **ii.** .5506 **e. i.** About 20 months. **ii.** About 29 months. **3.37** Almost all the differences are positive, indicating that the runners slow down. The histogram is positively skewed. A typical difference is about 150. About 2% of the runners ran the late distance more quickly than the early distance. **3.39** Histogram I: symmetric. Histogram II: positively skewed. Histogram III: bimodal and approximately symmetric. Histogram IV: bimodal. Histogram V: negatively skewed. **3.41 a.** The graft weight ratio tends to decrease as recipient body weight increases, but the relationship does not look linear. **3.43 b.** There is a positive relationship between the number of violations and the total fines. As the number of violations increases, so does the total fine. **3.45 c.** The graph suggests that elongation tends to increase as temperature increases until about 70°, and then elongation tends to decrease as temperature increases. **3.47** The plot shows a linear pattern. The two points that stand out in the plot look like they conform to the linear pattern seen in the rest of the data. **3.49 a.** No, there are two data points that have the same x value but different y values. **c.** Yes, as plasma cortisol concentration increases, there is a tendency for oxygen consumption to increase. **3.51 c.** There has been a steady increase in life expectancy for both men and women from 1940 to 2000. For both sexes, the life expectancy in 2000 is about 14 years higher than it was is in 1940, but the life expectancy for women is consistently about 6 years higher than that of men. **3.53** In both 2001 and 2002 the box office sales dropped in Weeks 2 and 6 and in the last two weeks of the summer. The seasonal peaks occurred during Weeks 4, 9, and 13. **3.55 b.** .4759 **3.57 a.** The number of transplants from a living relative has been increasing steadily from 1994 to 2001. **b.** There are many more related donors than unrelated donors used for kidney transplants. However, the use of unrelated donors is rising at a faster rate than that of related donors. **3.59 a.** For the first histogram, it appears that more frozen meals have a sodium content of about 800 mg. How-

ever, the second histogram suggests that sodium content is fairly uniform from 300 to 900 mg and then drops off above 900 mg. **3.61 a.** There appear to be two outliers. **c.** Both histograms are positively skewed. **3.63 b.** Positively skewed. **d. i.** .2979 **ii.** .4043 **iii.** .2978 **3.65 b.** .3647, .9176 **c.** Relative frequency.

CHAPTER 4
4.1 Mean = 970.7, median = 738. They are different because of a couple of outliers at the high end of the data set that influence the mean more than the median. **4.3** The supervisors will be paid the median: $4286. The mean (which is more favorable to the taxpayers) is $3968.67. **4.5 a.** Because the distribution is approximately symmetric, the mean and the median should be similar. **b.** $\bar{x}$ = 370.69, median = 369.50 **c.** The largest value could be increased without limit without affecting the value of the median. It could be decreased to 370 without affecting the value of the median. **4.7** 10% trimmed mean = 13.4375, 20% trimmed mean = 12.5833, $\bar{x}$ = 23.1, sample median = 13 **4.9** Statement 1: The fact that only half the homes in this county cost less than $278,380 is the definition of a median. Statement 2: The median is not the midpoint of the range. It is the price at which 50% of the homes cost less and 50% of the houses cost more than this price. Statement 3: If there are no houses that cost less than $300,000 and unless at least half the homes in this county cost exactly $300,000, Walker is correct; the median will be above $300,000. **4.11** x_5 = 32 **4.13** Median = 68, 20% trimmed mean = 66.17 **4.15** Median = 11; the median is not influenced by the unusually large value of 33. **4.19** $1,961,158 **4.21 a.** Standard deviation = $406.98, iqr = $370 **b.** Standard deviation = $186.236, iqr = $155 **c.** Moderately priced midsize cars. **d.** Moderately priced: $348.9; inexpensive: $298.36 **e.** On average, the repair cost is higher for a moderately priced midsize car than for an inexpensive midsize car. The variability (or spread) of the individual costs is greater for the moderately priced cars. **4.23 a.** Mean = 2965.2, standard deviation = 542.6, iqr = 602 **b.** Chocolate pudding: iqr = 602; catsup: iqr = 1300; less variability in chocolate pudding. **4.25** s^2 = 435.1868, s = 20.86 **4.27** The variance and standard deviation are not affected by adding or subtracting a number from each observation. **4.29 a.** Sample 1: $\bar{x}$ = 7.81, s = 0.39847. Sample 2: $\bar{x}$ = 49.68, s = 1.73897

b. Sample 1, CV = 5.10. Sample 2, CV = 3.50 **4.31** As with the dotplot, it is clear that there are two outliers, at 84 and 331. Although individual values cannot be shown by a boxplot, it is clear that 75% of the class accessed the web site 20 times or less during the first month. **4.33** Half the states in the continental United States have nitrous oxide emissions of less than 34,000 tons. There is one state that has a much larger quantity of emissions than any of the other states. The distribution looks positively skewed, meaning that most states have emission values at the lower end of the range. **4.35** The median score for creamy peanut butter is about 45, and the median score for crunchy peanut butter is about 51. The ranges appear to be equal. The data for creamy peanut butter are slightly skewed in the positive direction, whereas the data for crunchy butter are fairly symmetric. **4.37 a.** Excited delirium: median = 0.4, Q1 = 0.1, Q3 = 2.8, iqr = 2.7. No excited delirium: median = 1.6, Q1 = 0.3, Q3 = 7.9, iqr = 7.6 **b.** There are two mild and two extreme outliers for the excited delirium sample. There are no outliers in the no excited delirium sample. **4.39 a.** Approximately 68%. **b.** Approximately 16%. **4.41 a.** At least 75%. **b.** (2.90, 70.94) **c.** If the distribution of NO_2 values was normal, approximately 2.5% of the values would be less than −9.64 (2 standard deviations below the mean). Because NO_2 concentration cannot be negative, the distribution shape must not resemble a normal curve. **4.43** The z score is larger for the second exam, so performance was better on the second exam. **4.45 a.** Approximately 68%. **b.** Approximately 5%. **c.** Approximately 13.5%. **d.** Because the histogram is well approximated by a normal curve. **4.47** The proportion is at most .16. **4.49** One standard deviation below the mean is a negative number, and the number of answers changed from right to wrong cannot be negative. Therefore the distribution cannot be well described by a normal curve. From Chebyshev's Rule, the proportion who changed at least six answers from right to wrong is at most .106. **4.51 b. i.** 13.158 **ii.** 16.25 **iii.** 6.923 **iv.** 21 **v.** 11.842 **4.53 a.** Median = 20.88, Quartile 1 = 18.09, Quartile 3 = 22.20 **b.** iqr = 3.91. There are two outliers, 35.78 and 36.73. **c.** Most of the subjects have an oxygen capacity of between 20 and 25 ml/kg/min. There are two subjects with much higher oxygen capacities: over 35 ml/kg/min. **4.55 a.** $\bar{x}$ = 192.57, median = 189 **b.** $\bar{x}$ = 189.71, median = 189 (unchanged). **c.** 191, 7.14% trimmed mean. **d.** If the largest observation is 204, then the new trimmed mean = 185. If the largest observation is 284, the trimmed mean is the same as in Part (c). **4.57 a.** Lower quartile = 7.82, upper quartile = 8.01, iqr = 0.19 **b.** The boxplot indicates that the distribution is not symmetric, with a longer tail on the upper end. **4.59 b.** For the cancer group, $\bar{x}$ = 22.81 and median = 16; for the no-cancer group, $\bar{x}$ = 19.15 and median = 12. **c.** For the cancer group, s = 31.6553; for the no-cancer group, s = 16.99. There is more variability in the cancer sample values than in the no-cancer sample values. **4.61** It implies that the median salary is less than $1 million and that the histogram is positively skewed or that there are outliers with very large values. **4.63 b.** The mean and the median are not the same because the distribution is not symmetric. **c.** In the interval (0, 45.71). (Note: 0 is used for the lower endpoint because a score cannot be negative.) **4.65** Various graphical and numerical methods can be used for comparison. In general, the methods will suggest that milk volume for nonsmoking mothers is greater than that for smoking mothers. **4.67 a.** $\bar{x}$ = 22.15, s = 11.366 **b.** 19.4375 **c.** Upper quartile = 20.5, lower quartile = 18, iqr = 2.5 **d.** There are two mild outliers (25 and 28) and one extreme (69) outlier.

CHAPTER 5

5.1 a. Positive. **b.** Negative. **c.** Positive. **d.** No correlation. **e.** Positive. **f.** Positive. **g.** Negative. **h.** No correlation. **5.5 a.** r = .94. The correlation is strong and positive. **b.** Increasing sugar consumption does not cause or lead to higher rates of depression; there may be another reason that causes an increase in both. For instance, high sugar consumption may indicate a need for comfort food for a reason that also causes depression. **c.** The seven listed countries may not be representative of other countries. It may be that only these seven countries have a strong positive correlation between sugar consumption and depression rate and that other countries have a different type of relationship between these factors. It is therefore not a good idea to generalize these results to other countries. **5.7** Strong and positive. **5.9 a.** r = .335. There is a weak positive relationship between timber sales and the amount of acres burned in forest fires. **b.** No. Looking at the scatterplot, there does not seem to be much of a relationship at all. Even if there was, it wouldn't

necessarily mean that one caused the other. In fact, out of the six years with the highest timber sales, two of the years had the least amount of acreage lost as a result of forest fires. **5.11 a.** There appears to be a strong positive correlation. **b.** r = .9366 indicates a strong positive linear relationship. **c.** If x and y had a correlation coefficient of 1, all the points would lie on a straight line, but the line would not necessarily have slope 1 and intercept 0. **5.13** r = .39. There is a moderate positive linear relationship. **5.15** No. An r value of $-.085$ indicates, at best, a very weak relationship between support for environmental spending and degree of belief. **5.17 a.** There appears to be a weak negative relationship. **b.** $\hat{y}$ = 1082.24 − 4.691x **c.** Negative, yes. **5.19 a.** There appears to be a positive relationship between the percentages of public school students who were at or above proficient level in mathematics in the fourth grade in 1996 and in the eighth grade in 2000. There are two states with slightly unusual results: California and Utah. **b.** $\hat{y}$ = −3.14 + 1.52x **5.21 a.** Yes. As the carbonation depth increases, the strength of the concrete decreases. **b.** $\hat{y}$ = 24.39 − 0.275x **5.23** Slope = −4000 **5.25 a.** $\hat{y}$ = 59.910 + 27.462x **b.** 554.234 **c.** No, because the data used to obtain the least-squares equation were from steeply sloped plots, so it would not make sense to use these data to predict runoff sediment from gradually sloped plots. You would need to use data from gradually sloped plots to create a least-squares regression equation to predict runoff sediment from gradually sloped plots. **5.27 a.** The graph shows a moderately positive linear relationship between chest capacity and vital capacity. **c.** 0.113, 1.123 **d.** 4.9994 **e.** No. There are 2 data points whose x values are 81.8, but each has a different y value. **5.29 a.** 100 + 1.5(standard deviation). **b.** r = .67 **5.31 a.** $\hat{y}$ = −424.7 + 3.891x **b.** Both the slope and the y intercept are changed by the multiplicative factor c. **5.33 b.** There are three points with large residuals, those points with x values of 40, 50, and 60. **c.** Yes, the residuals for small x values and large x values are positive, whereas the residuals for the middle x values are negative. **5.35 b.** There is a curved pattern in the plot, suggesting that the linear model may not be appropriate. **5.37 a.** 15.4%. **b.** No, r^2 = 16%, so only 16% of the observed variation in first-year college grades can be attributed to the approximate linear relationship between first-year college

grades and SAT II score. The least-squares line does not effectively summarize the relationship between SAT score and first-year college grades. **5.39 a.** There does appear to be a positive linear relationship between rainfall volume and runoff volume. **b.** Slope = 0.827, intercept = −1.13 **c.** 65.03 **d.** For small values of x the residuals are relatively small in size. For the larger values of x the residuals are much larger, creating a fan-shaped pattern in the residual plot. This indicates that a linear relationship may not be appropriate. **5.41 a.** 184.31, residual = −19.31 **b.** r^2 = .963 **5.43** r^2 = .9512 This means that 95.12% of the variability in hardness can be explained by the linear relationship with elapsed time. **5.47 a.** No. **c.** Yes, the coefficient of determination between y' and x' is .976. This suggests that a least-squares line might effectively summarize the relationship between x' and y'. **d.** 2.47973 **e.** Yes. **5.49 a.** r = −.717 **b.** r = −.835 for the transformed data. **5.51 a.** Nonlinear. **b.** There is a definite pattern in the residual plot, confirming the nonlinear relationship. **5.53 a.** The plot does show a positive relationship. **b.** The plot appears straighter, but the variability of y increases as x increases. **c.** The plot appears to be as straight as the plot in Part (b), but the variability of the y values is smaller. **d.** The plot has a curvature in the opposite direction as the plot in Part (a). **5.55 a.** r = .9684 **b.** The plot resembles a graph of a quadratic equation and, despite a high correlation coefficient, does not resemble a linear relationship at all! **5.57 a.** $\hat{y}$ = 4027.083 − 577.895x **b.** For every 1 hr increase in finishing time, on average the myoglobin level decreased by 577.895 ng/ml. **c.** The least-squares equation yields a negative value for the estimated level of myoglobin when the finishing time is 8 hr. This is clearly unreasonable because myoglobin level cannot have a negative value. **5.59 a.** The predicted values are 40.4048, 42.6245, 49.2837, 53.7231, 54.8330, 56.4978, 58.1626, 58.7175, 62.6020, 68.1513; the residuals are −17.4048, 9.3755, 15.7163, 1.2769, −22.8330, 3.5022, 19.8374, 0.2825, −1.6020, −8.1513. **b.** SSResid = 1635.6833, r^2 = .2859 **5.61 b.** The plot of log(y) versus log(x) looks the straightest. The line for predicting y' = log(y) from x' = log(x) is $\hat{y}'$ = 1.61867 − 0.31646x'. When x = 25, predicted y = 15.0064. **5.63 a.** The plot does not suggest a linear relationship. **b.** 32.254 **d.** With the outlier deleted, r = .027; there is still little

evidence of a linear relationship.
5.65 a. $r = 0$ **b.** For example, $y = 1$ (any value > 0.973 will work). **c.** For example, $y = -1$ (any value < -0.973 will work). **5.67 b.** Based on the plot in Part (a) and Figure 5.34, a transformation going down the ladder on x or y is suggested. The transformation log(time) will produce a reasonably straight plot.

CHAPTER 6

6.1 a. In the long run, 1% of all people who suffer cardiac arrest in New York City will survive. **b.** 23 or 24 **6.3** Dependent; the probability that a U.S. adult experiences pain daily is higher if the adult selected is a woman.
6.5 They are dependent. **6.7** They are dependent. **6.9 a.** .7; assumed that whether Jeanie remembers any one errand is independent of whether she remembers any of the other errands. **b.** .5 **c.** .714286 **6.11 a.** The expert assumed that the positions of the two valves were independent. **b.** Larger, the actual probability is 1/12. **6.13 a.** .81 **b.** .19, .19 **c.** .0361, .9639 **d.** .993141 **e.** .926559 **6.15** The ultrasound predicts females with a success rate of $432/562 = 0.769$ and males with a success rate of $390/438 = 0.890$, suggesting that an ultrasound has a better success rate at predicting boys.
6.17 $60/96 = 0.625$ of current smokers view smoking as very harmful compared to $78/99 = 0.7878$ for former smokers and $86/99 = 0.8687$ for those who have never smoked. The proportion who view smoking as harmful is certainly less for current smokers.
6.19 a. .85 **b.** .19 **c.** .6889 **d.** .87 **6.21 a.** .3941 **b.** .1765 **c.** .0529 **d.** .3412 **6.23** The simulation results will vary from one simulation to another. The following probabilities were obtained from a computer simulation with 100,000 trials. **a.** .71293 **b.** .02201 **6.25** The simulation results will vary from one simulation to another. The approximate probability should be about .8504. **6.27 a.** The simulation results will vary from one simulation to another. The approximate probability should be about .6504. **b.** The decrease in the probability of on-time completion for Jacob made the biggest change in the probability that the project is completed on time. **6.29 a.** .0155 **b.** .0823 **c.** .1359 **d.** Assumption of independence is reasonable. **6.31 a.** .40 **b.** .81 **c.** .94 **6.33 a.** .00243 **b.** .00275

CHAPTER 7

7.1 a. Discrete. **b.** Continuous. **c.** Discrete. **d.** Discrete. **e.** Continuous. **7.3 b.** .4 **7.5 b.** $x = 0$, because its prob-

ability is the largest. **c.** .25 **d.** .55 **7.7 a.** .82 **b.** .18 **c.** .65, .27 **7.9 a.** The distribution for Supplier 1. **b.** The distribution for Supplier 2. **c.** Supplier 1, because the bulbs generally last longer and the lifetime is less variable. **d.** About 1000. **7.13 b.** .5, .125, .125 **c.** .375 **d.** .4375 **7.15 a.** Height = 2 **b.** 1/4 **c.** 7/16 **7.17 a.** .1003 **b.** .1003 **c.** .8185 **d.** .5 **e.** 1 **f.** .9390 **g.** .5910 **7.19 a.** .5398 **b.** .4602 **c.** .1469 **d.** .1469 **e.** .4577 **f.** .8944 **g.** .0730 **7.21 a.** 1.88 **b.** 2.33 **c.** -1.75 **d.** -1.28 **7.23 a.** 1.34 **b.** 0.74 **c.** 0 **d.** -1.34 **e.** They are negatives of one another. **7.25 a.** .119, .6969 **b.** .0021 **c.** .7019 **d.** Weights less than 1836 g or weights greater than 5012 g. **e.** The same answer is obtained. **7.27** The second machine. **7.29** 94.4 sec or less. **7.31 a.** .5, .5 **b.** .8185 **c.** Yes, the probability is about .0013. **d.** .0252 **e.** 47.4 or less words per minute. **7.33** Because this plot appears to look like a straight line, it is reasonable to conclude that a normal distribution provides an adequate description of the steam rate distribution. **7.35** Because the graph exhibits a curve, we cannot conclude that the variable *component lifetime* is adequately modeled by a normal distribution. **7.37** Although the graph seems to follow a straight line for values of DDT concentration below 45 and above 45, they do not appear to be the *same* straight line. A normal distribution may not be an appropriate model for this population. **7.39** The cube-root transformation is more symmetric. **7.41 b.** The histogram is positively skewed. **c.** A square-root transformation will result in a more symmetric histogram. **7.45** .0228, 0 **7.47 a.** No, only about 69.15% are shorter than 5 ft 7 in. **b.** About 69%. **7.49 a.** .7745 **b.** .1587 **c.** .3085 **d.** The largest 5% are those pH readings that exceed 6.1645. **7.51 a.** .8247 **b.** .0516 **c.** .6826 **d.** .0030 **e.** $P(x < 261) = .3783$.

CHAPTER 8

8.1 A statistic is computed from the observations in a sample, whereas a population characteristic is a quantity that describes the whole population. **8.3 a.** Population characteristic. **b.** Statistic. **c.** Population characteristic. **d.** Statistic. **e.** Statistic.
8.7 a.

t value	6	11	15	21	25	30
Probability	1/6	1/6	1/6	1/6	1/6	1/6

b. $\mu = 9$, $\mu_t = 18$
8.9

Statistic 1
Value	2.67	3.00	3.33	3.67
Probability	.1	.4	.3	.2

Statistic 2
Value	3	4
Probability	.7	.3

Statistic 3
Value	2.5	3.0	3.5
Probability	.1	.5	.4

8.11 a. 11.8 **8.13** Histograms for samples of size 10 would be centered in the same place but would be less spread out than for samples of size 5. **8.15** For $n = 36, 50, 100,$ and 400. **8.17 a.** $\mu_{\bar{x}} = 40$, $\sigma_{\bar{x}} = 0.625$, approximately normal. **b.** .5762 **c.** .2628 **8.19 a.** $\mu_{\bar{x}} = 2$, $\sigma_{\bar{x}} = 0.267$ **b.** $n = 20$: $\mu_{\bar{x}} = 2$, $\sigma_{\bar{x}} = 0.179$; $n = 100$: $\mu_{\bar{x}} = 2$, $\sigma_{\bar{x}} = 0.08$ **8.21 a.** .8185, .0013 **b.** .9772, 0 **8.23** .0026 **8.25** Approximately 0. **8.27 a.** 0.65, 0.1508 **b.** 0.65, 0.1067 **c.** 0.65, 0.0871 **d.** 0.65, 0.0675 **e.** 0.65, 0.0477 **f.** 0.65, 0.0337 **8.29 a.** $\mu_p = 0.07$, $\sigma_p = 0.0255$ **b.** Because $n\pi < 10$, the sampling distribution of p cannot be assumed to be approximated by a normal curve. **c.** If $n = 200$, μ_p does not change, but σ_p becomes 0.01804. **d.** Yes, $n\pi$ and $n(1 - \pi)$ are now both > 10 **e.** .0485 **8.31 a.** $\mu_p = 0.005$, $\sigma_p = 0.007$ **b.** Because $n\pi < 10$, the sampling distribution of p cannot be assumed to be approximated by a normal curve. **c.** $n \geq 2000$ **8.33 a.** $\pi = .5$: $\mu_p = 0.5$, $\sigma_p = 0.0333$; $\pi = .6$: $\mu_p = 0.6$, $\sigma_p = 0.0327$. In both cases the sampling distribution of p is approximately normal. **b.** $\pi = .5$: $P(p \geq .6) = .0013$; $\pi = .6$: $P(p \geq .6) = .5$ **c.** The probability for $\pi = .5$ would decrease and the probability for $\pi = .6$ would stay the same. **8.35 a.** .9744 **b.** Approximately 0. **8.37 a.** Approximately normal with a mean of 50 lb and a standard deviation of 0.1 lb. **b.** .9876 **c.** .5 **8.39** $\mu_{\bar{x}} = 52$ min, $\sigma_{\bar{x}} = 0.33$ min. **b.** Approximately 1, approximately 0. **8.41 a.** .0207 **b.** If $\pi = .4$ for your state, it would be very unlikely to observe such a sample.

CHAPTER 9

9.1 Statistic II, because it is unbiased and has a smaller variability than the other two statistics. **9.3** .252 **9.5 a.** 1.957 **b.** 0.159 **c.** 0.399 **9.7 a.** $\bar{x} = 20.6526$ **b.** $p = .2105$ **9.9** $p = .70$ **9.11 a.** 95% confidence level. **b.** $n = 100$ **9.13 a.** Yes. **b.** No. **c.** No. **d.** No. **e.** Yes. **f.** Yes. **g.** Yes. **h.** No. **9.15 a.** (0.2441, 0.4159) **b.** (0.1307, 0.2513) **c.** Increased confidence level in Part (a), increased sample proportion in Part (a). **9.17 a.** (0.891, 0.909) **b.** The sample was large ($np > 10$ and $n(1 - p) > 10$, which in this case, they are), and the respondents were randomly selected. **c.** All 18- to 24-year-olds in the nine different counties in the study.

9.19 $z = 2.06$. It was a 96% confidence interval. **9.21** Based on the sample data, we estimate that the percentage of all households that experienced some sort of crime during the past year is between 22% and 28%. The method used to construct this interval estimate has a 5% error rate. **9.23** (0.437, 0.523). With 90% confidence, we can conclude that the true proportion of students at the College of New Jersey who are binge drinkers is between 43.7% and 52.3%. **9.25** (0.586, 0.714). Because the lower bound is greater than 0.5, yes. **9.27** 385 **9.29 a.** 2.12 **b.** 1.80 **c.** 2.81 **d.** 1.71 **e.** 1.78 **f.** 2.26 **9.31** The interval (49.4, 50.6) is based on the larger sample size because the interval is narrower. **9.33 a.** (179.03, 186.97) **b.** (185.44, 194.56) **c.** The new FAA recommendations are above the upper bounds for both summer and winter. **9.35 a.** (52.07, 56.33) **b.** If we took many samples using this sampling method and obtained confidence intervals from each of the samples, 90% of them would capture the true mean airborne time for Flight 448. **c.** 10.57 A.M. **9.37 a.** (0.409, 0.591) **b.** No, even if the mean is nonzero, there could be some individual 0's in the population, meaning that there are some students who don't lie to their mothers. **9.39** (1.79, 2.61) **9.41** (19.51, 24.29) **9.43** (5.77, 28.23) **9.45** No. The sample size is small and there are two outliers in the data set, indicating that the assumption of a normal population distribution is not reasonable. **9.47** 385 **9.49** (0.741, 0.799) **9.51 a.** Because $np \geq 10$ and $n(1 - p) \geq 10$, we can justify the use of the large-sample confidence interval to estimate the population proportion. **b.** (0.024, 0.056) **9.53 a.** (14.21, 16.99) **b.** CBS: (10.51, 13.29); FOX: (10.31, 13.09); NBC: (9.61, 12.39). **c.** Yes. **9.55** (7.56, 11.84) **9.57** 246 **9.59** (9.976, 11.824) **9.61** 39.7277 **9.63** (5.704, 6.696)

CHAPTER 10

10.1 Hypotheses are always expressed in terms of a population characteristic. $\bar{x}$ is a statistic. **10.3** The burden of proof is placed on the contractor to show that the mean weld strength is greater than 100 lb/in.^{2}. **10.5** H_0: $\mu = 7.3$ versus H_a: $\mu > 7.3$ **10.7** H_0: $\pi = .6$ versus H_a: $\pi > .6$ **10.9** H_0: $\pi = .5$ versus H_a: $\pi > .5$ **10.11 a.** Type I error. **b.** Type I error. **c.** Type II error. **d.** Lowering the risk of a Type I error will increase the risk of a Type II error. **10.13 a.** Failed to reject H_0. **b.** Type II error. **c.** Because the null hypothesis is initially assumed to be true and is rejected only if there is

strong evidence against it, if we fail to reject the null hypothesis, we have not "proven" it to be true. We can say only that there is not strong evidence against it. **10.15 a.** Reject H_0. **b.** Type I error. **10.17 a.** A Type I error is returning to the supplier a shipment that is not of inferior quality. A Type II error is accepting a shipment of inferior quality. **b.** The calculator manufacturer would consider a Type II error more serious, because they would end up producing defective calculators. **c.** The supplier would consider a Type I error more serious because they would lose the profits from perfectly good printed circuits. **10.19 a.** The manufacturer claims that the defective rate is 10%. Because we would probably not complain if the defective rate was below 10%, we would want to determine whether the defective rate is greater than that claimed by the manufacturer. **b.** A Type I error is concluding that the defective rate exceeds 10% when, in fact, it doesn't. This would lead to false advertising charges against the manufacturer, who would not be guilty of false advertising. A Type II error is deciding that there isn't enough evidence to conclude that the defective rate exceeds 10% when, in fact, it does. This would lead to the manufacturer continuing to advertise with what is false advertising. **10.21 a.** H_0: $\pi = .02$ versus H_a: $\pi < .02$ **b.** A Type I error is changing to robots when in fact they are not superior to humans. A Type II error is not changing to robots when in fact they are superior to humans. **c.** Because a Type I error would result in a substantial monetary loss to the company and also in the loss of jobs for former employees, a small α should be used. **10.23** .001, .021, .047 **10.25 a.** .0808 **b.** .1762 **c.** .0250 **d.** .0071 **e.** .5675 **10.27** $z = -11.7$, P-value = 1, fail to reject H_0. **10.29** $z = 7.75$, P-value $\approx$ 0, reject H_0. **10.31** $z = 1.90$, P-value = .058, fail to reject H_0 (at $\alpha = .01$ and at $\alpha = .05$). **10.33 a.** $z = -1.56$, P-value = .0594, fail to reject H_0 (at $\alpha = .05$), reject H_0 (at $\alpha = .10$). **b.** $z = -2.48$, P-value = .0066, reject H_0 (at $\alpha = .05$ and $\alpha = .10$). **10.35** $z = 2.20$, P-value = .0139, fail to reject H_0. **10.37 a.** $z = 0.12$, P-value = .4522, fail to reject H_0. **b.** A Type I error would occur if we concluded that more than two-thirds of consumers preferred either baked or deep-fat-fried nuggets. **d.** No, P-value = .0694. **10.39** $z = 3.20$, P-value = .0007, reject H_0. **10.41 a.** .0358 **b.** .0802 **c.** .562 **d.** .1498 **e.** 0 **10.43 a.** .040 **b.** .003 **c.** .019 **d.** 0 **e.** .13 **f.** .13 **g.** 0 **10.45 a.** Reject H_0.

b. Reject H_0. **c.** Fail to reject H_0. **10.47 a.** Fail to reject H_0. **b.** Fail to reject H_0. **c.** Fail to reject H_0. **d.** H_0 would be rejected for any $\alpha > .002$. **10.49** $t = -3.89$, P-value $\approx$ 0, reject H_0. **10.51** $t = 1.0$, P-value = .187, fail to reject H_0. **10.53** $t = -14.75$, P-value $\approx$ 0, reject H_0. **10.55** $t = -5.05$, P-value $\approx$ 0, reject H_0. **10.57 a.** Because the boxplot is nearly symmetric with no outliers and the normal probability plot is reasonably straight, the t test is appropriate. **c.** $t = 1.21$, P-value = .256, fail to reject H_0. **10.59 a.** When the significance level is held fixed, increasing the sample size will increase the power of the test. **b.** When the sample size is held fixed, increasing the significance level will increase the power of the test. **10.61 a.** .1003 **b.** .2358 **d.** .7642, 1 **10.63 a.** $t = 0.47$, P-value $\approx$ .317, fail to reject H_0. **b.** $\beta \approx .75$ **c.** Power $\approx$ .25 **10.65 a.** $\beta \approx .06$ **b.** $\beta \approx .06$ **c.** $\beta \approx .21$ **d.** $\beta \approx 0$ **e.** $\beta \approx .06$ **f.** $\beta \approx .01$ **10.67** $z = 1.78$, P-value = .0375, fail to reject H_0 at $\alpha = .01$. **10.69** $z = 2.50$, P-value = .0124, reject H_0 at $\alpha = .05$; fail to reject H_0 at $\alpha = .01$. **10.71** $t = 3.86$, P-value $\approx$ 0, reject H_0. **10.73 a.** If the distribution was normal, the Empirical Rule would imply that about 16% of the values would be less than the value that is 1 standard deviation below the mean. Because the standard deviation is larger than the mean, this value is negative. Caffeine consumption cannot be negative, so it is not plausible that the distribution is normal. However, because the sample size is large ($n = 47$), the t test can still be used. **b.** $t = 0.44$, P-value $\approx$ .346, fail to reject H_0. **10.75** $t = -1.16$, P-value = .15, fail to reject H_0 at $\alpha = .01$. **10.77** $t = 5.68$, P-value = 0, reject H_0. **10.79** $\alpha = 0 < \beta < .3$.

CHAPTER 11

11.1 Approximately normal with mean 5 and standard deviation 0.529. **11.3** $t = -0.21$, P-value $\approx$ .842, fail to reject H_0. **11.5 a.** The sample sizes are too small, and we don't know whether the response distributions are approximately normal. **b.** The sample sizes are now large enough (>30) to use a two-sample t-test. **c.** $t = -2.21$, df = 79, P-value $\approx$ .032, fail to reject H_0. **11.7** For small prey: $t = 1.27$, df = 7, P-value $\approx$ .23, fail to reject H_0; for medium prey: $t = 1.83$, df = 10, P-value $\approx$ .102, fail to reject H_0; for large prey: $t = -18.08$, df = 12, P-value $\approx$ 0, reject H_0. **11.9 a.** Yes. **b.** Yes. **c.** No, you would need to know the sample means and standard deviations. **11.11 a.** $t = -6.56$, df = 201,

P-value ≈ 0, reject H_0. **b.** $t = -6.25$, df $= 201$, P-value ≈ 0, reject H_0. **c.** $t = 0.07$, df $= 201$, P-value $\approx .92$, fail to reject H_0. **11.13 a.** $t = 1.77$, df $= 37$, P-value $\approx .08$, fail to reject H_0. **b.** $t = -1.28$, df $= 29$, P-value $\approx .204$, fail to reject H_0. **c.** $t = -7.1$, df $= 332$, P-value ≈ 0, reject H_0. **11.15 a.** For any value of $\alpha > .212$. **b.** $t = 2.72$, df $= 473$, P-value $\approx .003$, reject H_0. **11.17** $t = 3.1$, df $= 113$, P-value $\approx .001$, reject H_0. **11.19 a.** $t = 2.93$, df $= 54$, P-value $\approx .006$, reject H_0. **b.** $t = 0.15$, df $= 68$, P-value $\approx .842$, fail to reject H_0. **c.** $(0.242, 1.458)$ **11.21** $t = 113.17$, df $= 45$, P-value ≈ 0, reject H_0. **11.23** $t = 3.30$, df $= 98$, P-value $\approx .001$, reject H_0. **11.25** $t = 2.97$, df $= 10$, P-value $\approx .007$, reject H_0. **11.27** $t = 2.94$, df $= 26$, P-value $\approx .92$, fail to reject H_0. **11.31 a.** $t = -1.15$, df $= 6$, P-value $\approx .853$, fail to reject H_0. **b.** No, the seven schools in the sample are not representative of all schools in California. **c.** No, there is an outlier in the sample distribution of the differences. **11.33** P-value $= .001$, reject H_0. **11.35 b.** $(-1.03, -0.63)$ **11.37 b.** $(-0.738, 1.498)$ **c.** $t = 0.764$, df $= 849$, P-value $= .445$, fail to reject H_0. **11.39** $t = -1.34$, df $= 26$, P-value $= .192$, fail to reject H_0. **11.41** $z = -1.67$, P-value $= .0475$, reject H_0. **11.43** $z = 4.22$, P-value ≈ 0, reject H_0. **11.45 a.** $(-0.0561, 0.0695)$ **b.** With 90% confidence, it is estimated that the true proportion of patients developing diabetes may be as much as .0695 more in the insulin group than in the control group; but it also may be as much as .0561 less in the insulin group than in the control group. **c.** Because 0 is in the interval, it is possible that there is no difference in the proportion of the patients developing diabetes in the two groups. The proposed treatment does not appear to be very effective. **11.47** $z = 2.36$, P-value $= .0182$, reject H_0. **11.49** The decision was based on an analysis of all the soldiers who served in the Gulf War (the population). Because a census was performed, no inference procedure was necessary. **11.51** $z = -4.09$, P-value ≈ 0, reject H_0. **11.53** $z = -1.84$, P-value $= .066$, fail to reject H_0. **11.55** $(-0.2053, 0.0777)$ **11.57** Rank sum $= 83$, P-value $> .05$, fail to reject H_0. **11.59 a.** Rank sum $= 48$, P-value $> .05$, fail to reject H_0. **11.61** Rank sum $= 63$, P-value $> .05$, fail to reject H_0. **11.63** The confidence interval indicates that the mean burning time of oak may be as much as 0.5699 hr longer than the mean burning time of pine but also that the mean burning time of oak may be as much as

0.4998 hr shorter than the mean burning time of pine. **11.65** $(-0.2743, -0.0823)$. The interval does not contain 0. The proportion of children with decayed teeth is lower for children who drink fluoridated water by somewhere between 0.0823 and 0.2743. **11.67 a.** $(-4.75, 22.75)$ **b.** $t = 0.1398$, df $= 18$, P-value $\approx .890$, fail to reject H_0. **c.** $t = 0.4462$, df $= 17$, P-value $\approx .331$, fail to reject H_0. **11.69** $t = 2.94$, df $= 21$, P-value $\approx .0039$, reject H_0. **11.71 b.** P-value $= .13$, fail to reject H_0. **c.** H_0: same, H_a: $\mu_1 - \mu_2 > 0$, $t = 1.13$, P-value: same. **11.73 a.** $t = 0.56$, df $= 95$, P-value $\approx .5795$ fail to reject H_0. **b.** $t = 0.1306$, df $= 92$, P-value $\approx .8964$, fail to reject H_0. **c.** $t = 2.4882$, df $= 811$, P-value $\approx .0148$, reject H_0. **11.75 a.** $z = -9.45$, P-value ≈ 0, reject H_0. **b.** $z = -4.60$, P-value ≈ 0, reject H_0. **c.** $(0.1333, 0.2005)$ **11.77** $t = 1.09$, df $= 42$, P-value $\approx .282$, fail to reject H_0. **11.79** $z = 5.59$, P-value ≈ 0, reject H_0.

CHAPTER 12

12.1 a. $.020 < P$-value $< .025$ (or P-value $= .023$), fail to reject H_0. **b.** $.040 < P$-value $< .045$ (or P-value $= .043$), fail to reject H_0. **c.** $.035 < P$-value $< .040$ (or P-value $= .0352$), fail to reject H_0. **d.** P-value $< .001$, reject H_0. **e.** P-value $> .10$ (or P-value $= .172$), fail to reject H_0. **12.3 a.** $X^2 = 19.0$, P-value $< .001$, reject H_0. **b.** If $n = 40$, the chi-square test should not be used because the expected cell count for one of the categories (nut type 4) would be less than 5. **12.5** $X^2 = 1.08$, P-value $> .100$ (or P-value $= .982$), fail to reject H_0. **12.7** $X^2 = 90.28$, P-value $< .001$, reject H_0. **12.9** $X^2 = 4.035$, P-value $> .100$ (or P-value $= .258$), fail to reject H_0. **12.11 a.** $X^2 = 4.63$, P-value $\approx .100$ (or P-value $= .0988$), fail to reject H_0. **b.** The analysis would not change. **12.13** $X^2 = 4.8$, P-value $> .100$ (or P-value $= .684$), fail to reject H_0. **12.15 a.** P-value $> .10$ (or P-value $= .844$), fail to reject H_0. **b.** P-value $> .10$ (or P-value $= .106$), fail to reject H_0. **12.17** $X^2 = 49.81$, P-value $< .001$, reject H_0. **12.19 a.** The proportions falling into each category appear to be dissimilar. For instance, only 17.62% of the subjects in the "never smoked" category consumed 1 drink per day, whereas 34.97% of those in the "currently smokes" category consume 1 drink per day. Similar discrepancies are seen for other categories as well. **b.** $X^2 = 980.068$, P-value $< .001$, reject H_0. **c.** The results are consistent with our observations in Part (a). **12.21** One cell has an expected count

of less than 5. The chi-square approximation is probably not a reasonable test to use. **12.23** $X^2 = 10.976$, P-value $< .001$, reject H_0. **12.25** $X^2 = 22.855$, P-value $< .0001$, reject H_0. **12.27** $X^2 = 1.978$, P-value $> .10$ (or P-value $= .372$), fail to reject H_0. **12.29** $X^2 = 8.684$, $.065 < P$-value $< .070$ (or P-value $= .069$), fail to reject H_0. **12.31** $X^2 = 6.621$, P-value $> .05$ (or P-value $= .085$), fail to reject H_0. **12.33** $X^2 = 9.84$, $.005 < P$-value $< .010$ (or P-value $= .0073$), reject H_0. **12.35 a.** $X^2 = 8216.5$, P-value $< .001$, reject H_0. **c.** $X^2 = 10.748$, P-value $> .100$ (or P-value $= .465$), fail to reject H_0. **12.37** $X^2 = 197.62$, P-value $< .001$), reject H_0. **12.39** $X^2 = 6.54$, $.035 < P$-value $< .040$ (or P-value $= .038$), reject H_0. **12.41** $X^2 = 103.87$, P-value $< .001$, reject H_0. **12.43** $X^2 = 75.35$, P-value $< .001$, reject H_0. Note: There are two cells with expected counts less than 5. If the last two rows of the table are combined, $X^2 = 73.03$, P-value $< .001$, reject H_0.

CHAPTER 13

13.1 a. $y = -5.0 + 0.017x$ **c.** 30.7 **d.** 0.017 **e.** 1.7 **f.** No, the model should not be used to predict outside the range of the data. **13.3 a.** 3.6, 4.3 **b.** .121 **c.** .5, .1587 **13.5 a.** 47, 4700 **b.** .3156, .0643 **13.7 a.** $r^2 = .6228$ **b.** $s_e = 1.5923$ **c.** $b = 3.4307$ **d.** 3.4426 **13.9 a.** $r^2 = .883$ **b.** $s_e = 13.682$, df $= 14$ **13.11 b.** $\hat{y} = -0.002274 + 1.246712x$. When $x = .09$, the predicted value is .1099. **c.** $r^2 = .436$. This means that 43.6% of the total variability in market share can be explained by the linear relationship with advertising share. **d.** $s_e = 0.0263$, df $= 8$ **13.13 a.** $\sigma_b = 0.253$ **b.** $\sigma_b = 0.179$, no. **c.** It would require 4 observations at each x value. **13.15 a.** $s_e = 9.7486$, $s_b = 0.1537$ **b.** $(2.17, 2.83)$ **c.** Yes, the interval is relatively narrow. **13.17 a.** $a = 592.1$, $b = 97.26$ **b.** $\hat{y} = 786.62$. Residual $= -29.62$ **c.** $(87.76, 106.76)$ **13.19 a.** $t = 2.8302$, P-value $= .008$, reject H_0; therefore we conclude that the model is useful. **b.** $(4.294, 25.706)$. Based on this interval, we estimate that the change in mean average SAT score associated with an increase of $1000 in expenditures per child is between 4.294 and 25.706. **13.21 a.** $\hat{y} = -96.6711 + 1.595x$ **b.** $t = 27.17$, P-value ≈ 0, reject H_0 and conclude that the model is useful. **c.** The mean response time for individuals with closed-head injuries is *1.48 times* as great as for those with no head injury. **13.23 a.** $t = 6.53$, P-value ≈ 0, reject H_0 and conclude that the model is useful. **b.** $t = 1.56$, P-value $= .079$, fail to reject H_0. **13.25** $t = -17.57$, P-value $\approx$ 0, reject H_0. **13.27** The standardized residual plot does not exhibit any unusual features. **13.29 b.** The plot is reasonably straight. The assumption of normality is plausible. **c.** There are two residuals that are unusually large. **13.31 a.** There are several large residuals and potentially influential observations. **13.33** The residuals are positive in 1963 and 1964 and then negative from 1965 through 1972. This pattern in the residual plot casts doubt on the appropriateness of the simple linear regression model. **13.35** If the request is for a confidence interval for β, the wording would likely be "estimate the *change* in the average y value associated with a 1-unit increase in the x variable." If the request is for a confidence interval for $\alpha + \beta x^*$, the wording would likely be "estimate the *average y* value when the value of the x variable is x^*." **13.37 a.** $(-0.4788, -0.2113)$ **b.** $(17.78, 21.78)$ **13.39 a.** $(6.4722, 6.6300)$ **b.** $(6.4779, 6.6973)$ **c.** Yes, because $x^* = 60$ is farther from $\bar{x}$. **13.41 a.** $\hat{y} = -133.02 + 5.919x$ **b.** $s_b = 1.127$ **c.** $t = 5.25$, P-value ≈ 0, reject H_0. **d.** 251.715 **e.** You should not use the regression equation to predict for $x = 105$ because it is outside the range of the data. **13.43 a.** $\hat{y} = 2.78551 + 0.04462x$ **b.** $t = 5.25$, P-value ≈ 0, reject H_0 and conclude that the model is useful. **c.** (3.671, 4.577) **d.** Because 4.1 is in the interval, it is plausible that a box that has been on a shelf 30 days will not be acceptable. **13.45 a.** $\hat{y} = 30.019 + 0.1109x$ **b.** (38.354, 43.864) **c.** (30.311, 51.907) **d.** The interval of Part (b) is for the mean blood level of all people who work where the air lead level is 100. The interval of Part (c) is for a single randomly selected individual who works where the air lead level is 100. **13.47 a.** Prediction interval for $x = 35$: (1.9519, 3.0931). Prediction interval for $x = 45$: (1.3879, 2.5271). **b.** The simultaneous prediction level would be at least 97%. **13.51** $t = 2.07$, P-value $\approx .036$, reject H_0. **13.53 a.** $t = -6.175$, P-value ≈ 0, reject H_0. **b.** No, $r^2 = .0676$. **13.55** $r = .574$, $t = 1.85$, P-value $\approx .1$, fail to reject H_0. **13.57** $t = 2.2$, P-value $= .028$, reject H_0. **13.61 a.** $t = -3.95$, P-value $= .003$, reject H_0 and conclude that the model is useful. **b.** $(-3.6694, -0.9976)$ **c.** (60.276, 70.646) **d.** Because $x = 11$ is farther from $\bar{x}$ than $x = 10$ is from $\bar{x}$. **13.63 a.** $t = -6.09$, P-value ≈ 0, reject H_0. **b.** (2.346, 3.844) **c.** According to the least-squares line, the predicted trail length when soil hardness is 10 is -2.58. Because trail length cannot be

negative, the predicted value makes no sense. Therefore, one would not use the simple linear regression model to predict trail length when hardness is 10. **13.65 a.** $t = 0.49$, P-value $= .624$, fail to reject H_0. **b.** (47.083, 54.099) **13.67** (18.0284, 22.2218) **13.69** A linear model is not appropriate for data sets 2, 3, and 4. **13.71 a.** $\hat{y} = 57.964 + 0.0357x$ **b.** $t = 0.023$, P-value ≈ 1, fail to reject H_0. **d.** The residual plot has a distinct curved pattern, which indicates that a simple linear regression model is not appropriate. **13.73 a.** $t = -0.06$, P-value $= .922$, fail to reject H_0. **b.** $t = 4.35$, P-value ≈ 0, reject H_0. **c.** $a = 3.0781$, $b = 0.01161$ **d.** (2.9778, 4.3392) **e.** (3.6197, 4.1619)

CHAPTER 14

14.1 A deterministic model does not have the random deviation component e, whereas a probabilistic model does contain such a component. **14.3** $y = \beta_0 + \beta_1 x_1 + \beta_2 x_2 + e$. An interaction term is not included in the model because it is given that x_1 and x_2 make independent contributions. **14.5 a.** 103.11 **b.** 96.87 **c.** $\beta_1 = -6.6$; 6.6 is the expected *decrease* in yield associated with a 1-unit increase in mean temperature when the mean percentage of sunshine remains fixed. $\beta_2 = -4.5$; 4.5 is the expected *decrease* in yield associated with a 1-unit increase in mean percentage of sunshine when mean temperature remains fixed. **14.7 b.** The mean chlorine content is higher for $x = 10$ than for $x = 8$. **c.** When the degree of delignification increases from 8 to 9, the mean chlorine content increases by 7. Mean chlorine content decreases by 1 when degree of delignification increases from 9 to 10. **14.9 c.** The parallel lines in each graph are attributable to the lack of interaction between the two independent variables. **d.** Because there is an interaction term, the lines are not parallel. **14.11 a.** $y = \alpha + \beta_1 x_1 + \beta_2 x_2 + \beta_3 x_3 + e$ **b.** $y = \alpha + \beta_1 x_1 + \beta_2 x_2 + \beta_3 x_3 + \beta_4 x_1^2 + \beta_5 x_2^2 + \beta_6 x_3^2 + e$ **c.** $y = \alpha + \beta_1 x_1 + \beta_2 x_2 + \beta_3 x_3 + \beta_4 x_1 x_2 + e$; $y = \alpha + \beta_1 x_1 + \beta_2 x_2 + \beta_3 x_3 + \beta_4 x_1 x_3 + e$; $y = \alpha + \beta_1 x_1 + \beta_2 x_2 + \beta_3 x_3 + \beta_4 x_2 x_3 + e$ **d.** $y = \alpha + \beta_1 x_1 + \beta_2 x_2 + \beta_3 x_3 + \beta_4 x_1^2 + \beta_5 x_2^2 + \beta_6 x_3^2 + \beta_7 x_1 x_2 + \beta_8 x_1 x_3 + \beta_9 x_2 x_3 + e$ **14.13 a.** Three dummy variables would be needed to incorporate a nonnumerical variable with four categories. **14.15 a.** $b_3 = -0.0996$ is the estimated change in mean maximal oxygen uptake associated with a 1-unit increase in "1-mi walk time" when the values of the other predictor variables are fixed. **b.** $b_1 = 0.6566$ is the estimated difference in mean maximal oxy-

gen uptake for males versus females (male minus female) when the values of the other predictor variables are fixed. **c.** $\hat{y} = 2.8049$, residual $= 0.3451$ **d.** $R^2 = .706$ **e.** $s_e = 0.397$ **14.17 a.** $.01 < P$-value $< .05$ **b.** P-value $> .10$ **c.** P-value $= .01$ **d.** $.001 < P$-value $< .01$ **14.19** $F = 24.41$, P-value $< .001$, reject H_0 and conclude that the model is useful. **14.21** P-value $< .001$, reject H_0 and conclude that the model is useful. **14.23 a.** $\hat{y} = 86.85 - 0.12297x_1 + 5.090x_2 - 0.07092x_3 + 0.001538x_4$ **b.** $F = 64.4$, P-value $< .001$, reject H_0 and conclude that the model is useful. **c.** $R^2 = .908$ means that 90.8% of the variation in the observed tar content values has been explained by the fitted model. $s_e = 4.784$ means that the typical distance of an observation from the corresponding mean value is 4.784. **14.25 a.** $\hat{y} = -511 + 3.06(\text{Length}) - 1.11(\text{Age})$. **b.** $F = 10.21$, $.001 < P$-value $< .01$, reject H_0 and conclude that the model is useful. **14.27 a.** $\hat{y} = -859 + 23.7(\text{minimum width}) + 226(\text{maximum width}) + 225(\text{elongation})$. **b.** Adjusted R^2 takes into account the number of predictors used in the model, whereas R^2 does not do so. **c.** $F = 16.03$, P-value $< .001$, reject H_0 and conclude that the model is useful. **14.29 a.** SSResid $= 390.4347$, SSTo $= 1618.2093$, SSRegr $= 1227.7746$ **b.** $R^2 = .759$. This means that 75.9% of the variation in the observed shear strength values has been explained by the fitted model. **c.** $F = 5.039$, $.01 < P$-value $< .05$, reject H_0 and conclude that the model is useful. **14.31** $F = 96.64$, P-value $< .001$, reject H_0 and conclude that the model is useful. **14.35** $\hat{y} = 35.8 - 0.68x_1 + 1.28x_2$, $F = 18.95$, P-value $< .001$, reject H_0 and conclude that the model is useful. **14.37 a.** $(-0.697, -0.281)$. With 95% confidence, the change in the mean value of a vacant lot associated with a 1-unit increase in distance from the city's major east–west thoroughfare is a decrease of as little as 0.281 or as much as 0.697, assuming that all other predictors remain constant. **b.** $t = -0.599$, P-value $\approx .550$, fail to reject H_0. **14.39 a.** $F = 144.36$, P-value $< .001$, reject H_0 and conclude that the model is useful. **b.** $t = 4.33$, P-value ≈ 0, reject H_0 and conclude that the quadratic term is important. **c.** (72.0626, 72.4658) **14.41 a.** The value 0.469 is an estimate of the expected change (increase) in the mean score of students associated with a 1-unit increase in the student's expected score, assuming that time spent studying and student's grade point average are fixed. **b.** $F = 75.01$,

P-value $< .001$, reject H_0 and conclude that the model is useful. **c.** (2.466, 4.272) **d.** 72.856 **e.** (61.562, 84.150) **14.43** $t = 2.22$, P-value $= .028$, reject H_0 and conclude that the interaction term is important. **14.45 a.** $F = 5.47$, $.01 < P$-value $< .05$, reject H_0 and conclude that the model is useful. **b.** $t = 0.04$, P-value $\approx .924$, fail to reject H_0 and conclude that the interaction term is not needed in the model. **c.** No. The model utility test is testing all variables as a group. The t test is testing the contribution of an individual predictor when used in the presence of the remaining predictors. **14.49 a.** $F = 30.81$, P-value $< .001$, reject H_0 and conclude that the model is useful. **b.** For β_1: $t = 6.57$, P-value $= 0$, reject H_0. For β_2: $t = -7.69$, P-value $= 0$, reject H_0. **c.** (43.0, 47.92) **d.** (44.0, 47.92) **14.51** One possible way would have been to start with the set of predictor variables consisting of all five variables along with all quadratic terms and all interaction terms and then to use a selection procedure, such as backward elimination, to arrive at the given estimated regression equation. **14.53** The model using the three variables x_3, x_9, and x_{10} appears to be a good choice. **14.55 a.** $F \approx 86$, P-value $< .001$, reject H_0 and conclude that the model is useful. **b.** The predictor added at the first step was smoking habits. At Step 2, alcohol consumption was added. No other variables were added to the model. **14.57 b.** 0.0567 **c.** Return on equity; yes. **d.** No. CEO tenure would be considered first. **e.** P-value $\approx .03$ **14.59** The best model, using the procedure of minimizing C_p, would use variables x_2 and x_4. These are not the same predictor variables as selected in Exercise 14.58. **14.63 a.** $F = 12.28$, P-value $< .001$, reject H_0 and conclude that the model is useful. **b.** Adjusted $R^2 = .5879$. **c.** (0.1606, 0.7554) **d.** x_9 would be considered first, and it would be eliminated. **e.** H_0: $\beta_3 = \beta_4 = \beta_5 = \beta_6 = 0$. None of the procedures presented in this chapter could be used. **14.65 b.** $F = 21.25$, $.001 < P$-value $< .01$, reject H_0 and conclude that the model is useful. **c.** $t = 6.52$, P-value ≈ 0, reject H_0. The quadratic term should not be eliminated. **14.67** Adjusted R^2 will be negative for values of R^2 less than 0.5.

CHAPTER 15
15.1 a. $.001 < P$-value $< .01$ **b.** P-value $> .10$ **c.** P-value $= .01$ **d.** P-value $< .001$ **e.** $.05 < P$-value $< .10$ **f.** $.01 < P$-value $< .05$ (using $df_1 = 4$ and $df_2 = 60$). **15.3 a.** H_0: $\mu_1 = \mu_2 =$

$\mu_3 = \mu_4$ versus H_a: At least two of the four μ_i's are different. **b.** $.01 < P$-value $< .05$, fail to reject H_0. **c.** $.01 < P$-value $< .05$, fail to reject H_0. **15.5 a.** The 1996 and 1997 boxplots are similar, but the 1998 boxplot values are higher. The boxplot for 1998 also has a longer tail on the upper end than the other two boxplots. **b.** $F = 6.83$, P-value $\approx .01$, reject H_0. **15.7** $F = 1.71$, P-value $> .10$, fail to reject H_0. **15.9** $F = 0.71$, P-value $> .10$, fail to reject H_0. **15.11** $F = 1.895$, $.05 < P$-value $< .10$, fail to reject H_0. **15.13** $F = 1.70$, P-value $> .10$, fail to reject H_0. **15.15** $F = 2.56$, $.05 < P$-value $< .10$, fail to reject H_0. **15.17** $F = 6.97$, $.001 < P$-value $< .01$, reject H_0. **15.19** Because the intervals for $\mu_1 - \mu_2$ and $\mu_1 - \mu_3$ do not contain 0, μ_1 and μ_2 are judged to be different and μ_1 and μ_3 are judged to be different. Because the interval for $\mu_2 - \mu_3$ contains 0, μ_2 and μ_3 are not judged to be different. Hence, statement (c) is the correct choice. **15.21** μ_1 differs from μ_2; μ_1 differs from μ_3; μ_1 differs from μ_5; μ_2 differs from μ_4; μ_3 differs from μ_4; μ_4 differs from μ_5.

Fabric	3	2	5	4	1
$\bar{x}$	10.5	11.63	12.3	14.96	16.35

15.23 No differences are detected. This conclusion is in agreement with the results of Exercise 15.7.
15.25

Group	Simulta-neous	Sequen-tial	Control
Mean	$\bar{x}_3$	$\bar{x}_2$	$\bar{x}_1$

15.27 The mean water loss when the plants are exposed to 4 hr of fumigation is different from all the other means. The mean water loss when the plants are exposed to 2 hr of fumigation is different from that for levels 16 and 0 but not for level 8. The mean water losses for duration 16, 0, and 8 hr are not different from one another. **15.29 a.** $F = 3.30$, $.01 < P$-value $< .05$, reject H_0. **b.** No significant differences are determined using the T-K method. **15.31 a.** See answer below. **b.** $F = 0.37$, P-value $> .10$, fail to reject H_0. **15.33** $F = 79.4$, P-value $< .001$,

reject H_0. **15.35 a.** Other environmental factors (amount of rainfall, number of days of cloudy weather, average daily temperature, etc.) vary from year to year. Using a randomized complete block helps control for variation in these other factors. **b.** $F = 15.03$, $.01 < P$-value $< .05$, reject H_0.
c.

Application Rate	1	2	3
Mean	138.33	152.33	156.33

15.37 $F = 0.868$, P-value $> .10$, fail to reject H_0. **15.39 b.** The lines are nearly parallel. The plot does not show evidence of an interaction between gender and type of odor. **15.41 a.** The plot does suggest an interaction between peer group and self-esteem. **b.** The change in the average response is greater for the low self-esteem group than it is for the high self-esteem group, when changing from low to high peer group. The authors are correct in their statement. **15.43** Test for interaction: $F = 2.18$, P-value $> .10$, fail to reject H_0. Test for main effect for size: $F = 4.00$, $.01 < P$-value $< .05$, fail to reject H_0. Test for main effect for species: $F = 4.36$, $.05 < P$-value $< .10$, fail to reject H_0. **15.45 a.** $F = 0.70$, P-value $> .10$, fail to reject H_0. **b.** $F = 20.53$, P-value $< .001$, reject H_0. **c.** $F = 2.59$, $.05 < P$-value $< .10$, fail to reject H_0. **15.47** Test for interaction: $F < 1$, P-value $> .10$, fail to reject H_0. Test for main effect of A: $F = 4.99$, $.01 < P$-value $< .05$, reject H_0. Test for main effect of B: $F = 4.81$, $.01 < P$-value $< .05$, reject H_0. **15.49 a.** Test for interaction: $F = 0.21$, P-value $> .10$, fail to reject H_0. **b.** Test for main effect of race: $F = 5.57$, $.01 < P$-value $< .05$, fail to reject H_0. **c.** Test for main effect of gender: $F = 1.89$, P-value $> .10$, fail to reject H_0. **15.51** Test for main effect of rate: $F = 76.14$, P-value $< .001$, reject H_0. Test for main effect of soil type: $F = 54.14$, P-value $< .001$, reject H_0.

Soil Type	Ramiha	Konini	Wainui
Mean	10.217	20.957	24.557

15.53 $F = 5.56$, $.001 < P$-value $< .01$, reject H_0. See top of next page.

Answer for Exercise 15.31a

Source of Variation	Degrees of Freedom	Sum of Squares	Mean Square	F
Treatments	2	11.7	5.85	0.37
Blocks	4	113.5	28.375	
Error	8	125.6	15.7	
Total	14	250.8		

Group	Apathetic	Conservative	Erratic	Strategic
Mean	2.96	4.91	5.05	8.74

15.55 a. $F = 6.24$, $.001 < P\text{-value} < .01$, reject H_0.
b.

Storage period	4	3	2	1
Mean	4.902	4.917	4.923	4.923

15.57 Test for interaction: $F = 0.44$, $P\text{-value} > .10$, fail to reject H_0. Test for main effect of oxygen: $F = 2.76$, $P\text{-value} > .10$, fail to reject H_0. Test for main effect of sugar: $F = 13.29$, $.001 < P\text{-value} < .01$, reject H_0.
15.59 Test for interaction: $F = 8.67$, $.001 < P\text{-value} < .01$, reject H_0; there is significant interaction, so no tests for main effects should be performed.
15.61 $F = 0.907$, $P\text{-value} > .10$, fail to reject H_0. **15.63** $F = 3.23$, $.01 < P\text{-value} < .05$, fail to reject H_0.

■ Index

Standard normal probabilities (cumulative z curve areas)

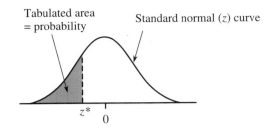

Tabulated area = probability

Standard normal (z) curve

z^* 0

z^*	.00	.01	.02	.03	.04	.05	.06	.07	.08	.09
−3.8	.0001	.0001	.0001	.0001	.0001	.0001	.0001	.0001	.0001	.0000
−3.7	.0001	.0001	.0001	.0001	.0001	.0001	.0001	.0001	.0001	.0001
−3.6	.0002	.0002	.0001	.0001	.0001	.0001	.0001	.0001	.0001	.0001
−3.5	.0002	.0002	.0002	.0002	.0002	.0002	.0002	.0002	.0002	.0002
−3.4	.0003	.0003	.0003	.0003	.0003	.0003	.0003	.0003	.0003	.0002
−3.3	.0005	.0005	.0005	.0004	.0004	.0004	.0004	.0004	.0004	.0003
−3.2	.0007	.0007	.0006	.0006	.0006	.0006	.0006	.0005	.0005	.0005
−3.1	.0010	.0009	.0009	.0009	.0008	.0008	.0008	.0008	.0007	.0007
−3.0	.0013	.0013	.0013	.0012	.0012	.0011	.0011	.0011	.0010	.0010
−2.9	.0019	.0018	.0018	.0017	.0016	.0016	.0015	.0015	.0014	.0014
−2.8	.0026	.0025	.0024	.0023	.0023	.0022	.0021	.0021	.0020	.0019
−2.7	.0035	.0034	.0033	.0032	.0031	.0030	.0029	.0028	.0027	.0026
−2.6	.0047	.0045	.0044	.0043	.0041	.0040	.0039	.0038	.0037	.0036
−2.5	.0062	.0060	.0059	.0057	.0055	.0054	.0052	.0051	.0049	.0048
−2.4	.0082	.0080	.0078	.0075	.0073	.0071	.0069	.0068	.0066	.0064
−2.3	.0107	.0104	.0102	.0099	.0096	.0094	.0091	.0089	.0087	.0084
−2.2	.0139	.0136	.0132	.0129	.0125	.0122	.0119	.0116	.0113	.0110
−2.1	.0179	.0174	.0170	.0166	.0162	.0158	.0154	.0150	.0146	.0143
−2.0	.0228	.0222	.0217	.0212	.0207	.0202	.0197	.0192	.0188	.0183
−1.9	.0287	.0281	.0274	.0268	.0262	.0256	.0250	.0244	.0239	.0233
−1.8	.0359	.0351	.0344	.0336	.0329	.0322	.0314	.0307	.0301	.0294
−1.7	.0446	.0436	.0427	.0418	.0409	.0401	.0392	.0384	.0375	.0367
−1.6	.0548	.0537	.0526	.0516	.0505	.0495	.0485	.0475	.0465	.0455
−1.5	.0668	.0655	.0643	.0630	.0618	.0606	.0594	.0582	.0571	.0559
−1.4	.0808	.0793	.0778	.0764	.0749	.0735	.0721	.0708	.0694	.0681
−1.3	.0968	.0951	.0934	.0918	.0901	.0885	.0869	.0853	.0838	.0823
−1.2	.1151	.1131	.1112	.1093	.1075	.1056	.1038	.1020	.1003	.0985
−1.1	.1357	.1335	.1314	.1292	.1271	.1251	.1230	.1210	.1190	.1170
−1.0	.1587	.1562	.1539	.1515	.1492	.1469	.1446	.1423	.1401	.1379
−0.9	.1841	.1814	.1788	.1762	.1736	.1711	.1685	.1660	.1635	.1611
−0.8	.2119	.2090	.2061	.2033	.2005	.1977	.1949	.1922	.1894	.1867
−0.7	.2420	.2389	.2358	.2327	.2296	.2266	.2236	.2206	.2177	.2148
−0.6	.2743	.2709	.2676	.2643	.2611	.2578	.2546	.2514	.2483	.2451
−0.5	.3085	.3050	.3015	.2981	.2946	.2912	.2877	.2843	.2810	.2776
−0.4	.3446	.3409	.3372	.3336	.3300	.3264	.3228	.3192	.3156	.3121
−0.3	.3821	.3783	.3745	.3707	.3669	.3632	.3594	.3557	.3520	.3483
−0.2	.4207	.4168	.4129	.4090	.4052	.4013	.3974	.3936	.3897	.3859
−0.1	.4602	.4562	.4522	.4483	.4443	.4404	.4364	.4325	.4286	.4247
−0.0	.5000	.4960	.4920	.4880	.4840	.4801	.4761	.4721	.4681	.4641

Tabulated area = probability

Standard normal (z) curve

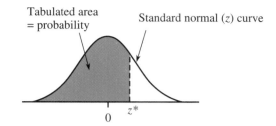

0   z*

z*	.00	.01	.02	.03	.04	.05	.06	.07	.08	.09
0.0	.5000	.5040	.5080	.5120	.5160	.5199	.5239	.5279	.5319	.5359
0.1	.5398	.5438	.5478	.5517	.5557	.5596	.5636	.5675	.5714	.5753
0.2	.5793	.5832	.5871	.5910	.5948	.5987	.6026	.6064	.6103	.6141
0.3	.6179	.6217	.6255	.6293	.6331	.6368	.6406	.6443	.6480	.6517
0.4	.6554	.6591	.6628	.6664	.6700	.6736	.6772	.6808	.6844	.6879
0.5	.6915	.6950	.6985	.7019	.7054	.7088	.7123	.7157	.7190	.7224
0.6	.7257	.7291	.7324	.7357	.7389	.7422	.7454	.7486	.7517	.7549
0.7	.7580	.7611	.7642	.7673	.7704	.7734	.7764	.7794	.7823	.7852
0.8	.7881	.7910	.7939	.7967	.7995	.8023	.8051	.8078	.8106	.8133
0.9	.8159	.8186	.8212	.8238	.8264	.8289	.8315	.8340	.8365	.8389
1.0	.8413	.8438	.8461	.8485	.8508	.8531	.8554	.8577	.8599	.8621
1.1	.8643	.8665	.8686	.8708	.8729	.8749	.8770	.8790	.8810	.8830
1.2	.8849	.8869	.8888	.8907	.8925	.8944	.8962	.8980	.8997	.9015
1.3	.9032	.9049	.9066	.9082	.9099	.9115	.9131	.9147	.9162	.9177
1.4	.9192	.9207	.9222	.9236	.9251	.9265	.9279	.9292	.9306	.9319
1.5	.9332	.9345	.9357	.9370	.9382	.9394	.9406	.9418	.9429	.9441
1.6	.9452	.9463	.9474	.9484	.9495	.9505	.9515	.9525	.9535	.9545
1.7	.9554	.9564	.9573	.9582	.9591	.9599	.9608	.9616	.9625	.9633
1.8	.9641	.9649	.9656	.9664	.9671	.9678	.9686	.9693	.9699	.9706
1.9	.9713	.9719	.9726	.9732	.9738	.9744	.9750	.9756	.9761	.9767
2.0	.9772	.9778	.9783	.9788	.9793	.9798	.9803	.9808	.9812	.9817
2.1	.9821	.9826	.9830	.9834	.9838	.9842	.9846	.9850	.9854	.9857
2.2	.9861	.9864	.9868	.9871	.9875	.9878	.9881	.9884	.9887	.9890
2.3	.9893	.9896	.9898	.9901	.9904	.9906	.9909	.9911	.9913	.9916
2.4	.9918	.9920	.9922	.9925	.9927	.9929	.9931	.9932	.9934	.9936
2.5	.9938	.9940	.9941	.9943	.9945	.9946	.9948	.9949	.9951	.9952
2.6	.9953	.9955	.9956	.9957	.9959	.9960	.9961	.9962	.9963	.9964
2.7	.9965	.9966	.9967	.9968	.9969	.9970	.9971	.9972	.9973	.9974
2.8	.9974	.9975	.9976	.9977	.9977	.9978	.9979	.9979	.9980	.9981
2.9	.9981	.9982	.9982	.9983	.9984	.9984	.9985	.9985	.9986	.9986
3.0	.9987	.9987	.9987	.9988	.9988	.9989	.9989	.9989	.9990	.9990
3.1	.9990	.9991	.9991	.9991	.9992	.9992	.9992	.9992	.9993	.9993
3.2	.9993	.9993	.9994	.9994	.9994	.9994	.9994	.9995	.9995	.9995
3.3	.9995	.9995	.9995	.9996	.9996	.9996	.9996	.9996	.9996	.9997
3.4	.9997	.9997	.9997	.9997	.9997	.9997	.9997	.9997	.9997	.9998
3.5	.9998	.9998	.9998	.9998	.9998	.9998	.9998	.9998	.9998	.9998
3.6	.9998	.9998	.9999	.9999	.9999	.9999	.9999	.9999	.9999	.9999
3.7	.9999	.9999	.9999	.9999	.9999	.9999	.9999	.9999	.9999	.9999
3.8	.9999	.9999	.9999	.9999	.9999	.9999	.9999	.9999	.9999	1.0000